Mark E. Jensen
Patrick S. Bourgeron
Editors

A Guidebook for Integrated Ecological Assessments

With 101 Figures

Springer

Mark E. Jensen
USDA Forest Service, Northern Region
200 East Broadway
Missoula, MT 59807, USA
mjensen@fs.fed.us

Patrick S. Bourgeron
University of Colorado
Institute of Alpine and Arctic Research
1560 30th Street
Boulder, CO 80309-0450, USA
patrick.bourgeron@colorado.edu

Library of Congress Cataloging-in-Publication Data
A guidebook for integrated ecological assessments / edited by Mark E. Jensen and Patrick S. Bourgeron.
p. cm.
Includes bibliographical references.
ISBN 0-387-98582-4 (hc : alk. paper) — ISBN 0-387-98583-2 (sc : alk. paper)
1. Ecological assessment (Biology) I. Jensen, Mark E. II. Bourgeron, Patrick S.
QH541.15.E22 G85 2001
577′.028′7—dc21 00-061265

Printed on acid-free paper.

Cover art: Front cover shows human use intensity model at 1-km search radius. Back cover shows human use intensity and predicted species richness. See complete Figure 29.3 on page 440 and also in color insert.

Production coordinated by WordCrafters Editorial Services, Inc., and managed by Lesley Poliner; manufacturing supervised by Joe Quatela.
Typeset by Matrix Publishing Services, Inc., York, PA.
Printed and bound by Edwards Brothers, Inc., Ann Arbor, MI.
Printed in the United States of America.

9 8 7 6 5 4 3 2 1

ISBN 0-98582-4 SPIN 10657809 hardcover
ISBN 0-98583-2 SPIN 10657817 softcover

Springer-Verlag New York Berlin Heidelberg
A member of BertelsmannSpringer Science+Business Media GmbH

Acknowledgments

This book was developed over a three-year period through a collaborative effort involving numerous federal agency, nongovernment organizations, and university personnel. The original idea and primary funding for this book was provided by the USDA, Forest Service, Washington Office, Ecosystem Management Staff Group, and the USDA, Forest Service Interregional Ecosystem Management Coordination Group. In particular, we wish to acknowledge the efforts of Dr. Christopher Risbrudt in identifying the purpose and need for the book and organizing funding for its completion. Additional funding was provided by the U.S. Environmental Protection Agency, Office of Research and Development.

An editorial board assisted the two senior editors in (1) developing initial section and chapter outlines; (2) identifying, contacting, and finalizing chapter content with authors; (3) directing the anonymous peer review process; and (4) assisting the authors with chapter revisions. Members of this board included Dr. J. Michael Vasievich (USDA Forest Service, North Central Station), Dr. Richard Everett (USDA Forest Service, retired), Gene Lessard (USDA Forest Service, retired), Dr. Russell Graham (USDA Forest Service, Rocky Mountain Research Station), and Dr. Bennett A. Brown (Animas Foundation). Additional technical assistance to the editorial board was provided by Dr. James Andreasen (U.S. Environmental Protection Agency, Office of Research and Development), Iris Goodman (U.S. Environmental Protection Agency, Office of Research and Development), Dr. John E. Estes (University of California at Santa Barbara), Dr. John Lehmkuhl (USDA Forest Service, Pacific Northwest Research Station), and Dr. Roland Redmond (University of Montana).

Primary work on the book began in May of 1997 with a weeklong workshop held in Denver, Colorado. At that time, most authors made presentations of the topics covered in each chapter. Discussion among editorial board members and authors resulted in final chapter outlines. Anonymous peer review of draft manuscripts included both a scientific review of the technical content and a management review of application appropriateness. Chapters were revised by the authors based on input from reviewers' comments. Dawn Wiser (USDA Forest Service, Pacific Northwest Research Station) and Melissa Hart (University of Montana) performed technical editing of the manuscripts. Judy Tripp (USDA Forest Service, Northern Region) made final modifications to the chapters to follow publisher specifications, and provided general administrative assistance during the development of the book.

To all previously named individuals we extend our gratitude and appreciation for their assistance in developing this book. Additionally, we wish to acknowledge the collaborative effort between the senior editors in completing this book. Authorship for this book does not imply that one senior contributed more than the other; both contributed equally. Instead, the authorship of this book represents the fact that Patrick S. Bourgeron lost the coin toss.

MARK E. JENSEN
Editor

PATRICK S. BOURGERON
Editor

January 2001

Contents

Contributors

JILL BARON
Natural Resource Ecology Laboratory
Colorado State University
Fort Collins, CO 80523, USA

PATRICK S. BOURGERON
Institute of Arctic and Alpine Research
University of Colorado
Boulder, CO 80309-0450, USA

C. KENNETH BREWER
USDA Forest Service
Northern Region
Missoula, MT 59807, USA

DANIEL G. BROWN
Department of Geography
Michigan State University
East Lansing, MI 48824-1115, USA

DAVID J. CAMPBELL
Geography and African Studies Center
Michigan State University
East Lansing, MI 48824-1115, USA

DANIEL E. CHAPPELLE
Department of Resource Development and Forestry
Michigan State University
East Lansing, MI 48824-1115, USA

NORMAN L. CHRISTENSEN, JR.
Nicholas School of the Environments
Duke University
Durham, NC 27708-0328, USA

DAVID T. CLELAND
USDA Forest Service
North Central Research Station
Rhinelander, WI 54501-9128, USA

THOMAS R. CROW
USDA Forest Service
North Central Research Station
Rhinelander, WI 54501-9128, USA

RAY FORD
Department of Computer Science
The University of Montana
Missoula, MT 59812, USA

MARIE-JOSÉ FORTIN
School of Resource and Environmental Management
Simon Fraser University
Burnaby, BC, Canada, V5A 1S6

JANET FRANKLIN
Department of Geography
San Diego State University
San Diego, CA 92182-4493, USA

CHRISTOPHER A. FRISSELL
Flathead Lake Biological Station
The University of Montana
Polson, MT 59860-9659, USA

IRIS A. GOODMAN
U.S. Environmental Protection Agency
Landscape Ecology Branch
Environmental Sciences Division
Las Vegas, NV 89119, USA

RUSSELL T. GRAHAM
USDA Forest Service
Moscow Forestry Sciences Laboratory
Rocky Mountain Research Station
Moscow, ID 83843, USA

RICK HAEUBER
Ecological Society of America
Washington, DC 20036, USA
Currently: U.S. EPA
Clean Air Markets Division
Washington, DC 20460, USA

MELISSA M. HART
Wildlife Spatial Analysis Lab
Montana Cooperative Wildlife Research Unit
The University of Montana
Missoula, MT 59812, USA

HOPE C. HUMPHRIES
Institute of Arctic and Alpine Research
University of Colorado
Boulder, Co 80309-0450, USA

MICHAEL HURLEY
M&M Environmental Enterprises
Boise, ID 83702, USA

MARK E. JENSEN
USDA Forest Service
Northern Region
Missoula, MT 59807, USA

ALAN R. JOHNSON
Department of Environmental Toxicology
Clemson University
Clemson, SC 29670, USA

VICKY C. JOHNSON
USDA Forest Service
Remote Sensing Applications Center
Salt Lake City, UT 84119, USA

HENRY M. LACHOWSKI
USDA Forest Service
Remote Sensing Applications Center
Salt Lake City, UT 84119, USA

LARRY A. LEEFERS
Department of Forestry
Michigan State University
East Lansing, MI 48823, USA

CHARLES A. LEVESQUE
Innovative Natural Resource Solutions
Antrim, NH 03440, USA

BAI-LIAN LI
Department of Biology
University of New Mexico
Albuquerque, NM 87131, USA

S. S. LIGHT
Environment and Agriculture Program
Institute for Agriculture and Trade Policy
Minneapolis, MN 55404, USA

EUNICE A. PADLEY
USDA Forest Service
Eastern Region
Milwaukee, WI 53203, USA

N. LEROY POFF
Department of Biology
Colorado State University
Fort Collins, CO 80523-1858, USA

WALTER D. POTTER
The University of Georgia
Artificial Intelligence Programs
Graduate Studies Research Center
Athens, GA 30602-7415, USA

H. MICHAEL RAUSCHER
USDA Forest Service
Southern Research Station
Bent Creek Research and Demonstration Forest
Asheville, NC 28805, USA

ROLAND L. REDMOND
Wildlife Spatial Analysis Lab
Montana Cooperative Wildlife Research Unit
The University of Montana
Missoula, MT 59812, USA

KIMBERLY ROLLINS
Department of Agricultural Economics and Business
University of Guelph
Guelph, ON, Canada, N1G 2W1

JAMES V. SCHUMACHER
Wildlife Spatial Analysis Lab
Montana Cooperative Wildlife Research Unit
The University of Montana
Missoula, MT 59812, USA

D. SCOTT SLOCOMBE
Department of Geography and Environmental Studies
Wilfrid Laurier University
Waterloo, ON, Canada, N2L 3C5

BRIAN M. STEELE
Department of Mathematical Sciences
The University of Montana
Missoula, MT 59812, USA

THOMAS J. STOHLGREN
Midcontinent Ecological Science Center
U.S. Geological Survey
Natural Resources Ecology Laboratory
Colorado State University
Fort Collins, CO 80523-1499, USA

S.J. TURNER
Department of Biological Sciences
St. Cloud State University
St. Cloud, MN 56301-4498, USA

CHARLES VAN SICKLE
19 Nottingham Drive
Candler, NC 28715, USA

J. MICHAEL VASIEVICH
USDA Forest Service
North Central Research Station
East Lansing, MI 48823, USA

STEPHANIE P. WILDS
Blue Star Consulting
Black Mountain, NC 28711, USA

PETER S. WHITE
Department of Biology
University of North Carolina at Chapel Hill
Chapel Hill, NC 27599-3280, USA

Introduction

Mark E. Jensen and Patrick S. Bourgeron

Ecological Assessments: What Are They?

Ecological assessments are a critical component of land management planning and regulatory decision making. Their scope and nature commonly differ due to the issues addressed; the discipline, agency, or audience involved; and the associated legislative or regulatory requirements (Lessard et al., 1999). In the United States, for example, the National Environmental Policy Act (NEPA) of 1969 directed that all federal lands be managed to "encourage productive and enjoyable harmony between man and his environment; to promote efforts which will prevent or eliminate damage to the environment and biosphere and stimulate the health and welfare of man; (and) to enrich understanding of the ecological systems and natural resources important to the Nation." To ensure compliance with these objectives, NEPA dictated that the environmental consequences associated with proposed management of federal lands be fully disclosed to the general public through appropriate ecological assessment documents (e.g., environmental impact assessments). In a similar manner, over 40 other countries have legislated the use of some type of ecological assessment as a prerequisite for effective environmental planning and land management (Robinson, 1992; Treweek, 1999).

In the United States, most ecological assessments have as their purpose either regulatory decision making or land management planning. Ecological risk assessment (U.S. Environmental Protection Agency, 1992, 1998; National Research Council, 1994) is a specific category of ecological assessment designed to meet regulatory mandates. Ecological risk assessment focuses on the relationships among one or more stressors and associated ecological effects to assist environmental regulatory decision making. These assessments are usually geographically restricted and their scope narrowly defined to satisfy the regulations(s) that triggered them; however, they also may be conducted at broader levels dependent on analysis needs (Hunsaker et al., 1990; Graham et al., 1991).

In contrast to regulatory-based ecological assessments, those based on land management planning objectives have traditionally followed methods of land evaluation that emphasize fundamental questions such as what is there, where is it, when is it present, and how does it function in determining optimal land use and required management actions for an assessment area (Klingebiel and Montgomery, 1961; Beek and Bannema, 1972; Olson, 1974; Food and Agriculture Organization, 1976, 1993; Zonneveld, 1988). Ecological assessments for land management planning are often tactical (project level) or strategic (regional or subregional level) in scope (Treweek, 1999). To date, most ecological assessments within the United States have been performed primarily at the project level (following NEPA guidelines) to evaluate environmental consequences associated with individual proposals or actions.

Because most project-level ecological assessments are limited in spatial scale and address a reduced set of issues, they often fail to meet the intent of NEPA (e.g., full disclosure of cumulative effects, relation of proposed action to other resources). This situation has placed increased importance on the development of strategic regional or subregional ecological assessments that provide context for more localized project-level assessments (Ortolano and Shepherd, 1995). Such strategic ecological assessments also provide a framework for proactive assessments of the impact of human activities and environmental changes at broad spatial scales and the placement of site-specific land-use impacts within an appropriate regional or national context (Thĕrivel et al., 1992; Thĕrivel and Rosario Pǎrtidorio, 1996; Thĕrivel and Thompson, 1996; Kessler, 1997).

Although ecological assessments can be conducted at a variety of spatial scales and may address one or many issues, the main intent of this book is to describe topics with particular importance to strategic integrated ecological assessments at a regional or subregional scale. In this introduction, we begin our discussion with a brief overview of the relation of ecological assessments to ecosystem management. In Section 2 of this introduction, we place emphasis on key ecological principles of ecosystem management and their implications to the assessment process. In Section 3, we present a general planning model and describe how ecological assessments are used in adaptive resource planning. In Section 4, we take the reader on a tour of the book. First we describe our logic for organizing chapters into generic issue group sections and related topics. We then present a general framework for ecological assessments to assist the reader in understanding how the chapters in this book relate to the procedural steps that should be considered in any integrated ecological assessment.

Relation of Ecological Assessments to Ecosystem Management

Ecosystem management is an evolving philosophy that many government agencies have adopted in the multiple-use, sustained-yield management of federal lands. The primary objective of this philosophy is to sustain the integrity of ecosystems for future generations while providing immediate goods and services to an increasingly diverse public. This objective can best be achieved through broad-level integrated ecological assessments that describe the maintenance or development of landscape patterns and processes that meet societal expectations within the limits of the land's ecological potentials. Landscape ecology, conservation biology, and adaptive management principles are critical components of this philosophy (Jensen et al., 1996; Lessard et al., 1999). In the following discussion, we review key concepts and objectives of ecosystem management to illustrate some of the underlying components that integrated ecological assessments must address if they are to have broad application to most land management or regulatory agencies.

There are about as many definitions of ecosystem management as there are groups that advocate its use (or limitations) as a basis for land-use planning (Slocombe, 1993; Christensen et al., 1996; Samson and Knoph, 1996; Boyce and Haney, 1997). According to Wagner (1995), "Definers commonly couch their definitions in terms of their own values, and it is the differences in individuals and group values that produce the definitional differences." For example, members of the environmental community commonly emphasize the values of preserving biodiversity, ecosystem health and integrity, and sustainability in their definitions of ecosystem management (e.g., Grumbine, 1994; Wilcove and Blair, 1995). Advocates of sustainable resource flows from public lands (e.g., industry) generally acknowledge the merits of these values; however, they also emphasize that the goals and objectives of ecosystem management must include economic and social needs because humans are also part of the ecosystem (Slocombe, 1993; Heissenbuttel, 1995).

Wagner (1995) suggests that ecosystem management is the "skillful manipulation of ecosystems to satisfy specified societal values." This definition is useful because it does not imply that optimization of biodiversity, ecosystem health and integrity, and commodity production be included in every ecosystem management effort. Instead, these values are articulated on a case by case basis, given the planning objectives and the laws and regulations that apply to a specific planning area. Management goals, therefore, determine the ecosystem(s) to be managed (i.e., the appropriate planning area), the goods and services, and the desired conditions (e.g., biodiversity, health and integrity, sustainability) of an ecosystem (Wagner, 1995).

Regardless of the specific definition of ecosystem management, three basic ecological principles should be considered in any ecological assessment (Haynes et al., 1996; Lessard et al., 1999; Bourgeron et al., 2001).

1. *Hierarchy theory:* Hierarchy theory provides a framework for characterizing ecosystem components and their linkages among different scales of ecological organization (O'Neill and King, 1998; Peterson et al., 1998). In multiscaled systems with hierarchical organization, higher levels provide the context or environment within which lower levels evolve; in other words, higher levels constrain the "behavior" of lower levels (Allen and Starr, 1982). In aquatic systems, for example, the characteristics of in-stream channel features are largely determined by the geoclimatic properties of the valley bottom and watershed settings within which they are nested (Maxwell et al., 1995). Knowledge of hierarchical scaling is important to the monitoring and characterization of ecological systems.

One major consequence of the hierarchical organization of ecological systems is that nonequilibrium dynamics or spatial heterogeneity at one scale can be translated into equilibrium at a higher level (O'Neill et al., 1986; Urban et al., 1987; Levin, 1992). Accordingly, landscape patterns persist within a hierarchical framework because a pattern may be stable at one level and not at another (Rahel, 1990). Therefore, ecological patterns should be analyzed at more than one spatial scale, and ecological assessments need to consider all appropriate levels of ecological organization (Urban et al., 1987; Baker, 1992; Levin, 1992).

2. *System dynamics:* Assessments of the spatial relations of ecosystem patterns cannot ignore the temporal dynamics of the processes that create them. It may be convenient for some modeling and planning efforts to assume equilibrium at a given spatial scale; however, it should be recognized that ecosystem dynamics may include succession along multiple pathways, discontinuities, and unexpected changes (called *surprise events* by Holling, 1978) and that ecosystem patterns and processes follow constantly changing environmental conditions (Holling, 1986; Kay, 1991; Constanza et al., 1993). Accordingly, ecological assessments must consider the complexity of ecosystem dynamics, and analysis protocols should be designed to account for this complexity.

3. *Limits to predictability:* Ecosystems exhibit complex nonlinear dynamics, hierarchical structuring, and multiple optimal operating points (most far from equilibrium), raising important questions about the predictability of ecosystem behavior (Costanza et al., 1993). Furthermore, predictability varies over spatial and temporal scales, with the characteristics of some events being predictable at one scale, but not at another. In some cases, a small fluctuation in an ecosystem component can lead to a large change in the functioning of the system as a whole, a process known as chaos (May, 1976). Because of the inherent limitations associated with predicting ecosystem behavior, an adaptive management process (Walters, 1986; Walters and Holling, 1990) should be used in implementing ecosystem management through integrated ecological assessments (Everett et al., 1994).

Adaptive ecosystem management assumes that uncertainties in current and future ecological conditions and societal values and expectations limit our ability to develop long-term ecosystem management strategies. Uncertainties arise from variability in biological and social systems, errors of estimation, bias in projection models, and ambiguous questions (Marcot, 1986). According to Walters and Holling (1990), "not only is the science incomplete, the ecosystem itself is a moving target, evolving because of the impacts of management, and the progressive expansion of the scale of human influences on the planet." For example, historical changes in management emphasis of Pacific Northwest forests—from protection of water rights and yields, to livestock grazing, to timber harvest, to threatened and endangered species, and, currently, to sustainable ecosystems—clearly demonstrate the transient nature of the preferences of society in the use of National Forest lands (Everett et al., 1994). Consequently, it is critical that an adaptive management approach be used in both ecological assessment and ultimately in land-use planning decision-making processes. A key feature of this approach is that partnerships must be developed between society, public land managers, resource management professionals, and scientists in the continual reshaping of management goals, objectives, and actions in response to changing socioeconomic information and evolving biological, physical, and societal conditions.

Relation of Ecological Assessments to Adaptive Resource Planning

Adaptive resource planning involves four basic components: monitoring, assessment, decisions, and implementation (Figure 1). In this section we present a general planning model for adaptive resource planning following the conceptual framework of Haynes et al. (1996) and principles of ecosystem management as articulated by various authors in Jensen and Bourgeron (1994).

Monitoring is the cornerstone of adaptive resource planning because it facilitates the identification of whether new ecological assessments, decisions, or strategies for implementation need to be made to existing regulatory or land-use planning documents (Figure 1). There are two distinct phases in a monitoring and evaluation program. Traditionally, the first phase monitors and evaluates the consequences of management actions and the progression toward desired future ecological conditions as described in land management decision documents. Noss and Cooperrider (1994) identify four types of monitoring that commonly fall under

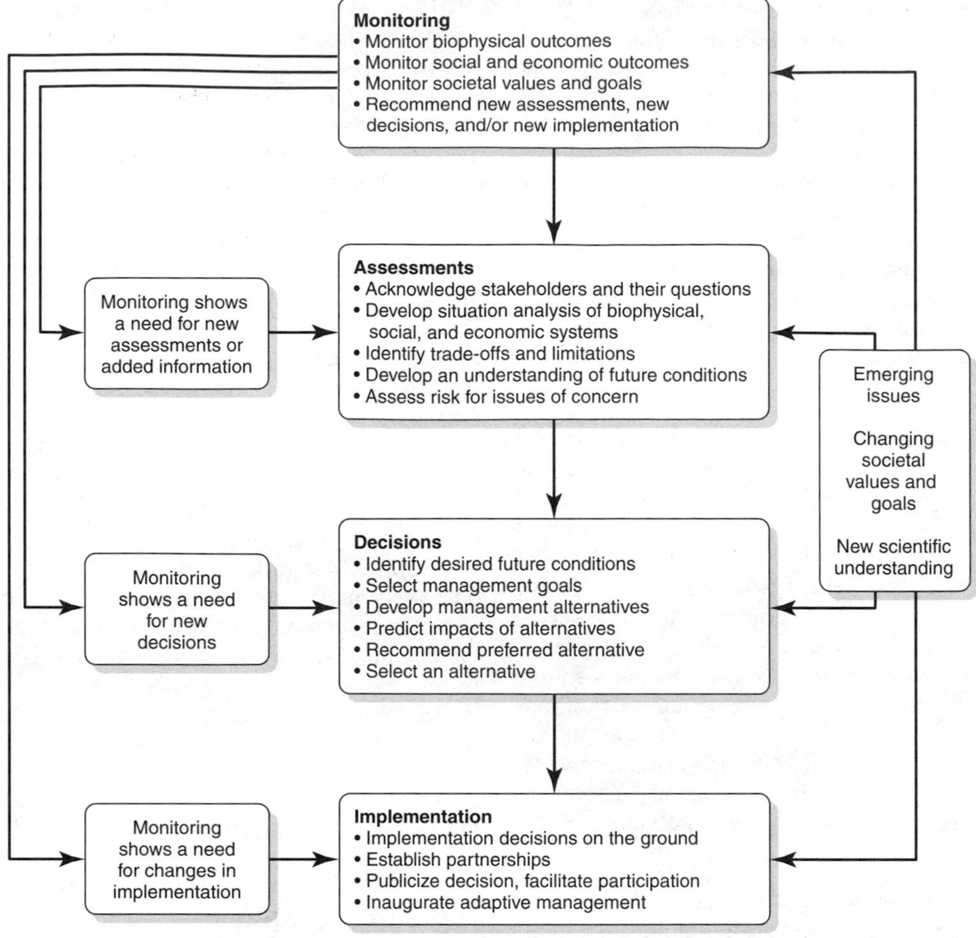

FIGURE 1. A general planning model for adaptive resource management (Haynes et al., 1996).

this phase: (1) implementation monitoring, which determines if a planned activity was accomplished, (2) effectiveness monitoring, which evaluates if an activity achieved its stated goal or objective, (3) validation monitoring, which determines to what degree assumptions and projections used in developing a decision document or ecological assessment are correct, and (4) baseline monitoring, which measures ecosystem patterns and processes that may be affected by management activities.

The second and less traditional phase of monitoring evaluates social needs in relation to changing societal values, new scientific understanding, and emerging issues (Figure 1). In this phase of monitoring, social values and needs are constantly assessed in relation to the validity of stated desired future ecological conditions. Such monitoring is critical, because societal values often change more rapidly than ecological conditions. Additionally, this phase of monitoring facilitates the incorporation of new knowledge into adaptive resource planning and the testing and modification of previous assumptions.

Integrated ecological assessments provide a much needed broader-level context for regulatory or land management planning decision documents (e.g., NEPA project-level environmental assessments; USDA, Forest Service forest plans; and USDI, Bureau of Land Management area plans). They are commonly conducted at regional or subregional levels to address issues that are not appropriately handled at finer assessment levels (e.g., site or landscape). Examples of such issues often include the viability of broadly distributed species, such as neotropical birds, grizzly bears, and salmonids; maintenance of mature forest structure (i.e., old growth); and maintenance of regional economies and societal landscape values.

Integrated ecological assessments are commonly conducted when major land management or regulatory decisions need to be reevaluated. The need for an ecological assessment is identified through ongoing monitoring efforts (Figure 1) based on one or more of the following criteria: (1) initial assumptions change as new knowledge is obtained, (2) social values and needs change, and (3) significant unexpected change occurs in important ecosystem patterns or processes (Lessard et al., 1999).

To best answer critical land management or regulatory questions, integrated ecological assessments often involve comprehensive descriptions of an area's past, present, and potential future conditions. They synthesize our knowledge of ecological systems by describing the biophysical and social limits of a system, the interrelation of its components, and the uncertainty and assumptions that underlie a given assessment effort. For federal land managers, ecological assessments are not decision documents, because they do not resolve issues or provide direct answers to specific problems. Instead, they provide the foundation for proposed additions or changes to existing land management direction, as well as the necessary information for policy discussions and decisions (Haynes et al., 1996).

The decision-making component of adaptive resource management (Figure 1) typically involves the articulation of a set of desired ecological conditions for an assessment area, evaluation of different alternatives for achieving desired conditions, and the selection of the preferred alternative. Such decisions are documented in appropriate NEPA documents, such as environmental impact statements (for proposed actions with significant environmental effects) and environmental assessments (for proposed actions with nonsignificant environmental effects). Although the term *condition* implies a static point in time, in reality, ecosystems are dynamic. Therefore, the desired ecological conditions (or more commonly the desired future conditions) for an assessment area must be described in measurable terms across multiple time periods. The use of ecological modeling and scenario planning in integrated ecological assessments can greatly reduce the complexity of describing desired future conditions in the decision-making process (Haynes et al., 1996; Lessard et al., 1999; Bourgeron et al., 2001).

Implementation is the process of turning plans and decisions into projects and practices on the ground, establishing mutual learning experiences for all involved groups, and realizing the projected outcomes of planning (Haynes et al., 1996). This component of adaptive resource management (Figure 1) is where inputs are transformed to outputs and where communities dependent on the flow of goods and services place emphasis (in other words, where "real work" occurs). Accordingly, principles of adaptive management must be utilized in this step of the planning process to ensure that decisions are informed; that they gain understanding, acceptance, and support by a wide audience; that they recognize their inherent uncertainty; and that they are capable of adjustment in the face of surprise.

A Tour of the Book

The chapters in this book range from overviews of basic ecological principles, to suggestions concerning ecosystem characterization and analysis, to systematic reviews of selected case studies. Our initial objective in designing this book was to provide a "how to" description of protocols for integrated ecological assessments. As we progressed in our efforts, however, it rapidly became apparent that there is no single correct set of protocol for ecological assessments; the diversity of issues, spatial scales, and ecosystem patterns and process relations that they must address is simply too great. Accordingly, we have organized this book into a more generic set of issues (sections) and topics (chapters) common to most integrated ecological assessments. In this respect, our book provides both theoretical and practical advice for future ecological assessments given specific land-use planning objectives. In this portion of the introduction, we first take the reader on a quick tour of the book to explain how we have ordered the chapters by topical sections. In the second part, we present a general procedural framework for integrated ecological assessments, with notes on how the chapters of this book fit the framework.

This book is comprised of eight sections and 35 chapters. In Section A (Basic Principles), we present four chapters that describe the fundamental principles of landscape ecology (Naveh and Lieberman, 1984; Forman and Godron, 1986; Golley, 1994), conservation biology (Hunter, 1991; Bourgeron and Jensen, 1994), and adaptive management (Holling, 1978; Walters, 1986; Walters and Holling, 1990; Everett et al., 1994) that should be considered in any integrated ecological assessment effort. These chapters emphasize the importance of hierarchy theory, system dynamics, and limits to ecosystem predictability to integrated ecological assessments and offer practical solutions as to how they may be addressed in future assessment efforts.

Section B (Information Management) contains five chapters that describe key components of information management in the ecological assessment process. These components include data acquisition (Chapter 5), sampling design (Chapters 6 and 7), data storage and maintenance (Chapter 8), and the integration of ecological assessment information (Chapter 9).

In Section C (Basic Technologies), we present three chapters that describe emerging technologies critical to integrated ecological assessments. These technologies include remote sensing (Chapter 10), geographic information systems (Chapter 11), and knowledge-based systems (Chapter 12). We believe these chapters are appropriately placed in this section of the book, because they represent fundamental technologies that are commonly utilized in most ecological analysis and characterization efforts described later in the book.

Basic concepts regarding analytical methods commonly used in ecological assessments are provided in Section D (Analytical Methods: Concepts). In this section we provide five chapters (13 through 17) that describe theoretical considerations that should be considered in the evaluation of ecological assessment data. In Section E (Analytical Methods: Applications), we provide four chapters (18 through 21) that provide practical advice concerning the implementation of suggestions presented in section D.

In Section F (Ecosystem Characterization), we present eight chapters that describe various methods and techniques that should be considered in describing the current conditions of terrestrial, aquatic, and human resources. These chapters cover three basic categories of information important to any ecosystem characterization effort: biophysical environment, ecological process, and dynamic ecosystem pattern description. Chapters 22, 24, and 27 provide suggestions concerning the identification and mapping of biophysical environments important to terrestrial, aquatic, and human resources, respectively. Ecological processes that influence terrestrial, aquatic, and human resource patterns are described in Chapters 23, 25, and 28, and the techniques for characterizing these resource patterns are provided in Chapters 23, 26, and 29.

In Section G (Case Studies), we present five chapters that describe various integrated ecological assessment efforts recently completed within the United States. These case studies range from relatively simple assessments with limited financial investments (Chapters 31 and 33) to very complex assessments with large financial investments (Chapters 30 and 34). These chapters were developed to suggest to the reader how the information contained in Sections A through F might actually be integrated into ecological assessments of varying scope, spatial extent, and complexity. To ensure that this objective was met, a similar format of topics was addressed in each chapter of this section. These topics include description of the process used in identifying the issues that were addressed; the strategy employed in creating partnerships and working groups; the scale, extent, and resolution of the assessment; information management; approaches used in ecosystem characterization; data analysis methods; techniques used in data integration and reporting; results of the assessment; implementation strategies; lessons learned from the assessment; and suggestions for future work.

The final section of this book (Implementation Strategies) contains one chapter that addresses various political, institutional, and social issues that need to be considered in the successful implementation of an integrated ecological assessment effort. Emphasis is placed on describing the role of ecological assessments in implementing ecosystem management or any other set of land-use objectives across mixed land ownerships and diverse public partnerships.

In the previous discussion, we described the manner in which we organized chapters of this book by topical sections. Integrated ecological assessments, however, often defy such convenient categorizations of issues (sections) and related topics (chapters). Therefore, we present a general framework for ecological assessment in Figure 2 to further assist the reader in understanding how the chapters of this book relate to basic procedural components of the assessment process. In the following discussion we describe these components, as well as the various book chapters that address them.

The first task in any integrated ecological assessment is to clearly define the scope of the assessment effort. This step is critical to successful completion of an assessment, because it establishes the nature, boundaries, and operational limits of the proposed effort. Key components of this task include identification of the issues that need to be addressed, the stakeholders that need to be involved, the proposed spatial boundaries of an assessment (which include integrated analysis units as well as

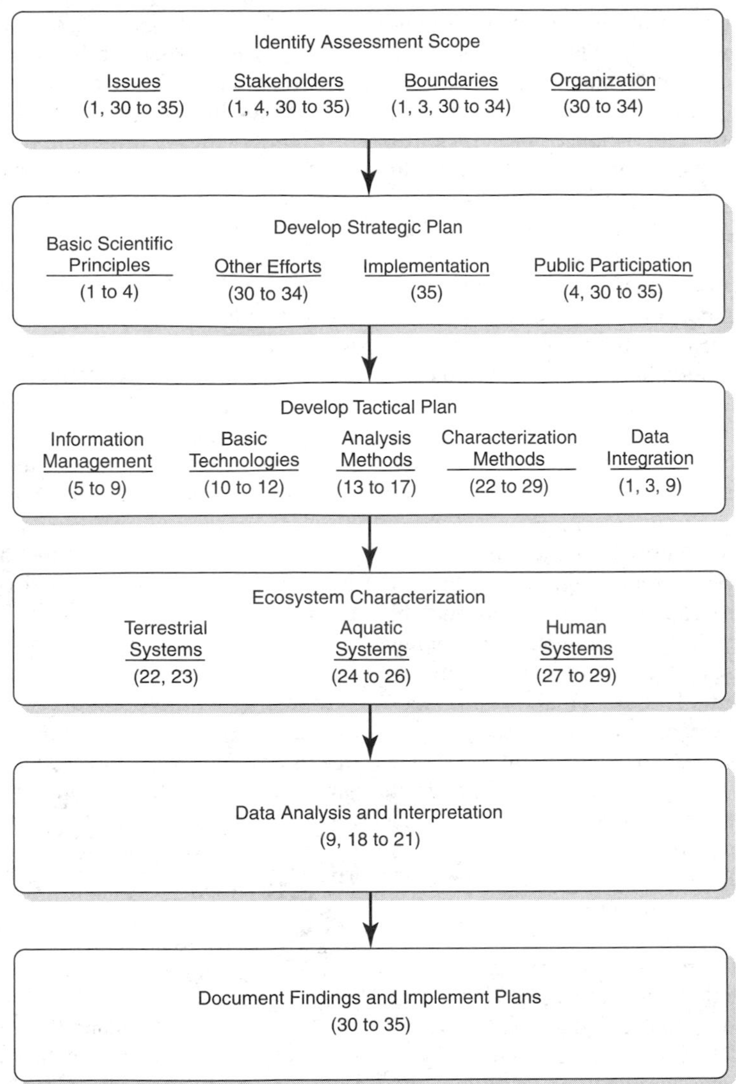

FIGURE 2. A procedural framework for integrated ecological assessments (numbers within parenthesis correspond to associated chapters in this book).

individual resource analysis areas), and an organization plan for group activities that considers appropriate technical expertise, administration, and budget. Paramount to any assessment effort is the concise articulation of the issues that need to be addressed (Chapters 1 and 30 through 35). Given the fact that most ecological assessment needs are identified by a relatively small group of organizations (e.g., federal land management or regulatory agencies), it is imperative that such groups consider the concerns of potentially affected interest groups or stakeholders early in the assessment process (Chapters 1, 4, and 30 through 35) to validate that the issues identified are appropriate to the general public. Once agreement on the basic issues to be addressed is reached, the spatial boundaries of the integrated assessment area (and common reporting units), as well as the spatial extent of resource-based characterization areas (which are often larger than the assessment area), need to be defined (Chapters 1, 3, and 30 through 34). The final step of this task involves the development of an organization plan for an assessment that describes the structure and roles of proposed resource area specialists and the general administrative and support needs (Chapters 30 through 34).

The second basic task of an integrated ecological assessment involves the development of a strategic plan that considers basic science principals, other related efforts, implementation, and public participation. The science principles of ecosystem management addressed previously in this discussion should be considered in any ecological assessment (Chapters 1 through 4). The challenge, therefore, is to implement these concepts into measurable terms, given the ecosystem patterns and processes considered to be important to a given assessment effort. Another key component of developing a strategic plan for an integrated ecological assessment involves a rigorous review of "lessons learned" from previous assessments. In Chapters 30 through 34, we present a variety of ecological assessment case study descriptions to suggest what did and did not work, given the scope of the project and the methods employed.

The primary purpose of an integrated ecological assessment is to provide context for more locally focused decision-making documents concerning land-use objectives and regulatory requirements (Haynes et al., 1996; Lessard et al., 1999). Accordingly, such assessments must consider the manner in which their findings will be implemented given federal, state, and local laws and regulations (Chapter 35). Additionally, the strategic plan of any ecological assessment should describe the approach that will be used to involve public participation in the assessment process (Chapters 4 and 30 through 35).

Once the scope of an assessment and a strategic plan for its completion are developed, a tactical plan is often required that considers information management, basic technologies, generic analysis methods, generic characterization methods, and data integration issues (Figure 2). Information management is often one of the most expensive (if not the most important) aspects of an ecological assessment. Included under this category are such issues as sampling design and appropriateness of data; assimilation of existing data sets; data storage, management, and dissemination; and integration of diverse data sources. We address each of these issues in Chapters 5 through 9 of this guidebook. A related topic to information management is basic technologies that have particular importance to ecological assessments. In Chapters 10 through 12, we present overviews concerning the use of remote sensing, geographic information systems, and decision support systems in ecological assessments to emphasize why these technologies should be considered in future analysis efforts.

A critical component of the tactical planning stage of any integrated ecological assessment involves an assessment of the analytical methods that should be used (Chapters 13 through 17) in the interpretation of basic information concerning terrestrial, aquatic, and human systems (Chapters 22 through 29). Additionally, plans should be developed that facilitate integration of different results from varied data sources (Chapters 1, 9, and 30 through 34).

The third and fourth steps of our general procedural framework for integrated ecological assessments (Figure 2) address the major work areas of an assessment: ecosystem characterization and analysis. For simplicity of presentation, we divided ecosystem characterization into three general groupings: terrestrial system techniques (Chapters 22 and 23), aquatic system techniques (Chapters 24 through 26), and human system techniques (Chapters 27 through 29). Recommendations concerning the analysis of this information are provided in Chapters 9 and 18 through 21.

The final step of our procedural framework for integrated ecological assessments involves the development of appropriate documents and implementation through the planning process. Papers that discuss these aspects of ecological assessments are provided in Chapters 30 through 35.

References

Allen, T. F. H.; Starr, T. B. 1982. *Hierarchy: perspectives for ecological complexity*. Chicago: University of Chicago Press.

Baker, W. L. 1992. The landscape ecology of large disturbances in the design and management of nature reserves. *Landscape Ecol.* 7:181–194.

Beek, K. J.; Bannema, J. 1972. *Land evaluation for agricultural land use planning—an ecological methodology*. Wageningen, The Netherlands: Department of Soil Sciences and Geology, Agricultural University.

Bourgeron, P. S.; Jensen, M. W. 1994. An overview of ecological principles for ecosystem management. In: Jensen, M. E.; Bourgeron, P. S., tech. eds. *Volume II: ecosystem management: principles and applications*. Gen. Tech. Rep. PNW-GTR-318. Portland, OR: U.S. Dept. Agric., For. Serv., Pacific Northw. Res. Sta.: 45–57.

Bourgeron, P. S.; Humphries, H. C.; Jensen, M. E.; Brown, B. A. (2001). Integrated regional ecological assessments and land use planning. In: Dale, V.; Haueber, R., eds. *Applying ecological principles to land management*. New York: Springer-Verlag.

Boyce, M. S.; Haney, A. 1997. *Ecosystem management: applications for sustainable forest and wildlife resources*. New Haven, CT: Yale University Press.

Christensen, N. L. 1996. The scientific basis for sustainable use of land. In: Diamond, H. L.; Noonan, P., eds. *The use of land.* Washington, DC: Island Press: 273–308.

Costanza, R.; Wainger, L.; Folke, C.; Maler, K. 1993. Modeling complex ecological economic systems. *BioScience* 43:545–555.

Everett, R.; Hessburg, P.; Jensen, M.; Bormann, B. 1994. *Volume 1: executive summary*. Gen. Tech. Rep. PNW-GTR-317. Portland, OR: U.S. Dept. Agric., For. Serv., Pacific Northw. Res. Sta.

Food and Agriculture Organization (FAO). 1976. *A framework for land evaluation*. 32. Rome: United Nations Food and Agriculture Organization.

Food and Agriculture Organization (FAO). 1993. *Guidelines for land use planning*. Rome: United Nations Food and Agriculture Organization.

Forman, R. T. T.; Godron, M. 1986. *Landscape ecology*. New York: John Wiley & Sons.

Golley, F. B. 1994. Development of landscape ecology and its relation to environmental management. In: Jensen, M. E.; Bourgeron, P. S., tech. eds. *Volume II: ecosystem management: principles and applications*. Gen. Tech. Rep. PNW-GTR-318. Portland, OR: U.S. Dept. Agric., For. Serv., Pacific Northw. Res. Sta.: 34–41.

Graham, R. L.; Hunsaker, C. T.; O'Neill, R. V.; Jackson, B. L. 1991. Ecological risk assessment at the regional scale. *Ecol. Appl.* 1:196–206.

Grumbine, R. E. 1994. What is ecosystem management? *Conserv. Biol.* 8:27–38.

Haynes, R. W.; Graham, R. T.; Quigley, T.M. 1996. *A framework for ecosystem management in the interior Columbia basin and portions of the Klamath and Great Basins*. Gen. Tech. Rep. PNW-GTR-374. Portland, OR: U.S. Dept. Agric., For. Serv., Pacific Northw. Res. Sta.

Heissenbuttel, A. E. 1995. Industry perspectives on ecosystem management: an evolution in forest science and policy. In: Everett, R. L.; Baumgartner, D. M., eds. *Ecosystem management in western interior forests*. Pullman, WA: Department of Natural Resource Sciences, Washington State University: 11–14.

Holling, C. S., editor. 1978. *Adaptive environmental assessment and management*. London: John Wiley & Sons.

Holling, C. S. 1986. The resilience of terrestrial ecosystems: local surprise and global change. In: Clark, W. M.; Munn, R. E., eds. *Sustainable development in the biosphere*. Oxford, UK: Oxford University Press: 292–320.

Hunsaker, C. T.; Graham, R. T.; Suter II, G. W.; O'Neill, R. V.; Barnthouse, L.W.; Gardner, R. H. 1990. Assessing ecological risk on a regional scale. *Environ. Manage*. 14:325–332.

Hunter, M. 1991. Coping with ignorance: the coarse-filter strategy for maintaining biodiversity. In: Kohn, K. A., ed. *Balancing on the brink of extinction—the endangered species act and lessons for the future*. Washington, DC: Island Press: 256–281.

Jensen, M. E.; Bourgeron, P. S., editors. 1994. *Volume II: ecosystem management: principles and applications*. Gen. Tech. Rep. PNW-GTR-318. Portland, OR: U.S. Dept. Agric., For. Serv., Pacific Northw. Res. Sta.

Jensen, M. E.; Bourgeron, P. S.; Everett, R.; Goodman, I. 1996. Ecosystem management: a landscape ecology perspective. *J. Amer. Water Resources Association* 32:208–216.

Kay, J. J. 1991. A nonequilibrium themodynamic framework for discussing ecosystem integrity. *Environ. Manage*. 15:483–495.

Kessler, J. J., editor. 1997. *Strategic environmental analysis, AIDEnvironment*. Amsterdam, The Netherlands: The Hague.

Klingebiel, A. A.; Montgomery, D. H. 1961. *Land capability classification*. Agricultural Handbook 210. Washington, DC: Soil Conservation Service, U.S. GPO.

Lessard, G.; Jensen, M.E.; Crespi, M.; Bourgeron, P.S. 1999. A general framework for integrated ecological assessments. In: Cordel, H. K.; Bergstrom, J. C., eds. *Integrating social sciences with ecosystem management: human dimensions in assessment policy and management*. Champaign–Urbana, IL: Sagamore: 35–60.

Levin, S. A. 1992. The problem of pattern and scale in ecology. *Ecology* 73:1942–1968.

Marcot, B. G. 1986. Concepts of risk analysis as applied to viable population assessment and planning. In: Wilcox, B. A.; Brussard, P. F.; Marcot, B. G., eds. *The management of viable populations: theory, applications, and case studies*. Standord, CA: Center for Conservation Biology.

Maxwell, J. R.; Edwards, C. J.; Jensen, M. E.; Paustian, S. J.; Parrott, H.; Hill, D. M. 1995. *A hierarchical framework of aquatic ecological units in North America*. NC-GTR-176. St. Paul, MN: U.S. Dept. Agric. For. Serv., North Central Exp. Sta.

May, R. 1976. Simple mathematical models with very complicated dynamics. *Nature* 261:459–467.

National Research Council. 1994. *Rangeland health: new methods to classify, inventory, and monitor rangelands*. Washington, DC: National Academy Press.

Naveh, Z.; Lieberman, A. 1984. *Landscape ecology: theory and application*. New York: Springer-Verlag.

Noss, R. F.; Cooperrider, A. Y. 1994. *Saving nature's legacy*. Washington, DC: Island Press.

Olson, G. W. 1974. *Land classifications*. Ithaca, NY: Cornell University.

O'Neill, R. V.; King, A. W. 1998. Homage to St. Michael; or, why are there so many books on scale? In: Peterson, D. L.; Parker, V. T., eds. *Ecological scale: theory and applications*. New York: Columbia University Press.

O'Neill, R. V.; DeAngelis, D. L.; Waide, J. B.; Allen, T. F. H. 1986. *A hierarchical concept of ecosystems*. Princeton, NJ: Princeton University Press.

Ortolano, L.; Shepherd, A. 1995. Environmental impact assessment. In: Vanclay, F.; Bronstein, D. A., eds. *Environmental and social impact assessment*. Chichester, UK: John Wiley & Sons: 3–31.

Peterson, G.; Allen, C. R.; Holling, C. S. 1998. Ecological resilience, biodiversity, and scale. *Ecosystems* 1:6–18.

Rahel, F. J. 1990. The hierarchical nature of community persistence: a problem of scale. *Amer. Naturalist* 136:328–344.

Robinson, N. 1992. International trends in environmental impact assessment. *Boston College Environ. Affairs Law Rev.* 19(3):591–621.

Samson, F. B.; Knopf, F. L. 1996. *Ecosystem management: selected readings.* New York: Springer-Verlag.

Slocombe, D. S. 1993. Implementing ecosystem-based management: development of theory, practice, and research for planning and managing a region. *BioScience* 4:612–622.

Thĕrivel, R.; Rosărio Partidărio, M. 1996. *The practice of strategic environmental assessment.* London: Earthscan.

Thĕrivel, R.; Thompson, S. 1996. *Strategic environmental assessment and nature conservation.* Peterborough, UK: English Nature.

Thĕrivel, R.; Wilson, E.; Thompson, S.; Heaney, D.; Pritchard, D. 1992. *Strategic environmental assessment.* London: Earthscan.

Treweek, J. 1999. *Ecological impact assessment.* Oxford, UK: Blackwell Science.

Urban, D. L.; O'Neill, R. V.; Shugart, H. H., Jr. 1987. Landscape ecology: a hierarchical perspective can help scientists understand spatial patterns. *BioScience* 37:119–127.

U.S. Environmental Protection Agency. 1992. *Framework for ecological risk assessment.* EPA/630/R-92/001. Washington, DC: Risk Assessment Forum, U.S. Environmental Protection Agency.

U.S. Environmental Protection Agency. 1998. *Guidelines for ecological risk assessment.* EPA/630/R-95/002F. Washington, DC: Risk Assessment Forum, U.S. Environmental Protection Agency.

Wagner, F. H. 1995. What have we learned? In: Wagner, F. H., ed. *Proceedings of the symposium: ecosystem management of natural resources in the Intermountain West,* April 20–22, 1994, Logan, Utah. Natural Resources and Environmental Issues, Logan, UT: College of Natural Resources, Utah State University, 5:121–125.

Walters, C. 1986. *Adaptive management of renewable resources.* New York: Macmillan.

Walters, C. J.; Holling, C. S. 1990. Large-scale management experiments and learning by doing. *Ecology.* 71(6):2060–2068.

Wilcove, D. S.; Blair, R. B. 1995. The ecosystem management bandwagon. *Tree* 10:345.

Zonneveld, I. S. 1988. Basic principles of land evaluation using vegetation and other land attributes. In: Kuchler, A. W.; Zonneveld, I. S., eds. *Vegetation mapping.* Boston: Kluwer Academic Publishers: 499–517.

Part 1
Basic Principles

1
An Overview of Ecological Assessment Principles and Applications

Mark E. Jensen, Norman L. Christensen, Jr., and Patrick S. Bourgeron

1.1 Introduction

Ecological assessments facilitate understanding of an area's past, present, and future conditions through comprehensive description of ecosystem patterns, processes, and functions (Lessard et al., 1999). They synthesize our knowledge of ecological systems and commonly describe the biophysical and social limits of a system, the interrelations of its ecosystem components, and the uncertainties and assumptions that underlie a given assessment effort. Ecological assessments are not decision documents because they do not resolve issues or provide direct solutions to specific policy questions. Instead, they provide the foundation for proposed additions or changes to existing land management plans or regulatory policies and are a critical component for implementing principles of ecosystem management in land management planning (Grossarth and Nygren, 1994; Morrison, 1994; Haynes et al., 1996).

Over the years, many different definitions and approaches have been used in ecological assessment efforts, depending on the discipline, agency, or audience involved and the associated legislative or regulatory requirements (Lessard et al., 1999). For example, earlier regulatory-based efforts in the United States commonly followed traditional risk assessment methodologies (National Research Council, 1983) and often focused on single resources and stressors over limited geographic scales. Over the past decade, these have evolved into *ecological risk assessments* (Risk Assessment Forum, 1992; Suter, 1993) that typically include a wider consideration of the risks to humans and ecosystem integrity associated with multiple stressors. These types of assessment are either predictive, focusing on a particular stressor and estimating possible future effects, or retrospective, focusing on a particular effect and subsequently identifying stressors that might have contributed to the effect (Rapport, 1992; Suter, 1993).

In contrast to regulatory-based ecological assessments, those based on land-use planning have largely followed traditional methods of land evaluation that hinge on fundamental questions (What is there? Where is it? When is it present? How does it function?) in determining optimal land uses and required management actions for an assessment area (Klingebiel and Montgomery, 1961; Beek and Bennema, 1972; Olson, 1974; Food and Agriculture Organization, 1976; Zonneveld, 1988).

Most government agencies collect different information (and at different spatial scales) in their ecological assessment efforts, largely because a clearly defined, unifying purpose for such information is lacking. Ecosystem management (Slocombe, 1993; Grumbine, 1994; Jensen and Everett, 1994; Keystone Center, 1996) provides this unifying purpose for ecological assessments: it represents a management philosophy that seeks to integrate scientific knowledge of ecological relationships within a complex sociopolitical and values framework toward the general goal of protecting native ecosystem integrity over the long term (Grumbine, 1994). Accordingly, we emphasize the linkages between ecosystem management philosophies (and related scientific principles) and ecological assessments throughout this chapter.

First, we present a brief overview of the evolution of ecosystem management philosophies within the United States. Goals and implementation strategies for ecosystem management are discussed and their relations to ecological assessments are briefly described. The basic ecological principles that support an ecosystem-based approach to land management are presented next, with particular emphasis on how the concepts of system dynamics, hierarchy theory, thresholds, and predictability apply to multiscale ecological assessments. The majority of this chapter is devoted to describing the basic properties of an ecological assessment that should be considered when such an assessment is used to implement ecosystem management objectives. We present a brief overview of the relation of purpose, issues, spatial scale, information needs, and analysis methodologies to ecological assessments in this section. This chapter concludes with a general discussion of the importance of information management to ecological assessments and suggested linkages between assessments and adaptive management. Recommendations concerning future research and management needs for ecological assessments are also provided.

1.2 Ecosystem Management and Its Relation to Ecological Assessments

1.2.1 Evolving Philosophies of Ecosystem Management

Although interest in and support for ecosystem management are relatively recent, many of its directives are deeply rooted in the work of resource managers and ecologists early in this century. Concerns regarding the sustainability of our natural resources are expressed in the writings of Marsh, Pinchot, and Clements. Certainly, sustainability is a core objective in so-called sustained yield management for forests, fisheries, and wildlife first promulgated in the United States over a century ago. That sustainability should be the overarching goal of land management was the central thesis of Aldo Leopold's "Land Ethic" (Leopold, 1949).

During the first six decades of this century, natural resource managers defined sustainable management within the context of ecological paradigms, such as equilibrium population dynamics and directional succession to stable climax communities. Even today, in the face of much contrary data, many models for sustained yield of wood fiber, wildlife, and fish stocks are based on the supposition that natural processes regulating competition, predation, and material cycling will produce stable equilibrium populations and communities. Furthermore, ecologists presumed that most of these natural processes operated at relatively small spatial and temporal scales and that the effects of episodic disturbances and climatic fluctuations could be discounted (Christensen, 1988).

This century began with a human population slightly over 1 billion. Five decades later, our numbers had barely doubled, and we were just beginning to appreciate the pervasiveness of our activities and their impacts on our globe. During most of this period, the markets for natural resources such as wood and wildlife were largely local and regional. Our transportation and communication systems certainly had not achieved their current continental and global extent. Thus, it could be argued that the need to include humans in resource management processes (as in protocols for ecosystem management; see below) was not nearly so great nor did it have the same meaning as today.

Ecological science during the second half of this century has taught us that notions of directional change and stable equilibria are, at best, naïve and, at worst, wrong-headed. Ecosystems are being influenced by periodic and stochastic events operating at a variety of time scales, and the behavior of critical elements at particular locations is heavily influenced by the character of the surrounding landscape. Furthermore, we have discovered that seemingly sustainable activities occurring at individual locations can accumulate across watersheds and landscapes to produce unsustainable outcomes in terms of resource production and collateral environmental impacts.

As we approach the new millennium, our numbers are nearing 6 billion. Humans now use or influence over 50% of Earth's primary production. Even the most remote and wild regions of Earth are now influenced by human activities. Perhaps more important from a management perspective, global communication is virtually instantaneous, and markets for most natural resources are truly global as well. Including and involving humans in natural resource management is not only more compelling, but also a far more complex and daunting challenge than it was during the first half of this century.

During the past several decades, a number of legislative initiatives have attempted to deal with these changing management challenges. For example, the Multiple Use Sustained Yield Act of 1960 has been the centerpiece of management in the USDA Forest Service for nearly four decades. Like ecosystem management, the intent was to reconcile conflicting values and demands on public lands. However, some would argue that, instead, the effect was that of institutionalizing such conflicts or that of setting a higher priority on conflict resolution than on sustainability (Fedkiw, 1999).

The 1969 National Environmental Policy Act (NEPA) fully acknowledges the tension between growing demands for natural resources and the ability to sustain these resources. Its goal is to "encourage productive and enjoyable harmony between man and his environment; to promote efforts which will prevent or eliminate damage to the environment and biosphere and stimulate the health and welfare of man; to enrich the understanding of the ecological systems and natural resources important to the nation." NEPA only applies to federal lands and actions significantly affecting the environment, in contrast to the 1973 Endangered Species Act (ESA), which protects threatened and endangered species on all lands and management activities. Although the ESA does acknowledge the importance of ecosystems, in reality, efforts have focused on management strategies for individual species, rather than the ecosystems that they inhabit. Ecosystem management has been advocated by many as a means of reconciling public conflict over the management of endangered species (e.g., Diamond and Noonan, 1987).

Formalization of the notion of ecosystem management began about ten years ago with the publication of a set of papers on park and wilderness management edited by Agee and Johnson (1988). Since then, an array of publications have attempted to clarify and define this concept (e.g., Overbay, 1992; Forest Ecosystem Management Assessment Team, 1993; Slocombe, 1993; Society of American Foresters, 1993; Eastside Forest Health Assessment Team, 1994 [Everett et al., 1994]; Grumbine, 1994; Jensen and Bourgeron, 1994; Wood, 1994; Christensen, 1996; Christensen and Franklin, 1996; Christensen et al., 1996; Boyce and Haney, 1997). Samson and Knopf (1996) have assembled an anthology of literature relevant to this issue.

Some have argued that the phrase "ecosystem management" is unfortunate because we do not manage ecosystems per se. Rather, our objectives and goals are inevitably oriented toward particular elements (e.g., species) or processes (e.g., hydrologic management) and, more often than not, our protocols involve managing the actions of humans. A more apt but cumbersome descriptor might be "management in an ecosystem context"—management that acknowledges the dynamic and spatially open character of ecosystems and the inevitability and complexity of humans as ecosystem components.

1.2.2 Goals of Ecosystem Management

Most government agencies (e.g., USDA Forest Service, USDI Bureau of Land Management) have adopted ecosystem management in their multiple-use, sustained-yield management of federal lands. The primary objective of this philosophy is to sustain the integrity of ecosystems (i.e., their function, composition, and structure) for future generations while providing immediate goods and services to an increasingly diverse public (Jensen and Everett, 1994). Grumbine (1994), through his review of the literature on ecosystem management, suggests that five goals are commonly recognized as essential to sustaining ecological integrity (Table 1.1). To realize these goals, we must integrate information on the land's ecological potential (goals 1 to 4) and on human values, technology, and economics (goal 5). The Keystone Center's National Policy Dialogue Group on Ecosystem Management, which conducted an extensive review and synthesis of regional ecosystem management initiatives across the United States (Keystone, 1996), also identified five central goals for ecosystem management (Table 1.1). Both sets of goals expressed similar intent, but with a different bent. Whereas Grumbine (1994) focused almost exclusively on ecological objectives, Keystone (1996) made the role of humans

TABLE 1.1. Goals of ecosystem management as stated by Grumbine (1994) and Keystone Center (1996).

Goal	Grumbine (1994)	Keystone Center (1996)
1	Maintain viable populations of all native species in situ.	Maintain ecosystem integrity.
2	Represent, within protected areas, all native ecosystem types across their natural range of variation.	Sustain biodiversity and ecosystem processes at a regional scale.
3	Maintain evolutionary and ecological processes (i.e., disturbance regimes, hydrological processes, nutrient cycles, etc.).	Sustain vibrant, livable, and economically diverse human communities.
4	Manage over periods of time long enough to maintain the evolutionary potential of species and ecosystems.	Incorporate distinct community and stakeholder values in the design and implementation of ecosystem management initiatives.
5	Accommodate human use and occupancy within these constraints.	Integrate the ecological, economic, and social goals of stakeholders in an ecosystem.

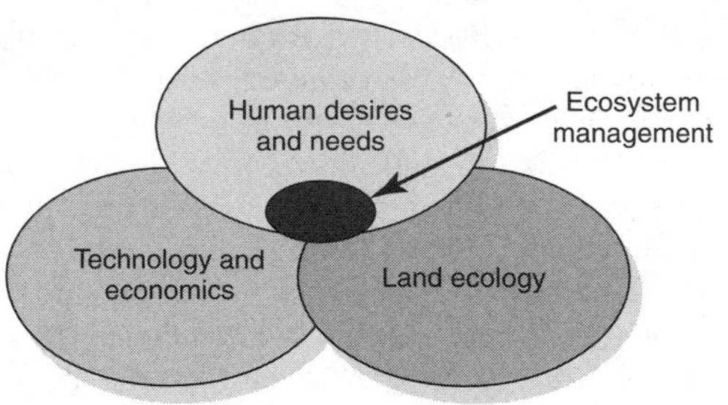

FIGURE 1.1. A conceptual framework for ecosystem management based on landscape ecology (Jensen et al., 1996).

more explicit. Accordingly, ecosystem management may be viewed as the optimal integration of societal desires and needs, ecological potential of the landscape, and economic plus technological considerations (Figure 1.1; Jensen et al., 1996).

1.2.3 Implementation of Ecosystem Management Goals through an Ecological Assessment Process

The Keystone Dialogue Group on Ecosystem Management (Keystone, 1996) identified eight steps that are critical to successful implementation of ecosystem management goals, and indeed to nearly any project. These include (1) recognizing and defining the problems or opportunities; (2) delineating boundaries; (3) identifying and involving participants; (4) establishing a common vision; (5) assessing ecological, economic, and social constraints and opportunities; (6) acquiring funding; (7) making decisions and implementing solutions; and (8) monitoring progress, evaluating impacts, and adapting based on new information.

In adaptive resource planning, ecological assessments are used to develop characterizations of the affected environment, as well as the purpose and need for action statements as required under NEPA (see Chapter 4). Because ecological assessments provide the foundation for subsequent decision making, the first six steps outlined by Keystone (1996) apply to their construction as well. In particular, the assessment of ecological, economic, and social constraints and opportunities is of obvious relevance. The Keystone Dialogue Group suggests that "Ideally, an assessment should consider (1) ecological, economic, and social structures, components, processes and functional linkages to other ecosystems; (2) current ecological, economic, legal, and social conditions and trends; (3) the conditions necessary to sustain ecosystem integrity; and (4) the effects of human activities on the ecosystem."

Slocombe (1993) further elaborates on these points in his review of strategies for implementing ecosystem management. Specifically, he suggests that the following points need to be considered in ecological assessments to support ecosystem management:

1. Recognition of goals and active management orientation.
2. Incorporation of stakeholder and institutional factors in the analysis.
3. Identification of people and their activities in the ecosystem considered.
4. Description of parts, systems, environments and their interactions.

TABLE 1.2. Questions regarding ecosystem sustainability to be answered by integrated ecological assessments.

- Do elements of structural diversity occur in conditions, amounts, and geographic patterns that contribute to natural species richness, ecological functions, and overall biological diversity (Keystone Center, 1991)?
- What are the patterns of diversity in nature, and what are their critical ecological and evolutionary determinants (Lubchenco et al., 1991)?
- How are the fundamental patterns and processes (e.g., connectivity and dispersion of habitats, disturbance recovery processes, and movement of individuals) that operate within each regional ecosystem maintained (Keystone Center, 1991)?
- How does fragmentation of the landscape affect the spread and persistence of populations (Lubchenco et al., 1991)?
- Biological communities occur in ecological conditions, frequency, amounts, area, and geographic distributions that, taken together, perpetuate their full range of diversity, ecological processes, species, and populations (Keystone Center, 1991). What factors govern the assembly of communities and ecosystems and the ways these ecosystems respond to various stressors? What patterns emerge from cross-system comparisons (Lubchenco et al., 1991)?
- What are the feedbacks between the biotic and abiotic portions of ecosystems and landscapes? How do climate, anthropogenic, and biotic processes regulate biogeochemical processes (Lubchenco et al., 1991)?
- How do patterns and processes at one spatial or temporal scale affect those at other scales (Lubchenco et al., 1991)?
- What are the consequences of environmental variability, including natural and anthropogenic disturbance, for individuals, populations, or communities (Lubchenco et al., 1991)?
- Can consumptive and nonconsumptive resources be produced in an environmentally sensitive manner at levels that ensure a high probability that long-term human and economic well-being can be maintained (Keystone Center, 1991)?

5. Holistic, comprehensive, and transdisciplinary considerations.
6. Delineation (definition) of the ecosystem naturally, for example, bioregionally instead of arbitrarily.
7. Consideration of different levels or scales of system structure, process, and function.
8. Description of system dynamics through ecological concepts such as stability and feedback.
9. Use of anticipatory, flexible research, and planning processes.
10. Consideration of ethics of quality, well-being, integrity, and resilience.
11. Recognition of systemic limits to action-defining and seeking sustainability.

Ecological assessments (conducted at national and regional scales) are essential if resource managers and stakeholders are to describe desired ecological conditions for the nation's major ecosystems and to set the proper ecological context for lands managed at finer scales (e.g., National Forests). By providing basic characterizations of ecosystems, ecological assessments may also help to identify needs for more specific (and traditional) issue-driven risk assessments. To meet these objectives, integrated ecological assessments must answer a number of questions regarding ecosystem sustainability (Table 1.2).

1.3 Basic Ecological Principles and Their Relation to Ecological Assessments

To address questions like those in Table 1.2, highly diverse data are required, including vegetation (plot-based and remotely sensed), maps of biophysical variables (soils, watersheds, landforms, etc.), and socioeconomic data. For a number of reasons, interpreting such data often proves to be complicated. First, many aspects of ecosystems remain enigmatic: defining interactions between ecosystem components and distinguishing between cause and effect relations are difficult tasks; the complexity of ecosystem dynamics is usually unknown; and the hierarchical arrangement of ecological systems is often poorly known, including scale dependencies of ecological processes and patterns. Furthermore, time is often limited, both for testing the hypotheses formulated in the development of protocols for sampling design and analytical methods and for validating and verifying the models used in an assessment.

These challenges to ecological assessments have long been recognized (Lessard et al., 1999). To develop a comprehensive method for dealing with these problems, several different conceptual frameworks (Hutchinson, 1953; Forman and Godron, 1986; Urban et al., 1987; Levin, 1992; Costanza et al., 1993) must be integrated in the articulation of basic ecological principles that underlie any ecological assessment effort. In this section, we provide a brief overview of some of the more important ecological principles that need to be considered in ecological assessments, synthesized from Lessard et al. (1999) and Haynes et al. (1996). The reader is encouraged to see Chapter 2 for a more complete discussion of ecological principles.

In any ecological assessment, the following four ecological principles (Haynes et al., 1996) should be considered:

1. Ecosystems are dynamic, evolutionary, and resilient.
2. Ecosystems can be viewed spatially and temporally within organization levels.
3. Ecosystems have biophysical, economic, and social limits.
4. Ecosystem patterns and processes are not completely predictable.

1.3.1 Ecosystems Are Dynamic, Evolutionary, and Resilient

Most ecological assessments emphasize the description of static ecosystem patterns, like existing vegetation; however, assessments of the spatial relations of ecological systems cannot ignore the dynamics of such patterns and the processes that create them. Ecological systems exhibit temporal changes along various developmental pathways that result in different types of organization. Processes that drive these changes are called *structuring processes.* Many conceptual models used in ecological assessments assume that all ecosystems reach an equilibrium or quasi-equilibrium (e.g., succession leads to climax). It has long been documented, however, that in reality ecosystems are rarely in equilibrium. Although it may be convenient for modeling and planning exercises to assume equilibrium at a given scale, it is impossible to ignore ecological system dynamics that exhibit multiple successional pathways, discontinuities, and surprises and that follow constantly changing environmental conditions (e.g., Holling, 1986; Kay, 1991; Costanza et al., 1993). Accordingly, ecological assessments must consider the complexity of ecosystem dynamics, and analysis pro-

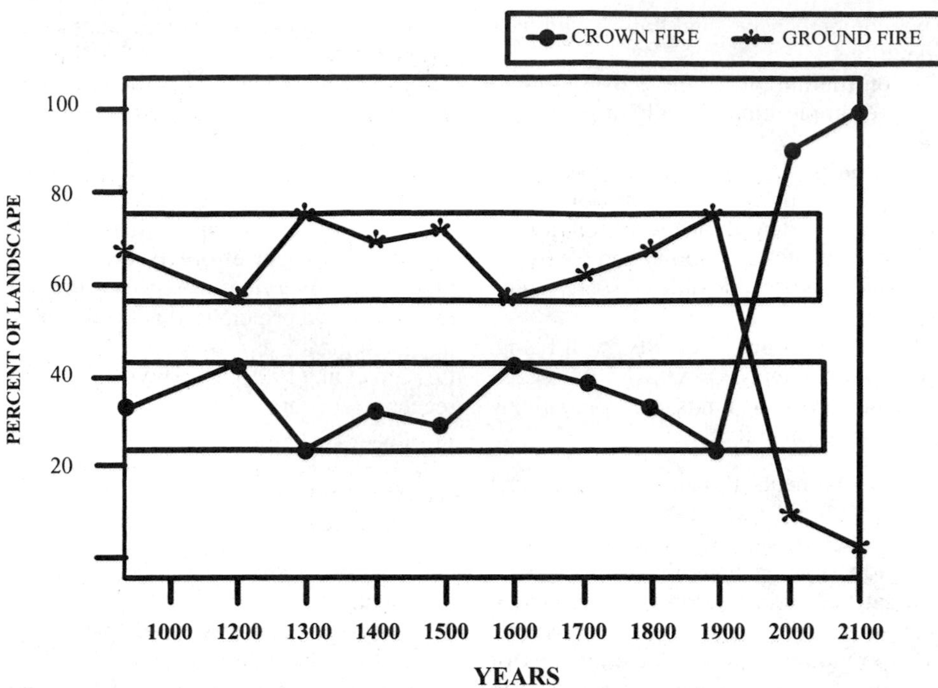

FIGURE 1.2. Characterization of historic variability for ground and crown fire regimes within selected dry Douglas fir forest environments of the Northern Rockies (Jensen et al., 1996).

tocols should be designed to account for this complexity.

Analysis of trends in ecosystem pattern and process relations is important if the temporal dynamics of systems are to be understood (Jensen et al., 1996). Trend analyses facilitate an understanding of both the nature of the ecosystem dynamics (stochasticity) and the periodicities, limits, and trends of system dynamics. Assessments of historical or "natural" variability are currently being conducted by many land management agencies in an attempt to improve this understanding (Swanson et al., 1994).

A typical assessment of historical variability is illustrated by the relative percentage of ground fires versus crown fires over time in selected dry Douglas fir forests of the Northern Rockies (Figure 1.2). In this example, both types of fire regimes displayed characteristic ranges until fire suppression efforts became effective in the mid-1900s. Following fire suppression, the ratio of ground to crown fires changed abruptly. Compared to the past 300 years, these forests are now outside their historical range of variability with respect to fire dynamics. Results of this sort do not necessarily imply that past patterns must be re-created. Instead, this type of assessment provides a larger context for understanding the temporal dynamics that have influenced the system (Swanson et al., 1994). Such assessments are commonly conducted to determine if present conditions have been experienced historically or not. If the answer is no, the ecological system of interest is in a state for which we have no information; therefore, our present knowledge and predictive abilities may not be useful for understanding system behavior, and present technology may be useless in preventing rapid large-scale change to the system (Hann et al., 1994).

The terms *ecological stability* and *resiliency* have been used in different ways for various characteristics of ecological systems. Kay (1991) suggests that stability be defined in the strict sense of the physical sciences (i.e., Lyapunov stability). This definition requires the quantification of ecosystem behavior and an explicit recognition that a number of different stable equilibrium points (or clouds) may exist for an ecosystem. Accordingly, if an ecosystem is displaced from equilibrium but remains within the cloud, it will return to the initial equilibrium point. Conversely, if the system is displaced outside the cloud, it will move to some new equilibrium state under this definition of stability. Resilience, on the other hand, is defined by Holling (1973) as the ability of an ecosystem to bounce back to its original state. For example, forests that undergo pest outbreaks as part of their natural cycle are unstable (populations oscillate strongly), but they are resilient (they bounce back). It is important to clarify what is meant by resilience and stability before ecosystem response to change can be assessed. In light of the nonlinear, hierarchical, and nonequilibrium characteristics of ecosystems, assessment protocols must be designed and interpretations made using consistent terminology.

1.3.2 Ecosystems Can be Viewed Spatially and Temporally within Organizational Levels

The concepts of scale and pattern are interwoven (Hutchinson, 1953; Levin, 1992). Complex ecosystem patterns, landscapes, and the multitude of processes that form them exist within a hierarchical framework (Allen and Starr, 1982; Allen et al., 1984; O'Neill et al., 1986). Scale dependency is very significant because the relationships between ecological processes (and the patterns that they create) change with spatial scale (Turner, 1990; Davis et al., 1991). In recent years, considerable attention has been directed to describing the formal hierarchical organization of ecological systems. As applied to ecological assessments, hierarchy theory provides a needed framework for the description of the components of an ecosystem and their scaled relations.

The following four tenets of hierarchy theory are required for an understanding of landscape patterns and their dynamics (Allen et al., 1984; O'Neill et al., 1986).

1. The whole–part duality of systems states that every component of a system, ecological or otherwise, is a whole and a part at the same time. For example, a forest (a whole) is made up of trees (the parts). However, at larger spatial scales, the forest is part of a regional landscape. In this case, the regional landscape is the whole and the forest becomes a part. The notion of whole–part duality is very important to the characterization of ecological systems because it must be clearly stated which role a component plays at a given scale of examination.
2. Patterns, processes and their interactions can be defined at multiple spatial and temporal scales. These scales need to be clearly identified in the assessment process and should be appropriate to the issues being addressed.
3. No single scale of ecological organization is correct for all purposes, although scientists often provide information and interpretations on ecological systems at a single scale or limited number of scales.
4. The definition of an ecological hierarchy (component patterns and processes) is dictated by the critical issues or policy decisions that need to be addressed in an ecological assessment.

Because a primary goal of ecosystem management is to maintain natural ecological patterns over time (sustainability), an understanding of pattern persistence is critical to most ecological assessments. One major consequence of the hierarchical organization of ecological systems is that nonequilibrium dynamics or spatial heterogeneity at one scale can be translated into equilibrium at a higher level (O'Neill et al., 1986; Urban et al., 1987; Levin, 1992). Accordingly, pattern persists within a hierarchical framework because a pattern may be stable at one level and not at another (Rahel, 1990). Therefore, ecological pattern should be analyzed at more than one scale (Rahel, 1990) and ecological assessments need to consider all levels of appropriate ecological organization (Baker, 1992; Levin, 1992; Urban et al., 1987). Examples of hierarchical organization within vegetation patterns and social systems are shown in Figure 1.3.

1.3.3 Ecosystems Have Biophysical, Economic, and Social Limits

In response to change, ecological systems may follow different pathways. Eventually, the processes that trigger change come into balance with the processes that lead to ecological organization. This state is called an *optimal operating point* (Kay, 1991) and may be used to define the limit of ecosystem development along one pathway. It has been suggested that ecosystems have more than one such point (or limit) (Holling, 1986; Kay, 1991) and that such limits are more appropriately described as "a set of points in state space whose membership changes over time" (Kay, 1991) or as "evolution at the edge of chaos" (Bak and Chen, 1991; Kauffman and Johnson, 1991; Costanza et al., 1993).

The concept that ecosystems may have multiple limits, some more stable than others, should be considered in the design of ecological assessments. For example, the climax community is the endpoint of ecological succession and as such exemplifies an optimum operating point (Kay, 1991). However, after a disturbance, succession may proceed along multiple pathways, and a given pathway may stop before the endpoint of succession is reached. Fur-

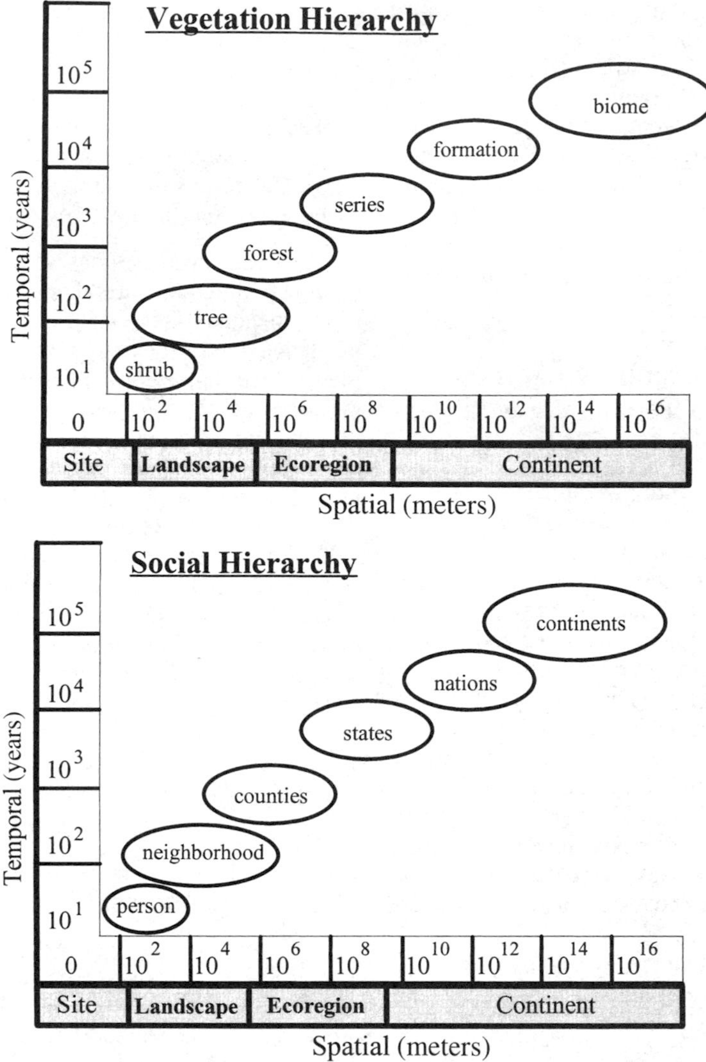

FIGURE 1.3. Hierarchical organization of vegetation patterns and social systems along characteristic temporal and spatial scales.

thermore, if there is a large-scale change in the climate or in the regional pool of plant species, new species may enter the system, resulting in new vegetation types (i.e., new optimum operating points). All ecosystems possess limits in their rate of production and accumulation of biomass (Kay, 1991); human populations must recognize that this translates to a limited ability to provide goods and services. Unfortunately, people often make demands on ecosystems that exceed their biophysical capacities (Robbins, 1982; Young and Sparks, 1985). This situation needs to be clearly addressed in any ecological assessment.

1.3.4 Ecosystem Patterns and Processes Are Not Completely Predictable

The fact that ecosystems exhibit complex, nonlinear dynamics; hierarchical structuring; and multiple optimum operating points, most far from equilibrium, raises important questions about the predictability of ecosystem behavior. Concerns about predictability and scale include the problem of data aggregation from the bottom up, for example, with maps of biophysical environments or maps depicting the regionalization of simulation model results (Running et al., 1989). One limitation of using aggregate data or models is that, by integrating many fine-level details, the data or models may not represent ecosystem behavior at the appropriate spatiotemporal scale (see discussion in Costanza et al., 1993, on aggregate models and ecological processes; also see Hofmann, 1991). Costanza and Maxwell (1994) discuss the relationship between level of resolution and predictability. They found that increasing resolution provides more information about the patterns in data, but also increases the difficulty of accurately modeling these patterns. These authors also found that predictability exhibits fractallike characteristics with decreasing resolution.

How are these findings relevant to ecological assessments? Knowing that predictability in explaining a particular pattern increases with resolution, whereas predictability in modeling the pattern decreases, an optimal resolution may be defined for a particular problem. For example, forest response to beetle outbreaks may exhibit chaotic dynamics at the level of a National Forest, but be more predictable at the continental level. Indeed, Sugihara and May (1990) found that measles epidemics show chaotic dynamics at the level of individual cities, but more predictable periodic dynamics for entire nations. This example reinforces the point that data used across spatial scales must be analyzed carefully in terms of a particular problem: there are limits to the predictability of natural phenomena, and assessment protocols should reflect this fact.

Predictability of ecosystems also is complicated when the processes that historically have created heterogeneity in landscape patterns are altered, resulting in new and unprecedented landscape patterns (see the previous discussion of fire regimes and Figure 1.2). Quantification of ecosystem stability and resiliency is extremely difficult under such conditions. Consequently, any predictions concerning highly altered ecosystems should be monitored closely under an adaptive management process (Lessard et al., 1999).

1.4 General Properties of an Ecological Assessment

1.4.1 Purpose

To reiterate, ecological assessments are an important component of any strategy for implementing ecosystem-based management (Slocombe, 1993; Jensen and Bourgeron, 1994). Furthermore, hierarchy theory suggests that ecosystems may be described at different spatial scales and that levels of ecosystem organization at broader scales bound the range of ecological properties that emerge at finer scales (Allen and Starr, 1982; Allen et al., 1984; O'Neill et al., 1986; Bourgeron and Jensen, 1994). Accordingly, ecological assessments should be conducted at multiple spatial scales (Table 1.3) to improve our understanding of the relations between different ecosystem components most appropriately addressed at different scales (e.g., vegetation biomes, plant communities, rare species).

Broad-scale ecological assessments facilitate better land management decisions by placing more localized ecosystem management strategies into proper context. Regional assessments of mature forest patterns, for example, can greatly assist the development of management plans for old-growth forest networks, which in turn can be used to help to direct local timber sale activities. Strategies of landscape design directed toward desired ecological condition objectives for management are best framed within a multiscale ecological assessment process. In this approach, desired ecological conditions are defined at different scales dependent on the specific issues at hand. Design objectives identified at broader scales are incorporated into management strategies for smaller ecosystems to ensure that activities at more localized levels are consistent with the larger plan. Multiscale ecological assessments enable decisions to be made from scale-appropriate information, and they place allocation and regulation decisions into the proper biophysical and social context, hence contributing to improved management decisions.

Ecological assessments are commonly conducted when major land management or regulatory decisions need to be reevaluated. The need for an ecological assessment should be identified through ongoing monitoring efforts based on one or more of the following changes:

1. *Initial assumptions change as new knowledge is obtained.* Implicit in this statement is an awareness that our understanding of ecosystem composition, structure, and function is, more often than not, incomplete. Given this fact, it is often necessary to revisit management decisions as our understanding of ecosystem interactions improves.
2. *Social values and needs change.* As new issues are identified by various sectors of the public, different scales of analysis are often required. Climate change, forest health, and biodiversity are examples of issues that have recently achieved high-profile status and often were not

TABLE 1.3. A generalized description of the scales used in ecological assessments.

Traditional planning scales[a]	Assessment scale	Typical analysis area size (sq. mi.)	Typical issues
	Continental	>10,000	Neotropical birds Climate change National timber supply
RPA[b]			
	Regional	1000–10,000	Widely distributed species (salmon, grizzly bear) Roadless areas Wildernesses River basin health
Regional guides			
	Subregional	10–1000	Old growth River subbasin health General fire management needs
Forest plans			
	Landscape	1–10	Timber volumes Watershed health Vegetation pattern (cover type and structural stage) Specific fire management needs
Area plans			
	Land unit	1	Soil compaction Specific vegetation management design
Project plans			
	Site	<1	Vegetation regeneration Small mammal activity

[a]A given level of planning requires complete spatial representation of information at the assessment level above, and subsampled information at the level(s) below.
[b]Resource Planning Act

adequately addressed in previous land allocation or regulation strategies. These issues commonly require that different spatial scales of assessment be utilized in the planning process; they tend to be broader in scope and require a larger geographic context for their resolution.

3. *Significant change occurs to the ecosystem.* When unexpected significant change occurs to an ecosystem, management goals and objectives for the system must be reevaluated. Such catastrophic changes are referred to as surprise events by Holling (1978) and can be better predicted through monitoring programs designed to account for uncertainty. Large-scale fire is an example of a significant ecosystem process that can greatly alter the vegetation patterns of an area. If such process–pattern relations are not explicitly addressed in planning documents, integrated ecological assessments may be triggered by such surprise events.

1.4.2 Information Needs

A wide variety of information can and must be used in an integrated ecological assessment. It is important, however, that the information utilized correspond directly to the purpose and need for the assessment. The first step in the assessment process, therefore, should be the identification of the major policy questions or issues to be addressed (see Table 1.2).

Once the policy questions or issues for the assessment are identified, they are further examined to select appropriate scales of analysis (Table 1.4). Climate change, for example, is a process that is meaningfully addressed at broader assessment scales. Grizzly bear viability assessments, by contrast, commonly assess conditions within the home range area of the species (i.e., the regional or subregional scale). Soil erosion assessments are commonly related to patterns of soil bodies as they occur on landforms, which are usually displayed at landscape or land unit levels of assessment.

Another critical step in the assessment process is the identification of scale-specific, measurable, and mappable features that relate to the issues being addressed. Techniques for pattern recognition (Jensen et al., 1996) are particularly useful because they require concise description of the hierarchical relations between the pattern of interest and the agents of its formation (i.e., biophysical environments, disturbance regimes, and biotic processes). For exam-

TABLE 1.4. Relation between ecological assessment scales and selected resource issues.

Assessment scale	Climate change	Grizzly bear viability	Soil erosion
Global	×		
Continental	×		
Regional	×	×	
Subregional		×	
Landscape			×
Land unit			×
Site			×

ple, in assessing grizzly bear viability, the current and historical patterns of grizzly bear distribution are described first in the assessment process. Other ecosystem pattern components thought to influence grizzly bear distribution, like existing vegetation and road networks, are then described at the regional or subregional scale. In the case of vegetation, coarse-scale maps of cover type groups (e.g., Douglas fir–ponderosa pine forests) could be used to describe existing patterns for this example. Changes in historical versus current vegetative patterns are then described in terms of the ecosystem processes that influence these patterns, such as large, infrequent, stand-consuming fires. Biophysical environment maps (e.g., geoclimatic maps of large landform areas) are commonly used to model ecosystem process–pattern relations in this type of assessment. This characterization facilitates the description of historical and current patterns and, in scenario planning, also allows a consistent methodology for predicting future patterns based on an understanding of the underlying biophysical environments and ecosystem processes.

Basic Types of Map Themes

Four types of maps are critical to most ecological assessment efforts (Jensen et al., 1996), including maps of the biophysical environment, maps of the existing or historical status of various ecosystem components (e.g., vegetation), resource interpretation maps, and coordinated planning unit maps.

Maps of the biophysical environment are used to describe terrestrial or aquatic units that behave in a similar manner, given their potential ecosystem composition, structure, and function (see Chapters 22 and 24). They delineate areas with similar response potential and resource production capabilities and are based on landscape components that display low temporal variability at a given map scale (e.g., regional climate, geology, landform). The differentiating criteria used in designing biophysical environment maps are also commonly selected to include those that exert primary control on the ecosystem patterns and processes of interest in an assessment (e.g., vegetation, flooding, fire). These maps commonly are used to describe how the landscape could look or function under historical or current ecosystem process regimes (e.g., fire and successional pathway relations), as well as under different management scenarios. They provide a semipermanent map theme that can be used to extrapolate ecosystem pattern–process relations from sampled areas to unsampled locales and are extremely useful when stratified sampling design strategies are desired for purposes of environmental monitoring (Bailey et al., 1994).

Unlike maps of the biophysical environment, existing or historical status maps describe ecosystem components or patterns (e.g., vegetation, species distribution, human settlement patterns) that commonly display large temporal variability at a given map scale. Accordingly, they provide dynamic snapshots of ecosystem trends and require ongoing monitoring and periodic update depending on the amount of change experienced in an assessment area. For example, the occurrence of large-scale fires in an area can greatly alter vegetative patterns; consequently, maps of existing vegetation would need to be revised or rebuilt following such an event. Species distributions, existing vegetation, roads, land use, fire occurrence, and human activity and values are all examples of dynamic information that needs to be spatially referenced in the assessment process. Interpretation of change in ecosystem status can be improved by overlaying such information on appropriate biophysical environment maps: assessments of ecosystem health or condition are most efficiently described by contrasting the existing condition of an area with other managed or unmanaged areas that occur on similar biophysical environments. The natural variability in pattern–process relations among sites is minimized by this approach; consequently, differences in observed condition may be correlated more directly to the different treatments imposed.

Resource interpretation maps are derived themes; they are commonly generated by overlaying existing ecosystem pattern maps, like vegetation, on similarly scaled biophysical maps, like geoclimatic settings or watersheds, to develop response units appropriate to the type of resource interpretation being made, such as big-game habitat. Multiple themes for existing patterns (e.g., roads and vegetation) are usually associated with a primary biophysical template (ecological unit) in spatially displaying the resource response units of concern. Because they are derived map themes, resource interpretation maps may be constructed as needed from more basic maps delineating bio-

TABLE 1.5. Hierarchical relations between assessment scales and various types of ecosystem delineation.

Assessment scale	Biophysical environments		Existing conditions	
	Terrestrial units[a]	*Aquatic units*[b]	*Vegetation units*[c]	*Social assessment units*
Global	Domain	Zoogeographic region	Class	Continent
Continental	Division	Zoogeographic subregion	Subclass	Nation
Regional	Province	River basin	Group	State
Subregion	Section subsection	Subbasin	Formation	County
Landscape	Landtype association	Watershed Subwatershed	Series	Community
Land unit	Landtype Landtype phase	Valley section Stream reach	Association	Neighborhood
Site	Ecological site	Channel unit	Species	Individual

[a]Cleland et al., 1997.
[b]Maxwell et. al., 1995.
[c]Driscoll et al., 1984.

physical and ecosystem patterns. Examples of resource interpretation maps that are commonly derived for many ecological assessments include species habitat suitability, ecosystem health or condition ratings, risk or hazard ratings, and management potential ratings.

Coordinated planning unit maps are used to delineate areas for specific planning or reporting needs at different spatial scales (e.g., national resource planning, area planning, and project planning). Maps of an assessment area's biophysical environments and existing conditions provide an initial template useful in delineating coordinated planning units. Social and political criteria are also incorporated when defining such map units to ensure multiagency, multiownership, and multigovernment collaboration in monitoring and land-use planning efforts. Coordinated planning units define an assessment areas's boundaries; however, information used to describe an area's biophysical and existing conditions often will extend beyond its boundaries. For example, large river basins (Columbia, Missouri) may be used to define coordinated planning units in regional-scale ecological assessments. The biophysical environment maps commonly used to describe potential vegetation relations at this scale (e.g., ecoregions) follow regional climate zones that do not coincide with the topographic divides used to delineate basins. Consequently, the area described by biophysical environments will often be larger than the coordinated planning unit area.

Scaled Relations of Information

It cannot be overemphasized that, depending on the issues or policy questions to be addressed, different scales of ecosystem characterization may be required (Table 1.5). A common requirement of most assessments is that continuous map coverage for ecosystem components be developed for all lands within the area under investigation. Such maps allow consistent descriptions of ecological relations across the assessment area; however, they may be too general for some needs. In these situations, stratified subsampling of ecological patterns is often required.

In the regional-scale ecosystem assessment of the Columbia Basin, for example, stratified subsampling of 10,000- to 30,000-acre watersheds was required to address differences between historical and current vegetation patterns across the basin (Quigley and Arbelbide, 1997). These watersheds were selected on a stratified-random sampling basis using section-level terrestrial biophysical mapping units. Results of this analysis—changes in vegetation cover type composition—were summarized at the river basin scale to display trends across the whole analysis area. This characterization facilitated general descriptions of ecosystem pattern and process relations at the landscape scale by displaying trends, obtained through subsampling, at coarser regional scales. Such information provides the context for landscape design activities (e.g., vegetation management) at finer scales through the identification of appropriate treatment areas and the types of activities that may be required to meet management objectives.

The scaled relations of information used in assessments also have major implications for the design of monitoring programs for detecting ecosystem change. Because inventory costs generally increase exponentially as the scale of description becomes finer, it is important to describe ecosystem pattern–process relations at the coarsest appropriate scale to reduce costs. Often this can be done by identifying the pattern of interest and moving up one scale in selecting the pattern that will be described in assessment or monitoring efforts. For example, if we are interested in the change of vegetation at the plant association level (e.g., Douglas fir–Idaho fescue or Douglas fir–snowberry), sampling strategies could be designed to detect change at the series level (e.g., Douglas fir forests) as an initial screening mechanism to identify potential areas of plant association change. The spatial area that could be described by this approach, for the same level of funding, is much larger than that which could be described continuously at the plant association level. In this example, areas that exhibit change at the series level would be prioritized for more detailed study to better understand plant association pattern relations in an assessment area. An approach of this sort follows the coarse-filter strategy for biodiversity conservation (Hunter, 1991) and is a practical way of applying pattern recognition theory (Jensen et al., 1996).

1.4.3 Analysis of Ecosystem Pattern–Process Relations

Ecological assessments require the synthesis of information across many disciplines (e.g., biological, social, economic, and physical sciences). If these assessments are to assist managers in implementing ecosystem-based strategies for land management, they must address both the biophysical and human components of ecosystems (Slocombe, 1993; Grumbine, 1994). Aggregation of information up and down scales must be used in combination with integration models to maximize the use of current information. The value of information will change through this aggregation process and should be well documented during the assessment process.

To facilitate improved ecosystem management, ecological assessments should provide descriptions of the following:

1. Current and historical composition, structure, and function of an ecosystem.
2. Abiotic and biotic events (including human actions) that contributed to the development of the current ecosystem condition.
3. Probable future scenarios that might exist under different types of management strategies (e.g., current management direction, maintenance of historical processes and patterns).

These basic components of an ecological assessment are briefly described next, following Lessard et al. (1999).

The fact that ecosystems are dynamic and evolutionary suggests that the current status of an ecosystem is a function of its evolutionary and ecological history. Accordingly, assessments of ecosystem conditions and management potential should describe existing and historical ecosystem composition, structure, and function in the context of specific biophysical and human processes associated with each time period. A characterization of existing conditions gives quantitative or qualitative evidence of current structure and function. Historical characterizations provide insight into the kinds, magnitudes, and rates of change in ecosystem structure and function. Historical characterizations also provide insight into possible future ecosystem development pathways by providing evidence of dominant disturbance types and regimes, typical environmental constraints, and variability of biotic patterns and processes (see Chapter 19).

Because there are real limitations to the productive capacities, structure, and functioning that might be achieved within an ecosystem at any time, ecological assessments should characterize plausible ranges of potential future ecosystem conditions with regard to climatic trends, biophysical environmental conditions, historical development, current structure, organization, and disturbance regimes. Additionally, recognizing that ecosystem patterns and processes are not always perfectly predictable, the accuracy and specificity of planning and management expectations should be in line with such uncertainty. Accordingly, desired conditions for planning should reflect desired dynamics, and management prescriptions should describe a quantifiable range of plausible conditions, rather than a single, narrowly defined target.

This range of desired conditions can best be described through scenario planning, a tool used to inform society about the implications and trade-offs in ecosystem management (Haynes et al., 1996). Scenario planning differs from traditional alternative evaluation methods of land-use planning, which describe the effects of different methods for achieving a desired state, because scenarios describe a set of possible futures given different management strategies and then explore their consequences. Scenarios project qualitative social and biological outcomes over time, allowing us to learn about the merits, pitfalls, and trade-offs of management choices and providing required information for decision makers. Examples of scenarios that might be considered in an ecological assessment include (1) continue existing management, (2) manage for minimal disturbance, (3) manage for historical disturbance processes, and (4) manage to maximize outputs of goods and services (Lessard et al., 1999). To construct these scenarios, ecological modeling is required so that the processes (both natural and anthropogenic) that influence landscape patterns can be described at different spatial and temporal scales. Displays of different landscape patterns at different periods of time assist the description of ecological consequences and resource outputs predicted for each management scenario. Such scenario projections are useful in identifying the desired future conditions for an assessment area and can significantly reduce the number of alternatives that need to be addressed in detail in subsequent NEPA-based planning.

1.5 Conclusion

"If you look where you're going, you will certainly end up where you're headed." This aphorism implies the need for an understanding of the landscape, as well as a sense of direction. Navigational metaphors abound in the ecosystem management and assessment literature; conceptual and analytical models may provide the maps, goal-directed monitoring the compass, and institutional flexibility the gyroscopes necessary to reach particular goals (Lee, 1993). At its best, an ecological assessment describes all these elements.

Change represents the most daunting challenge to sustainable management. Even in the absence of human influences, ecosystems are constantly changing; human actions have added a new source of complexity and unpredictability to natural change processes. Our understanding of the ways in which ecosystems function and humans affect these functions is constantly changing. To add even more complexity, our human expectations from ecosystems constantly change as well. Our desires are different from our parents and will be no less different from those of our children. Obviously, sustainability and the ecological assessments needed to support it must not be viewed as an endpoint, but as an evolving process.

1.6 Recommended Reading

Boyce, M. S.; Haney, A. 1997. *Ecosystem management: applications for sustainable forest and wildlife resources.* New Haven, CT: Yale University Press.

Christensen, N. L.; Bartuska, A.; Brown, J. H.; Carpenter, S.; D'Antonion, C.; Francis, R.; Franklin, J. F.; MacMahon, J. A.; Noss, R. F.; Parsons, D. J.; Peterson, C. H.; Turner, M. G.; Woodmansee, R. G. 1996. The scientific basis for ecosystem management. *Ecological Applications* 6:665–691.

Jensen, M. E.; Bourgeron, P. S., technical editors. 1994. *Volume II: ecosystem management: principles and applications.* PNW-GTR-318. Portland, OR: U.S. Department of Agriculture, Forest Service, Pacific Northwest Research Station. 376 p.

Keystone Center. 1996. *The keystone national policy dialogue on ecosystem management.* Keystone, CO: Keystone Center. 99 p.

Samson, F. B.; Knopf, F. L. 1996. *Ecosystem management: selected readings.* New York: Springer-Verlag.

Slocombe, D. S. 1993. Implementing ecosystem management. *BioScience* 43:612–622.

1.7 References

Agee, J. K.; Johnson, D. R., editors. 1988. *Ecosystem management for parks and wilderness.* Seattle, WA: University of Washington Press.

Allen, T. F. H.; Starr, T. B. 1982. Hierarchy: perspectives for ecological complexity. Chicago: University of Chicago Press.

Allen, T. F. H.; Hoekstra, T. W.; O'Neill, R. V. 1984. *Interlevel relations in ecological research and management: some working principles from hierarchy theory.* RM-GTR-110. Fort Collins, CO: U.S. Dept. Agric., For. Serv., Rocky Mountain Res. Sta.

Bailey, R. G.; Jensen, M. E.; Cleland, D. T.; Bourgeron, P. S. 1994. Design and use of ecological mapping units. In: Jensen, M. E.; Bourgeron, P. S., tech. eds. *Volume II: Ecosystem management: principles and applications.* PNW-GTR-318. Portland, OR: U.S. Dept. Agric., For. Serv., Pacific Northw. Res. Sta.: 95–106.

Bak, P.; Chen, K. 1991. Self-organized criticality. *Sci. Amer.* 264: 46.

Baker, W. L. 1992. The landscape ecology of large disturbances in the design and management of nature reserves. *Landscape Ecol.* 7:181–194.

Beek, K. J.; Bennema, J. 1972. *Land evaluation for agricultural land use planning: an ecological methodology.* Wageningen, Netherlands: Dept. Soil Sci. and Geol., Agricultural Univ.

Bourgeron, P. S.; Jensen, M. E. 1994. An overview of ecological principles for ecosystem management. In: Jensen, M. E.; Bourgeron, P. S., tech. eds. *Volume II: Ecosystem management: principles and applications.* PNW-GTR-318. Portland, OR: U.S. Dept. Agric., For. Serv., Pacific Northw. Res. Sta.: 45–57.

Boyce, M. S.; Haney, A. 1997. *Ecosystem management: applications for sustainable forest and wildlife resources.* New Haven, CT: Yale University Press.

Christensen, N. L. 1988. Succession and natural disturbance: paradigms, problems and preservation of natural ecosystems. In: Agee, J.; Johnson, D., eds.

Ecosystem management. Seattle, WA: University of Washington Press: 62–86.

Christensen, N. L. 1996. The scientific basis for sustainable use of land. In: Diamond, H. L.; Noonan, P., eds. *The use of land.* Washington, DC: Island Press: 273–308.

Christensen N. L.; Franklin, J. F. 1996. Ecosystem processes and ecosystem management. In: Christensen, N. L.; Simpson, D., eds. *Human activities and ecosystem function.* New York: Chapman-Hall, Inc.: 1–23.

Christensen, N. L.; Bartuska, A.; Brown, J. H.; Carpenter, S.; D'Antonion, C.; Francis, R.; Franklin, J. F.; MacMahon, J. A.; Noss, R. F.; Parsons, D. J.; Peterson, C. H.; Turner, M. G.; Woodmansee, R. G. 1996. The scientific basis for ecosystem management. *Ecol. Appl.* 6:665–691.

Cleland, David T.; Avers, Peter E.; McNab, W. Henry; Jensen, Mark E.; Bailey, Robert G.; King, Thomas; Russell, Walter E. 1997. National hierarchical framework of ecological units. In: Boyce, Mark S; Haney, Alan, eds. *Ecosystem management: applications for sustainable forest and wildlife resources.* New Haven, CT, and London: Yale University Press: 181–200.

Costanza, R.; Maxwell, T. 1994. Resolution and predictability—an approach to the scaling problem. *Landscape Ecol.* 9:47–57.

Costanza, R.; Wainger, L.; Folke, C.; Maler, K. 1993. Modeling complex ecological economic systems. *BioScience* 43:545–555.

Davis, F. W.; Quattrochi, D.; Ridd, M. K.; Lam, N. S.; Walsh, S. J.; Michaelsen, J. C.; Franklin, J.; Stow, D. A.; Johnannsen, C. J.; Johnston, C. A. 1991. Environmental analysis using integrated GIS and remotely sensed data: some research needs and priorities. *Photogramm. Eng. Remote Sensing* 57:689–697.

Diamond, H. L.; Noonan, P., editors. 1987. *The use of land.* Washington, DC: Island Press.

Driscoll, R. S.; Merkel, D. L.; Radlof, D. L.; Snyder, D. E.; Hagihara, J. S. 1984. *An ecological land classification framework for the United States.* Misc. Publ. 1439. Washington, DC: U.S. Dept. Agric., For. Serv.

Everett, Richard; Hessburg, Paul; Jensen, Mark; Bormann, Bernard. 1994. *Volume 1: executive summary.* PNW-GTR-317. Portland, OR: U.S. Dept. Agric., For. Serv., Pacific Northw. Res. Sta.

Fedkiw, J. 1999. *Managing multiple uses on National Forests, 1905–1995.* FS-628. Washington, DC: U.S. Dept. Agric., For. Serv.

Food and Agriculture Organization (FAO). 1976. A framework for land evaluation. *Soils Bulletin 32.* Rome: Food and Agriculture Organization.

Forest Ecosystem Management Assessment Team (FEMAT). 1993. *Forest ecosystem management: an ecological, economic and social assessment.* Report of the Forest Ecosystem Management Assessment Team. Washington, DC: U.S. Govt. Printing Off. 1993-793071.

Forman, R. T. T.; Godron, M. 1986. *Landscape ecology.* New York: John Wiley & Sons.

Grossarth, O.; Nygren, T. 1994. Implementing ecosystem management through the forest planning process. In: Jensen, M. E.; Bourgeron, P. S., tech. eds. *Volume II: Ecosystem management: principles and applications.* PNW-GTR-318. Portland, OR: U.S. Dept. Agric., For. Serv., Pacific Northw. Res. Sta.: 276–284.

Grumbine, R. E. 1994. What is ecosystem management? *Conserv. Biol.* 8:27–38.

Hann, W.; Keane, R. R.; McNicoll, C.; Menakis, J. 1994. Assessment techniques for evaluating ecosystem processes, and community and landscape conditions. In: Jensen, M. E.; Bourgeron, P. S., tech. eds. *Volume II: Ecosystem management: principles and applications.* PNW-GTR-318. Portland, OR: U.S. Dept. Agric., For. Serv., Pacific Northw. Res. Sta.: 237–253.

Haynes, R. W.; Graham, R. T.; Quigley, T. M. 1996. *A framework for ecosystem management in the interior Columbia basin and portions of the Klamath and Great Basins.* PNW-GTR-374. Portland, OR: U.S. Dept. Agric., For. Serv., Pacific Northw. Res. Sta.

Hofmann, E. E. 1991. How do we generalize coastal models to global scale? In: Mantoura, R. F. C.; Martin, J. M.; Wollast, R., eds. *Ocean margin processes in global change.* New York: John Wiley & Sons.

Holling, C. S. 1973. Resilience and stability of ecological systems. *Ann. Rev. Ecol. Syst.* 4:1–24.

Holling, C. S., editor. 1978. *Adaptive environmental assessment and management.* London: John Wiley & Sons.

Holling, C. S. 1986. The resilience of terrestrial ecosystems: local surprise and global change. In: Clark, W. M.; Munn, R. E., eds. *Sustainable development in the biosphere.* Oxford, UK: Oxford University Press: 292–320.

Hunter, M. 1991. Coping with ignorance: the coarse-filter strategy for maintaining biodiversity. In: Kohn, K. A., ed. *Balancing on the brink of extinction: the endangered species act and lessons for the future.* Washington, DC: Island Press: 256–281.

Hutchinson, G. E. 1953. The concept of pattern in ecology. *Proc. Natl. Acad. Sci. (USA)* 105:1–12.

Jensen, M. E.; Bourgeron, P. S., editors. 1994. *Volume II: Ecosystem management: principles and applications.* PNW-GTR-318. Portland, OR: U.S. Dept. Agric., For. Serv., Pacific Northw. Res. Sta.

Jensen, M. E.; Everett, R. 1994. An overview of ecosystem management principles. In: Jensen, M. E.; Bourgeron, P. S., tech. eds. *Volume II: Ecosystem management: principles and applications.* PNW-GTR-318. Portland, OR: U.S. Dept. Agric., For. Serv., Pacific Northw. Res. Sta.: 6–15.

Jensen, Mark E.; Bourgeron, Patrick; Everett, Richard; Goodman, Iris. 1996. Ecosystem management: a landscape ecology perspective. *J. Amer. Water Resour. Assoc.* 32:203–216.

Kauffman, S. A.; Johnson, S. 1991. Coevolution to the edge of chaos: coupled fitness landscapes, poised states, and coevolutionary avalanches. *J. Theor. Biol.* 149:467–505.

Kay, J. J. 1991. A nonequilibrium thermodynamic framework for discussing ecosystem integrity. *Environ. Manage.* 15:483–495.

Keystone Center. 1991. *Biological diversity on federal lands. Report of a keystone policy dialogue.* Keystone, CO: Keystone Center.

Keystone Center. 1996. *The keystone national policy dialogue on ecosystem management.* Keystone, CO: Keystone Center.

Klingebiel, A. A.; Montgomery, D. H. 1961. Land capability classification. *Agric. Handb. 210,* U.S. Dept. Agric., Soil Conserv. Serv. Washington, DC: U.S. Govt. Printing Off.

Lee, K. N. 1993. *Compass and gyroscope.* Washington, DC: Island Press.

Leopold, A. 1949. *A Sand County almanac.* New York: Oxford University Press.

Lessard, G.; Jensen, M. E.; Crespi, M.; Bourgeron, P. S. 1999. A general framework for integrated ecological assessments. In: Cordel, H. Ken; Bergstrom, John C., eds. *Integrating social sciences with ecosystem management: human dimensions in assessment policy and management.* Champaign–Urbana, IL: Sagamore Press: 35–60.

Levin, S. A. 1992. The problem of pattern and scale in ecology. *Ecology* 73:1942–1968.

Lubchenco, J.; Olson, A. M.; Brubaker, L. B.; Carpenter, S. R.; Holland, M. M.; Hubbell, S. P.; Levin, S. A.; MacMahon, J. A.; Matson, P. A.; Melillo, J. M.; Mooney, H. A.; Peterson, C. H.; Pulliam, H. R.; Real, L. A.; Regal, P. J.; Risser, P. G. 1991. The sustainable biosphere initiative: an ecological research agenda. *Ecology* 72:371–412.

Maxwell, J. R.; Edwards, C. J.; Jensen, M. E.; Paustian, S. J.; Parrott, H.; Hill, D. M. 1995. *A hierarchical framework of aquatic ecological units in North America.* NC-GTR-176. St. Paul, MN: U.S. Dept. Agric., For. Serv., North Central Exp. Sta.

Morrison, J. 1994. Integrating ecosystem management and the forest planning process. In: Jensen, M. E.; Bourgeron, P. S., tech. eds. *Volume II: Ecosystem management: principles and applications.* PNW-GTR-318. Portland, OR: U.S. Dept. Agric., For. Serv., Pacific Northw. Res. Sta.: 266–275.

National Research Council. 1983. *Risk assessment in the federal government: managing the process.* Washington, DC: National Academy Press.

Olson, G. W. 1974. Land classifications. *Search, Agriculture* 4(7). Ithaca, NY: Cornell University.

O'Neill, R. V.; DeAngelis, D. L.; Waide, J. B.; Allen, T. F. H. 1986. *A hierarchical concept of ecosystems.* Princeton, NJ: Princeton University Press.

Overbay, J. C. 1992. Ecosystem management. In: *Proceedings of the national workshop: taking an ecological approach to management.* April 27–30; Salt Lake City, UT. WO-WSA-3. Washington, DC: U.S. Dept. Agric., For. Serv., Watershed and Air Manage.: 3–15.

Quigley, Thomas M.; Arbelbide, Sylvia J., technical editors. 1997. *An assessment of ecosystem components in the interior Columbia basin and portions of the Klamath and Great Basins: volume 1.* PNW-GTR-405. Portland, OR: U.S. Dept. Agric., For. Serv., Pacific Northw. Res. Sta.

Rahel, F. J. 1990. The hierarchical nature of community persistence: a problem of scale. *Amer. Naturalist* 136:328–344.

Rapport, D. J. 1992. Environmental assessment at the ecosystem level: a perspective. *J. Aquat. Ecosystem Health* 1:15–24.

Risk Assessment Forum (RAF). 1992. *Framework for ecological risk assessment.* EPA630R92001. Washington, DC: U.S. Environ. Protect. Agency.

Robbins, William G. 1982. *Lumberjacks and legislators: political economy of the U.S. lumber industry* 1890–1941. College Station, TX: Texas A&M University Press.

Running, S. W.; Nemani, R. R.; Peterson, D. L.; Band, L. E.; Potts, D. F.; Pierce, L. L.; Spanner, M. A. 1989. Mapping regional forest evapotranspiration and photosynthesis by coupling satellite data with ecosystem simulation. *Ecology* 70:1090–1101.

Samson, F. B.; Knopf, F. L. 1996. *Ecosystem management: selected readings.* New York: Springer-Verlag.

Slocombe, D. S. 1993. Implementing ecosystem-based management: development of theory, practice, and research for planning and managing a region. *BioScience* 4:612–622.

Society of American Foresters. 1993. *Sustaining long-term forest health and productivity: report of a task force.* Bethesda, MD: Society of American Foresters.

Sugihara, G.; May, R. M. 1990. Applications of fractals in ecology. *Trends Ecol. Evolut.* 5:79–86.

Suter, G. W. 1993. *Ecological risk assessment.* Ann Arbor, MI: Lewis Publishers.

Swanson, F. J.; Jones, J. A.; Wallin, D. O.; Cissel, J. H. 1994. Natural variability: implications for ecosystem management. In: Jensen, M. E.; Bourgeron, P. S., tech. eds. *Volume II: Ecosystem management: principles and applications.* PNW-GTR-318. Portland, OR: U.S. Dept. Agric., For. Serv., Pacific Northw. Res. Sta.: 80–94.

Turner, M. G. 1990. Landscape changes in nine rural counties in Georgia. *Photogramm. Eng. Remote Sensing* 56:379–386.

Urban, D. L.; O'Neill, R. V.; Shugart, H. H., Jr. 1987. Landscape ecology: a hierarchical perspective can help scientists understand spatial patterns. *BioScience* 37:119–127.

Wood, C. A. 1994. Ecosystem management: achieving the new land ethic. *Renewable Resour. J.* 6–12.

Young, James A.; Sparks, Abbot, B. 1985. *Cattle in the cold desert.* Logan, UT: Utah State University Press.

Zonneveld, I. S. 1988. Basic principles of land evaluation using vegetation and other land attributes. In: Kuchler, A. W.; Zonneveld, I. S., eds. *Vegetation mapping.* Dordrecht, Netherlands: Kluwer Academic Publishers.

2
A Theoretical Framework for Ecological Assessment

S. J. Turner and Alan R. Johnson

2.1 Introduction

Ecological assessment of natural and human-managed systems need not be a shot in the dark or a hopeful guess about processes and impacts. There is a surprisingly large body of theory that can inform ecological assessment and the prediction of change for ecosystems. This body of theory can be used to structure the way that we investigate systems, the way we interpret the results of our investigations, and the management decisions that we make based on the assessment. It is critically important that we narrow any assessment to those things that specifically and coherently address the focal question of the assessment (e.g., forest health, biodiversity, sustainable economic development). We have theories that can help us to do so. Paramount among them are theories of hierarchical organization, landscape ecology, and system dynamics. These theories provide us with a conceptual framework within which we can address the most important and difficult questions facing ecological science and management. How do we understand the ecosystems that we manage so that we can make decisions that will sustain these systems for our future?

The first step in understanding ecosystems, as Ricklefs and Schluter (1993, p. 350) suggest, is to "recognize that ecology, evolution, geography, and history are different facets of a single set of processes and the patterns they generate. One cannot isolate any one system of a particular dimension from processes and structures at a smaller scale embedded within it or from those at a larger scale containing it." This recognition must be incorporated explicitly into the design and implementation of research for ecological assessment. This is the point of departure for this chapter. We stress the importance of identifying the appropriate observation and measurement criteria and the appropriate temporal and spatial scales for measurements. We also address the dynamics of change—how we can find out how ecological systems may be altered with time and how we can predict what may happen next. We begin with hierarchy theory and landscape ecology and move on to systems dynamics; we then use these concepts as a backdrop for making critical choices about ecological assessments.

2.2 Ecosystems as Hierarchies

2.2.1 Nestedness of Systems

Ecosystems are groups of interacting, interdependent parts (e.g., species, resources) linked to each other by the exchange of energy, matter, and information (Costanza et al., 1993). Ecosystem is a scale-independent concept, so an ecosystem can be of any size that the investigator chooses to study (Allen and Hoekstra, 1992). This scale independence is very important, because our management units and our questions do not always neatly fit into convenient boundaries. Ecosystems can also be profitably viewed as hierarchical in organization. For instance, integral to hierarchy theory is the idea of the whole–part duality (Koestler, 1967; Allen and Starr, 1982; Allen et al., 1987). Any part of the system being investigated is simultaneously embedded in the larger context of the whole ecosystem and itself

This work was supported in part by the U.S. Environmental Protection Agency through the Office of Research and Development and by the U.S. Department of Agriculture, Jornada Agricultural Research Station. The authors thank P. S. Bourgeron and the anonymous reviewers for helpful comments on the manuscript.

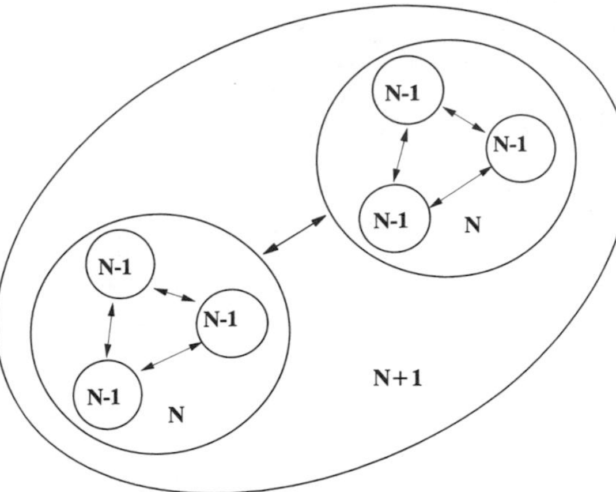

FIGURE 2.1. The hierarchical approach to ecosystem analysis suggests that the phenomenon of interest, level *N*, is created and maintained by mechanisms that are subsystems, level $N - 1$ of the system (*N*). The level $N + 1$ is the context of the level of interest and provides higher-order control of the system.

has important substructure. Clearly, a management unit we choose to study may be composed of subunits of soil or vegetation type, while the unit itself may be only part of a watershed.

Any ecosystem is a hierarchy of differently scaled structures and functions. The structure and function at a particular level (Figure 2.1) that we choose to observe, say a forest stand, has a particular spatiotemporal scale. This is the *N* level. The lower level ($N - 1$) is composed of subsystems, the components that together comprise the forest stand. Depending on the process under investigation, the relevant components might be patches or gaps or individual trees, all of which have smaller and quicker spatiotemporal dynamics and provide a mechanism for creation and maintenance of the stand. The higher level ($N + 1$) is larger in space than the forest stand, but changes more slowly. This larger, slower level might be the region, which exerts control on the forest stand through topography and mesoscale climate.

Hierarchy theory (O'Neill et al., 1986, 1989) applied to systems implies such upper-level control and lower-level mechanisms and provides an epistemology for measurement based on the scale of observation and level of analysis. Of course, the level of interest (*N*) and the $N - 1$, $N + 1$ levels are set by the problem being investigated and may not be the same hierarchy for all problems. Depending on the criterion employed, many different subsystems ($N - 1$) can be defined for any particular level (*N*). Observer perception determines the relevant levels (MacMahon et al., 1978). Characterizing the hierarchical organization of an ecosystem includes determining the spatial and temporal bounds of each pattern or process and the order in which patterns or processes are nested within the hierarchy (Urban et al., 1989). At any scale, we can observe many different types of things. What is included for analysis depends on what is important for the study.

The description of the system as a hierarchy of mechanisms ($N - 1$), phenomena (*N*), and controls ($N + 1$) helps in sorting though the complexity by isolating the dynamics of interest and structuring the research around important interactions (Johnson, 1996). This is the art. Leave out everything that is unimportant in sorting through the complexity. Find the correct criterion to address the problem. The hard part is to determine the correct level in the system, that which is truly important to the observer. It may be a single species or a biotic community or an ecosystem type. Choose the important level *N*. Then carefully, remaining within the same hierarchy, identify the $N + 1$ level of control and the $N - 1$ level of mechanisms. These should drive our investigations whether they are field tests or modeling, applied or theoretical (Turner, 1995).

2.2.2 Grain and Extent

Imagine casting a net into a sea populated with fish of widely varying sizes. The net will capture only a certain range of body sizes. Fish smaller than the mesh size can slip through, and fish larger than the net can hold will simply swim around the edge (O'Neill et al., 1986; Wiens, 1989; Ahl and Allen, 1996). The net represents our observational protocol, the mesh size is analogous to grain (the smallest features that can be retrieved from the data), and the size of the net determines the extent (the largest features captured by the observations). The concepts of grain and extent (Allen and Hoekstra, 1992) are particularly important because they set the primary limits on our perception of the system. We partially determine the grain and extent when we choose a system to study, and further specify grain and extent when we choose the protocols that we use to study the system. Will we choose to study a forest patch or watershed within a forest? How often will we measure a dynamic process? How will we analyze the data? These considerations again must be tied directly to the question addressed by the assessment and the levels in the system appropriate to the question. Each choice we make limits the information that we can expect to glean from the data. Do not expect the measurement of microbes to capture the dynamics of a whale.

Costanza and Maxwell (1994) used information theory to investigate the effect of spatial resolution (grain) on aspects of predictability. They use a raster-based representation of a landscape (e.g., a classified satellite image or a GIS layer) in which each pixel is assigned a unique state (e.g., land cover class). Raster pixels are equal-sized grid units of the image. Costanza and Maxwell (1994) used several images from the same time period with different resolutions so that the pixels in different images each represented a different amount of area on the ground. They also used images with constant resolution, but from different times. Two types of prediction are presented, which we will restate. The first is a type of spatial prediction within a scene (autopredictability): how well can we predict the state of a pixel given the states of its neighboring pixels? This prediction can be based on the images with different spatial resolutions. The second is a temporal prediction across scenes (cross-predictability): how well can we predict the state of a pixel in the current scene given its state in another scene with the same resolution? The general result is that, as resolution is increased or spatial grain in the data set is decreased (pixels represent less land surface), the ability to predict a pixel's state based on its neighbors increases. However, the ability to predict a pixel's state over time from one scene to the next decreases. This is intuitively reasonable. Landscapes usually display positive spatial autocorrelation, which decays with distance. Thus, as grain size decreases, each pixel represents a smaller area, and the states of neighboring pixels should be more strongly correlated, allowing greater spatial predictability. On the other hand, because pixels represent smaller areas, smaller-scale changes in state become evident, making prediction of changes from scene to scene more difficult (i.e., decreasing temporal predictability). Spatial resolution of the data should be chosen to yield an appropriate balance between spatial and temporal predictability. Just where that balance occurs will depend on the properties of the landscape and on the assessment objectives.

2.2.3 Continuous versus Discontinuous Change

Ecosystems, communities, populations, and organisms all change over time. Our measurements and our expectations determine whether we perceive change as continuous or discontinuous. A change is regarded as discontinuous if there is a sudden switch between distinct (qualitatively different) states. A bulldozer clears an area; a graph of vegetative cover versus time has a stair-step form. However, if a string of intermediate states is observed, the change is regarded as continuous. Measured year to year, livestock grazing may lead to a change in composition of the plant community; the graph is a curve. Continuity and discontinuity in ecological systems are both idealizations. The important issue is not whether a given change is *really* continuous or discontinuous, but how it can best be described, given our observations. The grain and extent of our observation set will determine whether a given change is regarded as continuous or discontinuous. A fine-grained observation set will be more likely to resolve intermediate states and therefore be described more naturally in terms of a continuous change. A coarser-grained observation set will more likely fit a description based on discontinuous transitions between discrete states. Observation sets with large extent will tend to yield evidence of discontinuous change, both because the larger spatial coverage or longer temporal duration is more likely to encompass changes of large magnitude and because observation sets with large extent usually will be relatively coarse grained (Ahl and Allen, 1996).

This can be illustrated by the altitudinal zonation of vegetation on a mountain. Standing back and looking at the mountain from afar, we can see distinct zones of vegetation, with discontinuous change from deciduous to coniferous forest, for example. Walking up the mountain, on the other hand, the transitions appear much more gradual, much more continuous, as purely deciduous stands give rise to mixed deciduous–coniferous stands and finally to purely coniferous stands. The difference in our perception stems from the scale of our observations. Viewing the mountain from a distance, we see a large extent (the entire altitudinal range from foot to peak), but our eye does not resolve fine-grained detail. Walking within the forest, our vision is limited in extent, but we see the fine-grained detail of individual trees in our local environment. Both perspectives can be useful. Thus, Whitaker (1956), in his classic study of the vegetation of the Great Smoky Mountains (Figure 2.2), showed cove hardwood forests and spruce–fir forests as distinct units on a mountain in one figure, while using gradient analysis to show a continuous gradation in another figure. The issue of continuous versus discontinuous change is a nonissue as we choose the appropriate description for the data set at hand.

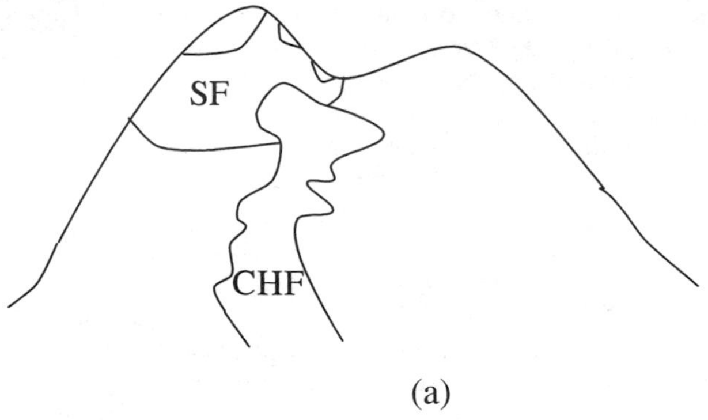

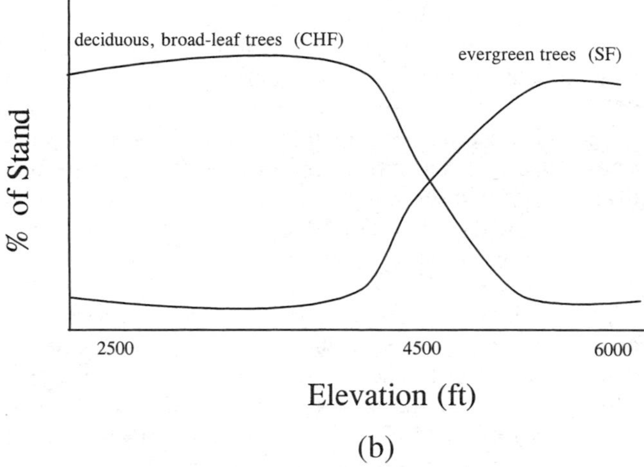

FIGURE 2.2. Two views of the transition from cove hardwood forest (CHF) to spruce–fir (SF) forest in the Great Smoky Mountains. A schematic representation of the patterns described by Whitaker (1956). (a) Discontinuous change between distinct patches on the mountainside viewed from a distance. (b) Continuous change apparent from quantitative data collected at sites spanning the transition zone.

Changing protocols will change the plot of the data, but it will not resolve the problem of which description is the most appropriate for the study (Ahl and Allen 1996). This decision must be made while choosing the appropriate grain and extent for the data and depends critically on the question to be answered.

2.3 Landscape Ecology

Landscape ecology is an integrative science that focuses on the way ecological systems are arrayed in space and through time. Landscape ecology has been described as a transdisciplinary, problem-solving science "bridging the gaps between bioecology and human ecology" (Naveh, 1995, p. 43). Landscape ecology covers such diverse topics as (1) elements in a park created for recreation, (2) the functional analysis of agricultural landscapes, (3) analysis of human flow through park and wilderness areas, (4) analysis of a beetle's-eye view of the world to understand fractal patterns (Wiens and Milne, 1989; Johnson et al., 1992), (5) understanding what reserve size means for maintaining species diversity (Margules et al., 1988; Nicholls and Margules, 1993; Shepherd and Brown, 1993), (6) how fire influences forest landscapes (Turner and Romme, 1994), and (7) the integration of human values into long-term monitoring of ecological processes (Turner, 1995).

2.3.1 Spatially Explicit Ecology

Although landscape ecology usually involves temporal considerations, the unifying theme of landscape ecology is the explicit consideration of space. For instance, Whitaker's (1972) work, which treated large spatial scales, is an example of landscape ecology even though it ignored temporal considerations. In contrast, the work of Lotka and of Volterra (Lotka, 1925; Volterra, 1926) cannot be considered landscape ecology because, while it considered temporal dynamics, their important work did not consider space.

Moreover, the spatial scale at which landscape ecologists work is often considerably broader than traditional ecology. Traditional plant ecology, for example, has focused inordinately on scales of 1 square meter and one growing season. However, the distribution, abundance, and processes affecting species at the community level are partly a function of the landscape in which they live. By acknowledging this interaction, a landscape study adds spatial heterogeneity to a population, community, or ecosystem study. Expanding the spatial scale can reveal how the distribution and abundance of species within a local community are influenced by the larger regional context in which it is embedded. As larger areas are considered, longer-term temporal processes become important determinants of ecological dynamics.

Understanding phenomena within their spatial context can provide important insight for management. The original efforts to save the Northern Spotted Owl *(Strix occidentalis caurina)* in the northwestern United States focused primarily on the total amount of reserved habitat without considering the ability of juveniles to disperse to suitable habitat within the landscape. Spatial models led researchers to conclude that the original patchy preserves were especially vulnerable to extinction and showed that the arrangement and size of habitat patches on the landscape were as important as the overall amount of habitat (Harrison et al., 1993). Thus, patch-level diversity at a larger scale can influence species diversity and species distributions at local and regional scales.

The consequences of adding broad-scale spatial heterogeneity and long-term temporal dynamics are profound. The spatial dimension allows for integration across studies from studies of small organisms to regional analysis. It allows for the analysis of impacts at scales different from those normally incorporated into the system. Expanding the time frame for temporal dynamics allows us to integrate history, both abiotic and biotic, and the activities and cultures of human beings into our analysis.

2.3.2 Landscape Structure, Function, and Change

Three properties of landscapes are important for assessment. First, landscapes have structure. That is, there are spatial relationships between elements. The relationship may be between trees and gaps in a forest. At a broader scale, there is a relationship between forest and agricultural patches in a landscape mosaic. Landscapes have spatial continuity between these elements, but they are not homogeneous. Landscapes exhibit spatial pattern and internal heterogeneity. It is this heterogeneity that is of interest. The continuity of the landscape suggests that it also has boundaries where landscape elements come together. Boundaries and barriers may prevent particular species from responding to changes by shifting between habitat types or sites, and this in turn may result in decreased diversity (Wiens et al., 1985; Forman and Moore, 1992; Vos and Opdam, 1993; Forman, 1995). Boundaries, natural and anthropogenic, are very important in designing research and monitoring for ecological assessment.

Second, landscapes also have function. That is, there are interactions among spatial elements. These may be flows of energy, materials, species, or genes among component elements of the landscape (Forman and Godron, 1986; Forman, 1995). These ecological functions are often important in a societal context. For instance, woodland corridors have been shown to act as conduits for many wildlife species. This may increase gene flow between interbreeding populations of important species and so help to maintain species diversity. A wetland or riparian zone may function as a filter for pollutants and maintain or increase water purity within the system (Forman, 1995).

Patterning and patchiness of both structure and function can be recognized at virtually every scale of investigation. Thus, landscapes may be of any size. In fact, we suggest that the notion of landscape is scale independent in the same sense as ecosystem is scale independent. This is a widely held view (Allen and Hoekstra, 1992); however, some researchers, among them Forman (1995), hold that a landscape "is a mosaic where the mix of local ecosystems or land uses is repeated in similar form over a kilometers-wide area" (Forman, 1995, p. 13). In this view, landscapes occupy a limited portion of the spectrum of ecological scales. We find that the scale-independent definition of landscapes is preferable because it can then be easily applied to any ecological property important for assessment (Turner, 1995). The questions and purpose of the assessment will determine the relevant scales.

Third, landscapes change. There is an alteration of structure and function of the ecological mosaic (the elements arrayed in space) over time. Clearly, change is a natural process, and natural changes in dynamical ecological systems are integrated from the smallest organisms to the watershed and regional levels. Change may be the normal turnover in species in a forest that changes in time or in space with the formation of gaps and the resulting successional dynamics. Alternatively, change may be anthropogenically induced, such as the clearing of an agricultural field or the harvesting of timber from a forest. Detection and analysis of change to the structure and function of the system are critical. Prediction turns on our ability first to identify change or trends in data and then on our ability to understand the meaning of these changes for ecosystem function at all levels in the hierarchy.

2.4 System Dynamics

Landscape change over time requires us to examine the theoretical basis for understanding ecosystem dynamics. The conceptual underpinnings of ecologists' understanding of ecological dynamics have shifted dramatically in recent decades. Earlier theories of ecosystems focused on equilibrium models, often expressed verbally in terms of the "balance of nature" (Egerton, 1973; Pimm, 1991; Wu and Loucks, 1995). Recent decades have seen the emergence of a number of alternative paradigms for describing ecological dynamics (Botkin, 1990; Pahl-Wostl, 1995; Wu and Loucks, 1995). We will not explore these theories in depth, but rather give a brief overview suggesting the breadth of the perspectives that may be applicable to the analysis of ecosystem or landscape dynamics. The reader should bear in mind that the applicability of these various theories depends on the spatiotemporal scale at which the system is considered. Different theories represent different ways of viewing the system. Thus the choice of which theoretical approach to apply to a given situation will depend on the phenomena of interest, as well as on the spatial and temporal scales.

2.4.1 Simple Dynamics

Consider a forest at a regional scale. At any point in time, some areas of the forest may be experiencing low endemic levels of pests, while others may experience epidemic outbreaks that kill trees. Part of the forest may be on fire, while another part may not have experienced fire for decades or centuries. Each of these are functional states of the forest that are happening right now at different points in space. These states may also be realized in one location at different times. There is potential for multiple equilibria (or states) of an ecosystem with the possibility of sudden (surprising, perhaps discontinuous) transitions from one state to another (e.g., Holling, 1986). Each of these equilibria acts as an attractor toward which the system moves over time. From a management perspective, the significance of these transitions varies. In some cases, certain equilibria (i.e., forest without fire or pests) may be deemed desirable, while others are considered disastrous. The management objective is to keep the system out of the undesired states and their domains of attraction. On the other hand, certain shifts between domains of attraction may be important in the long-term dynamics of natural systems.

Holling uses the term *resilience* to refer to the capacity of an ecosystem to shift to multiple states or domains in a manner that ensures the long-term adaptability and continued functioning of the system in the face of perturbations (Holling, 1986). In this context, proper management would allow transitions. Attempts to maintain a system in one desired state are likely to be unsuccessful in the long term. A system restricted to a particular attractor will eventually become fragile and will ultimately collapse when exposed to a perturbation for which the system lacks an adaptive response (Holling and Meffe, 1996). That is, when it is natural for an ecosystem to move back and forth between states such as fire–no fire, then it is a mistake to try to confine it to a single state, because the system will lose its ability to incorporate small disturbances into the normal function of the forest. Thus we have begun to rethink our management practice of fire suppression. If we had had these theories to consider at the point that managers initially decided to suppress fire in the Yellowstone ecosystem, we may have been able to anticipate the eventual catastrophic burn. The key point for assessment is to identify the scale at which the system is to be managed, along with the other scales of the system that management decisions may affect.

At times ecosystem dynamics converge not toward a fixed state (i.e., a mathematical equilibrium), but toward a fixed cycle of states. Predator–prey models often predict periodic fluctuations in species densities, and empirical data suggest that such behavior is a feature of some real-world systems (Pimm, 1991). Mathematically, the dynamics are governed by an attractor that exhibits cyclic behavior. Assessment strategies for systems with expected cyclic behavior will have to carefully consider the time scale of investigation relative to the time scale of the system's natural cycles (periodicities).

2.4.2 Chaotic Dynamics

Systems that display chaotic dynamics converge toward a strange attractor. Attractors are strange because they are a complex fractal, rather than the equilibrium point or cycle characteristic of simple dynamical systems such as we discussed above. Chaotic systems fluctuate in a complex and nonperiodic manner that can be described by fractal geometry. The prevalence of chaos in real-world ecosystems is still a matter of debate, although some data strongly suggest a chaotic component in the dynamics, for example, of insect population dynamics and childhood diseases (Schaffer and Kot, 1986a, b). Chaotic dynamics are apparent in certain nonlinear mathematical models of ecological systems (e.g., May, 1974; May and Oster, 1976; Gilpin, 1979; Hastings et al., 1993). These models suggest limitations to our ability to predict ecosystem dynamics. Chaotic systems display limited predictability due to extreme sensitivity to initial conditions. In other words, small uncertainties in the state of a system at time t (or small perturbations) can become amplified over time such that predictions of the expected state at future times become increasingly less accurate. On the other hand, although long-term prediction may be compromised, nonlinear forecasting of shorter-term events might well have applications in the assessment and management of ecological systems, although such applications have not been widely explored (Sugihara and May, 1990; Casdagli and Eubank, 1992). The major limitation is that these techniques generally require large quantities of precise data, which are difficult to obtain for most ecological variables.

2.4.3 Self-organized Criticality

A potentially valuable perspective is the emerging description of ecological dynamics in terms of *self-organized criticality*, in which ecosystems are poised at the edge of chaos, not displaying simple equilibrial or periodic behavior, but not displaying fully developed chaotic dynamics either (Bak et al., 1988; Bak and Chen, 1991; Kauffman, 1993). The systems are self-organized in the sense that the dynamics of the system itself drive it to the edge of chaos. For example, a forest contains trees that started to grow there (initial condition) and, over time, a fuel load is building up, so fire may remove the trees and create a clearing, from which the process begins anew. A characteristic feature of systems poised on the edge of chaos is fluctuations that vary in magnitude, displaying a power law relationship (linear in a log–log plot) between the magnitude of the fluctuation and its frequency of occurrence. Such power law relationships are evident in nature, such as the frequency–magnitude relation for earthquakes or the frequency–discharge relationship for floods (Malamud et al., 1996) and forest fires (Malamud et al., 1998). Analysis of extinction dynamics in the introduced avifauna of the Hawaiian islands provides evidence that ecological dynamics may also show such power law scaling (Keitt and Marquet, 1996). Certain models of evolutionary dynamics display self-organized critical behavior (Bak and Sneppen, 1993; Kauffman, 1993). Edge-of-chaos systems are sometimes described as transient dynamic systems (see Section 2.4.4) because they do not converge toward any attractor. If self-organized criticality proves to be a common feature of ecosystems, power laws should find general applicability in quantifying the frequency–magnitude relationships of various events. This would allow better prediction of the expected frequency of extreme events (floods, fires, droughts, etc.) and enhance our understanding of their role in the long-term dynamics of ecosystems.

2.4.4 Transient Dynamics

Mathematically, equilibria, periodic cycles, and chaos are all types of attractors that describe the ultimate behavior of the system. The concept of transient dynamics, on the other hand, focuses on the behavior of the system before an attractor is reached. If a forest is clear-cut, the transient dynamics are what occurs before the area settles into being a functioning forest again. Transient dynamics may be important in the analysis of ecosystems, because disturbances or changes in the constraints (e.g., fire frequency, climate) may prevent the system from reaching the attractor on the time scale of interest. Some systems may never converge to any attractor. Such systems may be said to have long or infinite transients, and they may display more open-ended behavior, showing greater adaptability in the face of perturbations or environmental fluctuations. History plays a significant role in these systems; their current response to a perturbation may be contingent on past events. Evolution may be an appropriate metaphor for systems that are constantly changing. Evolutionary innovation results in continuous change of species and morphologies through time. Technological systems may be another example—cultural use of resources keeps the system in flux. For instance, rubber was extracted from plant sources before World War II, but the pressure on these resources shifted with the development of synthetic rubber. People in tropical regions that depended on natural rubber for their livelihood had to shift to extraction of other resources. Assessment and analysis techniques have not yet been developed for long-transient dynamic systems, but we should keep these concepts in our consideration because they may offer a means to understand the continuing processes of change that are our management decisions. If we can develop knowledge about how one management decision may lead to new ecological phenomena (because we understand the ecological patterns and processes that it affects), we may be able to follow the thread of change in process to the necessity for a new management policy. From this point we may be able to better address the assessments required for sustainable development, for instance.

2.5 Theory Applied

With such a diversity of possible theoretical perspectives, the following question arises: Which one is the right one? There is no simple answer, because it depends crucially on the ecological phenomena of interest, as well as the scale. For instance, if we focus on net primary production (NPP) in the pelagic zone of a lake, a simple single-equilibrium model may suffice, because annual NPP is relatively constant year after year as determined by nutrient loading. The same lake, however, may display multiple states or attractors if we focus on the species composition of the phytoplankton that are the primary producers. There will be a sequence of dominant phytoplankton species over the course of a single year, and year to year variation in the timing of phytoplankton blooms and the dominant species involved may be great.

The temporal scale is important. For instance, community dynamics on time scales of a few generations for the species involved may be modeled as an approach to an attractor (e.g., a stable community composition), but if hundreds or thousands of generations are considered, evolutionary processes will make former attractors unstable—a dynamic that might be better represented by an edge-of-chaos system with infinite transients, rather than stable attractors.

Spatial scale is also important. A fire in a small forest stand will be perceived as a massive and sudden change in state. From a broader landscape perspective, there may be a shifting mosaic of burnt and unburnt forest stands, with the overall proportions remaining relatively constant over time. At the regional level, fire is not so much an external perturbation as an intrinsic part of the system dynamics. A simple equilibrium model may be sufficient to represent the relative constancy of mosaic composition over time; but if we focus on the fluctuations away from the long-term mean, there is evidence that forest fires display power law size–frequency relationships suggestive of self-organized criticality (Drossel and Schwabl, 1992; Malamud et al., 1998).

Hierarchy and landscape theories and techniques improve our abilities to conduct ecological assessments by providing us with methods that we can use to design research around important assessment issues. Figure 2.1 suggests that there is a focal level N toward which research, measurement, and monitoring should be directed. It also suggests that the research, whether modeling of ecological systems or field-based research, need only account for three levels, as suggested by Johnson (1996) and Turner (1995). Remember that functions at one level produce the structure of the next higher level (i.e., physiological processes grow and maintain the trees). Often we find that it is change or variability at the higher level that controls or constrains the system (i.e., climatic factors constrain forest expansion regionally and globally). The level at which the inquiry begins is arbitrary (Allen et al., 1987). This decision determines which observations, scales, and measurements are appropriate. N is defined uniquely for the question at hand; it may be defined by purely scientific interest or in response to ecological crisis and societal concerns.

Figure 2.3 is one (far from unique) conception of a phenomena and measurement criteria hierarchy. It could be any set of interesting and related phenomena. By related, we mean that there is a connection between the processes and functioning of one level and the expression of phenomena at a higher or lower level. More familiar are space–time diagrams in which structures, functions, systems, or other lists of ecologically interesting and related things are arranged along a diagonal of their inherent time and space scales. In this figure the time–space dynamics are combined in one line across the top, and the phenomena are arranged so that some of the possible measurement criteria can be associated with each phenomenon. Note that we have defined a slow–large level, which we call landscape. This is arbitrary. Note, too, that there are spatially arrayed processes that are appropriate criteria at most scales. For instance, measurement of spatial cover produced by a population of plants is amenable to landscape analytical techniques.

The phenomena along the diagonal—cell, organism, population, community, and ecosystem/landscape—may each be considered as a focal level N. If the population level is chosen as the level of investigative focus, then the level of the organism is the $N - 1$ level of mechanisms, and community is the $N + 1$ control level. Among the criteria that could be chosen to describe the population level are intraspecific competition effects, genetic diversity, spatial dispersion and biomass production of the population, and dispersal or invasion rate of propagules or offspring. Understanding of the

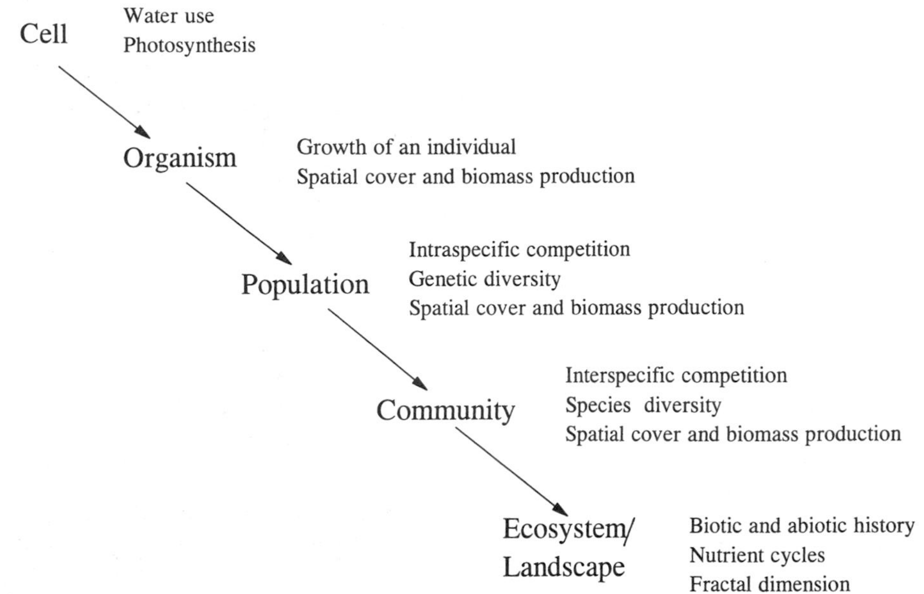

FIGURE 2.3. A hierarchy of phenomena and some of their associated measurement criteria. Intraspecific competition, genetic diversity, spatial cover, and biomass production are appropriate measurements for investigations at the population level.

mechanisms that create and maintain the population would come from the lower level of the single organism ($N - 1$), where reproduction and growth potential are important. The limits and context of the population will be found in the community ($N + 1$) interactions of interspecific competition, for instance, among various populations.

Assessment is not just about what is happening now. Often assessments are about predicting the consequences of an action. If assessment questions involve a specific anthropogenic impact, then the assessment plan should focus on the level at which the impact is targeted, and change must be followed through the scales of effect. Any impact at a particular hierarchical level (spatiotemporal scale) will change the constraints for lower (smaller, faster) levels. Repeated impacts at one level may interact to generate an effect on the dynamics at higher (larger, slower) levels.

2.6 Conclusions

Hierarchy theory, landscape ecology, and system dynamics provide ways to understand and sort through the complexity of the system so that investigations can focus on the information that will most clearly answer the assessment question. The level or levels in the system that are investigated will depend on the nature of the question and the type of answer required. A one-time assessment of the impact of a clear cut may be found by simply counting the trees and measuring the space cleared. The effect of the clear cut on the dynamics of forest ecosystem structure and function, however, will require the investigation of several levels in the system. Whatever the scale of investigation and the techniques chosen to conduct the investigation, vital questions can be answered by considering the system at three levels: mechanistic, focal, and control.

System dynamics, hierarchy, and landscape theories provide us with powerful ways to describe the behavior of systems once the assessment has been conducted. System dynamic descriptions and techniques may make complex systems predictable over a range of scales that is necessary for decisions based on ecological assessment. Importantly, system dynamics descriptions may provide the key method for integrating cultural history and economics with environmental systems. The system dynamical views presented here provide a basic template for predicting change in system behavior. Which view best describes the system being managed depends at least in part on the focus of the assessment and the goals of management.

2.7 References

Ahl, V.; Allen, T. F. H. 1996. *Hierarchy theory: a vision, vocabulary, and epistemology.* New York: Columbia University Press.

Allen, T. F. H.; Hoekstra, T. W. 1992. *Toward a unified ecology.* New York: Columbia University Press.

Allen, T. F. H.; Starr, T. B. 1982. *Hierarchy: perspectives for ecological complexity.* Chicago: University of Chicago Press.

Allen, T. F. H.; O'Neill, R. V; Hoekstra, T. W. 1987. Interlevel relations in ecological research and management: some working principles from hierarchy theory. *J. Appl. Syst. Anal.* 14:63–79.

Bak, P.; Chen, K. 1991. Self-organized criticality. *Sci. Amer.* 264(1):46–53.

Bak, P.; Sneppen, K. 1993. Punctuated equilibrium and criticality in a simple model of evolution. *Phys. Rev. Lett.* 71:4083–4086.

Bak, P.; Tang, C.; Wiesenfeld, K. 1988. Self-organized criticality. *Phys. Rev. A* 38:364–374.

Botkin, D. B. 1990. *Discordant harmonies: a new ecology for the twenty-first century.* Oxford, UK: Oxford University Press.

Casdagli, M.; Eubank, S. 1992. *Nonlinear modeling and forecasting.* Redwood City, CA: Addison-Wesley.

Costanza, R.; Maxwell, T. 1994. Resolution and predictability: an approach to the scaling problem. *Landscape Ecol.* 9:47–57.

Costanza, R.; Wainger, L.; Folke, C. 1993. Modeling complex ecological economic systems: toward an evolutionary, dynamic understanding of people and nature. *BioScience* 43:545–555.

Drossel, B.; Schwabl, F. 1992. Self-organized criticality in a forest-fire model. *Physica A* 191:47–50.

Egerton, F. N. 1973. Changing concepts of the balance of nature. *Quart. Rev. Biol.* 48:322–350.

Forman, R. T. T. 1995. *Land mosaics: the ecology of landscapes and regions.* Cambridge, UK: Cambridge University Press.

Forman, R. T. T.; Godron, M. 1986. *Landscape ecology.* New York: John Wiley & Sons.

Forman, R. T. T.; Moore, P. N. 1992. Theoretical foundations for understanding boundaries in landscape mosaics. In: Hansen, A. J.; di Castri, F., eds. *Landscape boundaries: consequences for biotic diversity and ecological flows.* New York: Springer-Verlag: 236–258.

Gilpin, M. E. 1979. Spiral chaos in a predator–prey model. *Amer. Naturalist* 113:306–308.

Harrison, S.; Stahl, A.; Doak, D. 1993. Spatial models and spotted owls: exploring some biological issues behind recent events. *Conserv. Biol.* 7:950–953.

Hastings, A.; Hom, C.; Ellner, S.; Turchin, P.; Godfray, H. C. J. 1993. Chaos in ecology: is mother nature a strange attractor? *Ann. Rev. Ecol. Syst.* 24:1–33.

Holling, C. S. 1986. The resilience of terrestrial ecosystems, local surprise and global change. In: Clark, W. C.; Munn, R. E., eds. *Sustainable development of the biosphere.* Cambridge, UK: Cambridge University Press: 292–317.

Holling, C. S.; Meffe, G. K. 1996. Command and control and the pathology of natural resource management. *Conserv. Biol.* 10:328–337.

Johnson, A. R. 1996. Spatiotemporal hierarchies in ecological theory and modeling. In: Goodchild, M. F.; Steyaert, L. T.; Parks, B. O.; Johnston, C.; Maidment, D; Crane, M; Glendinning, S., eds. *Integrating geographic information systems and environmental modeling: progress and research issues.* Ft. Collins, CO: GIS World: 451–456.

Johnson, A. R.; Milne, B. T.; Wiens, J. A. 1992. Diffusion in fractal landscapes: simulations and experimental investigation of Tenebrionid beetle movements. *Ecology* 73:1968–1983.

Kauffman, S. A. 1993. The origins of order: self-organization and selection in evolution. Oxford, UK: Oxford University Press.

Keitt, T. H.; Marquet, P. A. 1996. The introduced Hawaiian avifauna reconsidered: evidence for self-organized criticality? *J. Theor. Biol.* 182:161–167.

Koestler, A. 1967. *The ghost in the machine.* New York: Macmillan.

Lotka, A. J. 1925. *Elements of physical biology.* New York: Dover [reprinted in 1956 under the title *Elements of mathematical biology*].

MacMahon, J. A.; Phillips, D. L.; Robinson, J. V.; Schimpf, D. J. 1978. Levels of biological organization: an organism-centered approach. *BioScience* 28: 700–704.

Malamud, B. D.; Turcotte, D. L.; Barton, C. C. 1996. The 1993 Mississippi River flood: a one hundred or a one thousand year event? *Environ. Eng. Geoscience* 2:479–486.

Malamud, B. D.; Morein, G.; Turcotte, D. L. 1998. Forest fires: an example of self-organized critical behavior. *Science* 281:1840–1842.

Margules, C. R.; Nicholls, A. O.; Pressey, R. L. 1988. Selecting networks of reserves to maximize biological diversity. *Biol. Conserv.* 43:63–67.

May, R. M. 1974. Biological populations with nonoverlapping generations: stable points, stable cycles and chaos. *Science* 186:645–647.

May, R. M.; Oster, G. F. 1976. Bifurcations and dynamic complexity in simple ecological models. *Amer. Naturalist* 110:573–599.

Naveh, Z. 1995. Interactions of landscapes and cultures. *Landscape Urban Plan.* 32:43–54.

Nicholls, A. O.; Margules, C. R. 1993. An upgraded reserve selection procedure. *Biol. Conserv.* 64:165–169.

O'Neill, R. V.; DeAngelis, D. L.; Waide, J. B.; Allen, T. F. H. 1986. *A hierarchical concept of ecosystems.* Princeton, NJ: Princeton University Press.

O'Neill, R. V; Johnson, A. R.; King, A. W. 1989. A hierarchical framework for the analysis of scale. *Landscape Ecol.* 3:193–205.

Pahl-Wostl, C. 1995. *The dynamic nature of ecosystems: chaos and order entwined.* New York: John Wiley & Sons.

Pimm, S. L. 1991. *The balance of nature? Ecological issues in the conservation of species and communities.* Chicago: University of Chicago Press: 115–134.

Ricklefs, R. E.; Schluter, D. 1993. Species diversity: regional and historical influences. In: Ricklefs, R. E.; Schluter, D., eds. *Species diversity in ecological communities.* Chicago: University of Chicago Press: 350–363.

Schaffer, W. M.; Kot, M. 1986a. Chaos in ecological systems: the coals that Newcastle forgot. *Trend. Ecol. Evolut.* 1:58–63.

Schaffer, W. M.; Kot, M. 1986b. Differential systems in ecology and epidemiology. In: Holden, A.V., ed. *Chaos.* Princeton, NJ: Princeton University Press: 158–178.

Shepherd, S. A.; Brown, L. D. 1993. What is an abalone stock? Implications for the role of refugia in conservation. *Can. J. Fisheries Aquat. Sci.* 50:2001–2009.

Sugihara, G.; May, R. M. 1990. Nonlinear forecasting as a way of distinguishing chaos from measurement error in time series. *Nature* 344:734–741.

Turner, M. G.; Romme, W. H. 1994. Landscape dynamics in crown fire ecosystems. *Landscape Ecol.* 9:59–65.

Turner, S. J. 1995. Scale, observation and measurement: critical choices for biodiversity research. In: Boyle, T. J. B.; Boontawee, B., eds. *Measuring and monitoring biodiversity in tropical and temperate forests.* Jakarta: Center for International Forestry Research (CIFOR): 97–111.

Urban, D. L.; O'Neill, R. V.; Shugart, H. H., Jr. 1989. Landscape ecology. *BioScience* 37:119–127.

Volterra, V. 1926. Fluctuations in the abundance of a species considered mathematically. *Nature* 118:558–560.

Vos, C. C.; Opdam, P., editors. 1993. *Landscape ecology of a stressed environment.* London: Chapman & Hall.

Whitaker, R. H. 1956. Vegetation of the Great Smoky Mountains. *Ecol. Monogr.* 26:1–80.

Whitaker, R. H. 1972. *Communities and ecosystems.* New York: Macmillan.

Wiens, J. A. 1989. Spatial scaling in ecology. *Funct. Ecol.* 3:385–397.

Wiens, J. A.; Milne, B. T. 1989. Scaling of "landscapes" in landscape ecology, or, landscape ecology from a beetle's perspective. *Landscape Ecol.* 3:87–96.

Wiens, J. A.; Crawford, C. S.; Gosz, J. R. 1985. Boundary dynamics: a conceptual framework for studying landscape ecosystems. *Oikos* 45:421–427.

Wu, J.; Loucks, O. L. 1995. From balance of nature to hierarchical patch dynamics: a paradigm shift in ecology. *Quart. Rev. Biol.* 70:439–466.

3
Ecosystem Characterization and Ecological Assessments

Patrick S. Bourgeron, Hope C. Humphries, and Mark E. Jensen

3.1 Introduction

Ecological assessments are an important component of any strategy for making or reevaluating land management and regulatory decisions (see Chapters 1, 9, and 35; also Slocombe, 1993; Jensen and Bourgeron, 1994; Bourgeron et al., 1995). An important objective of ecological assessments is the identification, location, and description of the biotic and abiotic features of a landscape. Landscape features exhibit heterogeneity at a variety of scales (Turner et al., 1995). This heterogeneity is characterized by identifying relevant patterns and the processes that produce patterns in a landscape (Bourgeron and Jensen, 1994). Distinct patterns and processes occur at a variety of spatial and temporal scales of organization (see Chapter 2). For ecological assessments, an explicit understanding is needed of the scaled relationships of biological and biophysical characteristics from site to regional scales (Lessard, 1995; Lessard et al., 1999). Therefore, the characterization process is a multiscaled approach conducted within a hierarchical framework (Bourgeron and Jensen, 1994; Hann et al., 1994; Bourgeron et al., 1995; Jensen et al., 1996).

The emerging approach for land and resource management (e.g., O'Neill et al., 1997; Lessard et al., 1999) recognizes ecosystems as the context for assessment and monitoring of various commodities, species, and other resources (Lubchenco et al., 1991; Franklin, 1993; Mladenoff and Pastor, 1993; Jensen and Bourgeron, 1994; Stork et al., 1997). Such an ecosystem approach requires methods for characterizing the landscape attributes that influence ecosystem integrity and sustainability at various spatial scales (Zonneveld, 1988, 1989; Levin, 1992; Franklin, 1993; Mladenoff and Pastor, 1993; Bourgeron et al., 1994b, 1995; Bunce et al., 1996; Mladenoff et al., 1997). Regardless of their specific objectives and scope, all ecological assessments require the delineation of broad-scale land units to provide a context for the evaluation and management of commodities or elements of biodiversity; this is the process of regionalization (Omernik, 1987; Jensen et al., 1996; Bailey, 1995; Maxwell et al., 1995). Ecological assessments also require delineation of fine-scale ecological units to describe the basic landscape patterns and characteristics that are important for the issues at hand (Zonneveld, 1989, 1994).

This chapter presents general principles pertaining to two aspects of ecosystem characterization, classification and regionalization–mapping of ecosystem units, that are central to ecological assessments (Bailey, 1996; Rowe, 1996). As a consequence of the hypothesis that ecosystems can be characterized at different scales (hierarchically arranged), ecosystem units can be arranged in a hierarchy of sizes, can capture links between biotic and abiotic ecosystem components, and can link terrestrial and aquatic systems. Ecosystem characterization conducted for regional ecological assessments combines the principles of ecosystem theory and landscape ecology (see Chapter 2) with the methods developed in geography for describing tracts of land. This chapter includes discussion of issues of definition and prediction of ecosystem properties and extrapolation of results.

Patrick S. Bourgeron and Hope C. Humphries wish to acknowledge partial funding provided by a Science to Achieve Results grant from the U.S. Environmental Protection Agency ("Multi-scaled Assessment Methods: Prototype Development within the Interior Columbia Basin").

Ecosystem characterization is not itself a goal, but is a primary tool for ecological assessments and monitoring (see discussion of different approaches in Klijn, 1994; Sims et al., 1996; Grossman et al., 1999). The characterization process can be either static (e.g., physical land evaluation) or dynamic (e.g., see Chapters 18, 19, Chapter 23, and 26). However, all applications must somehow relate the patterns of interest to the agents of pattern formation (sensu Urban et al., 1987) or to the causes of ecological changes. Whereas ecosystems are characterized as functional units, ecological units must be delineated as their spatial expression (Bailey et al., 1994). These ecological units represent a stratification of landscapes for efficient data collection and analysis, as well as for presentation of results (e.g., Bunce et al., 1996).

3.2 General Description of Ecosystem Characterization

Ecosystems are groups of interacting, interdependent parts (e.g., species, resources) linked to each other by the exchange of energy, matter, and information. Such systems are considered complex because they are characterized by strong interactions among components and intricate feedback loops, as well as significant time and space lags, discontinuities, thresholds, and limits (Costanza et al., 1993). Ecosystems can be described at many different scales (see Chapter 2; also Levin, 1992), and therefore the spatial and temporal relationships of particular ecosystems or their components should be clearly defined. Describing ecosystems is a routine scientific endeavor that leads to the generation of classifications and maps (e.g., Klijn, 1994; Sims et al., 1996; Grossman et al., 1999).

Making predictions about future states of ecosystems under various scenarios requires elucidating the relationships between patterns and hypothesized causal factors, that is, processes or agents of pattern formation. Once a correlation or a cause–effect relationship between pattern and process is determined, predictions are made using (1) data summarization resulting from classification and/or mapping, (2) statistical or simulation models, or (3) a combination of these approaches. Many scientists establish a relationship between pattern and processes to make predictions, but the approach taken may lack rigor and standardization. Nevertheless, the formulation of recommended land and watershed management practices is accomplished by integrating predictions into proposed conservation and management actions.

Complex ecosystem patterns and the multitude of processes that form them exist within a hierarchical framework (see Chapter 2; also Allen and Starr, 1982; Allen et al., 1984; O'Neill et al., 1986). Scale dependency is very significant in the context of assessments conducted to understand relationships among patterns and processes, because such relationships change as spatial scale changes (Turner, 1990; Davis et al., 1991). In recent years, considerable attention has been directed toward describing the hierarchical organization of terrestrial and aquatic ecosystems (see Chapters 22 and 24; also Frissell et al., 1986; Amoros et al., 1987; Urban et al., 1987; Minshall, 1988, 1994; Naiman, 1988; Pringle et al., 1988; Wiens, 1989; Poff and Ward, 1990; Bayley and Li, 1992; Gardner et al., 1992; Holling, 1992; Lathrop and Peterson, 1992; Klijn, 1994; Sims et al., 1996; Grossman et al., 1999). As applied to landscape ecology, hierarchy theory provides a framework for characterizing the components of an ecosystem or set of ecosystems, as well as for detecting the linkages among different scales of ecological organization (see Chapter 2; also Allen and Starr, 1982; O'Neill et al., 1986; Urban et al., 1987).

Ecosystem characterization (Levin, 1992) is carried out to map and conduct landscape evaluation at scales ranging from the site to the continent. Ecological classifications are developed to delineate ecological units at multiple scales using criteria appropriate to the objectives of the project. Terrestrial ecological units are defined using terrestrial variables (e.g., ECOMAP, 1993; Bailey et al., 1994; Bailey, 1995). Aquatic–hydrologic units are defined using variables relevant to aquatic systems (e.g., Maxwell et al., 1995; Jensen et al., 1996). In all cases, boundaries are drawn around ecosystems to delineate the ecological units at appropriate ecological and planning scales (Omernik, 1987; Zonneveld, 1989; Maxwell et al., 1995; Jensen et al., 1996). Major ecological classification and mapping systems and their goals, objectives, and uses are presented in Klijn (1994), Sims et al. (1996), and Grossman et al. (1999).

Most ecological assessments and land management practices are concerned with patterns or static structures, but the dynamic nature of ecosystems cannot be ignored. It has long been documented that ecosystems are rarely in equilibrium. Nonequilibrium dynamics include succession along multiple pathways, discontinuities, and surprises as ecosystems track constantly changing environmental conditions (Holling, 1986; Kay, 1991; Costanza et al., 1993). Examples of dynamic processes that can lead to nonequilibrium behavior for terrestrial ecosystems (see Chapter 23) include fire (Christensen et al., 1989; Baker, 1992a, b; Turner et al., 1994) and grazing (McNaughton, 1985), and for aquatic systems (see Chapter 26) include flooding (e.g., Poff and Ward, 1989; Sparks, 1995) and fluvial geomorphologic processes (Ligon et al., 1995). As a consequence of the dynamic nature and hierarchical structure of ecosystems, there are likely to be limits to the predictability of natural patterns and processes at different scales (Costanza et al., 1993).

There are three general steps in conducting ecosystem characterization: (1) development of a hierarchical land classification of ecological units, including both a geographical or regional definition and a biological definition of the classes (Austin and Margules, 1986; Mackey et al., 1988; Belbin, 1993); (2) definition of the relevant ecological properties (e.g., processes) of these units; and (3) a method of allocating areas to the ecological units (mapping). Data resulting from ecosystem characterization can be used to assess the ecological conditions of land areas and to provide the framework for making regulatory and land management decisions (see Chapter 1; also see an example for regional conservation planning in Chapter 20).

Although much is known about ecosystem function and landscape dynamics, many questions need to be answered, including how to integrate population and landscape ecology, how to interpret the ecological effects of spatial patterns and changes in these patterns, and how to identify controls on ecological processes at different spatial scales (Turner et al., 1995). Ecosystem characterization provides a basis for answering many of these questions and for overcoming two major challenges to broad-scale ecological studies (Turner et al., 1995): assembling spatial databases over large areas (see Chapters 5 through 7) and offering alternative approaches to traditional experiments.

3.3 Delineating Ecosystem Units

Two main aspects of ecosystem characterization are classification and regionalization–mapping. Characterization shares with classification the fact that it is an abstraction (Zonneveld, 1994; Rowe, 1996). During the characterization process, we distinguish guiding principles, properties, and diagnostic characteristics (Zonneveld, 1994). Guiding principles are rules for choosing and calibrating properties and assessing the hierarchy of an ecological classification. These rules are derived based on intrinsic characteristics of the system (e.g., ecosystem function), relationships with the surroundings, or the genesis of the system, depending on the purpose of the characterization. The general guiding principle of characterization is the convergence of evidence (Zonneveld, 1994), which means that properties are chosen that result in convergence on an object (e.g., an ecological unit) or class of objects and lead to contrast of this object or class of objects with others. Properties are all the attributes of the object to be characterized; diagnostic characteristics for a class (or ecosystem) are selected from the suite of properties. Diagnostic characteristics should be properties of the entire system and not of the component parts (see Section 3.6), should be functional, and should correspond to the phenomenon of interest.

For example, the process of landform genesis is a logical guiding principle for geomorphological classifications, and the degree and form of slopes, sedimentation, and erosion pattern constitute properties. Slope degree, shape of slope, and horizontal configuration of slope types (pattern) are often used as diagnostic characteristics of geomorphological classifications. For vegetation classification, the development of stable successional stages can be distinguished as a guiding principle, and the vertical structure of the vegetation and floristic species composition are therefore good examples of measurable properties that can be used as diagnostic characteristics. This approach has also been used for soil, geological, and landscape classifications (see discussion in Zonneveld, 1994).

The degree of human influence in the landscape is important in selecting guiding principles and diagnostic characteristics. In landscapes that can be described as ranging from natural to seminatural, geomorphologic and climatic factors, in combination with pedogenic processes, may have the greatest constraining effect on ecosystem patterns. In contrast, land-use history may be the most important constraining factor in cultural landscapes (Haber, 1994; Zonneveld, 1994).

Mapping is a visual representation of the classification units. Much had been said about the merits of a priori versus a posteriori classifications during mapping efforts (e.g., see the classic work of Kuchler, 1973; also see Kuchler and Zonneveld, 1988). The mapping process forces a surveyor to confront all aspects of the landscape (Rowe, 1996). Although confusing samples (e.g., outliers) can be ignored in a classification framework, mapping necessitates the delineation of units over the entire surveyed area. The method for recognizing units, that is, the delineation of mapping unit boundaries, constitutes a set of de facto hypotheses about the compositional and structural differences lying on either side of the boundaries (Rowe, 1996). Differences in interpretation of features can be traced to the hypotheses, which can be tested (e.g., Uhlig and Jordan, 1996; see Section 3.4).

3.4 Boundaries

Boundaries are recognized by perceived changes in the relationships among attributes (e.g., vegetation, landform, climate) of ecological units. Boundary recognition may be difficult because of two problems: (1) the continuous or semicontinuous nature of most ecological attributes, which may be expressed as gradients, and (2) sampling considerations. The first problem leads to indistinct boundaries between ecological units; indeed, such boundaries are often termed *fuzzy*. Therefore, clear rules need to be formulated for the delineation of lines on a map derived from attributes. Three questions should be addressed (after Zonneveld, 1989):

1. What attributes determine the identity of the unit?
2. What attributes determine a unit's relevance for a particular purpose?
3. What attributes determine the patterns of spatial variability in the unit?

The first question should be answered by precisely defining the objectives of the characterization, which influences the choice of the criteria for classification and ecological unit mapping (see Section 3.6). Various analytical techniques can then be used to quantify the explanatory power of a map with respect to the patterns and processes of interest (see discussion in Chapter 22). The second question is answered by correlating a specific purpose with one or more of the attributes. For example, if assessment of erosion potential is the purpose of a project, slope is an important attribute. The third question deals with spatial heterogeneity among ecological units and becomes more important the higher the unit is in a hierarchy (Zonneveld, 1989, 1994).

The second problem of boundary recognition is rooted in the fact that ecosystem components (e.g., species) generally are not randomly distributed. For example, if an area contains n different biophysical environments that have sharp boundaries, each supporting a distinct set of plant species, a sufficiently large sample of randomly distributed adequately sized plots will generate n vegetation groupings after analysis, each characterized by a specific species combination (after Zonneveld, 1994). However, in such landscapes, which lack intermediate biophysical environments (ecotones), a purely random sampling design would be likely to include plots that overlap the boundary between two biophysical environments, describing a transitional condition that does not in fact exist. The size and shape of the plots sampled influence the degree to which this phenomenon affects ecosystem characterization. The pattern of distribution of ecosystems and their components affects the sampling design (see Chapters 6 and 7 for complementary views on sampling strategies for ecological assessments).

3.5 Characterization by Agglomeration or Subdivision: Bottom-up or Top-down Approach?

An important decision to be made when conducting ecosystem characterization is whether information should be integrated by agglomeration (also called *typification* or *bottom-up*) or by subdivision (also called *top-down*) of the data (Figure 3.1). The choice of the approach is important because the relationships between levels of ecological hierarchies are not symmetrical (see Chapter 2). Emergent properties of higher levels in the hierarchy cannot be predicted from the properties of lower levels. This phenomenon must be considered while attempting to extrapolate across scales within an ecological hierarchy (see discussion in Perera et al., 1996).

The bottom-up approach is nonspatial (i.e., independent from spatial constraints, which would not necessarily be identified; see Wessman, 1992). Bottom-up approaches are traditionally used for taxonomic classification systems and are inductive (Figure 3.1). There is no relationship to map scale. In a top-down approach, the context is developed (i.e., patterns are analyzed), and then function is predicted (i.e., process is inferred; Perera et al., 1996). Top-down approaches take into consideration spatial patterns and hierarchy (Zonneveld 1989, 1994) and identify those factors that constrain sublevels (Wessman, 1992). These approaches are traditionally used for biogeoclimatic classifications (see Chapter 22) and are deductive. The units defined may be complex mosaics of ecosystems (Figure 3.1). Starting from a site-specific ecosystem, the bottom-up approach leads to increasingly higher level taxonomic classes. In contrast, the top-down approach starts from large, ecologically heterogeneous areas

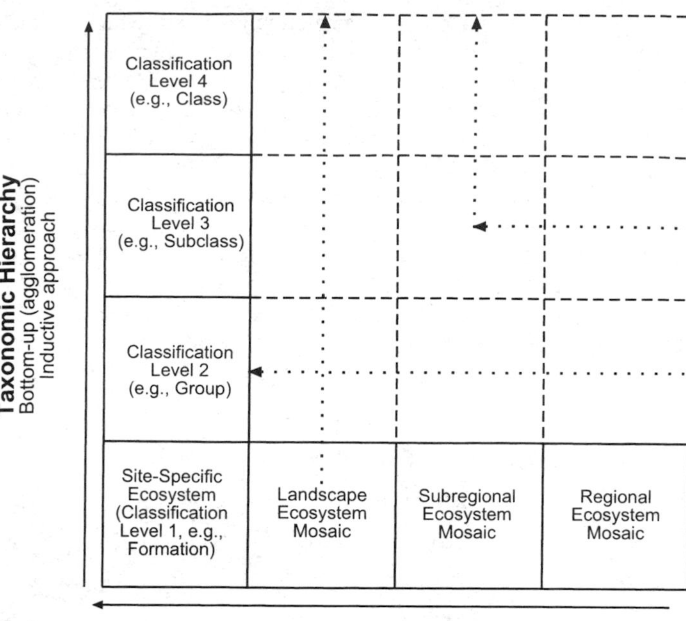

FIGURE 3.1. Relationship between spatial hierarchy produced with top-down (subdivision) approach and taxonomic hierarchy produced with bottom-up (agglomeration) approach. Examples of shared levels include the site-specific ecosystem level, classification level 1, and intersections of dotted lines.

and leads to site-specific ecosystems (Figure 3.1). Both hierarchies share a common level (e.g., site-specific ecosystem, classification level 1). The dotted lines in Figure 3.1 indicate that the shared level can be any level of the taxonomic and spatial hierarchies, depending on the purpose of the project (see further discussion in Zonneveld 1989, 1994).

3.6 Selection of Diagnostic Characteristics: What Process Should be Used?

Ecosystem characterization is contingent on the selection of patterns and processes, that is, the diagnostic characteristics of the system at hand. Although a single data layer (e.g., soil or vegetation) is sometimes used as a surrogate for all other attributes (see discussion in Chapter 22), characterization should be based on a combination of attributes. The phenomenon of interest in an assessment and the system in which the phenomenon occurs are determined by the objectives of the assessment (e.g., biodiversity of all ecosystems within the interior Columbia River basin). For example, if we are interested in the sustainable management and conservation of high-elevation subalpine fir forests of the Rocky Mountains, characterization focuses on the structure, composition, and function of these ecosystems. The following summarizes a discussion of ecological hierarchies and ecosystem characterization in Bourgeron and Jensen (1994). The ecological pattern of interest is the vegetation pattern, which can be resolved at six scales (Figure 3.2a), the individual ground cover plant, the individual mature canopy tree, the stand or community, the cover type, the physiognomic formation, and the biome. Each of these scales spans a particular spatial and temporal range. At each scale, the vegetation pattern produces patchiness. This patchiness can be related to the particular scales at which biotic processes, disturbances, and environmental constraints operate.

Biotic processes constraining Rocky Mountain high-elevation subalpine fir forests exhibit the scaled patterns shown in Figure 3.2b (Watt, 1947; Urban et al., 1987). Environmental constraints can be decomposed as shown in Figure 3.2c. Information about the scales associated with each constraint can be found in the literature; for example, detailed information on landforms has been published for the northern Rocky Mountains area (Donahue and Holdorf, 1990). Mitchell's (1976) climatic regions of the western United States provided the necessary information for Rocky Mountain climatic phenomena of interest. Finally, disturbances affecting vegetation can be arranged hierarchically for subalpine fir forests (Figure 3.2d) (Pickett et al., 1989). Information is readily available in the literature

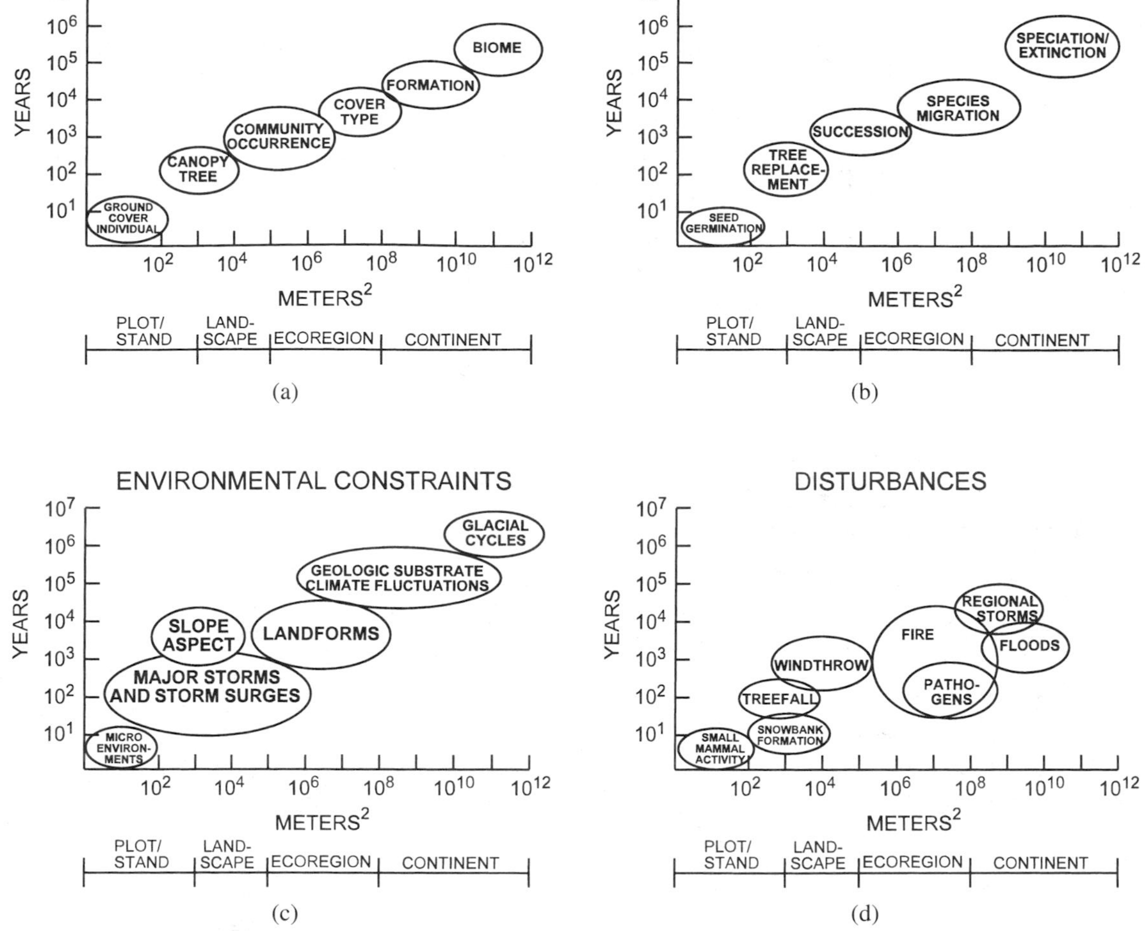

FIGURE 3.2. Spatial–temporal scaled patterns for (a) vegetation pattern, (b) biotic processes, (c) environmental constraints, and (d) disturbances (modified from Urban et al., 1987, and Bourgeron and Jensen, 1994).

concerning fire regimes and generalized successional models (Romme and Knight, 1982; Arno et al., 1985; Fischer and Bradley, 1987), snowbanks, pathogens, small mammal activities, and tree dynamics (Benedict, 1983).

An overlay of the four hierarchies depicted in Figure 3.2 provides a conceptual picture of ecological relationships at different scales (Figure 3.3). This process characterizes the composition, structure, and function of high-elevation subalpine fir ecosystems across a range of spatial and temporal scales. Therefore, vegetation pattern at different scales can be considered in light of the relevant biotic processes, disturbances, and environmental constraints, and hypotheses can be formulated concerning mechanisms that generate pattern.

The hierarchical approach described above provides a framework for characterizing ecosystems and identifying patterns and processes of interest at different scales (e.g., scales 1, 2, or 3 in Figure 3.3). Ecosystem classification and mapping can then be accomplished by integrating various land attributes (soil, vegetation, landform, hydrology, land use, etc.; see example in Figure 3.3). Variability within each level can be quantified to relate the emerging pattern to its causes and consequences. Correlational analysis usually provides an initial understanding of mechanisms generating patterns (Austin, 1985, 1991; Levin, 1992). For example, within the broad category of high-elevation subalpine fir forests, a definite substructure of spatial vegetation pattern is correlated primarily with moisture and temperature gradients, mineralizable nitrogen and phosphorus, and organic matter. Landforms interact with spatial patterns of ecosystem characteristics directly through control of nutrient cycling and water flow at a particular hierarchical level and indirectly through control of fire and wind

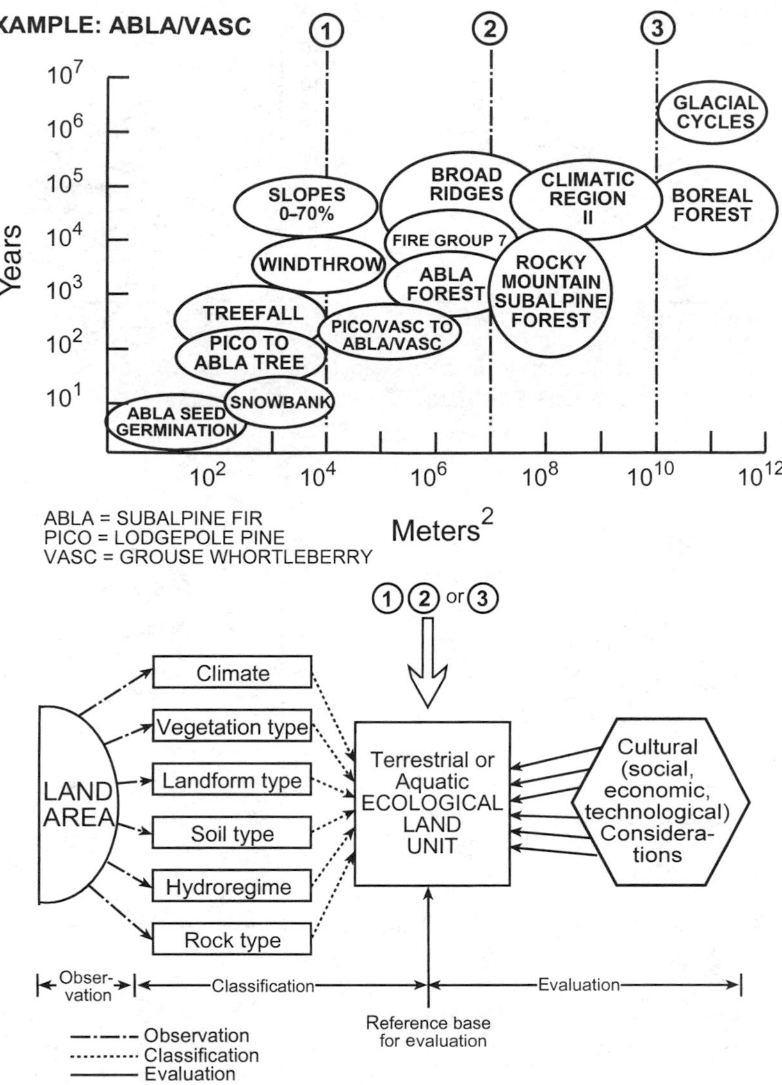

FIGURE 3.3. Hierarchical spatial–temporal representation of ecological relationships for high-elevation Rocky Mountain subalpine fir example (modified from Bourgeron and Jensen, 1994). Examples of different spatial scales, labeled 1, 2, and 3, characterized by means of observations, classification, and evaluation.

regimes at other levels (Swanson et al., 1988). The hierarchical approach leads to a listing of possible mechanisms (Levin, 1992) in a relevant spatial and temporal context (Urban et al., 1987).

3.7 Definition and Prediction of Ecosystem Properties

Much work in ecosystem science has concerned the use of ecosystem attribute information to predict various properties of interest. Three important properties in resource management and conservation planning are species distributions, species richness, and primary production. Prediction of these properties is part of the ecosystem characterization process (see Zonneveld, 1988, 1989, 1994). The most common and practical approach is to include values of the properties in the attribute file associated with mapping units (Bailey et al., 1994). However, doing so involves making two untested assumptions (Bourgeron and Jensen, 1994; Grossman et al., 1999): (1) the attribute is a clearly defined and predictable property of the characterized ecosystem, and (2) the attribute is invariant for that ecosystem over its known range of distribution. These assumptions may not be met.

The biotic component of an ecosystem is characterized by the link between distribution patterns of individual species, their occurrence in landscape features, and the distribution of landscape features (see discussion in Bourgeron et al., 1994a). Therefore, the distribution pattern of landscape features constrains the frequency and pattern of species distributions in the landscape. Attempts to characterize the distribution of species by using ecological classifications and mapping units can be successful only if the pattern of gradual change in individual species distributions is correlated with gradual changes in the environment at relevant scales. Ecosystem characterization should not assume that all occurrences of an ecological unit have identical properties with respect to a species requirement. Species distributions should be established with data that span the range of environmental variability over which they are distributed. Use of landscape classifications to predict species distributions include the Cherrill et al. (1995) model that explicitly orders ecological information into a hierarchical series of matrices representing the relationships between successive levels of spatial resolution and ecological organization. Despite limitations (e.g., poor prediction of scarce species, difficulties in testing landscape-scale models), this approach was successful for common species and provided insight into the consequences of land-use change. He et al. (1998) used a probabilistic algorithm to assign information from forest inventory sampling points and ecoregional boundaries to a satellite-based land cover classification map for regional forest assessment. New maps of tree species distributions and stand age were derived that reflected differences at the ecoregional scale. The inventory data provided important secondary information on age class and secondary species not available from the remotely sensed data.

Current theory regarding species diversity patterns has limited predictive power. Empirical relationships have been developed at one of three scales: global (Whittaker, 1972), regional (Pielou, 1979; Brown, 1984), or local (Grime, 1979; Woodward, 1987). Models relating diversity to disturbance (e.g., Huston, 1979, 1994) do not have site-specific predictive power. The cumulative impact of niche relations, habitat diversity, mass effects (the flow of individuals from favorable to unfavorable areas), and ecological equivalency (the fact that different species may be ecologically equivalent to each other) on diversity has been summarized in a multiscale context (Shmida and Wilson, 1985). Recent work has focused on relationships among hierarchical levels (e.g., Ricklefs, 1987; Caley and Schluter, 1997; Angermeier and Winston, 1998). The dependence of local diversity on regional patterns has been shown for some biota (Ricklefs, 1987), but not for others (Jackson and Harvey, 1989). Diversity at one scale may have complex relationships to structure, processes, and disturbances at other scales. Neilson et al. (1989) suggested that the prediction of local diversity patterns should be rooted in an understanding of the hierarchy of constraints imposed by regional and local factors, as well as of their interactions. The spatial and temporal context provided by multi-scaled ecosystem characterization should be useful for analysis of biotic diversity (Whittaker, 1972; Hoover and Parker, 1991).

Progress has been made in developing empirical relationships between diversity and environment using predictive statistical models. Useful relationships (Margules et al., 1987; Nicholls, 1991a, b; Austin et al., 1996) generally involve more than one environmental variable. These relationships should be derived from survey data and probably cannot be extended beyond the bounds of the data (Margules et al., 1987). However, site-specific predictions can be made for particular study areas. The use of hierarchical ecosystem characterization schemes could provide the appropriate stratification for extrapolating site-specific results.

Problems of scale interactions and model generalization beset predictions of ecosystem primary production. Spatially explicit models of biogeochemistry (see Chapter 18) have been developed using a combination of geographic information systems (GIS) and terrestrial regional ecosystem models (Houghton et al., 1983; Burke et al., 1990; Band et al., 1993), but their widespread use is limited by gaps in soil, climate, and vegetation databases (Stewart et al., 1989). Models like CENTURY (Parton et al., 1987) and BIOME-BGC (Running and Hunt, 1993; Hunt et al., 1996) explicitly link abiotic and biogeochemical factors with primary production and carbon storage (see also Schimel et al., 1990). The problem of predicting ecosystem processes is an active research field (Schimel et al., 1997). For example, VEMAP members (1995) compared simulations of net primary production (NPP) produced by three biogeochemistry models (CENTURY, BIOME-BCG, and TEM) for the conterminous United States under current climate conditions and a range of climate change scenarios. The models estimated similar continental-scale NPP values under current conditions. Schimel et al. (1991) summarized current activities and stressed

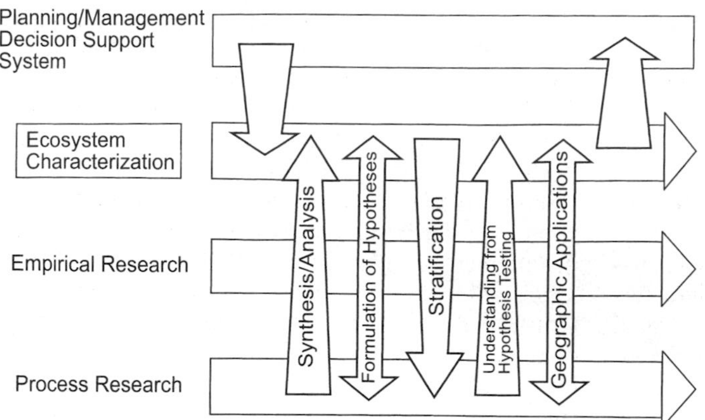

FIGURE 3.4. Integration of decision support systems, ecosystem characterization, empirical research, and process research (modified from Jones, 1990, and Uhlig and Jordan, 1996).

the need for ecosystem modeling based on a small set of critical environmental variables linked to the biotic component of ecosystems. Work is urgently needed to develop ecosystem models that include better information about the biotic components of ecosystems and ecosystem responses to land-use practices.

Useful theoretical frameworks are limited for predicting ecosystem properties, such as species distributions, species diversity, and primary production, over large areas. This scarcity of useful theory probably extends to most properties that are derived from individual ecosystem components. However, previous work has shown that empirical relationships can be derived from survey data that have been integrated into ecological classifications and mapping units. These relationships can be useful for ecological assessments.

Figure 3.4 (adapted from Jones, 1990, and Uhlig and Jordan, 1996) illustrates integration of ecosystem characterization, empirical research, and process research through spatially explicit methodologies to provide the input data needed for decision support systems (see Chapter 12). Planning and management inputs determine the need and the objectives for an ecosystem characterization effort. Such a characterization produces a synthesis of existing objective-specific information. Hypotheses can be generated about relationships of attributes and properties other than those used in the characterization process. Multiscale characterization schemes provide efficient stratification mechanisms for testing these hypotheses in subsequent research. Results derived from new work can be efficiently reported, extrapolated, and used for planning interpretation by the characterization scheme. The characterization scheme can be modified as research brings new understanding of the ecosystems (Uhlig and Jordan, 1996). The integration of remotely sensed data, GIS, and relational database management systems has led to the implementation of ecosystem characterization in a multithematic planning framework (see Chapter 12).

3.8 Impact of Grain and Extent of Data on Ecosystem Characterization and Extrapolation

Remotely sensed data (see Chapter 10) and GIS (see Chapter 11) are now commonly used for mapping and analysis of ecological data for ecological and conservation assessments (Davis et al., 1991; Cohen and Spies, 1992; Lachowski et al., 1992; Mladenoff and Host, 1994; Sample, 1994; Cohen et al., 1995; Kapner et al., 1995; Wolter et al., 1995; Host et al., 1996; Mladenoff et al., 1997; O'Neill et al., 1997; He et al., 1998, using the concepts and techniques of landscape ecology (Turner, 1989; Turner and Gardner, 1991; O'Neill et al., 1997). Although these tools facilitate the implementation of ecological assessments by making it easy to manipulate large amounts of data and conduct analyses over broad areas, their use raises issues concerning the grain (resolution) and spatial extent of the data (O'Neill et al., 1986, 1997), which can have complex consequences for landscape pattern analysis and interpretation (Turner et al., 1989a, b; Wickham and Riitters, 1995; Jelinski and Wu, 1996; Mladenoff et al., 1997; Wickham et al., 1997; Bourgeron et al., 1999). The impact of grain and extent on analyses is closely related to the problems of spatial variability and the ability to extrapolate results.

The spatial extent over which data are collected and analyses conducted affects the ecosystem characterization process. The assessment area is usually readily defined by the specific objectives of the study (e.g., Quigley and Arbelbide, 1997). However, the spatial scale for data collection and analysis may differ from the spatial scale of the assessment area, and the choice of a particular analysis scale may influence the results. In the Interior Columbia Basin Ecosystem Management Project, analyses of terrestrial and aquatic patterns were conducted either basinwide or, using data from restricted geographic locations, were conducted within provinces, sections, or subsections (Quigley and Arbelbide, 1997). For example, a vegetation structural stage data layer, important for determining landscape trends (Hann et al., 1997), was extrapolated from sampled to unsampled areas at coarse and midscales (Keane et al., 1996; O'Hara et al., 1996), without an assessment of the impact of the grain, extent, and spatial variability of the data on the results. Another effort to extrapolate the distribution of structural stages from an intensively studied site (Cohen and Spies, 1992) to a larger area and a broader range of ecological conditions showed clear advantages as well as serious limitations of the approach (Cohen et al., 1995). Therefore, it is critical to know the impact of the spatial scale of the data on characterization and analysis of landscape patterns.

Weaknesses in the use of GIS, remotely sensed data, and regional and landscape surveys, include (1) the lack of testing of the impact of the grain and extent of the data on the results, (2) the inability to predict the relative magnitude of spatial variability in different types of variables across the full range of conditions found in an area, and (3) the lack of standardized, repeatable methods for assessing the representativeness of individual study landscapes across multiple spatial scales and ecological conditions. Strengths include (1) the ability to take advantage of the remarkable advances in remotely sensed data, GIS, and modeling, (2) often lower costs than for intensive assessments of individual landscapes, and (3) easier implementation than ground-based surveys. This approach is included in the landscape characterization pursued by EMAP (O'Neill et al., 1994; Kapner et al., 1995).

3.9 Conclusions

Ecosystems exist at multiple scales, from global to local. They are defined by associations of biotic and abiotic components and the interactions among them. Each component has multiple attributes, but not all components are equally important in ecosystem characterization at all spatial scales. The challenges of ecosystem characterization include selecting the attributes and properties that best define ecosystems at all spatial scales, delineating boundaries between spatial representations of ecological units, defining and quantifying ecosystem properties, and extrapolating results to different geographic locations and across spatial scales.

The following principles should be followed during the ecosystem characterization phase of an ecological assessment.

1. Ecological classifications and mapping units must be defined according to precisely specified assessment goals.
2. Ecosystem patterns must be understood in terms of the processes and constraints generating them across their natural range of spatial and temporal scales.
3. Ecosystem characterization is influenced by the grain and extent of remotely sensed and ground-based data. The impact of data collected on the delineation of ecosystems and the range of spatial scales and geographic locations at which these data can be used should be evaluated.
4. The utility of ecosystem characterization depends on the formulation of testable hypotheses about ecological relations and on validating the extrapolation of results across geographic areas and spatial scales.
5. Ecosystem characterization usually is concerned primarily with spatial variability. Temporal variability and its interaction with spatial variability should also be investigated.

The following reviews and syntheses provide good sources of information and references on specific aspects of ecosystem characterization and on worldwide applications under different ecological, political, and socioeconomic conditions: Jensen and Bourgeron (1994), Klijn (1994), Sims et al. (1996), and Grossman et al. (1999).

3.10 References

Allen, T. F. H.; Starr, T. B. 1982. *Hierarchy: perspectives for ecological complexity.* Chicago: University of Chicago Press.

Allen, T. F. H.; Hoekstra, T. W.; O'Neill, R. V. 1984. *Interlevel relations in ecological research and management: some working principles from hierarchy theory.* RM-GTR-110. Fort Collins, CO: U.S. Dept. Agric., For. Serv., Rocky Mountain For. Range Exp. Sta.

Amoros, C.; Rostan, J. C.; Pautou, G.; Bravard, J. P. 1987. The reversible process concept applied to the environmental management of large river systems. *Environ. Manage.* 11:607–617.

Angermeier, P. L.; Winston, M. R. 1998. Local vs. regional influences on local diversity in stream fish communities of Virginia. *Ecology* 79(3):911–927.

Arno, S. F.; Simmermann, D. G.; Keane, R. E. 1985. *Forest succession on four habitat types in western Montana.* INT-GTR-177. Ogden, UT: U.S. Dept. Agric., Intermountain For. Range Exp. Sta.

Austin, M. P. 1985. Continuum concept, ordination methods and niche theory. *Ann. Rev. Ecol. Syst.* 16:39–61.

Austin, M. P. 1991. Vegetation theory in relation to cost-efficient surveys. In: Margules, C. R.; Austin, M. P., eds. *Nature conservation: cost effective biological surveys and data analysis.* Melbourne: Commonwealth Scientific and Industrial Research Organization: 17–22.

Austin, M. P.; Margules, C. R. 1986. Assessing representativeness. In: Usher, M. B., ed. *Wildlife conservation evaluation.* London: Chapman and Hall: 45–67.

Austin, M. P.; Pausas, J. G.; Nicholls, A. O. 1996. Patterns of tree species richness in relation to environment in southeastern New South Wales, Australia. *Aust. J. Ecol.* 21:154–164.

Bailey, R. G. 1995. *Description of the ecoregions of the United States.* Misc. Publ. 1391. Washington, DC: U.S. Dept. Agric., For. Serv.

Bailey, R. G. 1996. Multiscale ecosystem analysis. *Environ. Monit. Assess.* 39:21–24.

Bailey, R. G.; Jensen, M. E.; Cleland, D. T.; Bourgeron, P. S. 1994. Design and use of ecological mapping units. In: Jensen, M. E.; Bourgeron, P. S., eds. *Volume II: ecosystem management: principles and applications.* PNW-GTR-318. Portland, OR: U.S. Dept. Agric., For. Serv., Pacific Northw. Res. Sta.: 95–106.

Baker, W. L. 1992a. Effects of settlement and fire suppression on landscape structure. *Ecology* 73:1879–1887.

Baker, W. L. 1992b. The landscape ecology of large disturbances in the design and management of nature reserves. *Landscape Ecol.* 7:181–194.

Band, L. E.; Patterson, P.; Nemani, R.; Running, S. W. 1993. Forest ecosystem processes at the watershed scale: incorporating hillslope hydrology. *Agr. For. Meteorol.* 63:93–126.

Bayley, P. B.; Li, H. W. 1992. Riverine fishes. In: Calow, P.; Petts, G. E., eds. *The rivers handbook.* London: Blackwell Scientific: 251–281.

Belbin, L. 1993. Environmental representativeness: regional partitioning and reserve selection. *Biol. Conserv.* 66:223–230.

Benedict, N. B. 1983. Plant associations of subalpine meadows, Sequoia National Park, California. *Arctic Alp. Res.* 15(3):383–396.

Bourgeron, P. S.; Jensen, M. E. 1994. An overview of ecological principles for ecosystem management. In: Jensen, M. E.; Bourgeron, P. S., eds. *Volume II: ecosystem management: principles and applications.* PNW-GTR-318. Portland, OR: U.S. Dept. Agric., For. Serv., Pacific Northw. Res. Sta.: 45–67.

Bourgeron, P. S.; Humphries, H. C.; DeVelice, R. L. 1994a. Ecological theory in relation to landscape and ecosystem characterization. In: Jensen, M. E.; Bourgeron, P. S., eds. *Volume II: ecosystem management: principles and applications.* PNW-GTR-318. Portland, OR: U.S. Dept. Agric., For. Serv., Pacific Northw. Res. Sta.: 61–75.

Bourgeron, P. S.; Humphries, H. C.; Jensen, M. E. 1994b. Landscape characterization: a framework for ecological assessment at regional and local scales. *J. Sustain. For.* 2(3–4):267–281.

Bourgeron, P. S.; Jensen, M. E.; Engelking, L. D.; Everett, R. L.; Humphries, H. C. 1995. Landscape ecology, conservation biology, and principles of ecosystem management. In: Everett, R. L.; Baumgartner, D. M., eds. *Proceedings, ecosystem management in western interior forests,* May 3–5, 1994. Pullman, WA: Dept. Natural Resource Sciences, Washington State University: 41–47.

Bourgeron, P. S.; Humphries, H. C.; Barber, J. A.; Turner, S. J.; Jensen, M. E.; Goodman, I. A. 1999. Impact of broad- and fine-scale patterns on regional landscape characterization using AVHRR-derived land cover data. *Ecosystem Health* 5: 234–258.

Brown, J. H. 1984. On the relationship between the abundance and distribution of species. *Amer. Naturalist* 124:255–279.

Bunce, R. G. H.; Barr, C. J.; Clarke, R. T.; Howard, D. C.; Lane, A. M. J. 1996. Land classification for strategic ecological survey. *J. Environ. Manage.* 47:37–60.

Burke, I. C.; Schimel, D. S.; Yonker, C. M.; Parton, W. J.; Joyce, L. A.; Lauenroth, W. K. 1990. Regional modeling of grassland biogeochemistry using GIS. *Landscape Ecol.* 4:44–54.

Caley, M. J.; Schluter, D. 1997. The relationship between local and regional diversity. *Ecology* 78(1):70–80.

Cherrill, A. J.; McClean, C.; Watson, P.; Rushton, S. P.; Sanderson, R. 1995. Predicting the distributions of plant species at the regional scale: a hierarchical matrix model. *Landscape Ecol.* 10:197–207.

Christensen, N. L.; Agee, J. K.; Brussard, P. F.; Hughes, J.; Knight, D. H.; Minshall, G. W.; Peek, J. M.; Pyne, S. J.; Swanson, F. J.; Thomas, J. W.; Wells, S.; Williams, S. E.; Wright, H. A. 1989. Interpreting the Yellowstone fires of 1988. *BioScience* 39:678–685.

Cohen, W. B.; Spies, T. A. 1992. Estimating structural attributes of Douglas-fir/western hemlock forest stands from Landsat and SPOT imagery. *Remote Sens. Environ.* 41:1–17.

Cohen, W. B.; Spies, T. A.; Fiorella, M. 1995. Estimating the age and structure of forests in a multi-ownership landscape of western Oregon, USA. *Int. J. Remote Sens.* 16:721–746.

Costanza, R.; Wainger, L.; Folke, C.; Maler, K. 1993. Modeling complex ecological economic systems. *BioScience* 43:545–555.

Davis, F. W.; Quattrochi, D. A.; Ridd, M. K.; Lam, N. S. N.; Walsh, S. J.; Michaelsen, J. C.; Franklin, J.; Stow, D. A.; Johannsen, C. J.; Johnston, C. A. 1991.

Environmental analysis using integrated GIS and remotely sensed data: some research needs and priorities. *Photogramm. Eng. Remote Sensing* 57:689–697.

Donahue, J.; Holdorf, H. 1990. *Landforms for soil surveys in the northern Rockies.* Misc. Publ. 51. Missoula, MT: Montana For. and Conserv. Exp. Sta., School of Forestry, Univ. of Montana.

ECOMAP. 1993. *National hierarchical framework of ecological units.* Washington, DC: U.S. Dept. Agric., For. Serv.

Fischer, W. C.; Bradley, A. F. 1987. *Fire ecology of western Montana forest habitat types.* INT-GTR-223. Ogden, UT: U.S. Dept. Agric., For. Serv., Intermountain Res. Sta.

Franklin, J. F. 1993. The fundamentals of ecosystem management with application to the Pacific Northwest. In: Aplet, G.; Olson, J. T.; Johnson, N.; Sample, V. A., eds. *Defining sustainable forestry.* Washington, DC: Island Press: 127–144.

Frissell, C. A.; Liss, W. J.; Warren, C. E.; Hurley, M. D. 1986. A hierarchical framework for stream habitat classification: viewing streams in a watershed context. *Environ. Manage.* 10:199–214.

Gardner, R. H.; Turner, M. G.; O'Neill, R. V.; Lavorel, S. 1992. Simulation of the scale-dependent effects of landscape boundaries on species persistence and dispersal. In: Holland, M. M.; Risser, P. G.; Naiman, R. J., eds. *Ecotones: the role of landscape boundaries in the management and restoration of changing environments.* New York: Chapman and Hall: 76–89.

Grime, J. P. 1979. *Plant strategies and vegetation processes.* New York: John Wiley & Sons.

Grossman, D. H.; Bourgeron, P. S.; Busch, W. D. N.; Cleland, D.; Platts, W.; Ray, G. C.; Robins, C. R.; Roloff, G. 1999. Principles for ecological classification. In: Johnson, N. C.; Malk, A. J.; Sexton, W. T.; Szaro, R. C., eds. *Volume II. Ecological stewardship: a common reference for ecosystem management.* Amsterdam, The Netherlands: Elsevier Science: 353–393.

Haber, W. 1994. Systems ecological concepts for environmental planning. In: Klijn, F., ed. *Ecosystem classification for environmental management.* Dordrect, The Netherlands: Kluwer Academic Publishers: 49–67.

Hann, W.; Keane, R. E.; McNicoll, C.; Menakis, J. 1994. Assessment techniques for evaluating ecosystem processes, and community and landscape conditions. In: Jensen, M. E.; Bourgeron, P. S., eds. *Volume II: ecosystem management: principles and applications.* PNW-GTR-318. Portland, OR: U.S. Dept. Agric., For. Serv., Pacific Northw. Res. Sta.: 237–253.

Hann, W. J.; Jones, J. L.; Karl, M. G.; Hessburg, P. F.; Keane, R. E.; Long, D. G.; Menakis, J. P.; McNicol, C. H.; Leonard, S. G.; Gravenmier, R. A.; Smith, B. G. 1997. Landscape dynamics of the basin. In: Quigley, T. M.; Arbelbide, S. J., eds. *An assessment of ecosystem components in the interior Columbia Basin and portions of the Klamath and Great Basins: Volume II.* PNW-GTR-405. Portland, OR: U.S. Dept. Agric., For. Serv., Pacific Northw. Res. Sta.: 337–1055.

He, H. S.; Mladenoff, D. J.; Radeloff, V. C.; Crow, T. R. 1998. Integration of GIS data and classified satellite imagery for regional forest assessment. *Ecol. Appl.* 8(4):1072–1083.

Holling, C. S. 1986. The resilience of terrestrial ecosystems: local surprise and global change. In: Clark, W. M.; Munn, R. E., eds. *Sustainable development in the biosphere.* Oxford, UK: Oxford University Press: 292–320.

Holling, C. S. 1992. Cross-scale morphology, geometry, and dynamics of ecosystems. *Ecol. Monogr.* 62:447–502.

Hoover, S. R.; Parker, A. J. 1991. Spatial components of biotic diversity in landscapes of Georgia, USA. *Landscape Ecol.* 5(3):125–136.

Host, G. E.; Polzer, P. L.; Mladenoff, M. A.; White, M. A.; Crow, S. J. 1996. A quantitative approach to developing regional ecosystem classifications. *Ecol. Appl.* 6:608–618.

Houghton, R. A.; Hobbie, J. E.; Melillo, J. M.; Moore, B. I.; Peterson, B. M.; Shaver, G. R.; Woodwell, G. M. 1983. Changes in the carbon content of the terrestrial biota and soils between 1860 and 1980. *Ecol. Monogr.* 53:235–262.

Hunt, E. R., Jr.; Piper, S. C.; Nemani, R.; Keeling, C. D.; Otto, R. D.; Running, S. W. 1996. Global net carbon exchange and intra-annual atmospheric CO_2 concentrations predicted by an ecosystem process model and three-dimensional atmospheric transport model. *Global Biogeochem. Cycle.* 10:431–456.

Huston, M. A. 1979. A general hypothesis of species diversity. *Amer. Naturalist* 113:81–101.

Huston, M. A. 1994. *Biological diversity, the coexistence of species on changing landscapes.* Cambridge, UK: Cambridge University Press.

Jackson, D. A.; Harvey, H. H. 1989. Biogeographic associations in fish assemblages: local vs. regional processes. *Ecology* 70:1472–1484.

Jelinski, D. E.; Wu, J. 1996. The modifiable areal unit problem and implications for landscape ecology. *Landscape Ecol.* 11:129–140.

Jensen, M. E.; Bourgeron, P. S., editors. 1994. *Volume II: ecosystem management: principles and applications.* PNW-GTR-318. Portland, OR: U.S. Dept. Agric., For. Serv., Pacific Northw. Res. Sta.

Jensen, M. E.; Bourgeron, P. S.; Everett, R.; Goodman, I. 1996. Ecosystem management: a landscape ecology perspective. *J. Amer. Water Resour. Assoc.* 32:208–216.

Jones, R. K. 1990. Role of site classification in predicting the consequences of management on forest response. In: Dyck, W. J.; Mees, C. S., eds. *Impact of intensive harvesting on forest site productivity.* Proceedings, IEA/BE Workshop, March 1989. IEA/BE T6/A6 Report No. 2/FRI Bulletin No. 159. Rotorua, New Zealand: Forest Research Institute: 19–36.

Kapner, W. G.; Jones, K. B.; Chaloud, D. G.; Wickham, J. D.; Riitters, K. H.; O'Neill, R. V. 1995. *Mid-Atlantic landscape indicators.* Washington, DC: U.S.

Environ. Protect. Agency, Project Plan Ecological Monitoring and Assessment Program 620/R-95/003.

Kay, J. J. 1991. A nonequilibrium thermodynamic framework for discussing ecosystem integrity. *Environ. Manage.* 15:483–495.

Keane, R. E.; Morgan, P.; Running, S. W. 1996. *FIRE-BGC—a mechanistic ecological process model for simulating fire succession on coniferous forest landscapes of the Northern Rocky Mountains.* INT-RP-484. Ogden, UT: U.S. Dept. Agric., For. Serv., Intermountain Res. Sta.

Klijn, F., editor. 1994. *Ecosystem classification for environmental management. Ecology and environment.* Dordrecht, The Netherlands: Kluwer Academic Publishers.

Kuchler, A. W. 1973. Problems in classifying and mapping vegetation for ecological regionalization. *Ecology* 54:512–523.

Kuchler, A. W.; Zonneveld, I. S., editors. 1988. *Vegetation mapping.* Boston: Kluwer Academic Publishers.

Lachowski, H.; Maus, P.; Platt, B. 1992. Integrating remote sensing with GIS. *J. For.* 90:16–21.

Lathrop, R. G.; Peterson, D. L. 1992. Identifying structural self-similarity in mountainous landscapes. *Landscape Ecol.* 6:233–238.

Lessard, G. 1995. *A national framework for integrated ecological assessments.* Washington, DC: Federal InterAgency Working Group.

Lessard, G.; Jensen, M.; Crespi, M.; Bourgeron, P. 1999. A general framework for integrated ecological assessments. In: Cordel, H. K.; Bergstrom, J. C., eds. *Integrating social sciences with ecosystem management: human dimensions in assessment policy and management.* Champaign–Urbana, IL: Sagamore: 35–60.

Levin, S. A. 1992. The problem of pattern and scale in ecology. *Ecology* 73:1942–1968.

Ligon, F. K.; Dietrich, W. E.; Trush, W. J. 1995. Downstream ecological effects of dams: a geomorphic perspective. *BioScience* 45:183–192.

Lubchenco, J.; Olson, A. M.; Brubaker, L. B.; Carpenter, S. R.; Holland, M. M.; Hubbell, S. P.; Levin, S. A.; MacMahon, J. A.; Matson, P. A.; Melillo, J. M.; Mooney, H. A.; Peterson, C. H.; Pulliam, H. R.; Real, L. A.; Regal, P. J.; Risser, P. G. 1991. The sustainable biosphere initiative: an ecological research agenda. *Ecology* 72(2):371–412.

Mackey, B. G.; Nix, H. A.; Hutchinson, M. F.; MacMahon, J. P.; Fleming, P. M. 1988. Assessing representativeness of places for conservation reservation and heritage listing. *Environ. Manage.* 12(4):501–514.

Margules, C. R.; Nicholls, A. O.; Austin, M. P. 1987. Diversity of eucalyptus species predicted by a multivariable environmental gradient. *Oecologia (Berl.)* 71:229–232.

Maxwell, J. R.; Edwards, C. J.; Jensen, M. E.; Paustian, S. J.; Parrott, H.; Hill, D. M. 1995. *A hierarchical framework of aquatic ecological units in North America.* NC-GTR-176. St. Paul, MN: U.S. Dept. Agric., For. Serv., North Central For. Exp. Sta.

McNaughton, S. J. 1985. Ecology of a grazing ecosystem: the Serengeti. *Ecol. Monogr.* 55:259–294.

Minshall, G. W. 1988. Stream ecosystem theory: a global perspective. *J. N. Amer. Benthol. Soc.* 7:263–288.

Minshall, G. W. 1994. Stream–riparian ecosystems: rationale and methods for basin-level assessments of management effects. In: Jensen, M. E.; Bourgeron, P. S., eds. *Volume II: ecosystem management: principles and applications.* PNW-GTR-318. Portland, OR: U.S. Dept. Agric., For. Serv., Pacific Northw. Res. Sta.: 143–167.

Mitchell, V. L. 1976. The regionalization of climate in the western United States. *J. Appl. Meteorol.* 15:920–927.

Mladenoff, D. J.; Host, G. E. 1994. Ecological perspective: current and potential applications of remote sensing and GIS for ecosystem analysis. In: Sample, V. A., ed. *Remote sensing and GIS in ecosystem management.* Washington, DC: Island Press: 145–180.

Mladenoff, D. J.; Pastor, J. 1993. Sustainable forest ecosystems in the northern hardwood and conifer region: concepts and management. In: Aplet, G.; Olson, J. T.; Johnson, N.; Sample, V. A., eds. *Defining sustainable forestry.* Washington, DC: Island Press: 145–180.

Mladenoff, D. J.; Niemi, G. H.; White, M. A. 1997. Effects of changing landscape pattern and U.S.G.S. land cover data variability on ecoregion discrimination across a forest–agriculture gradient. *Landscape Ecol.* 12:379–396.

Naiman, R. J. 1988. Animal influences on ecosystem dynamics. *BioScience* 38:750–752.

Neilson, R. P.; King, G. A.; DeVelice, R. L.; Lenihan, J.; Marks, D.; Dolph, J.; Campbell, W.; Glick, G. 1989. *Sensitivity of ecological landscapes and regions to global climatic change.* EPA/600/3-89-073, NTIS-PB90-120-072/AS. Washington, DC: U.S. Dept. of the Interior.

Nicholls, A. O. 1991a. An introduction to statistical modelling using GLIM. In: Margules, C. R.; Austin, M. P., eds. *Nature conservation: cost effective biological surveys and data analysis.* Melbourne: Commonwealth Scientific and Industrial Research Organization: 191–201.

Nicholls, A. O. 1991b. Examples of the use of generalized linear models in analysis of survey data for conservation evaluation. In: Margules, C. R.; Austin, M. P., eds. *Nature conservation: cost effective biological surveys and data analysis.* Melbourne: Commonwealth Scientific and Industrial Research Organization: 54–63.

O'Hara, K. L.; Lathan, P. A.; Hessburg, P.; Smith, B. G. 1996. A structural classification for inland Northwest forest vegetation. *West. J. Appl. For.* 11(3):97–102.

Omernik, J. M. 1987. Ecoregions of the conterminous United States. *Ann. Assoc. Amer. Geogr.* 77(1):118–125.

O'Neill, R. V.; DeAngelis, D. L.; Waide, J. B.; Allen, T. F. H. 1986. *A hierarchical concept of ecosystems.* Princeton, NJ: Princeton University Press.

O'Neill, R. V.; Jones, K. B.; Riitters, K. H.; Wickham, J. D.; Goodman, I. A. 1994. *Landscape monitoring and assessment research plan.* Report 620/R-94/009. Washington, DC: U.S. Environ. Protect. Agency.

O'Neill, R. V.; Hunsaker, C. T.; Jones, K. B.; Riitters, K. H.; Wickham, J. D.; Schwartz, P. M.; Goodman, I. A.; Jackson, B. L.; Baillargeon, W. S. 1997. Monitoring environmental quality at the landscape scale. *BioScience* 47:513–519.

Parton, W. J.; Schimel, D. S.; Cole, C. V.; Ojima, D. S. 1987. Analysis of factors controlling soil organic matter levels in Great Plains grasslands. *Soil Sci. Soc. Amer. J.* 51:1173–1179.

Perera, A. H.; Baker, J. A.; Band, L. E.; Baldwin, D. J. B. 1996. A strategic framework to ecoregionalize Ontario. In: Sims, R. A.; Corns, I. G. W.; Klinka, K., eds. *Global to local ecological land classification.* Dordrect, The Netherlands: Kluwer Academic Publishers: 85–96.

Pickett, S. T. A.; Kolasa, J.; Armesto, J. J.; Collins, S. L. 1989. The ecological concept of disturbance and its expression at various hierarchical levels. *Oikos* 54: 129–136.

Pielou, E. C. 1979. *Ecological diversity.* New York: John Wiley & Sons.

Poff, N. L.; Ward, J. V. 1989. Implications of streamflow variability and predictability for lotic community structure: a regional analysis of streamflow patterns. *Can. J. Fisheries Aquat. Sci.* 46:1805–1818.

Poff, N. L.; Ward, J. V. 1990. Physical habitat template of lotic systems: recovery in the context of historical pattern of spatiotemporal heterogeneity. *Environ. Manage.* 14:629–645.

Pringle, C. M.; Naiman, R. J.; Bretschko, G.; Karr, J. R.; Oswood, M. W.; Webster, J. R.; Welcomme, R. L.; Winterbourn, M. J. 1988. Patch dynamics in lotic systems: the stream as a mosaic. *J. N. Amer. Benthol. Soc.* 7:503–524.

Quigley, T. M.; Arbelbide, S. J., editors. 1997. *An assessment of ecosystem components in the interior Columbia Basin and portions of the Klamath and Great Basins: Volume I.* PNW-GTR-405. Portland, OR: U.S. Dept. Agric., For. Serv., Pacific Northw. Res. Sta.

Ricklefs, R. E. 1987. Community diversity: relative roles of local and regional processes. *Science* 235:167–171.

Romme, W. H.; Knight, D. H. 1982. Landscape diversity: the concept applied to Yellowstone Park. *BioScience* 32:664–670.

Rowe, J. S. 1996. Land classification and ecosystem classification. In: Sims, R. A.; Corns, I. G. W.; Klinka, K., eds. *Global to local ecological land classification.* Dordrect, The Netherlands: Kluwer Academic Publishers: 11–20.

Running, S. W.; Hunt, E. R., Jr. 1993. Generalization of a forest ecosystem process model for other biomes, BIOME-BCG, and an application for global-scale models. In: Ehleringer, J. R.; Field, C., eds. *Scaling processes between leaf and landscape levels.* London: Academic Press: 141–158.

Sample, V. A. 1994. *Remote sensing and GIS in ecosystem management.* Washington, DC: Island Press.

Schimel, D. S.; Parton, W. J.; Kittel, T. G. F.; Ojima, D. S.; Cole, C. V. 1990. Grassland biogeochemistry: links to atmospheric processes. *Climatic Change* 17:13–25.

Schimel, D. S.; Kittel, T. G. F.; Parton, W. J. 1991. Terrestrial biogeochemical cycles: global interactions with the atmosphere and hydrology. *Tellus* 43(AB): 188–203.

Schimel, D. S.; Emanuel, W.; Rizzo, B.; Smith, T.; Woodward, F. I.; Fisher, H.; Kittel, T. G. F.; McKeown, R.; Painter, T.; Rosenbloom, N.; Ojima, D. S.; Parton, W. J.; Kicklighter, D. W.; McGuire, A. D.; Melillo, J. M.; Pan, Y.; Haxeltine, A.; Prentice, C.; Sitch, S.; Hibbard, K.; Nemani, R.; Pierce, L.; Running, S.; Borchers, J.; Chaney, J.; Neilson, R.; Braswell, B. H. 1997. Continental scale variability in ecosystem processes: models, data, and the role of disturbance. *Ecol. Monogr.* 67(2):251–271.

Shmida, A.; Wilson, M. V. 1985. Biological determinants of species diversity. *J. Biogeogr.* 12:1–20.

Sims, R. A.; Corns, I. G. W.; Klinka, K., editors. 1996. *Global to local ecological land classification.* Dordrect, The Netherlands: Kluwer Academic Publishers.

Slocombe, D. S. 1993. Implementing ecosystem-based management. *BioScience* 43(9):612–622.

Sparks, R. E. 1995. Need for ecosystem management of large rivers and their floodplains. *BioScience* 45:168–182.

Stewart, J. W. B.; Aselman, I.; Bouwman, A. F.; Desjardins, R. L.; Hicks, B. B.; Matson, P. A.; Rodhe, H.; Schimel, D. S.; Svensson, B. H.; Wassmann, R.; Whiticar, M. J.; Yang, W. 1989. Group report: extrapolation of flux measurements to regional and global scales. In: Andreae, M. O.; Schimel, D. S., eds. *Dahlem workshop on exchange of trace gases between terrestrial ecosystems and the atmosphere.* Berlin: John Wiley & Sons: 155–174.

Stork, N. E.; Boyle, T. J.; Dale, V.; Eeley, H.; Finegan, B.; Lawes, M.; Manokaran, N.; Prabhu, R.; Soberon, J. 1997. *Criteria and indicators for assessing the sustainability of forest management: conservation of biodiversity.* Working Paper No. 17. Jakarta: Center for International Forestry Research.

Swanson, F. J.; Kratz, T. K.; Caine, N.; Woodmansee, R. G. 1988. Landform effects on ecosystem patterns and processes. *BioScience* 38:92–98.

Turner, M. G. 1989. Landscape ecology: the effect of pattern on process. *Ann. Rev. Ecol. Syst.* 20:171–197.

Turner, M. G. 1990. Spatial and temporal analysis of landscape patterns. *Landscape Ecol.* 4:21–30.

Turner, M. G.; Gardner, R. H. 1991. Quantitative methods in landscape ecology: an introduction. In: Turner, M. G.; Gardner, R. H., eds. *Quantitative methods in landscape ecology: the analysis and interpretation of environmental heterogeneity.* New York: Springer-Verlag: 3–14.

Turner, M. G.; Dale, V. H.; Gardner, R. H. 1989a. Predicting across scales: theory development and testing. *Landscape Ecol.* 3:245–252.

Turner, M. G.; O'Neill, R. V.; Gardner, R. H.; Milne, B. T. 1989b. Effect of changing spatial scale on the analysis of landscape pattern. *Landscape Ecol.* 3:153–162.

Turner, M. G.; Hargrove, W. H.; Gardner, R. H.; Romme, W. H. 1994. Effects of fire on landscape heterogeneity in Yellowstone National Park, Wyoming. *J. Veg. Sci.* 5:731–742.

Turner, M. G.; Gardner, R. H.; O'Neill, R. V. 1995. Ecological dynamics at broad scales. *BioScience* 45:S29–S35.

Uhlig, P. W. C.; Jordan, J. K. 1996. A spatial hierarchical framework for the co-management of ecosystems in Canada and the United States for the upper Great Lakes region. *Environ. Monit. Assess.* 39:59–73.

Urban, D. L.; O'Neill, R. V.; Shugart, H. H., Jr. 1987. Landscape ecology: a hierarchical perspective can help scientists understand spatial patterns. *BioScience* 37:119–127.

VEMAP members. 1995. Vegetation/ecosystem modeling and analysis project (VEMAP): comparing biogeography and biogeochemistry models in a continental-scale study of terrestrial ecosystem responses to climate change and CO_2 doubling. *Biogeochem. Cycles* 9:407–438.

Watt, A. S. 1947. Pattern and process in the plant community. *J. Ecol.* 35:1–22.

Wessman, C. A. 1992. Spatial scales and global change: bridging the gap from plots to GCM grid cells. *Ann. Rev. Ecol. Syst.* 23:175–200.

Whittaker, R. H. 1972. Evolution and measurement of species diversity. *Taxon* 21:213–351.

Wickham, J. D.; Riitters, K. H. 1995. Sensitivity of landscape metrics to pixel size. *Int. J. Remote Sens.* 16:3585–3594.

Wickham, J. D.; O'Neill, R. V.; Riitters, K. H.; Wade, T. G.; Jones, K. B. 1997. Sensitivity of selected landscape pattern metrics to land-cover misclassification and differences in land-cover composition. *Photogramm. Eng. Rem. S.* 63:397–402.

Wiens, J. A. 1989. Spatial scaling in ecology. *Funct. Ecol.* 3:385–397.

Wolter, P. T.; Mladenoff, D. J.; Host, G. E.; Crow, T. R. 1995. Improved forest classification in the Lake States using multitemporal Landsat imagery. *Photogramm. Eng. Remote Sensing* 61:1129–1143.

Woodward, F. I. 1987. *Climate and plant distribution.* London: Cambridge University Press.

Zonneveld, I. S. 1988. Basic principles of land evaluation using vegetation and other land attributes. In: Kuchler, A. W.; Zonneveld, I. S., eds. *Vegetation mapping.* Boston: Kluwer Academic Publishers: 499–517.

Zonneveld, I. S. 1989. The land unit—a fundamental concept in landscape ecology, and its applications. *Landscape Ecol.* 3(2):67–89.

Zonneveld, I. S. 1994. Basic principles of classification. In: Klijn, F., ed. *Ecosystem classification for environmental management.* Dordrect, The Netherlands: Kluwer Academic Publishers: 23–47.

4 Adaptive Ecosystem Assessment and Management: The Path of Last Resort?

S. S. Light

Man's attitude to the world must be radically changed [emphasis added]. We have to abandon the arrogant belief that the world is merely a puzzle to be solved, a machine with instructions waiting to be discovered, a body of information to be fed into a computer in the hope that, sooner or later, it will spit out a universal solution.

—Vaclav Havel (1992)

The coming together of science, culture and spiritual activity is necessary for the fulfillment of human needs [emphasis added]. The world is presently in a difficult period of adjustment, and we urge people to be patient in judging what the longer-term future will be like. The complexity of this adjustment renders the future's possibilities very uncertain. But the role of science in this situation is essential for human survival.

—Ilya Prigogine (1984)

4.1 Introduction

Scientists understand that the Upper Mississippi River floodplain ecosystem evolved over eons of time through patterns of dynamic variability shaped by drought and flood, scour and deposition, channel formation, migration, and abandonment. Our scientists know that floodplain dynamics involving connectivity, hydroperiod, and energy distribution drive the river's ecosystems (Minnesota Department of Natural Resources, 1999). Yet basin-scale rehabilitation efforts on the Upper Mississippi River continue to focus on (1) single targets, optimizing habitat for relatively few highly valued species and (2) piecemeal environmental mitigation that shows little deference to the prime abiotic factors affecting biotic integrity: hydrodynamics and sedimentation.

Far from being an isolated incident, this dichotomy between our combined scientific and experiential understanding and the current state of management practice on the Upper Mississippi River is not unique. Experience suggests that this dichotomy is widespread within the natural resource and environmental management professional community. Professional scientists are frustrated by having their limited resources and energies squandered on mind-numbing meetings on where to locate the next dredge spoil pile while the ecological vitality of one of the world's great rivers slips slowly and possibly irreversibly beyond grasp. The inability for individuals and small groups of scientists to address root causes and positively influence institutional decision making such as on river degradation is driving some of our best people out of the profession or into early retirement.

The foundations of resource science and management are being rethought. The near century old vision of scientific management is being dismantled and it is not entirely clear what will replace it. Indisputably, understanding and managing nature as complex adaptive systems is the fundamental challenge resource that scientist and managers face as the 21st century approaches. The question is whether the current ways of organizing and using scientific knowledge can deliver the needed comprehension and action.

The belief that a scientifically rigorous technocratic elite could be empowered to solve our nation's environmental ills is over and done with. The unflagging faith in technocracy's ability to produce the next generation of add-on innovations that would "fix the problem" has backfired. We have been living under the fantasy that, if we could just put an additional program in place, or diffuse the next generation of technological innovation broadly enough, problems would go away. But the success that seemed assured has gradually proved unsettling, illusory. Competencies became traps, and would-be solutions have generated more unanticipated adverse social and ecological consequences. As the proliferation of agencies and programs accelerates, management institutions become increasingly disassociated from the health and resilience of natural systems and the vitality of human communities that they were charged with managing and serving, respectively.

Where once resource decisions enjoyed placid, straightforward decision environments, managing resources has grown fractious, turbulent, and congested. Instead of tackling the problems of changing the way that people think and act, society has opted for the path of least resistance—control technologies and subsequent add-on innovations that have left us in the long run bereft of constructive change. Indeed, resource management has slipped into an era of *no easy answers*, from which incremental and piecemeal attempts to get out only lead back in, only deeper into trouble. Instead of attempting to live with and profit from natural variability, we have chosen to rely on the science and technology of control and prediction instead.

Complexity encumbered with avarice, arrogance, and overreaching government programs has manifested itself in three distinct but interrelated ways: science, management, and policy formation. While management considerations relative to objectives and scales (temporal and spatial) were few, links among science, management, and policy appeared perfunctory and tractable. As the collapse of the "control at all costs" myth became undeniable, the shift to management of multiple objectives and scales became unavoidable, making the gaps between disciplines, constituencies, institutions, policy development, and implementation more striking and unruly. It would appear that we have reached the limits to control complexity through scientific reductionism, government bureaucracy, and special interests. In the words of Arthur (1990), "Complexity is a marvel when it evolves naturally and delivers powerful performance. But when we . . . allow it to go unchecked, it merely hampers. It is then that we need to discover new modes, the bold strokes that bring fresh simplicity to our organizations, or technology, or government, or lives."

Dealing with just the consequences of human action, the symptoms, will no longer do. The future lies in discovering new ways to work through the differences that separate people and nature and in learning new ways of thinking and acting, or the *path of last resort.* Focus must shift from just resource productivity to include sustainability as well. The leading edge of this transformation is already visible, and the profound change that it will exact from science, management, and public policy is substantial.

4.1.1 Science

To paraphrase Havel (1990), science has become adept at telling society about the many ways that we are destroying ourselves or abusing the planet, but utterly incapable of helping civilization avoid the disasters that threaten us. While our resource institutions excel in the science of parts (reductionism), the science of the integration of parts—dealing with the complex dynamic systems—is barely visible. The current generation of resource managers and scientists learned to consider "what is true, absolutely?" This classical view of science is dominated by "objectivism" and is conceptualized as the search for external truths associated with reality. Causation is seen as single and separable. The structure of inquiry focuses on fixed environment at a single scale and tries to eliminate uncertainty. So conceived, such an approach to science by definition appears beyond passion and governed by strict disciplinary perspectives. This is the type of scientific inquiry in which almost all our resource mangers and scientists frame their practice.

This is a rigorous science that offers volumes of information and fragments of insight, but limited relevance to the complex and conflicted situations in which our most stubborn resource issues exist. What is missing is science that serves citizens and community in the search to make sense of their world—a science of the integration of parts that can address the complex chaotic macro- and microsystem within the realm of human experience. Integrative science asks how we can make great sense of our world—where people's imaginations, intuitions, feelings, values, and beliefs are welcome, a science that is value variable and tailored to discovering ways to bring about greater harmony between people and nature, where humans learn to live and profit from natural variability.

When we looked for answers to relatively simple systems, reductionism worked well, especially in the laboratory. But in the messy real-world settings of natural resource management, where change occurs faster than we can comprehend, creating *collaboratories*—platforms for collective resource inquiry, use, and negotiation—seems better suited for building the commitment and local knowledge base needed to manage ecological complexity and uncertainty. A diversity of perspective and multiple modes of inquiry are more appropriate to identifying gaps in understanding and developing shared meaning on which collective action can be based. But few resource scientists as a community of scholars have yet to embrace the dream of unified science-based learning.

4.1.2 Management

Management is in an unenviable predicament. The public has lost confidence in government's capacity to accomplish desired ends. They sense an overall decline in performance simply based on the relatively narrow breadth and depth of public sector innovation when compared with that required by corporate and nonprofit sectors to stay competitive. Government rarely seems to fundamentally rethink the way it approaches management. The critique of resource management by Aldo Leopold (1949) over 50 years ago seems even more relevant today. Leopold was never enamored with the idea that more government was the answer to the conservation ills of the nation. He realized that government's influence for conservation was overestimated. Leopold saw society relegating "to government many functions eventually too large, too complex, or too widely dispersed to be performed by government." For him, there was no substitute for the ethical obligation on the part of the individual citizen. But this is not how our society has responded to increased complexity.

Instead, our penchant is for creating bureaucracies and then figuring out how to restructure, rename, and recast stale approaches and messages that seem to have as much consequence as shuffling deck chairs on a doomed ship. The results merely add costs for more information, coordination, and control. Real creativity, ingenuity, and breakthrough thinking are the casualty. One fundamental reason for this is that resource management agencies view workplace behaviors of the individual professional as filling a disciplinary niche, not as a boundary spanner or facilitator of change.

Behind it all is the gnawing sense that John Q. Public cannot comprehend the bureaucratic systems that we have created, let alone engage them. Increasingly, it appears that bureaucracy actually prohibits community from arising and from trying out new and creative solutions. In the Upper Mississippi River, the Corps of Engineers' presence is so pronounced, so entrenched in the status quo, and so indifferent to ecological concerns that some community-based conservationists appear to have given up hope of restoring the river's ecosystem, despite the fact that 50% of the upper river's floodplain is repairable.

4.1.3 Public Policy

Lobbyists and experts who are only surrogates for real interests and values increasingly dominate public policy formation. They get paid for keeping the policy pot sufficiently stirred to ensure their continued employment and that vested interests remain invested and advantaged. As competing scientific explanations for the more intractable problems proliferate, political powers survive by manipulating information to suit their ends. Public policy processes are so fragmented and fraught with misinformation that many citizens don't know what to believe. Others fight back by creating their own forums filled with more half-truth messages in a variety of media venues—talk shows, public hearings, Web sites, newspapers, and periodicals.

At every turn, the resource professional seems hampered to the point of immobility—unchecked complexity is in the saddle. The megastate created to house the nation's technical prowess seems incapacitated. Attempts to put natural systems into straightjackets of consistency through science, management, and public policy have failed. Fresh thoughts, real dialogue, and bold measures are needed. Improvements at the margin—tinkering—will not work. It is time to become adaptive, to fundamentally rethink resource assessment and management.

4.1.4 The Path of Last Resort?

Human efforts to subdue and control nature's dynamism over the past century have effectively decoupled social and ecological systems. The consequences have been the dramatic loss of biodiversity and the accumulation of evidence that nature's disturbance regimes continue to exceed human predictions and control strategies. Our resource science and management agencies have become proficient at documenting the price tag of our folly, but are incapable of discovering creative solutions that fundamentally learn how to live with and profit from the variability that is inherent in complex, adaptive biophysical systems. A new social culture must be implanted in our public policy processes; a culture based on public interest governed by informed skepticism and precautionary action toward nature. Resource policies must be viewed as hypotheses to be tested, evaluated, and continuously improved on. The presumption that resource policy is fail-safe must be deserted. Precautions must be embedded in practice to avoid the implementation of policy at temporal and spatial scales that are too large to be safe to fail. Through the unquestioned application of science to technological innovation, society has followed the path of least resistance. To reverse the ecological consequences that this approach has wrought, society must take the path of last resort—working with the differences that separate us from each other and nature and learning new ways of working and living together.

Adaptive ecosystem assessment and management (AEAM) is a philosophy and an approach to managing complex adaptive systems that offers some hope in addressing the fundamental challenges that face natural resource management in the coming century. Unfortunately, despite over 25 years of development as theory and practice, relatively little is known about it, even though the experience-based understanding for the need for such change is mounting among resource scientists and practitioners.

4.2 Purpose of the Paper

The adaptive environmental assessment (AEA) effort in the Everglades (1988–1993) stands as a testament to collaboration among scientists amid unrivaled political controversy (Davis and Ogden, 1994). But proving the value of AEA in principle and having this approach and way of thinking integrated into natural resource management are two very different challenges. Over the past 25 years, considerable effort has been quietly put into developing theory, conducting research, and implementing adaptive probes. Much learning about AEAM and how to implement it has occurred, primarily through hard-won experience. Much of the understanding gained from years of coupling grounded theory and thoughtful practice remains tacit and unexpressed, residing in the minds of those few who have done the work. The founding works (Holling, 1978; Clark and Munn, 1986; Walters, 1986) are out of print. Those most intimately engaged in practicing AEAM over the past quarter-century, despite best efforts, have not imparted their accumulated knowledge and understanding in comprehensive and systemic ways to current and future generations of resource professionals. This situation will change, but for present purposes a few concepts, principles, and the ethos pertaining to AEAM are offered in this chapter in the hope that they may in some small way contribute to guiding those interested in the pursuit of AEAM. Chapter 31 continues by summarizing in case study format how these principles were implemented in the Upper Mississippi River Basin adaptive environmental assessment.

4.3 Management Pathology

Someone once said, "For every complex problem there is a simple solution that just won't work." Over the past 50 years, resource management has exhibited a profound pathology. Examples from every resource discipline (fisheries, wildlife, forestry, waters) can be found (Gunderson et al., 1995). From a real-time management perspective, the syndrome in its advanced stages is characterized as "gridlock," "train wrecks," "perpetual crisis," or "whitewater." From a more systems-oriented perspective, the repetitious and cumulative set of symptoms surfaces as the phenomena of "fixes that backfire" (Senge, 1990). The central problem is that the implementation of simplistic solutions results in temporary alleviation of symptoms, followed by long-term adverse consequences that are often worse than the original problem. This pattern of decisions over time is seen as the endless pursuit of and redefinition of "the problem." For example, in the Everglades the management problem was successively redefined over 100 years as drainage, flood control, water supply, water quality, environmental mitigation, and ecological restoration. The cumulative effect of the layering of solutions one upon the other is the emergence of gridlock—institutional, technological, and political. All proposed solutions are seen as in conflict with each other, with none having a clear, decisive, competitive advantage as the sagas continue to unfold. Management in this sense has become pathological, a ritualistic repetition of steps designed to optimize one interest or value, as summarized in Gunderson et al. (1995). The pathology follows the following predictable sequence of events:

MANAGEMENT DECISION

1. Selection of single target
2. Full systems consequences ignored
3. Certainty assumed through control
4. Ecological consequences
5. Initial success followed by crisis
6. Attempts to get out of trouble lead back in only to deeper trouble
7. Loss of ecological structure and composition
8. Increasing vulnerability to subsequent shocks

4.3.1 Surprise and Pathology

In recent years, resource management has been shaken by a procession of external disturbances, discontinuities, innovations, and breakpoints. These dramatic changes have been manifested in technology, public opinion, life-styles, demo-

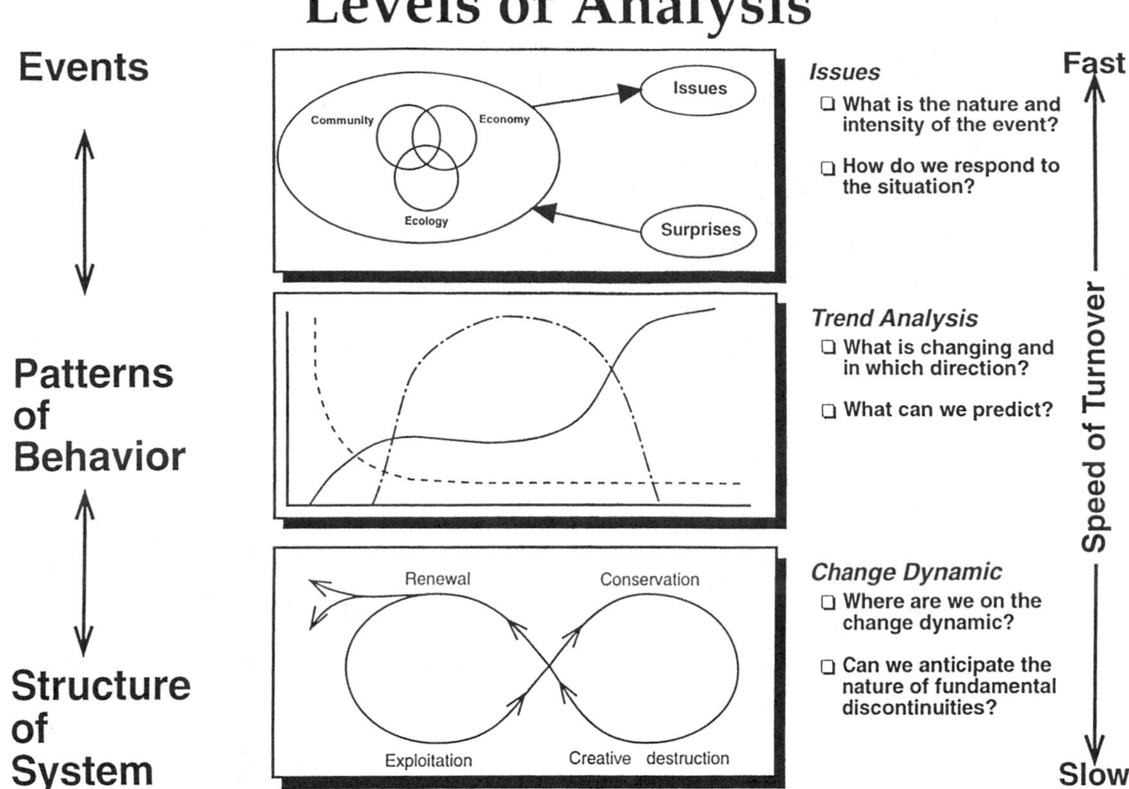

FIGURE 4.1. Levels of analysis.

graphics, habitats, biodiversity, and even global climate and nutrient cycles. Everything seems to be changing at once (i.e., natural systems, institutions, economies, and culture). Surprise is no longer the exception, it is the rule! Meanwhile, resource management remains largely trapped in a set of competencies that only lead into deeper and deeper trouble. Traditional approaches are consuming all the energy and resources, even as the power of these methods declines.

In the crush of technological and economic development of the past 150 years, the dominant culture of Western science and capitalism has sought ever more control over natural processes for various human benefits. The result is a profoundly fractured world view, collective psychoneurosis, and ecological crisis (Leopold, 1949; Wilbur, 1995). The Wise Use and Earth First movements, militia and back-to-the-land movements, animal rights activists, and group hysteria associated with the reintroduction of wolves are symptomatic of the cultural trauma our society is experiencing.

Time and again resource management decisions have placed restraints on nature that have resulted in adverse ecological consequences and long-term degradation. Short-term successes in timber production, flood control, water supply, and wildland protection have led to various forms of overexploitation, long-term oversimplification of ecological structure, and loss of diversity and ecological resilience (Regier and Baskerville, 1986).

To tackle this pathology requires us to shift from managing symptoms to exposing root causes. Why have fixes perpetually backfired? As Botkin (1990) stated, the search for solutions must penetrate the surface and probe the hidden order of things. We have failed to penetrate swarms of events and patterns of change to understand the logic of change—human and natural—that confounds attempts to resolving human and nature problems.

For the most part, resource managers tend to operate at two levels, those dealing with the issue and trend (Figure 4.1). But these levels provide only limited views and understanding of change, typically failing to capture the essence of major discontinuities in time and space. Resource management has tended to treat surprises as temporary digressions from business as usual. Were the 1988 fires of Yellowstone, the failures of stocking hatchery salmon in the Pacific Northwest, the outbreaks of spruce budworm in the Canadian boreal forests, and the algal blooms of Lake Okeechobee temporary digressions? Only the pursuit of what is unknown, uncertain, or fundamentally foreign to conventional wisdom will yield deeper understanding and reveal the pattern of change on which events and trends are based. While such an assertion may seem self-evident, even fatuous, this fundamental management pathology—the fixes that backfire—continues to be denied, overlooked, or ignored in hundreds, perhaps thousands, of resource settings. The structural barriers to overcoming the prevailing pathology are formidable and are discussed later in the chapter. Despite the obstacles, for those who are willing to traverse the treacherous terrain there is a clear alternative—adaptive ecosystem assessment and management—an alternative designed to recouple man and nature in ways that enhance resilience and knit futures that are economically productive and earth-honoring as well.

4.4 The Taproot: The Boundary-centered View

The philosophy of adaptive management is rooted in a radically new perspective of how ecosystems behave and are viewed (Holling, 1973). Holling reasoned that if ecosystems are profoundly affected by outside changes and continually confronted with the unexpected, the constancy of system behavior becomes less important than the persistence of relationships. Holling contrasted two very distinct views of natural systems:

1. An equilibrium-centered view (*stability*) based on the ability of systems to return to equilibrium state after a temporary disturbance.
2. A boundary-centered view (*resiliency*) based on the ability of a system to persist in the face of extremes disturbance to state variables, driving variables, and parameters.

Stability is used in the narrow sense of elasticity, the property that resists departure from equilibrium and maximizes the speed of return to equilibrium due to small disturbances. Stability deals with negative feedback and near steady states, while resilience focuses on the role of positive feedback, internally generated variability, and far-from-equilibrium behavior. At the heart of the equilibrium-centered view of the world, there is an assumption that all things can be explained and predicted. The implications of a boundary-centered view for management are profoundly different. Whereas stability emphasizes constancy and predictability and the prospect of regulation (control) to achieve maximum sustainable yield, a resilience perspective presumes the unexpected in the face of considerable ignorance about the consequences of future events and actions. Humility, not hubris, takes precedence.

The resilience viewpoint of ecosystems challenges the conventional wisdom on which traditional resource policy and economics have been based—economic efficiency, other things being equal, and steady-state conditions. As Holling wrote in 1973, "Strategies based upon these two different views of the world might well be antagonistic. . . . It is at least conceivable that the effective and responsible effort to provide maximum sustainable yield from a fish population or a nonfluctuating supply of water from a watershed (both equilibrium-centered views) might paradoxically increase the chance of extinctions."

For example, the concept of resilience differs significantly from the school of thought that emphasizes ecological health, stress, and endpoints. Ecosystems are moving targets; even when endpoints are initially conceptualized as a tool for serving short- and midterm management objectives, it is the nature of bureaucracy to canonize them rigidly over time. Health is an anthropocentric term. Dynamic processes such as fire, wind, flood, drought, and pestilence are only stressors when seen from a human perspective. Resilience instead focuses on the ecological functioning of such processes over temporal and spatial scales. It focuses on the capacity of all living systems to renew themselves not just in spite of, but *because* of, such processes. A resilience perspective thus shifts the emphasis from desired future *conditions* to desired future *behaviors*.

4.4.1 Far from Equilibrium Behavior

Since 1973, scientific advances in the disciplines of physics and chemistry, molecular biology, and ecology have amassed considerable support for the boundary-centered view based on greater understanding of behavior far from equilibrium in biotic and abiotic systems. "The dialogue between man and nature could be described as going from the concept of being in a static sense to the concept of being as becoming" (Prigogine, 1984). The shift to a boundary-centered view has focused greater attention on time, evolution, and complexity.

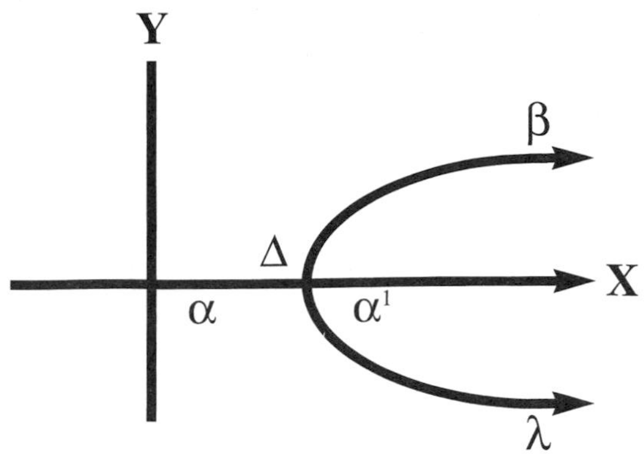

FIGURE 4.2. Simple bifurcation diagram.

A simple bifurcation diagram (Figure 4.2) illustrates the point. *X* represents the arrow of time and *Y* some medium. To the left of the bifurcation point, Δ is stable, but as time progresses past Δ, α becomes unstable. Past the bifurcation point, new stable structures β or λ appear. No additional information can predict which of the two branches will be selected. It is a unique and unpredictable event in a system that exhibits behavior far from equilibrium.

In summary, far-from-equilibrium conditions produce surprising features and radically new behavior not found in equilibrium or near-equilibrium worlds. Furthermore, systems far from equilibrium are much more sensitive to future disturbances. The bifurcation diagram demonstrates a perspective on the world with profound implications for management: ecosystems are dynamic and inherently uncertain, with the potential of multiple futures. While the act of management transforms ecosystems into new entities in order to create economic and social opportunity, the very success of that endeavor generates new classes of surprise and uncertainty.

4.5 The Adaptive Cycle: Of Time, Evolution, and Complexity

We must try harder to understand than explain. (Havel, 1992)

These eight words, spoken by the president of the Czech Republic, captured the essence of both the shift that is required of resource science and the challenges that face humanity in the 21st century. In a world of substance, precision is king, but in a world filled with pattern and context (boundary-centered view), accurately mapping the configuration of relationships is paramount. Current scientific predisposition favors a science of parts, substance, and precision. Technology has advanced to provide us with extraordinary analysis of detail complexity (e.g., geography information systems, sophisticated computer modeling), but such models are often underutilized in building understanding of complex

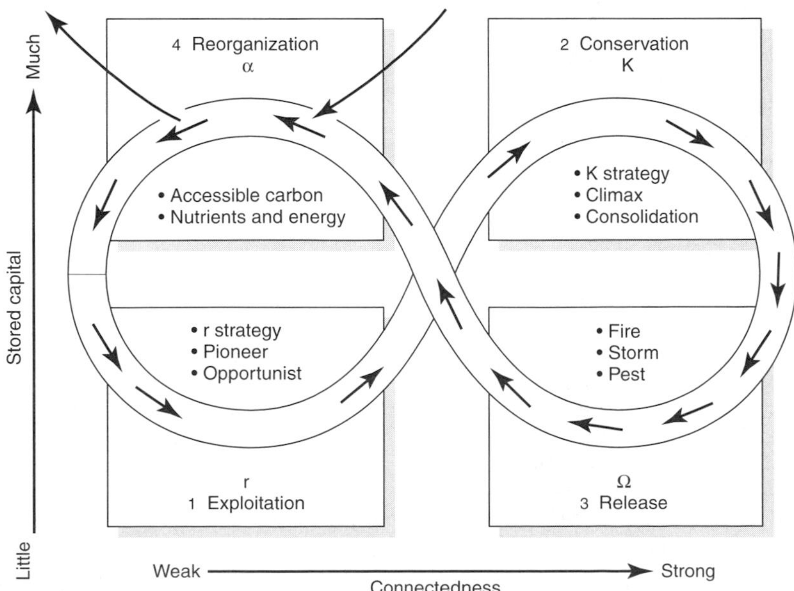

FIGURE 4.3. The adaptive cycle.

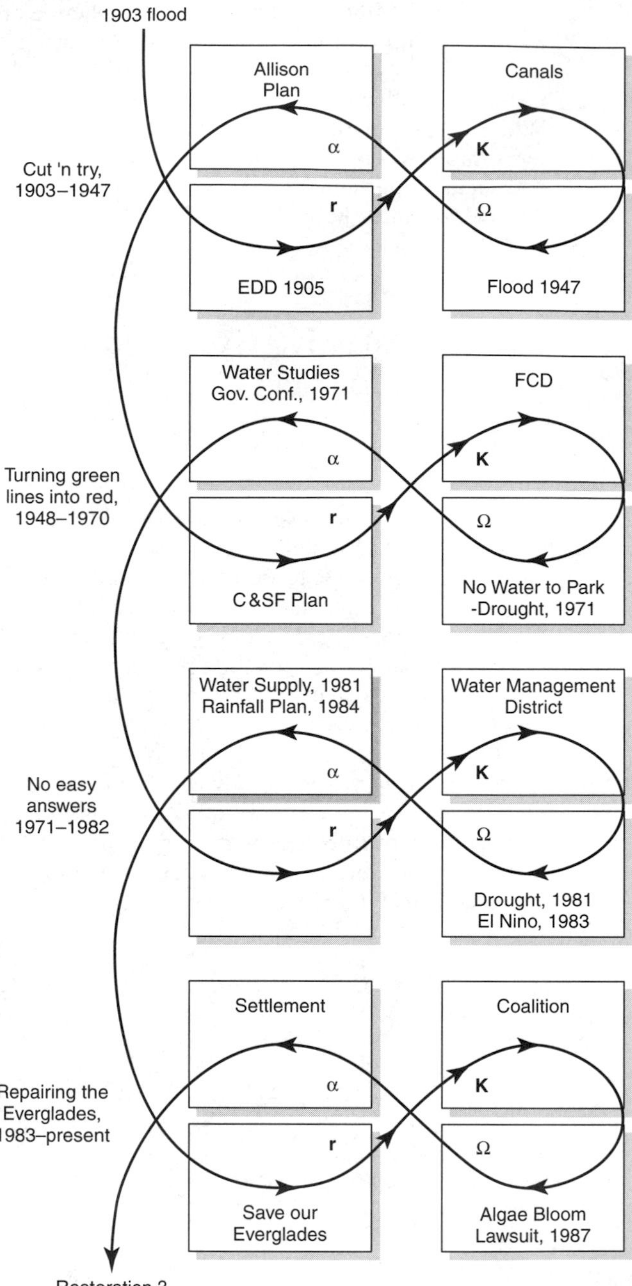

FIGURE 4.4. Coevolution of ecology and institutions in the Everglades.

adaptive systems. Dynamic complexity is about pattern—qualitative properties that fix the limits of change—and requires scientific methods of integrating and analyzing relationships among parts. To study complex adaptive systems, we need to try harder to understand patterns of behavior than to explain or predict limited manifestations of them. Faced with the difficulties of analyzing the problems of time, evolution, and complexity in systems far from equilibrium, Holling (1986) created a thought model of change called the *adaptive cycle* (Figure 4.3).

All living things go through the cycle of birth, death, and renewal, "the revolving fund of life" (Leopold, 1949). Holling's challenge was how to craft a heuristic that would capture change in time, states, and levels of complexity. As an example, the adaptive cycle (Holling, 1995) describes the process by which forest ecosystems grow and change in four distinct stages:

1. *Exploitation:* pioneer species rapidly colonize recently disturbed areas;
2. *Conservation:* growth of the forest through slow accumulation and storage of energy and matter (climax condition).
3. *Creative destruction:* fires in the forest consume biomass and release the nutrients.
4. *Reorganization:* soil processes prepare for the next generation of exploitation.

The model also attempts to capture the pace of change. From exploitation to conservation, the pace of change is slow; forests can take decades or centuries to mature. From conservation through release and renewal, however, the pace of change is very rapid; in hours a fire can undo a forest community that has been decades in the making.

"The secret of well being is simplicity, but the key to evolution is the continual emergence of complexity." This statement by Arthur (1990) describes the paradox of evolution and complexity that Holling has captured in the adaptive cycle. As forests grow, the interconnections become more tightly linked. But, just as the simple bifurcation diagram illustrates (Figure 4.2), dynamic systems evolve to points at which they become structurally unstable, or brittle, and a breakpoint is reached from which rapid change (creative destruction through renewal) in new directions occurs. The new forest that emerges may retain aspects of the old system, with new components thrown into the mix during reorganization, or it may be that no forest emerges at all. If nutrients are carried away and soil fertility is low, the next period of exploitation might tend for a time toward a grassland system. In answer to Arthur's riddle then, natural systems over time weave together both simplicity and complexity in an evolutionary dance.

Fire, floods, and storms are dynamic processes that are central to natural system renewal. Experience has shown that when these are eliminated long-term adverse ecological consequences are inevitable (Regier and Baskerville, 1986). The adaptive cycle helps to deepen our understanding of patterns of change, but does not necessarily promise to help us to predict how systems far from equilibrium will behave at any given time. Light et al. (1995) have demonstrated how ecosystems and institutions coevolve (Figure 4.4). This work suggests that this new understanding of institutions and ecosystems may help us rise above the crisis–response treadmill of the past to construct management systems that fit more adaptively within the constraints set by a cyclical, evolving nature.

The power of the adaptive cycle heuristic transcends the discipline of ecology. Analogs to the forest story containing components of the adaptive cycle can be found in many other fields of inquiry, for example, physics (Capra, 1975; Bohm, 1978, 1980), economics (Schumpeter, 1934; Arthur, 1990), business management (Mintzberg, 1978; Morgan, 1986; Hurst, 1995), anthropology (Douglas, 1986), literature (Goldstone, 1991; Fischer, 1996), psychology (Wilbur, 1995; Dorner, 1997), and religion (I Ching, Tielhard de Chardin, 1959). While this comes as no surprise—after all, the problems of time, evolution, and complexity are not confined to the science of ecology—the reductionist, positivist methods of traditional science and management continue to ignore these fundamental insights.

4.6 The Assessment and Management Challenge

Although scientific understanding of how ecological systems behave has advanced considerably, changes in management practice have lagged far behind (Holling, 1978; Clark and Munn, 1986; Walters, 1986; Gunderson et al., 1995). Economic success has been so convincing that, in many instances, the need and motivation to search for alternative ways of assessing and managing have been ignored, considered unnecessary. When apparent successes faltered, the principal option was to continue as before, only more rigorously (self-sealing beliefs). In the process, ecosystems have been driven further into less productive states and closer to potential irreversibilities (Regier and Baskerville, 1986). In addition, a new class of problems has become more apparent (Holling, 1995 paraphrased below). The new problems and their implications for assessment and management can be described as follows:

1. There are no simple cause and effect relationships. The problems are essentially systems problems in which aspects of behavior are complex and unpredictable and causes, although at times simple (when finally understood), are always multiple. Understanding of patterns of behavior, not "golden numbers," and integrated modes of understanding are needed to form new policies.
2. There are no universal stability regimes. The problems are fundamentally nonlinear in causa-

tion. They demonstrate multiple stable states and discontinuous behavior in both time and space. Simply relieving an ecological "stressor" won't solve the management problem.

3. There is no easy way out. Problems are increasingly caused by slow changes reflecting decadal accumulations of human influences on air and water and decadal or centuries-old transformations of landscapes. These slow changes cause sudden shifts in environmental variables that quickly and directly affect the health of people, productivity of renewable resources, and vitality of societies.
4. There can be no neat packaging of problems in time and space. The spatial extent of connections is intensifying so that the problems are now fundamentally cross-scale in nature in space as well as in time. More frequently, regional and national environmental problems now have their source at both local and global scales. Examples include the greenhouse gas accumulations, ozone depletion, and deterioration of biodiversity.
5. Time is not reversible. Both the ecological and social components of these problems have an evolutionary character. The problems are therefore not amenable to solutions based on knowledge of small parts of the whole, nor on the assumptions of constancy or stability of fundamental relationships, whether ecological, economic, or social. Policies must be focused on active adaptive designs and management actions that yield understanding as much as they do product.

4.7 The Adaptive Path: No Easy Answers

Instead of looking for saviors, we should be calling for leadership that will challenge us to face problems for which there are no simple painless solutions—problems that require us to learn new ways. (Heifetz, 1994)

Putting new ideas to work requires uprooting existing ways of thinking and acting. Implementing AEAM calls for a deep commitment to abiding change, not just in our skills and capabilities, but in our attitudes and beliefs, awareness, and sensibilities (Senge, 1990). Such profound shifts in the way we think and behave are extremely difficult. For practical and rational reasons, people resist "unlearning yesterday." Change is painful, threatening, inconvenient, and uncomfortable. It exposes incompetence and requires effort for new learning. The fundamental challenge for AEAM, or ecological policy design as it has sometimes been referred to, is fostering patterns of collective action based on our present limited understanding that remain open to learning how to discover, accept, and reflect on the transient solutions that will inevitably be found deficient. Learning is both painful and self-renewing—terrifying and reassuring. Implanting AEAM in our organizations will require overcoming deep structural barriers, including (1) institutions so attached to conventional methods as to drive out creativity, (2) threats to personal security or professional competence that break down commitment to change, (3) self-sealing beliefs that reify existing strategies, (4) limits to simplification, and (5) bureaucratic inertia (Hurst, 1995).

4.8 Principles for Adaptive Change

Adaptive ecosystem assessment and management aspires to create dialogue between humans and nature in which humans imagine ways to engage nature for human benefit by testing propositions (policy as hypothesis, management as experiment) on how to learn to live with and profit from (adapt to) the variability inherent in ecological, economic, and institutional systems and their interactions. Therefore, AEAM is a fundamentally new philosophy of management and not just a compilation of sundry methods (Baskerville, 1977).

Without an articulation of guiding principles and values, AEAM cannot be fully embraced and has no transformational power. What is offered next is a summary of evolving inferences about the way that the dialogue between people and nature should be conducted in the future. As with any propositions, they cannot be validated, only discredited in particular contexts. Some are roughhewn, while others have been polished through repeated application. Like any scientific model, they will be accepted not by faith, but through repeated testing.

- **Surprise is the rule, so plan to learn.** This is the soul of the adaptive philosophy; the realization that solutions are transient and will inevitably be found deficient makes learning by intent imperative. Building learning into organizational and institutional function is a challenge despite growing recognition in the management literature of its value. In place of a bottom line, public bureaucracies use rules and regulations, that is, adherence to procedure, as a measure of accountability and performance. Adaptive management creates a new bottom line and reasserts focus on outcomes, not just outputs. Learning requires that hypotheses and assumptions about how the natural world works and how human and natural systems interact can be made explicit. In doing so, accountability shifts away from mindless adherence to rules to an understanding of whether what we are doing is having the beneficial consequences intended.

• **Think and go slow to go fast.** The political dynamics that drive quick fixes simply lead to more unforgiving conditions for decisions, more fragile natural systems, more dependent industries, and more distrust from citizens (Holling, 1986). Analyze the pace at which change is occurring. More often than not, the variables that attract most attention cause rapid change and operate at small scales. Modern decision-making processes are beholden to the here and now. But life depends on a web of interrelationships, and all meaningful relationships take time and energy to build and sustain. Assessment should focus on long-term, slow changes in structural variables or system properties and challenge the assumptions on which current policy rests. Expand the domain of rationality beyond the *here and now* by reflecting on how the *long ago and far away* will influence how the future may unfold out *there and then.*

• **The devil is in the dynamics, not just the details.** Ecosystems are dynamic—constantly changing and inherently uncertain, with potential multiple futures—a moving target with no set endpoint. Focus on learning from past patterns and processes. Fire suppression in Yellowstone preserved scenic beauty for a time, until fire on an unanticipated scale was triggered by a slow moving dynamic. Drainage in the Red River Valley allowed for productive agriculture and development, until unprecedented snowfall and rains in 1997 reminded the world, with disastrous consequences, that the valley was in fact the ancient lakebed of the former Glacial Lake Agassiz.

• **Theory is not an intellectual luxury; it is a practical necessity.** Theory and practice must be tightly linked. Without one, the other withers. When theory or practice operates in a vacuum, precious time and resources are wasted, and unintended beneficial or adverse consequences go unexamined. Far too frequently, assessments and the decisions that follow are unlinked, and opportunities for learning are lost. Assessment needs an integrated scientific framework on which to build shared understanding that bridges with policy options. The lines that separate research, policy, and practice must eliminate work in tandem.

• **The goal of resource management is resilience.** We must maintain the capacity of living systems—ecological, cultural, and economic—to renew themselves and thrive, not just survive for a time. We must design ways of thinking and acting that will allow pattern (e.g., disturbance regimes) integral to ecosystem functioning to be restored. Resource assessments must communicate stories of renewal: how the slow (climate, geology) and fast (vegetation) variables are coupled (through hydrology and sedimentation) and evolve, revealing themselves as configurations over time that cannot be understood as snapshots. These stories must penetrate the surface of things and expose the deep logic that provides insight, not universal answers. Assessments should capture and convey the essence of change in readily understandable caricatures (see Section 31.5.1). Assessments must highlight uncertainties and search for flexibility, identifying management and policy options that eliminate or buffer negative relationships between ecological and economic objectives.

• **Embracing error is prerequisite to developing trust and understanding.** Human adaptation and resilience depend on unlearning the past. Learning new ways is predicated on understanding that yesterday's solutions are today's problems and tomorrow's blind spot. We must admit that we are captives of the moment, trapped in a world where change invariably outpaces comprehension, with little thought given to the lessons of the past or the emerging puzzles of the future. The result is that our professional competencies (i.e., doing what society has asked of resource scientists and managers) become increasingly short-lived, at risk, and untrustworthy. Too frequently, we resort to established patterns of activity and action that are economical and reassuring, but increasingly irrelevant and pathological. Facing tough ecological challenges requires inventing fundamentally new ways of thinking and acting and then communicating effectively and mobilizing with others to take action.

• **Truth lies at the intersection of competing explanations.** Inquiry and reflection are central to human endeavor. The era of science as the revealer of universal truths on which the future could be confidently predicted is over. Chaos, catastrophe, criticality, and fractal theories have exposed the inadequacies of a Newtonian, deterministic world view (not to mention the illuminating criticisms of generations of social scientists, historians, and philosophers). In its place is a world view that invites multiple and potentially competing modes of inquiry and lines of evidence, using the differences that separate people to learn new ways to work together. Truth is an approximation that is never at rest. Scientists must acknowledge a more humble role, leading from behind. They must stop acting like high priests; they are contributors to, rather than sole authors of, greater social and ecological understanding. The challenge is to reinvigorate and reempower societies by offering science as a culture for change, not a method for decision paralysis.

• **We know more than we think, but less than we understand.** The pat answer to most difficult resource questions is that we need more study. Rarely do we take full advantage of the breadth and depth of existing information to get a composite picture or integrated understanding of what we know or think we know. Integrative science brings together the knowledge of various disciplines and, with deference to local conditions, begins to explore the gaps between disciplines and lines of inquiry. Invariably, learning takes place and tacit understandings are made explicit.

• **Adaptive ecosystem assessment is an inductive and integrative approach to science.** There are two approaches to science, the science of parts and the science of the integration of parts. Dealing with a complex real-world setting requires development of scientific methods and tools that are well suited for putting puzzle pieces together, discovering and exploring gaps in understanding, and examining ways to combine experience, data, and intuition to develop a shared, but always incomplete, picture of the problem situation.

• **Adaptive ecosystem assessment strives for creative synthesis and breakthrough thinking.** Much of resource management is reactive, anemic, and uninspiring in nature. Management fails to concern itself with building adaptive capacity for creative futures that are uncertain. Adaptive ecosystem assessment encourages, facilitates, embraces human ingenuity, anticipating bursts of human insight that transcend conventional thinking and open new paths to action.

• **"You can't get there from here."** Institutions become trapped in the ritualistic framing of new ecological challenges in ways that work with less and less success. The new class of ecological problems (discussed earlier) requires the development of new methodological approaches that serve temporary platforms for managing the dynamics of science and democracy in deciding how to recouple social–ecological systems. The design of scientific inquiry must be more transdisciplinary, with the goal of consilience in mind.

• **Assessment and management are always aiming at a moving target.** Adaptive ecological assessments presume that knowledge of the combined social, economic, and ecological system that we are faced with is always incomplete; the system itself is constantly evolving, as is the imprint of management and the progressive expansion of the scale of human influences on the planet. Hence the actions taken by management must be ones that are renewed periodically by reassessment of the ever-shifting assumptions on which shared scientific understanding as well as the socially desired goals are founded.

4.9 Personal Ethos

In human endeavors in which people are collectively engaged, participants develop relationships and shared value systems. Such an observation would traditionally have been viewed as belonging in the realm of the history of science, largely irrelevant to the accumulation of scientific truths; however, AEAM recognizes that such relationships are not mere by-products, but explicit components of the learning process. Over the years those who have labored diligently and distinguished themselves in the field of AEAM have become a kind of family, an "institute without walls"—the Ralf Yorque Society. "The name was born as a product of the sometimes indelicate, perhaps naive, but always joyful, creative spirit of the group" (Holling, 1978). The collective code of norms has been partially articulated by Michael (1973, 1994):

1. *Embracing error.* Acknowledging specific uncertainties is the basis for building honesty and integrity, the stuff on which learning systems are based; it is also the operational precondition for building capacity to adapt. Institutions have a penchant for ignoring error and avoid dealing with it at almost any cost. Without embracing error, there is no learning, no change in ideas, attitudes, or behavior.
2. *Spanning boundaries.* Attempting to expose and address error even under the most constructive conditions can be very threatening to people. Thus it is necessary to develop trust and to reach out to who and what we are not, a way of being requiring considerable skill and patience. Boundary spanning is the negotiation of the no man's land that separates peoples and institutions. Akin to coordination and conflict resolution, it is a process of searching for ways to create new meaning and offers the potential to

transform relationships from vicious cycles to virtuous cycles, thus creating generative relationships that foster collaborative learning and creative synthesis.
3. *Coping with role ambiguity*. Learning and helping others to learn new ways will be perceived as not just pushing the envelope, but having gone over to the "other side." Michael says the feeling is like standing with both feet firmly planted in midair! Without considerable trust and respect among peers and by superiors, leaders of change can appear as "loose cannons," lacking discipline and having questionable loyalties. True leaders will emerge only from those with considerable personal mastery and integrity of character.
4. *Developing active listening skills*. Listening in an active way is essential, because unraveling systems problems requires going deeper. It requires a person to understand how what people think, feel, and act is linked to deeper belief systems and lived experience. Belief systems and natural systems are nested. Real leverage comes from searching below the surface of events and trends, probing for the logic of change. Active listening may be a hackneyed term, but next time you are exploring the depths of an ecological (physical, biological, cultural) problem, for every answer you think you have uncovered, ask "why" again—the causal chain of relationships will begin to reveal itself, grudgingly. Only the most tenacious listener will be rewarded.

4.10 Conclusion

Ideas lead change. Understanding how natural systems perform far from equilibrium and the design of precautionary management probes for societal benefit and learning forms the basis of adaptive ecosystem assessment and management. The AEAM is not about incremental change and patching holes in existing approaches to resource management; it is about revolutionizing the way that natural resource management is conducted. AEAM is not just about integrative science. At its core is the understanding that a new coming together of science, policy, culture, and ethical activity is necessary—consilience (Wilson, 1998)—a path of last resort, because deep learning is the only alternative.

The difficulty of this undertaking is considerable and the outcome uncertain. While the search for ways out of our current management pathology has begun in earnest, work is progressing slowly, with too much groping, too little insight, and, too many attempts to reinvent understanding that already exists. The field of natural resource management is in the midst of a protracted period of profound change. The question is not one of considering environmental values, but rather of understanding and benefiting from the invisible hands of nature at work on the landscape and how humans are an extension of this pattern. Leopold (1949) referred to this phenomenon of nature as the "revolving fund of life." The record of the past 150 years is one of humans evolving into a planetary force bent on harnessing nature to human aspirations for exploitation, growth, and development. But sustainable resource management is not simply a matter of honing our capacity to sustain ever-increasing yields from our products. Ecological sustainability will hinge on our capacity to learn, understand, live with, and profit from the variability that exists in ecological, economic, and community systems—resilience. A future must be envisioned that is predicated on the renewal of both economic and ecological systems that have the opportunity to fulfill their combined potential. This is not just a question of whether such a future is possible, although the role of science in the search for sustainable ecological and economic futures is clear. It is a question of resource scientists leading without authority, while taking responsibility for the whole, and creating collaborative platforms for science and democracy to flourish together.

4.11 References

Arthur, B. 1990. Positive feedback in the economy. *Sci. Amer.* 262:92–99.

Baskerville, G. L. 1977. *Adaptive management: a philosophy and a method.* Unpublished Collaborative Papers. Luxembourg, Austria: International Institute for Applied Systems Analysis.

Bohm, D. 1978. The implicate order: a new order for physics. *Process Stud.* 8:73–102.

Bohm, D. 1980. *Wholeness and the implicate order.* London: Routlege and Kegan Paul.

Botkin, D. B. 1990. *Discordant harmonies.* New York: Oxford University Press.

Capra, F. 1975. *The Tao of physics.* New York: Wildwood House.

Clark, W. C.; Munn, R., editors. 1986. *Sustainable development of the biosphere.* Cambridge, UK: Cambridge University Press.

Davis, S. M.; Ogden, J. C. 1994. Towards ecosystem restoration. In: Davis, S. M.; Ogden, J. C., eds. *Everglades: the ecosystem and its restoration.* Delray Beach, FL: St. Lucie Press: 769–796.

De Chardin, Teilhard P. 1959. *The future of man.* New York: Harper & Row.

Dorner, D. 1997. *The logic of failure.* Reading, MA: Addison-Wesley.

Douglas, M. 1986. *How institutions think.* Syracuse, NY: Syracuse University Press.

Fischer, D. H. 1996. *The great wave: price revolutions and the rhythms of history.* New York: Oxford University Press.

Goldstone, J. 1991. *Revolution and rebellion in the early modern world.* Berkeley, CA: University of California Press.

Gunderson, L. H.; Holling, C. S.; Light, S. S., editors. 1995. *Barriers and bridges to the renewal of ecosystems and institutions.* New York: Columbia University Press.

Havel, Vaclav. 1992. The end of the modern era. Op-Ed. *New York Times.* March 1, 1992.

Heifetz, R. 1994. *Leadership without easy answers.* Cambridge, MA: Harvard University Press.

Holling, C. S. 1973. Resilience and stability of ecological systems. *Ann. Rev. Ecol. Syst.* 4:1–23.

Holling, C. S. 1978. *Adaptive environmental assessment and management.* New York: John Wiley & Sons.

Holling, C. S. 1986. Resilience in ecosystems: local surprise and global change. In: Clark, W. C.; Munn, R., eds. *Sustainable development of the biosphere.* Cambridge, UK: Cambridge University Press: 292–317.

Holling, C. S. 1995. What barriers? What bridges? In: Gundersion, L. H.; Holling, C. S.; Light, S. S., eds. *Barriers and bridges to the renewal of ecosystems and institutions.* New York: Columbia University Press.

Hurst, A. 1995. *Crisis and renewal.* Cambridge, MA: Harvard Business Review Press.

Leopold, Aldo. 1949. *A Sand County almanac.* London: Oxford University Press.

Light, S. S.; Gunderson, L. H.; Holling, C. S. 1995. The Everglades: evolution of management in a turbulent ecosystem. In: Gunderson, L. H.; Holling, C. S.; Light, S. S., eds. *Barriers and bridges to the renewal of ecosystems and institutions.* New York: Columbia University Press.

Michael, D. N. 1973. *On learning to plan and planning to learn.* San Francisco: Jossey–Bass.

Michael, D. N. 1994. A letter to an aspiring policymaker. In: Senge, P.; Kleiner, A.; Roberts, C.; Ross, R.; Smith, B., eds. *Fifth discipline fieldbook: strategies and tools for building a learning organization.* Garden City, NY: Doubleday: 499–502.

Mintzberg, H. 1978. Patterns in strategy formation. *Manage. Sci.* 24(9):934–948.

Morgan, G. 1986. *Images of organization.* London: Sage Publications.

Prigogine, I. 1984. *Order out of chaos: man's new dialogue with nature.* New York: Bantam Books.

Regier, H. A.; Baskerville, G. L. 1986. Sustainable redevelopment of regional ecosystems degraded by exploitative development. In: Clark, W. C.; Munn, R., eds. *Sustainable development of the biosphere.* Cambridge, UK: Cambridge University Press: 75–103.

Schumpeter, J. A. 1934. *The theory of economic development: an inquiry into profits, capital, credit, interest and the business cycle.* Cambridge, MA: Harvard University Press.

Senge, P. 1990. *The fifth discipline: the art and practice of the learning organization.* Garden City, NY: Doubleday.

Walters, C. J. 1986. *Adaptive management of renewable resources.* New York: Macmillan.

Wilbur, K. 1995. *Sex, ecology, and spirituality; the spirit of evolution.* Boston: Shambhala Publications.

Wilson, E. O. 1998. *Consilience: the unity of knowledge.* New York: Knopf.

Part 2
Information Management

5
Data Acquisition

Thomas J. Stohlgren

5.1 Introduction

The science of data acquisition has changed a lot in the past two decades (Oppenheimer et al., 1974; Michener, 1986). In 1974, Austin (in Oppenheimer et al., 1974) marveled at the ability to view data and computer outputs directly on a television screen (cathode ray tube). It was a time when the acronym GIS referred to "general information system" (Oppenheimer et al., 1974, p. 232). In 1986, Klopsch and Stafford recommended the use of 8-inch single-sided, single-density or $5^1/_2$-inch double-sided, double-density diskettes for the storage of medium-sized data sets for the Long-Term Ecological Research (LTER) program. Kriging was touted as the best interpolation method for spatial analysis (Seilkop, 1986). There is little doubt that 10 or 20 years from now our current commonly used hardware (e.g., Unix-based workstations, Pentium-type personal computers, digitizing tablets), storage media (e.g., CD-ROM, $3^1/_2$-inch high-density diskettes, tape backups), and software (e.g., ARC/INFO geographic information systems, cokriging spatial analysis programs) will invoke a similar comedic response. However, some things haven't changed; there is a consistent, unrelenting commitment to improve data acquisition and management in the ecological sciences.

The challenges in the art and science of data acquisition are growing. Changing ecological paradigms, issues, threats, and increasing demands by society for accountability and management responsiveness encourage flexibility on the one hand; on the other, metadata standards, data transfer protocols, and demands for comparable information for multiple uses encourage increased standardization and rigidity. Data are expected to meet the needs of individual studies, along with other local, regional, national, and international needs, despite meager local funding (Stohlgren et al., 1995b). The push for interdisciplinary research involves more investigators and the simultaneous collection of data from multiple biological groups and environmental factors (Stohlgren et al., 1998). Investigations of ecological processes such as disturbance, competition, herbivory, nutrient cycling, and energy flux involve integrated data sets from the laboratory, field experiments, observational studies, and mathematical models. Spatially explicit data and geographic information system (GIS) analyses and modeling are expected outputs from many studies (Stohlgren et al., 1997c). Computing capabilities are increasing at a linear rate relative to data acquisition's exponential growth rate. Quality-checked data are expected to be "loaded on the Web" and accessible for public use in record-breaking time.

There is a perceived urgency for ecological assessments (see Chapter 1). Yet existing biotic inventories of natural landscapes commonly are poor (Stohlgren and Quinn, 1992; Stohlgren et al., 1995a, 1995c). The paradigms of ecosystem management and adaptive management are founded on the promise of a constant delivery of sound ecological data. Increased threats to ecosystems bring increased demands for reliable data and increased needs for ecological assessments.

The newest confounding factor to data acquisition is the issue of scale. The scale of inventory, monitoring, and research programs has changed from site by site assessments to the evaluation of landscape and regional changes in resource sustainability, land-use patterns, and biodiversity (National Research Council, 1990, 1994). It is becoming increasingly important to extrapolate plot-level information to landscape and regional scales (Scott et al., 1993; Stohlgren et al., 1997c, 1998). This, in turn, greatly influences the amount and complexity of ecological data by expanding the spatial and temporal scales of observations and the detail of each set of observations.

The purpose of this chapter is to highlight four major features of data acquisition in ecological assessments: (1) clearly articulated goals and objectives; (2) a commitment to preserving the integrity, longevity, and accessibility of the data for future unforeseen uses; (3) a detailed vision of how the data will be gathered, stored, summarized, statistically analyzed, displayed, and archived; and (4) an understanding of the quality and limitations of the data. Several successful survey, monitoring, and research programs have grappled with data acquisition issues for many years. The examples throughout this chapter are not exhaustive, and they have bias toward vegetation. Examples serve two major functions: to avoid "re-creating the wheel" and to avoid potential pitfalls by taking advantage of the hard-fought experiences of others.

5.2 Clearly Articulated Goals and Objectives

The first and most important step in data acquisition is to clearly articulate study goals and objectives (Oppenheimer et al., 1974; Krebs, 1989). Goals may be lofty, for example, to monitor changes in biological diversity, to detect the effects of acid deposition or global climate change, or to evaluate the processes influencing the structure and species composition of forests. The objectives are well focused. Goals are like a travel plan: we gather existing resources and theories, select a clear direction and trail, acknowledge the unknowns, and keep the goal in mind throughout the trip (Stohlgren, 1994).

Jones (1986) describes the process as "scoping" and "problem definition," by which general problems are reduced to specific ones, specific issues are identified, and priorities are set for specific data acquisition needs. Setting objectives requires a complete evaluation of existing data for spatial and temporal completeness, accuracy, and precision. This is usually no small task, but it is the only way to identify the types and levels of data that are needed.

Narrow objectives may eliminate the collection of extraneous data and preclude unrealistic expectations (MacDonald et al., 1991). As the objectives become clearly defined, we should be able to visualize data products (i.e., tables and figures) and assess the potential limits to which study results can be extrapolated spatially and temporally (see Berkowitz et al., 1989). Specific objectives also help in the selection of appropriate spatial scales, sampling designs, and field methodologies.

5.3 Commitment to Preserving the Integrity, Longevity, and Accessibility of the Data

Most ecologists would agree that a far greater commitment to data management is needed. This commitment begins with a long-term view of the value of well-collected data (Magnuson et al., 1991; Risser, 1991). It is a little frightening that only about 1.7% of ecological studies last at least five field seasons (Tilman, 1989). It is more frightening that the data from a smaller percentage of studies are probably archived, documented appropriately, and accessible for other uses. How can we change this?

A good rule of thumb is that about 20% to 25% of the budget for ecological assessments needs to be devoted to data and information management. Whether a centralized or distributed architecture is selected, several tasks must be fulfilled, including quality control and quality assurance for field data; data collection, transfer, storage, and archiving; data and metadata tracking; and statistical analyses (Shampine, 1993). One strength of the Long-Term Ecological Research program is its hiring of data managers, use of consistent hardware and software, early development of network-wide data standards, and attention to data synthesis (Stafford et al., 1986a, 1986b; Stafford, 1993).

Several other examples come to mind. For over 20 years, a program to monitor changes in forest structure and demography over time by establishing large (1 hectare or greater) *reference stands* in representative forest types has been underway (Hawk et al., 1978). Because each tree is tagged and mapped, large data sets amass quickly, with yearly or periodic observations on seedling establishment, tree growth, mortality, and pathogen effects (Franklin and DeBell, 1988; Riegel et al., 1988; Parsons et al., 1992). The program relies on full-time data managers and a very strong commitment to standardized information management, including data preservation.

The Smithsonian Institution/Man and the Biosphere Biological Diversity Program (SI/MAB) is establishing permanent vegetation plots in representative tropical forests (and in other Biosphere Reserve areas worldwide) to inventory and monitor changes in biological diversity (Dallmeier, 1992; Stohlgren, 1999). An extensive commitment to data management is evident: the program employs a full-time data manager, and the Smithsonian's own BIOMON software program (J. Comiskey, pers. comm.) was developed to store, analyze, and archive information. Other programs with an extensive information management program include the U.S. Department of Agriculture Forest Service Forest Health Monitoring Program (Burkman and Hertel, 1992) and Forest Response Program (Zedaker and Nicholas, 1990) and the U.S. Environmental Protection Agency's Environmental Monitoring and Assessment Program (EMAP; Messer et al., 1991; Palmer et al., 1991).

The challenge of obtaining and maintaining long-term support for data acquisition cannot be overstated (Stohlgren et al., 1995b). Funding sources appear more and more reluctant to support long-term projects, yet this support is necessary for continued consistent measurements and for maintaining data management systems and organizational infrastructures (Strayer et al., 1986; Franklin et al., 1990). Government agencies and others are downsizing the permanent research staff necessary to maintain continuity of field measurements over time and relying more on temporary staff. Regardless of the difficulties involved, the true measure of success in ecological studies should be the successful preservation and long-term use of the data.

5.4 How the Data Will Be Gathered, Stored, and Used

Throughout an ecological assessment, and especially in the initiation phase, investigators need a detailed vision of how the data will be gathered, stored, analyzed, and used (Risser and Treworgy, 1986). This involves careful planning, vision, and a strong commitment to information management. The planning phase includes the development of a detailed study plan (usually derived from the study proposal) with concomitant peer reviews and statistical reviews. The vision phase includes elaborate foresight of data and information products (e.g., data sets, GIS themes, summaries, graphics, mathematical models, and publications) from the data gathering stage to the analysis stage (discussed next). However, plans and vision are rendered useless without the obligatory commitment to information management.

5.4.1 Gathering Data

Although remotely sensed data are common, I will limit specific examples to data collected in the field. For example, in many forest monitoring studies, each tree is mapped, tagged, and identified, and the plots are recensused annually (Dallmeier et al., 1992). The 50-ha permanent plot on Barro Colorado Island, Panama (Hubbell and Foster, 1987), the North American Sugar Maple Decline Project (Miller et al., 1991), and the USDA Forest Service's monitoring programs continue to accrue and manage detailed data sets (Zedaker and Nicholas, 1990; Burkman and Hertel, 1992). It may be prudent to seek the advice of those with years of experience.

Several new tools are available to field ecologists to improve the efficiency of data collection. Global positioning systems help locate and relocate plots in rugged terrain and rangeland expanses. Electronic range finders improve the efficiency and accuracy of tree mapping. Palm-top computers are replacing data sheets, permitting immediate data entry and analysis. In the office, a GIS often provides the central framework for data management. A GIS can combine maps of biotic and abiotic information with historic data and new data. Perhaps the most underutilized capabilities of a GIS are in the unbiased selection of assessment and monitoring sites (Stohlgren and Bachand, 1997) and the predictive ecological modeling to support management decisions (e.g., Buckley et al., 1993). Spatially explicit, high-resolution GIS data are available from satellites and aerial photography. However, acquiring and using remotely sensed data require decisions of scale and resolution, the number and type of sensors, and whether the data are georeferenced or not (see Section 5.5).

5.4.2 Storing and Archiving Data

Data storage depends largely on the hardware and software used and the particular format desired. It is wise to store data sets in two different buildings, in two formats (one of them being ASCII), and on two media (e.g., tape and disks). The most important aspects of metadata (data describing data) are data set name, responsible organization, contact person and address, project codes, data form and format, key words, geographic coverage, scale, time span of data, accessibility, and additional comments about the data sets. This includes spatially explicit data that are incorporated into a GIS. Metadata should follow, as closely as possible, the Federal Geographic Data Committee's *Content Standards for Digital Geospatial Metadata* (1998). Adequate records of the process from data collection to preservation are every investigator's responsibility, including field technicians (data entry), data managers and statisticians, and principal investigators.

5.4.3 Analyzing and Using Data

Software packages have made it easier to perform many statistical operations, while making it more difficult to evaluate the assumptions and algorithms of individual tests. The primary responsibility of the investigator is to realize if the assumptions are met and to understand how the algorithms used influence the interpretation of the results. For example, some spreadsheet software programs have a mathematical function for the standard error of the mean for a population, rather than for a sample. This results in an underestimation of the true variability about the mean. Different kriging programs may produce wildly different "contour maps" from the same biological data depending on the algorithms that they use for interpolation (Legendre and Fortin, 1989; Stohlgren et al., 1997b).

Data often are simplified for display purposes. For example, vegetation classes may be lumped to accommodate storage or printer capabilities. It is important to maintain and link detailed records of data reduction procedures to map versions and their metadata. Maps and data made available on the Web should carry attached metadata.

5.4.4 Data Acquisition for Predictive Ecosystem Models for Decision Making

The real potential of data acquisition for ecological assessments lies in developing and validating predictive ecosystem models (Buckley et al., 1993; Stohlgren et al., 1997c). Here, various scenarios can be tested nondestructively. Conceptual and predictive models are an important step in avoiding the preventable degradation of biotic resources and populations of special concern. The models can also be powerful tools in planning, teaching, training, and public outreach. GIS-based ecosystem models provide an ideal tool for synthesizing existing information and making it useful to resource managers. For example, the spatially explicit ecosystem model Savanna (Coughenour, 1993) has been coupled to the ARC/INFO GIS system by the RAPiD/Arc development environment (Buckley et al., 1993) to investigate links between biodiversity and ecosystem function. Such models are in use in Elk Island National Park, Alberta, Canada (Buckley et al., 1993) and Yellowstone National Park (Coughenour and Singer, 1991).

5.5 Understanding the Quality and Limitations of the Data

The usefulness of data is dependent on their quality (Kanciruk et al., 1986). It is important to develop predetermined levels of accuracy and precision for field measurements (Messer et al., 1991; Palmer et al., 1991). Sources of error must be anticipated, evaluated, and minimized or eliminated. Besides systematic error (bias in measuring devices) and sampling error (bias in the methodology employed), long-term landscape-scale measurements may have considerable spatial variability and temporal variability; the "signal-to-noise ratio" may be very small.

5.5.1 Sampling Design Considerations

Despite the detailed data sets about particular sites, Berkowitz et al. (1989) stress that results from long-term study plots may be difficult to extrapolate because of (1) fundamental differences between the study system and surrounding areas; (2) unknowable boundary conditions or intrusion of unique events; (3) unknown uncertainty or bias in results due to poor replication; (4) the nature of underlying processes and phenomena (i.e., different perturbations may act on the system in different ways through time); and (5) inappropriate methods or poor data. This demonstrates that an important but underrated phase of the ecological assessment process is peer review of the experimental design. Hurlbert (1984) suggested that about 70% of ecological studies suffered design or statistical problems. Many of these problems, particularly pseudoreplication, are best alleviated early in a project. Often, in science, we give more attention to the review of manuscripts for publication than we do to the review of protocols and sampling designs. Rigorous plans for data analysis must be developed prior to data collection, understanding that some features of the sampling design may adapt and evolve in the first few years of the research program. That is, determining adequate sample sizes is an iterative process, recalculating variance and temporal and spatial replication needs as the study progresses. Befriend statisticians and sampling design experts.

The biggest design flaw in many ecological assessment programs is biased site selection (Stohlgren et al., 1998). Why are so many long-term study plots in flat terrain and close to roads? Krebs (1989, p. 202) defines this as "accessibility sampling," where "the sampling unit is restricted to units that are readily accessible," such as where samples of forest stands are taken only along roads. He also wrote that "judgmental sampling" occurs when "the investigator selects on the basis of experience a series of 'typical' sampling units—a botanist may select 'climax' stands of grassland to measure." The issue here is that these sampling approaches are rejected by statisticians because they cannot be evaluated by the theorems of probability (Krebs, 1989). Hence they are of limited use for extrapolating assessment results beyond the plots themselves (Stohlgren, 1994).

The long-term value of data sets also depends on the intensity and spatial pattern of sampling (Fortin et al., 1989; Stohlgren, 1994). Data from sparsely sampled sites may suggest no effect or trend where one exists. There is a growing appreciation for the need for adequate replication in long-term monitoring (Hinds, 1984; Likens, 1991; Stohlgren et al., 1995b). However, Kareiva and Anderson (1988) showed that (1) as plot size increased in ecological studies, the number of replicates decreased, and (2) once plot size reached about 3 m^2, the number of replicates was consistently less than 5.

Sample size determination is tricky business. Typically, appropriate sample sizes are determined after evaluating between-plot variance from initial field tests (Krebs, 1989, pp. 173–199). The appropriate sample size is then determined by corresponding the variance to a predetermined level of accuracy. Since disturbances of some kind or another frequently occur on most landscapes, studies designed to quantify disturbance effects must anticipate the need for large sample sizes. Interacting natural processes on an already heterogeneous landscape are further complicated by a range of species-specific and site-specific responses (Stohlgren et al., 1995b).

Sampling at multiple spatial scales is also becoming more commonplace because it is virtually impossible to inventory an entire landscape or region (Whittaker et al., 1979; Shmida, 1984; Stohlgren et al., 1998). Data acquisition for trend analysis must consider temporal and spatial variability simultaneously.

5.5.2 Accuracy Assessments of Mapped Data

It often proves difficult to assess the accuracy of maps. Field data, which are needed to assess map accuracy, are expensive (Kalkhan et al., 1995). Checking only large polygons on a map overestimates the accuracy of a map that contains many small polygons—the typical case. Accuracy assessments of satellite imagery are greatly improved by double-sampling techniques (Kalkhan et al., 1995) by which separate error matrices are developed between classifications on the satellite imagery and aerial photography and between aerial photography and ground observations. These error matrices provide information on the accuracy of map categories at various levels of resolution (Congalton, 1991; Stohlgren et al., 1997a). Other potential sources of map error include (1) sharp delineations of boundaries where gradients or broad ecotones exist (Stohlgren and Bachand, 1997), (2) exclusion of rare or heterogeneous habitats in the classification scheme (Stohlgren et al., 1997c), and (3) poor map resolution or a large minimum mapping unit (Stohlgren et al., 1997a). Regardless of the type of data, mapped or not, an understanding of the quality and limitations of the data will improve their usefulness.

5.6 Conclusion

I conclude with nine suggestions to improve the data acquisition process for ecological assessments.

1. *Clearly articulate goals and objectives.* Routinely review the objectives to see if they can achieve the goals. Modify objectives where necessary.
2. *Contact experienced groups with similar objectives.* Seek help before designing a complex plan for data acquisition. Others may prevent you from "re-creating the wheel" and help you to avoid mistakes.
3. *Work in teams.* Data acquisition and management are not a one-person job. Everyone involved in a project must help. Roles and responsibilities must be clearly defined.
4. *Commit the necessary funds early, and maintain them.* A good rule of thumb is that 20% to 25% (or more) of project funds should be allocated to data acquisition and management.
5. *Have a written plan.* The plan should include the goals and objectives, sampling protocols, ex-

ample data sheets, data dictionary and catalog, statistical analysis procedures, data archiving instructions, and metadata. Like any "Methods" section, the entire process (including analysis) must be repeatable.

6. *Understand and state the appropriate applications and limitations of the data.* This will improve the credibility of the data (and investigators) and may prevent inappropriate uses of the data.
7. *Data and documentation should be well archived.* Archived data should be in two locations, on two media, and in two formats. Metadata and quality assessment information should be attached.
8. *Improve access to your data.* Accessibility (on the Web and elsewhere) will ensure that the data do not disappear.
9. *Periodically review the data acquisition and information management process.* Keep the process updated and somewhat futuristic. A review of Oppenheimer et al. (1974) and Michener (1986) reminds us that the world of data acquisition is ever-changing.

5.7 References

Austin, T. 1974. Existing bodies of data—non-computerized. In: Oppenheimer, C. H.; Oppenheimer, D.; Brogden, W. B., eds. *Environmental data management.* New York: Plenum Press: 27–45.

Berkowitz, A. R.; Kolosa, K.; Peters, R. H.; Pickett, S. T. A. 1989. How far in space and time can the results from a single long-term study be extrapolated. In: Likens, G. E., ed. *Long-term studies in ecology: approaches and alternatives.* New York: Springer-Verlag: 192–198.

Buckley, D. J.; Coughenour, M. B.; Blyth, C. B.; O'Leary, D. J.; Bentz, J. A. 1993. *Ecosystem management model—Elk Island National Park: a case study of integrating environmental models with GIS.* Breckenridge, CO: Second International Conference on Integrating GIS and Environmental Modeling.

Burkman, W. G.; Hertel, G. D. 1992. Forest health monitoring. *J. For.* 90:2.

Congalton, R. G. 1991. A review of assessing the accuracy of classification of remotely sensed data. *Remote Sens. Environ.* 37:35–46.

Coughenour, M. B. 1993. *Savanna—landscape and regional ecosystem model: documentation.* Fort Collins, CO: Natural Resources Ecology Laboratory, Colorado State University.

Coughenour, M. B.; Singer, F. J. 1991. The concept of overgrazing and its application to Yellowstone's northern range. In: Keiter, R. B.; Boyce, M. S., eds. *The greater Yellowstone ecosystem.* New Haven, CT: Yale University Press: 209–230.

Dallmeier, F., editor. 1992. *Long-term monitoring of biological diversity in tropical forest areas: methods for establishment and inventory of permanent plots.* MAB Digest 11. Paris: United Nations Educational, Scientific, and Cultural Organization (UNESCO).

Dallmeier, F.; Taylor, C. M.; Mayne, J. C.; Kabel, M.; Rice, R. 1992. Effects of Hurricane Hugo on the Bisley Biodiversity Plot, Luquillo Biosphere Reserve, Puerto Rico. In: Dallmeier, F., ed. *Long-term monitoring of biological diversity in tropical forest areas: Methods for establishment and inventory of permanent plots.* MAB Digest 11. Paris: United Nations Educational, Scientific, and Cultural Organization (UNESCO): 47–72.

Federal Geographic Data Committee. 1998. *Content standards for digital geospatial metadata* (revised June 1998). FGDC-STD-001-1998. Washington, DC: Federal Geographic Data Committee.

Fortin, M. J.; Drapeau, P.; Legendre, P. 1989. Spatial autocorrelation and sampling design in plant ecology. *Vegetatio* 83:209–222.

Franklin, J. F.; DeBell, D. S. 1988. Thirty-six years of tree population change in an old-growth *Pseudotsuga-Tsuga* forest. *Can. J. For. Res.* 18:633–639.

Franklin, J. F.; Bledsoe, C. S.; Callahan, J. T. 1990. Contributions of the Long-Term Ecological Research Program. *BioScience* 40:509–523.

Hawk, G. M.; Franklin, J. F.; McKee, W. A.; Brown, R. B. 1978. *H. J. Andrews Experimental Forest reference stand system: establish and use history.* U.S. Dept. Agric., For. Serv. Bull. 12. Washington, DC: U.S. International Biosphere Program.

Hinds, W. T. 1984. Towards monitoring of long-term trends in terrestrial ecosystems. *Environ. Conserv.* 11:11–18.

Hubbell, S. P.; Foster, R. B. 1987. The spatial context of regeneration in a neotropical forest. In: Gray, A. J.; Crawley, M. J.; Edwards, P. J., eds. *Colonization, succession and stability.* Oxford, UK: Blackwell Scientific Publications.

Hurlbert, S. H. 1984. Pseudoreplication and the design of ecological field experiments. *Ecol. Monogr.* 54: 187–211.

Jones, K. B. 1986. The inventory and monitoring process. In: Cooperrider, A. Y.; Boyde, R. J.; Stuart, H. R., eds. *Inventory and monitoring of wildlife habitat.* Denver: U.S.D.I. Bureau of Land Manage., Serv. Center: 1–10.

Kalkhan, M. A.; Stohlgren, T. J.; Coughenour, M. 1995. An investigation of biodiversity and landscape-scale gap patterns using double sampling: a GIS approach. Vancouver, B.C.: In: *Proceedings of the ninth conference on geographic information systems:* 708–712.

Kanciruk, P.; Olson, R. J.; McCord, R. A. 1986. Data quality assurance in a research database: the 1984 national surface water survey. In: Michener, W. K., ed. *Research data management in the ecological sciences.* Belle W. Baruch Library in Marine Science Number 16. Columbia, SC: University of South Carolina Press: 193–208.

Kareiva, P. M.; Anderson, M. 1988. Spatial aspects of

species interactions: the wedding of models and experiments. In: Hastings, A., ed. *Community ecology. Lecture notes in biomathematics 77.* Berlin: Springer-Verlag: 35–50.

Klopsch, M. W.; Stafford, S. G. 1986. The status and promise of intersite computer communication. In: Michener, W. K., ed. *Research data management in the ecological sciences.* Belle W. Baruch Library in Marine Science Number 16. Columbia, SC: University of South Carolina Press: 115–124.

Krebs, C. J. 1989. *Ecological methodology.* New York: Harper & Row.

Legendre, P.; Fortin, M. J. 1989. Spatial pattern and ecological analysis. *Vegetatio* 80:107–138.

Likens, G. E., editor. 1991. *Long-term studies in ecology: approaches and alternatives.* New York: Springer-Verlag.

MacDonald, L. H.; Smart, A. W.; Wissmar, R. C. 1991. *Monitoring guidelines to evaluate effects of forestry activities on streams in the Pacific Northwest and Alaska.* Report EPA/910/9-91-001. Seattle, WA: Env. Protect. Agency.

Magnuson, J. J., Kratz, T. K.; Frost, T. M.; Bowser, C. J.; Benson, B. J.; Nero, R. 1991. Expanding the temporal and spatial scales of ecological research and comparison of divergent ecosystems: roles for LTER in the United States. In: Risser, P. G., ed. *Long-term ecological research: an international perspective.* SCOPE 47. New York: John Wiley & Sons: 45–70.

Messer, J. J.; Linthurst, R. A.; Overton, W. S. 1991. An EPA program for monitoring ecological status and trends. *Environ. Monit. Assess.* 17:67–78.

Michener, W. K., editor. 1986. *Research data management in the ecological sciences.* Belle W. Baruch Library in Marine Science Number 16. Columbia, SC: University of South Carolina Press.

Miller, I.; Lachance, D.; Burkman, W. G.; Allen, D. C. 1991. *North American sugar maple decline project: organization and field methods.* Gen. Tech. Rep. NE-154. Radnor, PA: U.S. Dept. Agric., For. Serv. Northeastern Field Exp. Sta.

National Research Council. 1990. *Forest research: a mandate for change.* Washington, DC: National Academy Press.

National Research Council. 1994. *Rangeland health: new methods to classify, inventory, and monitor rangelands. Committee on Rangeland Classification, Board of Agriculture.* Washington, DC: National Academy Press.

Oppenheimer, C. H.; Oppenheimer, D.; Brogden, W. B., editors. 1974. *Environmental data management.* New York: Plenum Press.

Palmer, C. J.; Ritters, K. H.; Strickland, J.; Cassell., D. C.; Byers, G. E.; Papp, M. L.; Liff, C. I. 1991. *Monitoring and research strategy for forests—Environmental Monitoring and Assessment Program (EMAP).* EPA/600/4-91/012. Washington, DC: U.S. Env. Protect. Agency.

Parsons, D. J.; Workinger, A. C.; Esperanza, A. E. 1992. *Composition, structure and physical and pathological characteristics of nine forest stands in Sequoia National Park.* U.S. Dept. Interior, Natl. Park Serv.; CPSU Tech. Rep. No. NPS/WRUC/NRTR 92/50. Davis, CA: University of California.

Riegel, G. M.; Green, S. E.; Harmon, M. E.; Franklin, J. F. 1988. *Characteristics of mixed conifer forest reference stands at Sequoia National Park, California.* U.S. Dept. Interior, Natl. Park Serv.; CPSU Technical Report No. 32. Davis, CA: University of California.

Risser, P. G., editor. 1991. *Long-term ecological research: an international perspective.* SCOPE 47. New York: John Wiley & Sons.

Risser, P. G.; Treworgy, C. G. 1986. Overview of research data management. In: Michener, W. K., ed. *Research data management in the ecological sciences.* Belle W. Baruch Library in Marine Science Number 16. Columbia, SC: University of South Carolina Press: 9–22.

Scott, J. M.; Davis, F.; Csuti, B.; Noss, R.; Butterfield, B.; Groves, C.; Anderson, H.; Caicco, S.; D'Erchia, F.; Dewards, T. C., Jr.; Ulliman, J.; Wright, R. G. 1993. GAP analysis: a geographic approach to protection of biological diversity. *Wildlife Monogr.* 123:1–41.

Seilkop, S. K. 1986. Management and analysis of spatially oriented data. In: Michener, W. K., ed. *Research data management in the ecological sciences.* Belle W. Baruch Library in Marine Science Number 16. Columbia, SC: University of South Carolina Press: 227–246.

Shampine, W. J. 1993. Quality assurance and quality control in monitoring programs. *Environ. Monit. Assess.* 26:143–151.

Shmida, A. 1984. Whittaker's plant diversity sampling method. *Israel J. Bot.* 33:41–46.

Stafford, S. G. 1993. Data, data everywhere but not a byte to read: managing monitoring information. *Environ. Monit. Assess.* 26:125–141.

Stafford, S. G.; Alabach, P. B.; Waddell, K. L.; Slagle, R. L. 1986a. *Data management procedures in ecological research.* In: Michener, W. K., ed. *Research data management in the ecological sciences.* Belle W. Baruch Library in Marine Science Number 16. Columbia, SC: University of South Carolina Press: 93–114.

Stafford, S. G.; Klopsch, M. W.; Waddell, K. L.; Slagle, R. L.; Alabach, P. B. 1986b. *Optimizing the computational environment for ecological research.* In: Michener, W. K., ed. Research data management in the ecological sciences. Belle W. Baruch Library in Marine Science Number 16. Columbia, SC: University of South Carolina Press: 73–92.

Stohlgren, T. J. 1994. Planning long-term vegetation studies at landscape scales. In: Powell, T. M.; Steele, J. H., eds. *Ecological time series.* New York: Chapman and Hall: 208–241.

Stohlgren, T. J. 1999. Measuring and monitoring biodiversity in nature reserves, forests, and grasslands in the United States. In: Aguirre-Bravo, C.; C.R. Franco, compilers. North American Science Symposium: *To-*

ward a Unified Framework for Inventorying and Monitoring Forest Ecosystem Resources. USDA Forest Service, Rocky Mountain Research Station Proceedings RMRS-P-12. Fort Collins, CO 80526.

Stohlgren, T. J.; Bachand, R. R. 1997. Lodgepole pine (*Pinus contorta*) ecotones in Rocky Mountain National Park, Colorado, USA. *Ecology* 78:632–41.

Stohlgren, T. J.; Quinn, J. F. 1992. An assessment of biotic inventories in western U.S. national parks. *Nat. Areas J.* 12:145–154.

Stohlgren, T. J.; Falkner, M. B.; Schell, L. D. 1995a. A modified-Whittaker nested vegetation sampling method. *Vegetatio* 117:113–121.

Stohlgren, T. J.; Binkley, D.; Veblen, T. T; Baker, W. L. 1995b. Attributes of landscape-scale, long-term studies: malpractice insurance for landscape ecologists. *Environ. Monit. Assess.* 36:1–25.

Stohlgren, T. J.; Quinn, J. F.; Ruggiero, M.; Waggoner, G. 1995c. Status of biotic inventories in U.S. National Parks. *Biol. Conserv.* 71:97–106.

Stohlgren, T. J.; Chong, G. W.; Kalkhan, M.A.; Schell, L. D. 1997a. Multi-scale sampling of plant diversity: effects of the minimum mapping unit. *Ecol. Appl.* 7:1064–1074.

Stohlgren, T. J.; Chong, G. W.; Kalkhan, M. A.; Schell, L. D. 1997b. Rapid assessment of plant diversity patterns: a methodology for landscapes. *Environ. Monit. Assess.* 48:25–43.

Stohlgren, T. J.; Coughenour, M. B.; Chong, G. W.; Binkley, D.; Kalkhan, M.; Schell, L. D.; Buckley, D.; Berry, J. 1997. Landscape analysis of plant diversity. *Landscape Ecol.* 12:155–170.

Stohlgren, T. J.; Bull, K; Otsuki, Y. 1998. Comparison of rangeland sampling techniques in the Central Grasslands. *J. Range Manage.* 51:164–172.

Strayer, D.; Glitzenstein, J. S.; Jones, C.G.; Kolasa, J.; Lichens, G. E.; McDonnell, M. J.; Parker, G. G.; Pickett, S. T. A. 1986. *Long-term ecological studies: an illustrated account of their design, operation, and importance to ecology.* Millbrook, NY: Institute of Ecosystem Studies Occasional Publication Number 2.

Tilman, D. 1989. Ecological experiments: strengths and conceptual problems. In: Likens, G. E., ed. *Long-term studies in ecology: approaches and alternatives.* New York: Springer-Verlag: 136–157.

Whittaker, R. H.; Niering, W. A.; Crisp, M. O. 1979. Structure, pattern, and diversity of a mallee community in New South Wales. *Vegetatio* 39:65–76.

Zedaker, S. M.; Nicholas, N. S. 1990. *Quality assurance methods manual for forest site classification and field measurements.* Document EPA/600/3-90/082. Corvallis, OR: U.S. Environ. Protect. Agency.

6 Sampling Design and Statistical Inference for Ecological Assessment

Brian M. Steele

6.1 Introduction

The success of ecological assessment depends on relevant and accurate information about the ecosystem or landscape under study. Data collected by sampling are a primary source of information about ecosystems and usually the only source of information that is specific to the ecosystem. The information value of sample data cannot be overestimated: these data directly reflect the processes and organisms constituting the system and are independent of human-held assumptions and theories. There are three general approaches to using sample data to describe ecosystems and ecosystem processes: (1) often important ecosystem components can be assessed by analyzing a few summary statistics; (2) ecosystem monitoring makes use of sample data for trend estimation and uses hypothesis testing for change detection; and (3) occasionally, ecological assessment involves the modeling of ecosystem processes, and sample data are used for model estimation and prediction.

Inference is the process of extrapolating from sample data to populations, and statistical inference is a methodology for minimizing the risk of drawing the wrong conclusions from sample data. Statistical inference can also quantify the risk incurred by extrapolation. Despite the risks involved, proper statistical inference based on good sample data is a highly effective technique for scientific inquiry and, more specifically, for ecosystem assessment. The quality of the data and the inference drawn from the sample are largely determined by the method of data acquisition; in particular, statistical inference is impossible without a statistically valid sample. Thus sampling design plays a key role in statistical inference and hence in ecological assessment.

In addition to providing statistically valid samples, sampling design promotes effective application of the scientific method to ecosystem assessment. Devising a sampling plan requires review of ecosystem knowledge, identification of assumptions regarding the ecosystem, and statement of unambiguous research questions. Scientific rigor is introduced into a sampling design by identifying a relatively short list of clear and concise questions regarding the ecosystem. For instance, if the objective is to investigate forest health, then the sample design defines a specific population of interest, such as forest stands located within a particular watershed. Formal inference requires unambiguous questions, such as "what proportion of the population is infected with white pine blister rust?" Statistical inference provides the basis for constructing an interval, for example, [0.23, 0.35], from the sample and ensures that the method succeeds in bracketing the true unknown proportion at a specified rate, for example, 95% of the time. Without a proper sampling design, a valid confidence interval cannot be constructed. Without statistical validity, the true unknown proportion will be bracketed at some other unknown rate.

Sampling designs are used to collect statistically valid samples efficiently. Useful criteria for assessing and comparing sampling designs are accuracy, cost, and feasibility. An efficient sampling design is one that produces accurate estimates or sensitive test statistics at low cost and effort. Efficiency is often difficult to achieve in ecological assessment because populations tend to consist of complex entities that are often widely dispersed and dissimilar. Furthermore, most ecosystems are dynamic and only partially, or poorly, understood. It is challenging to design an efficient and statistically valid sampling protocol for ecosystem assessment, and almost always problematic to determine the minimal sample size necessary for accurate estimates and sensitive tests. Sample design and the statistical analysis of sample data may benefit from the assistance or review of a trained statistician. In recognition of the diversity and complexity of sampling problems in ecological assessment, the objective of this chapter is to review relevant statistical concepts and discuss the general principles of sampling design so that researchers will be familiar with the issues and conversant in the language of sampling design and statistical inference.

6.2 Statistical Inference

We always risk drawing an incorrect conclusion when analyzing a population based on a sample, because not all population units are sampled. The goal of statistical inference is to minimize the risk of inferential errors and to provide a statement about the quality of the inference. Drawing on the previous example, a 95% confidence interval for the proportion of infected stands in a watershed is defined by its upper and lower bounds, U and L. A good sampling design will produce intervals of small width W, where $W = U - L$, and will ensure that the intervals are actually correct 95% of the time.

6.2.1 Populations and Probability Samples

A *population* is a collection of elements, objects, or organisms to which the findings of the study are to be extrapolated. The elements of the population are called *units* and the number of units in the population is denoted by N. Each unit provides a single measurement on a particular variable, attribute, or characteristic, although multiple variables may be measured on each unit. Multiple measurements on a single unit are often thought of as a single observation on a multivariate variable. We denote the variable of interest by Y, and a measurement of Y made on a population unit is denoted by y. For example, if we have a map identifying forest stands within a watershed, we may chose to define the population units to be the stands and define Y to be the triple: stand density, mean age of trees with d.b.h. $\geq$ 12 inches, and infection rate of trees with d.b.h. $\geq$ 12 inches. In principle, we can obtain a single measurement on this multivariate variable from each stand.

The elementary sampling setup assumes that N is finite and known so that a list of labeled population units can be formed. Other setups are also used, but this one is useful for its simplicity. In this spirit, suppose that the y variable is univariate, and let $y_1, y_2, \ldots, y_N$, denote the population of values. *Parameters* are descriptive characteristics of the population such as the population total, $T = \Sigma_{i=1}^{N} y_i$, the population mean, $\mu = T/N$, and the population standard deviation, $\sigma = \sqrt{\Sigma_{i=1}^{N} (y_i - \mu)^2/N}$. The population total is a particularly useful parameter for finite populations, because many other parameters are simple functions of the population total, and many estimators are functions of the sample total. For example, an estimator of μ is the sample mean, which is the sample total divided by the sample size.

6.2.2 Probability Samples

A sample is a subset of the population that has been selected by a researcher. A commonly used definition of a *probability sample* is one in which each population unit has a known, nonzero probability of being included in the sample (Hansen et al., 1953; Levy and Lemeshow, 1991). The probability that a population unit is included in the sample is called its *inclusion probability*. A *statistically valid* sample is a probability sample; the two terms are equivalent because probability sampling is necessary for statistical inference. A list of the population units with their associated inclusion probabilities is called a *sampling frame*. Once a population unit has been included in the sample, it is called a *sample unit*, and an *observation* is the value of the y variable measured on a sample unit.

There are two elemental types of sampling—with and without replacement. A sample selected without replacement is one in which every population unit can appear at most once in the sample, whereas sampling with replacement allows each unit to be selected more than once. Although the idea of including more than one observation from a population unit in the sample seems inefficient, it simplifies some sampling protocols. For instance, sampling with replication is appealing when estimating population size for a large animal by aerial survey, because it may be difficult to determine whether an animal has already been counted.

Simple Random Sampling

The simple random sample (SRS) design is the prototypical sampling design. In an SRS design that employs sampling without replacement, every subset of the population of size n has the same probability of being selected as the sample. Rather than listing and randomly selecting among all possible subsets, it is more practical to collect an SRS by successively selecting units from the population so that every unit has the same probability of being selected on each draw. On the first draw, the probability that a particular unit is selected is $1/N$; on the second draw, the probability that a particular unit (not selected on the first draw) is selected is $1/(N-1)$, and so on. For example, all possible samples of size 2 that can be selected from the population $P = \{0, 2, 3\}$ are $\{0, 2\}$, $\{0, 3\}$, and $\{2, 3\}$. Each sample must have a probability of $\frac{1}{3}$ of being selected in order that the sampling design be an SRS. Techniques for random selection of samples are discussed by Levy and Lemeshow (1991) and Thompson (1992).

6.2.3 Experimental Design

What is the difference between experimental and sampling designs? If a treatment is purposively imposed on the objects or organisms of interest, then the study and its design are experimental. In experimental studies, a probability sample is chosen from a population, and some of the sampling units are manipulated with the intent of affecting or eliciting a response from the units. Sampling designs do not manipulate sample units with this intent. For example, an experiment to test whether the presence of a suspected disease host (e.g., gooseberry) increases the infection rate of a pathogen (e.g., white pine blister rust) selects a probability sample from a population of stands. Then, some fraction of selected stands is randomly chosen to be cleared of gooseberry, and the remaining stands are left untreated. Measurement of the proportion of infected trees before and after clearing the stands provides data for a test of the hypothesis that removal of gooseberry does not affect the population infection rate, versus an alternative hypothesis stating that removal of gooseberry decreases the population infection rate. In this case, some of the sample units (i.e., stands) have been purposely manipulated by the researcher. In contrast, a sample design aimed at answering the same question collects a probability sample of stands, but no stands are cleared. An informative statistic that may be computed from the sample data is the sample correlation coefficient, measuring the degree of association between the proportion of infected trees and the areal cover of gooseberry. Both studies may provide convincing evidence that the rate of infection is positively associated with gooseberry coverage, but only the experiment provides the opportunity to obtain statistical evidence supporting the conclusion that gooseberry is a cause of infection in the population. The sampling design cannot rule out the possibility that gooseberry does not cause infection, but instead responds positively to an environmental variable that also favors the pathogen. For more information on statistical methods for environmental studies, see Eberhart and Thomas (1991) for a useful overview. Also, Petersen (1994) is a recent text on experimental design oriented toward biologists.

6.3 Random Sampling

The idea of a random sample must be understood to discuss statistical inference in more detail. A *random sample* is not a subset of population units, but a process by which a sample is selected from a population and a variable or attribute is observed on the sampled units. The nature of the process is random in the sense that the outcome of the sampling process is not known ahead of time. However, the distribution of possible outcomes of the process is determined by the population and the sampling design.

Let $\boldsymbol{Y} = (Y_1, \ldots, Y_n)$ denote a random sample. Once a sample is collected and the observations are recorded, then we have a realization of $\boldsymbol{Y}$, and the realized sample is denoted by $\boldsymbol{y} = (y_1, \ldots, y_n)$. For example, $\boldsymbol{Y}$ may represent the process of selecting an SRS of size 2 from the population $P = \{0, 2, 3\}$. A possible realization of $\boldsymbol{Y}$ is $\boldsymbol{y} = \{0, 2\}$. A *statistic* is a function of the random sample whose realizations are numerical. Because it is a function of a random process, a statistic is also a random variable. For instance, the sample mean is a random variable, and it may be expressed as $\overline{X}(\boldsymbol{Y})$, although the usual expression for the sample mean is $\overline{\boldsymbol{Y}}$.

Suppose that every possible sample is selected from a population, and a statistic S is computed using every one of the samples. The *expected value* of S, denoted by $E(S)$, is the weighted average of all its possible values; the weight associated with each value is the probability that the value is the realized value. For the population $P = \{0, 2, 3\}$, the population mean is $\mu = \frac{5}{3}$. An SRS specifies that each of the three subsets of size 2 listed in Section 6.2.2 has a probability of $\frac{1}{3}$ of being chosen. The possible realizations of the sample mean are 1, 1.5, and 2.5; each is equally likely. Hence the expected value of the sample mean is $(1 \times 1/3) + (1.5 \times 1/3) + (2.5 \times 1/3) = 5/3$.

6.4 Statistical Methods

Two rather different views of the population arise in sample design and statistical analysis. The first, and traditional, view is design based; it treats the population as a fixed and usually finite set. Commonly, the objectives of sampling are to describe the existing population units in terms of parameter estimates and confidence intervals. The second view is model based and is sometimes called a superpopulation view. A model-based view treats the population as a single realization of a stochastic process. In ecological terms, this is equivalent to viewing the ecosystem as dynamic rather than static. A model-based view is appropriate when the interest is in drawing inference not only about the population units in existence at the time of sampling, but also about population units that existed in the past or will exist in the future. A design-based view is sensible when sampling a forest stand with the intent of estimating basal area or volume, because the researcher is most likely interested in what exists at the time of sampling. A model-based view is more appropriate when the objective is to monitor a riverine ecosystem by sampling for changes in water chemistry or biotic components. In this situation, a model describing the distribution of chemical constituents is useful for distinguishing between natural variation and trends in average concentration.

The statistical analysis of model-based samples may differ from that of design-based samples; this difference affects sampling design. For instance, a routine objective of model-based sampling is to estimate a regression model describing a response variable as a function of a set of explanatory variables. A model-based design may be aimed at selecting a sample with observations regularly distributed across the range of the explanatory variable, because this design will tend to produce more accurate estimates and model predictions than a design that randomly samples the range. A model-based design may stratify the population based on the range of the explanatory variables and select an independent SRS from each stratum. This design will cover the range more fully than an SRS. A consequence of the model-based design is that some estimates may be biased when examined from the design view. For example, the sample mean of the response variable may be biased under the model-based design because subsamples have not been proportionally allocated to strata. Overton and Stehman (1995) provide further discussion relevant to natural resource sampling, and Thompson (1992) makes clear distinctions between the statistical methods appropriate for the two views.

Statistical methods for analyzing data obtained from both design- and model-based sampling can be differentiated from a much larger set of statistical methods in two principal aspects. First, the statistical analysis of sampling data often assumes a finite population. Second, and more importantly, many designs sample observations that are not independent. Because statistical analysis of designed samples is a large and difficult area, this chapter discusses only elementary estimators in the context of specific designs. The reader is referred to sampling design texts by Levy and Lemeshow (1991), Särndal et al. (1992), Schreuder et al. (1993), and Thompson (1992) for statistical analysis methods for data collected according to sampling designs. There are, however, general principles guiding the use of estimators, predictors, and hypothesis tests. The remainder of this section concentrates on these general principles, rather than specific methods. The finite-population, design-based view is adopted in the remainder of this chapter, because this view clearly departs from the conventional treatment of data as independent observations from infinite populations.

6.4.1 Bias, Precision, and Accuracy

Further discussion of sampling design requires a review of the properties of statistics, because sampling design affects the statistics used for inference and hence the quality of the inference. Three of these properties are bias, precision, and accuracy. Suppose that a statistic S is used to estimate a parameter ξ. The *bias* of S is the difference between its expected value and ξ; that is, the bias of S is $\mathrm{E}(S) - \xi$. If the expected value of S is ξ, then S is unbiased. An estimator need not be unbiased to be useful; for instance, the sample standard deviation is a biased estimator of the population standard deviation, but for practical applications the bias is small enough to be ignored.

The *precision* of an estimator is a measure of the repeatability of the estimator across many samples. Reliability and reproducibility are equivalent to precision. A precise estimator is reliable and reproducible in the sense that it produces similar estimates from different samples. Formally, the precision of an estimator is its *variance*, which is defined to be the expected value of the mean squared difference between the estimator and the expected value of the estimator; that is, the variance of S is $\sigma_s^2 = \mathrm{E}\{[S - \mathrm{E}(S)]^2\}$. Suppose that m possible values may be obtained from an estimator S when all possible samples are considered. If the set of possible values, or estimates, is denoted by $s_1, \ldots, s_m$, and p_i denotes the probability that S will take on the value s_i, then a computational formula for the variance is $\sigma_s^2 = \Sigma_{i=1}^m p_i\,[s_i - \mathrm{E}(S)]^2$. The standard deviation σ_s of S, or the *standard error* of S, is $\sigma_s = \sqrt{\sigma_s^2}$.

The *accuracy* of an estimator is a measure of closeness between the estimates and the parameter of interest. Accuracy reflects both bias and precision, because an estimator may be precise without being unbiased. An accurate estimator has low bias and high precision. Accuracy is usually defined mathematically to be the mean squared error (MSE) of the estimator, which is the expected value of the mean squared difference between the estimator and the parameter of interest; that is, $\text{MSE}(S) = \text{E}[(S - \xi)^2]$, where ξ is the parameter of interest. Mathematically, the MSE is the sum of the variance σ_S^2 and the squared bias; that is, $\text{MSE}(S) = \sigma_S^2 + [\text{E}(S) - \xi]^2$.

6.4.2 Confidence Intervals

It is rarely sufficient to provide an estimate without an accompanying measure of its reliability. For example, an estimate of the standard error of an estimator provides a measure of reliability. Confidence intervals are often more useful, however, because they are sets of candidate values for the parameter that are in agreement with the sample data. Confidence intervals are constructed so that there is a high probability of bracketing, or capturing, the true value of the parameter between the lower and upper limits. To illustrate the interpretation of confidence intervals, suppose that an estimate $\hat{\pi}$ of the proportion of infected stands π in a watershed is 0.29. A hypothetical 95% confidence interval for the true population proportion is [0.23, 0.35]. Thus, if the same sampling design were employed to sample the same population infinitely often and confidence intervals constructed from each sample, then 95% of these intervals would contain the true population proportion, no matter what the true value might be.

To clarify the interpretation of confidence intervals, suppose that π denotes the unknown population proportion. Let $L(\mathbf{Y})$ and $U(\mathbf{Y})$ denote the lower and upper confidence limits computed from the random sample $\mathbf{Y}$. Note that these limits are statistics and hence random variables. The confidence limits are constructed from $\mathbf{Y}$ so that the probability that the interval covers π is 0.95. Mathematically, this probability is expressed as $0.95 = \Pr[L(\mathbf{Y}) < \pi < U(\mathbf{Y})]$. What is sometimes confusing about confidence intervals is that the statement $0.95 = \Pr[0.23 < \pi < 0.35]$ is false, because π is not random and neither are the numbers 0.23 and 0.35. The statement $0.23 < \pi < 0.35$ is either absolutely true or absolutely false; there can be no probabilistic interpretation in classical frequentist statistics. The correct interpretation is that the *method* used to construct the interval will bracket π about 95% of the time, and *not* that there is a 0.95 probability that π is between 0.23 and 0.35.

A good sampling design will produce confidence intervals with narrow widths and achieve the stated confidence level.

6.5 Hypothesis Testing and p Values

Occasionally, the objective of sampling is to formally test a hypothesis about a population. Hypothesis testing is appropriate when the researcher identifies a theory, postulate, or hypothesis to be established or confirmed *before* the sample is collected. Testing a hypothesis that is suggested by a sample by using the sample itself greatly increases the risk of drawing an incorrect conclusion from the test.

A hypothesis test requires a statement of null and alternative hypotheses. The alternative hypothesis usually is the hypothesis of interest, and the null hypothesis represents the contrary to the alternative. For instance, suppose that it is thought that white pine blister rust is becoming more prevalent in a watershed, and the historical stand infection rate is believed to be 0.2 or less. If π represents the current unknown stand infection rate, then a useful null hypothesis is H_0: $\pi = 0.2$, and an appropriate alternative is H_1: $\pi > 0.2$. To test the null hypothesis, suppose that an SRS of stands is selected and the sample proportion of infected stands is observed to be 0.29. This value supports H_1 and contradicts H_0, but the strength of support is not obvious. A *test statistic* is a function of the sample that is used to measure how strongly the data contradict H_0. Strong evidence against H_0 lends credence to the contrary, as expressed by H_1. If the evidence is strong enough, then we reject H_0 and conclude that H_1 is true. If the evidence is not strong enough, then we conclude that the evidence is not strong enough to conclude that H_1 is true. The second conclusion is weak, because the hypothesis test is intended only to measure evidence against H_0; it does not measure the strength of evidence supporting H_0. Furthermore, we began with a preconceived notion that H_0 was false, and we are reluctant to give up the notion without evidence supporting its truth.

The *p value* is a quantitative measure of the strength of evidence against the null hypothesis. For example, if $Z(\mathbf{Y})$ is a test statistic, such as a Z statistic computed from the sample proportion (see Ostle and Mensing, 1975), then large values of the random variable $Z(\mathbf{Y}) = (\hat{\pi} - \pi)/\hat{\sigma}_{\hat{\pi}}$ correspond to large sample proportions and contradict H_0. Suppose that the observed data $\mathbf{y}$ yields a value of 1.96 when used to compute Z; that is, $Z(\mathbf{y}) = 1.96$. The approximate normality of $Z(\mathbf{Y})$ implies that the p value is approximately $\Pr[Z(\mathbf{Y}) \geq 1.96] = 0.025$, which says that the probability is approximately 0.025 that we would have observed a Z statistic as large or larger than 1.96 given that π were truly 0.2. The p value is the probability of obtaining as much or more evidence contradicting H_0 and favoring H_1, given H_0 were true. Hence a p value as small as 0.025 indicates that either H_0 is not true or an unlikely sample has been collected from a population with an infection rate of 0.2. Because we began with the notion that H_0 is false, we state that the data contradict H_0 and support H_1, but acknowledge the possibility that we may be wrong. These ideas can be formalized in terms of a decision rule. We reject H_0 when the p value is less than or equal to a predetermined significance level. A popular choice of significance level is 0.05, although there are often good reasons for using different values. The significance level is the probability that the test will incorrectly reject H_0 when H_0 is true. Rejecting H_0 when H_0 is true is called a Type I error. Hence the significance level is the probability of a Type I error. We may also fail to reject H_0 when H_0 is false and thus commit a Type II error. Type I errors are controlled by the significance level of the test. For example, if the significance level is chosen to be 0.0001, there is very little chance of committing a Type I error. Unfortunately, this strategy of minimizing Type I errors increases the risk of Type II errors. Type II errors are controlled by choosing an efficient sample design and ensuring that the sample size is adequate.

The use of hypothesis testing has drawn much criticism. Although the mathematics of hypothesis testing are elegant and the methodology works well in simple situations, the logic of hypothesis testing leaves much to be desired, as do the difficulties of interpreting multiple tests. In many instances, confidence interval construction can be used to accomplish the same general objectives as hypothesis testing. Hypothesis testing, confidence interval construction, and the relationships between the two are extensively discussed in applied statistics texts. Useful references are Ramsey and Schafer (1997), Ostle and Mensing (1975), and Snedecor and Cochran (1980).

6.5.1 Sample Size

A central topic in sampling design is that of choosing the sample size. Sample size strongly determines the accuracy of estimates and predictions, along with the Type II error rate of hypothesis tests. For example, suppose that an objective of the design is to estimate the population mean μ by calculating a point estimate and a confidence interval. The width of the confidence interval can be controlled in the sample design by controlling the size of the standard error of the estimator of μ. For an SRS design, the variance of the sample mean is $\sigma^2_{\bar{Y}} = (N - n)\sigma^2/(Nn)$, where σ^2 is the population variance. An approximate 90% confidence interval for the population mean has lower and upper limits $L(\mathbf{Y}) = \bar{Y} - 1.645\sigma_{\bar{Y}}$ and $U(\mathbf{Y}) = \bar{Y} + 1.645\sigma_{\bar{Y}}$ (Snedecor and Cochran, 1980). The width of the interval is $w = 3.29\sigma_{\bar{Y}} = 3.29\sigma\sqrt{(N - n)/(Nn)}$, and if we can provide an educated guess of the value of σ, then we can estimate the minimal sample size by solving for n.

Sample size calculations as simple and exact as the last example are rarely attainable for more complicated problems, such as estimating the relationship between a response variable and a set of explanatory variables. Approximations are sometimes possible. For example, the sample size for the regression problem can be approximated by calculating the sample size for a simpler problem, such as estimating the response variable mean using the sample mean. Provided that the explanatory variables are truly informative with respect to modeling the response variable, then the variance of the sample mean will be larger than the variance of the regression estimator of the response variable mean, given a set of explanatory variables. Then the estimated sample size for the simpler problem will be adequate for estimating the regression model. An alternative approach is to use Monte Carlo methods to simulate sample data and the distribution of the statistics of interest. Examples of sample size calculations for ecological monitoring are given by Kendall et al. (1992) and Lesica and Steele (1996).

6.6 Basic Sampling Designs

A unifying concept in sampling design is the Horvitz–Thompson estimator (Horvitz and Thompson, 1952; Overton and Stehman, 1995; Thompson, 1992). The Horvitz–Thompson estimator $\hat{T}$ gives an unbiased estimate of the population total T for *any* probability sampling design. In this section, we use the Horvitz–Thompson estimator to develop connections between the basic sampling designs and estimators of T and other population parameters. Estimators of many other parameters can be derived from $\hat{T}$; for example, $\hat{T}/N$ is an estimator of the population mean.

The general form of the Horvitz–Thompson estimator is simple to define. Let π_i denote the probability that the ith population unit will be included in the sample. Then the Horvitz–Thompson estimator of the population total T is

$$\hat{T} = \sum_{i=1}^{v} \frac{y_i}{\pi_i}, \tag{1}$$

where $y_1, \ldots, y_v$ are the distinct observations in the sample. If sampling occurs with replacement and an observation appears more than once, only the first appearance is used in the sum and counted toward v. To illustrate the Horvitz–Thompson estimator, simple random sampling without replacement yields an inclusion probability of $\pi_i = n/N$, and the Horvitz–Thompson estimator becomes $\hat{T} = (N/n) \Sigma_{i=1}^{n} y_i$, which is the usual sample mean multiplied by the number of population units N. An ecological example is that of estimating grass production. An SRS of plot locations can be selected from the population of plots, and the grass on each plot can be clipped, dried, and weighed. Mean production per square meter can be estimated by dividing $\hat{T}$ by the total number of square meters that were sampled.

The power of the Horvitz–Thompson estimator is demonstrated by the following example. Suppose that the objective is to estimate the number of birds in a fixed area by traversing a line transect and counting the number of birds that are flushed. The inclusion probabilities can be modeled as a function of the distance to the flushed birds (Hayne, 1949; Overton, 1969) according to the model $\pi_i = 2r_i/L$, where the ith bird will flush if the sampler is within a distance of r_i of the bird, and L is the transect length. Let y_i be 1 for all i; then the sample sum of the y_i's is the number of birds observed while traversing the transect. The Horvitz–Thompson estimator of the population size is then $\hat{N} = \Sigma_{i=1}^{v} y_i/\pi_i = (L/2) \Sigma_{i=1}^{v} 1/r_i$, where v is the number of flushed birds.

There are two important points to be made about this design. First, a bird is allowed to be observed more than once, which is important because it is often difficult to determine if the same bird has been flushed more than once. Second, we do not need to know the inclusion probabilities for all population units. This is important because we do not even know the number of units, much less the inclusion probabilities. In fact, the often made statement that inclusion probabilities must be known for all population units is stronger than necessary. Instead, the essential condition is that the inclusion probabilities must be known for all units that appear in the sample; in other words, the inclusion probabilities must be *knowable* for all population units (Overton and Stehman, 1995) and nonzero. The next three sections describe basic sampling designs used in ecological assessment.

6.6.1 Stratified Random Sampling

Stratified random sampling is a simple and effective method for improving on the efficiency of simple random sampling. The idea is to divide the population into a set of distinct strata, or subpopulations, so that the units within strata are more alike than units between strata. If stratification has been successful in this regard, then a smaller sample size is necessary to achieve the same accuracy level than if an SRS design were used. For example, a design aimed at assessing stream condition can be made more efficient if the population, or region of interest, is stratified according to watershed. Often, the principal reason for stratification is to obtain strata-specific estimates, for example, when strata correspond to administrative units. Stratified random sampling can be used to obtain both strata-specific and population-wide estimates. An equally important motivation for stratified random sampling is that the logistics of sampling are sometimes simplified, or the cost of sampling is reduced. For example, if a National Forest is to be sampled, it may be efficient to stratify on a district basis, because personnel located within each district can sample that area at a lower cost than centrally located personnel. Cost and feasibility are adequate reasons for stratified random sampling even if no gain in accuracy is anticipated. However, it is possible to degrade the quality of inferential methods if there are many strata that are not particularly different, and so it is best not to stratify unless there are clearly different strata or clear advantages in terms of feasibility or cost.

For stratified random sampling, the Horvitz–Thompson estimator is set up as follows. Suppose that there are N_r population units in the rth stratum. Stratified random sampling specifies that an SRS of n_r units be selected from the rth stratum. Suppose that the ith population unit is a member of the rth stratum. Then the inclusion probability for the ith unit is $\pi_i = n_r/N_r$. Equation (1) is the Horvitz–Thompson estimator of the population total. Stratum-specific estimates are based on estimates of the stratum total T_r for the rth strata. The Horvitz–Thompson estimate of T_r is $\hat{T}_r =$

$\Sigma_{i=1}^{n_r} y_i/\pi_i$ where the sum is only calculated from the y_i's from the rth stratum.

There are a variety of strategies for determining the stratum-specific sample sizes. For instance, proportional allocation specifies $n_r = nN_r/N$, where N_r is the number of population units in the rth stratum. Proportional allocation gives each population unit the same probability of being sampled. Stratum sample sizes can be allocated to minimize the variance of $\hat{T}$ if estimates of the within-stratum variances are available or to minimize the cost of sampling if stratum-specific costs and variances can be estimated. Levy and Lemeshow (1991) provide sample size formulas for these situations.

6.6.2 Systematic and Cluster Sampling

Systematic and cluster sampling designs are used when it is more efficient to sample clusters of units or to sample in regular patterns such as fixed intervals along a transect. Cluster sampling is based on the idea of choosing one sample unit randomly to start a cluster and several additional units nearby to fill out the cluster. Systematic sampling selects the sample units so that they occur in systematic patterns. In contrast to strata, which are comprised of population units close together with respect to the y variable, clusters are sets of sample units that are close together with respect to a different measure of distance, such as time or space. Cluster sampling is most efficient when the distance variable affects the logistics or cost of sampling, such as distance between plot locations when sampling vegetation. An important advantage of systematic sampling is that, when sampling an area, it is possible to distribute the sampling locations more uniformly than would be accomplished by simple random sampling.

Systematic and cluster sampling designs are connected by the property that only a few sample units are selected randomly, and the remainder of the sample is selected deterministically. The transects and the clusters are called *primary sampling units* and the cells within transect or within cluster are called *secondary units*. The primary units completely determine the secondary units, because once the primary units are selected no further randomization is involved in selecting sample units. The principal advantage of these designs is that the cost may be substantially less than the cost of an SRS of the same sample size. In contrast to stratified random sampling, there is little or no interest in the transects or clusters as distinct entities. In contrast, stratified random sampling is usually motivated by an interest in describing the strata individually. To maximize the efficiency of transect and cluster sampling designs, primary units should be defined in such a way that the variation of secondary units within transect, or within cluster, is as large as possible. Consequently, when sampling vegetation, it is preferable to orient the transects across, rather than along, important environmental gradients, such as elevation. Cluster sampling is inefficient in this regard for sampling vegetation, because adjacent plots tend to be similar.

To illustrate systematic sampling, suppose that vegetation is to be sampled on a steep mountainside by recording ocular estimates of species coverage on 100-m^2 plots. A lattice, or grid, with a 10-m cell size, is superimposed on a map of the mountainside, and the population of interest is defined to be all 10-m cells in the area. An SRS design selects n cells at random. This design may be inefficient because of the cost of traveling between sample units. Suppose that there are N lattice points along the base of the lattice, and we wish to choose every kth lattice point along the base to define a transect that will run uphill and perpendicular to the base of the mountainside. A systematic sample design selects $n = N/k$ transect starting points by choosing a random number between 1 and k. This random number defines the first transect starting point, and we use every kth lattice point after that to define another transect starting point. The same protocol is used to select a random starting *cell* for each transect. The first sample unit, or cell, on each transect is located between 1 and m cells from the

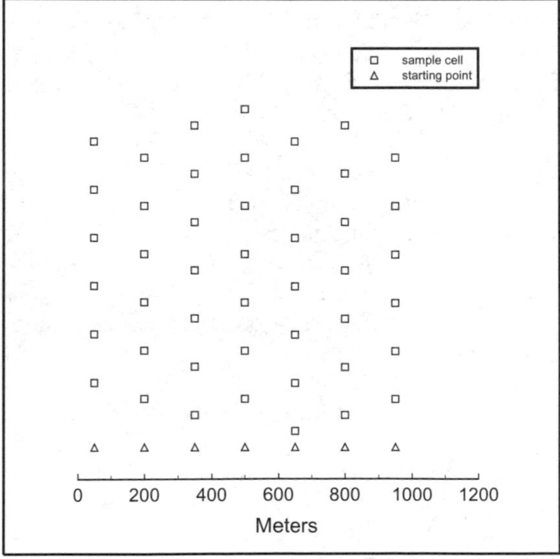

FIGURE 6.1. Plot layout from a systematic sampling design.

start, and the remaining cells are located at intervals of m cells from each other. Figure 6.1 shows plot locations from a systematic sampling design of this type. Under this design, the inclusion probabilities are $\pi_i = 1/(km)$ for each of the i cells in the lattice population. For example, if we choose every third lattice point to start a transect and every other cell on each transect, then the inclusion probabilities for the sample units are all equal to $\frac{1}{6}$.

A design that illustrates cluster sampling of the lattice population begins by choosing an SRS of n cells. Each cell in the SRS defines a cluster center, and the clusters are defined to be sets of $k > 1$ contiguous cells centered about the cluster center and the center cell. Figure 6.2 shows plot locations from a cluster sampling design of this type comprised of 9 clusters of $k = 5$ plots each. The inclusion probabilities are $\pi_i = nk/N$ for each of the N lattice cells.

Thompson (1992) gives a model-based example of systematic sampling of a river. Suppose that $n =$ 100 water samples are needed from a river to assess nutrient load. If the length of the river is 250 miles, then a systematic sampling design divides the river into intervals of 250/100 = 2.5 miles. A random starting point is chosen between river mile 0 and 2.5, and the river is sampled at successive intervals of 2.5 miles thereafter. For this and the previous mountainside example, there may be little advantage to treating the population as finite, because the plots and the river intervals are artificial constructs imposed by the researcher on an extremely large population. In cases like these, the motivation for using the finite population statistics is weak, and infinite population statistical methods are preferable.

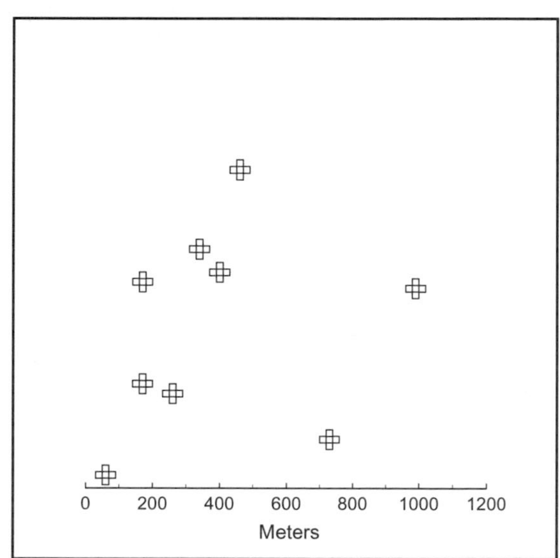

FIGURE 6.2. Plot layout from a cluster sampling design.

6.6.3 Stratified Systematic Sampling and Multistage Sampling

Stratified systematic sampling can also be useful for ecological assessment. Stratified systematic sampling is illustrated by a two-stage sampling design that partitions the population into a set of distinctive strata or subpopulations. Then each stratum is sampled according to independent, or unrelated, systematic sampling designs. Multistage sampling is useful when a population can be stratified according to a multiple-level hierarchy. For example, a landscape may be stratified according to major watersheds at the highest level of the hierarchy, then a second stratum is delineated by partitioning each watershed according to first-order streams. Multistage stratified random sampling selects an SRS at each stratum level of the hierarchy. Thus an SRS of watersheds is selected; then, within each selected watershed, an SRS of first-order stream basins is selected. Finally, an independent SRS is selected within each of the selected first-order stream basins. For example, we may divide each first-order stream into 100-m lengths and select an SRS from among the 100-m intervals to sample pH or nutrient load.

6.7 Variance of Estimators Derived from the Horvitz–Thompson Estimator

A critical component of statistical analysis is the estimation of standard errors. Section 6.6 discusses the estimator $\hat{T}$ of the population total and estimators derived from $\hat{T}$. This section discusses estimation of $V(\hat{T})$, the variance of $\hat{T}$, and the variance of estimators derived from $\hat{T}$. Calculation of $V(\hat{T})$ requires the second-order inclusion probabilities π_{ij}, $i = 1, \ldots, n, j = 1, \ldots, n, i \neq j$, where π_{ij} is the probability that *both* the ith and jth population units appear in the sample. For example, suppose that an SRS of size 2 is selected from the population $P = \{0, 2, 3\}$. The possible samples are {0, 2}, {0, 3}, and {2, 3}, and each sample has probability 1/3 of being the sample selected. Suppose that the first population unit is 0, and the second and third are 2 and 3. The first-order inclusion probabilities are $\pi_1 = \pi_2 = \pi_3 = n/N = 2/3$. The second-order inclusion probabilities are $\pi_{12} = \pi_{13} =$

$\pi_{23} = 1/3$, because there is only one sample in which each combination (0 and 2, 0 and 3, and 2 and 3) appear. In general, for an SRS, $\pi_{ij} = n(n - 1)/N(N - 1)$ (Thompson, 1992); for example, $\pi_{12} = 2(2 - 1)/(3 \cdot 2) = 1/3$.

For stratified random sampling, the second-order inclusion probability π_{ij} is $n_r(n_r - 1)/N_r(N_r - 1)$ if the ith and jth population units both belong to the rth stratum, and $\pi_{ij} = \pi_i \pi_j$ if the ith and jth population units belong to different strata. Here n_r is the number of units sampled from the rth strata, and N_r is the number of population units in the rth strata. For systematic sampling, $\pi_{ij} = \pi_i = 1/k$ if the ith and jth population units both appear in the sample; otherwise, $\pi_{ij} = 0$. For example, suppose that every fourth population unit is selected after a random starting unit has been determined. If population units are exactly 4 units apart, then they are guaranteed to appear simultaneously in the sample, and $\pi_{ij} = \pi_i = \pi_j = 1/4$. If the units are exactly 3 units apart, then they are guaranteed *not* to appear in the same sample, and $\pi_{ij} = 0$. Särndal et al. (1992) present second-order inclusion probabilities for other designs.

For *any* design, Thompson (1992) shows that the variance of $\hat{T}$ is estimated by

$$\hat{V}(\hat{T}) = \sum_{i=1}^{v} \left(\frac{1 - \pi_i}{\pi_i^2} \right) + 2 \sum_{i=1}^{v} \sum_{j>i} \left(\frac{1}{\pi_i \pi_j} - \frac{1}{\pi_{ij}} \right) y_i y_j.$$

The standard error of $\hat{T}$ is $\sqrt{\hat{V}(\hat{T})}$. The variances of estimators derived from $\hat{T}$ are estimated by functions of $\hat{V}(\hat{T})$ according to standard properties of the variance. For instance, the standard error of the sample mean is estimated by $\sqrt{\hat{V}(\hat{T})/v}$.

6.8 Nonprobability and Gradient Directed Transect Sampling

Probability sampling is sometimes forsaken in ecological assessment. Often the aim of sampling is to describe the range and extent of biotic communities across a landscape. In these situations, probability sampling may be inefficient, and sampling design objectives are oriented toward informal description, rather than formal statistical inference. Descriptive sampling designs are illustrated by the sampling protocols used by Cooper et al. (1987), Pfister et al. (1977), and Steele et al. (1981) to develop the forest habitat type classification systems for Montana and Idaho. These studies were largely observational in nature, and they dispensed with statistical validity in an effort to sample all environments without fully knowing the range of environments and plant communities before sampling commenced. The authors employed a sampling procedure termed by Mueller–Dombois and Ellenburg (1974) as "subjective, but without preconceived bias" to sample forest stands across large areas of remote and rugged mountains. Sampling was conducted by traveling forest roads and sampling whenever a forest stand was encountered that appeared representative of the local forest community. In contrast to the sampling design used by these authors, simple random sampling would be expected to sample representatively with respect to the areal coverage of forest communities. If an SRS had been obtained, unless sample sizes were substantially larger than those used by these authors, sample sizes may have been inadequate to describe some communities with limited distribution.

Gradient directed transect, or *gradsect*, sampling (Austin and Heyligers, 1989; Bourgeron et al., 1994) is a more useful nonprobability sampling design than Mueller–Dombois and Ellenburg's (1974) approach. Two objectives of gradsect sampling are to obtain complete coverage of the range of environmental conditions and biotic communities and to obtain information for constructing probability sampling designs. Gradsect sampling is used to sample regional-scale landscapes in a manner that distributes plots representatively with respect to major environmental gradients. This is accomplished by selecting sample locations to maximize variation between units with respect to major recognized environmental variables, such as elevation and moisture availability (Bourgeron et al., 1994). An assumption made of gradsect sampling is that the sample is representative of the landscape, and the validity of this assumption depends on the ability of the researcher to recognize the variables controlling the distribution of ecological communities. Because extrapolation, inference, and prediction hinge on the validity of this assumption, inference should be made with caution and accompanied by a clear statement of the conditional nature of the inference.

In some instances, probability sampling designs fail to accurately identify the population of interest. Also, researchers sometimes fail to sample the population according to design because of logistical problems that occur during the sampling process. A practical approach to these problems is to make a best effort at selecting a probability sample and admit that (1) the resultant estimators are nearly unbiased, rather than unbiased; (2) predictions are moderately precise, rather than highly precise; and (3) hypothesis tests nearly, rather than exactly, achieve a specified significance level. The difficulties of probability sampling for ecosystem assessment do not absolve researchers of the responsibility of striving for probability samples. Researchers must be knowledgeable about and use probability sampling when feasible, and be able to recognize the difference between a probability and a nonprobability sample. When probability sampling cannot be achieved, researchers should attempt to devise a sampling design that comes as close as possible to selecting a probability sample. This strategy defends against estimation bias and promotes efficiency in terms of the amount of information recovered from the sample for a given cost. The researcher is responsible for revealing the difficulties and failure of the design to his or her audience so that they may interpret the results objectively. Scientifically literate audiences usually are willing to consider an analysis based on nonprobability samples if provided with adequate explanations of the failure to obtain probability samples and if statistical inference is limited and conservative.

6.9 Advanced Sampling Techniques

In addition to the designs discussed earlier, advanced sampling techniques can be used to increase sampling efficiency. For example, adaptive sampling designs (Thompson, 1992) allow samplers to deviate from a sampling plan when a rare or unusual population unit is observed. Adaptive sampling allows the researcher to modify the sample design, while sampling is in progress, to increase the chances of observing another population unit with similar characteristics. It also provides a means for computing inclusion probabilities for all sample units. Consequently, the researcher is able to improve sampling efficiency and collect a statistically valid sample. Double sampling (Schreuder et al., 1993) is a method that reduces sampling costs by sampling a surrogate, or auxiliary, variable. Both the variable of interest and the surrogate are measured on some sample units; on the remainder, only the surrogate is measured at a reduced cost. Then sample units on which both are measured are used to predict the variable of interest for the surrogate-only sample units. For example, double sampling is sometimes used when estimating animal population size using aerial surveys. Ground visitation of a subsample of the aerial sample locations is used to correct the aerial only observations for undercounting (Thompson, 1992).

6.10 Example of a Large-scale Sampling Design

The National Surface Water Survey (NSWS) was initiated during the 1980s by the U.S. Environmental Protection Agency to investigate acidification of surface waters. The National Stream Survey (NSS) was charged with investigating the water chemistry of streams sensitive to acidic deposition. Objectives of the NSS were to detect and monitor regional patterns and trends in surface-water acidity and to analyze the relationship between observed patterns and trends in surface-water quality and atmospheric deposition (Ward et al., 1990). The population of interest was streams in southeastern and Mid-Atlantic states, and the population units were defined to be stream reaches appearing on 1:250,000-scale topographic maps (Kaufmann et al., 1988). A stream reach was defined as "a segment of stream between two confluences, or the segment between the origin of the stream and the first confluence if the stream was a headwaters reach" (Overton and Stehman, 1995, p. 261). Because of the scale and cost of the survey, a pilot study was used to assess the feasibility of the proposed sampling design (Ward et al., 1990). The pilot study sampled 54 stream reaches, and the data were used in conjunction with Monte Carlo simulation to compare and select statistical methods for trend detection of low-concentration chemicals.

The full sample design stratified the study area by region and subregion. Sampling units were selected by randomly locating a lattice over a subregion map. If a lattice point fell on a stream reach, then the stream reach was designated to be sampled (Overton and Stehman, 1995). There are several consequences of this design. Inclusion probabilities were proportional to stream reach area and unknown until the sample was selected. Once the sample units were selected into the sample, the inclusion probabilities for each sample unit were computed based on the number of lattice points, the area of each sample unit (a stream reach), and the total area of the subregion (Kaufmann et al., 1988). The sampling design did not use a sampling frame and hence avoided the cost and effort of constructing a sampling frame. (Recall that the purpose of the sampling frame is to list each population unit and its inclusion probability so that each unit can be sampled according to its inclusion probability. The use of the sampling frame in data analysis is restricted to identifying the inclusion probabilities for the sample units.) The sampling design yielded 445 stream reaches for sampling. Both the upper and lower ends were sampled in the spring to measure concentration of chemical constituents related to acidification, and these data provide a baseline against which surface-water quality trends and patterns can be measured through future sampling.

6.11 Conclusion

Sampling design begins with a statement of the research questions and study objectives. With the research questions in hand, a population and variables of interest can be identified. The population and variables should be relevant with respect to the study objectives, and they should be appropriate for addressing the research questions through statistical inference. One of the foremost difficulties in applying statistical inference for ecological assessment is that of defining the population of interest. Usually, limitations must be imposed on the original study objectives, because the population must be a precisely defined collection of units that can be sampled. The NSS example shows how a complex set of objectives regarding acidification of streams across large regions and over time can be addressed. The pivotal idea was that of identifying the population of interest to be a finite set of stream reaches covering the region. Then it was straightforward, but clever nonetheless, to select a stratified random sample by randomly positioning a lattice over each subregion map and selecting each reach located under a lattice point. Resources such as financing, personnel, and material for sampling must be identified once the population and variables have been determined. Then a sampling design and a methodology for data analysis can be proposed, given resource constraints. A proposal for data analysis methods should identify the parameters and estimators, confidence intervals, hypothesis tests, and predictive models to be used. At this stage, sample size calculations are necessary to ensure that the estimates and predictions will be sufficiently accurate and that the tests will be sufficiently sensitive. Sometimes simple sample size calculations are adequate, but large-scale sampling designs often necessitate Monte Carlo simulation studies.

Increasingly larger data sets are being used for ecosystem assessment as automated data collection methods become more widespread. A risk posed by massive data sets is the tendency to ignore statistical validity until data analysis commences, and then to proceed with data analysis even when data do not constitute a valid sample. Proper sampling design will produce data that are appropriate for their intended purpose and ensure that valid inference is drawn from the data. Statistical validity and proper sampling design are critical to nearly all quantitative ecological assessment efforts, no matter how many observations are collected. Researchers must be vigilant in their efforts to use sampling designs and expect the same of others.

6.12 References

Austin, M. P.; Heyligers, P. C. 1989. Vegetation survey design for conservation: gradsect sampling of forests in north-eastern New South Wales. *Biol. Conserv.* 50:13–32.

Bourgeron, P. S.; Humphries, H. C.; Jensen, M. E. 1994. General sampling considerations for landscape evaluation. In: Bourgeron, P.S.; Jensen, M.E., eds. *Volume II: ecosystem management: principles and applications.* Portland, OR: U.S. Dept. Agric., For. Serv., Pacific Northw. Res. Sta.: 109–120.

Cooper, S. V.; Neiman, K. E.; Steele, R.; Roberts, D. W. 1987. *Forest habitats of northern Idaho: a second approximation.* Ogden, UT: U.S. Dept. Agric., For. Serv., Intermountain Res. Sta.

Eberhart, L. L.; Thomas, J. M. 1991. Designing environmental field studies. *Ecol. Monogr.* 61:53–73.

Hansen, M. H.; Hurwitz, W. N.; Madow, W. G. 1953. *Sample survey methods and theory, volume 1: methods and applications.* New York: John Wiley & Sons.

Hayne, D. W. 1949. An examination of the strip census method for estimating animal populations. *J. Wildlife Manage.* 13:145–157.

Horvitz, D. G.; Thompson, D. J. 1952. A generalization of sampling without replacement from a finite universe. *J. Amer. Statist. Assoc.* 47:663–685.

Kaufmann, P. R.; Herlihy, A. T.; Elwood, J. W.; Mitch, M. E.; Overton, W. S.; Sale, M. J.; Messer, J. J.; Cougan, K. A.; Peck, D. V.; Reckhow, K. H.; Kinney, A. J.; Christie, S. J.; Brown, D. D.; Hagley, C. A.; Jager, H. I. 1988. *Chemical characteristics of streams in the Mid-Atlantic and southeastern United States, volume I: population descriptions and physio-chemical relationships.* EPA/600/3-88/021a. Washington, DC: U.S. Env. Protect. Agency.

Kendall, K. C.; Metzgar, L. H.; Patterson, D. A.; Steele, B. M. 1992. Power of sign surveys to monitor population trends. *Ecol. Appl.* 2:422–430.

Lesica, P.; Steele, B. M. 1996. A method for monitoring long-term trends: an example using rare arctic-alpine plants. *Ecol. Appl.* 6:879–887.

Levy, P. S.; Lemeshow, S. 1991. *Sampling of populations.* New York: John Wiley & Sons.

Mueller-Dombois, D.; Ellenberg, H. 1974. *Aims and methods of vegetation ecology.* New York: John Wiley & Sons.

Ostle, B.; Mensing, R. W. 1975. *Statistics in research,* 3rd ed. Ames, IA: Iowa State University Press.

Overton, W. S. 1969. Estimating the number of animals in wildlife populations. In: Giles, R.H., Jr., ed. *Wildlife management techniques.* Washington, DC: Wildlife Society: 403–455.

Overton, W. S.; Stehman, S. V. 1995. The Horvitz–Thompson theorem as a unifying perspective for probability sampling: with examples from natural resource sampling. *Amer. Statist.* 49:261–268.

Peterson, R. G. 1994. *Agricultural field experiments, design and analysis.* New York: Marcel Dekker.

Pfister, R. D.; Kovalchik, B. L.; Arno, S. F.; Presby, R. C. 1977. *Forest habitat types of Montana.* Ogden, UT: U.S. Dept. Agric., For. Serv., Intermountain Res. Sta.

Ramsey, F. L.; Schafer, D. W. 1997. *The statistical sleuth, a course in methods of data analysis.* Belmont, CA: Duxbury.

Särndal, C. E.; Swensson, B.; Wretman, J. 1992. *Model assisted survey sampling.* New York: Springer-Verlag.

Schreuder, H. T.; Gregoire, T. G.; Wood, G. B. 1993. *Sampling methods for multiresource forest inventory.* New York: John Wiley & Sons.

Snedecor, G. W.; Cochran, W. G. 1980. *Statistical methods*, 7th ed. Ames, IA: Iowa State University Press.

Steele, R.; Pfister, R. D.; Ryker, R. A.; Kittams, J. A. 1981. *Forest habitats of central Idaho.* Ogden, UT: U.S. Dept. Agric., For. Serv., Intermountain Res. Sta.

Thompson, S. K. 1992. *Sampling.* New York: John Wiley & Sons.

Ward, R. C.; Loftis, J. C.; McBride, G. B. 1990. *Design of water quality monitoring systems.* New York: Van Nostrand Reinhold.

7 General Data Collection and Sampling Design Considerations for Integrated Regional Ecological Assessments

Patrick S. Bourgeron, Hope C. Humphries, and Mark E. Jensen

7.1 Introduction

Large-scale data collection is the initial observational phase of integrated regional ecological assessments (IREAs). The data collection methodology determines to a large extent the accuracy and precision of all subsequent analyses, such as pattern recognition (Bourgeron and Jensen, 1994; Bourgeron et al., 1994a; Dale and O'Neill, 1999). The data analysis and interpretative phases of IREAs often tend to be emphasized by scientists over formal data collection procedures (see the case studies in Chapters 30 through 34; for a very formal approach to survey design, see the US-EPA Ecological Monitoring and Assessment Program, O'Neill et al., 1994; Kapner et al., 1995). The lack of formal, consistent data collection protocols for IREAs is unfortunate, because the quality of the characterization of a region and the quality of subsequent interpretations depend on the quality of the data, both in terms of the thoroughness of coverage and the type of information collected.

Several characteristics of IREAs make the use of classical sampling and survey design methodologies (see Chapter 6) difficult to implement at best (see discussion on aspects of the problem in Legendre and Legendre, 1998, p. 16): (1) IREAs usually take place in large areas (e.g., the 58 million ha of the interior Columbia River basin, ICRB), which may contain inaccessible locations; (2) they aim at characterizing the range of variability in the biotic and abiotic components of ecosystems, as well as biotic–abiotic interactions (e.g., plant communities in relation to environmental gradients) (see Chapter 22), regardless of their relative abundance (e.g., rare or common); (3) they aim at describing multiple components of ecosystems at multiple spatial scales.; (4) the data collected are generally used for multiple purposes; and (5) IREAs are often conducted under strict deadlines and budgetary and political constraints, which seriously limit the resources and time allocated to de novo sampling design and field survey (Treweek, 1999).

Therefore, the data collection effort of IREAs should address the following considerations: (1) the analysis of spatial (or temporal) structures is of primary interest; (2) spatial (or temporal) scale is a key concept; (3) as a result of the hierarchical structure of ecosystems (see Chapers 2 and 3), spatial heterogeneity is functional and not the result of some random, noise-generating process; (4) the recovery of all patterns, common and rare, and of their intrinsic range of variability is often prohibitively costly under the requirements of classical sampling designs; and (5) existing data usually have to be utilized because of time and resource limitations. The statistical methods and models used in IREAs also must include realistic assumptions about the spatial structuring of ecosystems (Legendre and Legendre, 1998).

In collecting data for IREAs, the emphasis on the multiscaled spatial heterogeneity of multiple ecosystem components contrasts with the purpose of classical sampling design, which is to determine unbiased estimates of the means of variables over entire populations (Austin and Heyligers, 1991; Legendre and Legendre, 1998; also see Chapter 6). A requirement of classical sampling design is the

Patrick S. Bourgeron and Hope C. Humphries wish to acknowledge partial funding provided by a Science to Achieve Results grant from the U.S. Environmental Protection Agency ("Multi-scaled Assessment methods: Prototype Development within the Interior Columbia Basin").

complete enumeration of the units from which a sample can be selected. However, such knowledge of the sampling frame may not exist for many ecosystem components. Furthermore, given the multiple purposes of IREAs, sampling is seldom directed at a single target, and an optimal sampling strategy may be hard to define (Bunce et al., 1996a). Finally, the scale of the information to be collected may vary from object to object within a sampling effort (e.g., sampling tree, grass, raptor, and invertebrate species).

The broad multiscale, multitheme nature of IREAs, in conjunction with the hierarchical structure of ecosystems, leads to four important questions regarding data collection for IREAs: (1) What are the general data requirements of IREAs? (2) What standard data collection procedures are appropriate? (3) What are the trade-offs among statistical theory, logistical practice, sampling efficiency, data for interpolation and extrapolation, and specific applications? (4) Can existing data be used alone or integrated with de novo data?

This chapter complements Chapters 5 and 6 by describing pragmatic, applied aspects of the data collection process for IREAs, including the use of and deviations from classical sampling design, as well as answers to the four questions listed above. The chapter contains discussions of the purpose of the data collection process and its impact on data attributes and other properties of data sets, the use of existing data, the need for representative data and the trade-off between statistical theory and application, the choice of sample sites, the detection of rarities, sample size and configuration, and interpolation and extrapolation from samples.

7.2 Purpose of Survey, Data Attributes, and Properties of the Data Collection Process

The first step in implementing a data collection scheme for an IREA is to clearly formulate the specific uses of the data in the context of the assessment (Green, 1979; Dale and O'Neill, 1999; Treweek, 1999). The specific objectives of an IREA define the data attributes and several other properties of the data sets. The data collection design can meet the purpose of an IREA only if its objectives are explicitly defined (Gauch, 1982; Austin, 1987; 1991b). The IREA objectives determine the applications for which the data will be used (Dale and O'Neill, 1999). These fall into seven generic types, each with its own, sometimes overlapping, data requirements: spatial variability characterization (see Chapters 3 and 13), temporal variability characterization (see Chapter 19), assessment of past and current ecological conditions (see Chapter 19), species and community distribution patterns, biodiversity characterization, linkages between terrestrial and aquatic systems (see Chapter 22), and inventory and monitoring.

Decisions about the sources of data are central to the development of a database for an IREA (see Chapters 1, 5, and 8). Data collection can be very costly when all aspects of the effort are considered: development of relational and spatial databases, quality control, standardization of data from different sources, and the like. Therefore, it is important to collect only data that are relevant to the issues at hand. Whether existing data, de novo data, or both are collected, a formal sampling design should guide the process.

The specific attributes of the data to be collected vary according to the objectives. For example, the data needed will differ depending on whether all ecosystems or only the forested ecosystems of a region are characterized; whether, if only forests are characterized, all successional stages or only ecosystems of a certain age (e.g., old growth) are considered; and whether the interest is in all taxa (plants or animals) or a subset of taxa (e.g., trees of a specific genus or granivorous birds).

IREAs involve spatial analysis of the patterns of interest because these patterns are structured by forces that have spatial components. The objectives define the spatial (and temporal) scale of several properties of a spatial sampling design (Wiens, 1989; Allen and Hoekstra, 1992; He et al., 1994; Legendre and Legendre, 1998; Dale and O'Neill, 1999):

1. *Grain*, the size of the elementary sampling unit. The resolution (Schneider, 1994) of an assessment is equal to the grain size of the sampling design.
2. *Sampling interval*, the average distance between neighboring sample units (i.e., the lag in time series analysis).
3. *Sampling extent*, the total area to be characterized by the assessment (see Chapter 1) or the duration of a time series. The sampling extent is termed the *range* by Schneider (1994).

The scales identified for the data collection process must satisfy the objectives of an assessment (see Chapter 1) and therefore address scales appropriate for the patterns and processes of interest (see Chapter 3). A single-objective assessment allows precise scaling of the specific components of a sampling design. However, in practice, IREAs have multiple objectives, which increases the difficulty of choosing the type of sampling design and its properties. The following guidelines should be followed wherever possible (Legendre and Legendre, 1998):

1. The sampling grain should be larger than a unit object (e.g., an individual) and the same as or smaller than the structure to be detected by the design (e.g., a patch).
2. The sampling interval should be smaller than the structure to be detected.
3. The sampling (characterization) extent should be the same as the total area covered by the patterns and processes of interest.

The sampling grain and extent define the observation window in spatial pattern analysis. No structure can be detected that is smaller than the grain or larger than the extent. The sampling (characterization) extent of an IREA may differ from the assessment area to effectively cover all patterns and processes of interest. In the Interior Columbia Ecosystem Management Project (ICBEMP), the characterization area was larger than the initial assessment area defined by the objectives (Hann et al., 1997; Jensen et al., 1997; Quigley and Arbelbide, 1997). As a consequence of the hierarchical structure of ecosystems (Dale and O'Neill, 1999; also see Chapter 2), more than one scale may be relevant to the patterns and processes of interest. For example, species diversity relationships change with changes in scale (Stoms, 1991; Stoms and Estes, 1993; He et al., 1994), and therefore diversity data should be collected in a manner that reflects this scale dependence. Data at multiple scales are often required to characterize the vegetation of an area (e.g., Franklin and Woodcock, 1997). In addressing the issue of scale, explicit consideration should be given to the need for data aggregation and its impact on the spatial properties of data sets (e.g., error propagation, sources of uncertainty).

7.3 Use of Existing Data

Existing data are utilized in IREAs either alone or integrated with de novo data for two reasons: (1) IREA objectives require the analysis of data on past conditions (e.g., for trend analysis), and (2) the cost and time involved in gathering all needed data by means of a de novo regional survey are likely to be prohibitive. Therefore, the kinds of data that are already available must be determined, along with their usefulness and cost effectiveness for a specific purpose, even if biases in such data exist (Austin, 1991a; Bourgeron et al., 1994a; Davis, 1995). Existing data may also be used to guide the sampling design for new surveys as needed. The use of existing data often involves compilation and collation of disparate and interdisciplinary data. Problems must be solved concerning differences in sampling dates, sampling design, data storage, spatial representation, spatial resolution, and data quality (Davis et al., 1991; Davis 1995). These differences may prevent the use of the data for some purposes. For example, some data may be suitable for hypothesis testing, while others may be suitable only for descriptive analyses.

However, existing data that are well organized in an integrated regional database (see Chapter 8) may be a cost-effective approach for IREAs. Many such collections of data have been compiled (e.g., the California Environmental Resources Evaluation System, CERES, discussed in Davis, 1995; also see Chapter 32: Southern Appalachian Assessment, Southern Appalachian Man and Biosphere (SAMAB), 1996; also see Chapter 34: ICBEMP, Gravenmier et al., 1997) or are being collected under general provisions for wide access (e.g., for digital data in the United States, U.S. Executive Office of the President, 1994; in the United Kingdom, the National Geospatial Database, Nanson et al., 1995).

7.3.1 Geographic Information Systems Data Layers

Geographic information systems (GIS) are important components of IREAs (see Chapter 11). One main advantage of GIS is the ability to use and interpret a variety of data sources (Treweek, 1999). Completed IREAs, such as the ICBEMP (see Chapter 34) and the Sierra Nevada Ecosystem Project (SNEP, 1996), are sources of many readily available regional and subregional data layers. The sheer number and variety of the data layers produced during a regional assessment make them appealing sources of information for analyses (e.g., more than 170 data layers and 20 associated databases for the ICBEMP). However, any use of data layers from multiple sources warrants caution because of varying scale, attributes, metadata completeness, and the like (Davis et al., 1991; Davis, 1995; Treweek, 1999; also see Chapter 11). During an IREA, problems may be encountered, such as incompatible map legends and different definitions of the same legend item, in combining multiple fine-scale maps of an attribute into a single coarse-scale data layer.

Existing data layers, such as maps of vegetation, soils, geology, climate, and terrain, can be used as initial surrogates for field data. Problems can arise if data layers have different scales, adding to the initial uncertainty and error rate inherent in each map. For example, inconsistencies in existing subregional vegetation maps acquired from multiple sources rendered them unsuitable for use in the ICBEMP (Hann et al., 1997). However, this approach has been used successfully by various investigators. Maps of land systems in Australia (Christian and Stewart, 1953) were incorporated into a database to describe the range of natural environments in the study region. The database was later used to construct a stratified sampling scheme to describe the range of natural variability contained in various areas in the region (Pressey and Nicholls, 1991).

7.3.2 Remotely Sensed Data

Remotely sensed imagery (see Chapter 10) is readily available over most of North America (e.g., the National Satellite Land Remote Sensing Data Archive at the USGS EROS Data Center, Holm 1997, Larsen et al., 1998). Use of remotely sensed data is a cost-effective approach for broad-scale mapping of land cover (e.g., the National Land Cover Characterization program at the USGS EROS Data Center, Van Driel and Loveland, 1996) and vegetation (e.g., the USGS GAP Analysis Program, or GAP, Scott et al., 1990, 1993; Scott and Jennings, 1997, 1998). In addition, remotely sensed imagery can provide a host of broad-scale measurements on the structure and function of regional ecosystems (Wessman and Asner, 1998). However, there are significant challenges in using information and producing maps from remotely sensed data (Estes and Mooneyhan, 1994) that pertain to the scale of measurement of the data (Davis et al., 1991; Wilkinson, 1996; Bian, 1997; Wessman and Asner, 1998).

There has been a trend toward acquiring ground and remotely sensed data over a range of scales (e.g., Running et al., 1989; Davis et al., 1991; Cao and Lam, 1997). Davis et al. (1991) concluded that some of the problems in the use of remotely sensed data relate to understanding the spatial scale dependence of patterns and processes. The search for multiscaled patterns and processes (Levin, 1992; Milne, 1994; Minshall, 1994; Turner et al., 1994) in turn requires the collection of ground data at the appropriate intensity and frequency. The spatial and temporal scales of remotely sensed data may differ from those of interest in particular cases (Treweek, 1999). Special attention is needed to understand the scaling relations of the phenomena of interest (e.g., see Mack et al., 1997, for deriving species–area relationships and Mladenoff et al., 1997, for discriminating among ecoregions, also see the wide-ranging series of approaches on the subject in Quattrochi and Goodchild, 1997) and to integrate remotely sensed and field data (e.g., Lobo et al., 1998; Wessman and Asner, 1998; Treweek, 1999).

7.3.3 Spatially Referenced Species Records

Another source of existing data is spatially referenced species records, such as herbarium and museum records. Databases that compile this kind of information are maintained over most of North America by state- and province-based Natural Heritage Programs or Conservation Data Centers and The Nature Conservancy (Jenkins 1985) using Biological Conservation Database (BCD) software. Herbarium and museum records for specific plant and animal species, plant communities, and/or ecosystems (including information on identification, location, ownership, and any relevant biological and management information) are collected and entered into the BCD for each state or province. Although the information in the database is descriptive, cost-effective uses can be made of the data for many purposes. The GAP programs (Scott et al., 1990, 1993; Scott and Jennings, 1997, 1998) combine such distribution information for selected animal species with vegetation maps to determine species–habitat relations. This exploratory analysis defines geographic areas that have suitable habitats for the species.

As part of SNEP, records of rare and endemic plant species were collected from a variety of sources, such as the California Native Plant Society's Inventory of Rare and Endangered Vascular Plants of California (Skinner and Pavlik, 1994), rare plant data maintained by the California Natural Diversity Database (California Department of Fish and Game, 1995), and rare plant information from the Nevada Heritage Program (Morefield and Knight, 1991; Morefield, 1994). Plant species distributions in the Sierra Nevada were attributed to coarse-scale river basins, counties, and topographic quadrangles for analyses (Shevock, 1996). The analysis was constrained by the fact that distribution information for many taxa was found to be incomplete due to limited field studies and collection of vouchered specimens.

7.3.4 Plot Data

Plot or transect data are commonly collected in routine biological surveys and monitoring programs. When incorporated into a standardized database, they provide a very cost effective tool for IREAs. A limitation in integrating different sources of existing survey data within a regional database is that the same biotic or abiotic information is rarely collected in each survey. However, if a minimum list of attributes common to all surveys is established, the data can be used for a variety of purposes (Austin, 1991a; Jensen et al., 1994). The lack of a common sampling design and differences in sample size may limit the use of these data. For example, over 37,000 vegetation plots were acquired for the ICBEMP and incorporated into ECADS, an integrated ecological data management and analysis system (Jensen et al., 1998). These plots provided data for many aspects of the ICBEMP, including deriving vegetation structural stages (O'Hara et al., 1996), evaluating regional-scale potential vegetation map accuracy (Reid et al., 1995), determining potential vegetation in subsample areas (Hann et al., 1997), and constructing predictive models of species and community distributions (Humphries and Bourgeron, in Jensen et al., 1997). However, not all 37,000 plots were suitable for addressing all objectives because of differing levels of data completeness, plot characteristics, and so on. Other examples of the use of regional plot databases for land management and conservation planning are provided by Austin et al. (1983, 1984, 1990, 1996) and Margules et al. (1987) in Australia. Such regional databases can also be used to determine which geographic areas and/or parts of environmental gradients have not been sampled. Systematic reexamination of existing data can prevent duplication of survey effort and make maximum use of resources. Disadvantages in the use of existing data (e.g., cost of establishing a regional database, limitation due to a minimum list of attributes at the regional scale) may often be offset by potential cost reduction and increasing effectiveness of new surveys and by providing an initial regional overview of phenomena of interest.

To assess the effect of grazing on rangeland in the Sierra Nevada for SNEP, existing Parker range condition and trend transects were acquired that had been repeatedly resurveyed over five decades in ten Sierra Nevada National Forests; the information was supplemented by resurveying 24 of the transects (Menke et al., 1996). Limitations in the use of Parker transect data were acknowledged, such as insensitivity of the method to changes in plant composition in areas with low plant cover and known species identification problems; but the transects comprised the only long-term widespread range information available for the Sierra Nevada (Menke et al., 1996).

Regional gradient analyses and modeling of plant species and community distributions have recently been conducted using regional plot databases assembled from a variety of existing sources. For example, Austin et al. (1990, 1996) analyzed individual *Eucalyptus* species and species richness responses to environmental variables using a 7200-plot data set covering 40,000 km^2 in southeast New South Wales, Australia. At a minimum, plots contained presence and absence of tree species. They also contained information on key environmental variables or were georeferenced accurately enough to attribute plots with such information. Similarly, Leathwick (1995, 1998) analyzed the response of New Zealand tree species to environmental variables using over 10,500 forest plots. Ohmann and Spies (1998) analyzed regional gradient relations of woody plant communities in Oregon based on a subset of a 10,000-plot data set acquired from USDA Forest Service and Oregon State University sources. Regional gradient analyses were conducted for vegetation and plant functional types in the ICRB using data from 19,000 plots (out of 37,000 plots acquired for the ICBEMP), which contained the appropriate information for the analyses (e.g., complete species cover and environmental data, appropriate plot size; Bourgeron et al., unpublished).

Regional plot databases can be integrated with GIS and remotely sensed data to derive detailed spatial environmental data for large areas. He et al. (1998) used Thematic Mapper (TM) data in conjunction with field inventory data (Forest Inventory and Analysis, FIA) (Hahn and Hansen, 1985; Hansen et al., 1992) in a GIS environment. They overlaid the satellite classification (adapted from Wolter et al., 1995) with an ecoregional classification (Host et al., 1996) in northwestern Wisconsin. The resulting classes represented the distributions of dominant tree species. FIA data were used to generate age-class distributions for dominant species and the distributions of associated species for each age class of the dominants. Finally, the age classes of the dominant and associated species were assigned to each pixel of the dominant species map. Results were used to assess forest patterns across regional landscapes and as input into LANDIS, a forest simulation model (Mladenoff et al., 1996; He and Mladenoff, 1999; also see Chapter 18), to examine forest landscape change over time.

7.4 Rationale for Representative Sampling

The primary goal of data collection for IREAs is to characterize as many ecological patterns of interest as possible. The constraints of most IREAs, which de facto limit the number of samples collected, require several possibly contradictory sampling design attributes: efficiency in the recovery of ecological patterns and processes, full representation of patterns and processes, geographic replication, logistical benefits, and cost effectiveness. Commonly used random sampling procedures do not necessarily have these attributes. The goal of the data collection phase of an IREA is acquiring data that are representative of the environments to be characterized under severely limited sample sizes. Sampling theory emphasizes randomization in order to provide the probability structure for statistical analysis or to give credibility to the statistical model used (Gillison and Brewer, 1985; see also Chapter 6). Simple random sampling does not capture the full range of variability of the biotic and abiotic components of ecosystems at regional scales unless the sampling intensity is very high (Pielou, 1974; Orlóci, 1978; Gauch, 1982).

The data collection process for IREAs should consider three aspects of pattern recognition: (1) the delineation of the pattern itself (e.g., a specific forested ecosystem); (2) the frequency and distribution of patches of the pattern (i.e., spatial distribution, number, and size of stands in a forested ecosystem) (Godron and Forman, 1983; Gillison and Brewer, 1985); and (3) the detection of boundaries or rates of change between samples (Fortin, 1994, 1999). In landscapes, patch frequency and distribution vary as a scale-sensitive function of environmental complexity and of the level of resolution of the ecological classifications used to characterize the pattern (Gillison and Brewer, 1985; Bissonette, 1997; Mladenoff et al., 1997; Hargis et al., 1998; Bourgeron et al. 1999). Landscape configuration variability should be analyzed in terms of the driving variables (the abiotic factors) controlling the biotic components of the ecosystem (Bourgeron et al., 1994b).

To obtain data representative of all patterns and processes, IREAs have made use of a series of stratified sampling schemes (SSs) (see Chapter 6). These schemes have been employed successfully to provide both accuracy and efficiency in the recovery of patterns and statistical validity over large heterogeneous areas with mostly unknown patterns. SSs used in IREAs fall into three broad categories: stratified random sampling, stratified semirandom sampling, and multistage stratified random and semirandom sampling. They have been used in terrestrial and aquatic ecosystems and to sample plant and animal species and communities.

7.4.1 Stratified Random Sampling

Stratified random sampling (SRS) (see Chapter 6) is used to stratify a region based on relevant data, such as abiotic variables (e.g., slope, aspect) or ecosystem units. Random sampling is then conducted within each stratum. The specific objectives of a project determine whether the number of samples in a stratum is proportional to its occurrence or is a fixed size. SRS has been used for a variety of biological assessments (e.g., Davis et al., 1990; Aspinall and Veitch, 1993; Walker et al., 1993). Stohlgren et al. (1997) used SRS as part of a methodology for rapid assessment of plant diversity patterns at landscape scales. The strata were six vegetation types of interest; the same number of samples was allocated to each type. Ecological classifications (see Chapter 22) have often been used as the basis for environmental stratification for sampling and monitoring. In Great Britain, the Institute of Terrestrial Ecology (ITE) land cover classification is used as the basis for stratification for national-scale surveys of land use and cover, as well as for monitoring and evaluation of land-use policies (e.g., Bunce et al., 1996b).

7.4.2 Stratified Semirandom Sampling

Stratified semirandom sampling (SSRS) is a variant of SRS. Stratification is conducted as for SRS, but randomly selected potential sites may be eliminated based on additional criteria. For example, Neave et al. (1996) conducted representative sampling of bird assemblages in a 7600-km^2 area of southern Australia containing open *Eucalyptus* forest. Forested sites were stratified by combinations of temperature, precipitation, and nutrient-supply classes, resulting in identification of 24 environmental domains. Potential sites for sampling bird species and their habitats were evaluated based on their suitability defined by road accessibility. Twenty-three of the domains were sampled with 165 1-ha plots, efficiently covering the range of environmental variability by sampling a very small fraction of the study area.

A methodology known as gradient directed transect (*gradsect*) sampling is a variant of SSRS. This approach, first described by Gillison and Brewer (1985), is based on the distribution of patterns along environmental gradients. The gradsect sampling design (Gillison and Brewer, 1985; Austin and Heyligers, 1989) is intended to provide a cost-effective description of the full range of biotic variability in a region by sampling along the full range of environmental variability present. Transects that contain the strongest environmental gradients in a region are selected in order to optimize the amount of information gained in proportion to the time and effort spent during a survey (Austin and Heyligers, 1989). In addition, sampling sites are deliberately located to minimize travel time. The method has been shown statistically to capture more information than standard designs (Gillison and Brewer, 1985).

Mackey (1993) conducted a vegetation survey in the wet tropics of Queensland, Australia, to examine prediction of rain forest structural characteristics from climate and soil parent material. A sampling design was required that would identify a minimum number of sites that provided both representation and replication. Stratification by climate zone, parent material, and topographic position, with replication of most variable combinations, resulted in a total of 61 plots for a preliminary analysis of rain forest structure and physiognomic attributes.

A fine-scale gradsect vegetation survey was conducted in semiarid woodlands of eastern Australia, in which gradsects were positioned along topographic gradients in three landscape types to measure vegetation, soil surface, and terrain attributes in contiguous 1-m^2 quadrats (Ludwig and Tongway, 1995). Gradsects ranging from 300 to 500 m in length were subjected to boundary analysis to characterize the spatial organization of each landscape at a range of scales.

Wessels et al. (1998) evaluated the effectiveness of the gradsect method for conducting animal surveys compared to a comprehensive habitat-specific survey in a 350-km^2 nature reserve in South Africa. Three 500-m-wide gradsects were positioned to incorporate the maximum variation in physiographic characteristics. The gradsect method performed well in detecting bird and dung beetle species compared to the more expensive and time-consuming habitat-specific method. The authors concluded that the gradsect method could be employed with confidence for faunal surveys in areas where little information is available about the vegetation. However, the gradsect method missed several species associated with small, isolated patches of vegetation habitat that were not identified by the environmental variables used to construct the gradsects. Therefore, vegetation information, where available, can refine and improve a gradsect sample design for animal surveys (Wessels et al., 1998).

7.4.3 Multistage Stratified Semirandom Sampling

Steele (see Chapter 6) discussed the usefulness of multistage stratified random sampling design in which a second level of stratification is imposed on a first level. This multistage approach has been widely used in biological assessments (e.g., Debinski and Brussard, 1992; Debinski and Humphrey, 1997). Helman (1983) and Austin and Heyligers (1989, 1991) have similarly expanded the gradsect methodology to include different levels of environmental stratification within each gradsect, thereby creating a multistage stratified semirandom sampling design (MSSRS). Their modified gradsect procedure utilizes a two-stage sampling design: (1) gradsects are selected and (2) adequate environmental stratification and replication are performed within gradsects.

The two-stage gradsect sampling design has been used to describe the rain forests of southern New South Wales, Australia (Helman, 1983) and for a mixture of eucalypt and rain forests in northern New South Wales (Austin and Heyligers, 1989, 1991). Similar MSSRS methodology was used to assess the conservation value of a 1300-km^2 area, the Gray Ranch, in southern New Mexico. The specific purpose of the survey was to characterize vegetation patterns and their associated floristic variability in relation to the range of environmental variability within the ranch (Engelking et al., 1994; Bourgeron et al., 1995), constrained by a two-week time limit for fieldwork by four surveyors. Stratifying variables (geology, elevation, and soil type) were chosen based on the results of a previous vegetation analysis. Taking access roads into consideration, a total of four gradsects was positioned to contain 49 out of the 55 biophysical environments (combinations of the stratifying variables) found at the ranch. Sampling of the biophysical environments and of the gradsects was proportional to their representation within the ranch. Geographic replication was provided by environmental overlap in the gradsects, as well as by dividing each of the two main gradsects into three segments. Actual sample location was chosen randomly for a particular physical environment within each segment.

Another variant of the MSSRS-gradsect design was utilized in designing a regional survey of forest-dwelling microchiropteran bats in New South Wales, Australia (Mills et al., 1996). Temperature and precipitation attributes of forested areas were selected for environmental stratification based on an understanding of the autecology of the region's bat species. First, three areas were chosen to represent climate strata. Second, four trapping sites were located within each stratum; each trapping site had two subsites. Sites and subsites were selected based on bat habitat characteristics, minimum distance of 1 km between sites, minimal disturbance and vehicle use, reasonable accessibility by four-wheel-drive vehicle, minimal traveling time among all sites in an area, and minimum distance from the edge of a stratum. In a pilot study, thirteen species were detected, including three listed as vulnerable and rare, out of the nineteen species expected to be present.

7.5 Evaluation of Sampling Strategies

The relative efficiency and cost effectiveness of various sampling strategies for biological inventories and IREAs have been compared by several authors. Austin and Adomeit (1991) used a simulated data set based on an actual landscape, with realistic biotic–environment relationships (Belbin and Austin, 1991). The authors provided informed guesses of sampling costs for the various methods. Random sampling (RS) recovered more species than systematic sampling or traditional transect sampling at intermediate cost. Transect sampling was found to have the highest costs and detected the fewest species. Gradsects established with various sampling rules generally detected more species than random sampling; some gradsect sampling rules achieved this at lower cost. The gradsect method allows placement of samples in logistically more accessible areas than other techniques such as systematic sampling or RS, thus improving cost effectiveness. Costs can also be reduced and effectiveness maximized when stratifying variables are carefully selected using existing information (Austin and Adomeit, 1991).

Neldner et al., 1995; Mills et al., 1996, and Neave et al., 1996 have also reported more effi-

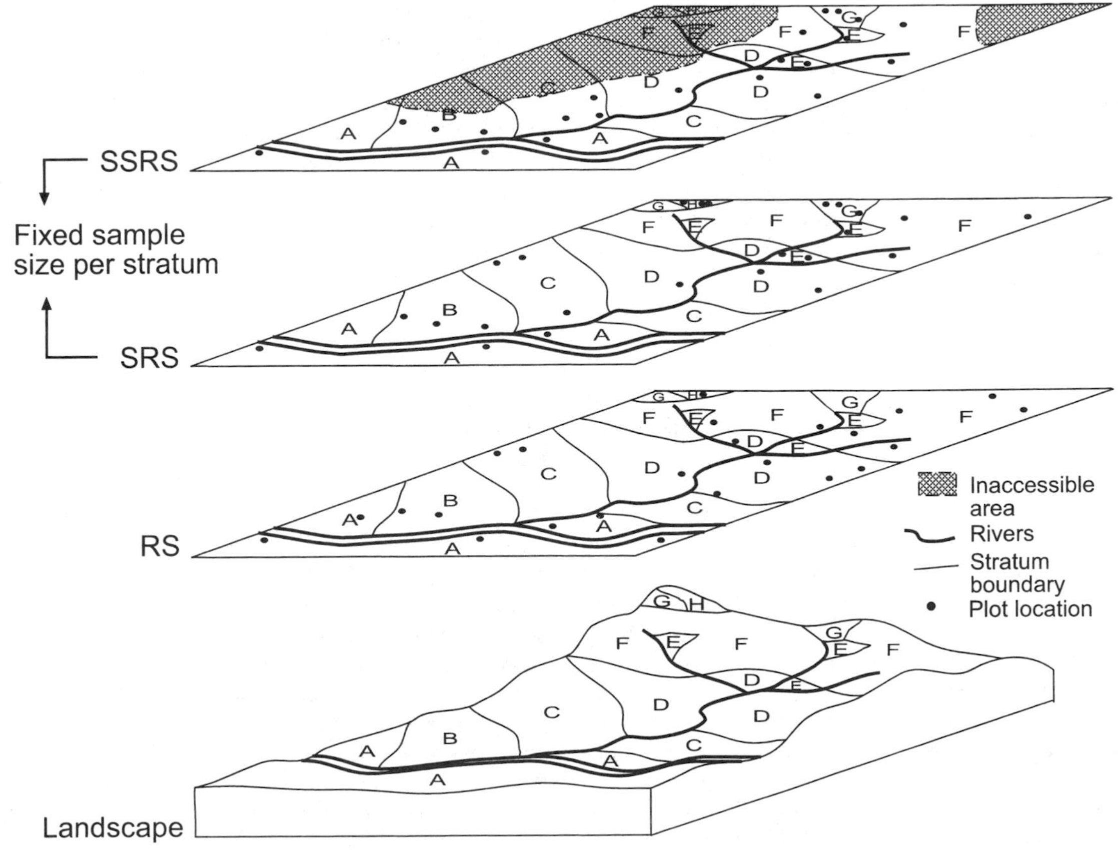

FIGURE 7.1. Three sampling methods applied to a hypothetical landscape containing eight environmental stratum (indicated with letters A through H) and a total of 24 plots. Stratified sampling methods shown for fixed sample sizes of three plots per stratum, regardless of stratum area. RS, random sampling; SRS, stratified random sampling; SSRS, semistratified random sampling.

TABLE 7.1. Characteristics of three sampling schemes.

Characteristics	Random sampling	Stratified random sampling	Stratified semirandom sampling (gradsect)
Represents region (as unbiased estimator)	Yes	Yes	No
Represents sampled area/transect only (as unbiased estimator)	NA	NA	Yes
Represents all aspects of variability in sample area	NA	NA	Yes
Represents all aspects of variability within region	No	No	No
Interpolation–extrapolation needed with limited number of samples	Yes	Yes	Yes
Cost/benefit (small/large areas)	High/Low	High/Medium	High/High
Efficiency in pattern recovery for large areas (low diversity/high diversity)	Medium/Low	High/High	High/High

cient, rapid representation of patterns using SSRS and MSSRS, including geographic replication, than could be attained using other sampling strategies. Neave et al. (1997) compared several sampling strategies to estimate the species richness of diurnal, terrestrial birds using Monte Carlo simulation in the southeastern region of Australia. The results of the study indicated that there was no apparent benefit in using SS designs (SRS, SSRS, and MSSRS) over RS for estimating bird species richness in the sampled area. The relatively poor performance of SS schemes was attributed to the limitations of the data used for stratification (e.g., incomplete biological knowledge about bird–environment relationships, inappropriate spatial resolution of the environmental variables). Another limitation of the simulation results may be the use of an existing bird survey as the sampling universe, which likely contained a nonrandom distribution of sites. Therefore, the effectiveness of SS schemes compared to RS may have been reduced if environmental variability was already accounted for by survey site locations. The authors concluded that a great benefit of SS schemes is their logistical efficiency and cost effectiveness.

A criticism of SSRS and MSSRS is that they may increase the effect of spatial autocorrelation among samples. This possibility may be reduced with appropriate stratification, geographic replication, and randomization (Austin and Heyligers, 1989, 1991; Stohlgren et al., 1997; Wessels et al., 1998). Differences between RS, SRS, and SSRS are represented in Figure 7.1. The landscape was classified into eight environmental strata (e.g., ecological units) with varying abundance (Figure 7.1), rare (E, G, and H), intermediate (B and C), and common (A, D, and F). Twenty-four samples were distributed over the landscape in each of the three sampling designs, with a fixed sample size of three per stratum for SRS and SSRS for clarity of presentation. SSRS differed from SRS in that the criterion of accessibility was used to eliminate portions of the landscape to provide cost effectiveness and logistical benefits. The results indicated that RS had the highest number of strata not represented (two rare strata).

Final decisions about the kind of sampling strategy to be followed should be made by taking into account the properties, advantages, and disadvantages of each scheme. Criteria used to compare different strategies should be selected in light of the specific applications of the data anticipated during an IREA. These criteria may include whether statistically unbiased regional estimators of the sampled objects are needed, the level of detail required to describe spatial variability, and trade-offs between statistical theory, efficiency, cost effectiveness, and other factors (Table 7.1).

7.6 Detection of Rarities

The problem of detecting rare ecological elements is an important topic in IREAs for three reasons. First, rarity is a widely used criterion in designing conservation strategies. The detection of rarities during the data collection phase of an IREA is a desirable feature that can help land managers meet legal requirements (e.g., threatened and endangered species management, designing representative research natural areas). Second, biotic–abiotic relationships need to be characterized to provide land managers with the ability to predict the response of an ecosystem to various management scenarios (Bourgeron et al., 1994b; also see Chapter 22). When rare ecological elements of a landscape are not identified, land units supporting them may be assigned to other known elements. This misidentification of a biotic–abiotic relationship may lead to erroneous ecosystem response predictions at the landscape level. Third, landscape configurations change over time. Ecosystems that have restricted extent today may become more extensive in the future. It is crucial to identify all segments of the ranges of biotic and abiotic variability, whether common or rare. Only then can all aspects of the natural variability at landscape and regional levels be considered for management and conservation planning.

Gillison and Brewer (1985) argued that gradsect design is more efficient than random and systematic design in recovering rare elements of an ecological pattern, because gradsects can recover significantly more patterns at finer scales of distribution, increasing the likelihood of locating rarities. Two other reasons make SSRSs and MSSRSs more likely to locate rarities. First, because physical environments are defined and mapped, less common environmental combinations are clearly identified and located (Austin and Heyligers, 1989, 1991; Bourgeron et al., 1994a; Engelking et al., 1994; Bunce et al., 1996a). Any biotic rarity associated with such environments will be included in a survey. Second, SSRSs and MSSRSs require intensive study of the range of environmental variability in an area, and this leads to a more thorough examination in the field. Hence the probability of locating rare ecological elements is increased. Mills et al. (1996) detected 3 rare bat species out of a total of 13 species recorded using MSSRS.

7.7 Sample Size and Configuration

Once sample sites have been selected, the size and configuration of samples become critical components of standardized sampling methods (Legendre and Legendre, 1998). Sampling at multiple scales within a site allows quantification of the influence of spatial scale on patterns such as local species diversity and provides better multiscaled analyses of community composition and diversity (Whittaker, 1977; Fortin et al., 1989; Podani et al., 1993; Reed et al., 1993; Stohlgren et al., 1995; Bellehumeur and Legendre, 1998). A number of nested survey designs have been used to measure spatially structured variation in a variety of organisms, including tropical trees and mollusks (Bellehumeur and Legendre, 1998; Legendre and Legendre, 1998). Results indicate that it is useful to collect samples at two or more different scales of observation, making possible the detection of nested structures (Bellehumeur and Legendre, 1998; Legendre and Legendre 1998). This approach can easily be integrated into SSRSs and MSSRSs. Finally, it would be advantageous to use preliminary data to test the impact of sample size and configuration on the results for a specific application. For example, Fortin (1999) showed that quadrat shape and orientation did not significantly affect the detection of boundaries in a forested landscape, although they might have such an effect in other data sets.

Some multiscale sampling methods, such as the 20 by 50 m Whittaker plots widely used in vegetation studies (Stohlgren et al., 1995), have flaws, such as suboptimal plot shapes, nested plots with different shapes at different sizes, and nonindependence of data where successively larger plots are superimposed on smaller plots, that reduce their effectiveness in detecting patterns (Stohlgren, 1994; Stohlgren et al., 1995). Stohlgren et al. (1995) introduced a modified-Whittaker vegetation sampling method that avoids these problems and is rapid and cost effective. The sampling method includes 1, 10, and 100-m^2 subplots within a 20 by 50 m (1000 m^2) plot, as in the Whittaker method, but all plots have the same rectangular shape and there is minimal plot overlap. The modified-Whittaker plot design produced significantly higher species diversity values than the Whittaker plot in forest and prairie vegetation in Colorado and South Dakota (Stohlgren et al., 1995). The modified-Whittaker method was also compared with several rangeland sampling methods in shortgrass, mixed grass, and tallgrass prairie vegetation in the central United States (Stohlgren et al., 1998). Parker transects, Daubenmire transects, and a large-quadrat design proposed by the USDA Agricultural Research Service all significantly underestimated total plant species richness and the number of native species compared with the modified-Whittaker method. In addition, the two transect methods missed half the exotic species and captured only 36% to 66% of the species detected by the modified-Whittaker method. The modified-Whittaker method was found to work equally well in sampling large or small vegetation patches (Stohlgren et al., 1997). The size of the plots can be adjusted (e.g., from 20 by 50 m to 10 by 25 m) to accommodate sampling in small patches (Stohlgren et al., 1997).

7.8 Interpolation and Extrapolation from Samples

Data collection efforts in most large areas are unlikely to be complete due to prohibitive costs. Therefore, it is often necessary to use models to interpolate or extrapolate the survey results to areas that have not been sampled. Predictive statistical techniques, such as generalized linear models (GLMs) (McCullagh and Nelder, 1989), generalized additive models (GAMs) (Hastie and Tibshirani, 1990), and classification and regression trees (CARTs) (Breiman et al., 1984), are powerful and flexible tools for generating quantitative relationships between species or communities and environmental factors (driving variables) using location-specific data sets (Nicholls, 1989, 1991a, b; Yee and Mitchell, 1991; Franklin 1998). Such models can be used to predict biotic distributions in unsampled areas if the environmental characteristics of these areas are known or can be derived from maps or simulation models. The reliability of spatial prediction is increased when the frequency of observations is evenly distributed across the environmental space (Nicholls, 1989).

Data collection should include explicit consideration of how well the full range of environmental factors is sampled when the use of statistical models for prediction to unsampled areas is anticipated. The SSRS and MSSRS schemes, including gradsects, with their purpose of covering the full range of biotic variability over the range of environmental variability, can generate the required data. Logistical constraints may make complete coverage unattainable, but the evenness of coverage determines the extent to which the prediction is an interpolation within the range of the data, rather than an extrapolation outside its range (Margules and Stein, 1989). The statistical properties needed to implement the data for a specific application will constrain the selection of a sampling strategy (see Chapter 6 and Section 7.4).

Another example of data extrapolation is He et al.'s (1998) integration of TM data and FIA data in a GIS environment (see Section 7.3.4) to represent the distribution of dominant tree species by age classes and their associated species.

7.9 Conclusions

Several attributes of IREAs restrict the use of classical sampling design methodologies: (1) the assessment areas are large and accessibility may be restricted in some locations; (2) data need to be collected for multiple components of ecosystems at multiple spatial scales; (3) data are used for multiple purposes; and (4) strict deadlines and budgetary and political constraints limit the resources and time allocated to de novo sampling design and field survey. IREA data collection must focus on the multiscale spatial heterogeneity of multiple ecosystem components, in contrast to the purpose of classical sampling design. In this chapter, we have reviewed and discussed four elements of data collection for IREAs: their general data requirements, appropriate standard data collection procedures, the use of existing data, and the trade-offs among statistical theory, logistical benefits, and efficiency. Identification of specific sampling strategies, including the number, shape, and configuration of samples, arises from the purpose of the IREA and from cost-effectiveness criteria. The following are general practical guidelines for the data collection phase of IREAs:

1. Define the scale and the purpose of the survey as clearly as possible. This will determine the grain, interval, and extent of the sampling design. The IREA characterization defined by the sampling extent may differ from the initial assessment area.
2. Review existing data and use them where possible for analysis and as templates for designing new surveys. Recognize that the gain realized by establishing databases for existing data may very well offset the cost.
3. Consider the potential of SS schemes, including SSRSs and MSSRSs, as cost-effective and efficient sampling designs. Systematic comparison of the pros and cons of simple RSs versus SSs should be conducted in the light of specific applications.
4. If an SS technique is used, match the scale of the stratifying variables with the scale and purpose of the IREA. Selection of inappropriate variables or scale of the data will diminish the efficiency of the design.
5. Choose sample size and configuration to detect nested spatial structures.
6. Assess the representativeness of the data collected and test the efficiency of the design, as appropriate.
7. Use suitable models (e.g., GLMs, GAMs, and CARTs) or other methods to extrapolate the collected data.

Any of the references on the topics discussed in this chapter can provide additional sources of information. Good comparisons and discussions of the various sampling strategies are found in Neldner et al. (1995) and Neave et al. (1997). Further details on nested designs and references are available in Fortin et al. (1989), Bellehumeur and Legendre (1998), and Legendre and Legendre (1998).

7.10 References

Allen, T. H. F.; Hoekstra, T. W. 1992. The integration of ecological studies—comment. *Funct. Ecol.* 6:118–119.

Aspinall, R.; Veitch, N. 1993. Habitat mapping from satellite imagery and wildlife survey data using a Bayesian modeling procedure in a GIS. *Photogramm. Eng. Remote Sensing* 59:537–543.

Austin, M. P. 1987. Models for the analysis of species response to environmental gradients. *Vegetatio* 69: 35–45.

Austin, M. P. 1991a. Vegetation: data collection and analysis. In: Margules, C. R.; Austin, M. P., eds. *Nature conservation: cost effective biological surveys and data analysis.* Melbourne: Commonwealth Scientific and Industrial Research Organization: 37–41.

Austin, M. P. 1991b. Vegetation theory in relation to cost-efficient surveys. In: Margules, C. R.; Austin, M. P., eds. *Nature conservation: cost effective biological surveys and data analysis.* Melbourne: Commonwealth Scientific and Industrial Research Organization: 17–22.

Austin, M. P.; Adomeit, E. M. 1991. Sampling strategies costed by simulation. In: Margules, C. R.; Austin, M. P., eds. *Nature conservation: cost effective biological surveys and data analysis.* Melbourne: Commonwealth Scientific and Industrial Research Organization: 167–175.

Austin, M. P.; Heyligers, P. C. 1989. Vegetation survey design for conservation: gradsect sampling of forests in north-eastern New South Wales. *Biol. Conserv.* 50:13–32.

Austin, M. P.; Heyligers, P. C. 1991. New approach to vegetation survey design: Gradsect sampling. In: Margules, C. R.; Austin, M. P., eds. *Nature conservation: cost effective biological surveys and data analysis.* Melbourne: Commonwealth Scientific and Industrial Research Organization: 31–36.

Austin, M. P.; Cunningham, R. B.; Good, R. B. 1983. Altitudinal distribution in relation to other environmental factors of several Eucalypt species in southern New South Wales. *Aust. J. Ecol.* 8:169–180.

Austin, M. P.; Cunningham, R. B.; Fleming, P. M. 1984. New approaches to direct gradient analysis using environmental scalars and statistical curve fitting procedures. *Vegetatio* 55:11–27.

Austin, M. P.; Nicholls, A. O.; Margules, C. R. 1990. Measurement of the realized qualitative niche: environmental niches of five Eucalyptus species. *Ecol. Monogr.* 60:161–177.

Austin, M. P.; Pausas, J. G.; Nicholls, A. O. 1996. Patterns of tree species richness in relation to environment in southeastern New South Wales. *Australia. Aust. J. Ecol.* 21:154–164.

Belbin, L.; Austin, M. P. 1991. ECOSIM—a simulation model for training in cost-effective survey methods. In: Margules, C. R.; Austin, M. P., eds. *Nature conservation: cost effective biological surveys and data analysis.* Melbourne: Commonwealth Scientific and Industrial Research Organization: 159–166.

Bellehumeur, C.; Legendre, P. 1998. Multiscale sources of variation in ecological variables: Modeling spatial dispersion, elaborating sampling designs. *Landscape Ecol.* 13:15–25.

Bian, L. 1997. Multiscale nature of spatial data in scaling up environmental models. In: Quattrochi, D. A.; Goodchild, M. F., eds. *Scale in remote sensing and GIS.* Boca Raton, FL: Lewis Publishers: 13–26.

Bissonette, J. A. 1997. Scale-sensitive ecological properties: Historical context, current meaning. In: Bissonette, J. A., ed. *Wildlife and landscape ecology.* New York: Springer-Verlag: 3–31.

Bourgeron, P. S.; Jensen, M. E. 1994. An overview of ecological principles for ecosystem management. In: Jensen, M. E.; Bourgeron, P. S., eds. *Volume II: ecosystem management: principles and applications.* PNW-GTR-318. Portland, OR: U.S. Dept. Agric., For. Serv., Pacific Northw. Res. Sta.: 45–67.

Bourgeron, P. S.; Humphries, H. C.; Jensen, M. E. 1994a. General sampling design considerations for landscape evaluation. In: Jensen, M. E.; Bourgeron, P. S., eds. *Volume II: ecosystem management: principles and applications.* PNW-GTR-318. Portland, OR: U.S. Dept. Agric., For. Serv., Pacific Northw. Res. Sta.: 109–120.

Bourgeron, P. S.; Humphries, H. C.; DeVelice, R. L.; Jensen, M. E. 1994b. Ecological theory in relation to landscape and ecosystem characterization. In: Jensen, M. E.; Bourgeron, P. S., eds. *Volume II: ecosystem management: principles and applications.* PNW-GTR-318. Portland, OR: U.S. Dept. Agric., For. Serv., Pacific Northw. Res. Sta.: 58–72.

Bourgeron, P. S.; Engelking, L. D.; Humphries, H. C.; Muldavin, E.; Moir, W. H. 1995. Assessing the conservation value of the Gray Ranch: rarity, diversity and representativeness. *Desert Plants* 11(2–3):1–68.

Bourgeron, P. S.; Humphries, H. C.; Barber, J. A.; Turner, S. J.; Jensen, M. E.; Goodman, I. A. 1999. Impact of broad- and fine-scale patterns on regional landscape characterization using AVHRR-derived land cover data. *Ecosystem Health* 5:234–258.

Breiman, L.; Friedman, J.; Olshen, R.; Stone, C. 1984. *Classification and regression trees.* Belmont, CA: Wadsworth.

Bunce, R. G. H.; Barr, C. J.; Clarke, R. T.; Howard, D. C.; Lane, A. M. J. 1996a. Land classification for strategic ecological survey. *J. Environ. Manage.* 47:37–60.

Bunce, R. G. H.; Barr, C. J.; Gillespie, M. K.; Howard, D. C. 1996b. The ITE land classification: Providing an environmental stratification of Great Britain. *Environ. Monit. Assess.* 39:39–46.

California Department of Fish and Game (CDFG). 1995. Special plants list. In: *Natural diversity data base.* Sacramento, CA: CDFG.

Cao, C.; Lam, N. S. N. 1997. Understanding the scale and resolution effects in remote sensing and GIS. In: Quattrochi, D. A.; Goodchild, M. F., eds. *Scale in remote sensing and GIS.* Boca Raton, FL: Lewis Publishers: 57–72.

Christian, C. S.; Stewart, G. A. 1953. *General report on survey of the Katherine–Darwin region, 1946. Land Research Series 1.* Melbourne: Commonwealth Scientific and Industrial Research Organization.

Dale, V. H.; O'Neill, R. V. 1999. Tools to characterize the environmental setting. In: Dale, V. H.; English, M. R., eds. *Tools to aid environmental decision making.* New York: Springer-Verlag: 62–93.

Davis, F. W. 1995. Information systems for conservation research, policy, and planning. *BioScience* 45:36–42.

Davis, F. W.; Stoms, D. M.; Estes, J. E.; Scepan, J.; Scott, J. M. 1990. An information systems approach to the preservation of biological diversity. *Int. J. Geogr. Inf. Syst.* 4(1):55–78.

Davis, F. W.; Quattrochi, D. A.; Ridd, M. K.; Lam, N. S. N.; Walsh, S. J.; Michaelsen, J. C.; Franklin, J.; Stow, D. A.; Johannsen, C. J.; Johnston, C. A. 1991. Environmental analysis using integrated GIS and remotely sensed data: some research needs and priorities. *Photogramm. Eng. Remote Sensing* 57:689–697.

Debinski, D. M.; Brussard, P. F. 1992. Biological diversity assessment in Glacier National Park, Montana: I. sampling design. In: McKenzie, D. H.; Hyatt, D. E.; McDonald, V. J., eds. *Proceedings of the international symposium on ecological indicators.* Essex, UK: Elsevier Publishing: 393–407.

Debinski, D. M.; Humphrey, P. S. 1997. An integrated approach to biological diversity assessment. *Nat. Areas J.* 17(4):355–365.

Engelking, L. D.; Humphries, H. C.; Reid, M. S.; DeVelice, R. L.; Muldavin, E. H.; Bourgeron, P. S. 1994. Regional conservation strategies: assessing the value of conservation areas at regional scales. In: Jensen, M. E.; Bourgeron, P. S., eds. *Volume II: ecosystem management: principles and applications.* PNW-GTR-318. Portland, OR: U.S. Dept. Agric., For. Serv., Pacific Northw. Res. Sta.: 208–224.

Estes, J. E.; Mooneyhan, D. W. 1994. Of maps and myths. *Photogramm. Eng. Remote Sensing* 60:517–524.

Fortin, M. J. 1994. Edge detection algorithms for two-dimensional ecological data. *Ecology* 75:956–965.

Fortin, M. J. 1999. Effects of quadrat size and data measurement on the detection of boundaries. *J. Veg. Sci.* 10:43–50.

Fortin, M. J.; Drapeau, P.; Legendre, P. 1989. Spatial autocorrelation and sampling design in plant ecology. *Vegetatio* 83:209–222.

Franklin, J. 1998. Predicting the distribution of shrub species in southern California from climate and terrain-derived variables. *J. Veg. Sci.* 9:733–748.

Franklin, J.; Woodcock, C. E. 1997. Multiscale vegetation data for the mountains of Southern California: spatial and categorical resolution. In: Quattrochi, D. A.; Goodchild, M. F., eds. *Scale in remote sensing and GIS.* Boca Raton, FL: Lewis Publishers: 141–168.

Gauch, H. G. 1982. *Multivariate analysis in community ecology.* Cambridge, UK: Cambridge University Press.

Gillison, A. N.; Brewer, K. R. W. 1985. The use of gradient directed transects or gradsects in natural resource survey. *J. Environ. Manage.* 20:103–127.

Godron, M.; Forman, R. T. T. 1983. Landscape modification and changing ecological characteristics. In: Mooney, H. A.; Godron, M., eds. *Disturbance and ecosystems: components of response.* New York: Springer-Verlag: 12–28.

Gravenmier, R. A.; Wilson, A. E.; Steffenson, J. R. 1997. Information system development and documentation. In: Quigley, T. M.; Arbelbide, S. J., eds. *Assessment of ecosystem components in the interior Columbia basin and portions of the Klamath and Great Basins: Volume II.* PNW-GTR-405. Portland, OR: U.S. Dept. Agric., For. Serv., Pacific Northw. Res. Sta.: 2011–2067.

Green, R. H. 1979. *Sampling design and statistical methods for environmental biologists.* New York: John Wiley & Sons.

Hahn, J. T.; Hansen, M. H. 1985. *Data bases for forest inventories in the North-central region.* NC-GTR-101. St. Paul, MN: U.S. Dept. Agric., For. Serv., N. Central Res. Sta.

Hann, W. J.; Jones, J. L.; Karl, M. G.; Hessburg, P. F.; Keane, R. E.; Long, D. G.; Menakis, J. P.; McNicol, C. H.; Leonard, S. G.; Gravenmier, R. A.; Smith, B. G. 1997. Landscape dynamics of the basin. In: Quigley, T. M.; Arbelbide, S. J., eds. *An assessment of ecosystem components in the interior Columbia basin and portions of the Klamath and Great Basins: Volume II.* PNW-GTR-405. Portland, OR: U.S. Dept. Agric., For. Serv., Pacific Northw. Res. Sta.: 337–1055.

Hansen, M. H.; Frieswyk, T.; Glover, J. F.; Kelly, J. F. 1992. *The eastwide forest inventory data base: user's manual.* NC-GTR-151. St. Paul, MN: U.S. Dept. Agric., For. Serv., N. Central Res. Sta.

Hargis, C. D.; Bissonette, J. A.; David, J. L. 1998. The behavior of landscape metrics commonly used in the study of habitat fragmentation. *Landscape Ecol.* 13:167–168.

Hastie, J. J.; Tibshirani, R. J. 1990. *Generalized additive models.* London: Chapman and Hall.

He, H. S.; Mladenoff, D. J. 1999. Dynamics of fire disturbance and succession on a heterogeneous forest landscape: a spatially explicit and stochastic simulation approach. *Ecology* 80:81–99.

He, F.; Legendre, P.; Bellehumeur, C.; LaFrankie, J. V. 1994. Diversity pattern and spatial scale: a study of a tropical rain forest of Malaysia. *Environ. Ecol. Stat.* 1:265–286.

He, H. S.; Mladenoff, D. J.; Radeloff, V. C.; Crow, T. R. 1998. Integration of GIS data and classified satellite imagery for regional forest assessment. *Ecol. Appl.* 8(4):1072–1083.

Helman, C. 1983. *Inventory analysis of southern New South Wales rainforest vegetation.* M.S. thesis. Biddeford, ME: University of New England.

Holm, T. M. 1997. National satellite land remote sensing data archive, technical note. *Photogramm. Eng. Remote Sensing* 63(10):1180.

Host, G. E.; Polzer, P. L.; Mladenoff, M. A.; White, M. A.; Crow, S. J. 1996. A quantitative approach to developing regional ecosystem classifications. *Ecol. Appl.* 6:608–618.

Jenkins, R. 1985. Information methods: why the heritage programs work. *Nature Conservancy* 35(6):21–23.

Jensen, M. E.; Hann, W.; Keane, R. E.; Caratti, J.; Bourgeron, P. S. 1994. ECODATA—A multiresource database and analysis system for ecosystem description and evaluation. In: Jensen, M. E.; Bourgeron, P. S., eds. *Volume II: ecosystem management: principles and applications.* PNW-GTR-318. Portland, OR: U.S. Dept. Agric., For. Serv., Pacific Northw. Res. Sta.: 192–205.

Jensen, M.; Goodman, I.; Brewer, K.; Frost, T.; Ford, G.; Nesser, J. 1997. Biophysical environments of the basin. In: Quigley, T. M.; Arbelbide, S. J., eds. *An assessment of ecosystem components in the interior Columbia basin and portions of the Klamath and Great Basins: Volume II.* PNW-GTR-405. Portland, OR: U.S. Dept. Agric., For. Serv., Pacific Northw. Res. Sta.: 99–314.

Jensen, M. E.; McGarvey, T.; Bourgeron, P.; Andreasen, J.; Goodman, I. 1998. ECADS—a multi-resource database and analytical system for ecosystem classification and mapping. In: *North American science symposium, toward a unified framework for inventory and monitoring forest ecosystem resources*; November 1–6, 1998, Guadalajara, Mexico.

Kapner, W. G.; Jones, K. B.; Chaloud, D. G.; Wickham, J. D.; Riitters, K. H.; O'Neill, R. V. 1995. *Mid-Atlantic landscape indicators. Project plan ecological monitoring and assessment program 620/R-95/003.* Washington, DC: U.S. Environ. Protect. Agency.

Larsen, D. M.; Faundeen, J. L.; Madigan, M. E.; Austad, J. K. 1998. *National satellite land remote sensing data archive Landsat multispectral scanner data CD-ROM. Proceedings of the conference on human interactions with the environment—perspectives from space.* August 20–22, 1996, Sioux Falls, SD. Bethesda, MD: Amer. Soc. Photogramm. Remote Sensing.

Leathwick, J. R. 1995. Climatic relationships of some New Zealand forest tree species. *J. Veg. Sci.* 6:237–248.

Leathwick, J. R. 1998. Are New Zealand's *Nothofagus* species in equilibrium with their environment? *J. Veg. Sci.* 9:719–732.

Legendre, P.; Legendre, L. 1998. *Numerical ecology.* Amsterdam: Elsevier Science.

Levin, S. A. 1992. The problem of pattern and scale in ecology. *Ecology* 73:1942–1968.

Lobo, A.; Moloney, K.; Chic, O.; Chiariello, N. 1998. Analysis of fine-scale spatial pattern of a grassland from remotely-sensed imagery and field collected data. *Landscape Ecol.* 13(2):111–131.

Ludwig, J. A.; Tongway, D. J. 1995. Spatial organization of landscapes and its function in semi-arid woodlands, Australia. *Landscape Ecol.* 10:51–63.

Mack, E. L.; Firbank, L. G.; Bellamy, P. E.; Hinsley, S. A.; Veitch, N. 1997. The comparison of remotely sensed and ground-based habitat area data using species-area models. *J. Appl. Ecol.* 34(5):1222–1229.

Mackey, B. G. 1993. A spatial analysis of the environmental relations of rainforest structural types. *J. Biogeogr.* 20:303–336.

Margules, C. R.; Stein, J. L. 1989. Patterns in the distributions of species and the selection of nature reserves: an example from *Eucalyptus* forests in south-eastern New South Wales. *Biol. Conserv.* 50:219–238.

Margules, C. R.; Nicholls, A. O.; Austin, M. P. 1987. Diversity of *Eucalyptus* species predicted by a multivariable environmental gradient. *Oecologia (Berl.)* 71:229–232.

McCullagh, P.; Nelder, J. A. 1989. *Generalized linear models.* New York: Chapman and Hall.

Menke, J. W.; Davis, C.; Beesley, P. 1996. Rangeland assessment. In: *Assessments and scientific basis for management options. Sierra Nevada Ecosystem Project: Final report to Congress Vol. III.* Davis, CA: Centers for Water and Wildland Resources, University of California.

Mills, D. J.; Norton, T. W.; Parnaby, H. E.; Cunningham, R. B.; Nix, H. A. 1996. Designing surveys for microchiropteran bats in complex forest landscapes—a pilot study from south-east Australia. *For. Ecol. Manage.* 85:149–161.

Milne, B. T. 1994. Pattern analysis for landscape evaluation and characterization. In: Jensen, M. E.; Bourgeron, P. S., eds. *Volume II: ecosystem management: principles and applications.* PNW-GTR-318. Portland, OR: U.S. Dept. Agric., For. Serv., Pacific Northw. Res. Sta.: 121–134.

Minshall, G. W. 1994. Stream–riparian ecosystems: rationale and methods for basin-level assessments of management effects. In: Jensen, M. E.; Bourgeron, P. S., eds. *Volume II: ecosystem management: principles and applications.* PNW-GTR-318. Portland, OR: U.S. Dept. Agric., For. Serv., Pacific Northw. Res. Sta.: 143–167.

Mladenoff, D. J.; Host, G. E.; Boeder, J.; Crow, T. R. 1996. LANDIS: a spatial model of forest landscape disturbance, succession, and management. In: Goodchild, M. F.; Steyaert, L. T.; Parks, B. O.; Johnston, C.; Maidment, D.; Crane, M.; Glendining, S., eds. *GIS and environmental modeling: progress and research issues.* Fort Collins, CO: GIS World: 175–179.

Mladenoff, D. J.; Niemi, G. H.; White, M. A. 1997. Effects of changing landscape pattern and U.S.G.S. land cover data variability on ecoregion discrimination across a forest–agriculture gradient. *Landscape Ecol.* 12:379–396.

Morefield, J. D. 1994. *Updated supplement to endangered, threatened, and sensitive vascular plants of Nevada.* Carson City, NV: Nevada Natural Heritage Program.

Morefield, J. D.; Knight, T. A. 1991. *Endangered, threatened, and sensitive vascular plants of Nevada.* Carson City, NV: Nevada Natural Heritage Program.

Nanson, B.; Smith, N.; Davey, A. 1995. What is the British National geospatial database? *Proceedings of the AGI 1995 conference; Birmingham.* London, UK: Association for Geographic Information: 1.41–1.45.

Neave, H. M.; Norton, T. W.; Nix, H. A. 1996. Biological inventory for conservation evaluation I. design of a field survey for diurnal, terrestrial birds in southern Australia. *For. Ecol. Manage.* 85:107–122.

Neave, H. M.; Cunningham, R. B.; Norton, T. W.; Nix, H. A. 1997. Preliminary evaluation of sampling strategies to estimate the species richness of diurnal birds using Monte Carlo simulation. *Ecol. Model.* 95:17–27.

Neldner, V. J.; Crossley, D. C.; Cofinas, M. 1995. Using geographic information systems (GIS) to determine the adequacy of sampling in vegetation surveys. *Biol. Conserv.* 73(1):1–17.

Nicholls, A. O. 1989. How to make biological surveys go further with generalised linear models. *Biol. Conserv.* 50:51–75.

Nicholls, A. O. 1991a. Examples of the use of generalized linear models in analysis of survey data for conservation evaluation. In: Margules, C. R.; Austin, M. P., eds. *Nature conservation: cost effective biological surveys and data analysis.* Melbourne: Commonwealth Scientific and Industrial Research Organization: 54–63.

Nicholls, A. O. 1991b. An introduction to statistical modelling using GLIM. In: Margules, C. R.; Austin, M. P., eds. *Nature conservation: cost effective biological surveys and data analysis.* Melbourne: Commonwealth Scientific and Industrial Research Organization: 191–201.

O'Hara, K. L.; Latham, P. A.; Hessburg, P.; Smith, B. G. 1996. A structural classification for Inland Northwest forest vegetation. *W. J. Appl. For.* 11(3):97–102.

Ohmann, J. L.; Spies, T. A. 1998. Regional gradient analysis and spatial pattern of woody plant communities of Oregon forests. *Ecol. Monogr.* 68(2):151–182.

O'Neill, R. V.; Jones, K. B.; Riitters, K. H.; Wickham, J. D.; Goodman, I. A. 1994. *Landscape monitoring and assessment research plan.* Report 620/R-94/009. Washington, DC: U.S. Environ. Protect. Agency.

Orlóci, L. 1978. *Multivariate analysis in vegetation research.* The Hague, The Netherlands: Dr. W. Junk.

Pielou, E. C. 1974. *Population and community ecology.* New York: Gordon and Breach.

Podani, J.; Czárán, T.; Bartha, S. 1993. Pattern, area and diversity: the importance of spatial scale in species assemblages. *Abstracta Botanica* 17:37–51.

Pressey, R. L.; Nicholls, A. O. 1991. Reserve selection in the Western Division of New South Wales: development of a new procedure based on land system mapping. In: Margules, C. R.; Austin, M. P., eds. *Nature conservation: cost effective biological surveys and data analysis.* Melbourne: Commonwealth Scientific and Industrial Research Organization: 98–105.

Quattrochi, D. A.; Goodchild, M. F., editors. 1997. *Scale in remote sensing and GIS.* Boca Raton, FL: Lewis Publishers.

Quigley, T. M.; Arbelbide, S. J., editors. 1997. *An assessment of ecosystem components in the interior Columbia basin and portions of the Klamath and Great Basins: Volume I.* PNW-GTR-405. Portland, OR: U.S. Dept. Agric., For. Serv., Pacific Northw. Res. Sta.

Reed, R. A.; Peet, R. K.; Palmer, M. W.; White, P. S. 1993. Scale dependence of vegetation–environment correlations: a case study of a North Carolina piedmont woodland. *J. Veg. Sci.* 4:329–340.

Reid, M. S.; Bourgeron, P. S.; Humphries, H. C.; Jensen, M. E. 1995. *Documentation of the modeling of potential vegetation at three spatial scales using biophysical settings in the Columbia River basin.* Unpublished report prepared for the U.S. Dept. Agric., For. Serv. Boulder, CO: Western Heritage Task Force, The Nature Conservancy. On file with: Interior Columbia Basin Ecosystem Management Project, Walla Walla, WA.

Running, S. W.; Nemani, R. R.; Peterson, D. L.; Band, L. E.; Potts, D. F.; Pierce, L. L.; Spanner, M. A. 1989. Mapping regional forest evapotranspiration and photosynthesis by coupling satellite data with ecosystem simulation. *Ecology* 70:1090–1101.

Schneider, D. C. 1994. *Quantitative ecology—spatial and temporal scaling.* San Diego, CA: Academic Press.

Scott, J. M.; Jennings, M. D. 1997. *A description of the national gap analysis program.* http://www.gap.uidaho.edu/gap/AboutGAP/GapDescription/Index.htm.

Scott, J. M.; Jennings, M. D. 1998. Large-area mapping of biodiversity. *Ann. Missouri Botanical Garden* 85(1):34–47.

Scott, J. M.; Csuti, B.; Smith, K.; Estes, J. E.; Caicco, S. 1990. Gap analysis of species richness and vegetation cover: an integrated conservation strategy for the preservation of biological diversity. In: Kohn, K. A., ed. *Balancing on the brink: a retrospective on the Endangered Species Act.* Washington, DC: Island Press.

Scott, J. M.; Davis, F.; Csuti, B.; Noss, R.; Butterfield, B.; Groves, C.; Anderson, H.; Caicco, S.; D'Erchia, F.; Edwards, T. C.; Ulliman, J.; Wright, R. G. 1993. Gap analysis: a geographic approach to protection of biological diversity. *Wildlife Monogr.* 123:1–41.

Shevock, J. R. 1996. Status of rare and endemic plants. In: *Assessments and scientific basis for management options. Sierra Nevada Ecosystem Project: Final report to Congress. Volume II.* Davis, CA: Centers for Water and Wildland Resources, University of California: 691–706.

Sierra Nevada Ecosystem Project (SNEP). 1996. *Final report to Congress, assessments and scientific basis for management options.* Davis, CA: Center for Water and Wildlands Resources, University of California.

Skinner, M. W.; Pavlik, B. M., editors. 1994. *Inventory of rare and endangered vascular plants of California. Special Publication 1*, 5th ed. Sacramento, CA: California Native Plant Society.

Southern Appalachian Man and the Biosphere (SAMAB). 1996. *The Southern Appalachian assessment summary report.* Atlanta, GA: U.S. Dept. Agric., For. Serv., Southern Region.

Stohlgren, T. J. 1994. Planning long-term vegetation

studies at landscape scales. In: Powell, T. M.; Steele, J. H., eds. *Ecological time series.* New York: Chapman and Hall: 209–241.

Stohlgren, T. J.; Falkner, M. B.; Schell, L. D. 1995. A modified-Whittaker nested vegetation sampling method. *Vegetatio* 117:113–121.

Stohlgren, T. J.; Chong, G. W.; Kalkhan, M. A.; Schell, L. D. 1997. Rapid assessment of plant diversity patterns: a methodology for landscapes. *Environ. Monit. Assess.* 48:25–43.

Stohlgren, T. J.; Bull, K. A.; Otsuki, Y. 1998. Comparison of rangeland vegetation sampling techniques in the Central Grasslands. *J. Range Manage.* 51:164–172.

Stoms, D. 1991. *Mapping and monitoring regional patterns of species richness from geographic information.* Ph.D. dissertation. Santa Barbara, CA: Department of Geography, University of California.

Stoms, D. M.; Estes, J. E. 1993. A remote sensing research agenda for mapping and monitoring biodiversity. *Int. J. Remote Sens.* 14(10):1839–1860.

Treweek, J. 1999. *Ecological impact assessment.* Oxford, UK: Blackwell Science.

Turner, M. G.; Gardner, R. H.; O'Neill, R. V.; Pearson, S. M. 1994. Multiscale organization of landscape heterogeneity. In: Jensen, M. E.; Bourgeron, P. S., eds. *Volume II: ecosystem management: principles and applications.* PNW-GTR-318. Portland, OR: U.S. Dept. Agric., For. Serv., Pacific Northw. Res. Sta.: 73–79.

U.S. Executive Office of the President. 1994. *Coordinating geographic data acquisition and access: the national spatial data infrastructure.* Executive Order 12906. Washington, DC: Executive Office of the President.

Van Driel, N.; Loveland, T. 1996. The U.S. Geological Survey's land cover characterization program. *Proceedings of the National Center for Geographic Information and Analysis (NCGIA) international conference/workshop on integrating GIS and environmental modeling*, January 21–25, 1996, Santa Fe, NM. Santa Barbara, CA: National Center for Geographic Information and Analysis. (CD-ROM)

Walker, D. A.; Halfpenny, J. C.; Walker, M. D.; Wessman, C. A. 1993. Long-term studies of snow vegetation interactions: a hierarchic geographic information system helps examine links between species distributions and regional patterns of greenness. *BioScience* 43:287–301.

Wessels, K. J.; Van Jaarsveld, A. S.; Grimbeek, J. D.; Van der Linde, M. I. 1998. An evaluation of the gradsect biological survey method. *Biodivers. Conserv.* 7:1093–1121.

Wessman, C. A.; Asner, G. P. 1998. Ecosystems and problems of measurement at large spatial scales. In: Pace, M. L.; Groffman, P. M., eds. *Successes, limitations, and frontiers in ecosystem science.* New York: Springer-Verlag: 346–371.

Whittaker, R. H. 1977. Evolution of species diversity on land communities. *Evol. Biol.* 10:1–67.

Wiens, J. A. 1989. Spatial scaling in ecology. *Funct. Ecol.* 3:385–397.

Wilkinson, G. G. 1996. A review of current issues in the integration of GIS and remote sensing data. *Int. J. Geogr. Inf. Syst.* 10(1):85–101.

Wolter, P. T.; Mladenoff, D. J.; Host, G. E.; Crow, T. R. 1995. Improved forest classification in the Lake States using multitemporal Landsat imagery. *Photogramm. Eng. Remote Sensing* 61:1129–1143.

Yee, T. W.; Mitchell, N. D. 1991. Generalised additive models in plant ecology. *J. Veg. Sci.* 2:587–602.

8
Ecological Data Storage, Management, and Dissemination

Ray Ford

8.1 Overview

The rapid evolution of computing, storage, and network technologies has changed the traditional rules for ecosystem data storage and dissemination. Some of the most profound changes have been changes in user expectations: in the post-Web world, both internal and external users expect to be able to not only download data sets electronically, but also to browse through well-organized catalogs of data holdings to make their download selections in a logical and systematic fashion. Thus it is not enough to simply fill disks or CD jukeboxes with gigabytes of data; an organization somehow must effectively build meaningful on-line catalogs that describe its holdings if it is to match the realities of data dissemination with increasingly sophisticated user expectations.

Continuing advances in information technology have dramatically improved our ability to collect ecological data sets, process them to derive new data sets, and disseminate results. In particular, dissemination technologies such as CD-ROMs and the World Wide Web (WWW) have made it much easier to move data from one site to another. Yet major problems still remain, which extend across the entire spectrum of project life cycles and project management. Instead of focusing on collecting data and preparing reports, a project manager now must also face the difficulties of selecting and integrating desirable data sets from a multitude of available options, organizing the project's own ever-increasing internal information resources, and dealing with numerous new options for data dissemination. Before beginning any study, the manager must determine what direct or background data sets are logically available and whether they can be acquired and easily used. Once data processing begins, it is now relatively easy to execute programs that derive new data sets. However, it remains difficult to effectively catalog all the derived data sets that result, much less to record and catalog all the processing attributes used in each derivation. At dissemination time, it is relatively easy to make CDs or tapes or to post Web-accessible data sets; it is not as simple to organize all the data and information about the processing methodology used to create and interpret the data so that recipients can effectively reuse what is received.

In short, moving from an environment in which data creation, processing, and storage are all very expensive to one in which the same aspects are all relatively cheap obviously changes the concerns of data managers. Having an abundant infrastructure for data creation eliminates some problems, but it also changes the characteristics of some old problems and creates new ones. This chapter will summarize some of the current problems facing ecological data managers, explain how these problems have evolved in parallel with advancing technology, and describe the options emerging as possible solutions.

8.2 Background Issues

To understand the current state of the art in ecological data management, it is important to recognize the key technological advances that have brought us to this point. To provide a framework for discussion of the important issues, we first identify the superset of information technology issues associated with overall project management. Subsequently, we then classify issues of major practical importance, both in historical and more modern technological terms.

Project management issues include, in the broadest sense, a wide range of scientific and practical issues associated with actually conducting an ecological study at medium to large scales. We assume that the primary purpose of such studies is to use computer processing to perform some type of analysis on existing digital data sets, which in turn produces new *derived data sets* and ultimately produces final results that are (or are based on) *digital data products*. From this base follows a set of information-related project management concerns, such as what input data sets will be used and how they will be collected, digitized if necessary, and spatially coregistered (i.e., have elements in each data set matched to common spatial points of reference). A related set of concerns exists for the project's output data products. Data processing implies an additional set of concerns, including what computer hardware and software are used, what processing parameters are used with that hardware and software, and how appropriate facilities are obtained, maintained, and scheduled. Thus the concerns in a modern ecological study of even modest size must include both those pertinent to ecological and scientific issues and those pertinent to data management and processing. The latter are more traditionally associated with the domain of the centralized computer center. However, technological advances allowing computing resources to be widely distributed have also shifted the responsibility of addressing data management concerns to the users of the distributed computing facilities.

The problems associated with maintaining general data repositories are not new. A wide variety of literature is associated with this topic, starting with a vast body of work describing how to construct a general-purpose database, typically in conjunction with a specific database modeling technique or commercial database system. Examples of this work include standard texts such as Kroenke (1998) and McFadden et al. (1999). Also pertinent here is work describing general quality metrics for data repositories (Ballow and Tayi, 1999). Another common approach is to look specifically at the data management issues associated with a particular type of data set. Work related to the storage and manipulation of image data sets is particularly widespread, ranging from standard texts (Russ 1995) to entire conference proceedings (Rosenholm and Osterlund, 1997).

Only recently has work appeared that focuses on the specific requirements of maintaining a variety of specialized data sets to support ecosystem modeling and management. Examples include reports from the Sequoia 2000 (Stonebraker, 1994; Frew, 1996) and GEMS (Bruegge et al., 1995) projects. Both are large-scale research projects involving the design of information systems to support a diverse collection of data sets, researchers, and ecosystem analysis projects. Sequoia 2000 is notable for its goals of utilizing high-speed network capabilities to support data sets and users distributed across a wide geographic area (Stonebraker, 1994; Frew, 1996). GEMS is notable for placing user requirements at the heart of its design, rather than implementation efficiency or efficacy (Bruegge et al., 1995). Although the software system architectures developed in projects such as these provide interesting examples of how such systems can be structured, neither provides much detail on the complex issues of data management faced by laboratories and agencies with large legacy data collections. More pertinent in this context are systems that address the specific problems faced in integrating existing data sets and databases with expanded, more ambitious goals in modeling and information dissemination. Examples here include two ecosystem information systems designed with both data managers and data clients in mind (Ford et al., 1994; Cowan et al., 1996). These systems focus on allowing data managers to organize their existing data collections by constructing new data indices, rather than requiring the conversion of massive amounts of data to a new database format. The systems also utilize Web facilities to provide access to data collections through these indices. The former system is discussed in detail in Section 8.5.

If we evaluate the most recent literature, looking specifically for evidence of how recent technological advances have forced data managers to view the data management and dissemination problem in different terms, several common themes emerge. As a starting point for this discussion, it is helpful to first divide the set of issues into subclasses that reflect their importance in historical terms.

8.2.1 Data Processing

An obvious special class of concerns in any project is that associated with the actual processing, focusing on the details of the computing hardware and software used to implement a particular form of data analysis. In historical terms, computational resources were relatively scarce. Software for spatial data analysis was both highly constrained and highly nonstandard, existing primarily in the form of locally written, nonportable code that was designed to be executed on locally available, highly constrained hardware. As constraints on both hardware and software have receded due to technological advances, ingrained assumptions about the importance, role, and difficulty of data processing must be reassessed.

8.2.2 Data Storage

Another special class of concerns involves the distinction between storing data sets *on line* during processing versus storing them in more permanent form *off line*. Traditionally, on-line storage was very limited, so off-line storage had to be used. Unfortunately, off-line storage systems and corresponding media were expensive and very error prone. Furthermore, conversion between on- and off-line forms was both highly constrained and centrally controlled, that is, by the central computer center that owned and operated this type of rare equipment, such as nine-track tape drives. Much of the focus in data storage was thus on the specialized techniques used at a particular site to move data sets between on-line and off-line forms. As with processing, storage technology has advanced rapidly, to the point where these traditional concerns have faded in importance.

8.2.3 Data Dissemination

The third class of concerns is that of the preparation of digital data products suitable for use outside the originating organization. Traditionally, data dissemination was more or less treated as a data storage issue, in the sense that the key feature of data exchange was that of producing off-line media in a portable form. This focus on producing data on appropriate media often obscured the need to attend to the inclusion of all the data actually needed to facilitate reuse. Also, organizations were generally slow to recognize that their own interests were best served by maintaining their data resources in a manner that promoted both internal and external reuse. Data dissemination concerns all too often have been thought of as an extra burden imposed by external governmental or support organizations. As media and format issues have faded in importance, the real issues in both internal and external dissemination have become more prominent: how does an organization manage its information resources to facilitate both internal and external use as a standard business practice, rather than as a special, extra requirement?

Each of these classes of traditional data management issues shares a common theme; in the past, the relevant information-processing technology was expensive and therefore relatively scarce, with the result that most projects operated in what was assumed to be an information-technology-poor environment. Today's situation is the result of a relatively rapid shift to an information-technology-rich environment; data access, storage, and processing facilities are very cheap compared to even a few years ago. As project staff have broken free from the constraints that limited the amount of data that they could collect, generate, and store, their information resources have exploded in size and complexity. Thus concern about the careful management of a small, scarce information resource must be refocused on the effective organization and management of a large, rapidly growing resource.

However, the explosive growth in data resources is not uniform across all types of data. For example, manually collected data sets have been and are likely to continue to be relatively scarce, small in spatial scope, expensive to collect, and highly variable in quality. Because these factors limit the amount of information that can be made available, it is fairly easy to deal with data organization and management issues for manually collected data sets. On the other hand, technological advances have made remotely sensed digital data sets readily available. Similar advances have also made access to vast hardware and software resources the norm. The availability of raw data and processing resources provides analysts with ample opportunities to process data sets in a variety of ways to derive new data sets. Thus data collections are increasingly dominated by remotely sensed and derived data sets. The problem in managing the overall information resource is how to include information about the methodology used in these types of automated data capture and derivation.

8.3 Technical Advances and Problem Evolution

What now seems clear is that a set of interrelated technical advances has essentially redefined the information management and access environment in which we all live. Not surprisingly, corresponding fundamental changes in the information environment are affecting ecological analysis. We have moved from an environment relatively poor in information technology to one in which it is hard to avoid becoming overwhelmed by the available information and opportunities to process it. This change has occurred so rapidly that we have had little time to reflect on how the underlying technological changes affect the assumptions and practices of the past. It would be easy to assume that advances in computing power, software capabilities, networking, and data collection somehow solve all the "hard" problems in ecological data management. In fact, the technological advances do solve some problems, but they also change the nature of others and create new ones as well. Only by looking at the issues that persist through radical technological change can we hope to identify the really fundamental issues.

By reclassifying project management concerns, we can focus on what might be called today's more technology driven approach to ecosystem study. In this new classification, we find some sets of issues that have grown in importance to demand their own specialized subclass and some that persist, but in greatly reduced importance. There are only a few key issues that persist in importance relatively unchanged by technological advances.

8.3.1 Data Assembly

Today we exist in an environment in which a huge volume of digital data is available and in which the critical decision is that of which data sets to assemble together, rather than how to overcome the inherent difficulties in manual data collection. New data collection is still important, but data collection problems are often dwarfed by our opportunity (and need) to choose between several, dozens, or even hundreds of existing data sets that could augment or complement a particular study. For example, compare the researcher who yearns for a single reliable data set of key weather-related values with one who has the option of selecting one of literally hundreds of such data sets being collected daily across dozens of regions. The practical problem of finding and reading the data shifts to a more methodological problem of why to select a particular data set. In addition, the availability of data sets with consistent measurements across a large space can cause significant shifts in project goals. Whereas researchers were once inherently constrained to relatively small study areas, they now can consider larger areas or multiscale studies that address ecological issues at scales ranging from small plots of land to watersheds, regions, and larger.

8.3.2 Processing and Storage

Constraints on processing and storage resources have not disappeared, but they have been relaxed by orders of magnitude so that familiar practical limits have essentially disappeared. Whereas the scarce computing resources of yesterday tended to be rigidly controlled and rationed out to users by central support organizations, nowadays virtually all project staff, from the most senior researchers and analysts to the most junior clerks and data-entry technicians, control their own powerful desktop computer. Once it took several days to months for staff members to collect data sets as inputs, digitize them appropriately, schedule processing time, create processing software, and determine the execution parameters that would be used to produce a single derived data set. Today a single staff member can rapidly derive dozens of such data sets on a desktop computer without ever leaving his or her desk. He or she can use network facilities to collect the inputs, standard software to do the processing, and local on-line disk storage to hold the results. The availability of local on-line storage is particularly significant; it has exploded to the point where it is now feasible to place tens of gigabytes of on-line storage on individual desktop computers. Removable media systems are still in use, but the reliable, standard, and inexpensive systems of today bear little similarity to those of the past. In particular, read–write media such as Zip and Jaz disks or read-only media such as CD-ROMs make it easy to replicate and distribute data sets. In fact, the devices themselves can be easily moved and distributed; many are designed specifically to be carried around in a backpack. Thus, although computing hardware, media, and software are still not free, the problem has shifted from finding a single reliable option to selecting between numerous affordable alternatives.

8.3.3 Networking and Dissemination

With advances in networking and on-line data storage, as exemplified by the capabilities of the World Wide Web, the distinctions among data sets prepared specifically for on-line processing, off-line storage, or off-line dissemination have rapidly crumbled. As information resources in even modestly sized labs grow, the parent organizations must recognize that the work that supports data dissemination to external users, typically viewed as extra, is in fact the core of what the organization must do to use its own data effectively. Vast stores of data can be kept on line, and modern networks permit essentially instantaneous delivery of data in standard formats to either internal or external users. Thus the problems of external media and data formats are virtually eliminated. What remains as the primary constraint for potential users, both internal and external, is determining the content and context for the available data sets and assuring their appropriateness for project goals.

Whether phrased in terms of processing methodology or content and context, the problem is that a complication arises. As the information resource expands, it becomes increasingly difficult for the owner of the information resource to handle the methodological context effectively, but it also becomes more important that this context be maintained. Implicit ways of handling this information, such as using file-naming conventions or "normal practices," break down over time with staff turnover and expanding information resources. It becomes essential to somehow make the context information explicit as an inherent part of the attendant data sets. The idea of maintaining data about data, or *metadata*, in explicit form is hardly new. What is still fairly new is the idea of making metadata about digital data sets an inherent, distinguishable, and accessible part of these data sets. Ideally, the metadata have characteristics such as the following. First, metadata creation should be an explicit part of the data derivation and data import processes so that modelers and data managers will understand the full impact of their most common data management actions (i.e., actions that create new data sets). Second, metadata should be easily distinguishable from the core data set and accessible on line in a manner that is at least as easy as accessing the data set itself. Finally, metadata should be used to build content-based indices that are searchable over the Internet in a manner analogous to the now-familiar Web search. Together, these three characteristics can guarantee that metadata are reliable, readily available for index and search engines, and usable as the basis for Web-like queries that are far more content specific and thus yield a much higher quality of hits than the simple-text indexing most common today.

The concept of maintaining some form of metadata is as old as the scientific method's explicit focus on experimental methodology, but the implementation of this concept in our information-technology-rich environment has been spotty at best. Too often the creation of metadata about digital data sets has been driven by specific dissemination requirements, rather than recognition on the part of the deriving organization that their own interests are best served if they appropriately, routinely, and carefully document the context in which derivations are performed. In the simplest cases, limited plans for dissemination may result in minimal *data annotation*, by which information is included only in the form of written reports about and summaries of the derived data. Note that such annotations are typically isolated from the data sets themselves. At an intermediate level are explicit techniques for metadata collection that involve both annotation and a certain amount of data organization; the annotation is again separate from the data, but the organization of the collection of data sets provides implicit information about derivation methodology. In contrast, at the most ambitious level is a system that automatically collects and electronically catalogs all metadata needed to document the processing used throughout the entire study. For example, this best-practice system would provide the complete set of records necessary to allow an independent group to duplicate a study and confirm or refute its results. There is clearly a wide gap between normal and best possible practices. The development of a system for automatic metadata collection, verification, and cataloging is essential; as rapidly as new data sets may be derived on line, any nonautomated scheme for collecting and cataloging metadata is unlikely to meet even modest dissemination goals.

8.4 The Solution, by Analogy

To make the problem of trying to reuse data that contain neither explicit nor implicit (organizational) metadata more concrete, consider a simple Web dissemination scenario. Suppose that a data manager builds a Web site with only a single short page (say 30 lines long) containing actual links to ten data sets. The single page per site, short textual description, and small number of data sets involved all combine to provide an implicit context for potential users that provides them with some information about what they actually get when they download the site's data. Contrast this with a data collection that has grown to the point where the Web site must describe and link to a collection of 1000 datasets. It is not feasible to simply list or show 1000 links on a page; furthermore, a simple, monolithic list of 1000 entities is clearly inadequate to allow a potential user to find any specific item of interest. Although a user can rapidly download any of the 1000 alternatives, his or her problem is that of determining which alternative to pick. That is, how can a potential user easily distinguish two or more data sets with names or short descriptions that are similar?

Creating an effective Web distribution for a larger set of items requires addressing two concerns. First, it is critical to provide some organization of the entities into smaller groups. Second, this organization must reflect key characteristics of the entities, as represented by one or more metadata characteristics that highlight similarities and/or differences. That is, when organizing a small information resource, as is typical in an information-technology-poor environment, it is easy to overlook the importance of details such as how (or if) the metadata are made available and how the metadata and data entities are organized. However, as each data set is added, both the complexity of the organization and the need to provide effective organization increase. At some point, simple, implicit, or monolithic management schemes no longer suffice and must be replaced by more carefully planned hierarchical schemes. Currently, most ecosystem research and development laboratories have recognized how radical advances in basic computing, networking, and storage technology can effectively remove the low-level issues in entity storage, processing, and dissemination. However, the same labs have generally been somewhat slower to recognize the attendant problems with data organization. The simple ability to compute faster or to post Web pages for public access does not solve the problem of data dissemination; instead, rapid proliferation of and public access to information make the need for coherent organization all the more critical.

To continue the analogy, note that the problem of effectively managing a rapidly growing information resource to facilitate internal and external use is hardly new. This is essentially the same problem that has been faced by libraries ever since technological advances made it easy to cheaply and reliably produce books. Librarians learned very early on that simple, monolithic lists of books did not represent an organized collection. Instead, they began to organize collections based on multiple indices of information collected about the books, resulting in the familiar by title, by author, and by subject indices that virtually every library uses today. The information in these indices represents metadata about the collection. Organizing these metadata in multiple indices is important because no single index meets all obvious access requirements of a disparate collection of users. Projects that have attempted to modernize and computerize libraries in recent years have expanded these indices and made them accessible on line, permitting potential users to rapidly discern what information is available and how it can be found. The parallels with data dissemination should be obvious. The first and most important goal should be to collect the metadata systematically, organize it into one or more easily searchable forms, and then, if possible, put the results on line. Even if it is not feasible or desirable to put the actual data sets on line, the primary goal in dissemination is to tell potential users exactly what is available. This goal is best met by providing convenient access to systematically indexed metadata.

To complete the analogy with traditional libraries, the amount of primary material that is available to libraries in electronic or on-line from other repositories is rapidly growing through subscription CD-ROMs, government publications, and electronically published journals. The experience in libraries indicates that having more source material available on line increases the dependency on and need for accessible on-line indexing, as well as the need to index on more content-specific keys.

The message for laboratories and organizations that use and produce spatial data sets should be clear: simply making data sets available on line is not enough to help potential users select the correct data set or to understand the implications of the selection in terms of key spatial attributes, such as data location, source scale, or sampling error. Furthermore, simply translating free-form textual descriptions to on-line form and then autoindexing the descriptive free text is not an effective way to document spatial data holdings. Just as libraries have devised specific indices (name, author, subject) to support the types of queries that dominate their use patterns, effective indexing schemes must be developed to support spatial data resources.

Several major initiatives are in progress to address issues of spatial data indexing on a large scale, along with several other smaller projects designed to address specific technical issues. The sections that follow first review one such major initiative, the Federal Geographic Data Committee's (FGDC) series of related programs to define standards for metadata content and format, to link individual spatial information repositories into recognized data clearinghouses, and to link the holdings of these clearinghouses into content-specific indices accessible to normal Web search engines. Then the evolution of a smaller, more specialized data repository construction tool, the University of Montana Ecosystem Information System (EIS) (Righter and Ford, 1994) is described. EIS provides an example of how customized repositories can adapt and evolve to retain viability in the current Web-dominated data dissemination environment.

As stated on their Web home page (http://www.fgdc.gov), the primary purpose of the FGDC is to coordinate the development of the National Spatial Data Infrastructure (NSDI) through policies, standards, procedures, and initiatives. The FGDC is a collaborative effort supported by 16 federal agencies, along with a variety of other state, local, and tribal governments, universities, and private companies. In keeping with the general goal of facilitating access to information from a range of potential users, the FGDC maintains a very active Web site that offers details on their various programs and initiatives.

Two related FGDC initiatives are particularly pertinent to the discussion of spatial data set dissemination. The first is an ongoing initiative to define standards for the metadata associated with spatial data sets. Like any major effort in standards definition, the FGDC metadata standards program is broad based, with participation from a wide collection of agencies and users. It is also an evolving process, with public comment on one version incorporated as feedback in the refinement of new versions. The first draft standard was released in 1994, and a second draft was released in 1998 (FGDC, 1998). The standard is very broad; it is designed to address metadata issues in data sets ranging from those derived through traditional field studies, to remotely sensed data sets, to data sets derived from complex computer simulations. As a result, it is neither simple nor particularly easy for the first-time user to grasp. For example, the standard describes over 300 different elements, over 100 of which are compound elements used to group more elementary items of information. Although many of these items are optional, the sheer size of the standard itself can discourage potential users.

Given the complexity that is inherent in any such standardization effort, it is essential that tools be developed in parallel with the standard that supports both the creation of the metadata and the effective use of metadata databases through standard electronic access tools. For simplicity, I lump both these aspects together in describing the second FGDC initiative of interest, the clearinghouse program. The purpose of this program is first and foremost to use modern networking technology, in the form of standard Internet software, to promote the dissemination of spatial data sets. Although it is desirable to make the data sets themselves available over the network whenever feasible, the key focus is on providing a standard type of data-set catalog that can be shared electronically over the Internet. With respect to the traditional library example discussed earlier, the idea is to create a flexible, multiindex catalog that can be used by a very broad range of data resource holders to describe their collections; it would be published freely and easily as part of normal Internet activity and effectively used by an even broader potential user community searching for data sets with particular characteristics.

The FGDC approach to meeting these very ambitious design requirements has evolved in concert with user expectations about how information should be located on the Internet, shaped by the rapid emergence of Web servers and browsers. The FGDC approach has four primary components. First is a database of indexable information, assumed to be metadata maintained on line in digital form and composed in accord with FGDC content standards. Second is a set of basic search protocols available to users, which are assumed to be the same standards used in modern libraries. One example here is the Z39.50 compliant indexing–querying systems that are standard in traditional library searches. Another example is the emerging pseudostandard represented by Web-based indexing–searching processes. Third, to demonstrate the viability and potential advantages of this approach, tools are provided to support an organization's ability to move through the entire process, from metadata creation, to indexing, to publishing the indices in standard Internet–Web accessible forms. Finally, simply making tools available is not enough; the effective use of the tools must be actively supported through training and demonstration programs for both data-set developers and users.

Generally, all these activities fall under the umbrella of the FGDC clearinghouse program. More specifically, the goal is to establish prominent sites whose collections are of significant public interest, hence assuring some user demand; whose managers have some commitment to support public access; and whose sites and collections can be linked and indexed according to Z39.50 and Web-index mechanisms. Concurrently, in an ongoing sequence of workshops, resource holders are shown the ways in which their collections can be described and advertised using such mechanisms. The hope is that, as the volume of information available in this manner grows, the public will become more aware of the quality of access that can be achieved for both the resource manager and the potential data user. For more details on activities related to the clearinghouse program, see the FGDC Web site noted earlier.

8.5 Ecosystem Information System

As a final example, we consider the ongoing evolution of other, more customized data-set classification mechanisms, as demonstrated by the University of Montana's Ecosystem Information System (EIS) (Ford et al., 1994; Righter and Ford, 1994). EIS allows spatial data-set managers to use object-oriented data design and specification techniques to

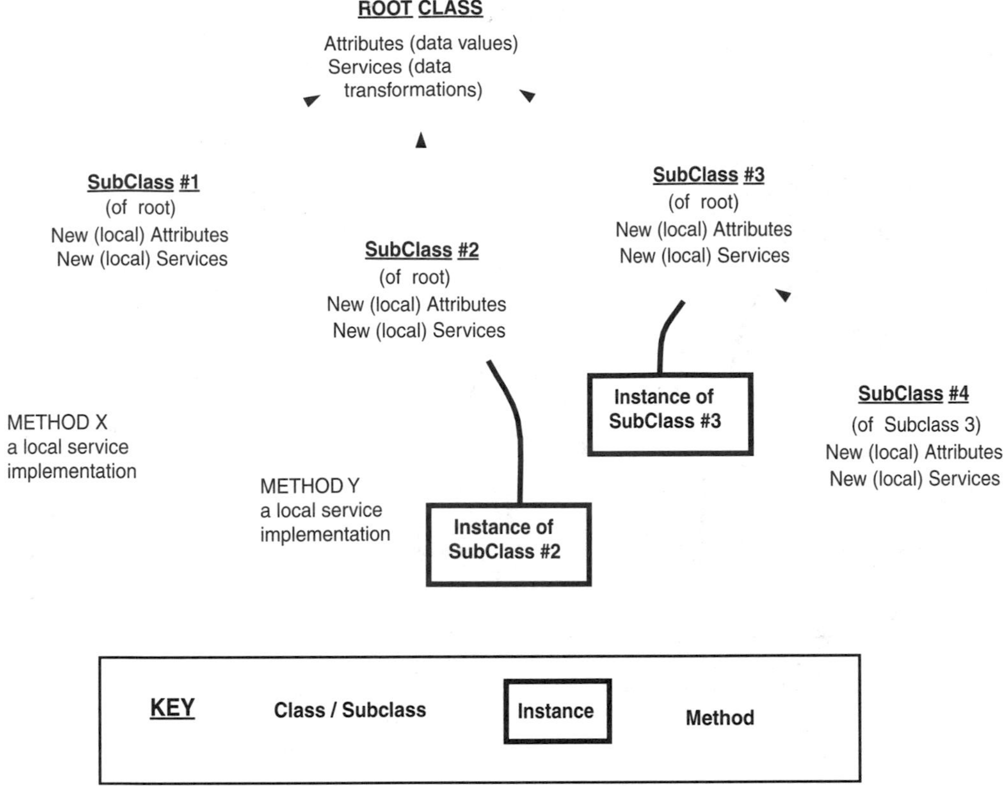

FIGURE 8.1. An object-oriented framework for data set–metadata hierarchical specifications.

build fairly traditional hierarchical classification schemes for their data holdings. Like the FGDC efforts, EIS makes metadata explicit and uses this information to organize data sets and publicize their existence on the Web. Whereas the FGDC efforts are based on use of a common, standard metadata framework, the EIS approach allows each organization (or subunits within a single organization) to devise unique classification hierarchies that best fit their data-set collections. However, EIS is similar to the FGDC efforts in that they both link the resulting framework to standard Web index and search techniques.

An example schematic for an EIS classification hierarchy is sketched in Figure 8.1. An EIS classification system first defines a set of *classes*, such that each class identifies a unique set of attributes shared by all data sets that fall within the class. Any class definition can be refined as its members evolve through the definition of *subclasses* that share all characteristics of the parent class, but also possess explicitly identified unique characteristics. Class attributes can describe either information components (data) or information-processing components in the form of data transformations called *methods*. A given class attribute can be *static*, meaning that each member of the class has the same data value or method, or *abstract*, meaning that the value or method is instance specific. Given a hierarchical framework of such class definitions, a collection of data sets is organized by identifying each data set as a member, or *instance*, of a particular class. The class hierarchy also provides a way to organize associated data-transforming programs, as instances of specific static or abstract methods. Finally, this technique allows various types of interclass relationships to be expressed, for example, by specifying that instances of a given class will always *contain* or *be associated with* one or more instances of another class. Booch provides a standard reference on object-oriented modeling techniques (Booch, 1994).

Figure 8.1 illustrates these techniques in general, showing a ROOT CLASS with SubClass #1, SubClass #2, and SubClass #3 as subclasses. SubClass #3 also has its own subclass, SubClass #4. Attributes and services common to all types of imagery are identified as properties of ROOT CLASS; at-

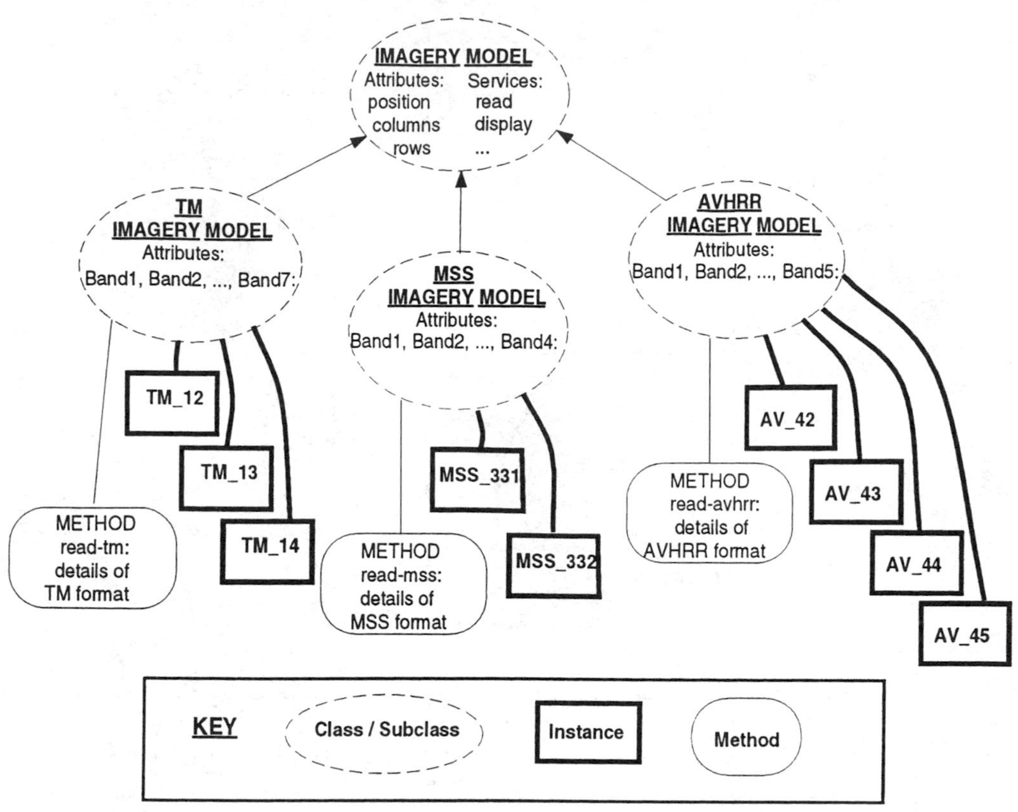

FIGURE 8.2. An example class hierarchy for remotely sensed imagery of various types.

tributes and services specific to different types of imagery are identified in the characteristics of the various subclasses. The class definitions form a framework that allows the specific characteristics of each data set to be deduced easily from the point of its attachment. Thus the characteristics of an instance of SubClass #2 are easily distinguishable from the characteristics of an instance of SubClass #1 or #3. Similarly, the specification of a service in a particular class is distinguished from its implementation in that class or a set of customized implementations in subclasses. For example, a common service can be defined in ROOT CLASS, yet implemented differently by method X for SubClass #1 and method Y for SubClass #2.

Figure 8.2 shows a more concrete example based on a simple framework describing various related but distinct types of remotely sensed imagery. The root class Imagery Model defines characteristics common to all types of remotely sensed imagery, such as attributes for the spatial location and size of the image, and services such as import and display. The subclasses TM Imagery Model, MSS Imagery Model, and AVHRR Imagery Model each define additional characteristics unique to these image types. With each of these subclasses are shown instances, or specific data sets, for which both local and inherited attribute values would be defined. Also attached to the subclasses are the specific implementations of methods, for example, the code to implement the common service to *read* an image would be customized for each type of data set to account for differing characteristics, such as number of bands, size, type, and layout of header information.

The attributes and services suggested in Figures 8.1 and 8.2 for imagery are obvious and match more or less directly with the information traditionally considered part of the data format. However, the object modeling technique is much more general than that and can also be used in the identification of the full range of metadata characteristics, whether encoded in the actual data-set header or not. The key to this approach is to incorporate the full range of metadata characteristics, independent of how this information is typically encoded. The implication is that, for a class of data sets whose set of identified characteristics is larger than the set of values encoded in a typical header file, an Import service must be broader than simply reading the header; for example, some other mechanisms must be used to obtain values for characteristics not defined within the header.

As conceived in the early 1990s, before Web tools dominated the Internet-based dissemination of information, EIS initially used basic facilities analogous to anonymous file transfer protocol (FTP) to support information exchange between designated EIS sites. However, by the time the implementation was actually stable enough for public distribution, it was clear that EIS would meet the fate shared by other niche data dissemination systems: no matter what advantages these systems might offer, the explosive growth in the popularity and ease of use of Web-based tools rapidly led the target user community to expect that information should be disseminated over the Internet with Web-based tools, not other customized software. As such, EIS development parallels other early efforts in building network-based data dissemination systems that were caught short by the rapid emergence of Web-based tools as de facto standards. Simply put, users quickly began to expect data to be found through the use of standard Web browsers and search engines, rather than with tools and interfaces customized for any other special-purpose system.

As a result, the designers of any custom network data repository now face three choices: simply abandon their efforts; continue to promote their system as an alternative to Web-based access; or reengineer their system to retain novel data classification and access components, but add some sort of standard Web-based access mechanism as a front end to the customized system. The last choice is obviously the most desirable, because it salvages the major part of the original effort. More importantly, this is also a way to provide access to information, such as the data repository's metadata, that is otherwise lacking in simple collections of Web pages or simple free text indexing. In the case of EIS, we have taken this approach of reengineering the interface. The newest version of EIS replaces the original, customized graphical user interface with a more familiar Web interface and uses Java mechanisms to link customized EIS databases and search mechanisms to standard Web-access techniques. This assures that user queries can be generated with familiar interfaces, yet be translated on the fly into appropriate EIS–Web searches. We also have been working to describe classification hierarchies consistent with FGDC metadata standards and for which FGDC indexing tools can also be used (Ford et al., 1997). In this respect, current EIS development attempts to complement, rather than compete with, both the FGDC clearinghouse program and the mass-market mechanisms in use on the Web. This complementary approach uses Web mechanisms for the basic interface associated with query formulation and display. It also uses Web-based protocols for information transfer. However, it interjects customized processing elements between the Web interface and final processing so that Web-formulated queries are directed to specific metadata-based indices, rather than generic Web indices. The use of specific metadata attributes adds significantly to the organization of the accessible information and provides considerable value to the searches that users formulate.

8.6 Summary and Conclusions

Our discussion traces the state of the art in data-set dissemination from the pre-Internet days, dominated by local issues of data format and storage, through early non-Web days, which saw the development of numerous customized access mechanisms for electronic data repositories, to the current Web-dominated environment. What has changed in recent years is the sort of mechanisms through which clients expect to be able to gain access to the information resources. What has not changed over time is the need for repository managers to effectively organize their ever-growing information resource collections. Thus the keys to maintaining effective systems for data dissemination are the following:

1. To accept standard Web facilities as the basic user interface and data-delivery element. It seems futile at this point to invest resources in offering software that competes with what has emerged as the de facto standard.
2. To focus local efforts first on building rich, content-specific descriptions of local repository contents. The key is to move beyond simple sets of Web pages into well-organized content-specific catalogs or indices based on the metadata associated with repository components.
3. For organizations with no preexisting catalogs and indices, to incorporate Web accessibility from the beginning as these catalogs and indices are built.
4. For organizations with significant existing investment in non-Web catalogs and indices, to focus attention on linking existing indices to Web-compatible forms, rather than converting or abandoning local systems. This allows an or-

ganization to make continued use of any prior investments in custom index development, particularly as represented in the training of staff in the development and use of local indices. Instead of investing resources in converting the indices to Web-based forms, which is expensive and time consuming, the organization can make more modest investments in developing dynamic *Web-link* processes that can build Web indices from custom indices and translate Web queries into custom queries. This helps avoid diverting resources from the all-important task of building the metadata database, yet guarantees that the result will still be accessible to both internal and external data clients.

8.7 References

Ballow, D.; Tayi, G. 1999. Enhancing data quality in data warehouse environments. *Commun. ACM* 42:73–78.

Booch, G. 1994. *Object-oriented analysis and design, with applications,* 2nd ed. Redwood City, CA: Benjamin/Cummings.

Bruegge, B.; Riedel, E.; Russell, A.; McRae, G. 1995. Developing GEMS: an environmental modeling system. *IEEE Comput. Sci. Eng.* 2:55–68.

Cowan, D.; Grove, T.; Mayfield, C.; Newkirk, R.; Swayne, D. 1996. An integrative information framework for environmental management and research. In: Goodchild, M., et al., eds. *GIS and environmental modeling: progress and research issues.* October 1993, Breckenridge, CO. Fort Collins, CO: GIS World Books: 423–427.

Federal Geographic Data Committee (FGDC). 1998. Content standard for digital geospatial metadata—2.0. Reston, VA: Federal Geographic Data Committee Secretariat, U.S. Geological Society. (Also available online at http://www.fgdc.gov.)

Ford, R.; Running, S.; Nemani, R. 1994. Large scale terrestrial ecosystem modeling. *IEEE Comput. Sci. Eng.* 1:32–44.

Ford, R.; Sweet, M.; Votava, P. 1997. An object-oriented database for cataloging, archiving, and disseminating spatial datasets and FGDC-compliant metadata. In: Rosenholm, D.; Osterlund, H., eds. *From producer to user: proceedings of 1997 ISPRS joint workshop*, October 1997. Boulder, CO: Int. Soc. Photogramm. Remote Sensing.

Frew, J. 1996. The Sequoia 2000 project. In: Goodchild, M., et al., eds. *GIS and environmental modeling: progress and research issues*. October 1993, Breckenridge, CO. Fort Collins, CO: GIS World Books: 69–72.

Kroenke, D. 1998. *Database processing: fundamentals, design, and implementation*, 6th ed. Upper Saddle River, NJ: Prentice Hall.

McFadden, F.; Hoffer, J.; Prescott, M. 1999. *Modern database management*, 5th ed. Reading, MA: Addison-Wesley.

Righter, R.; Ford, R. 1994. An object-oriented characterization of spatial ecosystem information. *Math. Comput. Modeling* 19:17–29.

Rosenholm, D.; Osterlund, H., editors. 1997. *From producer to user: proceedings of 1997 ISPRS joint workshop*. October 1997. Boulder, CO: Int. Soc. Photogramm. Remote Sensing.

Russ, J. 1995. *The image processing handbook*, 2nd ed. Boca Raton, FL: CRC Press.

Stonebraker, M. 1994. Sequoia 2000: a reflection on the first three years. *IEEE Comput. Sci. Eng.* 1:63–72.

9 Integration of Physical, Biological, and Socioeconomic Information

D. Scott Slocombe

9.1 Introduction

Ecological assessment is a process tool that supports a wide range of environmental management activities, from ecosystem-based management to impact assessment to ecological restoration. All these activities rely on an adequate information base for analysis, planning, and decision making; remember the old regional planners' rule "survey → analysis → plan." Environmental management activities are often linked in practice, but still may have different goals. Goals range from relatively specific and ecological, as in restoration, to broad and multifaceted, as in ecosystem-based management (cf. Slocombe, 1998). Different goals imply, to some degree, different kinds of needed information. But, even so, most environmental management activities draw on a very wide range of information from the biological, physical, and human (particularly mean social and economic) domains.

Integration of varied information is needed within ecological assessment for a range of reasons, some common and others depending on the contextual activity and its goals. It is a truism of environmental science and management that decisions based on partial knowledge usually create unexpected effects: integration of different kinds of knowledge can help to avoid this. Equally, in different ways, integration can support assessment, planning, and decision-making activities by helping to identify unexpected relationships within the ecosystem, fostering identification of a full range of possibilities, and providing full identification of the consequences in a range of realms of particular decisions (Table 9.1). This chapter focuses on the integration of basic information about the human, biological, and physical dimensions (or domains) of ecosystems at all scales to support the goals of assessment, planning, and management decision making. Emphasis is placed on assessment and integration of information as the basis of planning and management.

The precise meaning of integration is not a trivial question. In part, it is a technical or administrative task: putting information in common formats and within an integrated environmental information system so that information about different *things* can be overlaid and compared. Here such issues are discussed as steps on the way to a deeper meaning (see especially, Chapters 8, 11, and 12). The deeper meaning of information integration presents a conceptual challenge: finding ways to meaningfully link information about different dimensions of an ecosystem to provide new, fuller understanding of it, especially the connections and relationships between its biological, physical, and human dimensions (an entire *sociobiophysical* system: Grzybowski and Slocombe, 1988). A good place to start is Born and Sonzogni's (1995) characterization of integrated environmental management: comprehensive, interconnective, strategic, and interactive–coordinative.

Integrating information involves a threefold challenge: (1) integrating information about different kinds of things (e.g., geomorphology and human attitudes to development); (2) integrating information from different sources (e.g., scientists and citizens); and (3) integrating information that is in very different forms (e.g., quantitative data and qualitative survey results).

The quest for integration is not a new one. Many broadly environmental initiatives over the last hundred years have sought to integrate different aspects of the world around us, the views of different groups, and different kinds of information. These included early regional and watershed planning ef-

TABLE 9.1. Basic functions and data integration goals for representative environmental management activities.

	Basic functions	Data integration goal	Examples
Ecological assessment	Survey, analysis	Relating information on different ecosystem dimensions to produce more complete ecosystem understanding	Southern Appalachia Assessment, FEMAT, Bow Valley Task Force
Ecological restoration	Survey, design	Understanding ecological possibilities and human needs and desires for an ecosystem	Great Lakes Remedial Action Plans, Baltic Sea
Environmental impact assessment	Analysis, decision making	Understanding relationships among different ecosystem dimensions to better assess nature and significance of activity effects	Beaufort Sea EA, CARC Hudson's Bay Program,
Ecosystem-based management	Planning, decision making	Comparing varied human needs and expectation with ecological conditions to identify and pursue complex goals such as sustainability	Greater Yellowstone, Australian Alps, Adirondacks

FEMAT is Forest Ecosystem Management Assessment Team; CARC is Canadian Arctic Resources Committee.

forts in the 1920s and 1930s, integrated resource management from the 1960s onward, environmental impact assessment from the 1970s, adaptive environmental assessment and management from the late 1970s, and ecosystem management since the late 1980s (see, for example, Lang, 1986; Slocombe, 1993). But this kind of integration remains a challenge for all the technical, sociological, and philosophical reasons discussed in the following. Growing recognition of the need for and feasibility of integration is finally combining with a body of experience, tools, and methods to make integration more common and feasible.

In the next section, the basic conceptual challenge of information integration is defined in more detail. Then approaches to integration across disciplines and across space and time are discussed as the fundamentals on which most other approaches build; this is followed by a review of process, conceptual, and technical tools for integration. In conclusion, a summary is provided and current trends and their impact on data integration are assessed.

9.2 Data Sources and Integration Challenges

Integration presents three main challenges: integrating information about different things, in different formats, and from different sources. These are of course interrelated and share some common issues that will be highlighted at the end of this section; but first each will be examined in turn. Table 9.2 summarizes and partially distinguishes these challenges to information integration.

The most basic challenge is to integrate and relate information about different things, in this case an ecosystem's biological, physical, and human dimensions, or domains. In part, the challenge is a matter of integrating information from different disciplines, most obviously the natural and social sciences. Increasingly, however, ecological assessment and ecosystem-based management start with essentially multidisciplinary information about different domains. In addition, the amount of infor-

TABLE 9.2. Core issues for integrating different kinds of information.

Integration across domains	Integration across formats	Integration across sources
Identifying points of connection between domains	Different units Different accuracies and standards	Different variable definitions Different system characteristics considered significant
Certain–uncertain knowledge, predictive or not predictive, quantitative–qualitative	Different scales, resolution, and degrees of aggregation	Oral and written traditions Different views of cause and effect
Space and time variation generating new knowledge	Different hardware and software Different digital file formats	Information based on experience rather than experiment or formal study
Interdisciplinary communication		
Flexible, accessible products		Cross-cultural communication

mation is growing steadily, and much of it is increasingly complex. Consider, for example, environmental history information or ecological health assessments. The challenge here is relating this information not interdisciplinarily, but transdisciplinarily, as Erich Jantsch put it 25 years ago (Jantsch, 1971) and the ecological economists have more recently promoted (Costanza et al., 1991). Different domains and disciplines produce disparate kinds of knowledge. Most importantly, knowledge can be quantitative versus qualitative, historical versus ahistorical, specific–disaggregated versus general–aggregated, and certain or uncertain; to different degrees experimental, observational, theoretical, and/or experiential in origin.

Differences in format are often apparent between domains and disciplines. The most fundamental difference may be the division between quantitative and qualitative information. Also noteworthy are the presence or absence of spatial or temporal dimensions, issues regarding data volume, and issues regarding standards of certainty, reproducibility and statistical confidence. More specific differences may include technical and administrative definitions of variables collected, times and places collected, and formats for storing digital and paper time series and maps. Challenges to reconciliation vary accordingly. On the one hand, links must be made between qualitative and quantitative information, starting by accepting both as useful and valid in their own ways. More simply, approaches may require coordination and establishment of technical processes for format translation.

Differences also may stem from sources that embody fundamentally different assumptions and methods. The most common example is scientific knowledge and local, indigenous, or traditional knowledge (Inglis, 1993). Scientific knowledge is typically observational, quantitative, and experimentally testable, whereas traditional knowledge is typically historical, verbal, and mostly qualitative. Dealing with differences in source, we confront the basic issues around assumptions of cause and effect, repeatability, predictability, and quantifiability under which science has historically operated. These enter the fray in efforts as simple as integrating physical and biological knowledge, never mind social science. The fundamental challenges to reconciling source differences are as much cultural as technical and process tools at least as important as technical ones.

Thus the primary challenge is finding ways to integrate information in different formats, from different sources, so that information about different domains can be brought together in ways that a range of experts and decision makers find useful and acceptable—without making insupportable assumptions or fostering unjustifiable conclusions. This can entail a wide range of approaches: changing assumptions on the part of the analysts, developing information through participatory processes, coordinating processes for technical standardization, and combining conceptual and technical tools. All these are addressed next.

9.3 Information Integration Foundations

Two problems lie at the heart of the challenge to integrate information: integration across domains and disciplines and integration across space and time. Any substantial ecological assessment or ecosystem management exercise must achieve these forms of integration. Before addressing these kinds of integration more specifically, several prerequisites should be considered: defining integration products, understanding units and scales, and accepting both quantitative and qualitative information.

The type of product to emerge from information integration exercise depends on earlier definition of goals and activities (Table 9.3). Consider products of two sorts. One is a set of final products displaying information in particular ways or answering particular questions, for example, a simple atlas juxtaposing different kinds of information (Thomas et al., 1991) or a major technical report or environmental impact statement (Forest Ecosystem Management Association Team, 1993). Another type of product is an information system or set with which different users can work to create their own products and answer their own questions, whether through new analysis and map generation or simply through the ability to navigate nonlinearly

TABLE 9.3. Examples of information integration products categorized in terms of their customizability and linearity.

	Fixed products	Customizable products
Linear	Atlases, reports	Databases
Nonlinear	Hypertext documents, Web sites	GIS system, decision support systems

through, for example, hypertext links (e.g., Southern Appalachian Man and the Biosphere, 1996). The latter is more flexible and adaptable, although often harder and more expensive to develop and control.

Particularly with spatial data, but also with temporal data, it is increasingly possible to obtain very large volumes of data that complicate the data integration process. This may be unavoidable, given a large spatial or temporal scope; nonetheless it is worth being careful to match data to our needs. As the costs of computer storage, memory, and processing power fall, this becomes less of an issue; but data management remains a challenge.

At the core of working with information of different sorts and sources is managing and understanding the implications of different units and scales. There are no simple answers here. We must know our data and also understand issues of scale in spatial data, sampling in space and time, and the use of units. Standard texts in spatial and temporal data analysis, GIS, and quantitative methods address these issues (Haines-Young et al., 1993; Fotheringham and Rogerson, 1994; Schneider, 1994). In addition, we should have a good understanding of the assumptions, methods, and purposes under which the different kinds of data were collected. Information of this kind, termed metadata, is necessary to highlight the possible ways that data can be used—or should not be used.

Even more fundamental is the need to bridge the perceived chasm between quantitative and qualitative information. Later sections will offer some specific comments on linking the two. At this stage, the priority is to recognize the contributions of both, and especially those of qualitative information, which are too often denigrated or ignored. This means relating the different contributions of both kinds of information to assessment and management goals, also remembering that assessment and management take place in a context of human perception and action (cf. Burns et al., 1985; Slocombe, 1995). Hence ignoring qualitative data from human perceptions, recollections, and aspirations is a recipe for eventual failure. An important dimension of this is comparing and relating the various uncertainties in different kinds of qualitative data (e.g., Cleaves, 1995).

9.3.1 Integration across Domains and Disciplines

Beyond the prerequisites already noted, the most critical step for integrating information across domains is to adopt a systemic view of the system under study (e.g., Open Systems Group, 1981; Weinberg, 1975). This includes developing a conceptual model of the ecosystem, its subsystems, environment, flows, and hierarchies. Such a systems analysis provides a way to frame the assessment or planning problem, as well as a means of organizing information in a hierarchy of relationships and nested subsystems. It also requires assembling information from many domains for all but the most trivial systems and quantitative and qualitative information for any but narrowly defined purposes. A critical part of this step, and the next two as well, is identifying key variables, processes, functions, and boundaries. For examples of this, see Grzybowski and Slocombe (1988) and Slocombe (1990). Boundaries are an especially necessary starting point (although they may also be a product, e.g., in representativeness assessment or ecological mapping). It is necessary to establish, at least provisionally, both physical limits to the area of interest (usually based on multidisciplinary criteria) and disciplinary, informational limits to what is most relevant (usually based on the goals of the assessment and some initial understanding of the character of the study area).

The next step is to use specialized, domain-specific knowledge to identify points of interaction and connection between the system components and, where possible, to quantify interactions and develop a picture of dynamics and past, present, and even future behaviors. This almost always depends on detailed knowledge and experience of the system, at least some of it qualitative. Often it may help to focus on particular concepts or perspectives that foster identification of interrelationships in a way suited to our goals, as discussed later under conceptual tools.

Given the nature of the problems to be addressed and the need to integrate across disciplines, it is clear that multidisciplinary teams will often be needed to facilitate the process. Such teams have become reasonably common in the last 20 years in resource and environmental management. However, the mere creation of a team does not guarantee results and integration, any more than does commissioning a series of multidisciplinary research projects. Teams must be well chosen, focused, and skilled in integrative methods such as systems ideas, be open to integration and disciplines other than their specialty, and have the time and tools to do their job well (cf. Garcia, 1989; Giavelli, 1989). Some of these common tools are facilitated workshops, time for exploration and discussion, simulation modeling, and increasingly GIS and decision support systems, discussed further in

Sections 9.4 through 9.6 and elsewhere in this book.

Using time and technical tools for exploration, from GIS to decision support systems, still does little more than juxtapose or overlay information. Developing narrative histories, or pathway diagrams for the origins of possible impacts, provides opportunities for connections to be discovered and for possible implications to be explored. Few have explained this better than McHarg (1969) and Holling (1978). There are many forms of playing and experimenting, whether through qualitative scenarios or quantitative computer simulations, and maybe even some form of delphi analysis is critical to exploring the implications of the information. One rule is to use several. Another is that in the context of ecological assessment and ecosystem management it may be best to start with systems models, GIS, and simulation (cf. van der Weide, 1993; Patten, 1994; Russell, 1996), even if later efforts pursue more qualitative approaches.

Whatever the means, integration must yield a truly integrated product, a picture, or maybe even a story, depending on the problem and the type of information available. Simply compiling information, as in many an environmental impact statement or resource survey, is not enough. True integration means trading off quantitative and qualitative, certain and uncertain, information. We start by building a systems frame; solidifying it with quantitative, descriptive material from the natural or social sciences; and then relating and integrating qualitative and anecdotal data to flesh out the picture of possibilities, trends, dynamics, and implications of change. Olson (1987) offers a wonderful example of the use of largely qualitative information on ecological events to develop information for environmental management.

9.3.2 Integration across Space and Time

Integration across space and time entails both conceptual and technical challenges. Conceptual challenges include identifying the appropriate scale for integration, integrating information from different scales, and identifying and understanding processes at different scales. Technically, the critical challenges relate to standardizing and displaying data and maintaining metadata about different information.

Very often the purpose in integrating data across space and time is to identify patterns and correlations between events or structures or processes that can be related to domain-specific knowledge to yield management-relevant understanding (cf. Fresco and Kroonenberg, 1992). Examples include forest, vegetation, and biodiversity patterns or land use and water quality. Identifying patterns helps to develop a sense of ecosystem history, larger-scale dynamics such as path dependency, and constraints on future possibilities due to past changes.

This requires attention to problems of at least three different kinds. First is the need to identify and separate structures and processes at different scales and the ways they interact to produce macroscopic behaviors and patterns (e.g., Clark, 1987; Kirkby et al., 1996). The second is attending to issues of resolution and scale and the different patterns and emphases that may develop from attention at different scales (Costanza and Maxwell, 1994; Slocombe and Sharpe, 1997). And the third, related to the last, is choosing the appropriate scale for analysis and integration (Hoekstra et al., 1991; Dovers, 1995; Rowe, 1996).

Ideally, both spatial and temporal data will be integrated in the same system and analysis. In the case of a study of land use and water quality, for example, overall data on land-use change and point data on temporal change in water quality are linked through their location in the spatial database system. Fostering integration of spatial and temporal dimensions is critical to real understanding of ecosystem dynamics and the effects of activities and change. We seek, for example, to use hierarchy theory to identify multiple, significant scales or levels at which variables from one domain or subsystem appear in others (O'Neill, 1988; Fox, 1992; also see Chapter 2).

The practical challenge with spatial and temporal data is to get them into a common format and system so that they can be integrated through analysis and visualization. This is often primarily a technical question of compatibility. It also requires attention to resolution issues and comparability of data (metadata issues again). Ideally, all data will be spatially explicit, with standardized unit boundaries and resolution. Failing this, it is important to maintain awareness of differences in resolution and to use finer data to illustrate detail within the larger grain of the overall data system. In a spatial sense, this might mean preparing large-scale maps of sites of special interest within a landscape well understood at a smaller scale and, in a temporal sense, offering both long-term, low-resolution and detailed, short-term time-series data for systems that exhibit rapid short-term change within a longer cyclic pattern. Understanding the limitations of our data and the ways data were acquired and developed is critical. Some patterns will not be visible at large-scale resolution; others will not be visible at the small scale (see Chapters 6 and 7). A useful overall approach in the face of limited resources is to develop sufficient general, macroscopic understanding of a region to allow relatively few, focused site-specific studies to provide detailed information that can be generalized to other similar sites in the region.

9.4 Process Tools and Approaches

The degree of success achieved in integrating information depends on the process used or can at least be improved by a process that is well planned in advance. Advance planning is especially key when integration is to be part of a process that is generating much of the information, not simply collecting existing information. At root, information integration itself is a process—one of exchange and flow of information among different users, organizations, and computer systems. Figure 9.1 illustrates this and summarizes many of the components discussed in the following. The basic goals are planning, standardization, and coordination. Information needs and integration activities will change as we move through the planning and management cycle from recognizing to managing problems (e.g., Adriaanse et al., 1989).

Perhaps the first step in any information integration exercise is agreement on the goals of the exercise, both broadly and specifically. Broad goals include the purposes of integration in terms of key questions and activities to be supported; the scope in terms of domains, space, and time; and the overall personnel and other support for integration. These broad goals are likely to and probably should be addressed in a larger process of assessment or management. More specific goals include the number and type of users of the products, the nature and form of the products, how the process will be coordinated, and whether to opt for a centralized or decentralized process and system. Defining specific goals may be best done in a smaller, more focused process of organizing information users and providers. As with any technical, negotiated process, it is important that there be incentives for participation and agreement, some common ground for agreement, and ideally some financial support

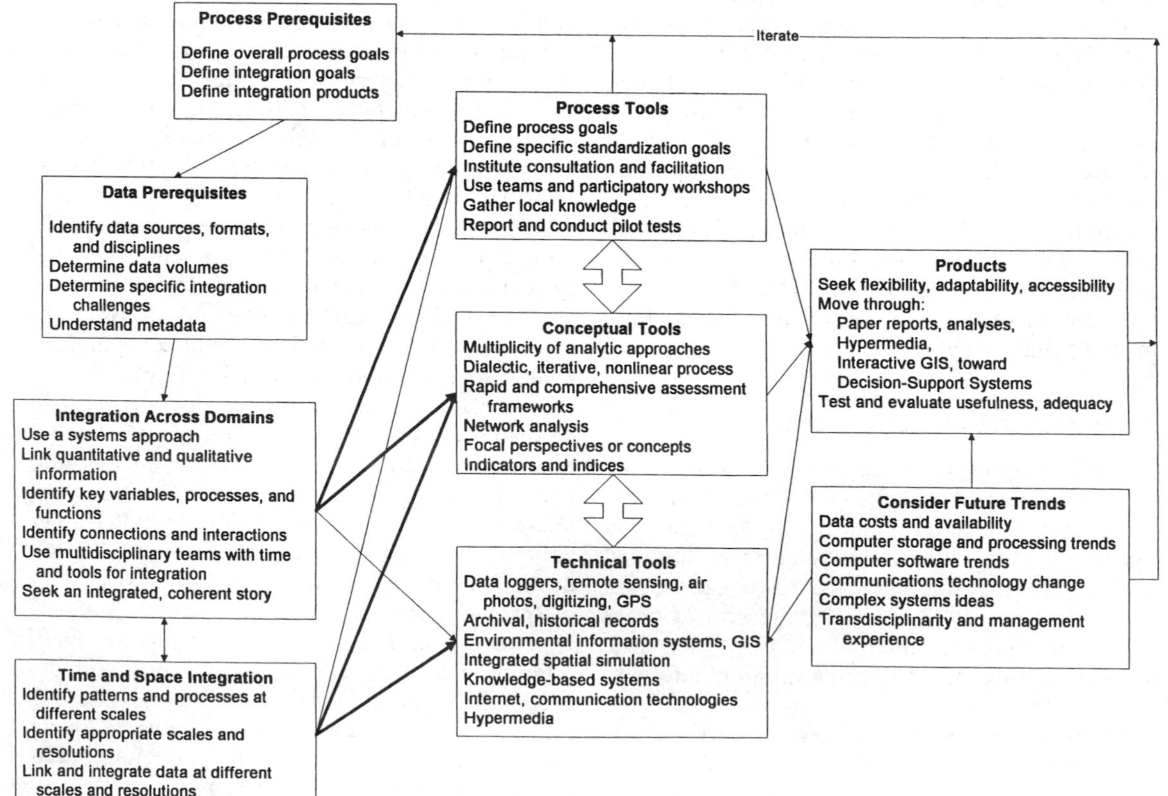

FIGURE 9.1. Overview of the information integration process.

for participation and implementation. A range of participation, negotiation, and conflict resolution skills and experience may be relevant to the initial phase of identifying goals and beginning the process (e.g., Susskind and Cruikshank, 1987; Cormick et al., 1996).

The specific goals lead to decisions on such matters as hardware and software to be used, standard data formats, and transfer and access protocols. It is critical that these be determined as early as possible, because so much depends on good choices here—and decisions are often irreversible once investments have been made in hardware and software. Choices must be consensual among all major partners; any partner's decision to go its own way can have major repercussions on integration later. While data format conversion is improving and there are at least a few widely used standards (e.g., Autocad's "dxf"), standardization can prevent many problems.

Guiding the process of integration is almost certainly a task for a dedicated interdisciplinary team—not too large, but with a range of expertise and links to all the major participants. Team members must have strong technical, disciplinary, organizational, and political skills. Making such teams work also involves clear mandates, strong senior support, appropriate tools, freedom to find what will work, and opportunities to make a difference in ways that provide incentives for participation and implementation.

Two useful, participatory components of the process can be identified. First is collection of local knowledge through participatory and consultative methods. This can serve as an excellent cross-check on scientific knowledge, add a diachronic dimension, and involve local people in the process (cf. Inglis, 1993). Second, ongoing reporting and pilot testing of integration products are important in order to determine their usability and utility. The whole information integration process needs to be adaptive and probably iterative.

9.5 Conceptual Tools and Approaches

The most basic conceptual tool is a systems approach, as highlighted previously. Next is an understanding of the conceptual process of information integration as nonlinear, adaptive, and iterative, by which we seek to make connections, develop insights, and find new ways of understanding typically separate bits of information. At one level, this is a matter of fostering creativity (e.g., de Bono, 1994) and trying many approaches to analysis (e.g., Wyatt, 1989). Beyond this, post-normal science (Funtowicz and Ravetz, 1993) ideas may also be useful in understanding the changing context of science, the participants, the nature of the information used, and the need for adaptive processes and management (see Chapter 4). One way of conceptualizing the process of information integration is as a dialectical process of analysis and synthesis of information moving toward transdisciplinarity. Synthetic transdisciplinary thinking is important, but so are analytic, descriptive inputs from the usual disciplines.

Complementary to the basic framework provided by systems analysis and models are approaches to identify system connections and linkages. Graphical, systems-oriented methods have been used in environmental impact assessment, including network diagrams and impact hypotheses (Shopley and Fuggle, 1984; Bisset, 1987; Kennedy and Ross, 1992). This is really just one variation of models, conceptual or otherwise, for the interaction of different parts of systems. More generally, it can often be useful to pick a concept or dimension of the system as a transdisciplinary starting point and work outward, multidisciplinarily, to build an integrated understanding. For example, landscape ecology or geomorphic processes in land planning (e.g., Wright, 1987), ecosystem functions (de Groot, 1992), environmental history (Moore and Witham, 1996), or ecological–economic–social connections may be useful (Svedin, 1991).

Of course, many frameworks also exist for identifying needed information, planning collection and categories, and organizing data collection in a range of environmental contexts such as land classification, state of the environment reporting, and environmental monitoring (e.g., Marshall et al., 1996; Rowe, 1996; also see review in Slocombe, 1992). The ABC Resource Survey methodology (Bastedo et al., 1984), Rapid Rural Appraisal (RRA) and Rapid Biodiversity Assessment frameworks are particularly relevant models.

Another more specific, technical approach is to develop sets of indicators or single, integrative, quantitative indices. The literature on indicators and indices is large and diverse (Kelly and Harwell, 1990; Victor et al., 1991; Indicators Evaluation Task Force, 1996). Their strengths lie in the ability to synthesize a wide range of information into a single, at least ordinal, number. There are, however, many concerns about how to define any single number, including choices of information to include, relative weightings, and overacceptance of the value of the number (remember Gross National

Product!). A small set of indicators may often be more informative and defensible. Karr's index of biotic integrity (1981) is a famous example based in aquatic faunal diversity. It can be integrated with other sorts of information, for example, land-use data, to yield more integrative knowledge and assessments (Steedman, 1988).

9.6 Technical Tools and Approaches

A wide range of tools is available for data acquisition, organization and analysis, understanding and communication, and learning and evaluation. In one way or another, all these are central to information integration; in fact, they may dominate to the point that other kinds of tools are sometimes forgotten. Many integration processes focus on acquiring and maintaining technical tools to the neglect of other more substantive issues.

There are many technical means of data acquisition, and each is a major area of research in its own right. We mention just a few. Data loggers that automatically record data on such variables as temperature or water level at predetermined intervals can provide large amounts of information at a high resolution. Remote sensing is increasingly used for acquisition of vegetation, land use, and other spatial information. Costs are high, except compared to field acquisition of the same data for large areas. Digitizing from existing maps can provide similar data for as far back as such maps go (which can be a long way in some places, e.g., 80 years in Ontario and 150 years in Italy). Digital photography is increasingly applicable in do-it-yourself remote sensing from low-altitude aircraft. Aerial photos have long provided a way of getting spatial information at lower cost than remote sensing, at least for smaller areas. Archival photos also have historical applications, for example, in tracking glacier retreat. Global positioning systems (GPS) are also an ever more available means of acquiring spatial data. Sample (1994) and Wilkie and Finn (1996) provide good overviews and examples of applications of most of these technologies; also see Chapters 5 and 10.

The most basic tool for data organization and analysis is an environmental database or information system (Günther 1998). When such a system is georeferenced in a geographical information system (GIS) and linked to spatial pattern analysis routines and includes historical data, it presents a powerful tool for integration and analysis (e.g., Kienast, 1993; Fotheringham and Rogerson, 1994; see also Chapters 13 through 15). Data analysis and visualization can be supported by tools from spreadsheets to scientific graphing and analysis packages to high-end visualization tools. No tools that are readily available incorporate all functions, yet most functions are necessary for a project of any size, with both spatial and temporal information, bibliographies, and qualitative data. In this context, it may be better to use a range of modestly priced tools, rather than trying to buy the very expensive top of the line product, which will still not be flexible enough (cf. Sharpe and Slocombe, 1995). Another option, increasingly common, may be networked, or distributed, environmental and conservation information systems that allow pooling of resources at different resolutions, domains, and extent (e.g., Davis, 1995).

Tools of growing importance for the process of integration, and especially the presentation of integrated information, are hypertext documentation, World Wide Web (WWW) sites, and interactive simulations. These allow the user to follow links in their own way, and in a nonlinear manner, depending on the questions of interest and the connections that make sense to them (cf., in a different context, Delany and Landow, 1994). While fostering integration in the development process, simulations also foster additional integration and understanding when they are widely used and experimented with in later phases (Holling, 1978; Hannon and Ruth, 1994). In addition to presenting integrated information to end users, Web sites may provide the means for teams to communicate, thus fostering consultative, cooperative work that would be harder with traditional paper documents and regular mail communication.

Finally, there are advanced tools for organizing data and facilitating, if not automating, learning and evaluation. Broadly, these embody efforts to integrate databases, GIS, and simulations in the first instance; and then to extend them with various kinds of expert, learning, and decision support (knowledge-based) tools (see Chapters 11, 12, and 18). Integrated GIS–simulation systems allow the extrapolation of existing spatial and temporal data, which is key to understanding a system. Many of these examples are driven by land and forest management problems for which predicting future spatial patterns is critical (e.g., Mackay et al., 1994; Miller, 1994; Sample, 1994). A related analytic tool is object-oriented analysis, which could complement standard systems analysis as well as computer-based tool development (e.g., Saarenmaa et al., 1994).

Knowledge-based systems seek to embed our knowledge of the world in a computer system that can help human users to define and answer questions, find and display relevant information, and in some cases develop prescriptions for action. Such systems usually entail efforts to link several of the above tools with special rule-based systems (*expert* systems: Saunders et al., 1992; Wright et al., 1993) that in some sense can reason about the information contained in them. This may or may not include spatial data in a GIS, data visualization tools, and simulations. Fedra (1995) provides a nice overview, and Baird et al. (1994), McClean et al. (1995), and Reynolds et al. (1996) provide relevant examples in fire management, land-use planning, and watershed analysis, respectively.

Complex multitool and knowledge-based systems can be useful on large projects where they will have very wide applications; however, they are not really necessary for useful information, and probably should not be developed before many of the other approaches discussed in this paper have been tried. The ultimate goal of information integration should be flexible, integrated systems that can handle, display, and interpret almost any kind of information. But systems also should be transparent and accessible and not consume resources to the extent that basic tasks like data collection, analysis, and consultation are neglected. All tools have their pluses and minuses, and the more complex the tool, the harder it is to remain aware of them (cf. Pickles, 1995, on GIS).

9.7 Trends and Opportunities

The most basic prerequisite for integration of biological, physical, and human information is its availability. Some technologies are making data more available, yet other trends, such as government cutbacks and cost recovery programs, are ending data collection or making data much more expensive to acquire. This is not a good trend from the perspective of ecosystem management, ecological assessment, or efforts to integrate diverse information.

Communication technologies such as the Internet and the WWW are making more information available and also providing a prime vehicle for the display of integrated information. The power of distributed computing in information systems was mentioned previously; the potential for linking networks of computers to increase computing power also has implications for simulation of complex, spatial systems (Alves, 1996). Posting information integration efforts on the web is a key opportunity to foster critical evaluation and progress. Other newer technologies, such as interactive electronic blackboards and videoconferencing, could further support the processes of integration in the future.

Rapid advances in computer processing power and fixed and movable storage media are making data integration more and more feasible. We could argue, however, that there is still a shortage of good integrated software tools for all the functions described here—at any price, never mind mass market prices. Future developments in knowledge-based tools will almost certainly have great impact in automating and organizing the volumes of information available and under development.

All these trends and technologies, as well as the asking of questions that require synthetic answers, are promoting better use of existing data, as well as targeted efforts to use simple tools such as workshops and simulation to reassess and integrate the data (e.g., Baskin, 1997). In addition, the development of complex systems ideas such as chaos, fractals, and self-organization is providing new paradigms and analytic tools for understanding complex systems (e.g., see Chapters 2, 14 and 15). Although their influence is still limited at a practical level, it will almost certainly grow (cf. Grzybowski and Slocombe, 1988; Slocombe, 1990, 1997; Jørgensen et al., 1992; Mainzer, 1996). Somewhat similarly, the growing interest in transdisciplinarity and integration has been catalyzed by systems and simulation ideas, as well as the need to understand the dynamics of complex systems. This interest itself is fostering more and more studies that seek to integrate understanding of change in a watershed or ecosystem (Smith, 1994; Frissel and Bayles, 1996). This alone will produce a good deal of learning and the experimentation on which it is based.

9.8 Conclusions

Integration from different domains, formats, and sources is a complex task. It often takes place with limited time and pressure to reach management decisions and faces debate over the validity, utility, and nature of the results. Equally, information integration faces technical and conceptual challenges. This chapter advocates an approach that combines advance planning to determine goals and standards with appropriate conceptual, process, and technical tools to develop useful, flexible information products. The overall process needs to be adaptive, iterative, and nonlinear and needs to be tailored to a particular information integration context (see Figure 9.1).

Certainly, it is important to use process and conceptual tools as well as technical ones, although their relative importance will vary depending on

	1	2	3	4	5	6	7	8	9	10	11	12	13	14	15	16	17	18
Integration across Domains																		
Quantitative–qualitative	x		x			x	x	x			x	x		x	x	x		
Identifying points of connection between domains			x	x	x	x		x	x	x	x				x	x	x	
Certain–uncertain knowledge			x	x				x		x		x			x	x		
Discovering new knowledge				x	x	x	x		x	x	x		x	x	x	x	x	x
Space and time variation	x		x		x	x	x	x						x	x			
Interdisciplinary communication	x		x	x	x	x	x	x	x	x	x	x		x	x	x		x
Flexible, accessable products	x	x					x					x	x	x		x	x	x
Integration across Formats																		
Different units		x												x	x			
Different accuracies and standards		x				x			x		x	x		x		x		
Different scales and resolution		x				x			x		x	x		x		x		
Different degrees of aggregation		x				x			x		x	x		x		x		
Different hardware and software		x							x		x	x		x				
Different digital file formats		x							x		x	x		x				
Data organization and availability	x	x										x				x	x	x
Integration across Sources																		
Oral and written traditions				x			x	x		x	x	x		x		x		
Different views of cause and effect			x				x	x	x			x			x	x		
Different variable definitions			x	x	x				x	x	x	x			x			
Different system characteristics considered significant	x			x	x	x	x	x	x	x	x	x		x	x	x	x	
Experiential information					x		x	x	x		x	x				x	x	
Cross-cultural communication					x		x	x	x	x	x	x		x		x	x	x
Usefulness of products	x	x			x					x		x	x	x		x	x	x

Key

1. Define process goals and products
2. Define standardization goals
3. Systems studies
4. Key variables, processes, functions
5. Conceptual starting points or foci
6. Identify patterns and processes at different scales
7. Multiplicity of analytic tools
8. Seek coherent narrative
9. Multidisciplinary teams
10. Dialectic, iterative, nonlinear process
11. Facilitated workshops
12. Consultation, negotiation
13. Reporting and pilot testing
14. Environmental information system, GIS
15. Simulation systems
16. Knowledge-based tools
17. Hypermedia
18. Internet, communication technologies

FIGURE 9.2. Indicative matrix of the applicability of selected tools for information integration and approaches to identified challenges to information integration.

context. A process with many stakeholders in an area with lots of data and good existing technical resources should probably emphasize process and conceptual tools. On the other hand, when there is little existing data and few stakeholders, greater priority may be put on technical data acquisition and organization tools to develop the information base. Many possible tools and approaches can foster integration of multidisciplinary information, as Figure 9.2 illustrates. A number of those listed are complementary and ought to be a part of most integration exercises. Others, notably some of the more advanced technical tools or large-scale facilitation and negotiation, may only be necessary in particular circumstances. Practically speaking, the choice of integration tools and effort should depend on the decision costs and alternatives—how great an expenditure can be justified economically and scientifically.

Clearly, the most widely used tools ought to be those with the potential to contribute to resolving many integration challenges; these include systems methods, simulation, teams, workshops, and knowledge-based tools. What is critical is how they are used, and earlier sections have sought to indicate useful ways. Making it all work together means using effective, committed teams to coordinate, facilitate, and integrate. But teams cannot work in isolation; they must have both support from above and the opportunity to build support from stakeholders and constituencies. And the whole process requires time, access to data, and sufficient funding. If integration is to get beyond lowest common denominator conclusions, public consultation and involvement are necessary.

Critical needs are access to information, better integrated tools, and more standardized data formats. More case studies of information integration, such as those in this volume, would assist the planning of ecological assessment projects. Critical factors in determining success include defining goals and products; adequate advance planning; fostering cooperation and standardization; choosing flexible, appropriate tools; using an appropriate transdisciplinary framework; and focusing on wholes, not parts (cf. Holling, 1989). A successful exercise in information integration is one that produces new knowledge and understanding of the system and new insights into appropriate management and the effects of human activities; that broadens and equalizes the base of understanding among stakeholders; and that creates flexible, adaptable products that can evolve and support future assessment and planning exercises. This is a tall order, rarely achieved as yet, but there is much experience in this area and great potential for rapid learning in the future.

9.9 Suggestions for Additional Reading

Baskin, Y. 1997. Center seeks synthesis to make ecology more useful. *Science* 275:310–11.

Born, S. M.; Sonzogni, W. C. 1995. Integrated environmental management: strengthening the conceptualization. *Environ. Manage.* 19(2):167–81.

Fedra, K. 1995. Decision support for natural resources management: models, GIS, and expert systems. *AI Applications* 9(3):3–19.

Günther, O. 1998. *Environmental information systems.* Berlin: Springer-Verlag.

Weinberg, G. M. 1975. *An introduction to general systems thinking.* New York: John Wiley & Sons.

9.10 References

Adriaanse, A.; Jeltes, R.; Reiling, R. 1989. Information requirements of integrated environmental policy experiences in the Netherlands. *Environ. Manage.* 13(3):309–315.

Alves, A. 1996. Parallel computing—windows style. *Byte*, May:169–170.

Baird, I. A.; Catling, P. C.; Ive, J. R. 1994. Fire planning for wildlife management: a decision support system for Nadgee Nature Reserve, Australia. *Int. J. Wildland Fire* 4(2):107–121.

Baskin, Y. 1997. Center seeks synthesis to make ecology more useful. *Science* 275:310–311.

Bastedo, J. D.; Nelson J. G.; Theberge, J. B. 1984. Ecological approach to resource survey and planning for environmentally significant areas: the ABC method. *Environ. Manage.* 8(2):125–134.

Bisset, R. 1987. Methods for environmental impact assessment: a selective survey with case studies. In: Biswas, A.K.; Geping, Q., eds. *Environmental impact assessment for developing countries.* Dublin: Tycooly Int.: 5–64.

Born, S. M.; Sonzogni, W. C. 1995. Integrated environmental management: strengthening the conceptualization. *Environ. Manage.* 19(2):167–181.

Burns, T. R.; Baumgartner, T.; DeVille, P. 1985. *Man, decisions, society: the theory of actor-system dynamics for social scientists.* New York: Gordon and Breach.

Clark, W. C. 1987. Scale relationships in the interactions of climate, ecosystems, and societies. In: Land K. C.; Schneider, S. H., eds. *Forecasting in the social and natural sciences.* Dordrecht: D. Reidel: 337–378.

Cleaves, D. A. 1995. Assessing and communicating uncertainty in decision support systems: lessons from an ecosystem policy analysis. *AI Applications* 9(3):87–102.

Cormick, G. W.; Dale, N.; Emond, P.; Sigurdson, S. G.; Stuart, B. D. 1996. *Building consensus for a sustainable future: putting principles into practice.* Ottawa, ON: National Round Table on the Environment and the Economy.

Costanza, R.; Maxwell, T. 1994. Resolution and predictability: an approach to the scaling problem. *Landscape Ecol.* 9(1):47–57.

Costanza, R.; Daly, H. E.; Bartholomew, J. A. 1991. Goals, agenda, and policy recommendations for ecological economics. In: Costanza, R., ed. *Ecological economics: the science and management of sustainability.* New York: Columbia University Press: 1–20.

Davis, F. W. 1995. Information systems for conservation research, policy, and planning. *BioScience Suppl.* 1995:S36–42.

de Bono, E. 1994. *Waterlogic.* Harmondsworth, UK: Penguin.

de Groot, R. S. 1992. *Functions of nature: evaluation of nature in environmental planning, management and decision making.* Amsterdam: Wolters-Noordhoff.

Delany, P; Landow, G. P., editors. 1994. *Hypermedia and literary studies.* Cambridge, MA: MIT Press.

Dovers, S. R. 1995. A framework for scaling and framing policy problems in sustainability. *Ecol. Econ.* 12:93–106.

Fedra, K. 1995. Decision support for natural resources management: models, GIS, and expert systems. *AI Applications* 9(3):3–19.

Forest Ecosystem Management Assessment Team (FEMAT). 1993. *Forest ecosystem management: an ecological, economic, and social assessment.* Washington, DC: U.S. Govt. Print. Off.

Fotheringham, A. S.; Rogerson, P., editors. 1994. *Spatial analysis and GIS.* Bristol, PA: Taylor and Francis.

Fox, J. 1992. The problem of scale in community resource management. *Environ. Manage.* 16(3):289–297.

Fresco, L. O.; Kroonenberg, S. B. 1992. Time and spatial scales in ecological sustainability. *Land Use Policy* 9(3):155–168.

Frissell, C. A.; Bayles, D. 1996. Ecosystem management and the conservation of aquatic biodiversity and ecological integrity. *Water Resour. Bull.* 32(2):229–240.

Funtowicz, S.; Ravetz, J. 1993. Science for the postnormal age. *Futures* 25(7):739–755.

Garcia, M. W. 1989. Forest Service experience with interdisciplinary teams developing integrated resource management plans. *Environ. Manage.* 13(5):583–592.

Giavelli, G. 1989. A rotational groups system approach for interdisciplinary research. *Environ. Manage.* 13(4):425–433.

Grzybowski, A. G. S.; Slocombe, D. S. 1988. Self-organization theories and environmental management: the case of South Moresby. *Environ. Manage.* 12(4): 463–478.

Günther, O. 1998. *Environmental information systems.* Berlin: Springer-Verlag.

Haines-Young, R.; Green, D.; Cousins, S., editors. 1993. *Landscape ecology and GIS.* Bristol, PA: Taylor and Francis.

Hannon, B.; Ruth, M. 1994. *Dynamic modeling.* New York: Springer-Verlag.

Hoekstra, T. W.; Allen, T. F. H.; Flather, C. H. 1991. Implicit scaling in ecological research. *BioScience* 41(3):148–154.

Holling, C. S., editor. 1978. *Adaptive environmental assessment and management.* Chichester: John Wiley & Sons.

Holling, C. S. 1989. Integrating science for sustainable development. *J. Bus. Admin.* 19(1/2):73–83.

Indicators Evaluation Task Force. 1996. *Indicators to evaluate progress under the Great Lakes water quality agreement.* Detroit, MI: International Joint Commission.

Inglis, J. T., editor. 1993. *Traditional ecological knowledge: concepts and cases.* Ottawa, ON: Canadian Museum of Nature, International Development Research Centre.

Jantsch, E. 1971. Inter- and transdisciplinary university: a systems approach to education and innovation. *Ekistics* 32:430–437.

Jørgensen, S. E.; Patten, B. C.; Straskraba, M. 1992. Ecosystems emerging: toward an ecology of complex systems in a complex future. *Ecol. Model.* 62:1–27.

Karr, J. R. 1981. Assessment of biotic integrity using fish communities. *Fisheries* 6:21–27.

Kelly, J. R.; Harwell, M. A. 1990. Indicators of ecosystem recovery. *Environ. Manage.* 14(5):527–545.

Kennedy, A. J.; Ross, W. A. 1992. An approach to integrate impact scoping with environmental impact assessment. *Environ. Manage.* 16(4):475–484.

Kienast, F. 1993. Analysis of historic landscape patterns with a geographical information system—a methodological outline. *Landscape Ecol.* 8(2):103–118.

Kirkby, M. J.; Imeson, A. C.; Bergkamp, G.; Cammeraat, L. H. 1996. Scaling up processes and models from the field plot to the watershed and regional areas. *J. Soil Water Conserv.* 51(5):391–396.

Lang, R., editor. 1986. *Integrated approaches to resource planning and management.* Calgary, AB: University of Calgary Press.

Mackay, D. S.; Robinson, V. B.; Band, L. E. 1994. A knowledge-based approach to the management of geographic information systems for simulation of forested ecosystems. In: Michener, W. K., et al., eds. *Environmental information management and analysis.* Bristol, PA: Taylor and Francis: 511–534.

Mainzer, K. 1996. *Thinking in complexity*, 2nd ed. Berlin: Springer-Verlag.

Marshall, I. B.; Smith, C. A. S.; Selby, C. J. 1996. A national framework for monitoring and reporting on environmental sustainability in Canada. *Environ. Monit. Assess.* 39:25–38.

McClean, C. J.; Watson, P. M.; Wadsworth, R. A.; Blaiklock, J.; O'Callaghan, J. R. 1995. Land use planning: a decision support system. *J. Environ. Plan. Manage.* 38(1):77–92.

McHarg, I. L. 1969. *Design with nature.* Garden City, NY: Natural History Press.

Miller, D. 1994. Coupling of process-based vegetation models to GIS and knowledge-based systems for analysis of vegetation change. In: Michener, W. K., et al., eds. *Environmental information management and analysis.* Bristol, PA: Taylor and Francis: 497–509.

Moore, E. H.; Witham, J. W. 1996. From forest to farm and back again: land use history as a dimension of ecological research in coastal Maine. *Environ. Hist.* 1(3):50–69.

Olson, S. 1987. Red destinies: the landscape of environmental risk in Madagascar. *Hum. Ecol.* 15(1):67–89.

O'Neill, R. V. 1988. Hierarchy theory and global change. In: Rosswall, T., et al., eds. *Scales and global change.* Chichester: John Wiley & Sons: 29–45.

Open Systems Group, ed. 1981. *Systems behaviour*, 3rd ed. London: Paul Chapman.

Patten, B. C. 1994. Ecological systems engineering: toward integrated management of natural and human complexity in the ecosphere. *Ecol. Modelling* 75/76:653–665.

Pickles, J., editor. 1995. *Ground truth: the social implications of geographic information systems.* New York: Guilford Press.

Reynolds, K.; Cunningham, P.; Bednar, L.; Olson, R.; Schmoldt, D.; Latham, D.; Miller, B.; Steffenson, J. 1996. A knowledge-based information management system for watershed analysis in the Pacific Northwest U.S. *AI Application* 10(2):9–21.

Rowe, J. S. 1996. Land classification and ecosystem classification. *Environ. Monit. Assess.* 39:11–20.

Russell, C. S. 1996. Integrating ecology and economics via regional modeling. *Ecol. Appl.* 6(4):1025–1030.

Saarenmaa, H.; Perttunen, J.; Vakeva, J.; Nikula, A. 1994. Object-oriented modelling of tasks and agents in integrated forest health management. *AI Applications* 8(1):43–59.

Sample, V. A., editor. 1994. *Remote sensing and GIS in ecosystem management.* Washington, DC: Island Press.

Saunders, M. C.; Coulson, R. N.; Folse, L. J. 1992. Applications of artificial intelligence to agriculture and natural resource management. In: Kent, A.; Williams, J. G., eds., *Encyclopedia of computer science and technology*, Vol. 25, Suppl. 10. New York: Marcel Dekker: 1–14.

Schneider, D. C. 1994. *Quantitative ecology: spatial and temporal scaling.* San Diego, CA: Academic Press.

Sharpe, B.; Slocombe, D. S. 1995. Building a GIS database for ecosystem modelling in the Grand River Basin, Ontario. In: Power, J. M.; Strome, M.; Daniel, T. C., eds. *Proceedings of decision support—2001.* Bethesda, MD: Amer. Soc. Photogramm. Remote Sensing: 846–859.

Shopley, J. B.; Fuggle, R. F. 1984. A comprehensive review of current environmental impact assessment methods and techniques. *J. Environ. Manage.* 18:25–47.

Slocombe, D. S. 1990. Assessing transformation and sustainability in the Great Lakes Basin. *GeoJournal* 21(3):251–272.

Slocombe, D. S. 1992. Environmental monitoring in protected areas: review and prospect. *Environ. Monit. Assess.* 21(1):49–78.

Slocombe, D. S. 1993. Environmental planning, ecosystem science, and ecosystem approaches for integrating environment and development. *Environ. Manage.* 17(3):289–303.

Slocombe, D. S. 1995. Understanding regions: a framework for description and analysis. *Can. J. Reg. Sci.* 18(2):161–178.

Slocombe, D. S. 1997. *Chaos, complexity and self-organisation—an annotated bibliography and review essay on management and design in complex systems from organisations to ecosystems.* Waterloo, ON: WLU Geography and Environmental Studies and Business.

Slocombe, D. S. 1998. Defining goals and criteria for ecosystem-based management. *Environ. Manage.* 22(4):483–493.

Slocombe, D. S.; Sharpe, B. 1997. *Scale, pattern, and interaction: toward assessing sustainability in the Grand River Basin, Ontario, Canada.* Manuscript. Originally Presented at Ecological Summit, Copenhagen, Denmark, August 1996.

Smith, C. L. 1994. Connecting cultural and biological diversity in restoring northwest salmon. *Fisheries* 19(2):20–26.

Southern Appalachian Man and the Biosphere (SAMAB). 1996. *The Southern Appalachian assessment.* Atlanta, GA: U.S. Dept. Agric., For. Serv., Southern Region. 5 vol.

Steedman, R. J. 1988. Modification and assessment of an index of biotic integrity to quantify stream quality in southern Ontario. *Can. J. Fisheries Aquat. Sci.* 45:492–501.

Susskind, L.; Cruikshank, J. 1987. *Breaking the impasse—consensual approaches to resolving public disputes.* New York: Basic Books.

Svedin, U. 1991. The contextual features of the economy–ecology dialogue. In: Folke, C.; Kåberger, T., eds. *Linking the natural environment and the economy: essays from the Eco-Eco Group.* Dordrecht: Kluwer: 3–29.

Thomas, G. L.; Backus, E. H.; Christensen, H. H.; Weigand, J. 1991. *Prince William Sound: Copper River: North Gulf of Alaska ecosystem.* Cordova, AK: Prince William Sound Science Centre and Copper River Delta Institute and Washington, D.C.: Conservation International.

van der Weide, J. 1993. A systems view of integrated coastal management. *Ocean Coast. Manage.* 21:129–148.

Victor, P. A.; Kay, J. J.; Ruitenbeek, H. J. 1991. *Economic, ecological, and decision theories: indicators of ecologically sustainable development.* Ottawa, ON: Canadian Environmental Advisory Council.

Weinberg, G. M. 1975. *An introduction to general systems thinking.* New York: John Wiley & Sons.

Wilkie, D. S; Finn, J. T. 1996. *Remote sensing imagery for natural resources monitoring.* New York: Columbia University Press.

Wright, J. R.; Wiggins, L. L.; Ravinder, K. J., editors. 1993. *Expert systems in environmental planning.* New York: Springer-Verlag.

Wright, R. L. 1987. Integration in land research for Third World development planning: an applied aspect of landscape ecology. *Landscape Ecol.* 1(2):107–117.

Wyatt, R. 1989. *Intelligent planning: meaningful methods for sensitive situations.* London: Unwin Hyman.

Part 3
Basic Techniques

10 Remote Sensing Applied to Ecosystem Management

Henry M. Lachowski and Vicky C. Johnson

10.1 Introduction

Ecosystems are complex and dynamic; often, they support many diverse and competing demands. Bailey (1996) states that "in simple terms, the ecosystem concept states that the earth operates as a series of interrelated systems within which all components are linked, so that a change in any one component may bring about some corresponding changes in other components and in the operation of the whole system." Managing ecosystems requires that we look at numerous phenomena and deal with information and analyses at multiple scales, whether geographic or temporal. The human dimension also must not be neglected. Effective management of ecosystems requires access to current and consistent geospatial information that can be shared by resource managers and the public. Geospatial information describing our land and natural resources comes from many sources and is most effective when stored in a geospatial database and used in a geographic information system (GIS). Information on the location and condition of current vegetation patterns is one of the key elements in ecosystem management. Remotely sensed data are primary sources for mapping vegetation. Furthermore, comparing images acquired several days, or several years, apart can assist in determining changes over time.

Several land management agencies have been using remote sensing and associated technologies for large area assessments, land and resource management plan updates, and specific land management activities, with emphasis on vegetation mapping and monitoring changes. Remote sensing is the general term for the field of study concerned with collecting and interpreting information about an object from a remote vantage point (Campbell, 1996). The platform can be anywhere, ranging from just above the surface of the object to several hundred miles in space. Examples of remotely sensed data include satellite imagery, aerial photographs, airborne video, and digital camera imagery.

Many types of remotely sensed data are in digital format; others can be readily digitized. Satellite imagery covers large areas and has the locational precision and spatial resolution to satisfy many natural resources mapping requirements. For most applications, satellite imagery is used in conjunction with a closer view provided by aerial photographs and digital camera images. The global positioning system (GPS), used in conjunction with imagery, helps users to determine precise locations and navigate in the field. This is important for linking data collected on the ground to data collected from airborne and satellite platforms (Rosenfeld and Thatcher, 1994; Cristofani, 1996). Ground data provide validation for the interpretation of airborne and satellite imagery and allow extrapolation of that interpretation to a broader area.

The ability to integrate remotely sensed imagery in a GIS is essential in order to derive the most benefits from these data. Vegetation and other layers derived from remote sensing, along with base layers such as roads, boundaries, water bodies, digital elevation models (DEMs), digital orthophoto quadrangles (DOQs), and other resource data layers, provide a solid geospatial data foundation. The Forest Service has identified several resource data layers known as *GIS core data* to support management of natural resources (Figure 10.1), but the applicability of these data is far from limited to one agency (GIS Core Data Team, 1997). These GIS core data are standardized layers and attributes that are suitable for addressing a wide variety of resource management questions.

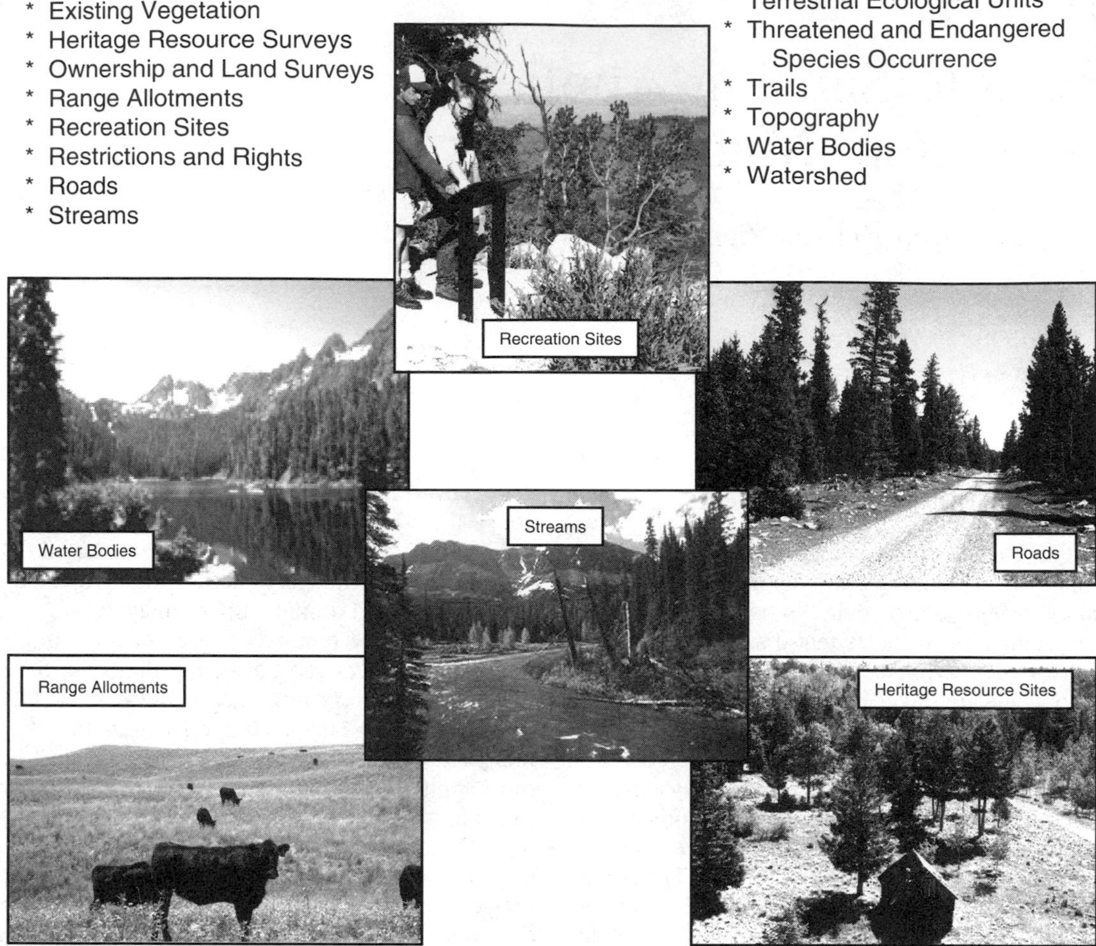

FIGURE 10.1. Example of GIS core data layers.

This chapter briefly covers concepts of remote sensing and discusses some of the many sources of remotely sensed data. We draw on experiences of various land management agencies, showing examples of applications of remote sensing at multiple scales, and providing a list of references that covers these topics in more detail.

10.2 Ecosystem Management and Remote Sensing: An Overview

10.2.1 Data Requirements for Ecosystem Management

Ecosystem management has been defined in many different ways (Agee and Johnson, 1988; Grumbine, 1994; Malone, 1998). Jack Ward Thomas, the former Chief of the Forest Service, defines ecosystem management as "not an end in itself, but a means to achieve sustainable conditions and provide wildlife and fish habitat, outdoor recreation, wilderness, water, wood, mineral resources, and forage for society while retaining the aesthetic, historic, and spiritual qualities of the land" (Thomas, 1999). To accomplish this, we need to understand ecological processes, current and past conditions of ecosystems, and the goals and needs of humans. Geospatial technologies, of which remote sensing is a part, can help us to gather, organize, and analyze information about ecosystems.

What kind of data do we need to manage ecosystems? A major part of ecosystem management is finding and understanding linkages, relationships, and thresholds within various landscapes and ecological systems. Map-based, or *geospatial*, information describing these linkages and relationships

TABLE 10.1 The national hierarchy of ecological units.

Planning and analysis scale	Ecological units	Purpose, objectives, and general use	General size range
Ecoregion Global Continental Regional	Domain Division Province	Broad applicability for modeling and sampling; large-area planning and assessment; international planning	Millions to ten thousands of square kilometers
Subregion	Section Subsection	Strategic, multiforest, statewide, and multiagency analysis and assessment	Thousands to tens of square kilometers
Landscape	Land-type association	Forest or area-wide planning and watershed analysis	Thousands to tens of hectares
Land unit	Land type Land-type phase	Project and management area planning and analysis	Hundreds to less than 10 hectares

helps us to put ideas into the right perspective, to visualize the linkages, and to communicate our common concerns and issues with others. Resource managers need information at various scales. These may range from very broad assessment areas, such as ecoregions, used for examining issues such as old-growth forests or transmigrational songbird nesting habitat to watersheds for examining hydrologic conditions and fisheries concerns to, finally, small land units with specific management concerns; here, finer-scale data are appropriate.

To facilitate information sharing among resource managers, the national hierarchical framework of ecological units was developed (Avers et al., 1993). This framework, shown in Table 10.1, is useful for defining boundaries for mapping ecological units and sharing information among the broad range of landowners and the public.

To gather data at reasonable cost and achieve organizational efficiency, we must agree to certain general scales that are likely to meet the majority of our requirements. Based on previous experience, several land management agencies acquire and manage basic information at the following general scales: 1 : 24,000, 1 : 100,000, 1 : 250,000, and 1 : 1,000,000. Many land management agencies and private organizations have a large amount of information available at these scales, especially at the 1 : 24,000 scale. A substantial portion of this information is derived from remotely sensed data.

10.2.2 Remote Sensing Contribution to Ecosystem Management

Remote sensing provides a very effective environmental overview for assessing landscapes or other land units. As a complement to field-collected data, it provides a wall-to-wall snapshot in time for assessing current conditions and trends. Figure 10.2 provides a general scenario of remote sensing platforms and of data at various scales.

Remotely sensed data are continuous over the landscape. However, users must be aware that the interpretation of imagery can be subjective and that sufficient ground reference data are needed to make reliable maps. Resource managers are facing a challenge to integrate alternative approaches of data collection, including (1) point data collection and (2) landscape data collection with remote sensing. Point data collection is a primary source for specific information about the environment and is not always suitable for landscape assessments. By integrating point data with landscape mapping from remote sensing, we can implement the powerful tool of stratified random sampling. In this approach, remote sensing provides the stratification of the landscape into homogeneous units, and point sampling provides specific data to characterize the strata (Lillesand and Kiefer, 1994).

The use of geospatial data, including remotely sensed data, is guided by government-wide federal standards established by the Federal Geographic Data Committee (FGDC, 1997a). These standards address the content, accuracy, and related parameters with which producers and users of data should be familiar and must comply to assure wide and unrestricted use of geospatial data. The Forest Service has recently prepared a set of guidelines and recommendations based on FGDC standards (USDA Forest Service Remote Sensing Advisory Team, 1998). These recommendations and examples, specifically for mapping vegetation cover, are summarized in Section 10.4. The emphasis is on vertical and horizontal integration of data for large area assessment and forest plan revisions.

Proper ecosystem management must look beyond arbitrary political or management boundaries

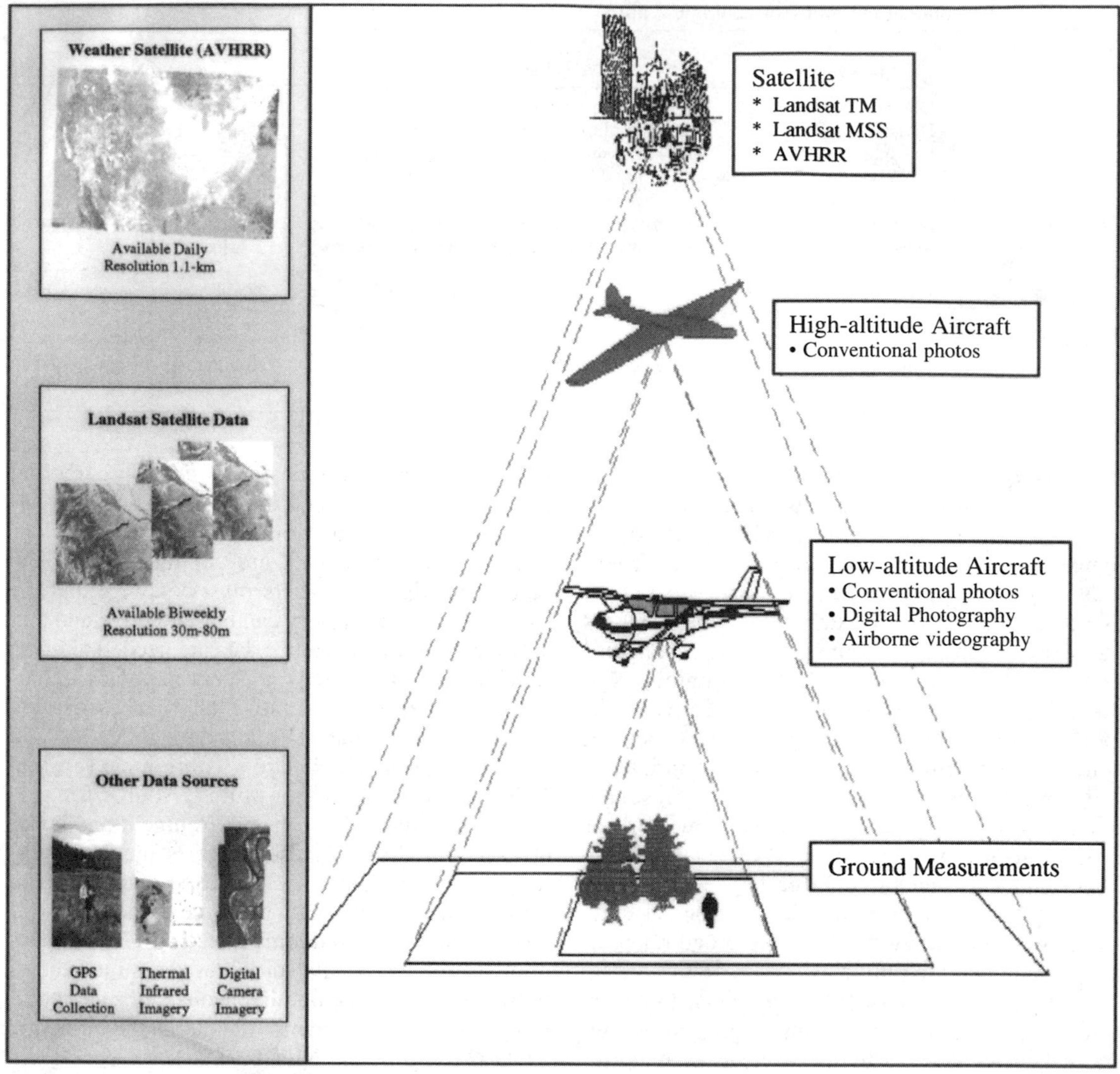

FIGURE 10.2. Acquisition platforms of remotely sensed data with potential applications of data at various scales.

since ecosystems rarely correspond to these types of boundaries. The Montreal Process Working Group (1994) was established with the specific purpose of developing and implementing a set of criteria and indicators (C&I) for sustainable forest management that transcends such boundaries. The C&I are tools for assessing national trends in forest conditions and management and provide a common framework for describing, monitoring, and evaluating progress toward sustainability at national and global levels. There are seven criteria and 67 indicators, many of which can be most effectively monitored by remote sensing. Brand (2001) summarizes progress to date and future direction of this monitoring initiative.

10.3 Sources of Remotely Sensed Data

Remote sensing is an evolving field in which continuous advances in technology have provided new tools for viewing Earth's resources. Data collection devices, known as sensors, have been refined over the last several decades. Besides photographic and video cameras, numerous digital satellite and aircraft sensors are used for mapping vegetation and other land cover features.

Digital imagery is the subset of all remotely sensed data that can be stored, analyzed, and displayed in digital form on a computer. Although imagery derived from camera and video sensors can be digitized, the satellite and some airborne sensors produce data that are already in digital form. Digital images are composed of a grid of picture elements, or pixels, representing reflectance information about objects. *Reflectance* is what allows us to distinguish features in an image. It is a measure of the amount and type of energy an object reflects (rather than absorbs or transmits). Reflected energy, modified by the atmosphere between the ground and the sensor, is called radiance. *Radiance* is what the sensor measures; it may differ from reflectance because of haze in the atmosphere. An in-depth discussion of this topic is presented in Chapter 2 of Jensen (1996).

10.3.1 Data Collection

Different sources of remotely sensed data vary in their ability to distinguish landscape features. In the following, we describe some of the most useful data sources for purposes of ecological assessment, with emphasis on digital imagery. The collection of primary data, creation of derived data in the form of vegetation maps, and detection of landscape change are addressed in turn.

Advanced Very High Resolution Radiometer (AVHRR)

AVHRR data are obtained through National Oceanic and Atmospheric Administration (NOAA) weather satellites. AVHRR imagery has a spatial resolution of 1.1 km by 1.1 km pixel size with five spectral bands. Broad area coverage is obtained with a swath width of 2700 kilometers. Temporal resolution is excellent, with repeat coverage over a given geographic area every 12 hours. Major applications for these data include vegetation discrimination, vegetation biomass, snow–ice discrimination, vegetation–crop stress, geothermal mapping, and fire mapping. This imagery has been used for fire detection and growth monitoring in Alaska and Spain (Kasischke and French, 1995; Pozo et al., 1997), forest classification of southeast Asia (Achard and Esteguil, 1995), and change detection in western Africa (Lambin, 1996) and to derive a relationship between AVHRR imagery and ecoregions in Utah (Ramsey et al., 1995).

Landsat Multispectral Scanner (MSS) and Landsat Thematic Mapper (TM)

The Landsat Program is the longest running program in the collection of multispectral, digital data of Earth's land surface from space. The temporal extent of the collection, characteristics and quality of Landsat data, and ability to collect new data directly comparable to that in the archive make Landsat data a unique resource for addressing a broad range of issues in ecosystem management. The July 1997 issue of *Photogrammetric Engineering & Remote Sensing* paid tribute to the 25-year history of the Landsat Program (French, 1997).

Landsat MSS imagery has a spatial resolution of 80 m, spectral resolution in four bands, revisits the same geographic location every 16 days, and covers an area of 185 by 170 km. The Landsat program routinely gathered digital MSS data from 1972 through 1992. The resulting data, spanning 20 years, can support evaluations of change in landscapes or land cover over a longer time frame than any available Earth observation system. MSS data are available through the EROS Data Center (EDC) working in cooperation with NASA. The North American Landscape Characterization (NALC) program has developed a data set of archived MSS images for the conterminous United States (Lunetta et al., 1998). The NALC data set, sometimes referred to as MSS triplicates, consist of at least three co-registered images acquired in the 1970s, 1980s, and early 1990s for all the conterminous United States.

Landsat TM imagery has 30-m spatial resolution with seven spectral bands. The satellite covers the same geographic location every 16 days, each scene covering 185 km by 170 km. The Multi-Resolution Landscape Characterization Consortium (Loveland and Shaw, 1996) was established to purchase TM imagery covering the conterminous United States. Partners in the consortium include USGS Gap Analysis Program, EPA Land Cover Program, NOAA Coastal Change Analysis Program, USGS National Water Quality Assessment Program, EPA/USGS North American Landscape Characterization, USGS EDC, and USDA Forest Service. Other ongoing large procurements of these data by federal agencies will allow easy and economical access to recent and future data.

Aerial Photography

Aerial photography at scales ranging from 1 : 12,000 to 1 : 24,000, sometimes referred to as *resource-scale* aerial photography, has been used for over 50 years as a tool for mapping and monitoring resource areas. Aerial photographs are typically acquired on a 5- to 10-year cycle on most federal lands. In recent years, aerial photography has become an important source of ancillary data for interpreting satellite images. Aerial photographs, with their high spatial resolution, provide a link between the resource data on the ground and the high spectral resolution of satellite imagery.

An interagency program called the National High Altitude Photography (NHAP) program was established in 1980 to build a uniform archive of aerial photos of the 48 conterminous states. Black-and-white and color-infrared photos were acquired at an approximate scale of 1 : 80,000 and 1 : 58,000, respectively. In 1987, the flight specifications were modified, as was the name: NHAP became the National Aerial Photography Program (NAPP). Black-and-white and color-infrared NAPP photos are 1 : 40,000 scale centered on quarters of 7.5-minute quadrangles. Many image processing software packages have modules that allow the user to analyze and manipulate digital copies of aerial photographs.

Airborne videography and digital camera systems provide powerful tools for ecosystem management. A camera can be mounted in a small airplane and imagery can be viewed as soon as the flight is complete (King, 1995). A wide range of spatial resolution can be obtained by varying the flying altitude and camera lenses. Digital camera imagery linked with GPS provides location information for each frame, allowing users to quickly locate specific areas on maps or images. Images can be georeferenced and used with other GIS layers. These systems are used for a variety of resource applications, including aerial surveys for forest pests (Knapp et al., 1998) and monitoring and mapping riparian areas (Bobbe et al., 1993; Bobbe and McKean, 1995) and as a high-resolution data source for satellite image classification (Slaymaker et al., 1996).

Digital orthophoto quadrangles (DOQs) combine the image characteristics of a photograph with the geometric qualities of a map. Wirth et al. (1994) describe the creation of DOQs as a process of scanning aerial photographs with each pixel of the image being corrected for relief displacement and camera orientation. DOQs are valuable data sources for updating road and existing land cover databases and for georeferencing other data for use in a GIS. However, the large file size associated with DOQs can make them cumbersome to use. A 7.5 minute quadrangle-based DOQ is approximately 160 Mbytes. Often multiple DOQs, requiring hundreds of megabytes or even gigabytes of hard disk space, need to be mosaicked together to cover a project area. The large file size should become less of an issue as larger hard drives, faster processors, and improved image compression techniques become available.

Radar Data

Radar systems form a unique addition to our capabilities for collecting remotely sensed data. The systems can operate during day or night and can penetrate clouds, thereby providing high-quality imagery of Earth's surface in situations where photographic or multispectral systems cannot be used (Hoffer et al., 1995). Radar imagery can be obtained from aircraft or satellite platforms. Kasischke et al. (1997) discuss the capabilities of radar systems for investigating terrestrial ecosystems. They organize applications into four broad areas: (1) classification and detection of change in land cover, (2) estimation of woody plant biomass, (3) monitoring the extent and timing of inundation, and (4) monitoring other temporally dynamic processes. Specific examples are provided for each application area. Imhoff et al. (1997) used radar imagery to discern structural differences relevant to bird habitat quality in the Northwest Territory, Australia.

Lidar Data

Compared to radar, lidar (light detection and ranging) is relatively new. This technology involves pulsing Earth's surface with lasers and collecting the reflected electromagnetic energy. By precisely timing the interval between the emission of the pulse and the detection of the reflected energy, distances and relative heights of objects can be determined, even for objects as small as 5 cm. Lidar is currently being used to produce precise digital terrain models; it also shows great promise as a method for estimating forest stand structure. Today, lidar sensors operate from airborne platforms (fixed-wing aircraft and helicopter), but in the year 2000 a satellite with three to five Lidar sensors was scheduled to become operational (Flood and Gutelius, 1997). Lidar was used by Means et al. (1999) to estimate forest stand characteristics such as height, basal area, and biomass in western Oregon. Similar forest stand characteristics were measured by Lefsky et al. (1999) in eastern Maryland.

10.3.2 Vegetation Characteristics Derived from Remote Sensing

Most remote sensing systems provide image data, as opposed to point data. Images are made of thousands or even millions of data, either single picture elements (pixels) or single grains of photographic emulsion. As such, these data are interpreted, not measured directly, by visual and computer-aided analysis. Interpretations are made using the data's collective brightness, texture, pattern, size, shape, shadow, and association (Avery and Berlin, 1992). Deriving information about vegetation and land cover is one of the primary applications of remotely sensed data. Several attributes can be derived from remotely sensed data that describe vegetation. Some characteristics are described in the following, followed by a discussion of accuracy assessment, without which a map of vegetation characteristics has limited utility.

Canopy Structure

Canopy structure describes the vertical structure of the vegetation, either single story or multistory. The vegetation of a single-story canopy is generally a uniform height, whereas a multistory canopy has overstory and understory components. A single-story canopy tends to have a smooth velvet texture in remotely sensed data, such as Landsat TM imagery, whereas a multistory canopy may have a rougher, irregular appearance.

Tree Size Class

Tree size class may be defined by the diameter of the trees that constitute the dominant or overstory canopy. To derive information on tree size class from remotely sensed data, we must assume that tree diameters are correlated with tree height, because diameters usually cannot be measured through remote sensing due to blockage by the overstory canopy. Typically, tree diameters are grouped into size classes, such as seedling, sapling, pole, and sawtimber, based on ground measurements. Size class determination requires substantially more field data than other attributes because of the complex relation between canopy reflectance and size class. Field data necessary to derive this attribute would include sampling of the diameters of the trees in the canopy across the full range of forest cover types. Problems associated with estimating tree size are discussed by Woodcock et al. (1994).

Percent Crown Closure

Percent crown closure refers to the area covered by the forest canopy. Determining percent crown closure can be complicated in multistory forest canopies because of shadows visible on remotely sensed data. In addition, low crown closures (less than 15%) are hard to estimate because they are spectrally very similar to bare ground. Percent crown closure is a characteristic that is more consistently estimated from remote sensing than on the ground due to the synoptic perspective provided by airborne and satellite sensors. Cohen and Spies (1992) assessed the utility of satellite data for analysis and inventory of stand attributes of Douglas fir–western hemlock forests in western Oregon and Washington. The authors concluded that of all the stand attributes evaluated the following were most reliably estimated using satellite data: standard deviation of tree sizes, mean tree size, tree density, a structural complexity index, and stand age. These attributes were for trees in the upper canopy layer of the sample stands.

Vegetation Composition

Species groups or associations are often used to define vegetation composition. The level of detail possible in determining vegetation composition is highly correlated to the spatial and spectral resolution of the remotely sensed imagery and the types being distinguished. Low spatial resolution data such as AVHRR can be used to identify very broad species groups. Landsat TM and aerial photos can supply more detailed species groups. Congalton et al. (1993) used satellite imagery and aerial photographs to produce a GIS database and map old-growth forest lands in western Oregon and Washington. GIS data layers derived from remotely sensed data included crown closure, size class–stand structure, species, and current-vegetation-type polygons.

Accuracy Assessments

Accuracy assessments are an essential part of all remote sensing projects. First, they enable the user to compare different methods and sensors. Second, they provide information regarding the reliability and usefulness of remote sensing techniques for a particular application. Finally, and most importantly, accuracy assessments support spatial data used in decision-making processes. Stehman and Czaplewski (1998) discuss the basic components and options available for conducting accuracy assessment of thematic maps constructed from remotely sensed data. Edwards et al. (1998) present an accuracy assessment methodology applied to a cover-type map developed for the Utah Gap Analysis. Further information on accuracy assessment can also be found in Congalton and Green (1998).

National Vegetation Classification System

The Federal Geographic Data Committee (1997a) approved a classification system to support the production of uniform statistics on vegetation resources at the national level. These standards include principles, definitions of important terminology, and the National Vegetation Classification System (FGDC, 1997b). The NVCS is a hierarchical classification system based on existing vegetation. The United Nations Educational, Scientific, and Cultural Organization (UNESCO) originally developed this system in the early 1970s, and The Nature Conservancy recently modified it for conservation planning. Several federal agencies are employing the NVCS for resource inventory, management, monitoring, and conservation. Some recent vegetation mapping activities and universal resource locators (URLs) to project Web sites follow.

1. The USGS Gap Analysis Program is mapping vegetation at the state level using Landsat TM imagery and the NVCS. (URL: http://www.calmit.unl.edu/gapmap/)
2. The U.S. Fish and Wildlife Service is testing the NVCS in a pilot project and hopes to classify the entire National Wildlife Refuge System. (URL: http://www.fws.gov/data/projcht2.html)
3. The National Park Service Vegetation Mapping Program, based on the NVCS, is currently underway in all national parks. (URL: http://biology.usgs.gov/npsveg/nvcs.html)
4. The Forest Service used the NVCS for the ecological assessment of the Pacific Northwest Region Columbia River Basin. (URL: http://www.icbemp.gov/)

10.3.3 Change Detection

Change detection, the process of monitoring changes in land cover over time, is a very important use of remotely sensed data. Mapping changes in land cover has many applications: updating vegetation layers; furnishing a historical context for land management planning; evaluating changes in patch characteristics relative to wildlife, water, recreation, and timber resource management; and understanding potential causes of change. Archives of remotely sensed data exist as far back as the 1930s for aerial photography and to the early 1970s for satellite imagery. For a general overview of change detection, refer to Jensen (1996). A variety of change detection techniques and applications exists and can be found in the literature. Collins and Woodcock (1997) assess several change detection techniques for mapping forest mortality in the Lake Tahoe Basin area. Lyon et al. (1998) compared vegetation indices from Landsat MSS data for their value for vegetation and land cover change detection over three time periods in Chiapas, Mexico. Muchoney and Haack (1994) used an automated approach to change detection to monitor forest defoliation in Virginia. Michener and Houhoulis (1997) also used an automated change detection process to monitor vegetation changes associated with flooding caused by Tropical Storm Albert in July 1994 in southwest Georgia. And Viedma et al. (1997) used satellite imagery to assess regrowth pattern of different species following a fire, to analyze the speed of recovery, and to estimate the rates of regrowth in Alicante, Spain.

10.3.4 Access to Remotely Sensed Data

Remotely sensed data are available from numerous sources. Listed next are the most commonly used repositories for aerial photography and satellite imagery.

Consolidated Farm Service Agency Aerial Photography Field Office
2222 West 2300 South, P.O. Box 30010
Salt Lake City, UT 84130-0010
(801) 975-3500
(URL: www.apfo.usda.gov)
Data available: Photography acquired by federal agencies that is less than 40 years old

National Archives and Records Administration
700 Pennsylvania Avenue, N.W.
Washington, DC 20408
(800) 234-8861
(URL: www.nara.gov)
Data available: Photography acquired by federal agencies that is more than 40 years old

EROS Data Center
47914 252nd Street
Sioux Falls, SD 57198-0001
(800) 252-4547 or (605) 594-6151
(URL: http://edcwww.cr.usgs.gov)
Data available: Satellite imagery and photography acquired by federal agencies that is less than 40 years old

Space Imaging / EOSAT
12076 Grant Street
Thornton, CO 80241
(800) 425-2997
(URL: www.spaceimaging.com)
Data available: Satellite and radar data and products

SPOT Image Corporation
1897 Preston White Drive

TABLE 10.2. Sample of new satellites in orbit or planned for launch in the next decade.

Country	Program	Date	Type[a]	Resolution in meters			No. of color bands	Stereo type[b]
				P	M	R		
France	SPOT 5B	2004	P&M	5	10		4	F/A
United States	EOS AM-2/L8	2004	P&M	10	30		7	
France	SPOT 5A	1999	P&M	5	10		4	F/A
United States	Landsat-7	1999	P&M	15	30		7	
United States	Space imaging	1998	P&M	1	4		4	F/A
Korea	KOMSAT	1998	P&M	10	10		3	F/A
United States/Japan	EOS AM-1	1998	M		15		14	F/A
European Space Agency	ENVISAT	1998	R			30		

[a]Type: M, multispectral; P, panchromatic; R, radar.
[b]Stereo type: F/A, fore and aft stereo.
Adapted from www.ccrs.nrcan.gc.ca/ccrs/tekrd/satsens/sats/asprse.html.

Reston, VA 22091-4368
(800) 275-7768
(URL: www.spot.com)
Data available: Satellite imagery and products

Access to remotely sensed data has been increasing rapidly as a result of new cooperative agreements among agencies to share data acquisition costs, remotely sensed data clearinghouses at various levels, and improved access to information through the Internet. Access to the Internet gives users the capability to preview satellite imagery acquired from several platforms, including Landsat. (See, for example, the Landsat 7 site at http://landsat7.usgs.gov). This allows users to assess the condition of the imagery (e.g., percent cloud cover), although problems may still exist on the imagery that were not apparent on the preview. Internet access also provides an easy method for ordering and transferring digital imagery.

Several new satellite sensors are scheduled to be launched within the next decade (Table 10.2). Once these plans come to fruition, users of remotely sensed data will have several new options. These sensors will allow for different landscape perspectives by providing imagery within different spectral, spatial, and temporal resolutions.

10.4 Remote Sensing and Ecosystem Management Examples

10.4.1 From Ecoregions to Land Units

Remotely sensed data can be used at many levels and scales of ecosystem management, from ecoregions to small project areas. Woodcock and Strahler (1987) as well as Atkinson and Curran (1997) provide an extended discussion of scale and spatial resolutions, respectively, as applied to remote sensing investigations. Remote sensing has been used in large-area assessments and for numerous forest planning efforts. Table 10.3 provides general guidelines for matching sensors to appropriate ecological levels. The following sections present examples for each of the levels, from ecoregion to land unit.

Ecoregion

Ecological units for the entire United States have been developed at this broadest level. Ecoregions are subdivided into domains, divisions, and provinces (Avers et al., 1993). Domains and divisions are defined primarily by climatic factors that are influ-

TABLE 10.3. Relationship of remotely sensed data to the national hierarchy of ecological units.

Ecological level	Remotely sensed data	Locational accuracy	Feature size	Scale
Ecoregion	AVHRR, MSS	1 km–100 m	100s of km^2	1 : 5,000,000 or smaller
Subregion	MSS, TM, photos[a]	100 m–30 m	10 ha	1 : 5,000,000 to 1 : 250,000
Landscape	TM, SPOT, photos[a]	30 m–10 m	1 ha	1 : 250,000 to 1 : 50,000
Land unit	SPOT, photos[b]	<10 m	<1 ha	1 : 50,000 or larger

[a]Small-scale aerial photos > 1:40,000.
[b]Medium-scale aerial photos, 1:40,000 to 1:15,000.

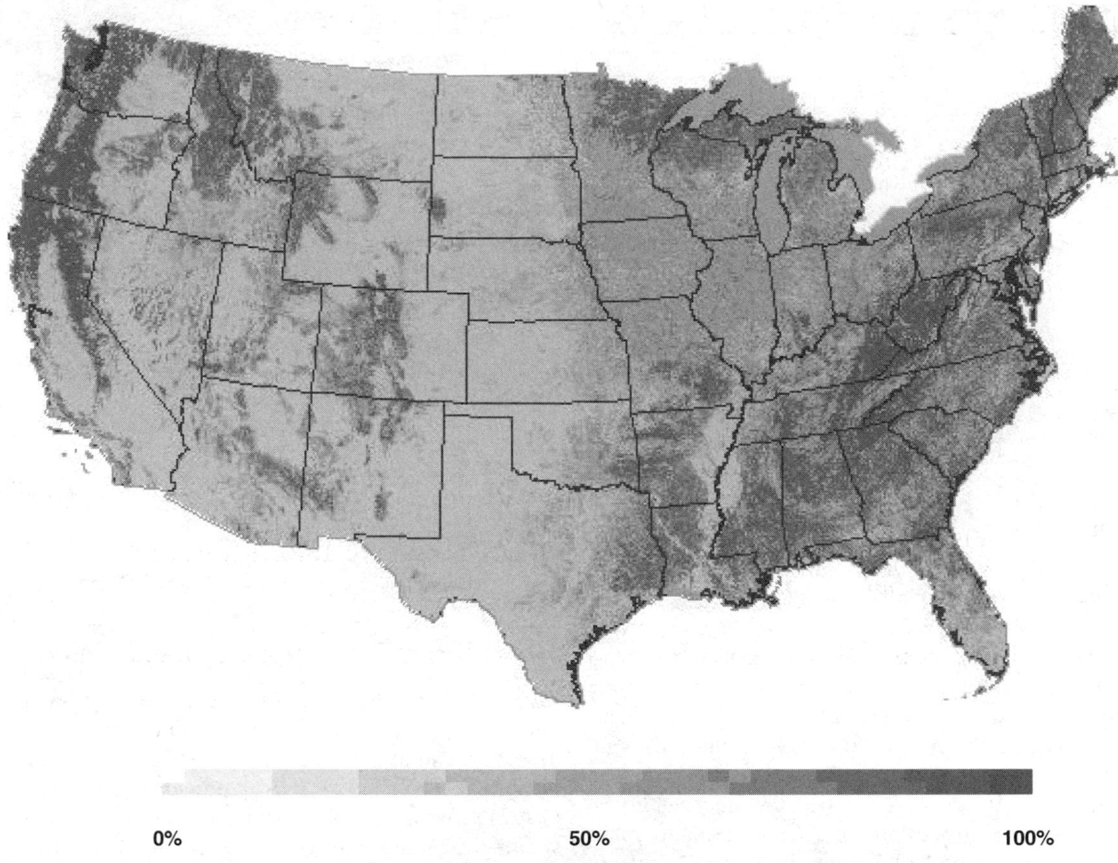

FIGURE 10.3. Percent forest cover area per square kilometer for the conterminous United States.

enced by latitude, continental position, and elevation. Vegetation macrofeatures determine where divisions are broken into provinces. AVHRR data are a good choice for assessing ecosystems at these levels. The AVHRR sensor collects multispectral data and has a 2400-km swath width and 1-km spatial resolution. Therefore, it can be used to evaluate vegetation and landforms throughout an ecoregion.

1993 Resource Planning Act Forest Type Group Map

Two forest cover maps for the conterminous United States have been derived from AVHRR data (Zhu and Evans, 1994). A forest density map was created by modeling the percent forest cover per square kilometer (Figure 10.3) (Zhu, 1994). A second map of forest-type groups depicting the spatial distribution of 22 forest-type groups was also produced for this project. These maps were made as part of the 1993 Resource Planning Act (RPA) assessment update (Powell et al., 1993). Forest Inventory and Monitoring (FIM) data from different Forest Service regions were linked to the images to identify regional vegetation patterns. For example, the FIM data showed a transition of pine to upland hardwoods in Alabama over the past two surveys. The forest-type group map depicted a related transitional zone between the loblolly–shortleaf type and the oak–hickory and oak–pine types. For more information, see the project's Web site (http://srsfia.usfs.msstate.edu/rpa/rpa93.htm).

Subregion

Next in the hierarchy are subregions, which are further divided into sections and subsections. Regional environmental factors, such as geomorphic processes, topography, and climate, influence the determination of subregions (Avers et al., 1993). Projects conducted over a number of National Forests or across a state are subregional. Sensors with midrange spatial resolution, such as Landsat TM, are appropriate for this level. Additionally, some small-scale, high-altitude aerial photos would be useful to map and analyze vegetation.

Vegetation Classification at the Subregional Level

The Classification and Assessment with Landsat of Visible Ecological Groupings (CALVEG) was developed to provide a classification system and mapping effort for the state of California to more adequately assess natural resources statewide. CALVEG uses a hierarchical classification that allows vegetation to be divided into broad units for region-wide or larger-area landscape planning. This classification system was designed to describe existing vegetation at various levels of detail (USDA Forest Service Regional Ecology Group, 1981). CALVEG mapped overstory size, structure, crown closure, and plant series associations. Plant series associations were designed to aggregate into lifeforms (i.e., shrub, conifer, deciduous, grassland). Maps were created using automated classification procedures based on multispectral Landsat TM imagery, field verification, aerial photography, and local knowledge.

Landscape

Landtype associations are mapped at the landscape level. General topography, geomorphic processes, surficial geology, type of soil, local climate, and potential natural communities determine landtype association boundaries (Avers et al., 1993). At the landscape level, forest or watershed assessments are possible using Landsat TM, SPOT, and aerial photos. The 185-km swath width and 30-m spatial resolution of Landsat TM imagery make these data useful for both subregional and landscape studies. The moderate swath width provides a synoptic view of a large area, and the mid-level spatial resolution supplies significant detail about vegetation or land cover conditions.

Aspen Change Detection and Mapping for Landscape Analysis in Southwestern Montana

The Beaverhead–Deerlodge National Forest conducted a landscape-level assessment of current and historical aspen populations (Lachowski et al., 1996). The study area, 186,000 hectares located in the Gravelly Mountain Range in southern Montana, was selected to analyze current aspen locations. A vegetation classification based on Landsat TM imagery yielded 15 vegetation classes. Further refinement was necessary to account for spectral confusion between willow and aspen stands. Aerial photos, digital camera images, and GIS modeling using elevation and slope were employed to reduce classification errors.

To evaluate aspen decline, historic aerial photos were compared to the Landsat TM classification over a 2400-hectare subsection of the study area. Aspen, aspen–conifer mix, and conifer cover types were interpreted from 1947 black-and-white aerial photos. The photointerpreted data were then digitized and overlaid onto the satellite classification to identify changes in aspen distribution. Successional change in conifer distribution was determined to be the primary cause of aspen decline (Figure 10.4).

Land Unit

Land units are the building blocks of landtype associations and are described as either landtypes or landtype phases (Avers et al., 1993). At this scale, local topography, rock and soil types, and vegetation determine unit descriptions and boundaries. Landtypes and landtype phases are used for managing and planning at the forest or project level. High-resolution remotely sensed data are combined with digital elevation models (DEMs), field data, large-scale aerial photos, and expert local knowledge to describe ecological units at this level.

Ecological Unit Mapping with Remote Sensing and GIS

The Bridger–Teton East Ecological Unit Inventory team inventoried ecological units on the western slopes of the Wind River Mountain Range in Wyoming. The inventory process integrated information on landform, geology, potential natural vegetation, and soils to map and characterize ecological units. Satellite imagery and DEMs within a GIS were used to expedite mapping and to stratify the study area for sampling into preliminary map units or landscape units. Repeating patterns were recognized across the landscape by grouping areas with similar ecological properties, thus further expediting the mapping process (Figure 10.5). Stratifying the study area helped to develop an effective strategy for field sampling. The boundaries of the map units and landscape units were refined through aerial photointerpretation and extensive field data collection. Products generated for this project included a classification scheme for ecological types, map units that comply with Natural Resource Conservation Service mapping standards, and data layers for the Bridger–Teton National Forest's GIS database.

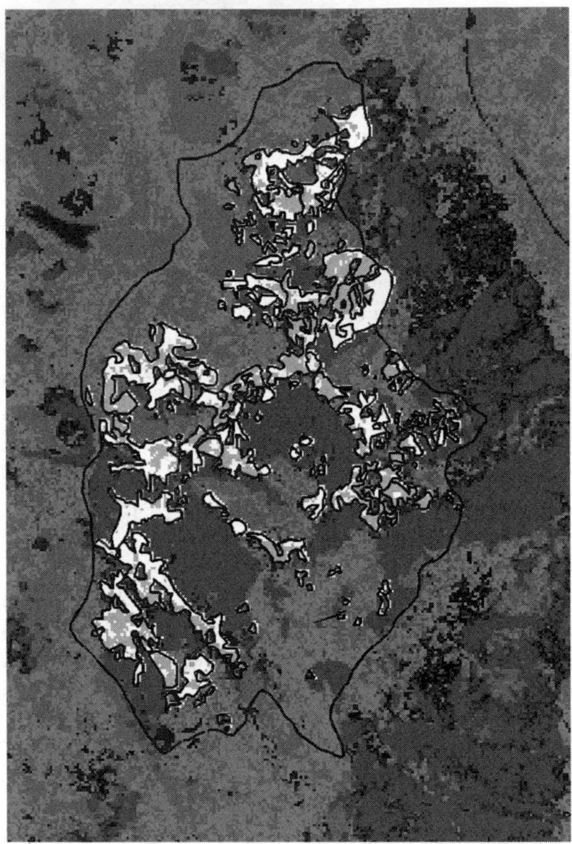

FIGURE 10.4. Map of change in aspen distribution between 1947 and 1992 as determined from remotely sensed data.

10.4.2 Disturbances to Ecosystems

Disturbance in an ecosystem can stem from many sources, some natural and others anthropogenic. These sources can include fire, wind, floods, heavy recreation use, and overgrazing. The timeline involved in these disturbances can range from hours to millennia. Acquired for multiple dates, remotely sensed data allow us to examine an area before, sometimes during, and after a disturbance occurs. Remotely sensed data have been successfully used to map fire extent and damage, changes occurring after a fire, and the spread of noxious weeds. Samples of these applications are described here.

Satellite imagery was used to map fire extent and intensity on the Payette National Forest in the summer of 1994 (Boudreau and Maus, 1996). A complex of three fires burned over 117,000 hectares. In the aftermath of the 1994 fires, significant changes occurred throughout the landscape. Analysis of these changes relative to the prefire vegetation type allowed a better understanding of spatial landscape qualities. Pre- and postfire vegetation maps were developed using a combination of Landsat TM imagery and high-altitude color-infrared aerial photography. These maps documented the vegetation that existed before the fire and the severity of the burn within those vegetation types after the fire. Boudreau and Maus (1996) list many other potential uses of the data compiled for this project.

The Burned Area Emergency Rehabilitation (BAER) Program gathers information on fire-induced watershed conditions after any fire that has burned more than 125 hectares (USDA Forest Service, 1995). Mapping burn intensity, in this instance referring to the fire's effect on the watershed, is a critical step in the survey, because it determines flooding potential and specific flood sources within the burned area. Inventory, analysis, and rehabilitation prescription must be completed before the first damaging rainstorms of the season. The use of digital color-infrared camera imagery integrated into GIS for mapping burn intensity was evaluated in 1996 on a fire on the

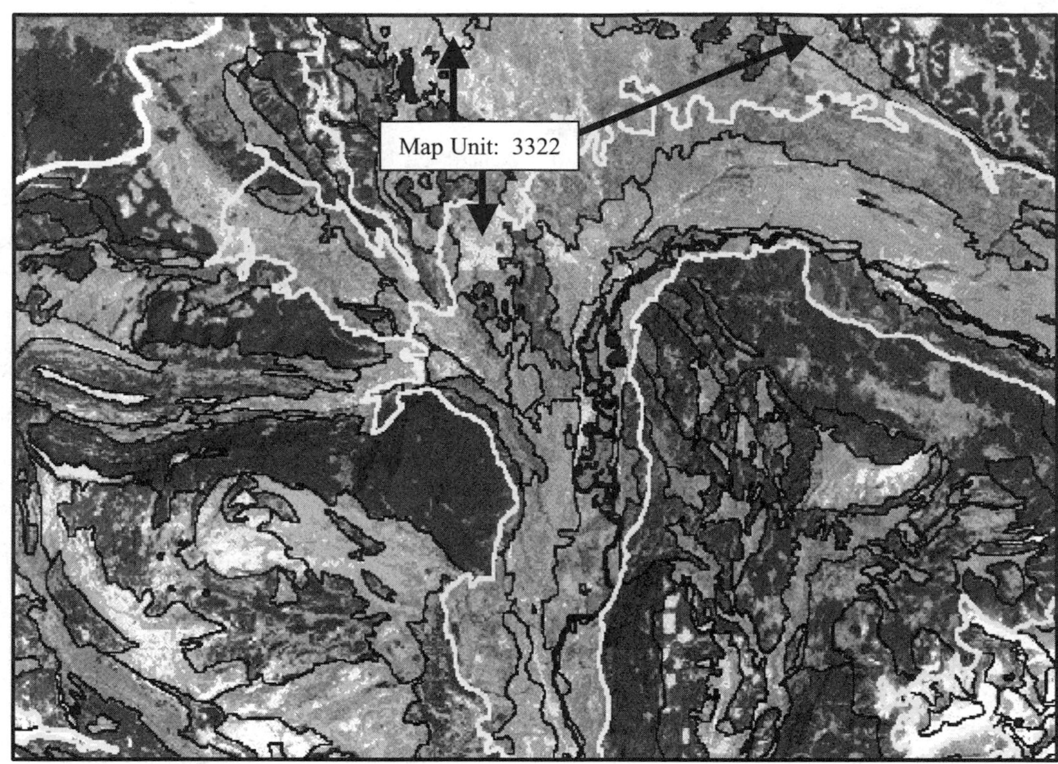

MAP UNIT: 3322

Mountain big sagebrush / Slender wheatgrass (ARTV2/ELTR7, Amesmont Family - ARCAV, Cambern Family ET - ARTV2/ELTR7, Guiser Family ET Complex)

This map unit consists of lateral and recessional glacial moraines with intermorainal swales, potholes, and kettles on moderately sloping undulating till plain. The ARTV2/ELTR7, Amesmont Family ET occurs on dry moraine sideslopes, till plains, and dry swales. The ARCAV2, Cambern Family ET occurs on moist positions including swales. The ARTRV2/ELTR7, Guiser Family ET occurs on dry moraine crests and backslopes.

MAP UNIT COMPOSITION

40 Percent - Mountain big sagebrush/slender wheatgrass (ARTV2/ELTR7, Amesmont Family ET)
25 Percent - Mountain silver sagebrush (ARCAV2, Cambern Family ET)
15 Percent - Mountain big sagebrush/slender wheatgrass (ARTRV2/ELTR7, Guiser Family ET)
20 Percent - Inclusions

PHYSICAL SETTINGS

Hierachy of Ecological Units: Union Pass Uplands Subsection of the Overthrust Mountains Section
General Location: Mosquito Lake
Landform: Pinedale recessional moraines and till plains, including intermorainal swales, kettles, and swales
Parent Material: Pinedale glacial till dominantly crystalline granitics and granodiorites with lesser amounts of sedimentary rock fragments.
Elevation: 8000 to 9000 feet **Annual Precipitation:** 21 to 24 in.
Slope: 0 to 20 percent **Areal Extent:** 3822 acres
Aspect: All

FIGURE 10.5. Map unit classification overlayed on Landsat TM imagery showing repeating occurrences and a description of Map Unit 3322. The black lines represent land units, the thin white lines represent landtype associations, and the thick white line in the upper-left corner represents subsections.

Mendocino National Forest in California (Lachowski et al., 1997; Hardwick et al., 1998). The display of digital data allowed faster identification of critical areas within the burn to focus the team's limited and expensive time. Mapping using digital camera imagery proved, in this instance, to be 20% more accurate than traditional mapping methods and saved an estimated $250,000 in rehabilitation costs.

Noxious weeds are transported by wildlife, livestock, wind, water, vehicles, and people to new sites and then become established in soil disturbed by construction, travel, or recreation. Once established, they are able to invade adjacent undisturbed plant communities, where they tend to crowd out native species. Many noxious weeds have site-specific requirements that can be modeled to identify areas susceptible to invasion and possibly the degree of susceptibility. The cost of data acquisition can be better allocated or even reduced by knowing if an area is susceptible to invasion and how susceptible it is. Many of the GIS data layers needed for susceptibility modeling come from remote sensing sources. A project in central Idaho demonstrated how a vegetation-cover-type map developed from Landsat TM and digital elevation models as well as other data layers could be used to generate a map of susceptible area (Varner et al., 1998).

10.5 Conclusions

Geospatial information describing our land and natural resources is the backbone of effective ecosystem management. The examples provided here illustrate just a few of the areas where geospatial information and specifically information originating from various remote sensors have been used to increase our knowledge of ecosystems. Information from remote sensing provides us with a unique vantage point that permits both broad-scale perspectives and project area analysis. Remote sensing not only offers a means for providing cost-effective information, but in many areas provides the only information available to study ecosystems. As new sensors with finer spatial and spectral resolution become available, the ability to interpret and process data from these sensors will follow. Image-processing software for remote sensing image analysis is becoming widespread and, with appropriate training, will be an important tool for managing ecosystems. Information derived from remote sensing serves many diverse applications in managing ecosystems. These applications range from monitoring forest sustainability through criteria and indicators at the national level to mapping and monitoring land cover at various ecosystem and land management levels.

10.6 References

Achard, F.; Estreguil, C. 1995. Forest classification of Southeast Asia using NOAA AVHRR data. *Remote Sens. Environ.* 54(3):198–208.

Agee, J. K.; Johnson, D. R. 1988. Introduction to ecosystem management. In: Agee, J. K.; Johnson, D. R., eds. *Ecosystem management for parks and wilderness.* Seattle: University of Washington Press: 3–14.

Atkinson, P. M.; Curran, P. J. 1997. Choosing an appropriate spatial resolution for remote sensing investigation. *Photogramm. Eng. Remote Sensing* 63(12): 1345–1351.

Avers, P. E.; Cleland, D. T.; McNab, W. H.; Jensen, M. E.; Bailey, R. G.; King, T.; Goudey, C. B. 1993. *National hierarchical framework of ecological units.* Washington DC: U.S. Dept. Agric., For. Serv.

Avery, T. E.; Berlin, G. L. 1992. *Fundamentals of remote sensing and airphoto interpretation*, 5th ed. Minneapolis, MN: Burgess Publishing Co.

Bailey, R. G. 1996. *Ecosystem geography*. New York: Springer-Verlag.

Bobbe, T.; McKean, J. 1995. Evaluation of a digital camera system for natural resource management. *Earth Observation Magazine*: 45–48.

Bobbe, T.; Reed, D.; Schramek, J. 1993. Georeferenced airborne video imagery. *J. For.* 91(8):34–37.

Boudreau, S. L.; Maus, P. A. 1996. An ecological approach to assess vegetation change after large scale fires on the Payette National Forest. In: Greer, J. D., ed. *Proceedings of the sixth biennial Forest Service conference on remote sensing*. Denver, CO: Amer. Soc. Photogramm. Remote Sensing: 330–339.

Brand, D. G. 2001. Criteria and indicators for the conservation and sustainable management of forests: progress to date and future directions. *New Zealand J. For.* An online version of the paper is located at: http://www.forest.nsw.gov.au/sfm/JCIPAPE.html.

Campbell, J. B. 1996. *Introduction to remote sensing*, 2nd ed. New York: Guilford Press.

Cohen, W. B.; Spies, T. A. 1992. Estimating structural attributes of Douglas-fir/western hemlock forest stands from Landsat and SPOT imagery. *Remote Sens. Environ.* 41(1):1–17.

Collins, J. B.; Woodcock, C. E. 1997. An assessment of several linear change detection techniques for mapping forest mortality using multitemporal Landsat TM data. *Remote Sens. Environ.* 56(1):66–77.

Congalton, R.; Green, K. 1998. *Assessing the accuracy of remotely sensed data, principles and practices.* Boca Raton, FL: Lewis Publishers.

Congalton, R.; Green, K.; Teply, J. 1993. Mapping old growth forests on national forest and park lands in the Pacific Northwest from remotely sensed data. *Photogramm. Eng. Remote Sensing* 59(4):529–535.

Cristofani, A. 1996. The earth in balance–maintaining Brazil's biodiversity. *GPS World* 7(6):20–30.

Edwards, T. C.; Moisen, G. G.; Cutler, D. R. 1998. Assessing map accuracy in a remotely sensed, ecoregion-scale cover map. *Remote Sens. Environ.* 63(1):73–83.

Federal Geographic Data Committee. 1997a. *Content standard for digital geospatial data*. Washington, DC: The Committee.

Federal Geographic Data Committee. 1997b. *National vegetation classification system.* FGDC-STD-005. Reston, VA: The Committee.

Flood, M.; Gutelius, B. 1997. Commercial implications of topographic terrain mapping using scanning airborne laser radar. *Photogramm. Eng. Remote Sensing* 63(4):327–329, 363–366.

French, W. D., publisher. 1997. *Photogramm. Eng. Remote Sensing* 63(7).

GIS Core Data Team. 1997. *Recommended core data standards for GIS application, report to the Ecosystem Management Corporate Team and Inter-Regional Ecosystem Management Coordinating Group.* Washington, DC: U.S. Department of Agriculture, Forest-Service.

Grumbine, R. E. 1994. What is ecosystem management? *Conserv. Biol.* 8:27–39.

Hardwick, P.; Lachowski, H.; Griffith, R.; Parsons, A. 1998. Burned area emergency rehabilitation project: an example of successful technology transfer. In: Greer, J. D., ed. *Proceedings of the seventh biennial Forest Service conference on remote sensing.* Nassau Bay, TX: Amer. Soc. Photogramm. Remote Sensing: 62–71.

Hoffer, R.; Maxwell, S.; Ochis, H. 1995. *Use of radar for forestry applications. Internal report.* Salt Lake City, UT: U.S. Dept. Agric., For. Serv., Remote Sensing Applications Center.

Imhoff, M. L.; Sisk, T. D.; Milne, A.; Morgan, G.; Orr, T. 1997. Remotely sensed indicators of habitat heterogeneity: use of synthetic aperture radar in mapping vegetation structure and bird habitat. *Remote Sens. Environ.* 60(3):217–227.

Jensen, J. R. 1996. *Introductory digital image processing.* Upper Saddle River, NJ: Prentice Hall.

Kasischke, E. S.; French, N. H. F. 1995. Locating and estimating the areal extent of wildfires in Alaska Boreal forests using multiple-season AVHRR NDVI composite data. *Remote Sens. Environ.* 51(2):263–275.

Kasischke, E. S.; Melack, J. M.; Dobson, M. C. 1997. The use of imaging radars for ecological assessments. *Remote Sens. Environ.* 59(2):141–156.

King, D. J. 1995. Airborne multispectral digital camera and video sensors: a critical review of system designs and applications. *Can. J. Remote Sensing* 21(3):245–273.

Knapp, K. A.; Disperati, A.; Sheng, Z. J. 1998. Evaluation and integration of a digital camera system into forest health protection programs in the Western United States, Southern Brazil, and Anhui Province, China. In: Greer, J. D., ed. *Proceedings of the seventh biennial Forest Service conference on remote sensing.* Nassau Bay, TX: Amer. Soc. Photogramm. Remote Sensing: 257–268.

Lachowski, H.; Wirth, T.; Maus, P.; Powell, J.; Suzuki, K.; McNamara, J.; Riordan, P.; Brohman, R. 1996. *Monitoring aspen decline using remote sensing and GIS, Gravelly Mountain landscape, Montana. Internal Project Report.* Salt Lake City, UT: U.S. Dept. Agric., For. Serv. Remote Sensing Applications Center.

Lachowski, H.; Hardwick, P.; Griffith, R.; Parsons, A.; Warbington, R. 1997. Faster, better data for burned watersheds needing emergency rehab. *J. For.* 95(6): 4–8.

Lambin, E. F. 1996. Change detection at multiple temporal scales: seasonal and annual variations in landscape variables. *Photogramm. Eng. Remote Sensing* 62(8):931–938.

Lefsky, M. A.; Harding, D.; Cohen, W. B.; Parker, G.; Shugurt, H. H. 1999. Surface lidar remote sensing of basal area and biomass in deciduous forests of eastern Maryland, USA. *Remote Sens. Environ.* 67(1):83–98.

Lillesand, T. M.; Kiefer. R. W. 1994. *Remote sensing and image interpretation*, 3rd ed. New York: John Wiley & Sons.

Loveland, T. R.; Shaw, D. M. 1996. Multiresolution landscape characterization: building collaborative partnerships. In: Scott, J. M.; Tear, T. H.; Davis, F. W., eds. *Gap analysis: a landscape approach to biodiversity planning*. Bethesda, MD: Amer. Soc. Photogramm. Remote Sensing: 83–89.

Lunetta, R.; Lyon, J. G.; Guindon, B.; Elvidge, C. D. 1998. North American landscape characterization dataset development and data fusion issues. *Photogramm. Eng. Remote Sensing* 64(8):821–829.

Lyon, J. G.; Yuan, D.; Lunetta, R. S.; Elvidge, C. D. 1998. A change detection experiment using vegetation indices. *Photogramm. Eng. Remote Sensing* 64(2): 143–150.

Malone, C. R. 1998. The federal ecosystem management initiative in the United States. In: Lemons, J.; Goodland, R.; Westra, L., eds. *Environmental stability: case studies on the prospects of science and ethics*. Dordrecht, The Netherlands: Kluwer Academic Publishers.

Means, J. E.; Acker, S. A.; Harding, D. J.; Blair, J. B.; Lefsky, M. A.; Cohen, W. B.; Harmon, M. E.; McKee, W. A. 1999. Use of large-footprint scanning airborne lidar to estimate forest stand characteristics in the western Cascades of Oregon. *Remote Sens. Environ.* 67(3):298–308.

Michener, W. K.; Houhoulis, P. F. 1997. Detection of vegetation changes associated with extensive flooding in a forested ecosystem. *Photogramm. Eng. Remote Sensing* 63(12):1363–1374.

Montreal Process Working Group. 1994. *Montreal Process Working Group* [on line]. Available: (http://www.dpie, gov.au/agfor/forests/montreal/international.html).

Muchoney, D. M.; Haack, B. N. 1994. Change detection for monitoring forest defoliation. *Photogramm. Eng. Remote Sensing* 60(10):1243–1251.

Nelson, R. F. 1983. Detecting forest canopy change due to insect activity using Landsat MSS. *Photogramm. Eng. Remote Sensing* 49(9):1303–1314.

Powell, D. S.; Faulkner, J. L.; Darr, D. R.; Zhu, Z.; MacCleery, D. W. 1993. *Forest resources of the United*

States, 1992. Gen. Tech. Rep. RM-234. Fort Collins, CO: U.S. Dept. Agric., For. Serv., Rocky Mt. For. Range Exp. Sta.

Pozo, D.; Olmo, F. J.; Alados-Arboledas, L. 1997. Fire detection and growth monitoring using a multitemporal technique on AVHRR mid-infrared and thermal channels. *Remote Sens. Environ.* 60(2):11–120.

Ramsey, R. D.; Falconer, A.; Jensen, J. R. 1995. The relationship between NOAA-AVHRR NDVI and ecoregions in Utah. *Remote Sens. Environ.* 53(3):188–198.

Rosenfeld, C. L.; Thatcher, T. 1994. On the move: GPS monitors Alaska's Surging Bering Glacier. *GPS World* 5(11):18–24.

Slaymaker, D. M.; Jones, K. M. L.; Griffin, C. R.; Finn, J. T. 1996. Mapping deciduous forests in southern New England using aerial videography and hyperclustered multi-temporal Landsat TM imagery. In: Scott, J. M.; Tear, T. H.; Davis, F. W., eds. *Gap analysis: a landscape approach to biodiversity planning*. Bethesda, MD: Amer. Soc. Photogramm. Remote Sensing: 87–101.

Stehman, S. V.; Czaplewski, R. L. 1998. Design and analysis for thematic map accuracy assessment: fundamental principles. *Remote Sens. Environ.* 64(3): 331–344.

Stoms, D.; Davis, F.; Cogan, C.; Cassidy, K. 1994. *Assessing land cover map accuracy for GAP analysis* [on line]. Available: (http://www.gap.uidaho.edu/gap/Publications/Publications.htm).

Thomas, J. W. 1999. Guiding principles and workshop overviews. In: Sexton, W.T.; Szaro, R.C.; Johnson, N.; Malk, A.J., eds. *Ecological Stewardship: A Common Reference for Ecosystem Management*. Oxford, UK: Elsevier Science.

USDA Forest Service. 1995. *Burned area emergency rehabilitation handbook*. GPO FSH 2509.13. Washington DC: Govt. Print. Off.

USDA Forest Service Regional Ecology Group. 1981. *CALVEG: a classification of California vegetation*. San Francisco: U.S. Dept. Agric., For. Serv., Region Five.

USDA Forest Service Remote Sensing Advisory Team. 1998. *Implementation of remote sensing for ecosystem management*. Salt Lake City, UT: U.S. Dept. Agric., For. Serv., Remote Sensing Applications Center.

Varner, V.; Lachowski, H.; Lake, L.; Anderson, B. 1998. Mapping and monitoring noxious weeds using remote sensing. In: Greer, J. D., ed. *Proceedings of the seventh biennial Forest Service conference on remote sensing*. Nassau Bay, TX: Amer. Soc. Photogramm. Remote Sensing: 182–193.

Viedma, D.; Melia, J.; Segarra, D.; Garcia-Haro, J. 1997. Modeling rates of ecosystem recovery after fire using Landsat TM data. *Remote Sens. Environ.* 61(3): 383–398.

Wirth, T.; Maus, P.; Lachowski, H.; Taylor, J.; Linquist, E. 1994. *Investigative report on digital orthophoto quadrangles using data from the Superior National Forest. Internal report.* Salt Lake City, UT: U.S. Dept. Agric., For. Serv., Remote Sensing Applications Center.

Woodcock, C. E.; Strahler, A. H. 1987. The factor of scale in remote sensing. *Remote Sens. Environ.* 21: 311–332.

Woodcock, C. E.; Collins, J. B.; Gopal, S.; Jakabhazy, V. D.; Li, X.; Macomber, S.; Ryherd, S.; Harward, V. J.; Levitan, J.; Wu, Y.; Warbington, R. 1994. Mapping forest vegetation using Landsat TM imagery and a canopy reflectance model. *Remote Sens. Environ.* 50: 240–254.

Zhu, Z. 1994. *Forest density mapping in the lower 48 states: a regression procedure*. Research Paper SO-280. New Orleans: U.S. Dept. Agric., For. Serv., Southern For. Exp. Sta.

Zhu, Z.; Evans, D. L. 1994. U.S. forest types and predicted percent forest cover from AVHRR data. *Photogramm. Eng. Remote Sensing* 60(5):525–531.

11 Geographic Information Science and Ecological Assessment

Janet Franklin

11.1 Introduction

Ecosystem management is carried out at a variety of spatial and temporal scales (Franklin, 1993a, 1993b; Kohm and Franklin, 1997). Ecological assessment (EA), however, especially ecosystem characterization, generally has a landscape-scale component and can hardly be conducted without the use of a geographic information system (GIS). Much of the biophysical and socioeconomic data that form the basis of EA are in digital map format (see Chapter 5), and carrying out EA requires that these map themes be overlaid, compared, analyzed, or used in a model (see Chapter 18), often in conjunction with nonspatial attribute data. Not long ago, GIS was a somewhat daunting new technology with a steep learning curve, but in the last decade it has become increasingly integrated into research, land-use and land management planning, ecosystem management, and related modeling at a variety of institutions and agencies. Therefore, this chapter will only briefly define GIS, point the reader to a number of excellent texts on the subject, and concentrate on how GIS is used in EA. Accordingly, the standard definition of GIS as a computer information system "for the capture, storage, retrieval, analysis and display of spatial data" (Clarke, 1995, p. 13) is enhanced by Goodchild's (1985, p. 36): "a GIS is best defined as a system which uses a spatial data base to provide answers to queries of a geographical nature."

The following sections discuss definitions of geographic information systems and science (11.2); institutional issues related to supporting a GIS for EA (11.3); data sources for GIS-based ecological assessments (11.4.1); effects of scale and accuracy of spatially explicit data on ecological assessment, analyses, and modeling (11.4.2 and 11.4.3); types of spatial data analysis that can support EA (11.5); and issues related to geographic information output (11.6).

The author is grateful to the anonymous reviewers and the technical editors for improving the quality of this chapter. I thank Patrick Bourgeron and Mark Jensen for inviting me to participate.

11.2 What Is a Geographic Information System?

Is GIS synonymous with GIS software? They are sometimes treated as if they were because the software systems are complex and specialized. However, a GIS would have to include the geographic database (digital maps, themes, or coverages—raster, point, line and polygon data), the software used to organize both the locational and attribute data (the information system) and to manipulate and query it, and the hardware required to capture or input the data (GPS, digitizer, scanner), manipulate and view it, and create hard- and soft-copy output. This paraphrases the textbook definition of a GIS, given above, and the components of a GIS are described in a number of excellent texts and references (Burrough, 1986; Star and Estes, 1990; Tomlin, 1990; Maguire et al., 1991; Star et al., 1997; Burrough and McDonnell, 1998; Clarke 1999). Also recommended is the National Center for Geographic Information and Analysis (NCGIA) GIS Core Curriculum Web site (http://ncgia.ncgia.ucsb.edu:80/giscc/) and web references therein for graphic introductions to GIS and spatial data analysis.

But is a GIS more than the sum of its parts? Burrough (1986, p. 7) asserted that the data in the GIS represent a model of the real world, and therefore the GIS serves as a test bed for studying environmental processes. Clarke (1999, p. 5) notes that GIS technology has changed our entire approach to spatial data analysis and, accordingly, Goodchild (1992a, p. 41) states that geographic information science should address "the generic issues that surround GIS technology, impeding its successful implementation, or emerge from an understanding of its potential capabilities." Subsequently, one flagship journal in the field changed its title from the *International Journal of Geographic Information Systems* to *IJGI Science*. GIS, like its sister technology, remote sensing, can be used with scientific or technological approaches to environmental problem solving (Curran, 1987).

Despite the growth of this literature on geographic information science, some GIS users—ecologists, land-use planners, and others—might profess (as one recently did to the author) that a GIS is simply one of the data management tools that they use, in the same category as spreadsheet software. However, a number of institutional issues related to GIS and methodological issues related to spatial data analysis distinguish it from other data storage support tools (spreadsheet and database software).

11.3 Institutional Issues

Institutional issues related to GIS are primarily related to the need for investments in software, hardware, and training. Low-cost or public-domain GIS software is available (GRASS, Idrisi, ArcView, for example), and some packages contain powerful tools for analyzing, or at least viewing, geographical data. High-cost software and hardware, however, are usually used to support activities requiring large amounts of data (geographical databases for large areas at high resolution) or sophisticated analyses (Morgan, 1987; Lauer et al., 1991; Clarke, 1999). Although until very recently a Unix-based (or comparable) workstation was needed to support GIS-based analysis, a high-end personal computer is now capable of supporting many of these activities (because of the increases in microcomputing power that have affected all computer users). Much of the available software requires substantial training for analysts who already have, or are concurrently acquiring, domain knowledge (geography, ecology, forestry, conservation biology, range management).

In addition to knowing how to use the programs or modules in the software package and write specialized applications programs, usually in a macro language, the GIS analysts in an organization must also understand (1) the nature of the input data, especially their source, currency, scale, and accuracy; (2) the fundamentals of geographic information science (Goodchild, 1992b), including coordinate geometry and topology, data structures and data models, algorithms and computational complexity, and spatial statistics and spatial analysis (Goodchild, 1985); and (3) the types of analyses that potentially can be applied to a particular problem. This is true whether the institution is a federal or other land management agency, a university research laboratory, or an environmental consulting firm, any of which could be involved in some capacity in an EA.

A number of other institutional issues are beyond the scope of this paper. These include the processes by which GIS technology is adopted by organizations and used in decision making (Lauer et al., 1991; Goodchild, 1992a; Obermeyer and Pinto, 1994) and ways of improving the efficacy of GIS in supporting management and planning decisions (Crossland et al., 1995; Zhu et al., 1998). Ethical and legal issues related to privacy (Curry, 1997), data ownership, and so forth, are also particularly relevant to GIS. Other issues will become increasingly important in the future; these include data security and access on computer networks, institutional policies regarding digital libraries and the global network, and cross-disciplinary community building.

11.4 Data Sources

Agencies and organizations involved in EA must also grapple with acquiring adequate data for their analytical needs. There are a growing number of sources of digital maps, as well as remotely sensed images and derived data products, notably those produced by the U.S. Geological Survey (USGS), and specialized mapping programs within land management and other agencies. The National Aeronautics and Space Administration's (NASA) Earth Observing System (EOS) and other programs promise expanding availability of global and regional remotely sensed imagery and data products. However, there is sometimes a misconception that much of the data required to conduct an ecological assessment or regional modeling effort already exist in an appropriate form and have only to be acquired, assembled, and analyzed. In fact, in most inventories and assessments that rely on digital spatial data, the majority of resources are still spent on developing the data, either because they are not available or because existing data are out of date or are not appropriate due to scale, accuracy, or the classification system used (Estes and Mooneyhan, 1994). On the other hand, existing digital imagery and maps are acquired much more easily now than they were even a few years ago because of access via the World Wide Web.

Stohlgren, in Chapter 5, notes that 20% to 25% of a budget for an EA must be devoted to data management. It could be argued, in the case of GIS data, that this figure is reasonable subsequent to acquisition of air photos or imagery or data development by field-based mapping, spatial interpolation, image interpretation, or modeling. Resources for developing the input data, if they are not already available, could equal or exceed by an order of magnitude the other costs of an EA. For example, detailed maps of vegetation that included canopy cover and crown size estimates for forested stands were produced for several million hectares of Forest Service lands in California; at 1 : 24,000 scale and 2-ha minimum mapping unit, the cost was about $0.30/ha (Franklin and Woodcock, 1997). On the other hand, interpolated climate surfaces might be produced for only a few cents per square kilometer. Other existing data, such as digital terrain models, may have been developed under federal programs and be available at low cost unrepresentative of the actual production costs of these data (Estes and Mooneyhan, 1994).

11.4.1 Data Sources

Digital spatial data essential to some forms of ecological assessment, especially at landscape or regional scales, include (1) remotely sensed data (satellite imagery, digital orthophotographs, aerial photography, or digital airborne camera data); (2) interpolated climate data; (3) digital elevation models; (4) digital cartographic data such as we find on a USGS 1 : 24,000 topographic map (transportation routes, hydrography, land ownership); and (5) thematic maps of environmental variables, including land use, landform, soils, geology, and vegetation (such as those compiled by the USGS, Forest Service, National Park Service, and Gap Analysis Program). Remotely sensed data were discussed in Chapter 10. In this chapter we will briefly discuss the other four types of data, especially with respect to the following: Were the spatial data created for the first time in a digital form (through remote sensing or by interpolation), or were they digitized from traditional maps?

Interpolated climate data are useful in ecosystem characterization (see Chapters 22 through 29) for predictively mapping vegetation or other ecosystem characteristics that are otherwise difficult to map (see Section 11.5) and as input to ecosystem models (discussed in Chapter 18). These may include interpolated surfaces of long-term climate (Hutchinson, 1987, 1996; Daly et al., 1994; Hutchinson and Gessler, 1994; Daly and Taylor, 1996; Hutchinson et al., 1996) or daily weather variables (Running et al., 1987; Glassy and Running, 1994; Running and Thornton, 1996). Some commercial data products are available; however, there is still considerable debate over the most appropriate methods for interpolation (Hutchinson, 1984; Hutchinson and Gessler, 1994). These climate surfaces can be generated for a specific region from climate station data by a climatologist with expertise in spatial analysis methods.

Digital elevation models (DEM), representing land surface topography, are essential for a number of purposes related to EA. Elevation and simple and complex terrain attributes derived from a DEM, such as slope and aspect, may be used directly in ecosystem characterization. More frequently, however, they are used to model the distributions of complex ecological or biophysical patterns and processes, such as site potential, vegetation type, plant or animal species distributions, potential solar insolation, and topographic moisture (Moore et al. 1991; McNab, 1993; Dubayah and Rich, 1995; reviewed in Franklin, 1995; Franklin et al., 2000). DEMs also are commonly used in ecosystem models, especially of hydrologically mediated processes (see Chapter 18). Although standard data products are available from the National Mapping Division of the USGS, these data are derived from existing topographic maps that may be more than 40 years old or are produced using photogrammetric methods that are prone to systematic and nonsystematic errors (Weibel and Heller, 1991; Estes and Mooneyhan, 1994). Although in the future greatly expanded sources of digital elevation models will include high-resolution stereo satellite imagery, the USGS products are currently the best or only source of data for most of the United States. Gridded (raster) products have, at best, 30-m cell size (resolution), whereas research indicates that DEMs of 10- to 20-m resolution are required to accurately depict topographic features and model topohydrological and other biophysical processes (Hutchinson et al., 1996; Hutchinson, 2000). A global digital elevation model with 30 arc second resolution, GTOPO30, is now available from USGS (http://edcwww.cr.usgs.gov/landdaac/gtopo30/ gtopo30.html).

In addition to the elevation contours, the other information found on USGS topographic quadrangle maps (hydrography, roads and water bodies, land ownership) is available from the USGS in vec-

TABLE 11.1. Geospatial data types and sources needed for developing vegetation data for ecosystem assessment.

Theme	Scale, description	Source	Use	Processing and analysis
Landsat Thematic Mapper imagery	30 m	NASA, EROS Data Center (via Forest Service)	Identify vegetation formations and land cover	Unsupervised classification, segmentation
Resource air photos	Variable; 1 : 12,000 to 1 : 30,000	Forest Service	Conduct field reconnaissance and label vegetation map features	Aerial photo interpretation
Digital elevation models (DEMs)	30-m, 7.5′ quadrangles	USGS (via Forest Service)	Model potential vegetation series	Mosaic, smooth, derive slope, aspect
Cartographic Feature Files (CFFs): land ownership, forest boundaries, hydrology, roads	Variable; usually 1 : 24,000	Forest Service (from USGS digital line graphs and various digitized sources)	Isolate study area, develop draft hard-copy maps for field checking	Overlay, georeference, edit
Existing digital thematic maps (forest plantations, previous vegetation maps)	Variable; usually 1 : 24,000	Forest Service	Label certain features in vegetation map (plantations)	Overlay, map editing

Example from a recently completed project to map existing vegetation on Forest Service lands in Southern California carried out at San Diego State University (Franklin and Woodcock, 1997; Franklin et al., 2000).

tor form as *digital line graphs* (DLGs). These represent fundamental data for an EA involving any type of cartographic modeling (discussed later) for which information such as the distribution of a resource by land ownership or in proximity to roads is needed. However, as with the raster DEMs discussed previously, these vector data are only as good as the maps from which they were derived (which may be outdated and contain errors) and the quality of the digital conversion process (never free from errors). For example, in our work in southern California, we found that the stream network represented in the DLG contained some errors in the location of streams when compared to georeferenced satellite imagery, but, more importantly, lacked information on lower-order (intermittent, seasonal) streams associated with important riparian ecosystems (Franklin et al., 2000). Owing to the importance of accurate digital maps of such basic features as land ownership and land use (wilderness, for example) for land management planning and EA, many federal, state, and regional agencies are beginning to develop their own data sets. These are built on DLGs or other widely available products, which are then mosaicked, edited, corrected, and updated.

One way to develop digital maps of more complex landscape features, such as soils, landforms, geology, or vegetation types, is to digitize existing maps. This is especially useful for small-scale relatively static features, such as geology, and for variables with source maps developed through tremendous expenditure of resources and input of expertise, such as soils. A problem with digitizing these types of maps is that they were probably developed using a traditional cartographic approach and a "communicative paradigm" (DeMers, 1991), through which the purpose of the map was to present a visual model of the spatial distribution of the phenomenon at a particular scale (Goodchild, 1988). The data used to develop this model were probably messy or noisy and have been smoothed using cartographic methods. However, in GIS-based analysis, we probably want to know the spatial distribution of the phenomenon with more precision and less smoothing, as well as with some representation of the spatial distribution of the variance of the phenomenon or uncertainty in the map (discussed later). Also, hard-copy source maps were produced at a particular map scale that can be generalized to a coarser scale in the GIS, but cannot be made finer without implying false precision.

Alternatively, over the last few decades, mapped data have increasingly been developed for the first time in digital form, especially using remotely sensed data. These include vegetation and land cover data at global and regional scales (Tucker et al., 1985; Townshend et al., 1987; Loveland et al.,

TABLE 11.2. Geospatial data types and sources needed for developing vegetation data for ecosystem assessment.

Theme	Scale, description	Source	Use	Processing and analysis
SPOT Panchromatic	10-m-resolution panchromatic satellite imagery, variable recent dates	SPOT Image Corporation (via BRD)	Base maps for on-screen digitizing vegetation boundaries	Contrast enhancement; on-screen digitizing GIS vegetation layer registered to SPOT
DEM	1 : 24,000 (for sampling, only a mosaic of 1 : 24,000 and 1 : 100,000 scale data of variable quality were available)	BLM, quality checked and mosaicked to "seamless" coverage	Allocate plot locations for field vegetation sampling; label vegetation polygons through predictive modeling	Mosaic; derive terrain variables (slope, aspect, topographic moisture index); stratify into terrain classes for sampling; use in modeling
Climate maps	1-km-resolution interpolated temperature and precipitation variables	University of California, Santa Barbara (via BRD)	"	Stratify into classes, overlay for sampling; use continuous variables in modeling
Geology map	1 : 6,000,000, digitized from preexisting state geology map	State of California, Teale Data Center (via BRD)	"	Aggregate into classes, overlay for sampling
Landform map	1 : 100,000, being developed in a concurrent project	Louisiana State University, Army Corps of Engineers (sponsored by DOD)	Label vegetation polygons through predictive modeling	Use in modeling vegetation labels

Example from the Mojave Desert Ecosystem vegetation mapping project at San Diego State University, coordinated by USGS Biological Resources Division (BRD) under sponsorship of the Department of Defense (DOD).

1991; Fuller et al., 1994; Stone et al., 1994; Zhu and Evans, 1994; Davis et al., 1995; Nemani and Running, 1997). It is only within perhaps the last five years that these data have been widely disseminated (particularly using the Internet) and that impediments to searching and browsing large spatial databases (Ehlers et al., 1991) are being overcome. However, current activities by a number of important land management agencies, including the Forest Service, Park Service, Bureau of Land Management, military, and state and regional entities, suggest that up to date, detailed, large-area digital maps of these themes and the ability to update them are critical to ecosystem management at the ecoregional scale. Consequently, improved methods for large-area mapping of properties of the vegetation, soil, wildlife habitat, and other aspects of the ecosystem are still an active area of research (e.g., see the review by Franklin, 1995, and Chapter 5).

For example, Tables 11.1 and 11.2 show, for recent and ongoing projects that the author has directed, the variety of data required to develop digital maps of existing vegetation for large regions. As mentioned earlier, in conjunction with the Forest Service in California, we developed fine-resolution (2-ha minimum mapping unit) maps of vegetation type and structure for the National Forests of southern and central coastal California, about 2 million ha (Table 11.1; see Franklin and Woodcock, 1997; Franklin et al., 2000). We are currently developing maps of vegetation alliances for about 5.6 million ha in the California portion of the Mojave Desert ecoregion, in conjunction with the Biological Resources Division of the USGS (Table 11.2). This is part of a larger Mojave Desert Ecosystem Science Program, a joint Department of Interior and Department of Defense enterprise (e.g., http://wrgis.wr.usgs.gov/MojaveEco/ and http://mojave.army.mil:90/Home/home.html). In each of these projects, diverse geospatial data are used as input to derive a complex variable or variables—vegetation composition and structure. These include the types of data discussed previously. The output, a digital vegetation database, is then used, often in conjunction with some of these same data sets, for higher-order applications (ecosystem management planning, EA). The lineage of some of the data sets that we received from our collaborators is complex, highlighting the importance of metadata. When evaluating the suitability of various data sets for our modeling purposes, one important criterion was its spatial scale or resolution.

11.4.2 Scale

Scale, both spatial and temporal, has been an important theme in science generally and ecology specifically (Allen and Star, 1982; Weins, 1989; Levin, 1992). Relevant to this chapter is the spatial scale or resolution (grain) and extent of geographical data used for studying Earth system processes (Meentemeyer, 1989; Lam and Quattrochi, 1992; Ehleringer and Field, 1993; Foody and Curran, 1994; Stewart et al., 1996; Quattrochi and Goodchild, 1997; van Gardingen et al., 1997).

Whereas Goodchild (1992b) noted that cartographic scale reflects a view of the database and is not a property of the database itself or of geographical reality, ecological processes may have characteristic spatiotemporal scales, and spatial databases may have been collected, modeled, or interpolated at a specific resolution. Therefore, we must exercise caution and good judgment in using spatial data to model an ecological distribution, drive a process model, or combine with data of greatly different resolution in a cartographic model (in which digital map layers are combined using Boolean operators). A global-scale soils database might be useless for predicting the distribution of a locally endemic plant in San Diego County or driving the CENTURY model for eastern Colorado (see Chapter 18). Generalized land ownership data for the state of California would not be used to design a habitat conservation area in the city of Escondido. A 1-meter-resolution DEM with centimeter vertical accuracy, developed from kinematic GPS (global positioning system) survey, would not be needed to model energy balance for a large watershed (Dubayah, 1994), but might be necessary to find suitable breeding habitat for an endangered bird nesting in a coastal salt marsh (Brewster, 1996).

11.4.3 Error and Accuracy

Spatial data accuracy has been a research focus in geographical information science for more than a decade (Goodchild and Gopal, 1989; Veregin, 1989; Lunetta et al., 1991). Issues of data and metadata quality are important to all types of scientific data management (Chapter 5), but some aspects of spatial data are unique in this regard (also see Section 11.6). One is that maps on paper or computer screens, with their sharp boundaries, interesting colors, and references to real places, seem to carry more authority than tables, pie charts, or histograms when viewed by land managers, decision makers, or the public (see Monmonier, 1996). This is in spite of the fact that they represent environmental phenomena, such as vegetation cover or water quality, with precision and accuracy that vary spatially and may not be fully specified. Furthermore, a map may be the output of a model (cartographic overlay, statistical, process, or other) built on several mapped input variables whose respective errors are propagated through that model.

Second, maps are models (representations) approximating complex spatial distributions (geographical reality sensu Goodchild, 1992b) that can vary continuously or with abrupt boundaries. These approximations are made with some uncertainty. Although it is highly desirable to quantify the level of uncertainty, as with confidence limits around an estimate, it can be very difficult with spatial data (Goodchild et al., 1992). But the important conceptual issue is that the "error" in digital maps of biophysical variables is not so much a cartographic "mistake" (Goodchild, 1988) as it is uncertainty in a spatially distributed estimate, akin to uncertainty or variance in other scientific measurements (Goodchild, 1994).

A number of methods, both simple and sophisticated, now exist for quantifying error or uncertainty in maps of continuous and categorical variables and in maps resulting from overlay operations (Goodchild et al., 1992; Goodchild et al., 1993; Heuvelink and Burrough, 1993; Veregin, 1995). Also, the effects of geographical data uncertainty on the output of spatially explicit process models can be explored through simulation and sensitivity analysis (Lodwick et al., 1990; Stoms et al., 1992; Journel, 1996). Error analysis or sensitivity analysis is an essential part of every EA that uses GIS data and modeling for spatial decision support, but one that is frequently overlooked.

11.5 Data Analysis

In the classic definition of GIS presented in the introduction of this chapter, data analysis was deemphasized. This accurately reflects the development of the technology and software, which in large part placed the challenges of data structures, format, storage, input, and output ahead of analysis (Goodchild, 1988; 1992a). It was (and is) often necessary to move data out of a GIS software package in order to execute statistical or process models. GIS and statistical software systems have lacked appropriate tools for statistical analysis of spatial data (Goodchild, 1992a; Anselin, 1993). However, several software packages for spatial statistical analysis now exist and are being more widely used (e.g., Anselin, 1999, and http://spacestat.com/).

The lack of integration of GIS and environmental models was a key topic of a series of international workshops and symposia (Goodchild et al., 1993; Goodchild et al., 1996; http://bbq.ncgia.ucsb.edu/conf/santa_fe/papers). Some have argued that GIS and applications software should be integrated (Parks, 1993), but it is getting much easier and more transparent to move data between software packages; this offers an increasingly viable option in spatial data modeling. The GIS may carry a tremendous overhead in terms of data storage (topology), and only a subset of the data may be required for an analysis. Spatially explicit environmental modeling is advancing on both fronts. Models are developed outside the GIS and spatial data are ported in and out with increasing ease (e.g., He and Mladenoff, 1999). Modeling tools that are integrated into a GIS are also being developed (e.g., see a dynamic modeling language embedded within a GIS, PCRaster, http://www.geog.uu.nl/pcraster).

11.5.1 Data Overlay and Map Algebra

The form of analysis that always has been well integrated into GIS software is map algebra or cartographic modeling—the arithmetic of combining map layers that are co-registered and of identical resolution (Burrough, 1986; Tomlin, 1990). Sometimes, viewing the spatial distribution of a variable (looking at a map) or the co-occurrence of two variables (map overlay) can be a very powerful source of information. Analogous to viewing a scatterplot in exploratory statistical data analysis, the locations of the intersection or union of several mapped variables can provide the information required for a land suitability analysis, as was recognized early on by McHarg (1969). Frequently, cartographic overlay is considered to be synonymous with analysis in GIS. However, digital spatial data can also be used in spatial extensions of statistical and process models and in analyses of spatial pattern.

11.5.2 Other Forms of Modeling in GIS

Of the multitude of potential types of spatial data modeling, we will discuss one that is particularly relevant to EA and mention several others that are covered in other chapters or that represent research challenges for the future. As discussed in Chapter 5, two important complementary uses of GIS that have yet to be fully realized for EA are stratified, unbiased, efficient selection of sites for field survey and predictive mapping as a method of extrapolating from those survey data. In addition to the references cited in that chapter, explicit methods for survey design were well articulated by Austin and Heyligers (1989, 1991), Austin and Adomeit (1991), and Cocks and Baird (1991). Statistical or other forms of (static) predictive modeling of ecological patterns (species distributions, soil properties, site potential) are a way to interpolate among expensive, sparse field survey data. This approach relies on the relationships between those ecological patterns and underlying environmental variables (climate, terrain, hydrology) that may be more easily mapped (reviewed by Elston and Buckland, 1993; Hunsaker et al., 1993; Franklin, 1995; Franklin et al., 2000).

Spatially explicit simulation models of ecological processes, supported by GIS, are discussed in Chapter 18. A current research issue in geographical information science, relevant to the integration of GIS and these process models, is the temporal data handling and modeling capabilities of GIS (for examples see Newell et al., 1992; Peuquet, 1994; Kraak et al., 1995; Egenhofer and Golledge, 1998).

11.6 Information Output

Discussed next are two issues related to spatial information output: data standards and visualizing scientific information. Data and metadata quality and standards are becoming increasingly important in the management of all types of scientific data including spatial data. The National Institute of Standards and Technology (1992) has adopted Federal Information Processing Standards for spatial data (Spatial Data Transfer Standards—SDTS). The Federal Geographic Data Committee (FGDC) "coordinates the development of the National Spatial Data Infrastructure (NSDI). The NSDI encompasses policies, standards, and procedures for organizations to cooperatively produce and share geographic data. The 16 federal agencies that make up the FGDC are developing the NSDI in cooperation with organizations from state, local and tribal governments, the academic community, and the private sector" (http://www.fgdc.gov/). The FGDC recently approved a revised Content Standard for Digital Geospatial Metadata (see the FGDC Web page). By executive order, data sets released by government agencies after 1994 must comply with these data and metadata standards. Federal standards will soon include (International Standards

Organization (ISO) standards so that data sets can be shared globally.

Ideas from the field of scientific visualization have implications for the ways that information output resulting from spatial data analysis with a GIS are displayed, perceived, and interpreted. Buttenfield (1996, p. 463) states that "GIS technology relies heavily on user visualization skills. . . . Poorly designed displays convey false ideas, and they bias analysis and interpretation. Many GIS users are not trained in graphical design, nor should they be." Butterfield suggests elements of map design and visualization methods that could help correct these problems for both static and animated map display (see also Hearnshaw and Unwin, 1994; MacEchren and Fraser Taylor, 1994).

11.7 Conclusion

It is difficult to envision an ecological assessment taking place without the support of a GIS. Consequently, several issues from geographical information science must be addressed in order for the geographical analyses that support an EA to be valid and defensible.

1. Those who are conducting GIS analysis in support of EA are most effective when they are trained in the fundamentals of both geographical information science and ecosystem management—and have the methodological and technical skills required to "drive" a GIS.
2. Although the availability of large spatial data sets to support ecosystem analysis has increased, this availability should never be assumed nor the cost of developing them underestimated. The appropriateness of the scale, precision, classification scheme, and other characteristics of existing digital maps to the task at hand should always be carefully considered.
3. Geographical analyses in support of EA can be very powerful tools, even when based on simple cartographic overlay modeling; however, the results of these analyses can be spurious or misleading when the effect of data uncertainty on model output is not calculated or simulated.
4. Two fruitful areas of further research and development are (a) closer integration of GIS and the ecological and resource management models used in EA and (b) developing better decision support and visualization tools that help decision makers to see the uncertainty in both the input data and the model outcome.

11.8 References

Allen, T. H. F.; Starr, T. B. 1982. *Hierarchy: perspectives for ecological complexity*. Chicago: University of Chicago Press.

Anselin, L. 1993. Discrete spatial autoregressive models. In: Goodchild, M. F.; Parks, B. O.; Steyaert, L. T., eds. *Environmental modeling with GIS*. New York: Oxford University Press: 454–469.

Anselin, L. 1999. Interactive techniques and exploratory spatial data analysis. In: Longley, P.; Goodchild, M. F.; Maguire, D.; Rhind, D., eds. *Geographical information systems: principles, techniques, management and applications*. New York: John Wiley & Sons: 253–266.

Austin, M. F.; Adomeit, E. M. 1991. Sampling strategies costed by simulation. In: Margules, C. R.; Austin, M. P., eds. *Nature conservation: cost effective biological surveys and data analysis*. East Melbourne, Australia: Commonwealth Scientific and Industrial Research Organization, 167–175.

Austin, M. F.; Heyligers, P. C. 1989. Vegetation survey design for conservation: gradsect sampling of forests in North-eastern New South Wales. *Biol. Conserv.* 50:13–32.

Austin, M. F.; Heyligers, F. C. 1991. New approaches to vegetation sample survey design: gradsect sampling. In: Margules, C. R.; Austin, M. P., eds. *Nature conservation: cost effective biological surveys and data analysis*. East Melbourne, Australia: Commonwealth Scientific and Industrial Research Organization: 31–36.

Brewster, A. E. 1996. *Utilizing geographic technologies to analyze the nesting habitat preferences of the Belding's Savannah sparrow*. Master of Arts thesis. San Diego, CA: Department of Geography, San Diego State University.

Burrough, P. A. 1986. *Principles of geographical information systems for land resources assessment*. Oxford, UK: Clarendon Press.

Burrough, P. A.; McDonnell, R. 1998. *Principles of geographical information systems (spatial information systems and geostatistics)*, New York: Oxford University Press.

Buttenfield, B. P. 1996. Scientific visualization for environmental modeling: interactive and proactive graphics. In: Goodchild, M. F.; Steyaert, L. T.; Parks, B. O., eds. *GIS and environmental modeling: progress and research issues*. Fort Collins, CO: GIS World Books: 463–467.

Clarke, K. C. 1995. *Analytical and computer cartography*, 2nd ed. Upper Saddle River, NJ: Prentice Hall.

Clarke, K. C. 1999. *Getting started with geographic information systems*, 2nd ed. Upper Saddle River, NJ: Prentice Hall.

Cocks, K. D.; Baird, L. A. 1991. The role of geographic information systems in the collection, extrapolation and use of survey data. In: Margules, C. R.; Austin, M. P., eds. *Nature conservation: cost effective biological surveys and data analysis*. East Melbourne,

Australia: Commonwealth Scientific and Industrial Research Organization: 74–80.

Crossland, M. D.; Wynne, B. E.; Perkins, W. C. 1995. Spatial decision support systems: An overview of technology and a test of efficacy. *Decis. Support Syst.* 14:219–235.

Curran, P. J. 1987. Remote sensing methodologies and geography. *Int. J. Remote Sens.* 8:1255–1275.

Curry, M. R. 1997. The digital individual and the private realm. *Ann. Assoc. Amer. Geogr.*: 681–699.

Daly, C.; Taylor, G. H. 1996. Development of a new Oregon precipitation map using the PRISM model. In: Goodchild, M. F.; Steyaert, L. T.; Parks, B. O., eds. *GIS and environmental modeling: progress and research issues*. Fort Collins, CO: GIS World Books: 91–92.

Daly, C.; Neilson, R. P.; Phillips, D. L. 1994. A statistical–topographic model for mapping climatological precipitation over mountainous terrain. *J. Appl. Meteorol.* 33:140–158.

Davis, F. W.; Stine, P. A.; Stoms, D. M.; Borchert, M. I.; Hollander, A. 1995. Gap analysis of the actual vegetation of California: 1. The southwestern region. *Madrono* 42:40–78.

DeMers, M. 1991. Classification and purpose in automated vegetation maps. *Geogr. Rev.* 81:267–280.

Dubayah, R. C. 1994. Modeling a solar radiation topoclimatology for the Rio Grande River Basin. *J. Veg. Sci.* 5:627–640.

Dubayah, R.; Rich, P. M. 1995. Topographic solar radiation for GIS. *Int. J. Geogr. Inf. Syst.* 9:405–419.

Egenhofer, M. J.; Golledge, R. G. 1998. *Spatial and temporal reasoning in geographic information systems*. New York: Oxford University Press.

Ehleringer, J. R.; Field, C. B., editors. 1993. *Scaling physiological processes: leaf to globe*. San Diego, CA: Academic Press.

Ehlers, M.; Greenlee, D.; Smith, T.; Star, J. 1991. Integration of remote sensing and GIS: data and data access. *Photogramm. Eng. Remote Sensing* 57:669–675.

Elston, D. A.; Buckland, S. T. 1993. Statistical modelling of regional GIS data: an overview. *Ecol. Model.* 67:81–102.

Estes, J. E.; Mooneyhan, D. W. 1994. Of maps and myths. *Photogramm. Eng. Remote Sensing* 60:517–524.

Foody, G. M.; Curran, P. J., editors. 1994. *Environmental remote sensing from regional to global scales*. Chichester, UK: John Wiley & Sons.

Franklin, J. 1995. Predictive vegetation mapping: geographic modeling of biospatial patterns in relation to environmental gradients. *Prog. Phys. Geog.* 19:474–499.

Franklin, J.; Woodcock, C. E. 1997. Multiscale vegetation data for the mountains of Southern California: spatial and categorical resolution. In: Quattrochi, D. A.; Goodchild, M. F., eds. *Scale in remote sensing and GIS*. Boca Raton, FL: CRC/Lewis Publishers: 141–168.

Franklin, J.; McCullough, P.; Gray, C. 2000. Terrain variables used for predictive mapping of vegetation communities in Southern California. In: Gallant J.; Wilson, J., eds. *Terrain analysis: principles and applications*. New York: John Wiley & Sons: 331–353.

Franklin, J. C.; Woodcock, C. E.; Warbington, R. 2000. Digital vegetation maps of forest lands in California: Integrating satellite imagery, GIS modeling, and field data in support of resource management. *Photogrammetric Engineering and Remote Sensing* 66:1209–1217.

Franklin, J. F. 1993a. The fundamentals of ecosystem management with applications in the Pacific Northwest. In: Aplet, G. H., et al., eds. *Defining sustainable forestry*. Washington, DC: Island Press: 127–143.

Franklin, J. F. 1993b. Preserving biodiversity: species, ecosystems, or landscapes? *Ecol. Appl.* 3:202–205.

Fuller, R. M.; Groom, G. B.; Jones, A. R. 1994. The land cover map of Great Britain: an automated classification of Landsat Thematic Mapper data. *Photogramm. Eng. Remote Sensing* 60:553–562.

Glassy, J. M.; Running, S. W. 1994. Validating diurnal climatology logic of the MT-CLIM model across a climatic gradient in Oregon. *Ecol. Appl.* 4:248–257.

Goodchild, M.F. 1985. Geographic information systems in undergraduate geography: a contemporary dilemma. *Operational Geographer* 8:34–38.

Goodchild, M. F. 1988. Stepping over the line: technological constraints and the new cartography. *Amer. Cartographer* 15:311–319.

Goodchild, M. F. 1992a. Geographical data modeling. *Comput. Geosci.* 18:401–408.

Goodchild, M. F. 1992b. Geographical information science. *Int. J. Geogr. Inf. Syst.* 6:31–45.

Goodchild, M. F. 1994. Integrating GIS and remote sensing for vegetation analysis and modeling: methodological issues. *J. Veg. Sci.* 5:615–626.

Goodchild, M. F.; Gopal, S., editors. 1989. *The accuracy of spatial databases*. London: Taylor & Francis.

Goodchild, M. F.; Guoqing, S.; Shiren, Y. 1992. Development and test of an error model for categorical data. *Int. J. Geogr. Inf. Syst.* 6:87–104.

Goodchild, M. F.; Parks, B. O.; Stayaert, L. T., editors. 1993. *Environmental modeling with GIS*. New York: Oxford University Press.

Goodchild, M. F.; Steyaert, L. T.; Parks, B. O., editors. 1996. *GIS and environmental modeling: progress and research issues*. Fort Collins, CO: GIS World Books.

He, H.; Mladenoff, D. J. 1999. Dynamics of fire disturbance and succession on a heterogeneous forest landscape: a spatially explicit and stochastic simulation approach. *Ecology* 80:81–99.

Hearnshaw, H. M.; Unwin, D. J., editors. 1994. *Visualization in geographical information systems*. New York: John Wiley & Sons.

Heuvelink, G. B. M.; Burrough, P. A. 1993. Error propagation in cartographic modelling using Boolean logic and continuous classification. *Int. J. Remote Sens.* 7:231–246.

Hunsaker, C. T.; Nisbet, R. A.; Lam, D. C. L.; Browder, J. A.; Baker, W. L.; Turner, M. G.; Botkin, D. B. 1993.

Spatial models of ecological systems and processes: the role of GIS. In: Goodchild, M. F.; Parks, B. O.; Stayaert, L. T., eds. *Environmental modeling with GIS*. New York: Oxford University Press: 248–264.

Hutchinson, M. F. 1984. *A summary of some surface fitting and contouring programs for noisy data*. Volume Consulting Report No. ACT 84/6. Canberra, ACT, Australia: Commonwealth Scientific and Industrial Research Organization, Division of Mathematics and Statistics and Division of Water and Land Resources.

Hutchinson, M. F. 1987. Methods for generation of weather sequences. Bunting, A. H., editor. *Agricultural environments: characterisation, classification and mapping*. Walingford: CAB International: 149–157.

Hutchinson, M. F. 1996. Thin plate spline interpolation of mean rainfall. In: Goodchild, M. F.; Steyaert, L. T.; Parks, B. O., eds. *GIS and environmental modeling: progress and research issues*. Fort Collins, CO: GIS World Books: 85–90.

Hutchinson, M. F.; Gallant, J. C. 2000. Digital elevation models and representation of terrain shape. In: Gallant J.; Wilson, J., eds. *Terrain analysis: principles and applications*. New York: John Wiley & Sons: 29–50.

Hutchinson, M. F.; Gessler, P. E. 1994. Splines—more than just a smooth interpolator. *Geoderma* 62:45–67.

Hutchinson, M. F.; Nix, H. A.; McMahon, J. P.; Ord K. D. 1996. The development of a topographic and climatic database for Africa. *Proceedings, Third International Conference on Integrating GIS and Environmental Modeling*, Santa Fe, NM, January 21–16, 1996. Santa Barbara, CA: National Center for Geographic Information and Analysis (http://www.ncgia.ucsb.edu/conf/SANTAFECD-ROM/main.html).

Journel, A. G. 1996. Modelling uncertainty and spatial dependence: stochastic imaging. *Int. J. Geogr. Inf. Syst.* 10:517–522.

Kohm, K. A.; Franklin, J. F. 1997. *Creating a forestry for the 21st century: the science of ecosystem management*. Washington, DC: Island Press.

Kraak, M. J.; Mueller, J.C.; Ormeling, F. 1995. GIS cartography: visual decision support for spatio-temporal data handling. *Int. J. Geogr. Inf. Syst.* 9:637–645.

Lam, N. S. N.; Quattrochi, D. A. 1992. On the issues of scale, resolution and fractal analysis in the mapping sciences. *Prof. Geogr.* 44:88–98.

Lauer, D. T.; Estes, J. E.; Jensen, J. R.; Greenlee, D. D. 1991. Institutional issues affecting the integration and use of remotely sensed data and geographic information systems. *Photogramm. Eng. Remote Sensing* 57:647–654.

Levin, S. A. 1992. The problem of pattern and scale in ecology. *Ecology* 73:1943–1967.

Lodwick, W. A.; Monson, W.; Svoboda, L. 1990. Attribute error and sensitivity analysis of map operations in geographical information systems: suitability analysis. *Int. J. Geogr. Inf. Syst.* 4:413–428.

Loveland, T. R.; Merchant, J. W.; Ohlen, D.; Brown, J. F. 1991. Development of a land cover characteristics database for the conterminous U.S. *Photogramm. Eng. Remote Sensing* 57:1453–1463.

Lunetta, R. S.; Congalton, R. G.; Fenstermaker, L. K.; Jensen, J. R.; McGwire, K. C.; Tinney, L. R. 1991. Remote sensing and geographic information system data integration: error sources and research issues. *Photogramm. Eng. Remote Sensing* 57:677–687.

MacEchren, A. M.; Fraser Taylor, D. R., editors. 1994. *Visualization in modern cartography*. Kidlington, Oxford, UK: Pergamon/Elsevier Science.

Maguire, D. J.; Goodchild, M. F;. Rhind, D. W., editors. 1991. *Geographical information systems: principles and applications*. New York: John Wiley & Sons.

McHarg, I. L. 1969. *Design with nature*. New York: John Wiley & Sons.

McNab, W. H. 1993. A topographic index to quantify the effect of mesoscale landform on site productivity. *Can. J. For. Res.* 23:1100–1107.

Meentemeyer, V. 1989. Geographical perspectives of space, time and scale. *Landscape Ecol.* 3:163–173.

Monmonier, M. S. 1996. *How to lie with maps*, 2nd ed. Chicago: University of Chicago Press.

Moore, I. D.; Grayson, R. B.; Ladson, A. R. 1991. Digital terrain modeling: a review of hydrological, geomorphologic and biological applications. *Hydrol. Process.* 5:3–30.

Morgan, J. M., III. 1987. Academic geographic information systems education: a commentary. *Photogramm. Eng. Remote Sensing* 53:1443–1445.

National Institute of Standards and Technology. 1992. FIPS PUB 173: *Spatial data transfer standards (SDTS)*. Washington, DC: U.S. Dept. Commerce.

Nemani, R.; Running, S. 1997. Land cover characterization using multitemporal red, near-IR, and thermal-IR data from NOAA/AVHRR. *Ecol. Appl.* 1:79–90.

Newell, R. G.; Theriault, D.; Easterfield, M. 1992. Temporal GIS B modeling the evolution of spatial data in time. *Comput. Geosci.* 18:427–433.

Obermeyer, N. J.; Pinto, J. K. 1994. *Managing geographic information systems*. New York: Guilford Press.

Parks, B. O. 1993. The need for integration. In: Goodchild, M. F.; Parks, B. O.; Stayaert, L. T., eds. *Environmental modeling with GIS*. New York: Oxford University Press: 31–34.

Peuquet, D. J. 1994. It's about time: a conceptual framework for the representation of temporal dynamics in Geographic Information Systems. *Ann. Assoc. Amer. Geogr.* 84:441–461.

Quattrochi, D. A.; Goodchild, M. F., editors. 1997. *Scale in remote sensing and GIS*. Boca Raton, FL: CRC/Lewis Publishers Inc.

Running, S. W.; Thornton, P. E. 1996. Generating daily surfaces of temperature and precipitation over complex topography. In: Goodchild, M. F.; Steyaert, L. T.; Parks, B. O., editors. *GIS and environmental modeling: progress and research issues*. Fort Collins, CO: GIS World Books: 93–98.

Running, S. W.; Nemani, R. R.; Hungerford, R. D. 1987. Extrapolation of synoptic meteorological data in mountainous terrain and its use for simulating forest evapotranspiration and photosynthesis. *Can. J. For. Res.* 17:472–483.

Star, J. L.; Estes, J. E. 1990. *Geographic information systems: an introduction*. Upper Saddle River, NJ: Prentice Hall.

Star, J. L.; Estes, J. E.; McGwire, K. C., editors. 1997. *Integration of geographic information systems and remote sensing*. Cambridge, UK: Cambridge University Press.

Stewart, J. B.; Engman, E. T.; Feddes, R. A.; Kerr, Y., editors. 1996. *Scaling-up in hydrology using remote sensing*. Chichester, UK: John Wiley & Sons.

Stoms, D. M.; Davis, F. W.; Cogan, C. B. 1992. Sensitivity of wildlife habitat models to uncertainties in GIS data. *Photogramm. Eng. Remote Sensing* 58:843–850.

Stone, T. A.; Schlesinger, P.; Houghton, R. A.; Woodwell, G. M. 1994. A map of the vegetation of South America based on satellite imagery. *Photogramm. Eng. Remote Sensing* 60:541–551.

Tomlin, C. D. 1990. *Geographic information systems and cartographic modeling*. Upper Saddle River, NJ: Prentice Hall.

Townshend, J. R. G.; Justice, C. O.; Kalb, V. 1987. Characterization and classification of South American land-cover types using satellite data. *Int. J. Remote Sens.* 8:1189–1207.

Tucker, C. J.; Townshend, J. R. G.; Goff, T. E. 1985. African land cover classification using satellite data. *Science* 227:369–375.

van Gardingen, P. R.; Foody, G. M.; Curran, P. J., editors. 1997. *Scaling up from cell to landscape*. Cambridge, UK: Cambridge University Press.

Veregin, H. 1989. *A taxonomy of error in spatial databases*. Santa Barbara, CA: National Center for Geographic Information and Analysis.

Veregin, H. 1995. Developing and testing of an error propagation model for GIS overlay operations. *Int. J. Geogr. Inf. Syst.* 9:595–619.

Weibel, R.; Heller, M. 1991. Digital terrain modeling. In: Maguire, D. J.; Goodchild, M. F.; Rhind, D. W., eds. *Geographic information systems. Volume 1: Principles*. Harlow, UK: Longman: 269–297.

Weins, J. A. 1989. Spatial scale in ecology. *Funct. Ecol.* 3:311–332.

Zhu, Z.; Evans, D. L. 1994. U.S. forest types and predicted percent forest cover from AVHRR data. *Photogramm. Eng. Remote Sensing* 60:525–531.

Zhu, X.; Healey, R. G.; Aspinall, R. J. 1998. A knowledge-based systems approach to design of spatial decision support systems for environmental management. *Environ. Manage.* 22:35–48.

12 Decision Support for Ecosystem Management and Ecological Assessments

H. Michael Rauscher and Walter D. Potter

12.1 Introduction

In the face of mounting confrontation and after almost 20 years of increasingly contentious public unhappiness with the management of National Forests, the USDA Forest Service officially adopted ecosystem management as a land management paradigm (Overbay, 1992). Other federal forest land management agencies, such as the USDI Bureau of Land Management, the USDI National Park Service, the USDI Fish and Wildlife Service, the USDC NOAA, and the Environmental Protection Agency, have also made the commitment to adopt ecosystem management principles (Government Accounting Office, 1994). Ecosystem management represents different things to different people. At the heart of the ecosystem management paradigm lies a shift in emphasis away from sustaining yields of products toward sustaining the ecosystems that provide these products (Thomas, 1995; Rauscher, 1999). The ecosystem management paradigm represents the latest attempt, in a century-long struggle between resource users and resource preservers, to find a sensible middle ground between ensuring the necessary long-term protection of the environment while protecting the right of an ever-growing population to use its natural resources to maintain and improve human life (Chase, 1995; Taylor, 1998). As the concept of ecosystem management evolves, debates over definitions, fundamental principles, and policy implications will probably continue and shape the new paradigm in ways not yet discernible.

The ecosystem management paradigm was adopted quickly. No formal studies were conducted to identify the consequences of the changes ushered in by this new approach, nor was any well-documented, widely accepted, organized methodology developed for its implementation (Thomas, 1997). Today, ecosystem management remains primarily a philosophical concept for dealing with larger spatial scales, longer time frames, and the requirement that management decisions must be socially acceptable, economically feasible, and ecologically sustainable. Because the definition and fundamental principles that make up the ecosystem management paradigm have not yet been resolved and widely accepted, the challenge is to build the philosophical concept of ecosystem management into an explicitly defined, operationally practical methodology (Wear et al., 1996; Thomas, 1997). Effective ecosystem management processes are urgently needed so that federal land managers can better accommodate the continuing rapid change in societal perspectives and goals (Bormann et al., 1993).

Ecosystem management represents a shift from simple to complex definitions of the ecosystems that we manage (Kohm and Franklin, 1997). It will require the development of effective, multiobjective decision support systems to (1) assist individuals and groups in their decision-making processes; (2) support, rather than replace, the judgment of the decision makers; and (3) improve the quality, reproducibility, and explicability of decision processes (Janssen, 1992; Larsen et al., 1997; Reynolds et al., 1999). The complexity of environmental dynamics over time and space, over-

The authors thank Don Nute, Geneho Kim, and Shanyin Liu of the University of Georgia Artificial Intelligence Center for their many years of hard work and wonderful cooperation in our long-term effort to build good decision support systems. We also thank Mark Twery and the entire NED development team for their hard work and great ideas.

whelming amounts of data, information, and knowledge in different forms and qualities, and multiple, often conflicting, management goals virtually guarantee that few individuals or groups of people can consistently make good decisions without powerful decision support tools (Janssen, 1992).

Both the ecosystem and the management subsystems of ecosystem management are part of an interlocking, nested hierarchy (Bonnicksen, 1991; Forest Ecosystem Management and Assessment Team, 1993; Kaufmann et al., 1994). Local management of ecosystems occurs within and is influenced by national and international social–economic–political systems. Similarly, local ecosystems operate within and are influenced by larger biophysical systems, such as ecoregions or biomes. Ecosystem management can and should occur at many scales: global–international, biome–national, ecoregion–multistate, or forest landscape–National Forest (GAO, 1994, p. 62). At the regional, national, and international scales, the ecosystem management decision process should render the mosaic of environmental issues manageable by (1) identifying and labeling the issues, (2) defining the problems, and (3) identifying who is causing the problems, who has responsibility for solving the problems, who are the stakeholders associated with the problems, and who will pay for finding and implementing solutions (Hannigan, 1995). The decision process at this macro scale should also coordinate solution efforts and supervise social and ecological system sustainability (Tonn et al., 1998). Decision support systems that operate on the multistate and national scale have been developed and tested in Europe and can serve as illustrations of what is needed for United States federal forest management (Van den Berg, 1996). There is as much work to be done in ecosystem management at the larger scales as there is at the local scale. At present, no one agency, committee, or other organized body in the United States manages ecosystems at the scale suggested by Bonnicksen (1991) and Tonn et al. (1998). It is not obvious that our society has addressed the need to manage ecosystems at the biome–national and ecoregional–multistate scale to cope with the complex environmental problems that we have created for ourselves at these scales (Caldwell, 1996). It seems reasonable that we should try.

For the purposes of this chapter, we consider three spatial scales for which clear, precise, practical ecosystem management processes are needed: (1) the regional assessment scale, (2) the forest landscape management scale, and (3) the project implementation scale. These three scales are connected by complex information linkages that make it difficult to treat them separately. Indeed, decision support systems (DSS) have been developed to operate at each of these scales, with special attention given to what is needed from the higher scale and what needs to be supplied to the lower one. This chapter reviews the state of ecosystem management decision support across all three scales. Although most of the examples are drawn from experiences with USDA Forest Service research and management, the principles discussed are generally applicable to all federal land management agencies.

12.2 The Decision-making Environment

The decision-making environment consists of the social, economic, political, and legal context in which any agency charged with ecosystem management operates. Ecosystem management itself is composed of two parts: an ecological subsystem and a management subsystem (Figure 12.1) (Bonnicksen, 1991). The ecological subsystem contains physical or conceptual objects such as trees, birds, deer, rivers, smells, sights, and sounds. Each object can be a resource if it has positive value in the minds of people or a pest if it has negative value. Otherwise, it is value neutral (Behan, 1997). A critical feature of this view is that, as goals change, objects can change in status from resource to value neutral or even to pest (Bonnicksen, 1991). For example, the white-tailed deer, once regarded as a sought-after resource, is now considered a pest in some forest ecosystems of the eastern United States. People need to be considered as part of the community of organisms that inhabit, use, or directly influence an ecosystem (Behan, 1997). Thus people, in their role as users, are part of the ecological subsystem, like trees, soil, and wildlife.

The management subsystem is defined as making decisions about and controlling ecosystems to achieve desired ends. People in the management role participate in ecosystem management in a very different manner. They are the risk takers, the setters of objectives, the judges of value, the substituters, in other words, the decision makers. It is useful to keep this distinction clearly in mind. Normally, many people participate in the management subprocess of ecosystem management. The specific values, goals, and constraints that characterize public preferences and needs may be identified through a group negotiation process involving

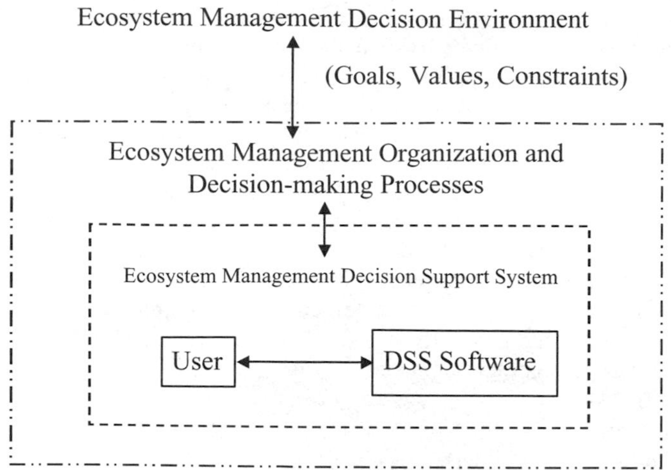

FIGURE 12.1. Decision environment for ecosystem management.

a variety of stakeholders and management decision makers (Figure 12.1). Management decision makers organize and lead the group negotiation process; they must ensure that the resultant goals are socially acceptable, legal, economically feasible, and ecologically sustainable. If ecosystem management decision support systems (EM-DSS) are available, all participants need to be able to rely on them at a reasonable level of confidence for relevant information and analyses. This group negotiation process is probably the most difficult part of ecosystem management (Bormann et al., 1993).

Most forest managers, administrators, and scientists are much more familiar with the structure and function of the ecological subsystem than with the management subsystem. This state of affairs is indicative of where we have put our attention and energy in the past. One new message for ecosystem management is that forest managers, administrators, and scientists need to rapidly redress this imbalance.

While biophysical scientists at the national and international scale struggle to understand local, regional, and global environmental systems, social and institutional scientists must struggle to understand how to sustain societal systems that will protect the ability of humans and nature to coevolve (Bormann et al., 1993; Tonn et al., 1998). Thus defining and understanding the nature of sustainable societies and the nature of sustainable ecosystems are equally important. Tonn and White (1996) describe sustainable societies as wise, participative, tolerant, protective of human rights, spiritual, collaborative, achievement oriented, supportive of stable communities, able to make decisions under uncertainty, and able to learn over time. One of the defining characteristics of a society will be how effectively it manages to sustain both itself and its ecosystems. Even a cursory review of history reveals numerous extinct civilizations that did not successfully sustain both society and ecosystem (Toynbee, 1946).

12.2.1 Major Elements of the Management Subsystem

The study of the management subsystem must include understanding the dynamics of public preferences, conflict management and resolution, and cost evaluation and containment as it relates to ecosystem management. Defining and understanding stakeholders and their preferences is an important part of ecosystem management (Garland, 1997). Stakeholder and general public preferences are volatile and sensitive to manipulation through the control of information transmitted through public media (Montgomery, 1993; Smith, 1997). Understanding the dynamics of social preferences and how they can be influenced over both the short and long term is a vital part of the ecosystem management process. Ecosystem management processes and the institutions that use them must be able to detect and accommodate rapid, and sometimes radical, changes in public preferences (Kohm and Franklin, 1997).

Successful social conflict management is as important as understanding stakeholder preference dynamics. Currently, “the dominant means of settling public land disputes have been either litigation or quasi-judicial administrative appeals. Such contentious methods of handling disputes expend much goodwill, energy, time, and money. These methods produce winners and losers, may leave fundamental differences unresolved, and potentially please few or none of the parties" (Daniels et al., 1993, p. 347). Decision makers need a fundamental understanding of the nature of environmental conflicts and disputes and how to use conflict-positive dispute management techniques effectively (Daniels et al., 1993). New approaches to managing the social debate surrounding ecosystem management, such as alternative dispute resolution (ADR) techniques (Floyd et al., 1996), should be evaluated, taught, and used. Adaptive management techniques are as applicable to the management side of ecosystem management as they are to the ecosystem side. They could be used to suggest a series of operational experiments that study actual public participation and conflict management activities to quickly determine what works and what does not (Daniels et al., 1993; Shindler and Neburka, 1997).

The ability of federal land managers to avoid gridlock is heavily dependent on stakeholder willingness to negotiate and ultimately agree on the goals for ecosystem management (Bormann et al., 1993). Unfortunately, people sometimes have preferences based on core values that are so strong and so conflicting that no solution is acceptable (Smith, 1997). To avoid societal gridlock, we must design and implement robust strategies that encourage voluntary conflict resolution among contentious stakeholders and explore other options leading toward a settlement if voluntary resolution is impossible. Such options might include binding arbitration, an agreed-upon delay in order to improve our data and knowledge about the ecosystem, or various other forms of conflict resolution.

Ecosystem management cost evaluation is a critical area for economists to study. As a general rule, increases in problem complexity increase the cost of finding satisfactory solutions (Klein and Methlie, 1990). Ecosystem management should accommodate limits on time, expertise, and money (Smith, 1997), because sustainable forest management is impossible if there are unsustainable social and economic costs (Craig, 1996). Documentation of costs should be prepared and made public, because few people know or appreciate the costs of efforts to solve complex ecosystem management problems. For example, the USDA Forest Service has spent approximately $2 billion, equal to 16% annually of the entire National Forest system budget, on planning since the National Forest Management Act was passed in 1976 (Behan, 1990). The additional costs of implementing ecosystem management prescriptions and monitoring and evaluating the results have not yet been estimated. Are we willing or able to marshal the funding to implement ecosystem management in a way that will ensure that federal forest managers can comply with the law and satisfy public preferences? The amount of money that could be spent on ecosystem management nationally may be extremely large, and identifying clear benefits may be difficult (Oliver et al., 1993).

In the last century of federal forestland management, timber harvesting has largely paid for multiple-use management activities. Many forecast that the level of timber harvesting under ecosystem management will greatly decline, while the cost of ecosystem management will greatly increase. Until managers evaluate the true costs and benefits, it will be difficult to determine whether the public is willing to pay for ecosystem management programs. In any case, a new and rational means of capital resource allocation will be required to fund the ecosystem management process adopted (Sample, 1990; Kennedy and Quigley, 1993; Oliver et al., 1993; Dombeck, 1997). Refusing to fund ecosystem management and opting for the "do nothing" alternative is likely to result in unacceptable future conditions. "Plant and animal species do not stop growing, dying, and burning; and floods, fires, and windstorms do not stop when all management is suspended" (Botkin, 1990). Nature does not appear to care, either about threatened and endangered species or about humans. People care and people must define goodness and badness. Nature will not do it for us. "Nature in the twenty-first century will be a nature that we make; the question is the degree to which this molding will be intentional or unintentional, desirable or undesirable" (Botkin, 1990). Making the nature that we want may be expensive. A good understanding of ecological economics will help society make rational choices.

12.2.2 Regional Ecological Assessments Viewed as a DSS

Recently, regional ecological assessments have been used to describe the large-scale context for ecosystem management and therefore can be considered decision support tools (Figure 12.1). Regional assessments have been large, collaborative interagency efforts, often with public stakeholder participation, that have taken 2 to 5 years and several millions of dollars to finish. The objectives for integrated ecological assessments are to provide (1) a description of current and historic composition, structure, and function of ecosystems; (2) a description of the biotic (including human) and abiotic processes that contributed to the development of the current ecosystem conditions; and (3) a description of probable future scenarios that might exist under different types of management strategies (Jensen et al., 1998). For examples of regional assessments, see the case studies provided in this guidebook.

Currently, precisely how these regional assessments fit into the ecosystem management process is unclear. One alternative would be to view regional assessments as DSS for ecosystem management at the ecoregion–multistate scale. In this capacity, the current objectives for assessments focus entirely too much on the ecosystem component of ecosystem management. The Southern Appalachian Assessment (SAMAB, 1996), for example, examines the social and economic activities of people within the region, but only in their roles as ecosystem members and users. The role of people as managers of ecosystems, including their role as managers of social–economic–political systems in the region, is largely ignored. To correct this deficiency, another list of objectives for regional assessments might include the following: (1) identify a set of regional scale goals and desired future conditions and compare these to the current conditions; (2) identify regional stakeholders, their preferences and values, and how they compare to the general public in the region; (3) identify the legal and political climate within which ecosystem management must function; (4) identify the regional costs and who will bear them, as well as the regional gains and who will reap them; and (5) identify the major problems, who is responsible for solving them, and who has supervisory responsibility to monitor progress and assure that a satisfactory solution is eventually reached.

12.3 Ecosystem Management Processes

The decision-making environment determines the goals, values, and constraints for the organization. Organizational policy then translates the mandates of the decision-making environment into specific decision-making processes. A decision-making process is a method or procedure that guides managers through a series of tasks, from problem identification and analysis to alternative design and finally alternative selection (Mintzberg et al., 1976; Clemen, 1996). Ideally, decision support systems should not be developed until the ecosystem management decision-making processes that they are to support have been articulated. In reality, both the decision processes and the software systems needed to support them are evolving simultaneously, each helping to refine the other.

First-generation ecosystem management processes have evolved from two sources: (1) academia, where several ecosystem management processes have been described at a general, conceptual level and their macrolevel structures and functions have been identified; and (2) federal forest managers at the field level, where numerous, local adhoc processes have been developed and tested under fire. The academic, high-level descriptions of ecosystem management processes do not supply adequate details to guide the development of decision support systems; also, they are theoretical, lacking adequate field testing to determine how they work in practice. The local, ad-hoc ecosystem management processes are too numerous for effective software-based decision support (approximately 400 ranger districts in the U.S. National Forest System each have their own process; other federal agencies have hundreds of additional field organizations), and few, if any, have been studied and described formally so that similarities and differences can be identified. Moreover, no particular process(es) have been widely accepted and implemented in federal forest management. We should devote as much creative attention to devising good ecosystem management decision processes as we do to assuring the quality of the decisions themselves (Ticknor, 1993). In this section, the major elements of a generic ecosystem management process are identified based on a synthesis of the literature.

Adaptive management, a continuing process of planning, monitoring, evaluating, and adjusting management methods (Bormann et al., 1993; FEMAT, 1993; Lee, 1993), provides a framework for describing a generic management process. The usefulness of adaptive management as an ecosystem management process is being field tested by the Forest Service in the Northwest and in other regions of the country (Shindler et al., 1996). Described at the most general level, adaptive management consists of four activities, *plan–act–monitor–evaluate,* linked to each other in a network of relationships (Figure 12.2). At each cycle, the results of the evaluation activity are fed back to the planning activity so that adaptive learning can take

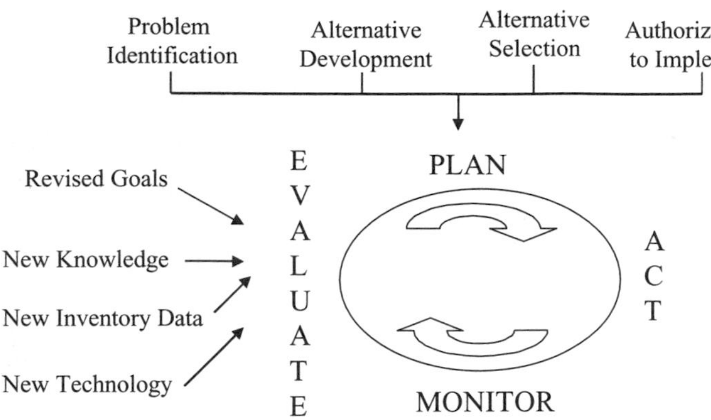

FIGURE 12.2. Adaptive management process for ecosystem management.

place. Without adding further detail to this definition, almost any management activity could erroneously be labeled adaptive management. In reality, adaptive management is a well-described, detailed, formally rigorous, and scientifically defensible management-by-experiment system (Walters and Holling, 1990). Baskerville (1985) prescribes a nine-step process for implementing adaptive management correctly. Adaptive management requires that a series of steps be followed for each of the four major activities described.

12.3.1 Plan

The Mintzberg et al. (1976) planning process can be viewed as a detailed description of the planning stage of the adaptive management process (Figure 12.3). Janssen (1992) argued that the planning stage of any decision process generally would need to be some variant of the Mintzberg et al. (1976) method. The planning stage consists of four steps: (1) problem identification, including goal selection, (2) alternative development, (3) alternative selection, and (4) authorization to implement the selected alternative (Figures 12.2 and 12.3). Each of these major steps can be decomposed into one or more phases (Janssen, 1992).

The problem identification step consists of two phases: (1a) recognition: identifying opportunities, problems, and crises, launching the decision process; and (1b) diagnosis: exploring the different aspects of the problem situation, identifying the goals, and deciding how to approach the problem. If the diagnosis phase, step 1b, is unnecessary, it

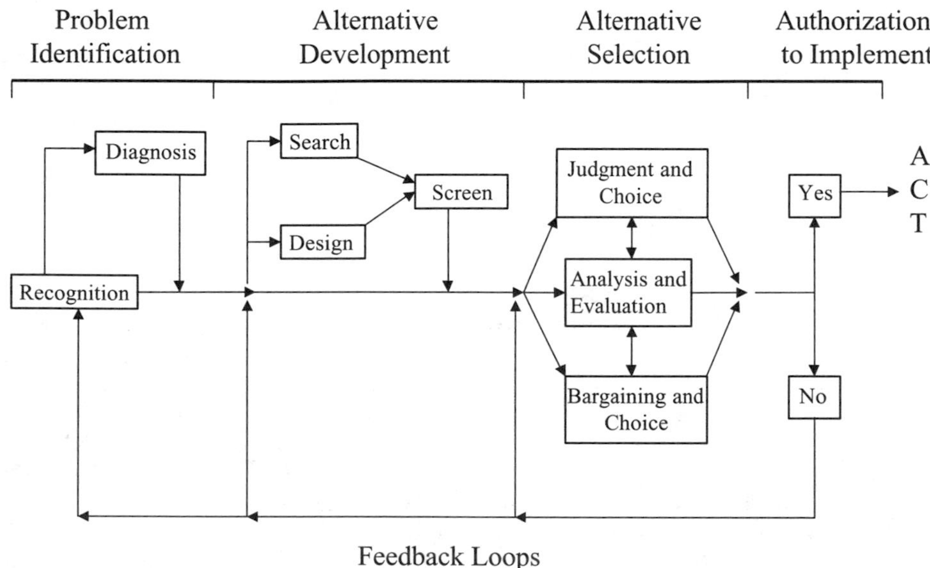

FIGURE 12.3. Detailed view of the planning activity in the adaptive management process for ecosystem management.

can be skipped (Figure 12.3). The next step, alternative development, has three phases: (2a) search: finding previously designed and tested solutions to the entire problem or to any of its parts; (2b) design: developing new alternatives; and (2c) screen: determining whether the number and quality of the alternatives found, developed, or both, provide an adequate range of choices for the selection step. The selection step also has three phases: (3a) analysis and evaluation: evaluating and understanding the consequences over space and time of each of the proposed alternatives and communicating these results clearly to the decision makers; (3b) judgment and choice: one individual makes a choice; and (3c) bargaining and choice: a group of decision makers negotiates a choice. The final step, authorization, may have two outcomes: (4a) authorization achieved: approval inside and outside the institutional hierarchy is obtained, marking the end of the planning process and the beginning of the implementation process; and (4b) authorization denied: evaluating the cause for denial and looping back to the appropriate part of the decision process to make another attempt at achieving authorization (Janssen, 1992).

A particular decision can take many pathways through these four steps, and iterative cycles are a normal part of how environmental decisions are actually made (Janssen, 1992). These cycles occur as the decision participant's understanding of a complex problem evolves and when alternative solutions fail to meet administrative, scientific, or political standards. Mintzberg et al. (1976) maintained that problems can be classified into seven types and that the solution cycle for each type can be mapped on Figure 12.3. Janssen (1992) illustrates this point by presenting and discussing the solution cycles of 20 actual environmental problems in the Netherlands, ranging from measures to reduce NH_3 emissions, to clean-up of a polluted site, to protecting forests from acid rain. The effectiveness of competing EM-DSSs may be evaluated by how many of the above phases are supported, how well they are supported, and whether the complex iterative cycling of real-world problems is supported (Janssen, 1992).

12.3.2 Act

The planning stage of adaptive management results in decisions about goals and constraints. The action stage determines how, where, and when to implement activities to achieve the goals and adhere to the constraints. Given a clear statement of management goals and objectives, the implementation stage creates testable adaptive management hypotheses, explicitly describes the assumptions supporting them, and generates an appropriate set of targeted actions (Everett et al., 1993). How each hypothesis is tested must be carefully and clearly documented (Walters and Holling, 1990; Lee, 1993; Kimmins, 1995).

12.3.3 Monitor

Documentation in the action stage is stressed because the monitoring stage often occurs months to years later, and the individuals who implemented the actions may not be involved in monitoring or subsequent evaluation. Documentation may be the only link between the two stages. The monitored results of experimental actions must also be recorded carefully and in detail so that a complete, understandable package exists for the evaluation stage.

This definition of the monitoring stage of adaptive management has immediate consequences. Which variables are monitored and when, how, and where they are monitored depend almost entirely on the hypotheses created in the action stage and on the type of actions deemed necessary to test those hypotheses. A unique *goal–hypotheses–action* sequence will probably need to be designed for each specific management unit. Similarly, each unit will probably have unique monitoring requirements to distinguish among the adaptive experimental hypotheses proposed for it. As a result, no general, broad-spectrum monitoring program can or should be designed to support adaptive management. Adaptive management means management by experiment. Management by experiment requires hypotheses that must be implemented and tested. Hence monitoring can only occur after the hypotheses have been designed and their tests devised so that it is clear what needs monitoring (FEMAT, 1993).

12.3.4 Evaluate

Finally, the documentation describing each adaptive management experiment must be analyzed and the results evaluated. Promising statistical methods have been identified (Carpenter, 1990), but using them requires considerable expertise. At the end of the adaptive management cycle, a written report should communicate the results publicly to stakeholders and managers, influencing future cycles of the planning activity of adaptive management (Everett et al., 1993). In fact, a metaanalysis of all adaptive experimental results should be compiled periodically and forwarded to the next higher planning level for corrective change leading to new actions.

Adaptive management, when implemented as defined by Walters and Holling (1990), FEMAT (1993), and Lee (1993), is a complex and challenging process. The adaptive management process is not a license to manipulate ecosystems haphazardly simply to relieve immediate sociopolitical pressure (Everett et al., 1993). Adaptive management must be applied correctly and rigorously as management by experiment if we are to achieve our stated goals. "Managing to learn entails implementing an array of practices, then taking a scientific approach in describing anticipated outcomes and comparing them to actual outcomes. These comparisons are part of the foundation of knowledge of ecosystem management" (FEMAT, 1993, p. II-87). The whole point of adaptive management is to generate change in the way ecosystem management is applied.

A number of institutional challenges must be addressed before adaptive management can make its expected positive contribution to the ecosystem management process (Lee, 1993). Adaptive management requires a greater level of expertise in statistical experimental design and analysis than other competing decision processes. Kessler et al. (1992) suggested that adaptive management requires close collaboration between forest managers and scientists. "Finding creative ways of conducting powerful tests without forcing staffs to do things they think are wrong or foolish is of central importance to the human part of adaptive management" (Lee, 1993, p. 113). Managers and the interested stakeholders must accept that adaptive management means making small, controlled mistakes to avoid making big ones. Keeping adaptive management unbiased may be difficult; research that has consequences is research with which managers or stakeholders may try to tamper or prevent altogether (Lee, 1993). The costs of properly monitoring results and documenting the entire managerial experiment are unknown (Smith, 1997). Nonetheless, adaptive management, supplemented by the Mintzberg et al. (1976) planning process, is an attractive candidate for an ecosystem management process at several operational scales. Despite much supportive rhetoric, the institutional and funding changes needed to implement adaptive management as an ecosystem management process for federal forestland management have not yet been widely accomplished.

Although the adaptive management concept appears to be the most well developed candidate for an operational ecosystem management process, others also should be investigated. Lindblom (1990), cited by Smith (1997), advocated a concept called *probing* as a candidate for an ecosystem management decision process. Probing is an informal process of observation, hypothesis formulation, and data comparison in which people of all backgrounds can engage. Jensen and Everett (1993) pointed to a method called a *land evaluation system* as another candidate for an ecosystem management process. The land evaluation system (Zonnefeld, 1988) has been used by the United Nations Food and Agricultural Organization and by the International Society for Soil Sciences in forestry land-use planning. Howitt (1978), cited by Allen and Gould (1986), offered a "simple" approach to dealing with decision processes for wicked problems that might be useful for ecosystem management. Rittel (1972) advocated a "second generation systems approach" to wicked problem solution based on the logic of arguments (Conklin and Begeman, 1987; Hashim, 1990). Problems and their consequences can be made understandable to individuals and groups by asking and answering crucial questions while diagramming the process using the formal logic of argumentation. Vroom and Jago (1988), cited in Sample (1993), suggested their *contingent decision process* may be used for problems like ecosystem management. Of course, any decision-making process used to implement ecosystem management in the United States must satisfy the requirements of the 1969 National Environmental Policy Act (NEPA) and the 1976 National Forest Management Act (NFMA).

Several formal, well-described candidates for an ecosystem management decision process have been introduced here. In addition, numerous local, ad-hoc decision processes have been developed and tested under fire in every ranger district in the U.S. National Forest System. Few case studies (e.g., Steelman, 1996) have been published, and not many evaluations (e.g., Shindler and Neburka, 1997) of the strengths and weaknesses of these informal decision-making processes have been conducted. Surely a concerted effort to study the existing formal and informal ecosystem management processes would result in some powerful candidates to implement ecosystem management. Because the adaptive management concept can also be used to improve our management systems, it may not be overly important which particular ecosystem management decision processes we choose. It is, however, critically important that we choose several and then use the adaptive management philosophy to test and improve them in real-life situations (Kimmins, 1991).

12.4 Decision Support Systems Defined

DSSs help managers make decisions in situations where human judgment is an important contributor to the problem-solving process, but where limitations in human information processing impede decision making (Rauscher, 1995). The goal of a DSS is to amplify the power of the decision makers without usurping their right to use human judgment and make choices. DSSs attempt to bring together the intellectual flexibility and imagination of humans with the speed, accuracy, and tirelessness of the computer (Klein and Methlie, 1990; Sage, 1991; Turban, 1993; Holsapple and Whinston, 1996).

A DSS contains a number of subsystems, each with a specific task (Figure 12.4). The first, and most important, is the subsystem composed of the decision maker(s). Decision makers are consciously diagrammed as part of the DSS because, without their guidance, there is no DSS. The group negotiation management subsystem helps decision makers to organize their ideas, formulate relationships surrounding issues and arguments, and refine their understanding of the problem and their own value systems (Jessup and Valacich, 1993; Holsapple and Whinston, 1996). Examples of group negotiation tools include the active response GIS system (AR/GIS) (Faber et al., 1997), the issue-based information system (IBIS) (Conklin and Begeman, 1987; Hashim, 1990), and various socioecological logic programming models (Thomson, 1993, 1996).

Group negotiation tools are used to construct issue-based argument structures using variants of belief networks to clarify the values and preferences of group members in the attempt to reach group consensus. For example, IBIS uses formal argument logic (the logic of questions and answers) as a way to diagram and elucidate argumentative thinking (Hashim, 1990). By asking and answering crucial questions, you can begin to better understand the problem and its solution set. Understanding the meaning of terms, and through them our thoughts, lies at the heart of collaborative management. Greber and Johnson (1991) illustrated how the malleable nature of many terms, in this case "overcutting," creates logically defensible differences of opinion that have nothing to do with a person's honesty or dishonesty in the argument. DSS should specifically deploy mechanisms by which the biological realities guide and, if appropriate, constrain the desires of the stakeholders (Bennett, 1996). For example, compromise is not acceptable for some issues. If the productive ca-

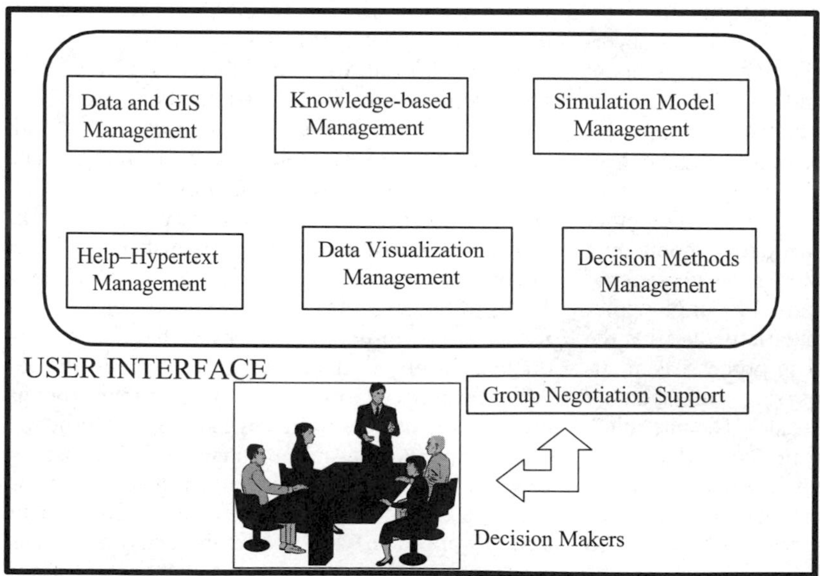

FIGURE 12.4. Major components of a generic decision support system.

pacity of an ecosystem is fixed, yet key stakeholders all want to extract a product from that ecosystem at a higher level, a compromise midway between the levels will be unsustainable.

The next major subsystem, spatial and nonspatial data management, organizes the available descriptions of the ecological and management components of ecosystem management. Data must be available to support choices among alternative management scenarios and to forecast the consequences of management activities on the landscape. There is a tension between the increasing number of goals that decision makers and stakeholders value and the high cost of obtaining data and understanding relationships that support these choices. Monitoring both natural and anthropogenic disturbance activities and disturbance-free dynamics of managed forest ecosystems are also extremely important if an EM-DSS is to accurately portray the decision choices and their consequences. Barring blind luck, the quality of the decision cannot be better than the quality of the knowledge behind it. Poor data can lead to poor decisions. It is difficult to conceive of prudent ecosystem management without an adequate biophysical description of the property in question.

The next four subsystems, knowledge-based, simulation model, help–hypertext, and data visualization management, deal with effectively managing knowledge in the many diverse forms in which it is stored, represented, or coded (see Rauscher et al., 1993, for more detail). Knowledge that is not language based is either privately held in people's minds or publicly represented as photographs, video, or graphic art. Language-based knowledge is found in natural language texts of various kinds, in mathematical simulation models, and in expert or knowledge-based systems. Data visualization software has been developed that can manipulate photographic, video, and graphic art representations of current and future ecosystem conditions. Data visualization software is beginning to be incorporated into EM-DSS on a routine basis to help decision makers see for themselves the likely impact of their decisions on the landscape.

In the last 20 years, an impressive amount of mathematical simulation software has been developed for all aspects of natural resource management. Schuster et al. (1993) conducted a comprehensive inventory of simulation models available to support forest planning and ecosystem management. They identified and briefly described 250 software tools. Jorgensen et al. (1996) produced another compendium of ecological models that incorporate an impressive amount of ecosystem theory and data. The simulation model management subsystem of the EM-DSS is designed to provide a consistent framework into which models of many different origins and styles can be placed so that decision makers can use them to analyze, forecast, and understand elements of the decision process.

Despite our most strenuous efforts to quantify important ecological processes to support a theory in simulation model form, by far the larger body of what we know can only be expressed qualitatively, comparatively, and inexactly. Most often this qualitative knowledge has been organized over long years of professional practice by human experts. Theoretical and practical advances in the field of artificial intelligence applications in the last 20 years now allow us to capture some of this qualitative, experience-based expertise into computer programs called expert or knowledge-based systems (Schmoldt and Rauscher, 1996). It is still not possible to capture the full range and flexibility of knowledge and reasoning ability of human experts in knowledge-based software. We have learned, however, how to capture and use that portion of expertise that the human expert considers routine. The knowledge management subsystem of the EM-DSS is designed to organize all available knowledge-based models in a uniform framework to support the decision-making process.

Finally, a large amount of text material exists that increases the decision maker's level of understanding about the operation of the decision support system itself, the meaning of results from the various modeling tools, and the scientific basis for the theories used. This text material is best organized in hypertext software systems. Hypermedia methodology supports a high degree of knowledge synthesis and integration with essentially unlimited expandability. The oak regeneration hypertext (Rauscher et al., 1997b) and the hypermedia reference system to the FEMAT report (Reynolds et al., 1995) are recent examples of the use of hypertext to synthesize and organize scientific subject matter. Examples of the use of hypertext to teach and explain software usage can be found in the help system of any modern commercial computer program.

The software subsystems of an EM-DSS described so far help decision makers to organize the decision problem, formulate alternatives, and analyze their future consequences. The decision methods management subsystem (Figure 12.4) provides tools and guidance for choosing among the alternatives, for performing sensitivity analysis to identify the power of specific variables to change the ranking of alternatives, and for recording the decisions made and their rationale. Many facets or dimensions influence the decision-making process. The rational–technical dimension, which concerns itself with the mathematical formulation of the methods of choice and their uses, is the one most often encountered in the decision science literature (Klein and Methlie, 1990; Rauscher, 1996). But there are others, including the political–power dimension (French and Raven, 1959; O'Reilly, 1983) and the value–ethical dimension (Brown, 1984; Klein and Methlie, 1990, p. 108; Rue and Byars, 1992, p. 61).

Decision makers might find themselves at any point along the political–power dimension bounded by a dictatorship (one person decides) on the one extreme and by anarchy (no one can decide) on the other. Intermediate positions are democracy (majority decides), republicanism (selected representatives decide), and technocracy–aristocracy (experts or members of a ruling class decide). Currently, three approaches seem to be in use at multiple societal temporal and spatial scales: management by experts (technocracy), management by legal prescription (republicanism), and management by collaboration (democracy) (Bormann et al., 1993). No one approach predominates. In fact, the sharing of power among these three approaches creates tensions that help to make ecosystem management a very difficult problem. In the context of ecosystem management, the value–ethical dimension might be defined on the one extreme by the preservationist ethic (reduce consumption and let nature take its course) and on the other by the exploitation ethic (maximum yield now and let future generations take care of themselves). Various forms of the conservation ethic (use resources, but use them wisely) could be defined between these two extremes. The rational–technological dimension is defined by normative–rational methods on the one hand and expert–intuitive methods on the other. Numerous intermediate methods also have been described and used (Janssen, 1992; Rauscher, 1996). The formal relationships among these dimensions affecting the decision process have not been worked out.

Informally, it is easy to observe decision-making situations where the political–power or value–ethical dimensions dominate the rational–technical dimension. Choosing an appropriate decision-making method is itself a formidable task (Silver, 1991; Turban, 1993) that influences both the design of alternatives and the final choice. Many EM-DSSs do not offer a decision method subsystem due to the complexity and sensitivity of the subject matter. Unfortunately, providing no formal support in EM-DSS for choosing among alternatives simply places all the burden on the users and may make them more vulnerable to challenges of their process and choice mechanisms.

12.5 Comparison of Existing Ecosystem Management DSS

Mowrer et al. (1997) surveyed 24 of the leading EM-DSSs developed in the government, academic, and private sectors in the United States. Their report identified five general trends: (1) while at least one EM-DSS fulfilled each criteria in the questionnaire used, no single system successfully addressed all important considerations; (2) ecological and management interactions across multiple scales were not comprehensively addressed by any of the systems evaluated; (3) the ability of the current generation EM-DSSs to address social and economic issues lags far behind biophysical issues; (4) the ability to simultaneously consider social, economic, and biophysical issues is entirely missing from current systems; and (5) group consensus-building support was missing from all but one system, a system that was highly dependent on trained facilitation personnel (Mowrer et al., 1997). In addition, systems that offered explicit support for choosing among alternatives provided decision makers with only one choice of methodology. The reviewers noted that little or no coordination had occurred between the 24 development teams, resulting in large, monolithic, stand-alone systems, each with a substantially different concept of the ecosystem management process and how to support it.

Different EM-DSSs appear to support different parts of the ecosystem management process. Table 12.1 lists 33 EM-DSSs, the 24 systems surveyed by Mowrer et al. (1997) plus 9 DSSs not included in that study. Nineteen of the 33 are labeled full-service EM-DSSs at their scale of operation because they attempt to be comprehensive EM-DSSs, offering or planning to offer support for a complete ecosystem management process. These EM-DSSs can be further classified by the scale of support that is their primary focus: regional assessments, forest planning, or project-level planning. The remainders, labeled functional service modules, provide specialized support for one or a few phases of the entire ecosystem management process. These service modules can be organized according to the type of functional support that they provide group

TABLE 12.1. A representative sample of existing ecosystem management decision support software for forest conditions of the United States arranged by operational scale and function.

Full-service EM-DSS		Functional service modules	
Operational scale	*Models*	*Function*	*Models*
Regional assessments	EMDS	Group negotiations	AR/GIS
	LUCAS[a]		IBIS[a]
		Vegetation dynamics	FVS
Forest-level planning	RELM		LANDIS
	SPECTRUM		CRBSUM
	WOODSTOCK		SIMPPLLE
	ARCFOREST		
	SARA	Disturbance simulations	FIREBGC
	TERRA VISION		GYPSES
	EZ-IMPACT[a]		UPEST
	DECISION PLUS[a]		
	DEFINITE[a]	Spatial visualization	UTOOLS/UVIEW
			SVS[a]
Project-level planning	NED		SMARTFOREST[a]
	INFORMS		
	MAGIS	Interoperable system architecture	LOKI
	KLEMS		CORBA[a]
	TEAMS		
	LMS[a]	Economic impact analysis	IMPLAN
		Activity scheduling	SNAP

[a]References for models not described in Mowrer et al. (1997): EZ-IMPACT (Behan, 1994); DECISION PLUS (Sygenex, 1994); IBIS (Hashim, 1990); DEFINITE (Janssen and van Herwijnen, 1992); SMARTFOREST (Orland, 1995); CORBA (Otte et al., 1996); SVS (McGaughey, 1997); LMS (McCarter et al., 1998); LUCAS (Berry et al., 1996).

negotiations, vegetation dynamics, disturbance simulation, spatial visualization, and interoperable system architecture.

12.5.1 Full-service EM-DSS

Regional Assessments

The Ecosystem Management Decision Support (EMDS) program is a software system specifically designed to support the development of ecological assessments, usually at regional or watershed scales. It provides a general software environment for building knowledge bases that describe logical relations among ecosystem states and processes of interest in an assessment (Reynolds et al., 1997). Once users construct these knowledge bases, the system provides tools for analyzing the logical structure and the importance of missing information. EMDS provides a formal logic-based approach to assessment analysis that facilitates the integration of numerous diverse topics into a single set of analyses. It also provides robust methods for handling incomplete information. A variety of maps, tables, and graphs provides useful information about which data are missing, the influence of missing data, and how data are distributed in the landscape. EMDS also provides support for exploring alternative future conditions. Finally, EMDS is general in application and can be used at the scale relevant to an assessment problem (Reynolds et al., 1997).

LUCAS is a multidisciplinary simulation framework for investigating the impact of land-use management policies (Berry et al., 1996). LUCAS has been used to support regional assessments of land-use change patterns as a function of social choices and regulatory approaches (Wear et al., 1996). LUCAS can be used to compare the effects of alternative ecosystem management strategies that could be implemented over an ecoregion of any size. These alternatives could be evaluated based on any number of social choice assumptions ascribed to private landowners (Wear et al., 1996). LUCAS could also be used to address the effects of land cover changes on natural resource supplies and local incomes. The advantage of an EM-DSS operating at the ecoregional scale is that regional decision-making activities and their consequences can be forecast with reasonable credibility.

Forest-level or Strategic Planning

Forest-level planning corresponds to the strategic planning process of decision science (Holsapple and Whinston, 1996). Many federal agencies, including the USDA Forest Service, have relied on linear programming systems of various kinds as the primary strategic planning decision support tool. In 1979, a linear programming, harvest scheduling model, FORPLAN, was turned into a forest-level planning, too, and until 1996 all national forest supervisors were required to use it as the primary analytical tool for strategic forest planning. After 17 years of increasingly fierce criticism that the normative, rational, optimization approach to decision analysis implemented by FORPLAN and its successor, SPECTRUM, was not adequate, the Forest Service finally removed its formal requirement to use FORPLAN/SPECTRUM (Stephens, 1996). The specifics of the arguments critical of FORPLAN/SPECTRUM as an analytical tool for forest planning are beyond the scope of this chapter and can be readily found in the following publications: Barber and Rodman (1990), Hoekstra et al. (1987), Shepard (1993), Liu and Davis (1995), Behan (1994, 1997), Kennedy and Quigley (1993), Howard (1991), Canham (1990), Morrison (1993), and Smith (1997).

Forest-level planning may be more successfully performed using soft, qualitative decision analysis formalisms than the hard, quantitative methods employed in rational, linear or nonlinear optimization schemes. Many other decision analysis formalisms exist (see Rauscher, 1996; Smith, 1997) along with the tools that make them useful and practical (Table 12.1). A number of these techniques may offer greater support for dealing with power struggles, imprecise goals, fuzzy equity questions, rapidly changing public preferences, and uneven quality and quantity of information (Allen and Gould, 1986). In particular, EZ-IMPACT (Behan 1994, 1997) and DEFINITE (Janssen and van Herwijnen, 1992) are well-developed and tested analysis tools for forest planning that use judgment-based, ordinal, and cardinal data to help users to characterize the system at hand and explore hidden interactions and emergent properties.

A forest plan should demonstrate a vision of desired future conditions (Jensen and Everett, 1993). It should examine current existing conditions and highlight the changes needed to achieve the desired future conditions over the planning period (Grossarth and Nygren, 1993). Finally, the forest plan should demonstrate that recommended alternatives actually lead toward desired future conditions by tracking progress annually for the life of the plan. The forest plan should be able to send accomplishable goals and objectives to the level of project implementation and receive progress reports that identify the changes in forest conditions that management has achieved. Ideally, all competitors in this class of EM-DSS should be objectively evaluated for their effectiveness in supporting these tasks, their ease of use in practice, and their ability to communicate their internal processes clearly and succinctly to both decision makers and stakeholders. Such an evaluation has not yet been conducted.

Project-level Implementation or Tactical Planning

"Forest plans are programmatic in that they establish goals, objectives, standards, and guidelines that often are general. Accordingly, the public and USDA Forest Service personnel have flexibility in interpreting how forest plan decisions apply, or can best be achieved, at a particular location. In addition, forest plans typically do not specify the precise timing, location, or other features of individual management actions" (Morrison, 1993, p. 284). EM-DSSs at the project level help to identify and design site-specific actions that will promote the achievement of forest plan goals and objectives. For example, a strategic-level forest plan might assign a particular landscape unit for management of bear, deer, and turkey, with minimum timber harvesting and new road construction and no clearcutting. The tactical project implementation plan would identify specific acres within this management unit that would receive specific treatments in a specific year. Project-level EM-DSSs have been developed to support the tactical-level planning process.

Project-level EM-DSSs (Table 12.1) can be separated into those that use a goal-driven approach and those that use a data-driven approach to the decision support problem. NED (Rauscher et al., 1997a; Twery et al., 2000) is an example of a goal-driven EM-DSS. Rauscher et al. (2000) present a practical, decision-analysis process for conducting ecosystem management at the project implementation level and provide a detailed example of its application to Bent Creek Experimental Forest in Asheville, North Carolina. Because management is defined to be a goal-driven activity, goals must be defined before appropriate management actions can be determined. It cannot be overemphasized that, without goals, management cannot be properly practiced (Rue and Byars, 1992). Goal-driven systems, such as NED, assist the user in creating an *explicitly defined goal hierarchy* (Rauscher et al., 2000).

A goal is an end state that people value and are willing to allocate resources to achieve or sustain (Nute et al., 1999). Goals form a logical hierarchy with the ultimate, all-inclusive goal at the top, subgoals at the various intermediate levels, and a special goal, which may be called a desired future condition, at the bottom (Saaty, 1992; Keeney and Raiffa, 1993). A *desired future condition* (DFC) is a goal statement containing a single variable measuring some observable state or flow of the system being managed (Nute et al., 1999). DFCs are the lowest level of the goal hierarchy. They are directly connected to the management alternatives being considered (Saaty, 1992). Furthermore, DFCs precisely define the measurable variables that each alternative must contain (Mitchell and Wasil, 1989). In other words, each DFC provides a measure of the degree to which any given ecosystem state, current or simulated future, meets the goal statement (InfoHarvest, 1996).

For example,

Goal: Freshwater Fishing Opportunities Exist IF
DFC(1): pH of all (> = 90%) freshwater lakes > 5 AND
DFC(2): popular game fish are plentiful AND
DFC(3): access to all (> = 90%) freshwater lakes is adequate.

Given that "plentiful" and "adequate" are further defined so that they are measurable, the three DFCs above define three attributes, that is, pH, popular game fish, and access to freshwater lakes, that each alternative under consideration must have. Otherwise, it will be impossible to determine whether the alternative can satisfy the goal "Freshwater Fishing Opportunities Exist." DFCs should be objective in nature. That is, there should exist a commonly understood scale of measurement, and the application of that scale to the ecosystem should yield roughly the same results no matter who makes the measurement. DFC(1) is objective in this sense. On the other hand, DFC(2) and DFC(3) are using subjective, binary scales for "plentiful" and "adequate." Research results indicate that subjectively developed scales can be reliably used by qualified professionals (Keeney and Raiffa, 1993, p. 40). An operationally practical DFC should (1) provide the appropriate information on theoretical grounds and (2) also be obtainable. In other words, a DFC should accurately define the lowest-level subgoal and be measurable in practice.

Unlike goals, which depend primarily on value judgments, defining how to achieve a goal with a set of DFCs depends primarily on factual knowledge (Keeney, 1992). A set of DFCs that defines a lowest-level goal is not generally unique (Keeney and Raiffa, 1993). There are usually alternative ways to define the same lowest-level goal, and no a priori tests exist to show that one way is better or worse than another. This disparity typically results from competing scientific theories or professional judgment. Consequently, it is important to document the justification for defining the lowest-level goal in any particular way.

Constraints, like goals, have a standard that is either met or not (Keeney, 1992). This standard is meant to screen unacceptable goals, objectives, alternatives, and management prescriptions from consideration. Requirements are equivalent to constraints, but are phrased differently. Logically, "you must" do something is exactly the same as "you must not" do something else. Permissions are the logical inverse of constraints. Permissions make it explicitly clear that certain goals, objectives, alternatives, and silvicultural prescriptions are allowed if they naturally surface in the management process. Permissions are not mandatory; if they were, they would be requirements.

The development of a goal hierarchy, the constraint network, and the specification of DFCs for a complex decision problem is more art than science. Although no step by step procedures are possible, some useful guidelines have been developed and summarized by Keeney and Raiffa (1993), Clemen (1996), and by InfoHarvest (1996).

Alternatives are the courses of action open to a decision maker for satisfying the goal hierarchy (Holtzman, 1989). In ecosystem management, each alternative contains a set of *action–location–time* triples that is intended to change the landscape so that goal satisfaction is improved. These triples, called *prescriptions*, embody the purposeful application and expenditure of monetary, human, material, and knowledge resources that define ecosystem management.

The design of alternatives, like the design of the goal hierarchy, is largely an art that relies heavily on decision science expertise, along with an expert-level understanding of forest ecosystem management (Klein and Methlie, 1990; Clemen, 1996). It is also very much an iterative process. High-quality decisions require the design of a set of promising, distinct alternatives to evaluate (Holtzman, 1989; Keeney, 1992). Given knowledge about (1) the current condition of the forest ecosystem; (2) the goals and DFCs; (3) the standards, guides, best management practices, and constraints in force at any given time; and (4) available management prescriptions, an experienced manager can craft alternatives that represent reasonable answers to the question of what state of organization we want for this forest ecosystem so that we can best meet the goals. The human mind is the sole source of alternatives (Keeney, 1992).

In contrast to goal-driven systems, INFORMS (Perisho et al., 1995; Williams et al., 1995) is a data-driven EM-DSS. Data-driven systems do not require the existence of an explicit goal hierarchy. Indeed, it is often the case that the only existing goals are *implicit goals* that reside in the private knowledge base of the manager. Data-driven systems begin with a list of actions that the user wants to explore and search the existing landscape conditions, as reflected in the system database, to find possible locations where these management actions can be implemented.

Both approaches have their strengths and weaknesses. Goal-driven systems tend to be rather prescriptive. They require the user to follow a certain sequence of events and force the user to make certain critical decisions in order to follow a predefined ecosystem management process. Data-driven systems allow users more freedom to craft their own process in an ad-hoc fashion; this provides great freedom of action, but places all the burden of knowing what to do and why to do it on the user. Goal-driven systems, by definition, tend to ensure that management actions move the landscape toward the specified desired future conditions by committing to a particular ecosystem management decision process. This reduces the utility of the EM-DSS to that set of decision makers who wish to use this particular decision process. On the other hand, data-driven systems offer no guarantee that the results of the sum of the actions have any resemblance to the desired future conditions as defined by the strategic objectives. Data-driven systems, however, do allow competent and knowledgeable decision makers maximum flexibility in the analyses that they perform and how they put them together to arrive at a decision. Hybrid goal- and data-driven systems may offer users the advantages of both approaches.

12.5.2 Functional Service Modules

The full-service EM-DSSs rely on specialized software service modules to add a broad range of capabilities (Table 12.1). Tools to support group negotiation in the decision process are both extremely important and generally unavailable and underutilized. AR/GIS (Faber et al., 1997) is the most fully developed software available for this function. IBIS, another group negotiation tool, is an issue-based information system that implements argumentation logic (the logic of questions and answers) to help users to formally state problems, understand them, clearly communicate them, and explore alternative solutions (Conklin and Begeman, 1987; Hashim, 1990). Vegetation dynamics simulation models, both at the stand and at the landscape scale, provide EM-DSSs with the ability to forecast the consequences of proposed management actions. Disturbance models simulate the effects of catastrophic events, such as fire, insect defoliation, disease outbreaks, and wind damage. Models that simulate direct and indirect human disturbances on ecosystems are not widely available. Although models that simulate timber harvesting activities exist, they provide little, if any, ecological impact analyses, such as the effect of extraction on soil compaction, on damage to remaining trees, or on the growth response of the remaining tree and understory vegetation. Models that simulate the impact of foot traffic, mountain bikes, and horseback riding on high-use areas are largely missing. Models that simulate climate change, nutrient cycling processes, acid-deposition impacts, and other indirect responses to human disturbance exist, but are rarely practical for extensive forest analyses. Stand- and landscape-level visualization tools have improved dramatically in the last few years. It is now possible, with relatively little effort, to link to and provide data for three-dimensional stand-level models such as SVS (McGaughey, 1997) and landscape-level models such as UVIEW (Ager, 1997) and SMARTFOREST (Orland, 1995).

12.6 Interoperability in Ecosystem Management DSS

Existing EM-DSSs (Table 12.1), with few exceptions, are islands of automation unable to easily communicate with each other. They have been written in different software languages, they reside on different hardware platforms, and they have different data access mechanisms and different component–module interfaces. For example, nongeographical databases may be written in Oracle, geographical information system (GIS) databases in ARC/INFO, knowledge bases in Prolog, and a simulation model in C or Fortran. Some execute only on a UNIX platform, others only in a Microsoft Windows environment. As a group, they have poorly developed mechanisms for achieving integrated operations with (1) other existing full-service EM-DSS; (2) the many available functional-service modules (Schuster et al., 1993; Jorgensen et al., 1996); (3) readily available, high-quality commercial software; or (4) new software modules that independent development groups are continually producing in their efforts to support ecosystem management.

Interoperability is the ability for two or more software components to cooperate by exchanging services and data with one another, despite the possible heterogeneity in their language, interface, and hardware platform (Heiler, 1995; Wegner, 1996; Sheth, 1998). Interoperable systems provide a software standard that promotes communication between components and provides for the integration of legacy and newly developed components (Potter et al., 1992; Liu et al., 2000). To date, efforts to achieve interoperability between EM-DSS modules have used ad hoc techniques yielding unique, point-to-point custom solutions. Although such unique solutions work, sometimes very efficiently, they are typically difficult to maintain and transfer to other developers because of their idiosyncratic nature. After evaluating several of the leading EM-DSSs, Liu et al. (2000) concluded that no comprehensive, theory-based interoperability standards currently exist for achieving integrated operations of EM-DSS.

In contrast, interoperability outside the ecosystem management domain has received extensive attention. There is heavy emphasis in the larger computer science field toward the construction of systems from preexisting components based on interoperability standards (Mowbray and Zahavi, 1995). Liu et al. (2000) evaluated four such approaches (see Liu, 1998, for details). The approaches addressed include CORBA (the Common Object Request Broker Architecture, 1997), DCOM (the Distributed Component Object Model, Microsoft 1995, 1996, 1997), intelligent agents (Finin et al., 1994; Genesereth and Ketchpel, 1994; Mayfield et al., 1996), and DIAS/DEEM (the Dynamic Information Architecture System/Dynamic Environmental Effects Model, Argonne National Laboratory 1995a, 1995b). These approaches encompass several different areas of computer science, including databases in which the emphasis is on interoperation and data integration, software engineering in which tool and environment integration issues dominate, artificial intelligence for which systems consisting of distributed intelligent agents are being developed and explored, and information systems.

With NED-1 (Rauscher et al., 1997a; Twery et al., 1999) and FVS (Teck et al., 1996, 1997) as example EM-DSSs, Maheshwari (1997) used CORBA and Liu et al. (2000) used DCOM to develop a framework for achieving integrated operations. CORBA and DCOM both provide standard specifications for achieving language interoperability and platform independence. They define their own interface standards to deal with peculiarities of legacy applications. They support distributed processing, object reuse, and the Internet. Both architectures are well documented, and the documentation materials are easily accessible to the public, on line as well as through books and journal articles. CORBA can be purchased from multiple vendors, and DCOM is shipped with Windows NT/98 or can be downloaded free for Windows 95. Based on their tests, Liu et al. (2000) concluded that DCOM was easier to learn and more productive than CORBA.

The proposed DCOM-based interoperability design was found to be general and to make no assumptions about the software applications to be integrated (Liu et al., 2000). Its standardized interface scheme enables integration of a variety of applications. It is an open framework in the sense that application components can be added and/or removed easily without drastically affecting the functionality of the whole system. DCOM comes with Windows NT and 98, so there is no up-front cost. The prototype worked smoothly and was totally transparent to the user. Although no performance tests were carried out, there appeared to be no performance degradation as a result of using DCOM (Liu et al., 2000).

The design, implementation, and maintenance of interoperable software architectures for EM-DSS are challenging activities. System integrators face computer science problems with different hardware platforms, software languages, compiler versions, data access mechanisms, module interfaces, and networking protocols (Mowbray and Zahavi, 1995). In addition, the ecosystem management arena contributes challenges such as different data sources, ecosystem management process visions, decision-making methods, and solution strategies. Future generations of EM-DSS must become more interoperable to provide the best possible support for ecosystem management processes.

12.7 Conclusions

Ecosystem management has been adopted as the philosophical paradigm guiding federal forest management in the United States. The strategic goal of ecosystem management is to find an acceptable middle ground between ensuring the necessary long-term protection of the environment while allowing an increasing population to use its natural resources for maintaining and improving human life. Adequately described and widely accepted ecosystem management processes do not yet exist, but a concerted effort to study the many formal and informal ecosystem management processes that do exist is yielding results. Several powerful candidate processes to support the practical implementation of ecosystem management have been identified to date and are undergoing evaluation.

The generic theory of decision support system development and application is well established. Numerous specific ecosystem management decision support systems have been developed and are evolving in their capabilities. Given a well-defined and accepted set of ecosystem management processes to support, along with adequate time and resources, effective EM-DSSs can be developed. However, major social and political issues present significant impediments to the efforts to make ecosystem management operational. A sociopolitical environment in which everyone wants to benefit and no one wants to pay incapacitates the federal ecosystem management decision-making process. The very laws that were adopted to solve the problem, the 1969 National Environmental Policy Act and the 1976 National Forest Management Act, have led to procedural paralysis at exponentially rising costs (Behan, 1990). Developing a workable ecosystem management process and the decision-making tools to support it is probably one of the most complex and urgent challenges facing forest ecosystem managers today.

To date, none of the available EM-DSSs has been found capable of addressing the full range of support required for ecosystem management (Mowrer et al., 1997). This is not surprising, because it is highly unlikely that a single DSS can provide adequate support for ecosystem management (Grossarth and Nygren, 1993; Mowrer et al., 1997), meaning that a familiarity with the entire range of available decision analysis methodology and related modeling tools is required. Many of the EM-DSSs introduced in this chapter hold great promise, but this promise has not been fully realized. A major reason for this situation is that system development has been primarily driven by technology, not by user requirements. The requirements to guide EM-DSS development are unknown or poorly defined, because the ecosystem management decision processes themselves have been inadequately identified and described. The frequently observed tendency to substitute technology for an inadequate or nonexistent ecosystem management decision-making process should be avoided because it is rarely satisfactory. Although formal evaluation procedures are available (see Adelman, 1992), few of the current EM-DSSs have undergone an unbiased, critical evaluation of their suitability for ecosystem management decision support. Such an evaluation is long overdue. In the final analysis, EM-DSS software should be evaluated using a simple question: Does the EM-DSS improve the decision maker's ability to make good decisions?

A number of more specific conclusions about the current state of EM-DSS appear obvious:

1. The complexity of environmental dynamics over time and space; the overwhelming amounts of data, information, and knowledge in different forms and qualities; and the multiple, often conflicting, management goals virtually guarantee that few individuals or groups can consistently make good decisions without adequate support tools.
2. Ecosystem management, by definition, is concerned with *both* ecological and management science. Yet much of the effort seems to concentrate on ecological issues to the detriment of equally important management issues. Understanding and developing good decision-making processes is as important for the success of ecosystem management as understanding ecological structure and function (http://biology.usgs.gov/dss/def.html).
3. Ecosystem management ought to be practiced at many different scales: forest landscape–national forest, ecoregion–multistate, biome–national, and global–international. This fact has not been as widely recognized in the United States as it has in Europe. In the United States, we have, however, scaled ecosystem management up to the ecoregional scale by developing regional ecological assessments as a management tool.
4. People play two roles in ecosystem management. First, they are part of the communities of organisms that interact with each other and their abiotic environment. Second, they are the risk takers, the setters of objectives, the judges of value; in other words, the decision makers. We need to understand human behavior and characteristics in both roles.
5. Defining and understanding the nature of sustainable societies and the nature of sustainable ecosystems are equally important. Human societies and ecosystems are inseparable and cannot be understood or managed without reference to each other.
6. The study of the management subsystem must include understanding the dynamics of public preferences, conflict management and resolution, and cost evaluation as it relates to ecosystem management.

In closing, it is appropriate to discuss how ecological assessments and EM-DSS may be mutually supportive. From a decision support point of view, ecological assessments have not been as useful as they might be. The ecological assessments undertaken so far have concentrated mostly on providing (1) a description of current and historic ecosystem composition, structure, and function and (2) a description of the biotic (including human) and abiotic processes that contributed to the development of the current ecosystem conditions. Although such descriptive emphasis on the ecological foundations of the region is necessary, it is not sufficient. This is another example of how the management side of ecosystem management is often all but ignored.

We suggest that ecological assessments can be more effective decision support tools if some or all of the following items are included in their scope:

1. Clarify, label, and define the major regional issues that concern people.
2. Explicitly identify and define the major socio–political–biological problems facing the region.
3. Identify what or who is causing the problems, who has responsibility for solving them, who the stakeholders are that are associated with the problem, and who will pay for finding and implementing solutions.
4. Clarify how ecosystem management efforts are to be coordinated at the regional scale and how progress in social and ecological sustainability is going to be identified and tracked.
5. Develop a set of regional-scale goals and desired future conditions to measure these goals in terms of regionally appropriate variables.
6. Describe probable future scenarios that might exist under different types of regional-scale management strategies.

Including such management-oriented items in an ecological assessment is likely to make it politically more sensitive. However, the resultant document would help to organize the entire ecosystem management process at lower scales, set the tone and parameters of the public and professional debate, and explicitly state the problems as well as the desired solutions for examination by everyone concerned. It would help to diffuse the impact of purely emotional points of view and bring the debate back to a more reasonable, rational arena. More specifically, such a regional assessment would provide solid direction, through the regional goals and desired future conditions, to forest-level ecosystem management planning and implementation.

12.8 References

Adelman, L. 1992. *Evaluating decision support and expert systems*. New York: John Wiley & Sons.

Ager, A. 1997. UTOOLS/UVIEW. In: Mowrer, H. T.; Barber, K.; Campbell, J.; Crookston, N.; Dahms, C.; Day, J.; Laacke, J.; Merzenich, J.; Mighton, S.; Rauscher, M.; Reynolds, K.; Thompson, J.; Trenchi, P.; Twery, M., eds. *Decision support systems for ecosystem management: an evaluation of existing systems*. Gen. Tech. Rep. RM-296. U.S. Dept. Agric., For. Serv., Rocky Mountain Res. Sta.: 144–148.

Allen, G. M.; Gould, E. M., Jr. 1986. Complexity, wickedness, and public forests. *J. For.* 84:20–23.

Argonne National Laboratory. 1995a. *The dynamic information architecture system: A high level architecture for modeling and simulation.* Chicago: Argonne National Laboratory, University of Chicago (http://www.dis.anl.gov:80/DEEM/DIAS/diaswp.html).

Argonne National Laboratory. 1995b. *DEEM high level architecture.* Chicago: Argonne National Laboratory, University of Chicago (http://www.dis.anl.gov:80/DEEM/hla.html).

Barber, K. H.; Rodman, S. A. 1990. FORPLAN: the marvelous toy. *J. For.* 88:26–30.

Baskerville, G. 1985. Adaptive management—wood availability and habitat availability. *For. Chron.* 61: 171–175.

Behan, R. W. 1990. The RPA/NFMA: solution to a nonexistent problem. *J. For.* 91:20–25.

Behan, R. W. 1994. Multiresource management and planning with EZ-IMPACT. *J. For.* 92:32–36.

Behan, R. W. 1997. Scarcity, simplicity, separatism, science—and systems. In: Kohm. K. A.; Franklin, J. F., eds. *Creating a forestry for the 21st century*. Washington, DC: Island Press: 411–417.

Bennett, A. J. 1996. Sustainable land use: interdependence between forestry and agriculture. In: *Caring for the forest: research in a changing world*, Congress Report, Volume I, IUFRO XX World Congress, August 6–12 1995, Tampere, Finland: 144–152.

Berry, M. W.; Flamm, R.O.; Hazen, B. C.; MacIntyre, R. M. 1996. The land-use change and analysis (LUCAS) for evaluating landscape management decisions. *IEEE Comput. Sci. Eng.* 3:24–35.

Bonnicksen, T. M. 1991. Managing biosocial systems: a framework to organize society–environment relationships. *J. For.* 89:10–15.

Bormann, B. T.; Brooks, M. H.; Ford, E. D.; Kiester, A. R.; Oliver, C. D.; Weigand, J. F. 1993. *A framework for sustainable ecosystem management*. Gen. Tech. Rep. PNW-331. Portland, OR: U.S. Dept. Agric., For. Serv., Pacific Northw. Res. Sta.

Botkin, D. B. 1990. *Discordant harmonies*. New York: Oxford University Press.

Brown, T. C. 1984. The concept of value in resource allocation. *Land Econ.* 60:231–246.

Caldwell, L. K. 1996. Science assumptions and misplaced certainty in natural resources and environmental problem solving. In: Lemons, J., ed. *Scientific uncertainty and environmental problem solving*, Cambridge, MA: Blackwell Science, Inc.: 394–421.

Canham, H. O. 1990. Decision matrices and weighting summation valuation in forest planning. *No. J. Appl. For.* 7:77–79.

Carpenter, S. R. 1990. Large-scale perturbations: opportunities for innovation. *Ecology* 71:2038–2043.

Chase, A. 1995. *In a dark wood: the fight over forests and the rising tyranny of ecology*. Boston: Houghton Mifflin Co.

Clemen, R. T. 1996. *Making hard decisions: an introduction to decision analysis*, 2nd ed. Pacific Grove, CA: Duxbury Press.

Conklin, J.; Begeman, M. L. 1987. GIBIS: a hypertext tool for team design deliberation. In: *Proceedings of "Hypertext '87,"* November 13–15, 1987. Chapel Hill, NC: 104–115.

Craig, L. E. 1996. Letter from the United States Senate, Committee on Energy and Natural Resources, Washington, D.C. to The Honorable Dan Glickman, Secretary of Agriculture. June 20, 1996.

Daniels, S. E.; Walker, G. B.; Boeder, J. R.; Means, J. E. 1993. Managing ecosystems and social conflict. In: Jensen, M. E.; Bourgeron, P. S., eds. *Eastside forest ecosystem health assessment: volume II ecosystem management: principles and applications*. Portland, OR: U.S. Dept. Agric., For. Serv., Pacific Northw. Res. Sta.: 347–359.

Dombeck, M. 1997. *Integrating science and decision-making: guidelines for collaboration among managers and researchers in the Forest Service*. FS-608. Washington, DC: U.S. Dept. Agric., For. Serv.

Everett, R.; Oliver, C.; Saveland, J.; Hessburg, P.; Diaz, N.; Irwin, L. 1993. Adaptive ecosystem management. In: Jensen; M. E.; Bourgeron, P. S., eds. *Eastside forest ecosystem health assessment: volume II ecosystem management: principles and applications*. U.S. Dept. Agric., For. Serv., Pacific Northw. Res. Sta.: 361–375.

Faber, B. G.; Wallace, W. W.; Croteau, K.; Thomas, V. L.; Small, L. R. 1997. Active response GIS: an architecture for interactive resource modeling. In: *Proceedings of GIS World '97, integrating spatial information technologies for tomorrow*, February 17–20, 1997, Vancouver, BC: 296–301.

Finin, T.; Fritzson, R.; McKay, D.; McEntire. R. 1994. KQML as an agent communication language. In: *Proceedings of the third international conference on information and knowledge management*. ACM Press: 1–8.

Floyd, D. W.; Germain, R. H.; ter Horst, K. 1996. A model for assessing negotiations and mediation in forest resource conflicts. *J. For.* 94(5):29–33.

Forest Ecosystem Management and Assessment Team (FEMAT). 1993. *Forest ecosystem management: an ecological, economic, and social assessment. Report of the Forest Ecosystem Management Assessment Team*. Washington, DC: U.S. Govt. Printing Off. 1993–793071.

French, J. R. P.; Raven, B. 1959. The bases of social power. In: Cartwright, D. ed. *Studies in social power*. Ann Arbor, MI: University of Michigan, Institute for Social Research: 150–167.

Garland, J. J. 1997. The players in public policy. *J. For.* 95(1):13–15.

Genesereth, M. R.; Ketchpel, S. P. 1994. Software agents. *Commun. ACM* 37:48–53.

Government Accounting Office (GAO). 1994. *Ecosystem management: additional actions needed to adequately test a promising approach. A report to congressional requesters*, GAO/RCED-94–111. Washington, DC: Govt. Acct. Off.

Greber, B. J.; Johnson, K. N. 1991. What's all this debate about overcutting? *J. For.* 89:25–30.

Grossarth, O.; Nygren, T. 1993. Implementing ecosystem management through the forest planning process. In: Jensen, M. E.; Bourgeron, P. S., eds. *Eastside forest ecosystem health assessment: volume II ecosystem management: principles and applications*. U.S. Dept. Agric., For. Serv., Pacific Northw. Res. Sta.: 291–299.

Hannigan, J. 1995. *Environmental sociology: a social constructionist perspective*. London: Routledge.

Hashim, S. H. 1990. *Exploring hypertext programming*. Blue Ridge Summit, PA: Windcrest Books.

Heiler, S. 1995. Semantic interoperability. *ACM Comput. Surv*. 27:271–273.

Hoekstra, T. W.; Dyer, A. A.; LeMaster, D. C., editors. 1987. *FORPLAN: an evaluation of a forest planning tool*. Gen. Tech. Rep. RM-140. U.S. Dept. Agric., For. Serv., Rocky Mountain Res. Sta.

Holsapple, C. W.; Whinston, A. B. 1996. *Decision support systems: a knowledge-based approach*. Minneapolis/St. Paul, MN: West Publishing Co.

Holtzman, S. 1989. *Intelligent decision systems*. Reading, MA: Addison-Wesley.

Howard, A. F. 1991. A critical look at multiple criteria decision making techniques with reference to forestry applications. *Can. J. For. Res*. 21:1649–1659.

Howitt, W. S. 1978. The value of the simple approach. *Interfaces* 8:64–70.

InfoHarvest. 1996. *Criterium decision plus: user's guide*. Seattle, WA: InfoHarvest.

Janssen, R. 1992. *Multiobjective decision support for environmental management*. Dordrecht, The Netherlands: Kluwer Academic Publishers.

Janssen, R.; van Herwijnen, M. 1992. *Definite: decisions on a finite set of alternatives*. Dordrecht: The Netherlands: Institute for Environmental Studies, Free University Amsterdam, Kluwer Academic Publishers.

Jensen, M. E.; Everett, R. 1993. An overview of ecosystem management principles. In: Jensen, M. E.; Bourgeron, P. S., eds. *Eastside forest ecosystem health assessment: volume II ecosystem management: principles and applications*. U.S. Dept. Agric., For. Serv., Pacific Northw. Res. Sta.: 9–18.

Jensen, M. E.; Crespi, M.; Lessard, G. 1998. A national framework for integrated ecological assessments. In: Cordel, H. K.; Bergstrom, J. C., eds. *Integrating social sciences with ecosystem management: human dimensions in assessment policy and management*. Champaign–Urbana, IL: Sagamore Press.

Jessup, L. M.; Valacich, J. S., editors. 1993. *Group support systems: new perspectives*. New York: Macmillan.

Jorgensen, S. E.; Halling-Sorensen, B.; Nielsen, S. N. 1996. *Handbook of environmental and ecological modeling*. Boca Raton, FL: Lewis Publishers.

Kaufmann, M. R.; Graham, R. T.; Boyce, D. A., Jr.; Moir, W. H.; Perry, L.; Reynolds, R. T.; Bassett,

R. L.; Hehlhop, P.; Edminster, C. B.; Block, W. M.; Corn, P. S. 1994. *An ecological basis for ecosystem management*. Gen. Tech. Rep. RM-246. U.S. Dept. Agric., For. Serv., Rocky Mountain For. and Range Exp. Sta.

Keeney, R. L. 1992. *Value focused thinking: a path to creative decision-making*. Cambridge, MA: Harvard University Press.

Keeney, R, L.; Raiffa, H. 1993. *Decisions with multiple objectives: preferences and tradeoffs*. Cambridge, MA: Cambridge University Press.

Kennedy, J. J.; Quigley, T. M. 1993. Evolution of Forest Service organizational culture and adaptation issues in embracing ecosystem management: In: Jensen, M. E.; Bourgeron, P. S., eds. *Eastside forest ecosystem health assessment: volume II ecosystem management: principles and applications*. U.S. Dept. Agric., For. Serv., Pacific Northw. Res. Sta.: 19–29.

Kessler, W. B.; Salwasser, H.; Cartwright, C. W., Jr.; Chaplan, J. A. 1992. New perspectives for sustainable natural resources management. *Ecol. Appl.* 2(3):221–225.

Kimmins, J. P. 1991. The future of the forested landscapes of Canada. *For. Chron.* 67:14–18.

Kimmins, J. P. 1995. Sustainable development in Canadian forestry in the face of changing paradigms. *For. Chron.* 71:33–40.

Klein, M.; Methlie, L. B. 1990. *Expert systems: a decision support approach*. Reading, MA: Addison-Wesley.

Kohm, K. A.; Franklin, J. F. 1997. Introduction. In: Kohm, K. A.; Franklin, J. F., eds. *Creating a forestry for the 21st century*. Washington, DC: Island Press: 642.

Larsen, D. R.; Shifley, S. R.; Thompson III, F. R.; Brookshire, B. L.; Dey, D. C.; Kurzejeski, E. W.; England, K. 1997. 10 Guidelines for ecosystem researchers: lessons from Missouri. *J. For.* 95(4):4–9.

Lee, K. N. 1993. *Compass and gyroscope: integrating science and politics for the environment*. Washington, DC: Island Press.

Lindblom, C. E. 1990. *Inquiry and change: the troubled attempt to understand and shape society*. New Haven, CT: Yale University Press.

Liu, G.; Davis, L. S. 1995. Interactive resolution of multi-objective forest planning problems with shadow price and parametric analysis. *For. Sci.* 41:452–469.

Liu, S. 1998. *Integration of forest decision support systems: a search for interoperability*. Master's thesis. Athens, GA: University of Georgia.

Liu, S.; Potter, W. D.; Rauscher, H. M. 2000. Using DCOM to support interoperability in forest ecosystem management decision support systems. *Comput. Electron. Agric.* 27:335–354.

Maheshwari, S. S. 1997. *A CORBA and Java based object framework for integration of heterogeneous systems*. Master's thesis. Athens, GA: University of Georgia.

Mayfield, J.; Labrou, Y.; Finin, T. 1996. Evaluation of KQML as an agent communication language. In: Wooldridge, M.; Muller, J. P.; Tambe, M., eds. *Intelligent agents volume II—proceedings of the 1995 workshop on agent theories, architectures, and languages*. Lecture notes in artificial intelligence. New York: Springer-Verlag.

McCarter, J. B.; Wilson J. S.; Baker, P. J.; Moffett, J. L.; Oliver, C. D. 1998. Landscape management through integration of existing tools and emerging technologies. *J. For.* 96(6):17–23.

McGaughey, R. J. 1997. Visualizing forest stand dynamics using the stand visualization system. In: *Proceedings of the 1997 ACSM/ASPRS annual convention and exposition*, April 7–10, 1997. Seattle, WA: Amer. Soc. Photogramm. Remote Sensing 4:248–257.

Microsoft. 1995. *The component object model specification*. Draft Version 0.9 (http://premium.microsoft.com/msdn/library/specs/tech1/d1/s1d139.htm).

Microsoft. 1996. *The component object model: technical overview* (http://www.microsoft.com/oledev/olecom/).

Microsoft. 1997. *Distributed component object model protocol—DCOM 1.0* (http://premium.microsoft.com/msdn/library/techart/msdn_dcomprot.htm).

Mintzberg, H.; Raisinghani, D.; Theoret, A. 1976. The structure of unstructured decision processes. *Admin. Sci. Quart.* 21:246–275.

Mitchell, K. H.; Wasil, E. A. 1989. *The analytical hierarchy process, applications, and studies*. New York: Springer-Verlag.

Montgomery, C. A. 1993. Socioeconomic risk assessment and its relation to ecosystem management. In: Jensen, M. E.; Bourgeron, P. S., eds. *Eastside forest ecosystem health assessment: volume II ecosystem management: principles and applications*. U.S. Dept. Agric., For. Serv., Pacific Northw. Res. Sta.: 325–335.

Morrison, J. 1993. Integrating ecosystem management and the forest planning process. In: Jensen, M. E.; Bourgeron, P. S., eds. *Eastside forest ecosystem health assessment: volume II ecosystem management: principles and applications*. U.S. Dept. Agric., For. Serv., Pacific Northw. Res. Sta.: 281–290.

Mowbray, T. J.; Zahavi, R. 1995. *The essential CORBA: systems integration using distributed objects*. New York: John Wiley & Sons.

Mowrer, H. T.; Barber, K.; Campbell, J.; Crookston, N.; Dahms, C.; Day, J.; Laacke, J.; Merzenich, J.; Mighton, S.; Rauscher, M.; Reynolds, K.; Thompson, J.; Trenchi, P.; Twery, M. 1997. *Decision support systems for ecosystem management: an evaluation of existing systems*. RM-GTR-296. U.S. Dept. Agric., For. Serv., Interregional Ecosystem Manage. Coord. Group, Decision Support System Task Team, Rocky Mountain For. and Range Exp. Sta.

Nute, D,; Nath, S.; Rosenberg, G.; Verma, B.; Rauscher, H. M.; Twery, M. J. 1999. Goals in decision support systems for ecosystem management. *Comput. Electron. Agric.*

Oliver, C. D.; Knapp, W. H.; Everett, R. 1993. A system for implementing ecosystem management. In: Jensen, M. E.; Bourgeron, P. S., eds. *Eastside forest ecosystem health assessment: volume II ecosystem management: principles and applications*. U.S. Dept. Agric., For. Serv., Pacific Northw. Res. Sta.: 377–388.

O'Reilly, C. A. 1983. The use of information in organizational decision making: a model and some propositions. In: Cummings L. L.; Shaw, B. M., eds. *Res. Org. Behav*. 5:103–139.

Orland, B. 1995. SMARTFOREST: a 3-D interactive forest visualization and analysis system. In: Power, J. M.; Strome, M.; Daniel, T. C., eds. *Proceedings of the decision support—2001 conference*, September 12–16, 1994, Toronto, ON: 181–190.

Otte, R.; Patrick, P.; Roy, M. 1996. *Understanding CORBA: the common object request broker architecture*. Upper Saddle River, NJ: Prentice Hall.

Overbay, J. C. 1992. Ecosystem management. In: *Proceedings of the national workshop: taking an ecological approach to management*, April 27–30, 1992, Salt Lake City, UT. WO-WSA-3. Washington, DC: U.S. Dept. Agric., For. Serv., Watershed and Air Manage. Branch.

Perisho, R. J.; Oliveria, F. L.; Forrest, L.; Loh, D. K. 1995. *INFORMS-R8—A tool for ecosystem analysis*. College Station, TX: STARR Lab, Texas A&M University.

Potter, W. D.; Byrd, T.A.; Miller, J. A.; Kochut, K. J. 1992. Extending decision support systems: the integration of data, knowledge, and model management. *Ann. Oper. Res*. 38:501–527.

Rauscher, H. M. 1995. Natural resource decision support: theory and practice. *AI Applications* 9(3):1–2.

Rauscher, H. M. 1996. Decision making methods for ecosystem management decision support. In: *Caring for the forest: research in a changing world*, Congress Report, Volume II, IUFRO XX World Congress, August 6–12, 1995, Tampere, Finland: 264–273.

Rauscher, H. M. 1999. Ecosystem management decision support for federal forests in the United States: A review. *For. Ecol. Manage*. 114:173–197.

Rauscher, H. M.; Alban, D. H.; Johnson, D. W. 1993. Managing the global climate change scientific knowledge base. *AI Applications* 7(4):17–40.

Rauscher, H. M.; Kollasch, R. P.; Thomasma, S. A.; Nute, D. E.; Chen, N.; Twery, M. J.; Bennett, D. J.; Cleveland, H. 1997a. NED-1: a goal-driven ecosystem management decision support system: technical description. In: *Proceedings of GIS World '97, integrating spatial information technologies for tomorrow*, February 17–20, 1997, Vancouver, BC: 324–332.

Rauscher, H. M.; Loftis, D. L.; McGee, C. E.; Worth, C. V. 1997b. Oak regeneration: a knowledge synthesis. *The Compiler* 15(1): pages 51–52, insert, 3 disks, 3.8 megabytes; 748 chunks; 742 links (electronic).

Rauscher, H. M.; Lloyd, F. T.; Loftis, D. L.; Twery, M. J. 2000. A practical decision-analysis process for conducting ecosystem management at the project level on national forests of the United States. *Comput. Electron. Agric.*, 27:195–226.

Reynolds, K. M.; Rauscher, H. M.; Worth, C. V. 1995. *A hypermedia reference system to the forest ecosystem management assessment team report and some related publications*. Gen. Tech. Rep. PNW-GTR-357. U.S. Dept. Agric., For. Serv., Pacific Northw. Res. Sta.

Reynolds, K.; Saunders, M.; Miller, B.; Murray, S.; Slade, J. 1997. An application framework for decision support in environmental assessment. In: *Proceedings of GIS World '97, integrating spatial information technologies for tomorrow*, February 17–20, 1997, Vancouver, BC: 333–337.

Reynolds, K.; Bjork, J.; Riemann, H. R.; Schmoldt, D.; Payne, J.; King, S.; Moeur, M.; DeCola, L.; Twery, M.; Cunningham, P.; Lessard, G. 1999. Decision support systems for ecosystem management. In: Szaro, R.; Sexton, W.; Johnson, N.; Malk, A., eds. *Ecosystem management: a reference guide to concepts and implementation*. Washington, DC: Island Press.

Rittel, H. 1972. On the planning crisis: systems analysis of the first and second generations. *Bedriftokonomen* 8:390–396

Rue, L. W.; Byars, L. L. 1992. *Management: skills and application*. Homewood, IL: Richard C. Irwin.

Saaty, T. L. 1992. *Multicriteria decision making—the analytical hierarchy process*. Pittsburgh, PA: RWS Publications.

Sage, A. P. 1991. *Decision support systems engineering*. New York: John Wiley & Sons.

Sample, V. A. 1990. *The impact of the federal budget process on national forest planning*. New York: Greenwood Press.

Sample, V. A. 1993. A framework for public participation in natural resource decisionmaking. *J. For*. 91(7): 22–27.

Schmoldt, D. L.; Rauscher, H. M. 1996. *Building knowledge-based systems for natural resource management*. New York: Chapman & Hall.

Schuster, E. G.; Leefers, L. A.; Thompson, J. E. 1993. *A guide to computer-based analytical tools for implementing national forest plans*. Gen. Tech. Rep. INT-296. Ogden, UT: U.S. Dept. Agric., For. Serv., Intermountain Res. Sta.

Shepard, W. B. 1993. Ecosystem management in the Forest Service: political implications, impediments, and imperatives. In: Jensen, M. E.; Bourgeron, P. S., eds. *Eastside forest ecosystem health assessment: volume II ecosystem management: principles and applications*. U.S. Dept. Agric., For. Serv., Pacific Northw. Res. Sta.: 31–37.

Sheth, A. 1998. Changing focus on interoperability in information systems: from system, syntax, structure to semantics. In: Goodchild, M. F.; Egenhofer, M. J.; Fegeas, R.; Kottman, C.A. eds. *Interoperating geographic information systems*. Dordrecht, The Netherlands: Kluwer Academic Publishing.

Shindler, B.; Neburka, J. 1997. Public participation in forest planning: 8 attributes of success. *J. For*. 95:17–19.

Shindler, B.; Steel, B.; List, P. 1996. Public judgments of adaptive management. *J. For*. 94:4–12.

Silver, M. S. 1991. *Systems that support decisionmakers: description and analysis*. New York: John Wiley & Sons.

Smith, G. R. 1997. Making decisions in a complex dynamic world. In: Kohm K. A.; Franklin, J. F. eds. *Creating a forestry for the 21st century*. Washington, DC: Island Press: 419–435.

Southern Appalachian Man and the Biosphere (SAMAB). 1996. *The southern Appalachian assessment summary report*. Report 1 of 5. Atlanta, GA: U.S. Dept. Agric., For. Serv., Southern Region.

Steelman, T. A. 1996. Public participation in national forest management: a case study of the Monongahela National Forest, West Virginia. In: *Proceedings of the IEEE 1996 international symposium on technology and society, technical expertise and public decisions*, June 21–22, 1996. Princeton, NJ: Princeton University Press: 226–230.

Stephens, J. 1996. A new era in planning analysis tools. In: *Analysis notes*. Ft. Collins, CO: U.S. Dept. Agric., For. Serv., WO/Ecosystem Manage. Analysis Center: 6:1–2.

Sygenex. 1994. *Criterium decision plus: the complete decision formulation, analysis, and presentation for Windows. User's guide*. Redmond, WA: Sygenex.

Taylor, R. 1998. Forest fighter. *Government Executive*, December: 23–26.

Teck, R.; Moeur, M.; Eav, B. 1996. Forecasting ecosystems with the forest vegetation simulator. *J. For.* 94: 7–10.

Teck, R.; Moeur, M.; Eav, B. 1997. *The forest vegetation simulator: A decision-support tool for integrating resources science* (http://www.fs.fed.us/ftproot/pub/fmsc/fvsdesc.htm).

Thomas, J. W. 1995. *The forest service program for forest and rangeland resources: a long-term strategic plan*. Draft 1995 RPA Program. Washington, DC: U.S. Dept. Agric., For. Serv.

Thomas, J. W. 1997. Foreword. In: Kohm, K. A.; Franklin, J. F. eds. *Creating a forestry for the 21st century: The science of ecosystem management*. Washington, DC: Island Press.

Thomson, A. J. 1993. Paradigm green: AI approaches to evaluating the economic consequences of changing environmental viewpoints. *AI Applications* 7:61–68.

Thomson, A. J. 1996. Asimov's phsychohistory: vision of the future or present reality? *AI Applications* 10:1–8.

Ticknor, W. D. 1993. Sustainable forestry: redefining the role of forest management. In: Aplet, G. H.; Johnson, N.; Olson, J. T.; Sample, V. A. eds. *Defining sustainable forestry*. Washington, DC: Island Press: 260–269.

Tonn, B.; White, D. 1996. Sustainable societies in the information age. *Amer. Sociol.* 27:102–121.

Tonn, B.; English, M.; Travis, C. 1998. *Frameworks for understanding and improving environmental decision making*. ORNL/NCEDR-7. Oak Ridge, TN: Oak Ridge National Laboratory.

Toynbee, Arnold J. 1946. *A study of history*. New York: Oxford University Press.

Turban, E. 1993. *Decision support and expert systems: management support systems*. New York: Macmillan.

Twery, M. J.; Rauscher, H. M.; Bennett, D. J.; Thomasma, S. A.; Stout, S. L.; Palmer, J. F.; Hoffman, R. E.; DeCalesta, D. S.; Gustafson, E.; Cleveland, H.; Grove, J. M.; Nute, D. E.; Kim, G.; Kollasch, R. P. 2000. NED-1: integrated analysis for forest stewardship decisions. *Comput. Electron. Agric.*, 27:167–193.

Van den Berg, J. C. J. M. 1996. *Ecological economics and sustainable development: theory, methods, and applications*. Cheltenham, UK: Edward Elgar.

Vroom, V. H.; Jago, A. G. 1988. *The new leadership: managing participation in organizations*. Upper Saddle River, NJ: Prentice Hall.

Walters, C. J.; Holling, C. S. 1990. Large-scale management experiments and learning by doing. *Ecology* 71:2060–2068.

Wear, D. N.; Turner, M. G.; Flamm, R. O. 1996. Ecosystem management with multiple owners: landscape dynamics in a southern Appalachian watershed. *Ecol. Appl.* 6:1173–1188.

Wegner, P. 1996. Interoperability. *ACM Comput. Surv.* 28:285–287.

Williams, S. B.; Roschke, D. J.; Holtfrerich, D. R. 1995. Designing configurable decision-support software: lessons learned. *AI Applications* 9:103–114.

Zonnefeld, I. S. 1988. Basic principles of land evaluation using vegetation and other land attributes. In: Kuechler, A. W.; Zonnefeld, I. S., eds. *Vegetation mapping*. Dordrecht, The Netherlands: Kluwer Academic Publishers.

Part 4
Analytical Methods: Concepts

13 Elements of Spatial Data Analysis in Ecological Assessments

Patrick S. Bourgeron, Marie-José Fortin, and Hope C. Humphries

13.1 Introduction

Virtually any aspect of an ecological assessment (EA) is likely to involve the topic of space, including its striking effect on landscapes and the distribution of human populations. For example, a spatially explicit approach is needed to address two policy questions common to many EAs. What is required for maintaining the long-term productivity of ecosystems? What is the impact of maintaining current management scenarios on, for example, major social issues or the maintenance of rural communities and their economies in a given area? Essential tasks of EAs also involve the explicit consideration of space, such as in combining information from various geographic areas and multiple scales. Most measurements of large-scale phenomena, such as the effect of regional carbon and nitrogen cycles, hydrologic regimes, changes in land-use patterns, and demographics, among many others, carry the imprint of spatial variability and scaling. Therefore, explicit consideration of all aspects of space (e.g., spatial variability and its corollary, spatial scaling) is increasingly a central concern in the design and implementation of EAs, whether in map creation or incorporation in predictive modeling (see Chapters 3 and 18; also, Haining, 1990; Ritchie, 1997).

EAs focus on environmental and social systems in which spatial heterogeneity is functional and not the result of random, noise-generating processes (see Chapter 2; also, Kolasa and Rollo, 1991; O'Neill et al., 1991; Legendre and Legendre, 1998). EA data are therefore inherently spatially and temporally structured, likely at more than one scale (see Chapter 2; also, Davis et al., 1991; Schneider, 1994; Bradshaw, 1998; Pahl-Wostl, 1998). The characterization of physical, ecological, and social systems depends on the particular spatial, temporal, and organizational perspectives resulting from addressing specific objectives (see Chapter 3). Therefore, it is essential to understand how patterns and processes vary in space and across geographic scales (King, 1997; Hammer, 1998; Legendre and Legendre, 1998; Pahl-Wostl, 1998). Spatial variability and scaling interact with each other to produce the spatial arrangement of patterns and processes (see Chapters 2 and 3). As a result, three questions must be addressed during the design phase of an EA (Haining, 1990). (1) What expressions of spatial variability in the area of interest should be examined? (2) What components of the assessment must be spatially explicit? (3) At what scale do processes act on phenomena of interest?

The main purpose of this chapter is to address the first two questions concerning quantification of spatial variability. The third question is treated in Chapters 13 and 14, in which several methods for examining scaling properties in spatial data analysis (fractal geometry, percolation theory, and fuzzy statistical and modeling approaches) are presented. Here we discuss and review concepts, issues, and approaches to spatial analysis and thereby provide a basis for developing a sound inductive approach to data analysis in a spatial context during the different stages of an EA. This chapter is organized

Patrick S. Bourgeron and Hope C. Humphries wish to acknowledge partial funding provided by a Science to Achieve Results grant from the U.S. Environmental Protection Agency ("Multi-scaled Assessment Methods: Prototype Development within the Interior Columbia Basin").

into two main sections: (1) general considerations and issues in spatial analysis and (2) an overview of approaches in spatial analysis. It is not the intent of this chapter to describe at length all possible techniques and how they relate to each other; this information is available in other publications, which are cited in the relevant sections. The scope of this chapter is to describe approaches appropriate for some of the questions of interest in the spatial analysis of EA data.

The basic methodologies for spatial data analysis presented in this chapter apply to both environmental and social sciences for three main reasons. First, EAs consider linked environmental and social systems. Second, data are generally observational rather than experimental in EAs and therefore present similar issues in their analysis and interpretation, regardless of whether they are environmental or social. Third, environmental and social analyses are often directed at similar spatial scales and data structures, so methods applicable to the spatial analysis of one type of data should also be applicable to other types.

13.2 General Considerations and Issues in the Spatial Analysis of EA Data

The analysis of spatial patterns has become increasingly prevalent in the investigation of environmental and social data (Haining, 1990), because most phenomena of interest are structured by forces that have spatial components (Bell et al., 1993; Legendre and Legendre, 1998). For example, ecologists examine the spatial patterns of single species or assemblages of species to understand the mechanisms that control their distributions; soil scientists study the movement of water, a driver of soil genesis and landscape evolution, to understand the causal agents of pedogenic variability and many practical aspects of soil behavior; social scientists study the settlement patterns of human populations and the impact of societal values to understand the extent to which human populations shape and are shaped by their environment. In this section, we discuss general considerations and issues that may arise in analyzing spatial data. Specific information regarding the sampling and storage of such data is provided in Chapters 5 to 8 and 11. A full presentation of the subject matter treated in this section is found in Haining (1990).

13.2.1 Importance and Nature of Spatial Structure in EA Data

Spatial structure in EA data sets has three main sources: measurement error, continuity effects (e.g., spatial heterogeneity), and space-dependent processes. Determining which of these sources is present in any given data set has implications for the analysis of EA data, indicating the types of spatial patterns that should be expected and the types of models that best represent such patterns (Haining, 1990).

Measurement errors have several origins. First, instrument precision can alter our ability to accurately quantify spatial structure at a given scale. Second, such errors can be generated by observer-dependent bias. For example, if a large region is divided into subregions for surveying purposes, each with a different survey team, even the smallest team-dependent bias can generate discontinuities at subregion boundaries (Milne, 1959; Haining, 1990). Third, remotely sensed data (see Chapter 10) can also suffer from spatial dependency effects induced by instruments. Although rapid progress is being made in hardware improvement and in modeling satellite data (see Chapter 10; also, Quattrochi and Goodchild, 1997), it is still important to carefully consider this type of error in EA data sets (Craig, 1979; Labovitz and Masuoka, 1984). Finally, measurement errors can propagate themselves from one scale to another (Heuvelink, 1998).

Three important related concepts to consider in the spatial analysis of EA data are continuity effects, homogeneity, and heterogeneity. Spatial continuity of patterns is due to the spatial continuity in the events (i.e., processes) that generates the patterns and are responsible for their spatial distribution, including boundaries between patterns, pattern shape, and the scale at which patterns are generated (see Chapter 2; also, King, 1997). Spatial variation in the distribution of a pattern or process is described using the two opposing concepts of homogeneity (i.e., the absence of variation), and heterogeneity (i.e., composed of different parts; see Kolasa and Rollo, 1991; Dutilleul and Legendre, 1993; Legendre and Legendre, 1998). The analysis of sharp boundaries (Fortin et al., 1996) between contrasting patterns (i.e., spatial heterogeneity) is a central aspect of EA spatial analysis.

Four important types of processes may generate spatial structure in an environmental or social attribute (Haining, 1990): (1) *diffusion*, a general term describing a process in which some attribute (e.g., commodity, gene flow, disturbance) is taken up by a fixed population; (2) exchange and transfer (e.g., income transfer, flow of nutrients); (3) interactions, in which events at one location influence and are influenced by events at other locations (e.g., pricing patterns across retail sites reflecting underlying competitive interactions among retailers, distribution of individuals of different species as a function of interspecific competition); and (4) dispersal or spread (e.g., migration of human and animal populations, seed dispersal). Diffusion and spread are differentiated by Haining (1990); *diffusion* is defined as the dispersal of an attribute through a fixed population, whereas *spread* refers to the dispersal of the population itself. These processes are often included in models of spatial distribution (e.g., Haining, 1990, in the social sciences; Pearson and Gardner, 1997, Tilman and Kareiva, 1997, in ecology; see also Chapter 18).

13.2.2 Issues in Spatial Data Analysis

Analytical Issues

Analytical issues encountered in spatial analysis of EA data fall into three broad categories (Table 13.1): (1) sampling, (2) numerical summary and characterization of the spatial properties of the data, and (3) analysis of multivariate data sets. The first analytical issue, the design of spatial sampling schemes to fit the goals of an EA (see Chapter 1), is an important consideration, whether with existing or de novo data (see Chapter 7), including accuracy assessment of remotely sensed data (see Chapter 10). The second analytical issue includes the numerical description of the spatial properties of map data. EAs often use maps as a basis for evaluation and scenario planning (see Chapters 3, 22, and 24); therefore, it is important to study the spatial arrangement of map values and to devise summary measures that characterize spatial patterns and their properties at different scales (e.g., Burrough, 1995).

For example, characterization of map data provides the basis for (1) constructing additional smoothed and interpolated maps (Webster, 1985); (2) defining the spatial attributes of models, such as the spatial variation in precipitation in hydrological models (e.g., TOPMODEL in RHESSYS; see Chapter 18) or in fire regimes in vegetation models (e.g., LANDIS; see Chapter 18); (3) comparing different maps; and (4) comparing map surfaces at different points in time. The third analytical issue stems from the fact that the issues addressed in an EA (see Chapter 1) generally require the analysis of data describing two or more variables in a region. Multivariate analysis of spatial data includes measures of association between variables and regression models such as higher-order trend surface analysis (Casetti and Jones, 1987; Anselin, 1988; Haining, 1990).

TABLE 13.1. Analytical and practical issues in the spatial analysis of ecological assessment data.

Analytical issues
Spatial sampling
Numerical summary and characterization of the spatial properties of map data
Analysis of multivariate data sets
Practical issues
Conceptual models and inference frameworks for spatial data
Modeling spatial variation
Statistical modeling of spatial data and:
Dependency in spatial data
Spatial heterogeneity
Spatial distribution of data points and boundary effects
Assessing model fit
Distributions
Extreme data values
Model sensitivity to the areal system
Size–variance relationships in homogeneous aggregates

Practical Issues in Spatial Data Analysis

There are three broad categories of practical issues in the analysis of spatial data for EAs (Table 13.1): choosing conceptual models and inference frameworks, modeling spatial variation as a function of attribute properties, and analyzing spatially referenced point and area data. The choice of an inference framework for the analysis of spatial data is a fundamental issue, because the assumptions of classical inference theory (i.e., that the data are the outcome of some well-defined experiment) and its implications (e.g., the stationarity assumption; see Section 13.3.1) may not apply to EA spatial data. Exploration of alternatives to classical theories of inference is an active field of research and includes the Bayesian approach (Leamer, 1978), which allows an explicit mixing of prior information with current data, and tests based on data randomization (e.g., bootstrapping, permutation tests; see Diaconis, 1985; Manly, 1997), among others.

The second practical issue, the choice of a model of spatial variation, is influenced by the geographic attributes of the data and the study region, which constrain how data analysis proceeds. It is important to note that, whereas temporal and spatial data share a number of similarities, they also differ in

TABLE 13.2. Examples of spatial attribute considerations for observations and region prior to data analysis.

1. Observations	From a continuous surface, patches internally continuous or discrete points
	Size of observations relative to region (e.g., points vs. areas)
	Size of observations relative to driving processes
	Sample vs. exhaustive coverage
	If sample, type of distribution (e.g., random, systematic)
	Within-observation heterogeneity
2. Region	Size of region relative to driving processes
	Type of boundary with other regions (e.g., natural vs. artificial)
	Within-region heterogeneity, including type of heterogeneity, presence of subregions, interior discontinuities (natural or human-caused)
	Relationships of subregions within region
	Differences and similarities in subregion responses
3. Across scale spatial relationships	External
	Influence on spatial variation of effects originating outside the region
	Relationship of each sample, site, or area to other regions
	Internal
	Scale relationships among samples, sites, or areas within region
	Scale relationships between region, subregion, and samples, sites, or areas within region
	Intra-site:
	Influence of site-specific effects

Source: Haining, 1990.

significant ways, which may prevent the use of the same models for both types of data. For example, both types of data are ordered (e.g., observations cannot be assumed to be independent in time and space), but the ordering of temporal data is unidirectional (or one-dimensioned), that is, largely constrained to the dependence of the present on the past. In contrast, spatial data are omnidirectional (or two-dimensioned and all-direction), and their structure is usually complex due to a multiplicity of paths and patterns of interactions (Haining, 1990). Table 13.2 summarizes some of the considerations that should be given to the spatial attributes of data and the study region prior to data analysis. Among the most important considerations are boundary conditions (e.g., type of boundary, its geometry), discontinuities (e.g., geological fault lines, administrative boundaries), and areal characteristics (e.g., scale relations). The latter depend on the nature of the objective addressed by the analysis.

The third practical issue to address in the analysis of spatial data for EAs is related to the statistical modeling of spatial data. Table 13.1 summarizes the various points that need to be considered in analyzing spatial data. Dependency in spatial data, that is, the amount of information carried by each observation that is duplicated by other observations in the data set, leads to a relative loss of information compared to independent observations. Model parameters may differ among subregions (Cliff et al., 1975), possibly reflecting the influence of processes operating at larger scales. Therefore, stratification of a region may be necessary (Balling, 1984; Semple and Green, 1984), although some environmental stratification schemes may be inappropriate for some analyses (see the discussion of specific applications of terrestrial environmental schemes in Chapter 22). If the data have uneven spatial coverage, model fitting for interpolation will be disproportionately influenced by those parts of the region where there are more observations; isolated points in undersampled areas will have high leverage, which is undesirable if there is any error in these points. Boundaries may present problems, in particular if they are artificial or if strong effects originate from outside the study region on those observations that are close to the boundary.

The process of assessing model fit includes estimation of parameters, significance testing, and analysis of the spatial properties of model residuals (i.e., examining the residuals for spatial autocorrelation). The presence of spatial structure in the distribution of the residuals is usually an indication of failure to account for important elements of the problem addressed by the analysis. It is imperative to examine the distribution of the data, because transformations may often be required to approximate normality in the spatial structures under investigation (Cressie and Read, 1989). Also critical is screening for extreme values that may represent sampling units that differ from the rest of the sampled population, rather than arising from errors in the data (Cox and Jones, 1981; Fortin, 1999a). Such observations have high leverage in the fit of models. Diagnostic procedures, such as examination of residuals and leverage, are used to assess the influence of individual observations (Wrigley, 1983; Austin et al., 1990).

Observations comprising areal units create a distinct set of problems for spatial data analysis in

TABLE 13.3. General framework for statistical analysis of spatial data.

Exploratory data analysis methods: Detecting data structure, hypothesis generation, data model specification
  Descriptive statistics
  Model evaluation methods
Confirmatory data analysis methods: Statistical inference and model development and testing
  Parametric methods: strict distributional assumptions
  Robust methods: data arise from set of possible distributions
  Nonparametric methods: no distributional assumptions

Source: Haining, 1990.

terms of the choice of the analysis method, such as trend surface or nearest-neighbor analyses (Ripley, 1984). Problems that may need to be addressed include grid orientation and origin, irregular areal units, arbitrary aggregation of areal units, number of units, and the effects of boundaries and geometric structure of the study area. Finally, the relationships between differently sized areas and variance among areas should be investigated. An example of a question to be answered concerns whether a small number of observations should be considered as reliable and given as much weight as a large number of observations (Haining, 1990).

13.2.3 Statistical Framework for Spatial Data Analysis

The spatial analysis of EA data involves an interplay between two distinct phases of statistical analysis: exploratory data analysis (EDA) and confirmatory data analysis (CDA) (sensu Haining, 1990; also see Hoaglin et al., 1983, and Table 13.3). EDA is the initial analytical phase during which the intrinsic characteristics of a data set (e.g., patterns, structure) are identified. The main purpose of EDA is to describe the properties of data sets and to formulate hypotheses. Because the first stages of data analysis are conducted with raw data, EDA must use methods to identify outlier values (Haining, 1990). For multivariate data sets, EDA includes the identification of variable relationships (scatterplots of y against x) or the detection of scale-level (variance–mean) relationships.

Formulation of hypotheses in EDA implies the need for testing them, which is provided by CDA. CDA, similar to traditional statistical inference, involves tests of significance, estimation, and prediction. CDA also includes sensitivity and residual analyses. Hoaglin et al. (1983) characterize EDA as the flexible search for evidence and CDA as the evaluation of the available evidence. EDA is particularly important for modeling EA spatial data because some formal CDA procedures may be difficult to implement or may have unknown reliability. However, when underlying distributional assumptions are violated, CDA uses robust methods, that is, methods that employ a robust estimator (Hoaglin et al., 1983, 1985; Hampel et al., 1986; Haining, 1990).

Figure 13.1 shows a framework for spatial data analysis in which either EDA and CDA can be used, depending on the specific objectives and data sets. Step 1 requires EDA, but all other steps can be conducted with EDA or CDA. The choice of a given framework is a function of the data and their properties, the purpose of the study, and the judgment of the analyst.

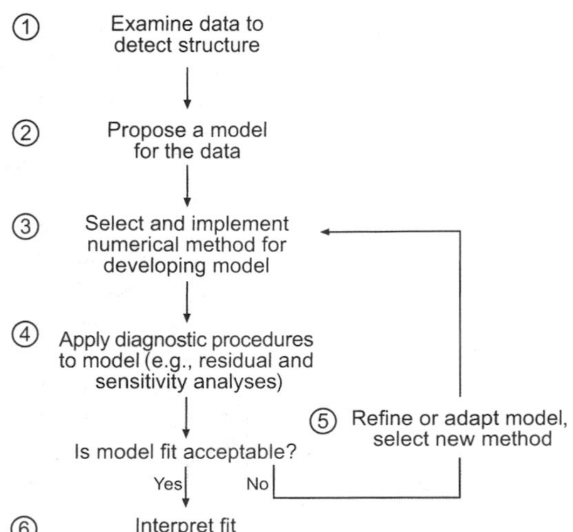

FIGURE 13.1. Framework for spatial data analysis. See text for further explanation.

13.3 Overview of Approaches in Spatial Analysis

13.3.1 Definitions and Assumptions

All spatial analysis approaches are concerned with quantifying and characterizing the spatial structure or pattern in the distribution of variables at their scale(s) of observation (Haining, 1990; Legendre and Legendre, 1998). As discussed in Section 13.2.1, the observed spatial structure in a data set may be due to different sources. Therefore, a useful description of approaches for spatial analysis must begin with clear definitions of basic terms, because inconsistent or imprecise use of terminology may lead to the appearance of conflict among the available methods.

Spatial autocorrelation refers to the lack of independence among values of a variable of interest due to error introduced by the effects of a process operating within a zone of spatial influence (Legendre and Fortin, 1989; Legendre, 1993; Legendre and Legendre, 1998). In contrast to spatial autocorrelation, spatial dependency results from the influence on a variable of interest of one or more explanatory variables that exhibit spatial structuring (Legendre and Legendre, 1998). In practice, it is difficult to determine the origin of spatial structure (see Section 13.3.2). In some complex statistical models, spatial structure is expressed as a combination of spatial autocorrelation in a response variable, the effects of explanatory variable(s), plus explanatory variable autocorrelation structure (Cliff and Ord, 1981; Griffith, 1988; Legendre and Legendre, 1998).

Implicit in statistical analyses is the assumption that the statistical properties of spatially correlated variation are the same over the entire study area. This assumption, known as the hypothesis of stationarity, is necessary for applying statistical methods to spatial analysis (Burrough, 1995). However, in spatial analyses, the stationarity assumption is usually replaced by second-order or weak stationarity, in which the mean is constant, covariance depends only on sampling interval, and the variance is finite and constant (Legendre and Legendre, 1998). The spatial and nonspatial definitions of stationarity differ in the recognition, for the spatial case, that the variance includes a covariance term that depends on sample spacing (Burrough, 1995). For spatial data, a third, relaxed form of stationarity, the intrinsic assumption (Legendre and Legendre, 1998), may be used. This assumption specifies that the differences between successive points must meet the stationarity hypothesis, rather than the values of the points themselves; that is, it considers only the increments of the values of the regionalized variables. This condition must be met in the use of surface pattern methods, such as the correlogram and semivariance (see Section 13.3.2) (Burrough, 1995; Legendre and Legendre, 1998).

There are four aspects of nonstationarity to consider: nonnormality in the distribution of a variable, nonstationarity of the mean, nonstationarity of the variance, and anisotropy (variation in the value of the variable as a function of direction). Nonnormal distributions are common in EA data and require transformation to approximate normality if possible. Nonstationarity of the mean is corrected for by filtering out long-range trends. Nonstationarity of the variance results in a meaningless expression for autocorrelation, because a constant variance is required (Burrough, 1995). In this case, using semivariance rather than covariance is suggested, because semivariance does not require stationarity of the variance (see discussion of the intrinsic assumption). Anisotropic variation occurs when there is strong directional control on a variable of interest. Detection is necessary because anisotropy affects the interpretation and applicability of the analyses. All techniques discussed in this section are sensitive to one or more of these aspects of nonstationarity and to extreme values that can bias the quantification of spatial patterns in the data.

The quantification and characterization of spatial structure can be accomplished using two distinct categories of methods, spatial statistics and landscape metrics. Spatial statistics quantify spatial structure based on sampled data, whereas landscape metrics characterize the geometric and spatial properties of mapped data. Therefore, spatial statistics estimate the spatial values of sampled variables, and landscape metrics characterize the properties of spatially homogeneous units (patches) or mosaics of patches (Fortin, 1999b). The two categories of techniques are discussed in Sections 13.3.2 and 13.3.3.

13.3.2 Spatial Statistics

Three families of spatial statistics, point pattern methods, surface pattern methods, and line pattern analysis, are used in spatial analysis, based on the type of data and objective of the analysis. Point pattern analysis is concerned with the spatial distribution of all individual objects (points) in a study area. Its main objectives are to quantify the geographic distribution of the objects, test for randomness, and describe the type of pattern and scale. The second family of methods, surface pattern analysis, applies to the study of spatially continuous variables. It is based on the concept of regionalized variables (Matheron, 1965), which are assumed to be distributed as a continuous surface (Oden et al., 1993; Legendre and McArdle, 1997). This family of methods includes trend surface analysis, a special case of traditional multiple-regression techniques (Turner et al., 1991; Bur-

TABLE 13.4. Spatial statistics: objectives, data types, families, and methods of analysis.

Objective	Data type	Family	Methods
Spatial structure description	Quantitative	Surface pattern	Correlograms (Moran's I, Geary's c) Semivariogram
	Qualitative	Surface pattern	Correlogram (Join–Count)
		Point pattern	Ripley's K, Ripley's K_{12}
Mapping (interpolation)	Quantitative	Surface pattern	Trend surface analysis, kriging, spline correlogram
Testing for the presence of spatial autocorrelation	Qualitative	Point pattern	Ripley's K, Ripley's K_{12}
		Surface pattern	Correlogram (Join–Count)
	Quantitative	Surface pattern	Correlograms (Moran's I, Geary's c), Mantel test, semivariogram–kriging
Analysis of the source of variation	Quantitative	Surface pattern	Partial Mantel test, partial CCA

Source: Legendre and Fortin, 1989; Legendre and Legendre, 1998; Fortin, 1999a.

rough, 1995). Trend surface analysis may be applied to one-, two-, or three-dimensional space (Legendre and Legendre, 1998). It is the simplest and oldest method for producing smoothed maps and will not be discussed further in this section; a review of the topic is found in Haining (1990). Surface pattern methods also include the correlogram and semivariance methods presented later. The third family of methods, line pattern analysis, developed by geographers, is concerned with the study of networks and connections among points. In line pattern analysis, graph theory is the main analytical approach used to optimize pathways based on networks of points (Upton and Fingleton, 1985). In this section, we describe only point and surface pattern methods (Table 13.4).

Spatial statistics are categorized as first- or second-order statistics. First-order statistics test whether an overall or large-scale spatial trend (i.e., the mean spatial trend) differs significantly from a random pattern (Fortin, 1999b). Tests explicitly distinguish randomness (absence of spatial structure) from nonrandom patterns (e.g., clumping or regularity). They include variance-to-mean, clumping index, Green's index, Lloyd's index, Morosita's index, and nearest-neighbor statistics. First-order statistics have a limited ability to discriminate among some spatial patterns, for example, distinguishing a trend from a patch (Fortin 1999b). For this reason, they are not discussed further in this section. Good reviews of these methods are found in Pielou (1977), Ripley (1987), Turner et al. (1991), and Dale (1999), among others.

In contrast to first-order methods, second-order statistics (Table 13.4) were developed to quantify small-scale pattern intensity (i.e., magnitude, degree) and scale (i.e., spatial extent). Second-order methods measure squared deviations from the mean (Bailey and Gatrell, 1995) and consequently are sensitive to nonstationarity (Legendre and Legendre, 1998; Fortin, 1999b). The sampling design (location and number of samples) and sample unit size (grain) affect the intensity and type of spatial autocorrelation identified (Fortin et al., 1989; Fortin, 1999b).

Table 13.4 summarizes the use of second-order techniques based on the objectives and type of data. It should be noted that some techniques can be used to address different objectives, and the same family of techniques can be used with different types of data. Detailed discussions of these techniques are found in Legendre and Fortin (1989), Legendre (1993), Legendre and Legendre (1998), and Fortin (1999b).

Point Pattern Methods

Ripley's K (Ripley, 1981; Upton and Fingleton, 1985; Fortin, 1999b) is a second-order univariate statistic that quantifies the spatial pattern intensity and range of qualitative point data (i.e., points in space such as x–y coordinates) that correspond to the locations of discrete objects (e.g., individual plant stems). This statistic requires mapping all objects in the study plot. Specifically, the number of objects is counted that lie in an area within a spatial lag (i.e., distance) of a randomly chosen object. Ripley's K is the overall mean of the number of objects within an area described by a circle whose radius is a given spatial lag. It is important to note that the statistic is cumulative, quantifying all objects from zero distance up to a given spatial lag. Its significance is determined using a randomization test. The univariate Ripley's K can be extended to bivariate patterns and is then known as Ripley's K_{12} (Ripley, 1981; Upton and Fingelton, 1985; Cressie, 1991). Positive values of the statistic indicate a positive interaction between the two variables, whereas negative values indicate a negative interaction (e.g., spatial segregation).

Surface Pattern Methods

Correlogram-based Methods

A correlogram is a graph in which spatial autocorrelation values are plotted against a measure of distance among sites or samples. Correlograms (Cliff and Ord, 1981) can be produced for qualitative data (Join–Count) or quantitative data (Moran's *I*, Geary's *c*, Mantel test, among others) and for single variables (Join–Count, Moran's *I*, Geary's *c*) or multivariate data (Mantel test). In all cases, a test of significance is available for each individual value plotted (Legendre and Legendre, 1998).

When the data are qualitative (categorical), the Join–Count statistic can be calculated (Upton and Fingleton, 1985). The Join–Count statistic tests whether neighboring samples are more likely to be in the same categorical state or in different states than would be expected for a random process. Spatial correlograms based on the Join–Count statistic are graphed using neighbor networks (Upton and Fingleton, 1985), rather than the Euclidean distances that are calculated for the surface pattern methods discussed later (Moran's *I*, Geary's *c*, Mantel test, and semivariogram-based methods). The significance of the Join–Count statistic is obtained by computing a standard normal deviate, using a two-sided test to detect both positive and negative spatial autocorrelations. Originally developed for binary data, the method has been extended to multicategorical data (Cliff and Ord, 1981; Upton and Fingleton, 1985).

For quantitative variables, spatial autocorrelation is measured using either Moran's *I* or Geary's *c* (Cliff and Ord, 1981). Moran's *I* is the degree of correlation between the values of a variable as a function of spatial locations. It is related to the Pearson's correlation coefficient, in that it represents the deviations between the values of a variable and its mean. Like Pearson's correlation coefficient, Moran's *I* varies between -1 (negative autocorrelation) and $+1$ (positive autocorrelation), with a value of 0 in the absence of spatial autocorrelation. In contrast, Geary's *c* is a distance-type function that measures the difference (distance) among values of the variable at nearby locations. It varies from 0 (positive autocorrelation) to some unspecified larger value close to 2 (strong negative autocorrelation). Both coefficients can be graphed against distance to produce a correlogram. To interpret the correlogram, coefficients can be tested for significance using the Bonferroni method (Oden, 1984), but the test requires limiting the correlogram to a small number of distance classes. To alleviate this problem, the spline correlogram technique (Bjørnstad and Falck, 1997) has been suggested.

The Mantel test (Legendre and Fortin, 1989) quantifies the degree of relationship between two distance matrices based on data sets (e.g., environmental variables and species abundance data) obtained at the same sampling locations. The normalized Mantel statistic, which is the sum of all products between corresponding elements of the distance matrices, behaves like a product–moment correlation coefficient, *r*, varying from -1 to $+1$. When one of the distance matrices contains the Euclidean distances between the sample locations, the resulting Mantel test is a measure of the overall spatial pattern of the data. Significance is assessed by using a randomization test under the null hypothesis of no relationship between the two distance matrices. The observed Mantel test is expected to have a value near the mode of a reference distribution obtained by randomization of the data.

Semivariogram-based Methods

Methods based on the analysis of the semivariogram, developed by geological engineers and soil scientists, quantify spatial pattern and allow modeling of this pattern by interpolation (Cressie, 1991; Rossi et al., 1992). Interpolation uses only the data that are spatially dependent to calculate estimated values of a variable, in contrast to regression-based methods (e.g., trend surface analysis), which use coefficients calibrated over the entire study region (see the discussion in Burrough, 1995). The semivariance is defined as one-half the variance in the differences between the values of a variable at two locations. The plot of semivariance against sampling interval or distance is called a *semivariogram*. Semivariance is relatively easy to calculate and is robust to deviation in nonstationarity of the variance; therefore, semivariograms have been favored over correlograms as a basis for predictive modeling of regionalized variables (Burrough, 1995). For a second-order stationary process, the semivariance levels out at some distance in the semivariogram. This level is called the *sill*. The distance at which the sill appears is the distance within which sample points are spatially dependent. Beyond that distance, the samples can be treated as independent. The semivariogram does not always pass through the (0, 0) point. This indicates that there is residual random variation, called the *nugget variance*, that is not spatially correlated. The semivariogram

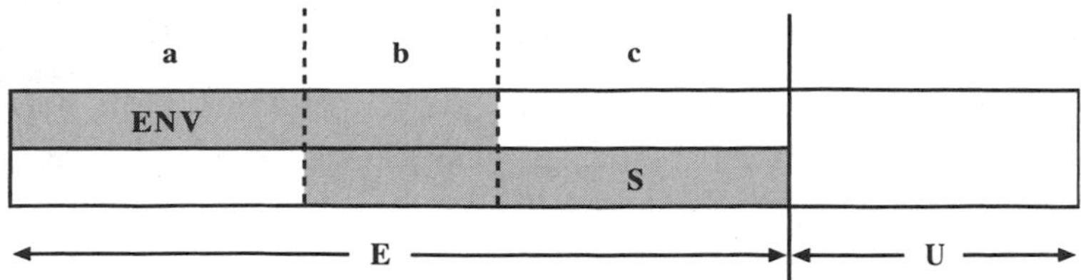

Fraction	Example of factors that can be used in interpretation
[a]	Nonspatially structured component of biotic or environmental factors
[b]	Spatially structured component of biotic or environmental factors Spatial autocorrelation in B and ENV
[c]	Spatially structured biotic or environmental factors not included in the analysis Spatial autocorrelation in B
U	Factors not included in an analysis and nonspatially structured at scale of study Random variation, measurement error

FIGURE 13.2. Partitioning the variation in a response variable, B, between explained (E) and unexplained (U) variation (after Borcard et al., 1992; Legendre and Legendre, 1998). Explained variation is further partitioned into environmental (ENV) and spatial (S) components and into fractions [a], [b], and [c]. See text for further description of fractions.

may take a variety of shapes (e.g., Haining, 1990, p. 97) as expressed by different models (e.g., spherical, linear, exponential), which form the basis for interpolation (Burrough, 1995; Legendre and Legendre, 1998). Traditionally, semivariance statistics are not tested for significance; Legendre and Legendre (1998) suggest that they could be tested with a variation of Geary's c under the condition of second-order stationarity.

Kriging (Matheron, 1965; David, 1977; Journel and Huijbregts, 1978) is an interpolation method that uses the information contained in the semivariogram. It is considered to be an optimum method in the sense that it is designed to provide the best linear unbiased estimate (BLUE) of the average value of a variable at a given point. Kriging is also an exact interpolator; that is, interpolated values coincide with the observed values at the data points. The estimated standard errors of the interpolated values provide the basis for optimal sampling design by identifying where sampling intensity should be increased or decreased (Burrough, 1995). Reviews by Haining (1990), Burrough (1995), and Legendre and Legendre (1998) are solid starting points for obtaining further information on the use of semivariance and interpolation methods.

Understanding the Source of Variation

Two challenges are encountered in conducting spatial analysis: (1) spatial structures may be a source of false correlations (Legendre and Legendre, 1998), and (2) the relationships between two variables may be due to a third, unmeasured variable or simply to their joint co-occurrence (Legendre, 1993; Legendre and Legendre, 1998; Fortin, 1999b). For these reasons, Legendre and Legendre (1998) ask the following question as a basis for understanding the origin of variation in spatial data: Is there a significant amount of correlation between the response and explanatory variables, other than some common spatial structure that may not have a relationship of cause to effect? To answer this question, the relationships among spatially distributed variables need to be tested.

Figure 13.2 (after Borcard et al., 1992; Legendre, 1993; Legendre and Legendre, 1998) represents a conceptual model of the origin of variation in a response variable (B) in a spatial data set. For example, B could represent biotic data (e.g., plant species composition) analyzed in conjunction with an environmental data set. In this model, variation in the response, B, is first partitioned into explained (E) and unexplained (U) components. Explained

TABLE 13.5. Landscape metrics: methods, description of characteristics, and factors influencing metric values.

System property	Description	Landscape metrics	Percent representation of map classes	Aggregation of classes into patches	Frequency distribution of patch size	Spatial distribution of patches
Nonspatial	Composition	Proportion	X			
		Richness	X			
		Evenness		X		
		Dominance		X		
Spatial	Configuration	Size		X	X	
		Shape		X		
		Edge		X		
		Perimeter–area ratio		X	X	
		Density	X			
		Connectivity		X	X	X
		Fractal dimension		X		X
		Contagion	X	X	X	X

Source: Hargis et al., 1997; Gustafson, 1998. See text for further explanation.

variation can then be partitioned into environmental (ENV) and spatial (S) components. By analyzing the overlap between these two components, variation can be partitioned further into three fractions. Fraction [a] corresponds to the portion of the variation remaining after the spatial structure common to both environmental and spatial explanatory variables (fraction [b]) has been removed. If [b] is large, there is support for the hypothesis that the apparent variable correlations are false; that is, there is no significant correlation among the variables other than a common spatial structure. Depending on the objectives, there may be interest in any of the fractions [a], [b], [c], or in U. For example, [c] may be caused by environmental variables that were not included in the analysis or by historical events or processes that may be masked by some dominant environmental effect (Borcard and Legendre, 1994). Analyzing [c] allows the generation of hypotheses about the processes responsible for the observed residual spatial pattern. This correlative approach is part of EDA (Section 13.2). Two methods for partitioning variation have been proposed and widely used: the partial Mantel test and partial constrained ordinations.

The partial Mantel test (Smouse et al., 1986; Legendre and Troussellier, 1988) was designed to test the effect of a geographic variable (i.e., distance) on two other variables. This test is the multivariate equivalent of a partial correlation coefficient (Sokal and Rohlf, 1995). The degree of relationship between two matrices is computed after the effect of a third matrix has been removed. The question of whether there is a significant correlation between the two matrices, other than a common spatial structure, can then be answered. A drawback is that the test of significance is conducted on all variables simultaneously; therefore, determining the specific source of variation is not possible.

A method to implement the partitioning of variation into specific sources was proposed by Borcard et al. (1992) using partial redundancy analysis (RDA) or canonical analysis of correspondence (CCA) (ter Braak, 1988; Borcard et al., 1992; Legendre, 1993; Palmer, 1993). Partial RDA and CCA quantify the relative contributions of several categories of variables; randomization tests are used to assess their significance (ter Braak, 1990). Specific hypotheses can be tested, further supporting models of causal relationships (Legendre and Legendre, 1998). Figure 13.2 summarizes some of the possible interpretations of the various fractions of variation. These interpretations can help formulate better models for further testing.

13.3.3 Landscape Metrics

In response to the growing demand for measurement and monitoring of regional landscape-level patterns and processes (e.g., the U.S. Environmental Protection Agency's Environmental Monitoring and Assessment Program: see Overton et al., 1990; Hunsaker et al., 1994), a family of metrics, known as landscape metrics (Baker and Cai, 1992; Li and Reynolds, 1995; McGarigal and Marks, 1995; Riitters et al., 1995; Gustafson, 1998), has recently been developed to take advantage of the increasing availability of categorical map data derived from aerial photographs or satellite images. The geometric and spatial properties of discrete data mapped into patches (i.e., spatially homogeneous entities) can be analyzed using two families of landscape metrics that characterize different properties of the study area (Table 13.5).

The first family of landscape metrics includes simple indices that describe composition (e.g., proportion or number of patch classes, also known as richness), a nonspatial characteristic. The second family describes configuration, a spatial property, using patch-based metrics (e.g., area, size, perimeter, shape, fractal dimension). These landscape metrics quantify four phenomena (Table 13.5; Hargis et al., 1997): proportional representation of each patch class, aggregation of each class into patches, frequency distribution of patch sizes, and spatial distribution of patches. Hargis et al. (1997, pp. 232–233) summarize the influence of these phenomena on landscape patterns as follows. The proportional representation of each class determines which class comprises the landscape matrix. The degree of aggregation affects the size, shape, and perimeter of each patch. The frequency distribution of patch sizes creates landscape texture. The spatial distribution of patches determines their pattern of clumpiness or dispersion. Landscape metrics are generally applied to thematic (categorical) maps, which assume within-class homogeneity, an assumption that may not always be met. Reviews of properties, applications, and limitations of landscape metrics can be found in Riitters et al. (1995), Hargis et al. (1997, 1998), and Gustafson (1998).

13.4 Conclusions

An important emphasis of this chapter is the need for identifying methods of spatial analysis that:

1. Make appropriate assumptions for addressing the objectives of the analysis.
2. Can identify and/or be resistant to data characteristics (e.g., extreme values, outliers, deviation from strict distributional assumptions).
3. Are appropriate for the spatial attributes of the study area (e.g., distribution of observations, the type of environmental stratification, influence of within- and between-region boundaries).

In practice, it is likely that arbitrary decisions may be made about how to represent spatial characteristics or about some properties of the data. Therefore, alternative methods, variables, and the like, should be considered in an analysis and the effects of such changes interpreted (Haining, 1990). However, EA spatial analysts should be aware of the fact that, given the number of possible analyses that can be conducted on a data set, the synthesis of multiple interpretations may be a very difficult task. A final word of caution concerns the volume of spatial data that is common in EAs. Haining (1990) suggests that the critical issue is not how many analyses are conducted on a given data set, but how much information is provided by the data.

13.5 References

Anselin, L. 1988. *Spatial econometrics: methods and models*. Dordrecht, The Netherlands: Kluwer Academic Publishers.

Austin, M. P.; Nicholls, A. O.; Margules, C. R. 1990. Measurement of the realized qualitative niche: environmental niches of five *Eucalyptus* species. *Ecol. Monogr.* 60:161–177.

Bailey, T. C.; Gatrell, A. C. 1995. *Interactive spatial data analysis*. Essex: Longman Scientific & Technical.

Baker, W. L.; Cai, Y. 1992. The r.le programs for multiscale analysis of landscape structure using the GRASS geographical information system. *Landscape Ecol.* 7:291–302.

Balling, R. C. 1984. Classification in climatology. In: Gaile, G. L.; Willmott, C. J., eds. *Spatial statistics and models*. Dordrecht, The Netherlands: Reidel: 81–108.

Bell, G.; Lechowicz, M. F.; Appenzeler, A.; Chandler, M.; DeBlois, E.; Jackson, L.; Mackenzie, B.; Preziosi, R.; Schallenberg, M.; Tinker, N. 1993. The spatial structure of the physical environment. *Oecologia* 96: 114–121.

Bjørnstad, O. N.; Falck, W. 1997. An extension of the spatial correlogram and the x-intercept for genetic data. In: Bjørnstad, O. N. *Statistical models for fluctuating populations—patterns and processes in time and space*. Ph.D. dissertation, University of Oslo. 9 p.

Borcard, D.; Legendre, P. 1994. Environmental control and spatial structure in ecological communities: an example using oribatid mites (Acari, Oribatei). *Environ. Ecol. Stat.* 1:37–61.

Borcard, D.; Legendre, P.; Drapeau, P. 1992. Partialling out the spatial component of ecological variation. *Ecology* 73:1045–1055.

Bradshaw, G. A. 1998. Defining ecologically relevant change in the process of scaling up: implications for monitoring at the "landscape" level. In: Peterson, D. L.; Parker, V. T., eds. *Ecological scale: theory and applications*. New York: Columbia University Press: 227–249.

Burrough, P. A. 1995. Spatial aspects of ecological data. In: Jongman, R. H. G.; ter Braak, C. J. F.; van Tongeren, O. F. R., eds. *Data analysis in community and landscape ecology*. Cambridge, UK: Cambridge University Press: 213–251.

Casetti, E.; Jones, J. P., III. 1987. Spatial aspects of the productivity slow down: an analysis of U.S. manufacturing data. *Ann. Assoc. Amer. Geogr.* 77:76–88.

Cliff, A. D.; Ord, J. K. 1981. *Spatial processes: models and applications*. London: Pion.

Cliff, A. D.; Haggett, P.; Ord, J. K.; Bassett, K.; Davies, R. B. 1975. *Elements of spatial structure*. Cambridge, UK: Cambridge University Press.

Cox, N. J.; Jones, K. 1981. Exploratory data analysis. In: Wrigley, N.; Bennet, R. J., eds. *Quantitative geography*. London: Routledge & Kegan Paul.

Craig, R. G. 1979. Sources of variation in LANDSAT autocorrelation. In: *Proceedings of the international symposium on remote sensing of the environment*, Ann Arbor, MI: 1517–1524.

Cressie, N. A. C. 1991. *Statistics for spatial data*. New York: John Wiley & Sons.

Cressie, N.; Read, T. R. C. 1989. Spatial data analysis of regional counts. *Biometrical J.* 6:699–719.

Dale, M. R. T. 1999. *Spatial pattern analysis in plant ecology*. Cambridge, UK: Cambridge University Press.

David, M. 1977. Geostatistical ore reserve estimation. In: *Developments in geomathematics, 2*. Amsterdam: Elsevier Science.

Davis, F. W.; Quattrochi, D. A.; Ridd, M. K.; Lam, N. S.-N.; Walsh, S. J.; Michaelsen, J. C.; Franklin, J.; Stow, D. A.; Johannsen, C. J.; Johnston, C. A. 1991. Environmental analysis using integrated GIS and remotely sensed data: some research needs and priorities. *Photogramm. Eng. Remote Sensing* 57:689–697.

Diaconis, P. 1985. Theories of data analysis: from magical thinking through classical statistics. In: Hoaglin, D. C.; Mosteller, F. M.; Tukey, J. W., eds. *Exploring data tables*. New York: John Wiley & Sons: 1–36.

Dutilleul, P.; Legendre, P. 1993. Spatial heterogeneity against heteroscedasticity: an ecological paradigm versus a statistical concept. *Oikos* 66:152–171.

Fortin, M.-J. 1999a. Spatial statistics in landscape ecology. In: Klopatek, J. M.; Gadne, R. H., eds. *Landscape ecological analysis: issues and applications*. New York: Springer-Verlag: 253–279.

Fortin, M.-J. 1999b. Effects of quadrat size and data measurement on the detection of boundaries. *J. Veg. Sci.* 10:43–50.

Fortin, M.-J.; Drapeau, P.; Legendre, P. 1989. Spatial autocorrelation and sampling design in plant ecology. *Vegetatio* 83:209–222.

Fortin, J.-J.; Drapeau, P.; Jacquez, G. M. 1996. Statistics to assess spatial relationships between ecological boundaries. *Oikos* 77:51–60.

Griffith, D. A. 1988. Estimating spatial autoregressive model parameters with commercial statistical packages. *Geogr. Anal.* 20:176–186.

Gustafson, E. J. 1998. Quantifying landscape spatial pattern: what is the state of the art? *Ecosystems* 1:143–156.

Haining, R. 1990. *Spatial data analysis in the social and environmental sciences*. Cambridge, UK: Cambridge University Press.

Hammer, R. D. 1998. Space and time in the soil landscape: the ill-defined ecological universe. In: Peterson, D. L.; Parker, V. T., eds. *Ecological scale: theory and applications*. New York: Columbia University Press: 105–140.

Hampel, F. R.; Ronchetti, E. M.; Rousseeuw, P. J.; Stahel, W. A. 1986. *Robust statistics*. New York: John Wiley & Sons.

Hargis, C. D.; Bissonette, J. A.; David, J. L. 1997. Understanding measures of landscape pattern. In: Bissonette, J. A., ed. *Wildlife and landscape ecology*. New York: Springer-Verlag: 231–261.

Hargis, C. D.; Bissonette, J. A.; David, J. L. 1998. The behavior of landscape metrics commonly used in the study of habitat fragmentation. *Landscape Ecol.* 13: 167–168.

Heuvelink, G. B. M. 1998. *Error propagation in environmental modeling with GIS*. London: Taylor and Francis.

Hoaglin, D. C.; Mosteller, F.; Tukey, J. W. 1983. *Understanding robust and exploratory data analysis*. New York: John Wiley & Sons.

Hoaglin, D. C.; Mosteller, F.; Tukey, J. W. 1985. *Exploring data tables, trends and shapes*. New York: John Wiley & Sons.

Hunsaker, C. H.; O'Neill, R. V.; Jackson, B. L.; Timmins, S. P.; Levine, D. A.; Norton, D. J. 1994. Sampling to characterize landscape pattern. *Landscape Ecol.* 9:207–226.

Journel, A. G.; Huijbregts, C. J. 1978. *Mining geostatistics*. London: Academic Press.

King, A. W. 1997. Hierarchy theory: a guide to system structure for wildlife biologists. In: Bissonette, J. A., ed. *Wildlife and landscape ecology*. New York: Springer-Verlag: 185–212.

Kolasa, J.; Rollo, C. D. 1991. Introduction: the heterogeneity of heterogeneity—a glossary. In: Kolasa, J.; Pickett, S. T. A., eds. *Ecological heterogeneity*. New York: Springer-Verlag: 1–23.

Labovitz, M. L.; Masuoka, E. J. 1984. The influence of autocorrelation in signature extraction: an example from a geobotanical investigation of Cotter Basin, Montana. *Int. J. Remote Sens.* 5:315–332.

Leamer, E. E. 1978. *Specification searches: ad hoc inference with non experimental data*. New York: John Wiley & Sons.

Legendre, P. 1993. Spatial autocorrelation: trouble or new paradigm. *Ecology* 74:1659–1673.

Legendre, P.; Fortin, M.-J. 1989. Spatial pattern and ecological analysis. *Vegetation* 80:107–138.

Legendre, P.; Legendre, L. 1998. *Numerical ecology*. Amsterdam: Elsevier Science B.V.

Legendre, P.; McArdle, B. H. 1997. Comparison of surfaces. *Oceanol. Acta* 20:27–41.

Legendre, P.; Troussellier, M. 1988. Aquatic heterotrophic bacteria: modeling in the presence of spatial autocorrelation. *Limnol. Oceanogr.* 33:1055–1067.

Li, H.; Reynolds, J. F. 1995. On definition and quantification of heterogeneity. *Oikos* 73(2):280–284.

Manly, B. F. 1997. *Randomization, bootstrap and Monte Carlo methods in biology*, 2nd ed. London: Chapman and Hall.

Matheron, G. 1965. *Les variables régionalisées et leur estimation—une application de la théorie des fonctions aléatoires aux sciences de la nature*. Paris: Masson.

McGarigal, K.; Marks, B. 1995. *FRAGSTATS: spatial analysis program for quantifying landscape structure*. PNW-GTR-351. Portland, OR: U.S. Dept. Agric., For. Serv., Pacific Northw. Res. Sta.

Milne, A. 1959. The centric systematic area sample treated as a random sample. *Biometrics* 15:270–297.

Oden, N. L. 1984. Assessing the significance of a spatial correlogram. *Geogr. Anal.* 16:1–16.

Oden, N. L.; Sokal, R. R.; Fortin, M.-J.; Goebl, H. 1993. Categorical wombling: detecting regions of significant change in spatially located categorical variables. *Geogr. Anal.* 25:315–336.

O'Neill, R. V.; Gardner, R. H.; Milne, B. T.; Turner, M. G.; Jackson, B. 1991. Heterogeneity and spatial hierarchies. In: Kolasa, J.; Pickett, S. T. A., eds. *Ecological heterogeneity*. New York: Springer-Verlag: 85–96.

Overton, W. S.; White, D.; Stevens, D. L., Jr. 1990. *Design report for EMAP environmental monitoring and assessment program*. EPA/600/3-91/053. Corvallis, OR: U.S. Environmental Protection Agency, Environmental Research Laboratory.

Pahl-Wostl, C. 1998. Ecosystem organization across a continuum of scales: a comparative analysis of lakes and rivers. In: Peterson, D. L.; Parker, V. T., eds. *Ecological scale: theory and applications*. New York: Columbia University Press: 141–170.

Palmer, M. W. 1993. Putting things in even better order: the advantages of canonical correspondence analysis. *Ecology* 74:2215–2230.

Pearson, S. M.; Gardner, R. H. 1997. Neutral models: Useful tools for understanding landscape patterns. In: Bissonette, J. A., ed. *Wildlife and landscape ecology*. New York: Springer-Verlag: 215–230.

Pielou, E. C. 1977. *Mathematical ecology*, 2nd ed. New York: John Wiley & Sons.

Quattrochi, D. A.; Goodchild, M. F., eds. 1997. *Scale in remote sensing and GIS*. Boca Raton, FL: Lewis Publishers.

Riitters, K. H.; O'Neill, R. V.; Hunsaker, C. T.; Wickham, J. D.; Yankee, D. H.; Timmins, S. P.; Jones, K. B.; Jackson, B. L. 1995. A factor analysis of landscape pattern and structure metrics. *Landscape Ecol.* 10:23–39.

Ripley, B. D. 1981. *Spatial statistics*. New York: John Wiley & Sons.

Ripley, B. D. 1984. Spatial statistics: developments 1980–3. *Inter. Statistical Rev.* 52:141–150.

Ripley, B. D. 1987. Spatial point pattern analysis in ecology. In: Legendre, P.; Legendre, L., eds. *Developments in numerical ecological*. Berlin: Springer-Verlag: 407–429.

Ritchie, M. E. 1997. Population in a landscape context: sources, sinks, and metapopulations. In: Bissonette, J. A., ed. *Wildlife and landscape ecology*. New York: Springer-Verlag: 160–184.

Rossi, R. E.; Mulla, D. J.; Journel, A. G.; Franz, E. H. 1992. Geostatistical tools for modeling and interpreting ecological spatial dependence. *Ecol. Monogr.* 62:277–314.

Schneider, D. C. 1994. *Quantitative ecology—spatial and temporal scaling*. San Diego, CA: Academic Press.

Semple, R. K.; Green, M. B. 1984. Classification in human geography. In: Gaile, G. L.; Willmott, C. J., eds. *Spatial statistics and models*. Dordrecht, The Netherlands: Reidel: 55–79.

Smouse, P. E.; Long, J. C.; Sokal, R. R. 1986. Multiple regression and correlation extensions of the Mantel test of matrix correspondence. *Syst. Zool.* 35:627–632.

Sokal, R. R.; Rohlf, F. J. 1995. *Biometry—the principles and practice of statistics in biological research*, 3rd ed. New York: W. H. Freeman.

ter Braak, C. J. F. 1988. Partial canonical correspondence analysis. In: Bock, H. H., ed. *Classification and related methods of data analysis*. Amsterdam: Elsevier Science: 551–558.

ter Braak, C. J. F. 1990. *Update notes: CANOCO version 3.10*. Wageningen, The Netherlands: Agricultural Mathematics Group.

Tilman, D.; Kareiva, P. 1997. *Spatial ecology: the role of space in population dynamics and interspecific interactions*. Princeton, NJ: Princeton University Press.

Turner, S. J.; O'Neill, R. V.; Conley, W.; Conley, M. R.; Humphries, H. C. 1991. Pattern and scale: statistics for landscape ecology. In: Turner, M. G.; Gardner, R. H., eds. *Quantitative methods in landscape ecology: the analysis and interpretation of environmental heterogeneity*. New York: Springer-Verlag: 17–49.

Upton, G. J.; Fingleton, B. 1985. *Spatial data analysis by example, volume 1: point pattern and quantitative data*. New York: John Wiley & Sons.

Webster, R. 1985. Quantitative spatial analysis of soil in the field. *Advances Soil Science* 3:1–70.

Wrigley, N. 1983. Quantitative methods: on data and diagnostics. *Progress Human Geo.* 7:567–577.

14
Applications of Fractal Geometry and Percolation Theory to Landscape Analysis and Assessments

Bai-Lian Li

14.1 Introduction

Landscapes are ecologically distinct because the flows of resources, organisms, pollutants, and sediment through space are controlled to some extent by the connections between points on the landscape. Landscape composition and pattern influence the nature and magnitude of ecological processes at a variety of spatiotemporal scales (O'Neill et al., 1988; Wiens, 1992; Li and Archer, 1997). This situation results from a set of biophysical constraints and processes, including geologic history, soils, topography, and climate. Changes in the distribution and pattern of ecological resources (e.g., woodlands, rangeland, streams and wetlands) and human activities can alter fundamental ecological processes, including the flows and balances of water, nutrients, energy, and biota. These changes in ecological processes can, in turn, influence many aspects of the environment valued by society. Characterizing and assessing landscapes at multiple spatiotemporal scales enables managers, ecologists, and designers to identify the causes of ecological change according to the scale at which the relevant ecological forces operate. Management decisions can also be made once the ecological implications of the location, juxtaposition, and the flow of resources, pollutants, and species are known.

Many thanks to Steve Archer, Craig Loehle and Hsin-i Wu for collaboration on these research subjects during the last eight years. Mike Fuller and How Passell provided editorial assistance. This work was partially supported by the U.S. National Science Foundation (BSR-91-09240, DEB-93-06679, and DEB-94-11976), DOE/Sandia National Laboratories (BE-0229), the University of New Mexico, and USDA Forest Service–Northern Region. This is Sevilleta LTER publication no. 136.

For two decades or more, spatial (landscape) analysis has been dominated by a style of model building that has sought high predictive understanding in numerical terms, but has paid little attention to the geometry of spatial form. Mandelbrot's (1983) concept of a fractal, one of the fast-moving research fronts coupled with concepts of complexity, criticality, and self-organization, extends our usual ideas of classical geometry beyond those of point, line, circle, and so on, into the realm of the irregular, disjoint, and singular. Fractals represent many kinds of patterns, including density, diversity, dendritic stream networks, geometrical shapes, mountainous terrain, and size distributions of islands (Mandelbrot, 1983). It has the potential to provide us with a new way to understand and analyze such natural spatial phenomena, which are not smooth, but rough and fragmented to self-similarity or self-affinity at all scales. Percolation represents the simplest model of a disordered system. Percolation transition is a simple example of a geometrical phase transition phenomenon (Stauffer, 1985; Sahimi, 1994). The percolation model, together with fractal analysis, can provide useful insight to assess landscape dynamics (Loehle et al., 1996).

Landscapes are spatially heterogeneous, and the structure, function, and change of landscapes are themselves scale dependent. Heterogeneity of environmental resources, succession, and disturbance result in landscape patches of diverse size, shape, type, and ecotone (or boundary) characteristics. The patch characteristics may be important factors in ecological diversity, stability, and function. The geometric features of heterogeneity, multiple scales, and self-similarity–affinity are characteristic of a variety of patch spatial patterns in landscapes. More generally, there are pansymmetries of biological spatial–temporal structure in nature, which are the pansystem holography in space and the pansystem repetition in time (Li, 1986). Recent studies have included measures of the fractal geometry of landscape patterns and percolation-based phase transitions in ecological literature (Burrough, 1981, 1983a, b; MacKay and Jan, 1984; Frontier, 1987; Gardner et al., 1987; Goodchild and Mark, 1987; Milne, 1988, 1992; Palmer, 1988, 1992; Wiens and Milne, 1989; Loehle, 1990; Warner and Fry, 1990; Bartoli et al., 1991; Williamson and Lawton, 1991; Young and Crawford, 1991; Zeide, 1991; Li et al., 1992; Montgomery and Dietrich, 1992; Berntson, 1994; Loehle et al., 1996; Milne et al., 1996; Harte et al., 1999).

This chapter will summarize the state of the art and introduce several updated developments in analysis and description of landscape patterns and landscape dynamics under Mandelbrot's fractal and percolation framework, with an emphasis on this author's current research results and with a personal view. A fractal analysis of southern Texas savanna landscape dynamics and a percolation model for the forest–prairie ecotonal phase transitions in Kansas are included. Future directions of fractal geometry and percolation theory in ecological assessment applications are also discussed.

14.2 What Are Fractals?

A line is a one-dimensional object, a square is two-dimensional, and a cube is three-dimensional. Most objects in nature are much more complicated geometric figures. It should not be surprising that the dimension of a tree, say, is not integral. Fractal geometry theory tells us that most objects in nature have fractional dimensions, for example, 1.5 for a tree. *Fractals* are conceptual objects showing structures at all spatial scales, with a scale-dependent self-similarity (Mandelbrot, 1977). The shape of fractals is nonrectifiable, consisting of an infinite sequence of clusters within clusters or waves within waves. Perhaps we have seen that snow crystals all have the same pattern, each part of a branch being similar to itself. In rectifiable objects, increasingly accurate measurements based on successive scale reductions give series converging to a limit: the true extent of the object. By contrast, in fractals the same procedure generates infinite series, according to the relationship $N(\epsilon) \propto \epsilon^{-D}$, where $N(\epsilon)$ is a number measure corresponding to the scale unit ϵ and D is the fractal dimension. The length of the object is then $L(\epsilon) \propto \epsilon^{1-D}$ and $D > 1$. The length diverges as $\epsilon \rightarrow 0$. In a volume of Euclidean dimension, E, the volume occupied by an object of fractal dimension D, is given by $V(\epsilon) \propto \epsilon^{E-D}$ (Mandelbrot, 1983). This parameter exceeds the topological dimension d of the object and is generally not an integer, but less than the space dimension of the object, that is; $d < D < d + 1$. For example, the fractal dimension of Koch's snowflake is $D = \log 4/\log 3 = 1.2618$ (see Figure 14.1). For all rivers, the fractal dimension of the mainstream calculated individually by changing coarse-graining level falls in the range from 1.1 to 1.3, with a mean value of 1.2, which can be derived from the best-known empirical law, Hack's law. [This law asserts that the relation between the length L (km) of the mainstream and area A (km^2) of the drainage basin is $L \propto 1.89\,A^{0.6}$. We can rewrite it as $A^{1/2} \propto L^{1/1.2}$, and the fractal dimension of the mainstream can be seen to be 1.2.) Taylor (1986) suggested that a set should be called a fractal if these different computations

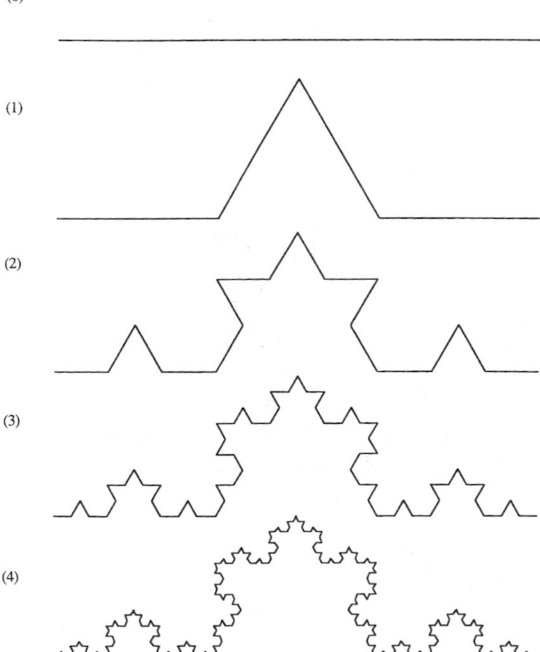

FIGURE 14.1. Koch's snowflake curve. Koch's curve can be constructed by taking out the middle third of a line segment and inserting two segments equivalent to the one that was removed. They are inserted to make an equilateral triangle with the removed segment. Therefore, at every iteration of the construction procedure, the length of the perimeter is multiplied by 4/3, which means that it diverges to infinity. A more complex Koch's curve can be considered as the diffusion fronts of ecological interfaces between two biome transition zones.

all lead to the same value for the index, which we then call the *dimension* of the set.

Fractals are characterized by so-called symmetries or pansymmetries (Li, 1986; Li et al., 1992), which are invariance under dilations or contractions. Hence the best fractals are those that exhibit the maximum of invariance. A fractal invariant under ordinary geometric similarity is called *self-similar* (Mandelbrot, 1983). Self-similar has two meanings. We can understand *similar* as a loose, everyday synonym of *analogous*. But there is also the strict textbook sense of *contracting similarity*. It expresses that each part is a linear geometric reduction of the whole, with the same reduction ratios in all directions. Self-similarity cannot be compatible with analyticity. Random fractals are self-similar only in a statistical sense, and to describe them it is more appropriate to use the term *scale invariance* than self-similarity. More recent developments have extended, in particular, to include *self-affine*, in which the reductions are still linear but the reduction ratios change in different directions.

Fractals have been used to study nonlinear spatial and temporal phenomena (such as D as a measure of complexity), but they can also be extended to abstract objects developing in a phase space, such as models of dynamic complex systems. Size-frequency distributions describing structured systems can also have a fractal dimension. There are several methods of measuring the fractal dimension, which include changing coarse-graining level (i.e., box-counting methods, information fractals, etc.), using the fractal measure relations (i.e., perimeter–area–volume methods), using the correlation function (i.e., autocorrelations, semivariograms, etc.), using the distribution function (i.e., hyperbolic distribution), and using the power spectrum (i.e., Fourier transformation, filters, wavelets, etc.) (Li et al., 1992). Although the theoretical origins of fractals in measure theory may seem abstruse, the basic ideas of fractal geometry are extremely simple and intuitive, and we can begin to work with them very quickly.

Fractal dimensions can be positive, negative (Mandelbrot, 1990), complex (Pietronero and Tosatti, 1986), fuzzy (Feng et al., 1991), and multifractals (Mandelbrot, 1989). Generally, there are three properties of fractal forms: heterogeneity, self-similarity, and the absence of a characteristic scale of length. These geometric features are also characteristic of patch patterns in landscape. The fractal dimension D has been shown to be a useful way to characterize the geometric structure of a number of these patchy spatial patterns (Milne, 1988).

The fractal concept is also useful for characterizing certain aspects of landscape patch dynamics. Consider a complex process of landscape patch change that cannot be expressed in terms of a simple characteristic rate, but instead is regulated by a self-similar or self-affine mechanism in time. The multiplicity in time scales will be reflected in a power spectrum with a broad profile of responses. The fractal scaling between variations on different time scales will lead to a frequency spectrum having an inverse power-law distribution. The fractal analysis of cluster-phase dynamics in southern Texas savanna landscape in Section 14.5 is an example.

14.3 Current Fractal Applications in Landscape Ecology

Recently, Milne (1990), Sugihara and May (1990), and Li (2000) reviewed fractal applications in ecological research and landscapes. Some interesting aspects associated with spatial patterns and landscape dynamics, combined with the author's current research in this field, are introduced.

14.3.1 Perimeter Versus Area Relationships

Krummel et al. (1987) used fractal models to show that patch shape varies with patch size. The relationship is generalized to fractal patches by the equality between area and patch length, $A = \beta L^D$, where A = patch area, L = patch perimeter, β = constant, and D = fractal dimension. By using data of patch area and patch perimeter, and relationships of $\log A = \log \beta + D \log L$, we can easily estimate the fractal dimension D. They found a marked ($p < 0.001$) discontinuity in D, with $D = 1.20 \pm 0.02$ at small scales and $D = 1.52 \pm 0.02$ at large scales. The discontinuity occurred at areas of around 60 to 70 hectares. Their result is interpreted to indicate that human disturbances predominate at small scales, making for smoother geometry and lower D, while natural processes (e.g., geology, distributions of soil types) continue to predominate at larger scales.

In general, the area of fractal patches can be expressed as a function of perimeter raised to an exponent. Similar studies can be found in DeCola (1989), Rex and Malanson (1990), and Haslett (1994).

14.3.2 Information Fractals of Landscape Diversity

The information fractal dimension, D_I, is a generalization of the box-counting method that takes into account the relative probability of the patchy types that cover the landscape. The information fractal dimension as the natural measure dimension has been used in calculating the dimensionality of strange attractors (Farmer et al., 1983; Grassberger and Procaccia, 1983; Ruelle, 1989).

The information fractal dimension is given by

$$D_I = \lim_{\epsilon \to 0} \frac{H(\epsilon)}{\text{Ln}(1/\epsilon)}$$

where $H(\epsilon)$ is a given Shannon function (Shannon, 1948),

$$H(\epsilon) = -\sum_{i=0}^{n(\epsilon)} p_i(\epsilon) \text{ Ln } p_i(\epsilon),$$

where p_i is the probability of observing the ith patch element measured using samples of ϵ units in size. For very complicated landscapes (e.g., n-phase mosaics of different vegetation types in a natural landscape), we can extend the above formula to a generalized hierarchical diversity function, for example, Pielou's hierarchical diversity index (Pielou, 1977) to count the total patch diversity in the landscape.

In practical applications, we could use the following relationship to estimate D_i, that is, the diversity $H(\epsilon)$ will vary ϵ according to

$$H(\epsilon) = H(0) - D_I \text{ Ln } \epsilon,$$

where D_I is the lower bound to the Hausdorff Besicovich dimension or information fractal dimension. In many cases, the lower bound is numerically identical to it (e.g., linear regression estimation). Comparing with the fractal dimension D from the box-counting method, in general, $D \geq D_I$ (Farmer et al., 1983). If all boxes have equal probability of the patchy types, $D = D_I$.

In the author's research about fractals in *point* patch patterns, it was found that information fractal dimensions varied with the different degree of randomness parameters (R) of Clark and Evans (1954). For instance, when $R > 1$ indicates regularity, we have fractals, $1.92 \leq D_I \leq 2$, where we used point patterns based on Wu et al. (1987). Figure 14.2 shows different information fractal dimensions for a regular point pattern, random point pattern, random clumped point pattern, and aggregated clumped point pattern. Because a fractal dimension is scale invariant, it provides us with a new index to measure point patch patterns and diversity. Information fractal dimensions can also be used in quantifying landscape habitat diversity (Loehle and Wein, 1994), nongeometric ecological

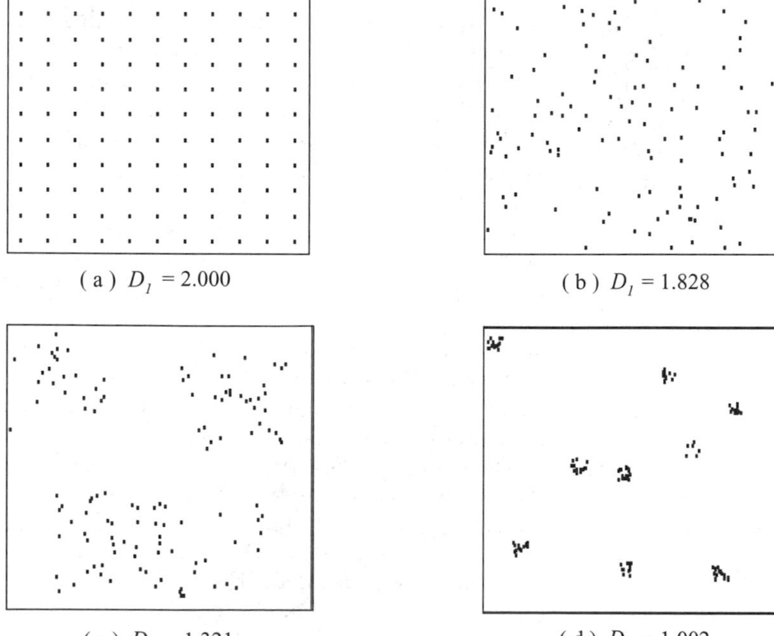

FIGURE 14.2. Information fractal dimensions (D_I) in different point patterns: (a) regular point pattern; (b) random point pattern; (c) random clumped point pattern; and (d) aggregated clumped point pattern.

properties such as permeability (Loehle and Li, 1996b), and uncertainty in ecological systems (Li, unpublished manuscript).

14.3.3 Patch Hierarchical Scaling

Different observational scales capture different aspects of structure, and these transitions are signaled by shifts in the apparent dimension of an object. This latter fact suggests an interesting application of fractals as a method for distinguishing hierarchical size scales of patches in nature, such as how to determine boundaries between hierarchical levels and how to determine the scaling rules for extrapolating within each level.

Bradbury et al. (1984) examined the possibility of hierarchical scaling in an Australian coral reef. They used the dividers method in transects across the reef to determine whether D depends on the range of length scales. They found that three ranges of scale correspond nicely with the scales of three major reef structures: 10 cm corresponds to the size of anatomical features within individual coral colonies; 20 to 200 cm corresponds to the size range of whole, adult living colonies; and 5 to 10 m is the size range of major geomorphologic structures. This showed that the shifts in D at different scales appear to signal where the breakpoints occur in the hierarchical organization of reefs.

In our recent study on simple patch patterns, change of fractal dimensions seems to predict hierarchical scales of patch size and structure in nature, both of different mean grain densities within patch and spatial patterns between patches. Since the change in the fractal dimension may tell us something about the underlying physical and biological processes, D could be used as a scaling indicator of patch phase transition and may help us decide on the appropriate scale of ecological experiment, assessment, and management.

14.3.4 Fractal Spatial Patterns and Modified Brownian Dynamics

There are simple relationships between persistence measured by the parameter H in modified Brownian diffusion models and fractal exponents D. Hastings et al. (1982) and Mandelbrot (1983) have discussed how fractal exponents may be incorporated into diffusion processes, as a scaling factor for normalizing increments in space and time. They find that D may be used as an index of succession in circumstances where simple patch-extinction models are reasonable. Sugihara and May (1990) consider that it would be interesting to follow up these provocative anecdotes with careful studies to determine to what extent D computed from snapshots can be used as an index of physiological state or persistence of patches in time and how such persistence may relate to the spatial scales involved. Recent studies have shown that many of nature's seemingly patch shapes can be effectively characterized and modeled as random fractals based on generalizations of fractional Brownian motion (Voss, 1988). The variance of the increments of a process (i.e., the semivariogram) is required to be independent of scale; the condition of independence of scale is satisfied by such generalizations of fractional Brownian motion. Extending the ideas in fractal correlation analysis of patchy systems is addressed in Li (2000).

14.3.5 Multiple-scale Sampling and Data Analysis

Biotic and metric scale-dependent structures exist in landscapes. Biotic scale dependence originates from the different responses of organisms to the abundance of a resource. Metric scale dependence results when physical processes produce statistically similar aggregations of abiotic quantities, such as water, temperature, or minerals (Milne, 1990). From a fractal point of view (Mandelbrot, 1977, 1983), we consider a sampling space dimension D_s and a specific ecological phenomena dimension D_p embedded in a space of dimension E. If $D_s + D_p > E$, we will obtain a nonzero measure; that is, a specific sampling space dimension D_s used in space of dimension E can only detect such phenomena of different dimension $D_p > E - D_s$. Lovejoy et al. (1986) argued that measuring network inhomogeneity by the fractal dimension raises new problems concerning the detectability of sparse phenomena. They suggested a new criterion for evaluating measuring networks: to detect geophysical phenomena, not only must a network have sufficient spatial resolution, but it must also have sufficient dimensional resolution. Because our sampling space is at multiple scales, of course, such sampling is multiple-scale sampling.

Theoretically, data from different scales have their corresponding probability distributions. Statistical fractals show us that the sum of a large number of identically distributed random variables has self-similarity with the distribution as each of the contributors to the sum. The class of distributions having this property is called the *Levy distributions* (Mandelbrot, 1977, 1983). This result implies that when we blend data from different sources we need to find a suitable transformation to form a stable distribution. There are some statistical methods to deal with data from unequal probability sampling, for example, the weighted distribution method (Patil and Taillie, 1989). Further discussion about this issue can be found in Li (2000).

14.4 Fractal Fragmentation of Habitat into Patches

The fragmentation of the habitat into discrete patches is a topic of concern in relation to biodiversity conservation and resource management efforts (Harris, 1984; Pimm and Gilpin, 1989; Loehle and Li, 1996a). It is an important feature of landscapes. Landscape habitat is fragmented by joints of natural processes and human and natural disturbances. Fragmentation is thought to play a dominant role in determining the size-frequency relation for astrophysics (Brown et al., 1983). There are a variety of ways to represent the size-frequency distribution of habitat fragments or patches. Li (2000) uses a simple power-law relation to define the fractal distribution for quantifying the habitat fragmentation processes and gives a hypothesis about the fragility or vulnerability of landscape habitat based on the result of Turcotte's renormalization group approach to the problem of fragmentation (Turcotte, 1986).

Using the renormalization group approach to scale invariant problems with fractal distributions, Turcotte (1986) proposes two renormalization group models to the fragmentation problem. The models yield a fractal behavior for fragmentation, but give different values for the fractal dimension. He indicates that the fractal dimension is a measure of the fracture resistance of the material relative to the process causing fragmentation. His conclusion can be used as our hypothesis about the fragility or vulnerability of habitat fragmentation. That is, a more fragile or vulnerable landscape habitat may be associated with a smaller fractal dimension. The results of Krummel et al. (1987) support our hypothesis. Their results imply that landscape habitats are more sensitive to human disturbances than to natural processes. This may be useful for managing and conserving the ecological habitat.

Several other studies in fractal measures of habitat fragmentation for quantifying human impact and describing various habitat structures or patchy landscape features can be found in Krummel et al. (1987), Milne (1988), Palmer (1988, 1992), Williamson and Lawton (1991), Haslett (1994), Loehle and Wein (1994), and Li (2000).

14.5 Fractal Analysis of Southern Texas Savanna Landscape

Recent trends toward increased woody plant abundance in temperate and tropical grasslands and savanna in recent history have been reported worldwide. We have little knowledge of the rates, dynamics, patterns, or successional processes involved. To determine the long-term patterns and dynamics of vegetation patch-to-patch interactions in a southern Texas subtropical savanna landscape, we defined different fractals and fractal relationships to describe (1) cluster growth relationships, (2) changes in the size and shape of clusters, (3) degree of coalescence or fragmentation, and (4) spatial pattern shifts of different types of vegetation clusters during succession. These fractals and their relationships included fractal kinetics of aggregation processes, area-perimeter fractals, patch-size distribution fractals, self-affine fractals, correlation fractals, and the information fractals introduced above.

The Rio Grande Plains of southern Texas and northern Mexico offer distinct examples of processes involved in the physiognomic conversion of grassland and savannas to woodlands. The process begins when the leguminous shrub, *Prosopis glandulosa* (mesquite), establishes in herbaceous zones. The seeds of *Prosopis* are widely dispersed by livestock and establish readily on sites where fire and competition from grasses have been reduced by grazing. As the *Prosopis* plant grows, it modifies soils and microclimate and facilitates the ingress and establishment of other shrubs. These patches of woody vegetation that form around the *Prosopis* nucleus enlarge over time as new plants appear and existing plants grow. As the canopies of the woody patches develop, light attenuation increases such that herbaceous production in well-developed patches is only 20% of that in herbaceous zones. As new patches are initiated and existing patches expand, coalescence will eventually occur and a grassland will have become a woodland. See Archer (1995) and Li (1993, 1995) for the details.

TABLE 14.1. Local growing fractals of woody patches at study sites from black and white aerial photographs in southern Texas.

Period	Fractal dimension	Roughness
1941–1960 (dry)	1.0819	0.4603
1960–1983 (wet)	1.1406	0.6328

TABLE 14.2. Anisotropic characteristics of correlated growth of woody patches.

Year	Fractal dimension (roughness) in *X* direction	Fractal dimension (roughness) in *Y* direction
1941	1.1478(5.7231)	1.2250(5.2634)
↑ Dry ↓		
1960	1.1588(5.6976)	1.1189(5.3163)
↑ Wet ↓		
1983	1.1526(5.7051)	1.1189(5.9614)

TABLE 14.4. Degree of coalescence and fragmentation in southern Texas Savanna landscape.

Vegetation type	Fractal dimension	Roughness
Herbaceous	1.4061	0.6505
Pioneer cluster	1.3120	0.6096
Mature cluster	1.2908	0.5930
Coalesced clusters	1.2592	0.5889
Woodland	1.1189	0.5128

The results in Tables 14.1 through 14.5 have shown that the fractal method looks very favorable and is straightforward for understanding and identifying landscape processes and successional mechanisms (Li and Archer, unpublished manuscript).

14.6 Landscape Phase Transitions and Percolation Theory

Study of phase transitions in ecology has included the application of catastrophe theory to pest outbreak (Ludwig et al., 1978; Casti, 1982) and forest ecosystem exploitation (Gatto and Rinaldi, 1987; Loehle, 1989); application of percolation theory to landscape ecology (Gardner et al., 1987; Milne et al., 1996), forest fire (MacKay and Jan, 1984), and epidemic spread (Grassberger, 1983, 1985); and application of nonlinear stability theory to non-equilibrium phase transitions of vegetation landscape in a southern Texas savanna (Li, 1995). These researches indicate that developing an ecological phase transitions theory is useful for understanding ecosystem dynamics. In particular, this approach may benefit determination of ecotonal changes, because ecotones are transitional areas between adjacent ecological systems or vegetation types. Ecotones occur under two types of conditions: (1) steep gradients in physical environmental variables that directly affect key ecological processes and the distribution of organisms, and (2) threshold or nonlinear responses to gradual gradients in the physical environment that cause large changes in ecosystem dynamics and the distribution of dominant species (Gosz, 1992; Risser, 1995). Identifying the thresholds of phase transitions at an ecotone is important for assessing ecotone landscape dynamics.

A phase is a state of a macroscopic system that is qualitatively different in its characteristics from other states of the same system. A phase transition is a transition from one phase state of a system to another. Vegetation, for example, is often described in physiognomic terms by the dominant life-form: trees (forest), shrubs (shrubland), and grasses (grassland) and combinations of these (woodland, savanna). Vegetation types can be considered ecological phases, and transformations from one type to another can be considered ecological phase transitions. A system that is poised between two or more phases is an *ecotone*.

Phase transitions are characterized by a fundamental change in an order parameter (Uzunov, 1993). They can be discontinuous (first-order phase

TABLE 14.3. Changing complexity of the size and shape of all vegetation patches.

Year and climate condition	Fractal dimension
1941	1.8732
↑ Dry ↓	
1960	1.8337
↑ Wet ↓	
1983	1.9377

TABLE 14.5. Spatial pattern shifts of southern Texas savanna landscape during succession.

Year	Fractal dimension	Roughness
1941	1.7905	5.7130
↑ Dry ↓		
1960	1.8282	5.5968
↑ Wet ↓		
1983	1.9130	6.2221

transition) and continuous (second-order phase transition). The point of phase transition is usually called the *critical point*. Strictly speaking, it is not necessary that a transition be realized just at a point. For example, an ecotone can be considered a phase transitional area or zone on [a, b].

This section will introduce a simplest phase transition model; the percolation model, together with fractal analysis, to assess ecotonal landscape dynamics (Loehle et al., 1996). Percolation processes were first developed by Flory (1941) and Stockmayer (1943) to describe how small branching molecules react and form very large macromolecules. This process is very similar to the expansion of forest or shrubs in grassland (Li et al., 1992). In mathematical literature, percolation was introduced by Broadbent and Hammersley (1957). They originally dealt with the concept of the spread of hypothetical fluid particles through a random medium. The terms *fluid* and *medium* were viewed as totally general: a fluid can be liquid, vapor, heat flux, infection, fire or forest spreading, and so on. The medium where the fluid is carried can be the pore space of a subsurface, an array of trees, grassland, or the universe. Percolation provides an intuitively appealing and transparent model for dealing with the unruly geometries that occur in many random media. The percolation threshold is a prototypical phase transition that occurs as we vary the richness of interconnections present (a generalized density or composition). For a detailed introduction to percolation models and theory, see Stauffer (1985) and Sahimi (1994). In ecology, percolation models have been applied to the spread of epidemics (Mollison, 1977, 1986; Grassberger, 1983, 1985; Cox and Durrett, 1988), forest fire (MacKay and Jan, 1984), microbial transport (Li et al., 1996), and ecotone and landscape ecology (Gardner et al., 1987; Milne et al., 1996).

Milne et al. (1996) used critical densities or percolation thresholds of percolation models to identify woodland ecotonal phase transitions. In percolation theory, the numerical value of every percolation quantity for any percolation probability p depends on the microscopic details of the system, such as its configuration or neighborhood. But near the bond or site percolation threshold p_c, most percolation quantities obey scaling laws that are largely insensitive to the network structure and its microscopic details. However, the percolation threshold p_c still varies with configuration or neighborhood. Percolation theory tells us that the topological exponents including fractal dimension near p_c are completely universal (Stauffer, 1985; Sahimi, 1994). They are independent of the microscopic details of the system and depend only on the dimensionality of the system. This discovery allows us to directly use fractal dimensions to detect spatial phase transitions of an ecotone, regardless of the spatial configuration of the object of concern.

Based on percolation theory and computer experiment simulation of general random percolation transitions (Stauffer, 1985; Stinchcombe, 1990; Sahimi, 1994), the relation between critical exponents and a dimension of a fractal occurring in percolation is $FD = d - \beta/v$, where d is the Euclidean dimension, β and v are the critical exponents, and $(-\beta/v)$ can be regarded as the length-scaling exponent. Based on currently accepted values of critical exponents (Stauffer, 1985; Sahimi, 1994), we have fractal dimensions, $\mathrm{FD}(p = p_c) = 1.8958$, $\mathrm{FD}(p > p_c) = 2$, and $\mathrm{FD}(p < p_c) = 1.56$. Here let us consider the forest spread as a dynamic percolation process. Forest percent cover is percolation probability. When forest cover is below the percolation threshold p_c, the landscape is dominated by tallgrass prairie. When forest cover is above p_c, the forest invasion can spread through the entire landscape, which landscape becomes forest dominant. On the basis of this intuitive argument, we suggested that fractal dimensions of forest spatial distributions could be used to characterize ecotones in landscapes and should range between 1.56 and 1.8958. This result can also be used to monitor ecotonal movement and responses to climate change.

Spatial phase transitions of an ecotone between tallgrass prairie and forest are complex. In general, if environmental or management conditions change in ways that are beneficial for one of the adjacent ecosystems, patch size is likely to increase in this system, and it is likely to invade habitats previously unsuitable in the adjacent system (Risser, 1995). Let us assume that 70% of the driving forces (or controlling factors and constraints) support forest spread. We could simply calculate the new fractal dimension of this critical transition in the two-dimensional lattice as

$$\mathrm{FD}_{70\%} = 1.8958 - (1.8958 - 1.56) \times (1 - 0.7) = 1.7951.$$

And if there are 80% driving forces, we have $\mathrm{FD}_{80\%} = 1.8286$. We could use the critical fractal dimension to detect a phase transition at a forest–tallgrass prairie ecotone. Now we can state our hypothesis in this study; that is, if the fractal dimension of forest cover is above 1.7951 or 1.8286, the forest invasion at the forest–prairie ecotone may spread to the entire system. We tested this hypothesis based on actual and simulated data of forest

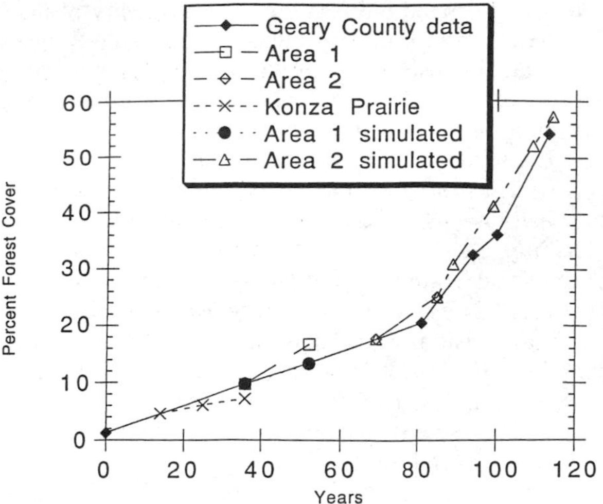

FIGURE 14.3. Long-term trends in forest area after fire suppression. Data of Bragg and Hulbert (1976) for Geary County and of Knight et al. (1994) for Konza Prairie are compared with simulated spread. To test for similar trends with time, data from Fort Riley and Konza Prairie are overlaid on the year at which their initial percent cover matches that in Geary County. This procedure allows rates of spread in an area to be compared with the initial year in which Geary County forest cover matches the particular site. Forest spread was slow and steady until about 20% forest cover was reached, at which point the rate increased (from Loehle et al., 1996).

spread at Fort Riley in eastern Kansas (Loehle et al., 1996). Using the above results, the predicted critical point of about 18.5% forest cover was very close to the observed result and might represent a phase transition at the forest–prairie ecotone (see Figure 14.3). Therefore, the self-accelerating response of forest in this case is due to spatial patterns created by the spreading trees that tend to accelerate the growth and invasion process after a critical point is reached. For details, see Loehle et al. (1996).

14.7 Summary and Discussion

The concept of fractal geometry provides us with new insights on analyzing and quantifying the spatial variability of spatial patterns and landscape dynamics. Fractal analysis as a tool for addressing problems of scale and hierarchy allows ecologists to view landscape patch patterns and dynamics at multiple spatial and temporal scales and thereby achieve predictability in the face of complexity; it also suggests that patch landscape properties are a function of the scales of measurement and that traditional concepts of stationarity and averaging in stochastic approaches may not capture all the heterogeneity. In addition, statistical fractals offer a viable way to analyze discontinuous, inhomogeneous processes in natural systems. Percolation theory can tell us whether a system is macroscopically connected or not. This macroscopic connectivity is of fundamental importance to many phenomena involving random media (Sahimi, 1994). Moreover, universal scaling laws near the percolation threshold tell us which aspects of a given dynamic system are important in determining its macroscopic properties and which aspects are not relevant, and we therefore do not have to collect certain data and build very complex models; a simplified model is enough.

Future directions in ecology will be strongly influenced by methodological advance, especially technologies imported from other disciplines (Wiens, 1992). Fractal geometry and percolation theory promise to play a very important role in building a spatially explicit ecology, because these approaches have obvious advantages in describing the following three related contexts: geometric, temporal (dynamical), and statistical. They also provide a bridge to concentrate chaos theory, fractal analysis, percolation model, wavelet analysis, scaling analysis, and spectral analysis into a spatiotemporal integrated methodology (Li, 1998, 2000). Future application of fractals, percolation, and their underlying nonlinear space–time dynamics, together with high-speed computational technology, will continue to bring ecological, physical, and mathematical sciences together for work on real landscape assessment problems that were formerly thought to be outside some of the artificially set ranges in each field.

14.8 References

Archer, S. 1995. Tree-grass dynamics in a *Prosopis*–thornscrub savanna parkland: reconstructing the past and predicting the future. *EcoScience* 2:83–99.

Bartoli, F.; Philippy, R.; Doirisse, M.; Niquet, S.; Dubuit, M. 1991. Structure and self-similarity in silty and sandy soils: the fractal approach. *J. Soil Sci.* 42:167–185.

Berntson, G. M. 1994. Root systems and fractals: how reliable are calculations of fractal dimensions? *Ann. Bot.* 73:281–284.

Bradbury, R. H.; Reichelt, R. E.; Green, D. G. 1984. Fractals in ecology: methods and interpretation. *Mar. Ecol. Prog.* Series 14:295–296.

Bragg, T. B.; Hulbert, L. C. 1976. Woody plant invasion of unburned Kansas bluestem prairie. *J. Range Manage.* 29:19–24.

Broadbent, S. R.; Hammersley, J. M. 1957. Percolation processes: crystals and mazes. *Proc. Camb. Phil. Soc.* 53:629–641.

Brown, W. K.; Karpp, R. R.; Grady, D. E. 1983. Fragmentation of the universe. *Astrophys. Space Sci.* 94:401–412.

Burrough, P. A. 1981. Fractal dimensions of landscapes and other environmental data. *Nature* 294:241–243.

Burrough, P. A. 1983a. Multiscale sources of spatial variation in soil: I. application of fractal concepts to nested levels of soil variations. *J. Soil Sci.* 34:577–597.

Burrough, P. A. 1983b. Multiscale sources of spatial variation in soil: II. a non-Brownian fractal model and its application in soil survey. *J. Soil Sci.* 34:599–620.

Casti, J. 1982. Catastrophes, control and the inevitability of spruce budworm outbreaks. *Ecol. Model.* 14:293–300.

Clark, P. J.; Evans, F. C. 1954. Distance to nearest neighbour as a measure of spatial relationships in populations. *Ecology* 35:445–453.

Cox, J. T.; Durrett, R. 1988. Limit theorems for the spread of epidemics and forest fires. *Stoch. Proc. Appl.* 30:171–191.

DeCola, L. 1989. Fractal analysis of a classified landsat scene. *Photogramm. Eng. Remote Sensing* 55:601–610.

Farmer, J. D.; Ott, E.; Yorke, J. A. 1983. The dimension of chaotic attractors. *Physica* 7D:153–180.

Feng, D.; Lin, M.; Jiang, C.; Liu, X. 1991. Fuzzy fractals: theory and application. *Proc. fractal theory and its application.* Wuhan: Central China University of Technology Press: 116.

Flory, P. J. 1941. Molecular size distribution in three dimensional polymers (I–III). *J. Amer. Chem. Soc.* 63:3083–3100.

Frontier, S. 1987. Applications of fractal theory to ecology. In: Legendre, P.; Legendre, L., eds. *Developments in numerical ecology.* Berlin: Springer-Verlag: 335–378.

Gardner, R. H.; Milne, B. T.; Turner, M. G.; O'Neill, R. V. 1987. Neutral models for the analysis of broad-scale landscape pattern. *Landscape Ecol.* 1:19–28.

Gatto, M.; Rinaldi, S. 1987. Some models of catastrophic behavior in exploited forests. *Vegetatio* 69:213–222.

Goodchild, M. F.; Mark, D. M. 1987. The fractal nature of geographic phenomena. *Ann. Assoc. Amer. Geogr.* 77:265–278.

Gosz, J. R. 1992. Ecological functions in biome transition zones: translating local responses to broad-scale dynamics. In: Hansen, A. J.; di Castri, F., eds. *Landscape boundaries: consequences for biotic diversity and landscape flows.* New York: Springer-Verlag: 55–75.

Grassberger, P. 1983. On the critical behavior of a general epidemic process and dynamic percolation. *Math. Biosci.* 63:157–172.

Grassberger, P. 1985. On the spreading of two-dimensional percolation. *J. Physics A* 18:L215–L219.

Grassberger, P.; Procaccia, I. 1983. Measuring the strangeness of strange attractors. *Physica D* 9:189–208.

Harris, L. D. 1984. *The fragmented forest.* Chicago: University of Chicago Press.

Harte, J.; Kinzig, A.; Green, J. 1999. Self-similarity in the distribution and abundance of species. *Science* 284:334–336.

Haslett, J. R. 1994. Community structure and the fractal dimensions of mountain habitats. *J. Theor. Biol.* 167: 407–411.

Hastings, H. M.; Pekelney, R.; Monticciolo, R.; Kannon, D. V.; Monte, D. D. 1982. Time scales, persistence and patchiness. *Biosystems* 15:281–289.

Knight, C. L.; Briggs, J. M.; Nellis, M. D. 1994. Expansion of gallery forest on Konza Prairie research natural area, Kansas, USA. *Landscape Ecol.* 9:117–125.

Krummel, J. R.; Gardner, R. H.; Sugihara, G.; O'Neill, R. V.; Coleman, P. R. 1987. Landscape patterns in disturbed environment. *Oikos* 48:321–324.

Li, B. 1986. Pansystems analysis: a new approach to ecosystem modelling. *Ecol. Model.* 32:227–236.

Li, B. 1993. A theoretical framework for analysis of cluster-phase dynamics. *Bull. ESA* 74:329.

Li, B. 1995. Stability analysis of a nonhomogeneous Markovian landscape model. *Ecol. Model.* 82:247–256.

Li, B. 1998. Wavelet-based multiscale analysis in landscape ecology. *Proceedings of 1998 joint statistical meeting*, August 9–13, 1998, Dallas, TX: 283.

Li, B. 2000. Fractal geometry applications in description and analysis of patch patterns and patch dynamics. *Ecol. Model.* 132:33–50.

Li, B.; Archer, S. 1997. Weighted mean patch size: a robust index for quantifying landscape structure. *Ecol. Model.* 102:353–361.

Li, B.; Wu, Y.; Wu, J. 1992. Patchiness and patch dynamics: II. description and analysis. *Chinese J. Ecol.* 11:28–37.

Li, B.; Loehle, C.; Malon, D. 1996. Microbial transport through heterogeneous porous media: random walk, fractal, and percolation approaches. *Ecol. Model.* 85:285–302.

Loehle, C. 1989. Forest-level analysis of stability under exploitation: depensation responses and catastrophe theory. *Vegetatio* 79:109–115.

Loehle, C. 1990. Home range: a fractal approach. *Landscape Ecol.* 5:39–52.

Loehle, C.; Li, B. 1996a. Habitat destruction and the extinction debt revisited. *Ecol. Appl.* 6:784–789.

Loehle, C.; Li, B. 1996b. Statistical properties of ecological and geologic fractals. *Ecol. Model.* 85:271–284.

Loehle, C.; Wein, G. 1994. Landscape habitat diversity: a multiscale information theory approach. *Ecol. Model.* 73:311–329.

Loehle, C.; Li, B.; Sundell, R. C. 1996. Forest spread and phase transitions at forest–prairie ecotones in Kansas, U.S.A. *Landscape Ecol.* 11:225–235.

Lovejoy, S.; Schertzer, D.; Ladoy, P. 1986. Fractal characterization of inhomogeneous geophysical measuring networks. *Nature* 319:43–44.

Ludwig, D. D.; Jones, D.; Holling, C. S. 1978. Qualitative analysis of insect outbreak systems: the spruce budworm and the forest. *J. Anim. Ecol.* 44:315–332.

MacKay, G.; Jan, N. 1984. Forest fires as critical phenomena. *J. Phys.* A 17:L757–L760.

Mandelbrot, B. B. 1977. *Fractals: form, chance and dimension*. New York: W. H. Freeman.

Mandelbrot, B. B. 1983. *The fractal geometry of nature*. New York: W. H. Freeman.

Mandelbrot, B. B. 1989. Fractal geometry: what is it, and what does it do? *Proc. Roy. Soc. London A* 423:3–16.

Mandelbrot, B. B. 1990. Negative fractal dimensions and multifractals. *Physica A* 163:306–315.

Milne, B. T. 1988. Measuring the fractal geometry of landscapes. *Appl. Math. Comput.* 27:67–79.

Milne, B. T. 1990. Lessons from applying fractal models to landscape patterns. In: Turner, M. G.; Gardner, R. H., eds. *Quantitative methods in landscape ecology*. New York: Springer–Verlag: 199–235.

Milne, B. T. 1992. Spatial aggregation and neutral models in fractal landscapes. *Amer. Naturalist* 139:32–57.

Milne, B. T.; Johnson, A. R.; Keitt, T. H.; Hatfield, C. A.; David, J.; Hraber, P. 1996. Detection of critical densities associated with piñon–juniper woodland ecotones. *Ecology* 77:805–821.

Mollison, D. 1977. Spatial contact models for ecological and epidemic spread. *J. Roy. Statist. Soc. B* 39: 283–326.

Mollison, D. 1986. Modelling biological invasions: chance, explanation, prediction. *Phil. Trans. Roy. Soc. London B* 314:675–693.

Montgomery, D. R.; Dietrich, W. E. 1992. Channel initiation and the problem of landscape scale. *Science* 255:826–830.

O'Neill, R. V.; Milne, B. T.; Turner, M. G.; Gardner, R. H. 1988. Resource utilization scales and landscape pattern. *Landscape Ecol.* 2:63–69.

Palmer, M. W. 1988. Fractal geometry: a tool for describing spatial patterns of plant communities. *Vegetatio* 75:91–102.

Palmer, M. W. 1992. The coexistence of species in fractal landscapes. *Amer. Naturalist* 139:375–397.

Patil, G. P.; Taillie, C. 1989. *Probing encountered data, meta analysis and weighted distribution methods*. CSEES Technical Report No. 89-0101: 1–46.

Pielou, E. C. 1977. *Mathematical ecology*. New York: John Wiley & Sons: 291–311.

Pietronero, L.; Tosatti, E., editors. 1986. *Fractals in physics*. Amsterdam: North-Holland.

Pimm, S. T.; Gilpin, M. E. 1989. Theoretical issues in conservation biology. In: Roughgarden, J.; May, R. M.; Levin, S. A., eds. *Perspectives in ecological theory*. Princeton, NJ: Princeton University Press: 287–305.

Rex, K. D.; Malanson, G. P. 1990. The fractal shape of riparian forest patches. *Landscape Ecol.* 4:249–258.

Risser, P. G. 1995. The status of the science examining ecotones. *BioScience* 45:318–325.

Ruelle, D. 1989. *Chaotic evolution and strange attractors*. Cambridge, UK: Cambridge University Press.

Sahimi, H. 1994. *Applications of percolation theory*. London: Taylor & Francis.

Shannon, C. 1948. A mathematical theory of communication. *Bell Tech. J.* 27:379–423, 623–656.

Stauffer, D. 1985. *Introduction to percolation theory*. London: Taylor & Francis.

Stinchcombe, R. B. 1990. Fractals, phase transitions and criticality. In: Fleischmann, M.; Tildesley, D. J.; Ball, R. C., eds. *Fractals in the natural sciences*. Princeton, NJ: Princeton University Press: 17–33.

Stockmayer, W. H. 1943. Theory of molecular size distribution and gel formation in branched chain polymers. *J. Chem. Phys.* 11:45–55.

Sugihara, G.; May, R. M. 1990. Applications of fractals in ecology. *TREE* 5:79–86.

Taylor, S. J. 1986. The measure theory of random fractals. *Math. Proc. Camb. Phil. Soc.* 100:383–406.

Turcotte, D. L. 1986. Fractals and fragmentation. *J. Geophys. Res.* 91:1921–1926.

Uzunov, D. I. 1993. *Introduction to the theory of critical phenomena: mean field, fluctuations and renormalization*. Singapore: World Scientific.

Voss, R. F. 1988. Fractals in nature: from characterization to simulation. In: Peitgen, H.; Saupe, D., eds. *The science of fractal images*. New York: Springer-Verlag: 21–70.

Warner, W. S.; Fry, G. 1990. Evaluating small-format photogrammetry for forest and wildlife surveys: Euclidean vs. fractal geometry. *For. Ecol. Manage.* 31: 101–108.

Wiens, J. A. 1992. Ecology 2000: an essay on future directions in ecology. *Bull. ESA* 73:165–170.

Wiens, J. A.; Milne, B. T. 1989. Scaling of "landscapes" in landscape ecology, or, landscape ecology from a beetle's perspective. *Landscape Ecol.* 3:87–96.

Williamson, M. H.; Lawton, J. H. 1991. Fractal geometry of ecological habitats. In: Bell, S. S.; McCoy, E. D.; Mushinsky, H. R., eds. *Habitat structure*. London: Chapman and Hall: 69–86.

Wu, H.; Malafant, K. W. J.; Penridge, L. K.; Sharpe, P. J. H.; Walker, J. 1987. Simulation of two-dimensional point patterns: application of a lattice framework approach. *Ecol. Model.* 38:299–308.

Young, I. M.; Crawford, J. W. 1991. The fractal structure of soil aggregates: its measurement and interpretation. *J. Soil Sci.* 42:187–192.

Zeide, B. 1991. Fractal geometry in forestry applications. *For. Ecol. Manage.* 46:179–188.

15
Fuzzy Statistical and Modeling Approach to Ecological Assessments

Bai-Lian Li

15.1 Introduction

Increasing attention on the extreme sensitivity of ecological systems to environmental insults has been changing the traditional view that humans are the most sensitive species. Ecological risk assessment, especially the relatively unexplored area of applied ecotoxicology, has been developed to meet this need. However, characterization, quantification, estimation, and prediction of ecological risks at multiple scales are often very difficult. Valuation of ecosystem components and scaling from the laboratory toxicity bioassay or intensive investigations of single sites (or relatively small geographic areas) to population, community, ecosystem or landscape level are also ill-defined. Some uncertainty is unavoidable in ecologists' assessment and prediction about ecological systems, simply because uncertainty emerges whenever information pertaining to the situation is deficient in some respect. It may be incomplete, imprecise, fragmentary, not fully reliable, vague, contradictory, or deficient in some other way. In these situations, unexpected risks and/or environmental changes may result from decisions that must be made. There are various information deficiencies resulting in different types of uncertainty. Traditionally, the only well-developed mathematical apparatus for dealing with uncertainty in ecological risk assessment has been probability theory (Suter and Barnthouse, 1993). However, the probabilistic approach alone cannot represent uncertainties attached to systems for which some deterministic dynamical characteristics are unknown or deliberately ignored, as well as uncertainties attached to their mathematical model. We have recognized that uncertainty is a multidimensional concept. Which of its dimensions are actually manifested in a description of an ecological situation is determined by the mathematical theory employed. The challenge now is to introduce and/or develop new methods to address these concerns.

This chapter describes a fuzzy statistical and modeling approach to ecological risk assessment under uncertainty to improve risk-based decision making for protection of our environment/ecosystem. This new method (maybe relatively new for ecologists, but not for mathematicians and systems engineers) integrates utilization of knowledge or judgment of experts together with statistics of the available vague data and imprecise information so as to provide the modeling basis for a predictive ecological risk assessment at multiple scales. Some potential applications of the fuzzy statistical and modeling approach to ecological assessment are also introduced with the intent to begin the process of bringing the concepts and principles of fuzzy sets (logic) and systems into mainstream ecological assessments.

Although multivalued logic was developed early in this century, the development of fuzzy set theory by Lotfi Zadeh of the University of California at Berkeley in 1965 marks the turning point of its development from an academic venue to modern application (Zadeh, 1965). The central idea is that members of a set may have only partial membership; that is, there is gray between the black and

This paper is based on my presentation at the VI International Congress of Ecology, August 21–26, 1994, Manchester, UK. Several major changes and improvements have been included in this version. This work was partially supported by the U.S. National Science Foundation (BSR-91-09240, DEB-93-06679, DEB-94-11976, and INT-95-12739), DOE/Sandia National Laboratories (BE-0229), the University of New Mexico start-up fund, and USDA Forest Service-Northern Region. This is Sevilleta LTER publication no. 137.

white of true and false. Zadeh referred to this gray area as "fuzzy", which was an inauspicious choice of terms for Western science and engineering. But, once we realize that fuzzy logic and set theory (and even fuzzy thinking!) are quantitative ways of characterizing intrinsic ambiguity, rather than of substituting imprecision for precision, the shorthand term *fuzzy* becomes acceptable in science and engineering. Despite much thought, no better term has yet emerged, even though fuzzy thinking does not sound like something intelligent people might do. Already the application of this theory to the development of fuzzy systems for commercial electronic products and system controllers in Japan has been remarkable (e.g., Williams, 1992). Fuzzy mathematics itself is not fuzzy. It is a mathematical theory and method for dealing with fuzzy phenomena in nature.

Allen and Hoekstra (1992) have described the conceptual approaches and potential of fuzzy systems analysis for applied ecology. Salski et al. (1996) and this author (Li, 1996) have also edited two special issues on fuzzy modeling in ecology for the international journal *Ecological Modelling*. (For people who would like to know more about fuzzy theory, the easier readable book without too much mathematical treatment by Kosko, 1993, is a good starting point.) It would be most valuable for us to begin to explore the horizons of this new approach for ecological assessment applications.

15.2 Uncertainty and Paradigm Shift Regarding Uncertainty

When we deal with real-world problems (e.g., ecological risk assessment), we can rarely avoid uncertainty. At the experimental level, uncertainty is an inseparable companion of any measurement and observation, resulting from a combination of resolution limits of measuring instruments and inevitable measurement errors. At the cognitive level, it emerges from the vagueness and ambiguity inherent in natural languages. At the social level, uncertainty results not only from the inevitable incompleteness of shared meanings obtained by people through social interaction, but it is also created and maintained by people for different purposes. For example, people may have different views in sustainable development indices on how to incorporate social, societal, resource management, and scientific perspectives into the assessment process in order to develop criteria that distinguish good (nominal), marginal, or poor (subnominal) ecological condition for various resources.

Uncertainty is thus fundamental to human beings at all levels of their interaction with the real world. The treatment of uncertainty can be time consuming and difficult. Policymakers often tend to ignore it. Instead, they may consider decision parameters to be single valued and deterministic or solicit expert opinion and trust it implicitly as the best estimate. It is increasingly recognized, however, that attitudes toward uncertainty have been undergoing a significant change in this century. In science, this change has been manifested by a transition from the traditional attitude, according to which uncertainty is a plague that should be avoided by all means, to an alternative attitude, according to which uncertainty is fundamental to science and its avoidance is often counterproductive. Two phases of this transition can be clearly recognized, each having the characteristics of a paradigm shift in the sense introduced by Kuhn (1996). When statistical mechanics was accepted, by and large, by the scientific community as a legitimate area of science early in this century (Prigogine, 1997), the traditional attitude toward uncertainty was for the first time revised. Uncertainty became recognized as useful, or even essential, in certain scientific inquiries. However, this recognition was strongly qualified: uncertainty was conceived solely in terms of probability theory.

A mathematical method to quantify uncertainty was first proposed by Hartley in 1928. This method was based on classical set theory. Later, Shannon (1948) proposed a measure for uncertainty based on probability theory. This measure is called the *entropy measure*. Until recently, entropy measure was the only measure for quantifying uncertainty. It is generally agreed that an important turning point in the evolution of the modern concept of uncertainty was the publication of a seminal paper by Zadeh (1965), even though some ideas presented in the paper were envisioned by the philosopher Max Black in 1937. Zadeh (1965) introduced a theory whose objects, fuzzy sets, are sets with imprecise boundaries. Membership in a fuzzy set is not a matter of affirmation or denial, but rather a matter of degree. The importance of Zadeh's paper was that it challenged the adequacy of classical set theory and probability theory as frameworks for expressing uncertainty. By allowing imprecise boundaries, fuzzy sets acquire the capability to express concepts of natural language that are inherently vague. They also acquire the capability to bridge, in whatever crude way, mathematics and empirical reality (this is really needed for our ecological assessment!).

The concept of a fuzzy set represents a basic mathematical framework for dealing with vagueness, nonspecificity, dissonance, and confusion in general. The novel theories of uncertainty, subsumed under the categories of fuzzy sets (Klir and Folger, 1988), rough sets (Slowinski, 1992), and fuzzy measures (Wang and Klir, 1992), as well as their various combinations (Dubois and Prade, 1988), have emerged over the relatively short period of the last three decades. On the basis of these studies, three principles, (1) the principle of minimum uncertainty, (2) the principle of maximum uncertainty, and (3) the principle of uncertainty invariance, have been formulated (Klir, 1991). The first principle can help us to simplify complex systems without information loss and conflict-resolution problems; the second can help us to deal with optimization and ampliative reasoning (this is reasoning in which conclusions are not entailed in the given premises); and the third enables us to convert results obtained in one theory into equivalent representations in the other theories.

These new theories of uncertainty provide useful tools for assessing ecological health. In general, sources of uncertainty in ecological assessment include the following:

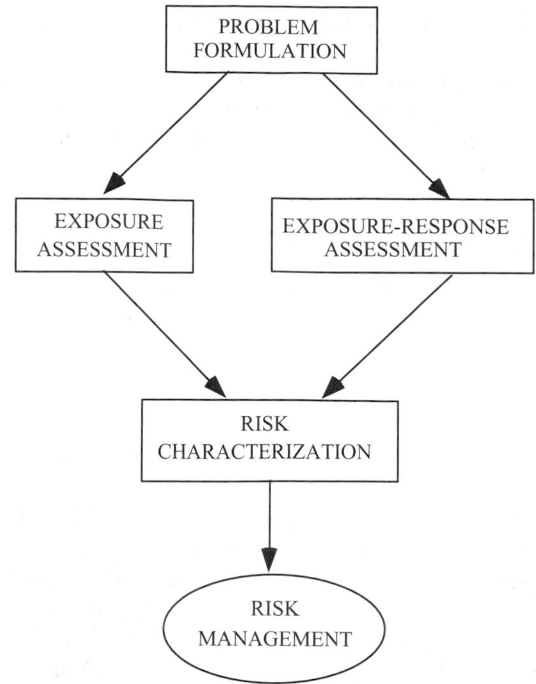

FIGURE 15.1. A general framework for ecological risk assessment.

1. *Complexity*: for example, linkages between species in ecosystems are complicated, indirect effects are difficult to predict, and nonlinear interactions are normally dominated in social-economic-natural complex ecosystems (we do not know where the threshold is and when we may go beyond the threshold).
2. *Variability and heterogeneity*: spatial scales, temporal scales, and "organism scales" of heterogeneous ecological systems are not well defined; spatiotemporal cross-scale dynamics are essential.
3. *Random variations*: Einstein doubted that God was tossing dice, but, in reality, the laws of nature have a stochastic character. This is not because we are not exactly sure about something or because we cannot calculate some things exactly. Rather, it is because the idea of probability is inherent in the very nature of things. Random variation is a very important uncertainty in our ecological assessment.
4. *Errors of measurement and estimation*: no measurement or observation is perfect, and estimation based on such measurement or observation has error unavoidably.
5. *Lack of knowledge* about the system that we are going to assess or predict.

We believe that the theory of fuzzy sets holds great promise for elucidating and resolving significant ecological problems, including those related to environmental decision making. Fuzzy logic and systems theory is useful for both theoretical and applied ecology. Wherever we have been forced by the dictates of binary logic to draw artificially sharp boundaries in ecology, we can now draw more realistic distinctions in terms of fuzzy sets. Organisms can exist for varying lengths of time in conditions that are lethal, because that lethal boundary is fuzzy, not the rigid tolerance limits of the textbooks. The boundaries of fundamental and realized niches are fuzzy, not precise. Species can be members of communities to a partial degree, expressed as a fuzzy membership function. Decisions are made because conditions are "just about right," not because some magically precise number has been quantitatively calculated or measured with extreme accuracy. Many ecological controversies have persisted for decades because a methodology for expressing intrinsic ambiguity has been lacking. Some of these controversies may now be resolvable, or at least more aptly stated, with fuzzy sets and fuzzy logic. Because many uncertainties are involved in every step of ecological risk assessment and management (see Figure 15.1) (Norton et al., 1992), fuzzy mathematics will provide an appropriate framework for the quantitative modeling of such complex systems.

15.3 Basics of Fuzzy Mathematics

Fuzzy sets as formulated by Zadeh (1965) are based on the simple idea of quantifying the degree of "belongingness" of an element with respect to a subset. Assume that the symbol S represents the entire set, in other words, a universe of discourse. In classical set theory, given a subset A of S, each element $x \in S$ satisfies either x belongs to A, or does not belong to A. The subset A is accordingly represented by a function $f_A : S \rightarrow \{0, 1\}$, where

$$f_A = \begin{cases} 1 & \text{if } x \in A \\ 0 & \text{if } x \notin A. \end{cases}$$

The function f_A is called the characteristic function of subset A.

Fuzzy sets are introduced by generalizing the characteristic function f_A. Let μ_B be a function defined on S whose values are in the unit interval [0, 1]; that is, μ_B: $S \rightarrow [0, 1]$, where B is a label identifying the function. We call the label B a fuzzy set when we have a certain interpretation of this label. Let us give an example.

Let S be the set of all nonnegative integers. We interpret S to be the set that includes landscape vegetation cover under grazing and dry conditions (Wu et al., 1996). Let us consider if we can define a subset B that implies "good vegetation cover." By experience, the cover above 70% is good, whereas the cover at about 60% is not so good, but not too bad either. There is no definite criterion that separates good and not good vegetation covers. Now we define a function μ_B that corresponds to the concept of "good vegetation cover." That is, for $x \in S$, μ_B is defined to show the degree of relevance of the cover percentage x to the concept "good." Thus $\mu_B(70) = 1$ indicates that 70% cover is good, while $\mu_B(30) = 0$ indicates that 30% cover is not good. For all other cover percentages, we might use

$$\mu_B(x) = \begin{cases} 1 & x \geq 70 \\ (x - 50)/20 & 50 < x < 70. \\ 0 & x \leq 50 \end{cases}$$

Thus a fuzzy set B is a grade of relevance of elements to a concept represented by the label B. In terms of the function $\mu_B(x)$, it is a generalization of the characteristic function for an ordinary (crisp) subset.

Mathematically, B is nothing but a label attached to the function μ_B, but an interpretation as in the above example enables us to call it a fuzzy set, a subset without a clearly defined boundary. The function μ_B is the membership function for the fuzzy set B. If $\mu_B(x) = 1$, then x certainly belongs to the fuzzy set B; if $\mu_B(x) = 0$, then x does not belong to B at all; if $0 < \mu_B(x) < 1$, then the belongingness of B is ambiguous. If $\mu_B(x_1) > \mu_B(x_2)$, then the relevance of x_1 to the concept represented by B is greater than the relevance of x_2.

We have another way to interpret the label B. A fuzzy set B defined by a membership function μ_B is interpreted as a monotone family of crisp (nonfuzzy) sets. Let α be a real parameter in the unit interval ($\alpha \in [0, 1]$). An α-cut of a fuzzy set B denoted by B_α is defined to be a crisp set:

$$B_\alpha = \{Sx|\ \mu_B(x) \geq \alpha, \qquad x \in .$$

An α-cut is sometimes called an α-level set. Thus a fuzzy set is interpreted as a family of α-cuts: $\{B_\alpha\}_{\alpha \in [0, 1]}$.

Every crisp set can be regarded as a fuzzy set, since the characteristic function of a crisp set is regarded as its membership function. In particular, the entire set S is a fuzzy set whose membership function is $\mu_s(x) = 1$ for all $x \in S$. In the same way, the empty set $\varnothing$ is a fuzzy set with $\mu_\varnothing(x) = 0$ for all $x \in S$.

The membership function is the basic idea in fuzzy set theory; its value measure degrees to which objects satisfy imprecisely defined properties. To manipulate fuzzy sets, it is necessary to have operations that enable us to combine them. There are many ways to define the fuzzy set operations (e.g., Klir and Folger, 1988), and it is not within the scope of this chapter to discuss them all. Here we discuss only the basic fuzzy set operations. Consider a finite universe $S = \{x_1, x_2, \ldots, x_n\}$ and let $A \subseteq S$ be a fuzzy set. Its membership values are expressed by a simplified notation:

$$A = \mu_A(x_1)/x_1 + \mu_A(x_2)/x_2 + \cdots + \mu_A(x_n)/x_n = \Sigma\ \mu_A(x_i)/x_i.$$

Note that here the symbol + does not refer to ordinary addition.

For example, let $S = \{1, 2, 3, 4, 5\}$ and consider a fuzzy set A given by $\mu_A(1) = 0$, $\mu_A(2) = 0.5$, $\mu_A(3) = 0.8$, $\mu_A(4) = 1$, and $\mu_A(5) = 0.2$. Using the preceding notation, A would be specified by

$$A = \frac{0}{1} + \frac{0.5}{2} + \frac{0.8}{3} + \frac{1}{4} + \frac{0.2}{5}.$$

Equality of two fuzzy sets is defined by the equality of the membership functions. That is, for two fuzzy sets $A, B \subseteq S$,

$$A = B \Leftrightarrow \mu_A(x) = \mu_B(x), \qquad \forall x \in S.$$

Inclusion of fuzzy sets is defined by the inequality of their membership functions:

$$A \subseteq B \Leftrightarrow \mu_A(x) \leq \mu_B(x), \qquad \forall x \in S.$$

Union and intersection of two fuzzy sets are defined by the maximum and minimum, respectively.

Union: $A \cup B$: $\mu_{A \cup B}(x) = \max[\mu_A(x), \mu_B(x)]$.

Intersection: $A \cap B$: $\mu_{A \cap B}(x) = \min[\mu_A(x), \mu_B(x)]$.

The complement of a fuzzy set A, which is denoted by A^c, is defined as follows:

Complement: A^c: $\mu_{A^c}(x) = 1 - \mu_A(x)$.

For example, let $S = \{1, 2, 3, 4, 5\}$ again and assume that fuzzy sets A and B are given by

$$A = 0/1 + 0.5/2 + 0.8/3 + 1/4 + 0.2/5,$$
$$B = 0.9/1 + 0.4/2 + 0.3/3 + 0.1/4 + 0/5.$$

Then

$$A \cup B = 0.9/1 + 0.5/2 + 0.8/3 + 1/4 + 0.2/5,$$
$$A \cap B = 0/1 + 0.4/2 + 0.3/3 + 0.1/4 + 0/5,$$
$$A^c = 1/1 + 0.5/2 + 0.2/3 + 0/4 + 0.8/5.$$

Many extensions of the above min–max definition to the fuzzy set operations can be found in Zimmermann (1991) and many other texts. These fuzzy set operations can be used to aggregate different ecological indicators during ecological risk analysis. Applications of fuzzy mathematical methods in ecology have been published by many authors (e.g., Li, 1986, 1996; Li and Zhou, 1987; Banyikwa et al., 1990; Bardossy et al., 1990a; Keesman and van Straten, 1990; Lindsey et al., 1992; McBratney et al., 1992; Bardossy, 1994; Feoli and Zuccarello, 1994; Salski et al., 1996).

15.4 Applications of the Fuzzy Modeling Approach to Ecological Assessment

Because fuzzy sets are a generalization of a classical set theory, the embedding of conventional models into a larger setting endows fuzzy models with greater flexibility to capture various aspects of incompleteness or imperfection in whatever information and data are available about a real process. For example, we developed a semiarid grazing ecosystem simulation model combining fuzzy imprecision with probabilistic uncertainty to characterize climate–plant–herbivore interactions (Wu et al., 1996); Cheng et al. (1996) used fuzzy systems analysis and conventional dynamic programming to optimize biological control of a greenhouse spider mite–cucumber system; Singer and Singer (1993) presented a fuzzy set formulation of the phenomenological equations of nonequilibrium chemical kinetics, which could be easily adopted to deal with nonequilibrium ecological "reactions" of multiple species; and Levary (1990) established fuzzy-logic-based system dynamics methodology for modeling, simulating, and analyzing real-life systems with imprecise and vague variables or events.

In general, we can use fuzzy models in any of the following circumstances: (1) the modeled process is too complicated and its exact description is not known; (2) many factors are not quantified, or are impossible to quantify, and they are characterized only verbally; (3) more exact specifications cannot be employed and, hence, our research would be gratuitous; (4) the exact description is too time consuming (the use of natural language is advantageous because it follows human thinking in the best way, and the necessary time for fuzzy model construction is very small); (5) the exact model construction might need some experiments (this might not, however, be possible either due to economic reasons or due to the nature of the process itself); and (6) the fuzzy model is more objective than verbal description based on a purely intuitive approach.

The general methodology for the fuzzy model construction can be summarized as follows:

1. Description of the problem
2. Factor identification, definition, and classification as dependent or independent
3. Analysis of all factors from the viewpoint of their dependence
4. Determination of the factors that will be processed as linguistic variables and determination of the scales of linguistic values
5. Determination of the universes, that is, determination of the kind of elements (numbers, intervals) and their extent (see the above example on membership functions of vegetation cover)
6. Transformation of linguistic values into fuzzy sets, that is, the estimation of membership functions
7. Formulation of the fuzzy model using conditional statements and/or its mathematical elaboration

This process should be understood to be iterative. It means that we must repeat some steps with modifications to reach satisfactory results.

15.4.1 Fuzzy Classifications

In fuzzy modeling techniques, many methods can be easily used for ecological data analysis (e.g., see Zimmermann, 1991; Bandemer and Nather, 1992). Among them, fuzzy classification (or fuzzy clustering or pattern recognition) is the most commonly used fuzzy data analysis in clustering of ecotoxicological data (Frederichs et al., 1996), vegetation analysis (Banyikwa et al., 1990), Landsat TM data classification for soil moisture estimates (Lindsey et al., 1992), GIS-based soil mapping (McBratney et al., 1992; Odeh et al., 1992; Zhu et al., 1996), the impact of winter climate on crop (Chen et al., 1992), downscaling from GCMs to local climate (Bardossy, 1994), ecological site classification (Ayoubi et al., 1991; El-Shishiny and Ghabbour, 1991; Owsinski and Zadrozny, 1991), and land suitability analysis (Hall et al., 1992).

Fuzzy classification or clustering is the search for structure in incomplete and uncertain data. Given any finite data set X of objects, the problem of clustering in X is to assign object labels that identify natural subgroups in the set. Because the data are unlabeled, this problem is often called *unsupervised learning*, the word learning here referring to learning the correct labels for good subgroups. In turn, it should be clear that good subgroups implies that we have a question or problem in mind that finding subgroups of some kind will answer, help describe the process, and so forth. The objective is to partition X into a certain number (c) of natural and homogeneous subsets, where the elements of each set are as similar as possible to each other and at the same time as different from those of the other sets as possible. The number (c) can be fixed beforehand or may result as a consequence of ecological or mathematical constraints. Because the technique of clustering is unsupervised (i.e., we do not have any information on the class structures or the labeled samples of classes), clustering algorithms attempt to partition X based on certain assumptions and criteria. The output partition, which includes both number of classes (c) and the class membership structures, depends on the criteria that are used to control the clustering algorithm.

Many algorithms have been developed to obtained hard clusters from a given data set. Among these, the c-means algorithms and their generalizations, the ISODATA (Iterative Self-Organizing Data Analysis Techniques A) clustering methods, are probably the most widely used. The c-means algorithms assume that c is known, whereas c is unknown in the case of ISODATA algorithms. By allowing algorithms to assign each object a partial or distributed membership in each of the c clusters, fuzzy clustering offers special advantages over conventional algorithms. Sometime we refer to fuzzy classification as supervised learning. The reference cited in the first paragraph basically used fuzzy c-means algorithms.

Let us look at a hypothetical example, habitat comparison and evaluation. Let us assume that we have five different habitats $U = \{x_1,x_2,x_3,x_4,x_5\}$ with four different indices for evaluation. Based on some criteria, experts may use 1 to 5 (excellent) to rank these indices. Now we assume that we have

$$\begin{aligned}
x_1 &= (x_{11},x_{12},x_{13},x_{14}) = (5,5,3,2)\\
x_2 &= (x_{21},x_{22},x_{23},x_{24}) = (2,3,4,5)\\
x_3 &= (x_{31},x_{32},x_{33},x_{34}) = (5,5,2,3)\\
x_4 &= (x_{41},x_{42},x_{43},x_{44}) = (1,5,3,1)\\
x_5 &= (x_{51},x_{52},x_{53},x_{54}) = (2,4,5,1)
\end{aligned}$$

We can use simple distance measure to construct membership function as follows:

$$r_{ij} = \begin{cases} 1, & \text{if } i = j \\ 1 - 0.1\sum_{k=1}^{4}|x_{ik} - x_{jk}|, & \text{if } i \neq j \end{cases} \quad i,j = 1, \ldots, 5$$

to establish the fuzzy similarity matrix:

$$R = \begin{bmatrix} 1 & 0.1 & 0.8 & 0.5 & 0.3 \\ & 1 & 0.1 & 0.2 & 0.4 \\ & & 1 & 0.3 & 0.1 \\ & & & 1 & 0.6 \\ & & & & 1 \end{bmatrix}$$

Mathematically, to have partitioning, we need to obtain the fuzzy equivalent matrix through, $R \rightarrow R^2 \rightarrow R^4 \rightarrow R^8$, that is,

$$R^* = R^4 = R^8 = \begin{bmatrix} 1 & 0.4 & 0.8 & 0.5 & 0.5 \\ & 1 & 0.4 & 0.4 & 0.4 \\ & & 1 & 0.5 & 0.5 \\ & & & 1 & 0.6 \\ & & & & 1 \end{bmatrix}$$

If we set the α-level equal to 0.6, we have three classes, $\{x_1,x_3\}$, $\{x_2\}$, and $\{x_4,x_5\}$; when you change the α-level, you can have different classes. Based on these results, we can say that habitats 1 and 3, and 4 and 5 have very similar quality, respectively; but habitat 2 has its own unique condition.

The work done by Hall et al. (1992) provides a good example for using the fuzzy classification technique to analyze the land-use suitability of the Cimanuk watershed in northwest Java, Indonesia. It is not too difficult to extend their work to general ecological landscape assessment by using an adaptive hierarchical fuzzy *c*-means clustering algorithm and its generalization (e.g., the ISODATA methods). We can use multitier data, quantitative or qualitative.

15.4.2 Adaptive Fuzzy Modeling Approach

An adaptive fuzzy system is a fuzzy logic system equipped with a training algorithm, where the fuzzy logic system is constructed from a collection of fuzzy IF-THEN rules, and the training algorithm adjusts the parameters of the fuzzy logic system based on numerical input-output pairs. Conceptually, adaptive fuzzy systems combine linguistic and numerical information in the following way. Because fuzzy logic systems are constructed from fuzzy IF-THEN rules, linguistic information in the form of fuzzy IF-THEN rules can be directly incorporated; on the other hand, numerical information in the form of input-output pairs is incorporated by training the fuzzy logic system to match the input-output pairs. Adaptive fuzzy systems can be viewed as fuzzy logic systems whose rules are automatically generated through training. There are two strategies for combining numerical and linguistic information by using adaptive fuzzy systems:

1. Use linguistic information to construct an initial fuzzy logic system, and then adjust the parameters of the initial fuzzy-logic-system-based numerical information. The final fuzzy logic system is, therefore, constructed based on both numerical and linguistic information.
2. Use numerical information and linguistic information to construct two separate fuzzy logic systems, and then use a fuzzy weighted average to obtain the final fuzzy logic system.

Many training methods have been proposed in the fuzzy research community, including back-propagation, orthogonal least squares, a table-lookup scheme, nearest-neighborhood clustering, and evolutionary learning. Among them, neuro-fuzzy data analysis has gained great attention recently (Ichihashi and Turksen, 1995).

Li and Yeung (1994) have considered using an adaptive fuzzy data modeling framework for characterization of subsurface contamination. Similarly, this approach can be used in data analysis in ecological assessment. Characterizing a contaminated subsurface involves a complex interaction between subsurface contaminants and the soil-water system. The composition and properties of contaminants are significant factors in the development of such interaction relationships. However, basic chemical-physical data, the concentration data, and the toxicological data are often unavailable. The existence of a large number of subsurface sites suspected of being contaminated requires us to obtain a relative estimation among these sites with respect to their hazard potential. We use the relative estimation of hazard as a basis for deciding about the urgency of carrying out a remedial measure.

We have developed an adaptive fuzzy modeling framework as a new approach to the characterization of subsurface on the basis of consideration of the relative estimation of hazard. This system involves several different data analysis methods (Figure 15.2), including a pure fuzzy logic system (IF-THEN rules) and fuzzy classification linking a training (or learning) algorithm (back-propagation algorithm) that have been developed recently in the fuzzy research community; for example, see Wang (1994). This system combines linguistic and numerical information from expert knowledge and field sampling data. In data preprocessing and algorithmic data analysis, we use traditional statisti-

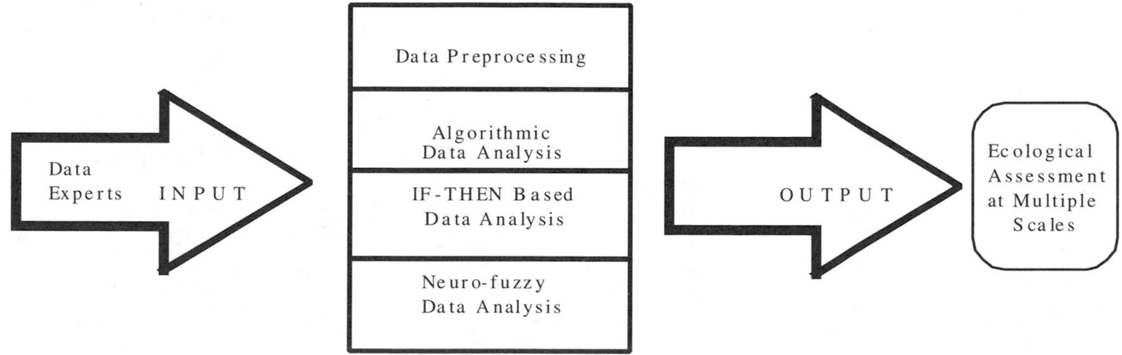

FIGURE 15.2. The procedure of data analysis in an adaptive fuzzy modeling framework for characterizing potentially subsurface contaminants (modified from Li and Yeung, 1994).

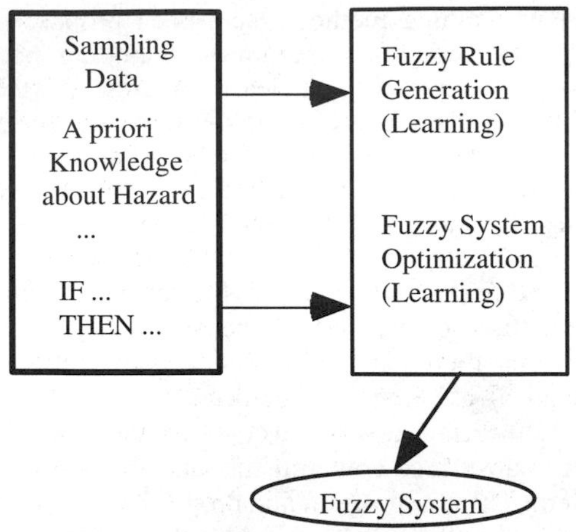

FIGURE 15.3. A neuro-fuzzy data analysis as an important part of our adaptive fuzzy modeling approach to characterization, quantification, estimation, and prediction of ecological risks (from Li and Yeung, 1994).

cal data exploration methods, fractal and wavelet approach, geostatistics, and fuzzy clustering to sort and analyze any available numerical information. The IF-THEN rule-based data analysis is used to quantify expert knowledge and experience. In our approach, we apply the modeling procedure of Sugeno and Yasukawa (1993) for our multiinput and single-output system. Neuro-fuzzy data analysis is an important step in our adaptive fuzzy modeling approach to characterizing subsurface contaminants that helps us to understand geophysical mechanisms, extracting knowledge and generating rules in fuzzy systems based on incomplete knowledge and uncertain sampling data (see Figure 15.3). A detailed introduction to these methods can be found in Wang (1994) and Ichihashi and Turksen (1995). Combining GIS and data mining technique, the adaptive fuzzy modeling approach should provide a more flexible method to assess ecological systems.

15.4.3 Fuzzy Comprehensive Assessment Model and Others

Mathematically, fuzzy relation calculus provides a foundation for us to develop a fuzzy comprehensive assessment model. Such a model is very important for our integrated ecological assessment, because so many important factors or indices with very complicated interactions needed to be considered. It might be easier to define binary relations among these indices, but it would be difficult for us to understand multiple relations at the same time. A fuzzy relational equation provides a way to deal with such problems, that is, transferring binary relations to fuzzy equivalence and ordering relations. Here we can define a fuzzy relation R in $X_1 \times X_2$ by the ordered pair

$$R = \{[(x_1,x_2),\ \mu_R(x_1,x_2)]\},\ x_1 \in X_1,\ x_2 \in X_2.$$

This can be generalized to dimensions.

Given a product space $X \times X$ on which two fuzzy relations R_1 and R_2 are defined, the composition of R_1 and R_2, denoted by $R_1 \circ R_2$, is a fuzzy subset of the product space with membership function given by

$$\mu_{R_1 \circ R_2}(x_1,x_2) = \sup_v \min[\ \mu_{R_1}(x_1,v),\ \mu_{R_2}(v,x_2)],$$

where the supremum (or maximum) is taken over all v in X. In the preceding habitat evaluation example, we used composition of relations to obtain the fuzzy equivalent matrix.

In real ecological assessment, we can derive a comprehensive assessment model from a single index evaluation matrix with any given weight by using a fuzzy relational equation. Such an assessment model can be hierarchical, multidimensional, and very flexible. For example, Li (1986) combined economic and ecological indices using a fuzzy relational equation and then solved the equation to obtain an optimal biological control threshold for cotton aphids. Because the membership functions are dimensionless, we can easily link social, economic, and environmental indices through their membership functions and fuzzy relations.

A fuzzy weighted average is another way to aggregate multidimensional ecological indicators. In reality, the rating criteria of ecological assessment and their corresponding importance weights could be evaluated in fuzzy numbers. Thus the value of the assessment variable is also a fuzzy number, which is the fuzzy weighted average of criteria rating.

To generalize the fuzzy weighted average, let A_1, $A_2, \ldots, A_n$ and $W_1, W_2, \ldots, A_n$ be the fuzzy numbers defined on the universes $X_1, X_2, \ldots, X_n$ and $Z_1, Z_2, \ldots, Z_n$, respectively. If f is a function which maps from $X_1 \times X_2 \times \ldots \times X_n \times Z_1 \times Z_2 \times \ldots \times Z_n$ to the universe Y, then the fuzzy weighted average y is

$$y = f(x_1, x_2, \ldots x_n,\ w_1, w_2, \ldots, w_n) = \frac{w_1x_1 + w_2x_2 + \ldots + w_nx_n}{w_1 + w_2 + \ldots + w_n}.$$

Let μ_B be the membership function of the fuzzy image B of $A_1, A_2, \ldots, A_n$, $W_1, W_2, \ldots, A_n$ through *f*. Then, by the extension principle in fuzzy mathematics,

$$\mu_B(y) = \max_{\substack{x_i \in X_i,\ W_i \in Z \\ i=1,2,\ldots,n \\ y=f(x_1,\ldots,x_n)}} \{\min[\mu_{A_1}(x_1), \ldots, \mu_{A_n}(x_n), \mu_{w_1}(w_1), \ldots, \mu_{w_n}(w_n)]\}$$

where μ_{A_i} and μ_{W_i} are the membership functions of fuzzy number A_i and W_i, respectively, $i = 1,2, \ldots, n$. Dong and Wong (1987) provided a way to calculate them.

Many other fuzzy modeling techniques can be used for ecological assessment, for example, fuzzy approach to expert data analysis (Khurgin and Polyakov, 1986; Bardossy et al., 1993), fuzzy spatial analysis (Leung, 1985), fuzzy urban design assessment model (Kumar and Goel, 1994), fuzzy kriging (Diamond, 1989; Bardossy et al., 1990b), fuzzy rule-based GIS technology (Saint-Joan and Desachy, 1995), and fuzzy cellular automata (Baldwin et al., 1993).

15.5 Conclusion

Integrated ecological assessment at multiple scales is complex and multidimensional. Uncertainty is a major issue in all applications of risk assessment, but it presents a particular problem for ecological risk assessment due to the inherent variability of biological and ecological systems. Errors in the information used for assessments can arise from errors in measurements (e.g., toxicology index, environmental concentrations); extrapolation from species to species or from one set of conditions to another and in other ways in which expert judgment is used; and the assumptions and arbitrary thresholds that are made at some stages. There is nothing to be gained from ignoring the potential inaccuracies that these errors may cause or the substantial risks of misjudging the scale of problems if they are not examined, even imperfectly. Increasing awareness of ecological issues has emphasized the need for improved ecological risk assessment methodology. This chapter briefly introduced a fuzzy statistical and modeling approach as a potential tool for integrated ecological risk assessment. The significance of this approach is that it can generalize conventional assessment models and, more importantly, it provides new perspectives, a higher degree of flexibility, and more appropriate approximations to reality. Although the phenomena analyzed are vague, the analysis, though not necessarily quantitative, is precise. While the present scope and depth of the fuzzy ecological assessment modeling is limited, with more researchers taking an active role in its development and application, its increased impact on complex ecological assessment and constructing ecological indicators is a strong possibility.

15.6 References

Allen, T. F. H.; Hoekstra, T. W. 1992. *Toward a unified ecology*. New York: Columbia University Press.

Ayoubi, B. A.; Beninel, F.; Le Calve, G. 1991. Comparison of some techniques of multidimensional scaling in soil fauna ecology. *Appl. Stochastic Models Data Analysis* 7:263–271.

Baldwin, J. F.; Martin, T. P.; Zhou, Y. 1993. Fuzzy cellular automata. In: Bouchon-Meunier, B., ed. *Uncertainty in intelligent systems*. Amsterdam: Elsevier Science: 235–245.

Bandemer, H.; Nather, W. 1992. *Fuzzy data analysis*. Dordrecht: Kluwer.

Banyikwa, F. F.; Feoli, E.; Zuccarello, V. 1990. Fuzzy set ordination and classification of Serengeti short grasslands, Tanzania. *J. Vegetation Sci.* 1:97–104.

Bardossy, A. 1994. Downscaling from GCMs to local climate through stochastic linkages. In: Paoli, G., ed. *Climate change, uncertainty, and decision-making*, Inst. for Risk Res.: 33–46.

Bardossy, A.; Bogardi, I.; Duckstein, L. 1990a. Fuzzy regression in hydrology. *Water Resources Res.* 26: 1497–1508.

Bardossy, A.; Bogardi, I.; Kelly, W. E. 1990b. Kriging with imprecise (fuzzy) variograms. *Math. Geol.* 22: 63–79.

Bardossy, A.; Duckstein, L.; Bogardi, I. 1993. Combination of fuzzy numbers representing expert opinions. *Fuzzy Sets Systems* 57:173–181.

Black, M. 1937. Vagueness. *Phil. Sci.* 4:427–455.

Chen, S.; Jiang, A.; Domroes, M. 1992. Studies on the impact of winter climate on rubber and wheat cultivation in the mountains of southern China, applying a fuzzy cluster analysis. *Int. J. Biometeorology* 36: 159–164.

Cheng, Z.; Horn, D. J.; Lindquist, R. K.; Taylor, R. A. J. 1996. Fuzzy analysis for a greenhouse spider mite management system. *Ecol. Modelling* 90:111–121.

Diamond, P. 1989. Fuzzy kriging. *Fuzzy Sets Systems* 33:315–332.

Dong, W. M.; Wong, F. S. 1987. Fuzzy weighted averages and implementation of the extension principle. *Fuzzy Sets Systems* 21:183–199.

Dubois, D.; Prade, H. 1988. *Possibility theory*. New York: Plenum Press.

El-Shishiny, H.; Ghabbour, S. I. 1991. A fuzzy interpretation for an ecological site classification case. *Appl. Stochastic Models Data Analysis* 7:257–261.

Feoli, E.; Zuccarello, V. 1994. Naivete of fuzzy system spaces in vegetation dynamics? *Coenoses* 9:25–32.

Friederichs, M.; Franzle, O.; Salski, A. 1996. Fuzzy clustering of existing chemicals according to their ecotoxicological properties. *Ecol. Modelling* 85:27–40.

Hall, G. B.; Wang, F.; Subaryono, 1992. Comparison of Boolean and fuzzy classification methods in land suitability analysis by using geographical information systems. *Environ. Planning* A 24:497–516.

Ichihashi, H.; Turksen, I. B. 1995. Neuro-fuzzy data analysis and its future directions. In: *Proceedings of 1995 IEEE International Conference Fuzzy Systems*, Yokohama, Japan, IEEE Press: 1919–1925.

Keesman, K.; van Straten, G. 1990. Set membership approach to identification and prediction of lake eutrophication. *Water Resources Res.* 26:2643–2652.

Khurgin, J. I.; Polyakov, V. V. 1986. Fuzzy approach to the analysis of expert data. *Systems Analysis Modeling Simulation* 3:29–35.

Klir, G. J. 1991. Measures and principles of uncertainty and information: recent developments. In: Atmanspacher, H.; Scheingraber, H. eds. *Information dynamics*. New York: Plenum Press: 1–14.

Klir, G. J.; Folger, T. 1988. *Fuzzy sets, uncertainty, and information*. Upper Saddle River, NJ: Prentice Hall.

Kosko, B. 1993. *Fuzzy thinking: the new science of fuzzy logic*. New York: Hyperion.

Kuhn, T. S. 1996. *Structure of scientific revolutions*, 3rd ed. Chicago and London: University of Chicago Press.

Kumar, S.; Goel, S. 1994. Fuzzy sets in urban design. *Int. J. Systems Sci.* 25:1727–1741.

Leung, Y. 1985. Basic issues of fuzzy set theoretical spatial analysis. *Papers of the Regional Sci. Assoc.* 58: 35–46.

Levary, R. R. 1990. Systems dynamics with fuzzy logic. *Int. J. Systems Sci.* 21:1701–1707.

Li, B.-L. 1986. Fuzzy decision model for optimal thresholds of biological control in cotton aphid. *J. Fuzzy Math.* 6:35–40.

Li, B.-L., ed. 1996. Fuzzy modelling in ecology. *Ecol. Modelling* 90:109–186.

Li, B.-L.; Yeung, A. T. 1994. An adaptive fuzzy modeling framework for characterization of subsurface contamination. In: *Proceedings of 1994 1st Int. Joint Conference of North American Fuzzy Information Processing Society Biannual Conference, the Industrial Fuzzy Control and Intelligent Systems Conference, and NASA Joint Technology Workshop on Neural Network and Fuzzy Logic*, San Antonio, TX: IEEE Press: 194–195.

Li, B.-L.; Zhou, C. 1987. Optimization model in wheat white smut (*Erysiphe garmins* DC. *f. sp. tritici* Marchal) control. *J. Biomath.* 2:154–156.

Lindsey, S. D.; Gunderson, R. W.; Riley, J. P. 1992. Spatial distribution of point soil moisture estimates using Landsat TM data and fuzzy-c classification. *Water Resources Bull.* 28:865–875.

McBratney, A. B.; De Gruijter, J. J.; Brus, D. J. 1992. Spacial prediction and mapping of continuous soil classes. *Geoderma* 54:39–64.

Norton, S. B.; Rodier, D. J.; Gentile, J. H.; van der Schalie, W. H.; Wood, W. P.; Slimak, M. W. 1992. A framework for ecological risk assessment in the EPA. *Environ. Toxicology Chemistry* 11:1663–1672.

Odeh, I. O. A.; McBratney, A. B.; Chittleborough, D. J. 1992. Fuzzy-c-means and kriging for mapping soil as a continuous system. *Soil Sci. Soc. Am. J.* 56:1848–1854.

Owsinski, J. W.; Zadrozny, S. 1991. Ecological site classification: an application of clustering. *Appl. Stochastic Models Data Analysis* 7:273–279.

Prigogine, L. 1997. *End of certainty: time, chaos, and the new laws of nature*. New York: Free Press.

Saint-Joan, D.; Desachy, J. 1995. A fuzzy expert system for geographical problems: an agricultural application. In: *Proceedings of 1995 IEEE International Conference on Fuzzy Systems*, Yokohama, Japan, IEEE Press: 469–476.

Salski, A.; Franzle, O.; Kandzia, P. eds. 1996. Fuzzy logic in ecological modelling. *Ecol. Modelling* 85: 1–98.

Shannon, C. E. 1948. The mathematical theory of communication. *Bell Systems Tech. J.* 27:379–423, 623–656.

Singer, D.; Singer, P. G. 1993. Fuzzy chemical kinetics: an algorithmic approach. *Int. J. Systems Sci.* 24:1363–1376.

Slowinski, R. ed. 1992. *Intelligent decision support: handbook of applications and advances of the rough sets theory*. Boston: Kluwer.

Sugeno, M.; Yasukawa, T. 1993. A fuzzy-logic-based approach to qualitative modeling. *IEEE Trans. Fuzzy Systems* 1:7–31.

Suter, G.; Barnthouse, L. 1993. Assessment concepts. In: Suter, G.W. II, ed. *Ecological risk assessment*. Boca Raton, FL: Lewis Publishers: 21–47.

Wang, L. 1994. *Adaptive fuzzy systems and control*. Upper Saddle River, NJ: Prentice Hall.

Wang, Z.; Klir, G. J. 1992. *Fuzzy measure theory*. New York: Plenum Press.

William, T. 1992. Fuzzy logic is anything but fuzzy. *Computer Design* 31:113–126.

Wu, H.; Li, B-L.; Stoker, R.; Li, Y. 1996. A semi-arid grazing ecosystem simulation model with probabilistic and fuzzy parameters. *Ecol. Modelling* 90:147–160.

Zadeh, L. A. 1965. Fuzzy sets. *Information and Control* 8(3):338–353.

Zhu, A.; Band, L. E.; Dutton, B.; Nimlos, T. J. 1996. Automated soil inference under fuzzy logic. *Ecol. Modelling* 90:123–145.

Zimmerman, H.-J. 1991. *Fuzzy set theory and its applications*, 2nd ed. Boston: Kluwer.

16 Measuring Ecosystem Values

Kimberly Rollins

16.1 Introduction

Campbell (Chapter 28) provides an overview of models that characterize societal and ecosystem interactions and dynamics. He emphasizes the advantages of using a variety of different models to characterize these interactions in order to gain a better understanding of their nature. In his overview, he provides a brief introduction to economic models. This chapter describes an economic approach to defining and quantifying ecosystem values. An important reason for quantification of ecosystem values is to provide input to social decision-making processes intended to address some of the negative and unintended side effects that have developed between the economy and the ecosystem. The measurement of ecosystem values is not an end in itself, but rather a means in a process to develop policies that would benefit both society and the environment. This larger context for valuation complements Campbell's focus on interactions that affect dynamic processes.

Chappelle (Chapter 21) provides a description of economic impact analysis. Economic impact analysis is useful in public policy situations when specific geographic regions are targeted for economic development attention due to poor employment conditions, sluggish economic growth, and other indications of underdevelopment relative to the rest of the economy. Natural resource development issues are often closely tied to local community development and economic stability. However, economic impact analysis is not intended to measure the value of the ecosystem itself to society, and its use for this purpose can have serious and potentially irreversible consequences.

It is sometimes mistakenly believed that the results of an economic impact analysis of a change in ecosystem use are equivalent to the value of the change in use. Strictly speaking, impact assessment quantifies the changes in indicators of economic activity within the boundaries of a well-defined region, as a result of a specific change in economic activity in one sector. The economic impacts from a given change in a specific resource sector are related to the complexity of the local economy, the linkages within the economy, the dependence of the economy on the resource sector, and other elements that are dependent on the structure of the local economy. While the resulting indicators may be measurable in dollar units, they are not equivalent to measures of the economic value of ecosystem services to society. Rollins and Wistowsky (1997) give an example of how using impact analysis estimates as a proxy for economic value could seriously underestimate the value of a wilderness area and bias policy decisions in favor of development rather than preservation.

The problem is that economic impact analysis measures expenditures associated with an economic activity, rather than the value of the good or service that is supplied. In reality, many ecosystems provide great economic value to society without requiring expenditures in a marketplace. In its focus on expenditures, impact analysis measures the costs associated with use of environmental resources. Using expenditures as a proxy for ecosystem benefits is not justifiable in economic terms and, in practice, can lead to a bias in favor of extractive high-cost uses of an ecosystem over low-impact ecosystem benefits. The key to a correct approach is to measure the benefits generated by ecosystems, not the cost of maintaining and using ecosystems.

Numerous published works define economic value in the context of ecosystem values (Randall and Stoll, 1983; Randall and Peterson, 1984; Cummings et al., 1986; Randall, 1987; Mitchell and

Carson, 1989; Johnson and Johnson, 1990; Braden and Kolstad, 1991; Freeman, 1993; Pearce, 1993; Tietenberg, 1996; Smith, 1997). Many authors explain the relevance of defining specific categories of economic values for natural amenities, such as use and nonuse values, existence values, bequest values, and option values. Literature comprising thousands of publications over the last 10 years focuses on methodological issues encountered in economic valuation. Carson et al. (1993) reference over a thousand nonmarket valuation studies using one technique, contingent valuation, alone. It is expected that Environment Canada's Environmental Valuation Reference Inventory (EVRI), an on-line, continually updated bibliographic reference tool, will catalogue several thousand environmental valuation studies from around the world (Environment Canada, 1998).

However, simultaneous with the development of the theory and methods of environmental valuation, others question economic valuation on a variety of grounds, suggesting that more research is needed to better understand how to use estimates of ecosystem value. Some authors question whether certain methods are up to the task of estimating nonmarket economic values (AERE, 1992; Boyle and Bergstrom, 1992; Smith, 1992; Arrow et al., 1993; Hausman, 1993). Others point out that, while the methods may be reliable in theory, many applications of the methods are sloppy and ill-informed, producing estimates of ecosystem benefits that are of questionable use (Sarker and McKenney, 1992; Arrow et al., 1993). Still others consider the philosophical implications of the differences between economic value, which is anthropocentric in nature, and other definitions, which attempt to identify intrinsic values apart from human derivations of value (Rescher, 1980; Kneese and Schulze, 1985; Van De Veer and Pierce, 1986; Norton, 1987; Sagoff, 1988). The result is that research on methods and theory is ongoing while, in practice, valuation measurements are used to inform day to day decision making.

Rigorous questioning and discourse are integral parts of the academic process. The process has refined and greatly improved the development of valuation theory and methods and our understanding of how measures of ecosystem value can be used. Many of those who most vociferously question aspects of ecosystem valuation also concur that in the real world individuals, firms, and governments make decisions daily that could alter ecosystems. The incentives driving many of these decisions are economic. Decision criteria include perceptions of net changes in value that occur under different alternatives. But, in fact, in many instances the full range of economic values that could be affected by a given decision may include ecosystem values that are not revealed in markets. Without an attempt to properly identify and quantify these changes in values, the decision process remains incompletely informed. Thus, while the academic discourse indicates that there is room for continued research in the area and that this research will likely change the details of how future valuation studies are to be performed and interpreted, valuation is already a powerful tool in ecosystem management.

The fact is that environmental degradation and resource misallocation problems exist precisely because the full range of ecosystem benefits to society now and in the future is not fully represented in the market economy. With respect to their use of environmental resources, individuals and firms acting in their own interests do not always have incentives to take those actions that would be most beneficial for society as a whole. In the terminology of economics, a mismatch between individual interests and social interests indicates that the market, or environmental policy, has failed. This failure to take into account the full value of ecosystems can potentially be corrected through the use of carefully designed economic instruments and policies that alter people's options and incentives. Quantitative estimates of the value of ecosystem services are often necessary in order to develop and justify the implementation of land-use regulations, taxes, subsidies, permit systems, user fees, liability laws, and other policies to encourage wise use and conservation of ecosystem resources.

This chapter provides a conceptual understanding of what ecosystem valuation attempts to measure as defined in economic theory. A major goal is to provide insight into the factors that determine and influence the economic value of ecosystems and their services. This chapter describes the policy contexts in which it is desirable to quantify ecosystem values, summarizes techniques available for measuring nonmarket ecosystem values, and provides relevant examples and references for further information. This chapter is conceptual in nature and written for the professional who has limited formal economic training. It is intended to be complementary to others in this volume.

Section 16.2 justifies the need to estimate ecosystem values to make information available that the market cannot provide to policymakers, governments, firms, and individuals who would prefer that the full value of environmental services be incorporated into economic decision making. Estimation of nonmarket ecosystem values, in particular, is necessary to develop policies that can re-align individual with social interests, thereby correcting for market failures.

Section 16.3 describes the concept of economic value. The definition is built up by demonstrating how value is revealed by trade between members of a simplistic world with two regions. The example is simply to introduce the reader to the notion that, under specific conditions, the market theoretically results in the highest-valued allocations of goods and services in society. The example is overly simple to introduce the reader to the concepts of marginal value, marginal cost, demand, and supply. The section ends by acknowledging that, in reality, the idealized conditions that must hold in order for markets to fulfill their promise fail for a great many situations commonly encountered in services that are generated by natural environments.

Section 16.4 describes several categories of economic values of ecosystems and describes how alternative uses of an ecosystem may not prevent the flow of various sources of value, albeit they may be at altered levels. The goal of economic analysis of alternative uses of an ecosystem is to first identify all values that may be altered among the alternative uses. Those that would be expected to remain unaltered need not always be measured in order to make an informed policy decision. The key is to identify which values would change and by how much. Section 16.5 outlines the major methods of nonmarket valuation used today. Finally, two examples are given to illustrate the importance of the policy context, in terms of achieving useful and reliable estimates or ecosystem valuation.

16.2 Why Measure Ecosystem Values?

Individuals, firms, and public sector agents regularly make decisions that involve explicit and implicit trade-offs among alternative uses of environmental resources. The results of these decisions imply an underlying structure of incentives that affect how people choose to interact with their environment. When choices made by individuals or firms are likely to cost society more than it can gain, the market has failed. Similarly, policy failure is said to occur when environmental policies and management guidelines cost society more than it can gain. Policy can be developed to alter economic incentives that influence individual and firm behavior and to bring individual and social interests more closely into alignment. Measurements of the extent to which economic markets and policies fail are useful to develop regulations and economic instruments to effectively alter incentive structures and thereby protect environmental resources from overexploitation.

16.2.1 Examples of Situations for Which Nonmarket Values Inform Decisions

Three commonly encountered situations illustrate how quantitative measurements of ecosystem values can assist in decision making. The first is to guide policy decisions under conditions of market failure. The second is to rank the societal desirability of alternative uses of publicly owned lands. The third is the need to set a protocol for determining the value of ecosystem damages in liability suits, which in turn creates the incentive structure for resource users to practice caution to prevent damages. In each type of situation, policy action is indicated to resolve economic problems.

Correction of Market Failures

In terms of economic theory, many environmental problems can be said to result from market failure. Market failure occurs whenever any of the basic assumptions necessary for markets to function properly is not met. These assumptions include perfect information and a well-defined system of property rights. The nature of many environmental resources is such that these basic assumptions are not met. Most notably, market failure can occur when the flow of benefits from an ecosystem is not fully represented within the market. Consider the benefits of a standing forest in providing oxygen, old-growth habitat, and watershed protection. The value of these services may not be incorporated into the decisions to harvest the forest and redevelop the land for other purposes. The benefits that do not flow directly to the owners of the forest are not realized in the market, while the benefits to the owners of harvest and land conversion are realized in the marketplace. In this case, the omission of some of the values of the standing forest may lead private decision makers to choose an option that would cost society as a whole more than the owners of the forest would gain. A correction of this problem might involve setting up incentives for private owners to maintain forest lands. Tax rebates for woodlots are one example of such an incentive mechanism. For such a mechanism to be most effective, some estimate of the value of woodlots to society is needed. Otherwise, the incentive mechanism may result in too much or too little forestland being withdrawn from harvest and developed.

Market failure also occurs when an individual's decision to use environmental resources imposes costs on other people that are not fully taken into account in market transactions. An example of market failure in the case of ecosystem benefits is groundwater recharge and purification functions of wetlands. The beneficiaries of these services do not pay for the full value of wetlands when they use water and thus do not have the incentive to conserve water efficiently. The owner of a wetland does not receive the full value of the water services provided by the wetland, and it may thus appear to the private landowner that the wetland is of greater value filled in and developed.

The value of a wetland to society is determined by a number of interacting factors, including substitutes for the ecosystem goods and services, the costs of substitutes, the existing level of services, and the types and costs of complementary goods and services. To be complete, the full economic value of a wetland includes benefits that are received external to any markets. Groundwater recharge and purification benefits from wetlands are not purchased in the marketplace. The water users who receive these benefits may not have paid for them in a market, and therefore the full value of these services is not reflected in a market price for water. However, water purification and supply is clearly of value to society. The value of water services that are provided to the wetland owner alone are likely to be a very small fraction of the total value of water services to the large number of people who depend on the aquifer serviced by the wetland.

Market failure can result in overexploitation of environmental resources, thereby reducing the welfare of society now and in the future. Market failures can cause a misallocation within the same time period, wherein certain groups in society gain at the expense of others. In the case of the wetland example, the landowner may choose to develop the wetland. If the societal value of the wetland's ecological functions in maintaining water quality and quantity is greater than the added value of the land after development, then clearly society has lost an amount that is greater than the private landowner has gained. Other market failures can cause intertemporal misallocations or resource overexploitation that benefits one generation at the expense of another. Over time, many wetlands may be developed. At first, it may be that there are so many wetlands that the loss to society is minimal in terms of the gain in agricultural acreage. The current generation of landowners gain, but the loss to future generations in terms of reduced water quality may be greater than the short-term gain to landowners. Market failure indicates that ecosystem resources are being wasted and that society would be better off if the full complement of costs and benefits was incorporated into resource use decisions. This could be done with a variety of economic instruments, such as taxes, permits, or subsidies. But if many ecosystem costs and benefits are nonmarket in nature, then the first step in designing policy would require identifying these values and attempting to estimate some measure of their magnitudes.

Full Information for Benefit Cost Analysis

When policy is recommended that would alter the use of an ecosystem, economic activity is altered as well. Often large-scale government-sponsored projects have major impacts on economic activity and ecosystem function. Large-scale dam and irrigation projects around the world are examples. The process of identifying potential projects and choosing whether or not to proceed requires the comparison of present-valued net benefits (benefits minus costs) with and without the project or among alternative projects. Benefit–cost analysis has become an increasingly important tool for public sector economic decisions. Properly done, benefit–cost analysis can provide criteria to help to determine the trade-off between the benefits of a proposed policy and the costs in terms of foregone economic benefits. Most benefit–cost analysis texts include extensive sections outlining the importance of properly measuring quantitative estimates of nonmarket ecosystem values. Increasingly, environmental impact assessments are required to include benefits and costs of the impacts of proposed projects. For example, the Canadian Environmental Assessment Act defines *environmental effects* that are covered by the act to include "any change that the project may cause in the environment, including any effect of any change on health and socio-economic conditions." Accordingly, it is becoming more common to see environmental assessments in Canada include an analysis of economic costs and benefits of nonmarket impacts.

While it is beyond the scope of this chapter to describe benefit–cost analysis, it is important to note that economic valuation is necessary in order to provide required data for any nonmarket costs or benefits associated with a proposed policy. The quality of a benefit–cost analysis will depend on the quality of the valuation estimates. If nonmarket costs and benefits of environmental impacts of a project are omitted, the implicit assumption is that the value of these environmental impacts is zero. If a project is turned down on the basis of environmental impacts, but these impacts have not been quantified, the implicit assumption is that the costs of the impacts outweigh any potential gains. This may be a difficult position to support without any valuation of nonmarket impacts. Most benefit–cost analysis texts include extensive sections outlining the importance of properly measuring quantitative estimates of nonmarket ecosystem values.

Liability Laws

Full liability implies that any agent who is deemed responsible for an environmental insult must pay the full costs of the loss. Identifying and calculating the full costs is important, but not just after the fact as a means to calculate a penalty. The most important role of valuation is *before* any person or firm makes a conscious decision to use production methods that threaten ecosystems. As the potential liability increases for loss of ecosystem services, the incentive to avoid such losses increases to those who would be legally liable. Thus, while liability laws themselves provide an incentive to avoid losses, it is the recognition of the potential magnitude of a claim that indicates how much precaution will be taken. Smith and Kopp (1993) provide a comprehensive summary of the role of natural resource valuation in the context of damage assessment and liability laws.

16.2.2 Is It Useful to Measure Ecosystem Values for All Interactions between the Economy and the Environment?

A measure of the total value of environmental amenities may not be feasible and, more importantly, measures of total value are often not necessary in most specific environmental policy contexts. Many human impacts on the environment do not cause concern. The majority of forestry, agriculture, and land-use practices, for example, do not result in situations in which public agencies are called on to intervene. Why is it that in some circumstances it is important to determine the value of ecosystem services, but in other circumstances the impact of human activity on the ecosystem is not a contentious issue?

The value of ecosystem services is of little policy concern when markets function well and all costs and benefits are properly accounted for, because the resources are already being used in their most highly valued uses. When we consider the full range of interactions between society and the environment, we can see that the market works reasonably well as an allocative mechanism. In other words, as long as current land uses are not creating unaccounted for costs for other people now or in the future, then the net benefits that are produced as a result of current uses are most likely greater than the benefits of ecosystem services that are foregone. We do not consider it imperative to measure a value for these foregone services simply because we know that they are probably lower than the value of the current uses. By definition, it is not necessary to measure values since all resources are already in their highest-valued use. Indeed, in this so-called perfect world, the concept of economic policy or natural resource policy is anathema, since it implies altering the functioning of an already well functioning system.

The three examples described previously, market failure, benefit–cost analysis, and damage assessment for liability rulings, share the feature that they describe situations in which decisions made in the market alone do not incorporate the full value of ecosystems. Thus determination of ecosystem values is one step in a larger process to attempt to develop policy responses to internalize the full values of ecosystems into economic decision making. The market is an effective and low-cost means to distribute resources throughout society. The market is so effective precisely because it is able to reveal the value of all goods and services relative to one another and relative to the preferences of all people, provided that it is functioning well. If the market is not functioning well and all ecosystem values are not incorporated, then the market values and true economic values deviate. This is the focus of much economic analysis of environmental problems.

The next section describes in more detail the concept of economic value. The main purpose of the discussion is to illustrate that economic value exists apart from markets. Markets serve as a highly effective allocative mechanism and, in so doing, reflect economic values of commodities traded. But the economic values exist without markets. The problem is that without being able to observe market activities it is much more difficult to reveal the true economic value of a good or service. So, while market prices can reveal value, without markets, analysts need some other proxy for value. The section is intended to demonstrate that the notion of a person's *willingness to pay* for a good or their *willingness to accept* compensation for the loss of this good can be used as a proxy for measuring economic value. The section also demonstrates two concepts that are important to economic value. The first is the difference between total value and marginal value, and the second is the difference between benefits, costs, and the notion that value is related to *net benefits*.

16.3 Concept of Economic Value

Fisheries, forests, water, clean air, soil, and other environmental resources provide significant stocks and flows of goods and services that are valuable to society. Activities of individuals, firms, and governments affect the conditions of these stocks and flows. In many cases, it is possible to quantify in physical units the resource stocks and flows and the impacts of human activities on them. The loss of 50 hectares of wetlands in conversion to agricultural cultivation is an example of a physical measure of an ecosystem impact from human activity.

Economic value refers to the ability of any good or service to satisfy needs and desires of people. To say that an object has value is to suggest that it contributes to the well-being of some person. Not every person is in a position to receive equal levels of well-being from the same good or service. We can start with a simple example to illustrate that economic value arises from the satisfaction of human needs and preferences. The example illustrates that the concept of economic value is relevant whether or not money is used to facilitate exchange. Imagine two regions; one is water abundant and the other is water poor. The arid region has a number of other goods that are of value to the water-rich region, such as minerals. Each person in the water-abundant region has more water than they can use.

It can easily be seen that, in a world in which people are able to trade among themselves, water has value even for the person who has no need for it, since it can be exchanged for something else that this person does value. The resource has value in use and, because of trade, in exchange. It is important to note that the potential for exchange generates, for a person who controls access to a resource, an incentive to conserve the resource, even if they have no need of it themselves. As long as another person is willing to trade for it, then there is a powerful incentive to steward and protect the value of the resource for current or future exchange value. In this way, overall patterns of scarcity, both over space and time, play a role in the economic value of a given resource.

There would be no exchange between the regions if the cost to transport a unit of water were higher than the maximum value any one person has for that unit. The cost of transport is more complex than the simple notion of the engineering costs of building a pipeline. It may be, for instance, less costly to conserve what water is accessible in the arid region and to purify nonpotable water than to bear the costs in terms of time, effort, and resources necessary to access the water of the other region. In this case, purification of nonpotable water is an imperfect substitute for access to the clean water of another region. Conservation of existing reserves is also a substitute for access to the other region's water. The availability and cost of substitutes affect the willingness of residents to bear the cost of transporting water. This in turn affects the exchange value of a unit of water for the people of the water-rich region. In this way, the availability and cost of substitutes influence the value of a unit of ecosystem goods and services.[1]

If the cost of transporting water is less than the maximum amount someone is willing to give to gain access to the water, then the water would likely be transferred from one region to another. Any technological innovation that reduces costs of transporting water would affect the value of a unit of water. As costs of transport are reduced, the costs of water purification and conservation become higher relative to the costs of accessing the water of the other region.

Both regions would trade water for other items that the people in the water-abundant region value until the point at which trade would not increase any further. This point would be reached because, as water becomes relatively more abundant in the arid region and less abundant in the water-rich region due to exchange, the use and exchange values

---

[1]How much water overall would be used if some benevolent donor gave the arid region a lump sum to pay for a pipeline? If the cost of accessing additional water is subsidized, then the full cost of supplying water to the region is not taken into account in any subsequent arrangements for trade. It is likely that the incentive to conserve the existing water resources of the arid region will be reduced and the local supply used at a faster rate, because the residents can anticipate less expensive access to the water-rich region's supplies. In a sense, the donation to cover the cost of supplying the water appears to lessen the water-scarcity problem, and the value to local residents of a unit of water appears to be reduced. This "subsidy" has effectively appeared to reduce the scarcity problem. Subsidies in the real world represent the expenditure of funds that have been transferred from one area of the economy to another. So, while the water-scarcity problem may appear to be temporarily mitigated, in the longer run the water will be used at a faster rate, and the other use of the funds that made up the subsidy are ruled out. The subsidy has bought time, but actually worsened a scarcity problem.

for both regions become equalized. At some point it is inevitable that the value of one more unit of water is the same for both parties. After that point, it is not worthwhile to continue to exchange water for other goods. The fact that this equilibrium point exists does not imply, however, that the physical quantities of water are the same for each region at the equilibrium point. The point at which exchange stops is where the unit value of one more unit of water is the same for each region. The region with the higher costs would likely have less water, while the region with lower costs would have more. The point is that trade would ensue with each unit of water commanding a lower and lower price until the two parties are not willing to trade for another unit of water. Thus the value of water and the quantity of water that each region eventually has are simultaneously determined in this situation.

One factor that determines how much water will flow between regions and, simultaneously, the value of the last unit of water exchanged is the level of overall wealth that each region has to start. Suppose that the water-rich region has absolutely no other resources but water, while the arid region is rich in many resources. The value of a unit of water would be very different in this case than it would be in a situation where all other resource endowments are similar between regions. That is, the value of the water relative to all other potential goods in exchange is a function of the initial distribution of resources across all people in society. If we were to suddenly redistribute all resource endowments, all exchange values would be altered, and the value of any resource relative to all others will be changed. If the water-rich region is poorly endowed with any other resources, while the arid region is heavily endowed, the economic value of water will change to reflect the ability to pay of the group that wants more water. In general, the economic value of any good or service in society is determined in part by the overall pattern of the initial endowments of wealth. These endowments determine the relative values of all resources in terms of their exchange values. If a war or government policy redistributes wealth, the relative economic values of all goods and services can be affected.[2]

Finally, the relative value of a unit of water is also determined by the ability of individuals who control access and wish to gain access to interact with one another. If people of the arid region and the water-rich region have little information about each other's reserves, then there is an initial cost associated with any potential exchange. The value of a unit of water may be very different within each of the two regions, and the situation would likely stay that way until the initial cost of gathering information about alternatives is low, relative to the use value of water within the arid region, or low relative to the potential exchange value of water for the water-rich region. The costs of acquiring information to facilitate exchange thus also affect the value of ecosystem resources.

Another way in which information costs affect the value of ecosystem services is by new information that directly affects the use value of the services for society. For example, suppose that the people in the water-rich region were not previously aware that the surface water that they divert to the canals that transport water to the arid region is connected to groundwater stocks. Suppose that new scientific evidence suggests that diversions will eventually lead to land subsidence and loss of wetlands that purify groundwater and provide habitat for waterfowl and fish. Since the people of the region value waterfowl and fish and do not wish to invest in a water treatment plant to substitute for wetland services, they may revise their preferences regarding water. The value of the water for their own needs becomes greater due to the new information about the role that water plays in their overall level of well-being.

Finally, information can affect the value of ecosystem services in terms of an individual's awareness that the potential for new information might in the future affect our understanding of the role the ecosystem plays in our well-being. As long as the possibility exists that new information might be forthcoming in the future that would alter people's choices now, then there is economic value in waiting to decide at some future time whether or not to engage in an activity that may irreversibly alter ecosystem functions. This is technically a value for information; but since the information has value in how it may affect ecosystem values, it is typically attributed to the resource itself. Economists have defined this as quasi-option value.

This example is intended to illustrate the complexity of the concept of economic value and the dependence of economic value on the underlying social structure. This example shows that the value of a unit of water for each region thus depends on the following:

[2] This issue is recognized as being particularly relevant when government policy redistributes endowments that may represent a fairly large proportion of any person's overall endowment. The result is that the original sets of relative prices and the subsequent sets of relative prices for goods and services are not strictly comparable using standard efficiency criteria. It is easy to see this in terms of the adage "a dollar is worth more to a poor person than to a rich person."

1. Preferences and needs of the people in society
2. Initial patterns of endowment and scarcity of water
3. Distribution of initial endowments of other items of value for exchange purposes
4. Costs of transporting water
5. Costs of substitutes such as conservation and purification
6. Information costs of facilitating transactions
7. Possibility of new information that could change the understanding of society in terms of how the ecosystem contributes to well-being

The economic value of any ecosystem service is determined by similar factors. Any method that attempts to determine value in a way that is consistent with the economic value of any other good and service to society must be defensible as a proxy that can take into account all these features. This defensibility of valuation methods has spawned innumerable debates and published papers. Before turning to methods of valuation, however, the remainder of this section will review a few basic economic concepts that are important to understanding economic value and what valuation methods attempt to measure.

16.3.1 Concept of Marginal Value

Demand

The previous example described a situation in which the value of a unit of water changed over the course of trading between regions. When people of the arid region had little water, they were willing to give up a great amount of other resources to gain access to one more unit of water. As trade progressed between the two regions, the value of one incremental unit of water changed for the two parties in a predictable manner. The value of one unit of water increased for the water-rich region as they gave up more and more water and decreased for the arid region as they gained access to more and more water. Trade was predicted to cease when the formerly arid region had enough water so that the additional use value of one more unit, less the cost of transporting it, was exactly equal to the value of that one unit of water to the people of the formerly water-rich region.

The notion of scarcity is also relevant within the framework of a single individual. The more of an item that an individual already has in her possession for personal use, the less she is likely to value an additional unit. One additional unit of any good or service is termed the *marginal* unit. The concept of the marginal unit always implies that there already are a specific number of units prior to the marginal unit. In terms of economic value, the number of units prior to the marginal unit is a critically important reference point. The marginal value of a resource is the value of the marginal unit. Given that an individual already has access to more water than she needs, the marginal value of 1 cubic meter of water is low. However, given that an individual has access to no clean water whatsoever, then his value for one unit, the first unit of water he would receive, is probably very high.

The notion of a *demand curve* is simply the graphical representation of the value of a single unit of a good or service, given that individuals already have access to successive quantities of the same good or service. Each point on Figure 16.1 illustrates the demand for one unit of water, given the quantity already available. The horizontal axis represents the quantity of water already available. The vertical axis measures the value of 1 cubic meter of water. Therefore, the point *A* represents a high marginal value for 1 cubic meter of water due to the smaller quantity already available. The point *B* represents a lower value for 1 cubic meter, given the greater quantity already available. The law of demand states that the locus of points that trace out marginal value of a good or service over quantities that are already available is downward sloping: as a person acquires more and more of a good, the value of an additional unit is progressively smaller and smaller. The demand curve traces out how marginal value changes with the scarcity of a good or service.

While a point on the demand curve indicates the marginal value, the *total value* is the area under the demand curve up to a specified quantity. Often the measurement of interest, for environmental pol-

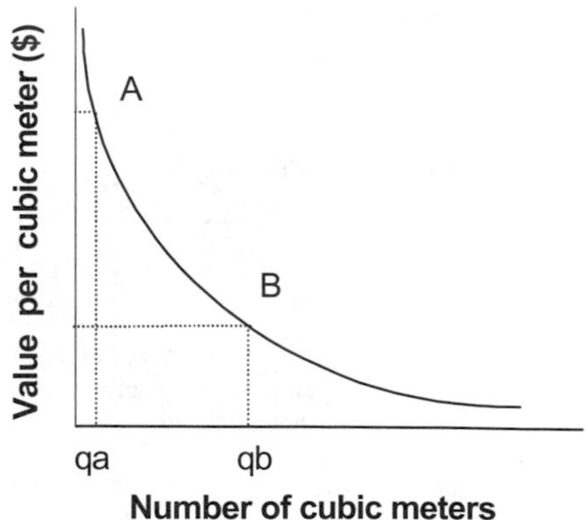

FIGURE 16.1. Marginal value of a cubic meter of water.

icy reasons, is not the total value of a particular quantity of a resource, but rather the *change* in total value when some action has altered the quantity or quality of the resource in question. This change in total value can be calculated from the marginal value curve as the area below the curve, between the two quantities. Referring to Figure 16.1, if the starting point is quantity *A* and the change results in quantity *B*, the value of the change is the area below the demand curve between quantities *A* and *B*. This measure of the change in total value would be relevant for a cost–benefit analysis of a proposed irrigation project that would affect the flow of water to a region, for example.

Supply

As we saw in the example used in the first part of this section, the marginal value of water is affected by the cost of supplying the water. In this example the cost of supply to the arid region is actually composed of two parts. The first is the cost of transport and other direct costs of delivery that depend on the amount of water used. The second is the opportunity cost to the water-rich region of the use value for each unit of water. Essentially, every unit they give up in trade is no longer available for their own use now or in the future. The foregone benefits are opportunity costs. As water becomes relatively more and more scarce to the people of the water-rich region, their marginal opportunity costs increase with each unit given up. The marginal cost of a unit of water is composed of both types of costs, direct costs of use and opportunity costs.

Marginal costs per unit of water can be traced out in a supply curve, analogously to the demand curve. Figure 16.2 illustrates how the marginal cost

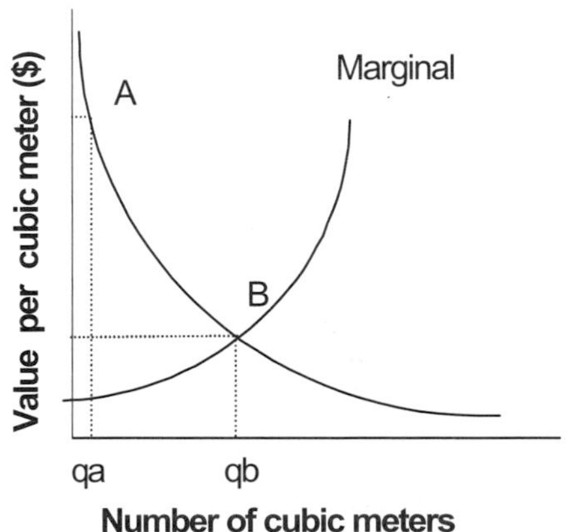

FIGURE 16.2. Marginal cost and marginal benefit.

to supply water to the arid region increases with each unit supplied. The marginal cost curve takes into account all costs, including the opportunity cost of the suppliers. The point at which the marginal value of one unit is equated between the two regions is at the point where the demand and supply curves intersect. This is the point at which the marginal value, less the cost of transportation, for the arid region is the same as the marginal value of water for the water-rich region.

Thus far, we have described how economic value for a marginal unit of water is determined in a situation in which two regions are able to trade with one another. The sum of the total value of water over both regions could theoretically be calculated as the point at which neither region would gain from any further trade. Each region in Figure 16.2 receives exactly the same marginal value from a unit of water at point *B*, so the incentives to continue to trade are exhausted; beyond this point the sum of the total value for the resource over both regions would decrease. At this point the value of the resource is maximized over all individuals in both regions. For this reason, economists refer to this point as a socially optimal allocation, because the value of the resource is at its maximum for all users.

Revelation of Value by Markets

If all markets actually incorporated all costs and all benefits of ecosystems, then market interactions of suppliers and those who demand resources would take all costs into account. Economic values of all resources relative to all others would be revealed in terms of the relative marginal values that are achieved in equilibrium, the point at which it does not increase value to anyone by trading further. The prices of all goods would reflect marginal values less marginal costs. Resulting allocations of resources among society would thus be those allocations that produced the greatest net benefits overall. In this sense, it is said that markets reveal value. In reality, not all costs and benefits are always included in market transactions that affect ecosystems. In this case, the market prices cannot possibly be consistent with the true economic value of ecosystems to society.

Our example has demonstrated in a simplistic manner a textbook concept of economic value. Many assumptions, however, lie beneath the surface. These assumptions concern certain properties of the resources in question. For instance, it is implicitly assumed that the water does not convey benefits to people who do not consume it, but who may share in a claim to the benefits of the resource. For example, some individuals in the water-rich region may receive value from the flow of benefits that they receive from wetlands, which would be endangered by declining groundwater levels once trade with the arid region starts. If they have a legitimate claim to these benefits and this claim is not taken into account in the water trading, then the losses to these individuals are said to be external to the market for water trading. Problems caused by external costs, public goods, poorly defined property rights, and other factors can result in situations for which trade cannot assure a socially optimal allocation of resources. In short, the value of natural amenities is not as great as it could be and the markets that do exist result in allocations that do not reflect the true value of the resources to society. One motivation for economic valuation of ecosystems is to attempt to correct for these market failures. We shall return to a discussion of causes of market failure in Section 16.6. But first we focus on the difference between market values and nonmarket values.

Our example described trade between two regions that resulted in the water resource attaining its highest value for both regions. The result was achieved because all assumptions necessary for trade fully represent the values that all individuals within each region are assumed to hold. As long as markets are perfect they reveal the value of goods and services that are traded in this manner. The market price is the marginal value that is equated for all individuals. It is said that the marginal value is *revealed* as the price that attains from the market process. Total values can then be calculated as the area beneath the demand curve and net value as the difference between total value and total costs. The net value, or net economic benefits, is represented by the triangle beneath the demand curve and above the supply curve. Net value is greatest at the point where marginal value is equal to the marginal cost. This basic result of microeconomic theory underlies the preoccupation that economists have for allowing free markets to determine the price and quantities of goods, including many supplied by the environment. The notion is that, as long as the markets are working correctly, that is, all values and all costs are fully incorporated into all decision making, the resulting allocation will be the one for which the value to society is highest.

However, in many cases, ecosystems and ecosystem services present situations in which the assumption that all costs and all values are completely incorporated into all decisions is not valid. In these cases, markets do not reveal the true value of goods that are traded. Also, goods and services that are not traded may provide substantial benefits that are not revealed at all.

For many ecosystem benefits, there are no markets, and where markets do exist, they often are unable to fully express the full value of the resources to society. If there are no markets for these environmental amenities, then can we still say that they have economic value? The simple answer is yes. Although the value of goods and services may be revealed in market transactions, markets are not a prerequisite for economic value. Recall, economic value arises when any good or service contributes to the well-being of society. If we are left better off knowing that a particular ecosystem is left intact so that species diversity can continue in an undisturbed state there, then we receive benefits from that knowledge. We can express the value of these benefits in terms of sacrifices that we may be willing to make to receive them. For example, we may be willing to live with fewer conveniences if we know that in return we would be assured that the ecosystem would be left intact. We may even be able to quantify the limit of exactly how far we would be willing to go in terms of sacrificing other conveniences. And because many of these other goods and services that we would be willing to sacrifice are valued in markets, we could express this limit in dollar terms. In this case, the dollar units we could use to reveal our willingness to pay to protect the ecosystem are representative of the sacrifice that we are willing to make in terms of other goods and services that we could receive with the same financial resources. We may be willing to forego \$500 per year in order to assure that the undeveloped public lands at the end of our block are not converted to residential lots. However, beyond \$500 per year, we would feel that other uses for the resources that the money represents would bring greater value, at the margin, than the knowledge that the habitat of that ecosystem had been preserved.

The problem is that nonmarket-valued ecosystem services would most likely be undersupplied if markets alone are relied on for allocation of ecosystems services. This would result in an overall diminution of the potential economic value of the ecosystem for society. Many of the benefits associated with preservation of natural ecosystems stem from nonmarket values. In this way, it could be predicted that uncorrected market activity would likely be biased toward extractive uses, rather than preservation of ecosystems. Thus many of the efforts to value ecosystems aim to identify and then quantify ecosystem values associated with preservation and other nonmarket benefits.

Dollar Units Provide the Convenience of Common Units of Measurement

In many contexts, a combination and comparison of physical quantities using different physical units may not provide enough input to decision making. How do we go about comparing the loss of 50 hectares of wetland to the gain in agricultural production from the farmland that was created? Do we compare the number of waterfowl nests that are lost, plus the number of recreational waterfowl hunting days plus the cubic meters per year of groundwater recharge foregone with the bushels of corn per year gained? What about the additional person-hours per year of employment gained by bringing the additional acreage into production or the addition to farm income?

The measurement units that are used to quantify the physical stocks and flows may not convey a consistent measure of the value of the resources to society. The value of an additional total allowable catch of 15,000 tons per year for menhaden would be quite different from the same amount of cod, a species that demands a higher price in consumer markets. The value of several thousand cubic meters per month of water recharging an aquifer is likely to be much greater in an arid, heavily developed region than in a region that is less heavily populated or with greater water reserves. The value of any additional recharge to an aquifer is likely to be greater in a region with no alternative sources for potable water than in a region with many alternative surface water sources. Measures of value take into account the interaction between physical properties, uses by society, scarcity of the resource in question relative to its own reserves, and the availability and cost of suitable substitutes.

The use of money facilitates trade in a complex social environment. The value of money depends on the value of the goods and services that it represents. Individuals' values for resources are revealed indirectly by the amounts of money that they are willing to pay for given market environment. Money is a medium for exchange in a complex market environment. It allows for a common unit to compare values of a myriad of different items and services, which are measured in a wide variety of physical units that are typically not directly comparable.

Physical Units of Ecosystem Impacts to be Valued Must be Appropriate to the Policy Context

Before we can ask how to measure ecosystem values, we first must understand what is measured and why and anticipate how this information is to be used. Typically, the notion of measuring ecosystem values arises in response to specific policy options or proposals. Thus it is often the case that the necessary policy-relevant information is not provided by answering the question, "What is the value of groundwater and surface protection services provided by this watershed?" but rather, "What would be the net change in value of water protection services effected by the proposed policy to limit clear-cutting in this watershed?" These are two very distinctly different questions, and the information obtained from each is not substitutable in terms of informing public policy.

Before any economic analysis of value can be achieved, the statement of the problem must include a precise definition of the ecosystem good or service that is being valued. A measure of ecosystem value must be policy relevant. Under what circumstance might these services be altered? Economic valuation requires a precise definition of how a proposed change in the ecosystem will affect clearly identified services received by society. Those who receive the value or bear the cost if the ecosystem service is altered must be identified. The next section focuses on the steps that lead up to applying valuation methods.

16.4 Nonmarket Valuation of Ecosystem Goods and Services

The goal of the remainder of this chapter is to describe circumstances that contribute to successful measurement of ecosystem values. This section briefly discusses considerations that need to be taken into account before any decision to conduct some form of ecosystem valuation is made. The purpose of the discussion is to describe issues that are encountered before a valuation study is set up. A large number of articles and books provide greater detail on these and related topics (e.g., Mitchell and Carson, 1989; Johnson and Johnson, 1990; Freeman, 1993). While most nonmarket valuation methods are, in theory, consistent with the economic concept of value, the application of methods and the usefulness of resulting value estimates can be problematic. Five basic steps that must be taken before any decisions can be made about the specific methods used in a valuation study are reviewed next.

The first step is to identify which ecosystem goods and services might be affected by the proposed policy. These may differ for a given type of ecosystem, depending on the nature of the interactions between the specific ecosystem and the human population. Thus generic lists of ecosystem services by type of ecosystem are only of very cursory use at this stage. For any given policy setting, a full accounting of relevant ecosystem goods and services must be compiled.

The second step is to determine how the units in which physical impacts are measured can be translated into the units that are relevant for human preferences and decision making. Often, it is conceptually easier for scientists to describe impacts of human activities on ecosystems in the physical units that are used to measure relationships within the ecosystem. However, these units may not be consistent with the units that individuals and firms are compelled to consider when deciding whether to undertake actions that would have an impact on the environment.[3] Since the goal of ecosystem valuation is to use estimates to develop policies and instruments to change incentive patterns and to encourage people to make decisions that are consistent with long-term interests of society, the units of analysis need to be relevant to human options and choices.[4] Heimlich et al. (1998) provide an excellent example for the case of wetlands in which they identify physical impacts, then identify the units that are appropriate for measuring how the physical impacts translate into human values, and then identify which human activities will be affected and by how much. They categorize and review nonmarket valuation estimates for wetlands from existing literature at this level.

Third, the scope of changes induced by the policy must be well understood. It is necessary to determine the extent of the potential changes in quantity or quality of the goods and services that might be affected by a policy. It is generally not useful to measure the total value of ecosystem services, unless the proposed policy would result in total loss of these services. Of greater value for informing decisions are measures of marginal changes in value that result from specific policy actions. Thus a given baseline level of the provision of the services must be established, along with the potential changes from this baseline. For the majority of situations in which ecosystems may be altered but not destroyed, valuation must be based on the marginal changes that would be expected to occur. A large source of bias can occur if a valuation estimate that is based on the total value of an ecosystem service is expressed as an average (by dividing over the number of units provided, for example) and then used as a proxy for a marginal value. If a marginal value is required, then the study design needs to assure that a marginal value is measured. For example, Rollins et al. (1997b) estimate the value of maintaining water quality for municipal drinking water services. The policy context required that they estimate the value of an incremental change in water quality. More precisely, they estimated the willingness to pay of users to maintain the level of quality to be consistent with health guidelines. A measure of total value would have provided information that would have been of little practical use in their policy context.

Fourth, the relevant population for each of the identified ecosystem values must be properly identified to avoid biased valuation estimates. The relevant population can vary, depending on which ecosystem services are considered. In the example of the public forestland taken out of commercial use, any potential changes in recreational use patterns may affect mostly local populations. The change in recreational benefits generated by the policy would depend on the size of the population in the region where the forest is located, the number of available substitutes for recreational land, and the change in the quality of the recreational experience. However, any changes in habitat that affect threatened or endangered species might be valued by a much larger population, most of which may not live in the region.

Fifth, the choice of valuation method must be appropriate to the circumstances of the proposed policy. For example, most travel cost models are not appropriate for determining marginal changes in the quality of a site for recreational purposes, but they work well for determining the economic value of creating a new site or the loss of totally eliminating a site. On the other hand, direct methods, which include contingent valuation and choice experiments, work well for marginal changes. Only direct methods are available for measuring existence, or passive use, values; it is simply not possible to use any kind of indirect method, such as a travel cost model for this purpose. While it is com-

[3]The issue of reconciling measurement units to bridge human values and ecosystem impacts can seem especially difficult in very complex ecosystems. This is one area of research that could receive more attention.

[4]The intention here is not to minimize the importance of measuring the significance of a physical impact on the overall function of a complex ecosystem, when some of these impacts or changes in overall function may not have an obvious role in human values. There are other ways that the term *ecosystem values* may be used, which refer purely to physical impacts. However, these notions of value are different from the definition here. More importantly, in terms of the goal of understanding and altering incentives that influence human interactions with ecosystems, a purely physical definition is of little use.

mon for the naïve to make sweeping statements that one method is generically better than another, the fact is that more well-informed practitioners are aware that the best method depends on the situation.

To summarize, to minimize problems that have affected using valuation to inform policy decisions, it is necessary to consider many details prior to making a commitment to any specific valuation procedure. Five of these details are identified as follows:

1. Identifying goods and services that may be affected by the proposed policy
2. Translating units of physical impacts into the units that are relevant for human preferences and decision making over these goods and services
3. Determining the potential scope of changes induced to these goods and services by the proposed policy
4. Identifying the relevant population for each of the identified ecosystem values
5. Assuring that the choice of valuation method is appropriate to the circumstances of the proposed policy

16.4.1 Categories of Values

To address the five previous points, it can be useful to refer to a checklist of categories of values. Not every policy situation will have values that fall in all categories. These categories are outlined next. Randall (1987) and Sarker and McKenney (1992), among others, provide more detailed explanations of the notion of such a total valuation framework as a starting point to establish the full range of economic values affected by a given policy.

Market Values

The values of environmental goods and services that are bought and sold in markets, such as wood, are accounted for if these markets are well-functioning markets. This is because quantities produced and consumed and prices of forest products relative to their next best alternatives are all jointly determined. Through this process, the price per unit for a forest product reveals what, on average, a typical consumer is willing to pay for one unit of the product, given how much that consumer already has consumed. Notice that the concept is "the last unit." The value of any good depends in part on the relative scarcity of the good, relative to other goods and services that might be purchased with the same financial and physical resources.

Production costs include wages paid to labor, capital tied up in equipment and infrastructure, variable costs of other inputs to production, the costs of hauling logs to mills, costs of harvesting timber and replanting forestlands, and the trees themselves. The proportion of the market value that is attributed to each input, calibrated in terms of one unit of input, is the marginal product of this input. Thus the value of labor to produce paper is termed the *marginal value product of labor*. In a well-functioning economy, the marginal value of an input is equal to the cost of the input to the production process. Thus the wage rate for mill employees reflects the value of labor to the production process. In a well-functioning market, therefore, it is only necessary to observe the price and quantity produced of the final end product to the consumer to determine the value of all the inputs to society. All inputs values are embodied in the final market price. It would be erroneous to measure the market value of building lumber and add to this the value of the wages paid to builders, to sawmill workers, and to timber forest workers, and so on. Such an approach would grossly overestimate the value of the forest product in the market.

However, if the market is not functioning properly, it can be determined that one or more of the potential costs of producing forest products may not be fully incorporated into market prices for forest products. If labor were not paid a wage that fully represented the value of its contribution to the final product, then the price of the final product would be too low. This is not likely to happen in the case of a mobile input like labor or capital, because these inputs would move to a different sector where they would receive a higher return. The mechanism that allows this mobility is the marketplace and the concept that a productive input will move to its highest-valued use. In the ideal, the market assures that all inputs are used in their most-valued use to society. But in natural resource sectors and many industrial sectors that use environmental inputs, the market alone may not adequately serve to allocate all inputs and outputs in the way that leaves society most well off. If inputs are not valued at their highest-valued use at the margin, then the full costs of production are not accounted for, the price is too low, and the amount produced for the market is too high. Thus, if water quality or air quality is an input to the production process that is not paid its full cost, then the produced products will have a market price that does not embody the true costs of these environmental inputs. The market price will be too low, and too much of the manufactured product will be produced and sold in markets, relative to a well-functioning market.

When it is determined that a market is well functioning, the contribution of ecosystem goods and services to the market prices and quantities can be used for the value of those goods and services. However, if markets do not fully take these into account, non-market valuation methods must be used.

Nonmarket Values

There are a variety of reasons why many ecosystem values are not expressed in terms of well-functioning markets. The property rights over ecosystem goods and services may be poorly developed, leading to open-access problems of the type described by Hardin's Tragedy of the Commons (Hardin and Badin, 1977). Often it is the nature of the resources themselves that make it difficult to exclude users. The atmosphere, waterways, scenic views, and other resources can be classified as nonexcludable public goods. The nonexclusive nature of these resources, in part, justifies public ownership and regulation of these types of goods as a means to prevent open-access overexploitation.

Many environmental amenities are nonrival in nature. This means that the use of the ecosystem service by one person does not diminish the amount that remains available for other people. It can easily be shown that the full value of nonrival goods is not fully revealed in markets, and thus in market prices and quantities, and that the market alone would underprovide them. Thus goods and services that are nonrival exhibit nonmarket values. Some values are not exhibited by the market simply because people hold value in knowing that ecosystems will continue to exist in a given state into perpetuity. People would be willing to express this value in terms of trade-offs that they would be willing to make to assure that the ecosystems are not altered, but there is no market available in which to express this willingness to accept trade-offs. The following categories are generally used to refer to the types of nonmarket values that flow from ecosystems.

Use Value

Use value refers to the benefits derived from direct contact with an ecosystem or its services. Recreational activities, ecosystem inputs to production processes, watershed protection, and other activities that involve direct or indirect contact with an ecosystem result in a use value. An indirect use value involves direct contact at some point in the process of creating the good or service, which directly produces benefits. Photos of scenic vistas on calendars, photos, books, and television shows about wildlife and other similar goods are examples of indirect use values of ecosystems.

Consumptive (Rival Goods) and Nonconsumptive (Nonrival Goods) Use Values

Benefits derived from use values may or may not involve consuming the good in question. If, in order to receive an ecosystem benefit, the level of benefits for other people is necessarily reduced, then it said that the benefit is consumptive, or rival. Examples of ecosystem benefits that are rival include hunting and extractive activities. Nonconsumptive or nonrival goods are those for which use by any one person does not appreciably reduce the use by any other person. Once the good exists, it provides a flow of benefits for many people at once. The addition of a marginal user does not reduce the marginal benefit for anyone else. The significance of nonrival goods is that there is a tendency for people to free-ride on the amount provided or protected by other people. Therefore, the full value of the good is not revealed in the marketplace, and the market alone may accord for too little provision of these goods. This characteristic of nonrival goods is the basic reason for public ownership of them. Another term for nonrival goods is *public goods*. As an economic definition, a public good is not synonymous with public ownership.

Option Value

In many circumstances, people may not be able to anticipate exactly when or even if they will use the resource in question. However, they receive value in knowing that the option exists, if they wish to exercise it. In this sense, the benefits that they gain aside from the actual use of the ecosystem are expressed as an option value. This concept is similar to a person's willingness to pay more for an insurance premium than the expected value of the event of an accident that the insurance would cover. The excess is the value to the person of avoiding the risk. An option value is the value to a person of avoiding to have to determine at any given time whether or not they will or will not use a resource in the future. It is the value in putting off the decision. A quasi-option value is a related concept. If the decision about use is to be put off until the future specifically because the potential user is awaiting more information about the benefits of future use, then the information itself has value. Thus quasi-option value is the value of putting off a decision specifically in anticipation of new information that will reduce uncertainty about the benefits of use in the future. Option and quasi-option values are difficult to measure. However, the concepts are especially significant in the case of potential decisions that could result in ecosystem losses that are of uncertain probability of occurring, of unknowable cost to society, and not reversible. Examples of projects that would threaten populations of endangered species readily come to mind. The probability that any one decision will result in loss may be unknowable, the potential cost to society of this loss may be unknowable, and, once the loss occurs, the decision is not reversible.

Nonuse (Existence) Value

Existence benefits flow to people who value the knowledge that an ecosystem continues to exist in a specific state into perpetuity. If the ecosystem were to be compromised so that it was permanently changed, then the full value of the ecosystem to society may be diminished.

16.4.2 It May Not be Necessary to Include All Values in an Assessment of a Proposed Policy Change

Any one action that affects an ecosystem could result in changes in several different categories of values. Depending on the nature of the policy question, however, it may not be necessary to measure all potential changes in values. For example, a policy might affect consumptive and nonconsumptive use values, as well as existence values and option values. Suppose that it can be determined on the basis of a subset of use values alone that the decision would cost more than it could generate in market benefits. Then, for policy needs, the number of goods and service changes that are valued with nonmarket valuation methods is sufficient. It is simply not necessary to apply methods to more categories of values.

16.5 Techniques to Measure Nonmarket Economic Values

A number of nonmarket valuation techniques have been developed. Since other authors, such as Smith (1992), offer overviews and appraisals of methodologies, this section will very briefly summarize the major categories of methods. In general, methods are divided into two classes: stated preference approaches and revealed preference approaches.

16.5.1 Stated Preference (Direct) Approaches

Stated preference approaches are also known as direct methods for nonmarket valuation. These methods allow people to state a value that directly relates to the specific environmental changes that would be incurred by the policy in question. Because the methods are used prior to a policy action, these methods require the use of hypothetical scenarios. The advantage of direct methods is that the valuation study can be developed to very closely approximate the same marginal changes that a given policy would generate. The two basic types of stated preference methods are contingent valuation and choice experiments. Contingent valuation is more widely recognized and used for nonmarket valuation. Choice experiments have evolved from marketing tools based on conjoint analysis and have only very recently been modified so that they can be used to estimate willingness to pay for specific attributes of nonmarketed environmental amenities. The new generation of choice experiments is based on random utility theory and can provide welfare measures that may be consistent with economic theory, although their use for this purpose is somewhat experimental at this time. Contingent valuation is the only method accepted by the U.S. government for measurement of existence values.

One criticism of stated preference methods is that it is not possible to determine whether people really would react to a real situation in the same manner in which they react to a hypothetical situation in an experimental context. This criticism has been the subject of a large body of research that appears to suggest, overall, that a well-designed scenario, questionnaire, information materials, and statistical procedures can minimize bias. In addition, a well-designed study can include a number of internal consistency tests to detect biases and other anomalies that would suggest that responses are not consistent with economic theory (Arrow et al., 1993; Diamond, 1996; Rollins and Lyke, 1998).

A tendency toward professional resistance within the economics profession against stated preference methods arises because they run counter to the tendency of empirical economists to develop predictive models using the notion of *revealed preference*. In the case of nonmarket valuation, revealed preference models are indirect models that use observable behavior in actual markets, which are in some way connected to the value that is the target of valuation. Many economists counter this argument by suggesting that revealed preferences are difficult to measure because an extrapolation must be made from the observation of particular choices made by the consumer to general conclusions about preferences. This implies that the researcher must be able to hypothesize the functional form and underlying theory that connect the observed market behavior with the unobservable and still hypothetical welfare change that would result from the policy scenario under consideration. As well, direct methods allow the researcher to hypothesize assumptions about the respondents' preferences and then test these assumptions to verify a respondent's motives. Revealed preference methods (also known as indirect methods, because they indirectly estimate demand for a nonmarket good based on some observed demand for a related market good) do not easily allow for these tests. Finally, arguments in favor of direct methods include the fact that there is a real distinction between private market goods and public goods, which make revealed preference methods harder to apply, especially in the case of public goods, which are often of national rather than local scope.

Contingent Valuation

The contingent valuation method (CVM) uses carefully worded surveys to gather primary data that is analyzed by an appropriate econometric model to predict what people would be willing to pay (or willing to accept in compensation) for a specified change in the price or the quality of an environmental amenity. The CVM has been shown to work especially well for recreational use values for goods with which respondents have had prior experience. The term *contingent* refers to the fact that the valuation of the good is contingent on the hypothetical assumption of a plausible market and method of payment for the good. A simple CVM study might consist of a detailed description of a good, followed by a question asking the respondent whether he or she would be willing to pay a given price for provision of the good. The yes or no responses to a range of prices can be used to establish a cumulative probability distribution of a representative individual's probability of answering yes. The area beneath is the expected value of a yes response and thus the average willingness to pay (WTP) for the environmental change in question. This example is based on a dichotomous choice model CVM, since respondents are given the option to say yes or no to given prices. Contingent valuation can be accomplished with other models as well, including those that ask follow-up questions after the first response.

The benefit to individuals is measured in terms of their maximum willingness to pay to achieve the change, under the assumption that the respondents would not be willing to pay more than the benefit that they would receive. Statistical models are used to estimate mean WTP for an individual from the survey sample. These estimates can then be aggregated over the relevant population to determine an estimate of benefits generated by a nonmarket amenity. The interested reader is referred to Mitchell and Carson (1989) for a discussion of the theoretical underpinnings of CVM and practical suggestions on how to perform CVM. Freeman (1993) provides a good review of the economic theory regarding the use of WTP to measure individual and aggregate benefits of nonmarket amenities.

As of 1993, no less than 1500 studies and papers using the CVM were documented by Carson et al. (1993), many of them applied to recreational use value. Since then, much of the theoretical and methodological advances have focused on applying CVM to measure nonuse values. One of the most well-known examinations of the CVM was that carried out by the National Oceanic and Atmospheric Administration (NOAA). At the time of the *Exxon Valdez* spill at Prince William Sound, Alaska, in March 1989, the U.S. government had considered CVM and valuation estimates from CVM as acceptable to determine compensation levels in liability cases.

The Exxon Corporation, which was responsible for the *Exxon Valdez* oil spill, hired two consulting economists to expose any reasons why CVM should not be used to estimate lost passive use values.[5] The CVM was, and still is, considered the only method acceptable for measurement of nonuse values. The stakes were high, since estimates of lost nonuse values would greatly increase the amount that the corporation could be held liable for. The stakes involved with the state of Alaska's lawsuit, consulting income, the attack on the academic literature supporting the theoretical basis of CVM, and academic reputations, precipitated a heated debate that divided the discipline. At the time, it was not uncommon to find Exxon representatives in attendance at academic conferences ready to criticize any application of CVM. The potential for public spectacle substantially boosted attendance at these sessions as the lines between theory, practice, academics, and politics became blurred.

The U.S. government reacted by creating the NOAA panel to review the critiques of CVM and

[5]They used the term *passive use values* to refer to existence values and nonuse values.

to make recommendations about whether estimates based on the method were valid and reliable enough to be used for applied work and damage assessments in liability claims. Kenneth Arrow and Robert Solow, two Nobel laureates in microeconomic and macroeconomic theory respectively, headed the panel. Since neither had been involved in CVM studies previously, they were considered objective chairs. After a lengthy review, the panel concluded the "contingent valuation method can produce results that appear to be consistent with assumptions of rational choice" (Arrow et al., 1993, p. 4604). They asserted that "contingent valuation studies convey useful information (that) can produce estimates reliable enough to be the starting point of a judicial process of damage assessment including lost passive use values" (Arrow et al., 1993, p. 4610). Thus the federal government continues to accept CVM today for assessing lost environmental values in damage liability cases in the United States.

While the hearings focused on the theoretical basis of CVM, it was also clear that methodological issues made a significant impact on the reliability of CVM. The panel recommended that, although estimates of existence values derived through CVM studies can, in theory, be reliable enough, the method must be conscientiously applied to meet rigorous standards. Otherwise, the numbers CVM produces could be essentially meaningless. The panel noted that, since no other generally accepted means of determining existence values exists, CVM takes on an "added importance in the light of the impossibility of validating externally the results of CV studies" (Arrow et al., 1993, p. 4603). For this reason, a system of internal checks on the findings is especially important to build into any study. The aftermath of the NOAA hearings resulted in a set of recommendations for future CVM studies. And while current research appears to indicate that some of these recommendations may be open to debate as methods are refined, it has become standard for CVM studies performed since then to explicitly address these recommendations.

Choice Experiments

Choice experiments are a relatively new valuation method, but are based on the same theoretical economic and behavioral principles regarding choice as contingent valuation. In terms of the modeling approaches used for each method, contingent valuation can be regarded as a simple version of a choice experiment. Since both methods are based on random utility models, some researchers have successfully combined contingent valuation and choice experiment response data to enhance the interpretive value of statistical findings (Boxall et al., 1996).

Choice experiments were developed from conjoint analysis, which is used in marketing to determine the attributes of a product likely to result in the largest market share for the product. Conjoint analysis is based on surveys that required potential consumers to rank alternative options in terms of combinations of features in a good or service. The difference between conjoint and choice experiments is that conjoint analyses ask a respondent to reveal their preferences by ranking alternatives. A choice experiment presents the respondent with several paired groups of selected profiles of a good. Each profile is composed of several attributes. For each pair, the respondent chooses which of the two options is preferred. Since the respondent is given several sets of paired profiles, it is possible to statistically generate the worth of each attribute to the individual. From these "part-worths," it is possible to calculate the value of any potential combination of attributes. Thus the value of any combination of attributes is defined as the sum of the partial values of each of the component parts. This is a rather simplistic description of a choice experiment.

This area of current valuation research potentially holds much promise. One particularly interesting use of choice experiments is in the development of joint models using both choice experiments and revealed preference models, such as in Adamowicz et al. (1997), and using contingent valuation, such as in Boxall et al. (1996). In both of these joint approaches, the use of choice experiments appears to enhance the value of the information from the method with which it is paired.

Contingent valuation focuses on the willingness to pay (or accept compensation for) a specific quality or quantity change for one well-specified good, whereas a choice experiment defines the good in terms of a number of characteristics and then focuses on the trade-offs between combinations of characteristics. Choice experiments have most commonly been used in marketing studies to determine what combination of characteristics of a new product would receive the greatest share of the market for the product. The technique is most well suited for goods that are very well defined and very familiar to consumers. For example, a choice experiment is particularly well suited to determine what bundle of options for long-distance telephone services are best for specific markets or to determine what bundles of cable TV channels should be targeted to specific markets. The nature of a choice experiment is to make a good generic, in a sense, by defining it as being composed of any combination of attributes that the consumer most prefers. On the other hand, one feature of a unique good is that it cannot exist in any other form. The value of the good lies in preserving it as it is. For this reason, the use of choice experiments for unique environmental goods or for existence values may be inconsistent with how a person views a unique good.

For environmental goods and services that are not unique, such as many forms of recreational fishing, boating, hiking, camping, and other experiences, choice experiments may prove to be exceptionally valuable since they allow for large numbers of combinations of site attributes in one experimental design. Because they are designed to be somewhat generic, the results of choice experiments may be well suited to apply to many sites.

16.5.2 Revealed Preference, or Indirect, Methods

Indirect methods rely on observing how people behave in markets that are indirectly related to the ecosystem good or service that is to be valued. Some difficulties with indirect methods are that they imply that it is possible to model precisely how the behavior in the observed markets is related to the specific policy impacts on the ecosystem goods and services. Such modeling sometimes requires heroic assumptions about individual utility functions and other details that are generally unknowable with certainty. A large number of empirical papers focus on the sensitivity of models used in indirect methods to alternative specifications of utility functions, the value of time, and other assumptions. It is generally a good idea to build into any study using indirect methods analyses that test the sensitivity of results to alternative specifications. Another limitation of indirect methods is that it may not be possible to construct a one-to-one correspondence between the activity that is valued and the value of the economic changes induced by the proposed policy. An advantage of many indirect methods is that often the necessary data can be collected from information that is already available from secondary sources. In contrast, direct methods require implementation of surveys to collect primary data.

Travel Cost Models

There are a number of types of travel cost models, ranging from basic zonal travel cost models to more sophisticated random utility travel cost models. The basic concept for all travel costs models is that the costs associated with travel time and distance to a potential recreation site in part determine how likely an individual is to visit that site. However, there is a great deal of variety in the types of models. Travel cost models allow for inclusion of substitute sites and can be designed to model how a person chooses from among multiple sites with different characteristics. Thus travel cost models are suited for valuation of recreational use of ecosystems and have been widely used to value recreational experiences, including fishing, hunting, hiking, boating, and swimming (e.g., see Caulkins et al., 1986; Englin and Mendelsohn, 1991; Boxall, 1995).

The majority of travel cost models are best suited for estimating the value of adding a new site, rather than for valuing marginal quality changes in existing sites. Well-designed studies do a reasonable job at predicting visitation at sites, provided substitute sites are properly accounted for in the model. Thus, if the policy issue under consideration involves estimating the value of and predicted number of visits to a proposed new recreational site, a travel cost model is likely to work well at a reasonable cost. However, if the policy context is one that explicitly involves decisions that will cause changes in the quality of existing sites, then travel cost models may not be as sensitive as stated preference models or some combination of a stated preference and travel cost model. Freeman (1993) provides a detailed summary of the types of travel cost models and the advantages and disadvantages of each.

Hedonic Pricing

Hedonic pricing is often used to value environmental amenities (or disamenities such as pollution) in the context of residential home markets (e.g., Rosen, 1974; Brown and Rosen, 1982; Palmquist, 1984, 1991). Hedonic wage models are used to value the contribution of environmental amenities and disamenities to explain wage differentials (e.g., see Cropper and Arriga-Salinas, 1980; Smith and Gilbert, 1984; Hoehn et al., 1987). The basic principle is that, along with various features of a house, such as square feet of living space, number of rooms, and proximity to schools, environmental amenities also contribute to purchase consumer decisions. Thus a home in a desirable location may carry a premium, which is related to the value of natural amenities. The value of these amenities can in theory be teased out of market data on housing purchases if the data sets are rich enough to include houses with similar features over a range of different levels of the environmental amenity in question. As with other indirect methods, a hedonic model can be developed using secondary data. For the model to be successful, the data must include enough variations of combinations of characteristics. The price of a particular model as a function of the quantities of its various characteristics is a hedonic price function. The partial change in the hedonic price function with respect to a change in one characteristic is the implicit price of that characteristic. Thus a change in the hedonic price function with respect to a change in the level of water quality would be interpreted as the additional cost necessary for a consumer to purchase a unit of water quality.

Hedonic pricing results can be difficult to estimate and to interpret, because the comparisons across purchase decisions assume that the differences in house prices can be attributed to different levels of the amenity. Since the comparisons are interpersonal, there are potentially any number of reasons why consumers may choose to make the purchases that they do. The preference functions of different individuals may not be comparable, but the analyst is not able to observe sufficient information about other factors that may influence an individual's choice. Freeman (1993) reviews many of the problems encountered with estimation of hedonic models.

16.5.3 Benefits Transfers

A third type of valuation method is the benefits transfer. In reality, a benefits transfer is not a proper valuation method, but rather a procedure that uses valuation estimates from other study sites to apply to a given policy site. One reason for performing a benefits transfer is to make a first-pass estimate on the potential range of values that might be attributed to a specific policy proposal in order to determine whether a full valuation study is warranted. Few economists now suggest that estimates from benefits transfer are reliable enough to inform a final decision for which the potential gain or loss of nonmarket values might be very great or in situations in which the decisions are not reversible (Downing and Ozuma, 1996; Kirchoff et al., 1997). Rollins and Ivy (1997) provide a practical example of the steps involved in a benefits transfer approach to assessing the nonmarket values of a policy proposal to construct a dam and irrigation system in a prairie and rangeland ecosystem in North America.

The concept of benefits transfer evolved from an initially naïve view that valuation measures of a given ecosystem-based good or service are applicable to situations other than the one in which the estimates were measured. This idea was implicitly adopted by noneconomists who knew enough about nonmarket values to realize the significance of omitting them from an analysis, but not enough about the underlying concept of economic value to understand that, in general, estimates are not transferable from one policy scenario to another. At a time when budgets for environmental impact assessments and cost–benefit analyses were weighted toward engineering studies and the notion of nonmarket valuation was still considered experimental, a benefits transfer approach may have been considered one of the few means to attract attention to the importance of nonmarket values. Over time, benefits transfers came to be seen as a fast and inexpensive substitute for nonmarket valuation.

Academic economists have generally agreed that benefits transfer is not valuation and is an inferior means to estimate nonmarket values. However, most are also aware that benefits transfers will continue to be used in applied policy, due to the speed and cost relative to the cost and time involved for proper valuation studies. Therefore, much research focuses on developing criteria to determine the types of circumstances for which benefits transfer is more or less likely to be a satisfactory alternative to valuation (AERE, 1992; Atkinson et al., 1992; Boyle and Bergstrom, 1992; Brookshire and Neill, 1992; Desvousges et al., 1992; Loomis, 1992; McConnel, 1992; Walsh et al., 1992; Smith, 1993; Boyle et al., 1994; Kask and Shogren, 1994; Pearce et al., 1994; Hagler Bailly Consulting, Inc., 1995; Loomis et al., 1995; VandenBerg et al., 1995; Bergstrom, 1996; Rollins and Ivy, 1997; Desvousges et al., 1999). One common conclusion from most of these studies is that there is no one set protocol for conducting benefits transfer. That is, there can be no cookbook approach because the circumstances vary so widely from one policy situation to another.

16.5.4 Choosing the Technique to Fit the Situation

There are often obvious circumstances that suggest the most suitable technique for a given valuation problem. For example, contingent valuation is the only accepted method available for measuring existence values. We may rule out a benefits transfer approach to provide information for a benefit–cost analysis of a proposed policy decision if the possible cost of an error is very great relative to the potential costs savings from performing a benefits transfer instead of a primary valuation study. However, in many situations the choice of method is not always obvious. In general, revealed preference methods, such as travel cost models, have the advantage that they are often less expensive to carry out than are stated preference methods.

16.6 Examples

Two examples of nonmarket valuation studies with the policy scenarios that they were intended to inform are given next. Two common themes in these examples are the importance of the policy context of the valuation problem and the importance of a well-planned experimental design. The policy context defines the precise good that is to be valued and the units and ranges of use that are relevant for policy alternatives. Planning requires time in the initial phases. However, as these examples illustrate, attention to design enhances the reliability and validity of estimates.

The first example is the measurement of the value of wilderness canoeing in Ontario Provincial Parks. The estimates were expected to provide information for a number of policy decisions. These potential policy actions included increasing the quotas on wilderness canoe routes to increase revenues from this activity as a way to address budget shortfalls, increasing the backcountry user fees, and increasing the area of wilderness-zoned areas in Ontario Parks. The second example is a study that measured existence values of creating new national parks in extremely remote regions. These parks would not be developed for recreation or visitation, but instead for ecosystem preservation. The interesting feature of this example is that a total of ten parks is required to fulfill the park system goal, whereas four were specifically proposed in the policy proposal. The study was designed to measure the value of the four parks relative to the goal of creating ten parks. The results indicate that the marginal value of a new park decreased as more parks were added to the system. This would be consistent with a perception among the public that as more and more areas were devoted to preservation, at the margin, additional protected areas would be less scarce than other potential uses of the same financial and capital public resources.

16.6.1 Example 1: Valuation of Wilderness Canoeing in Ontario

In 1993, the Ontario Ministry of Natural Resources was faced with large budget cuts and increasing demand by recreational users of provincial lands. One method discussed to meet an increasing shortfall was the relaxation of daily quotas to wilderness areas. It seemed reasonable that, if there was increasing demand, allowing more users per unit area would generate more user fees per unit area. The possibility of increasing the fees while maintaining or even further restricting quotas was considered to be politically difficult. One problem with this view was that the quest for increased operating revenues obscures the true economic value of the recreational resources. Past work on the effects of congestion on the value of recreational experiences suggests that by relaxing quotas the value per recreational user day may be expected to decline. Thus consumer surplus per trip day, or the value of individual benefits over and above the expenses for a trip, may decline. So, while more user days are experienced, the overall value per user day declines. It is possible that greater overall value might be achieved by increasing user fees per user day and maintaining or even further restricting day use quotas for wilderness areas.

A second question relates to the reliability of the estimated value of a trip day for trips of different duration. Numerous studies of recreational values of natural ecosystems report average values of a trip day. These values are often multiplied by the average number of user days at a site for an estimate of the recreational value of a site. This question is relevant when considering how to optimally apply a quota to a wilderness recreation area. Is it better to allocate a quota of user days per season by reducing the number of trips, and allowing trips of longer duration or to increase the overall number of people who may take trips, but limit the days per trip? In the latter case, a larger number of people per season are able to enjoy the area, but their trips may be shorter than they wish. If the marginal value per trip day for a single trip is constant, then each day is worth the same as every other day. In this case the answer to the allocational question may be an issue of how the benefits of the resource are to be distributed over potential users. On the other hand, if the marginal value of a trip day declines for days during a trip, then limiting the length of trips and allowing a larger number of shorter trips may increase the overall value of a site.

Rollins et al. (1997a) designed a study to address both willingness to pay for alternative crowding conditions and over trips of different durations. They determined that the marginal value for a trip day for wilderness canoeing in three of Ontario's most popular backcountry provincial parks was about $60 per user day, on average. However, when broken down the value on a day by day basis, the highest-valued days were the first three days and then the value per day dropped off considerably. This would support rationing backcountry use in part by limiting the duration of individual trips, given the particular attributes of the site.

The same study was also designed to determine how much recreational wilderness users were willing to pay per user day for wilderness canoe trips under actual crowding conditions that they experienced in the three Ontario wilderness destinations. The same respondents were split into two samples for a second set of questions. The first half were asked their willingness to pay for the same trip under the condition that they saw half as many other groups per day as they actually experienced. The other half were asked their willingness to pay for the same trip if they could expect to see double the number of other groups as they experienced. The actual crowding levels included a range of no other groups per day to about ten groups per day, on average. The willingness to pay estimates were compared over these ranges. Although expectations over what levels of crowding were expected prior to the trip were highly correlated with willingness to pay, it was still apparent that, all else equal, recreational users were willing to pay for reduced levels of crowding. So, overall, users were willing to pay on average about \$60 per day for existing crowding levels, and they were willing to increase that amount if guaranteed fewer encounters with other users. At the same time, the existing daily backcountry user fee was \$4.25 per day. The results indicated that user fees could be increased substantially before many existing users would choose to no longer visit the areas. This suggests that one possible strategy to increase value to users and also generate revenues for the province to use to maintain these areas, might be to increase user fees, limit trip duration, and reduce the existing quota to guarantee fewer encounters with other users. This strategy is very different from one of maintaining existing fee levels, but increasing overall user days by reducing quotas.

16.6.2 Example 2: Existence Values for Creation of New National Parks in the Northwest Territories

This example illustrates the importance of precisely defining the good to be valued and of understanding that the marginal value of a good depends on the context in which the good is offered and the number already available. The simple policy scenario was to determine the value to Canadians of creating four new parks to protect tundra ecosystems in the Northwest Territories. Parks Canada had already spent more than a decade in the planning stages of each of the four proposed parks, and the agency was very close to making its final recommendations. In situations with previous new parks, the economic analysis consisted almost exclusively of economic impact analysis of the effect of park creation on local economies. However, since these parks would be created for preservation of habitat and tourism would be minimal, at best, the value of each of the four new parks, as measured by impact analysis, appeared to be very low. Indeed, an impact analysis of the impact of the alternative to new park creation, mineral exploration and mining, would almost assuredly result in greater economic impacts. Clearly, the methods that Parks Canada had found adequate for previous situations did not appear to be suitable for the four proposed new parks.

The problem was that the policy makers at Parks Canada who had been relying on impact analysis were mistaken in their belief that impact analyses benefits measures were equivalent to measures of the values of new parks. As described in an earlier section of this chapter, impact analysis measures expenditures associated with an activity or the area associated with the supply curve. These new parks would actually cost very little to maintain, since no infrastructure would be developed for tourism and the lands were already in government ownership. On the other hand, a true measure of the economic value of each new park could only be determined by the area associated with the demand for the park. Total willingness to pay for the new parks, summed over all Canadians, would be an appropriate measure of the economic value of the parks. A key feature of the Northwest Territories parks is that they would not be developed for recreational use, but rather for protection of habitat. Thus the main category of value for these parks is the existence value that people hold for the knowledge that these examples of tundra ecosystems would remain intact for the indefinite future. The one nonmarket valuation method accepted for existence values is contingent valuation. Thus the choice of method was clear.

The study resulted in the first estimate of an existence value for a Canadian National Park. Parks Canada initially requested that we concentrate our efforts specifically on the four proposed parks in the Northwest Territories. However, as we learned more about the policy context, it was clear that the actual environmental good to be valued might not be as well defined as it appeared to be. The four parks were part of a larger system of National Parks. The system as a whole was defined to be complete when each of Canada's 39 distinct Natural Regions was represented by an example that was considered large enough to sustain most of the natural processes characteristics of the Natural Region. At the time, 29 of the 39 Natural Regions were already represented by National Parks, and there were 10 remaining to be included in the system. The four proposed parks would each be representative of four Natural Regions that were not yet represented.

This policy context is significant in terms of the definition of the good. The relevant good that the average Canadian resident values may be the features of the four specific parks, or the good may be the contribution that the park makes to the completion of the goal to represent each of the 39 Natural Regions within the National Park System. The relevant good may be some combination of the two. If a survey instrument focused exclusively on the features of the specific parks, outside of the context of the 39 Natural Regions and the fact that the park in question represented the thirtieth Natural Region in the system, then the good might be misspecified. It may be that individuals who do not care to know the details of specific parks do place a value on the knowledge that remnant examples of distinct natural regions are preserved within the system.

Additionally, the fact that there were still ten Natural Regions to incorporate into the system might be significant, because the marginal value of the thirtieth park, in terms of completing the system, may be different from the marginal value of the park 32 or park 38, or park 39. Economic theory could support either increasing or decreasing marginal values for additional parks, depending upon the situation. For example, if completion of the system generated great value for people, park 39 may be worth more to people than the park 30, all else equal. On the other hand, the additional value of creating successive new parks in very remote places, when compared with the demand for other social programs and public goods, may become less and less, as more of the Natural Regions are represented. It may be that the specific features of each park are of less importance than the number of natural regions that are already represented in the system. If this was the case, then it was feasible that the value of each of the four proposed parks might depend on the order in which they would be created. That is, if the marginal value of each successive park added to the system were declining, then park A may appear to be of greater value if it was the thirtieth park created, rather than if it were the thirty-third park created. Since we could not rule out this possibility, we designed a valuation study that asked each respondent to value one, two, four, or ten parks. The experimental design called for the sample to be split many times so that the four parks were rotated. The goal was to estimate the value for each new park in various combinations of new park creation.

As Rollins and Lyke (1998) demonstrate, it turned out that the order mattered more than the specific ecological features of the individual parks. That is, the demand for new park creation followed a diminishing marginal value relationship, which was quite pronounced. Individual Canadian households were willing to pay, on average, about $104, in terms of a one-time payment, for the thirtieth park (the first of the four Northwest Territory parks), regardless of which of the four parks was created first. The difference between the Natural Region 30 and Natural Region 31 was about $60 per household, which was statistically significant. However, the difference between Natural Regions 33 and 39 was not statistically significant. These results illustrate, first, that the definition of the good may not be as clear as we would first assume and, second, that the concept of marginal value is extremely relevant in a policy context. The economic value of a given ecosystem to society is relative to the quantity and quality of similar ecosystems that already exist and relative to other needs that compete for the same financial resources. Park A is worth relatively less to people if Parks B and C have already been created, even though they are different parks. This is one of many reasons why we may not generally assume that it is possible to use an ecosystem value that has been estimated in one context to serve as a proxy for other contexts.

16.6.3 Common Pitfalls

Several of the most common pitfalls that plague nonmarket valuation of ecosystem goods and services include goods defined in a way that is not relevant for the policy context, attempting to measure total values when marginal values are needed, confusing average values for marginal values, improper bases for comparison of values, confusion with regard to the precise definition of the good, and using a valuation estimate from one site to serve as a simple proxy for other sites and circumstances. This discussion has touched on several economic concepts that explain some of these pitfalls. A common reason for most improper uses of ecosystem valuation is a lack of understanding of the limitations of the underlying economic theory and methods.

On the other hand, the potential to misuse economic valuation estimates of ecosystems may exist among some who are well informed about economic theory and methods. The economic approach values environmental resources in terms of their perceived value to humans. Yet we live in a world of great uncertainty and limited information about how environmental amenities interact with one another and about humanity's reliance on ecosystems. Probability distributions can be assigned and measures of risk can be calculated to help to describe the effects of uncertainty on the values of ecosystems. However, extreme caution should be used whenever using valuation information in decision making that would compromise a rare ecosystem in a way that is potentially irreversible.

16.7 References

Adamowicz, W.; Swait, J.; Boxall, P.; Louviere, J.; Williams, M. 1997. Perceptions versus objective measures of environmental quality in combined revealed and stated preference models of environmental valuation. *J. Environ. Econ. Manage.* 32(1):65–84.

Arrow, K; Solow, R.; Portney, P. R.; Leamer, E. E.; Radner, R.; Schuman, H. 1993. Report of the NOAA Panel on contingent valuation. *Federal Register* 58:4601–4614.

AERE (Association of Environmental and Resource Economics). 1992. *Benefits transfer: procedures, problems, and research needs. Workshop proceedings.* Snowbird, Utah, June 3–5, 1992.

Atkinson, S. E.; Crocker, T. D.; Shore, J. F. 1992. Bayesian exchangeability, benefit transfer, and research efficiency. *Water Resour. Res.* 28(3):715–722.

Bergstrom, J. C. 1996. *Current status of benefits transfer in the U.S.: a review.* Faculty Series 96-9. Atlanta, GA: Department of Agricultural and Applied Economics, University of Georgia.

Boxall, P. C. 1995. The economic value of lottery-rationed recreational hunting. *Can. J. Agr. Econ.* 43(1):119–31.

Boxall, P. C.; Adamowicz, W. L.; Swait, J.; Williams, M.; Louviere, J. 1996. A comparison of stated preference methods for environmental valuation. *Ecol. Econ.* 18(3):243–253.

Boyle, K. J.; Bergstrom, J. C. 1992. Benefit transfer studies: myths, pragmatism, and idealism. *Water Resour. Res.* 28(3):657–663.

Boyle, K. J.; Poe, G. L.; Bergstrom, J. C. 1994. What do we know about groundwater values? Preliminary implications from a meta analysis of contingent valuation studies. *Amer. J. Agr. Econ.* 76(5):1055–1061.

Braden, J. B.; Kolstad, C. D., editors. 1991. *Measuring the demand for environmental quality.* New York: Elsevier/North-Holland.

Brookshire, D. S.; Neill, H. R. 1992. Benefit transfers: conceptual and empirical issues. *Water Resour. Res.* 28(3):651–655.

Brown, J. M; Rosen, H. S. 1982. On the estimation of structural hedonic price models. *Econometrica* 50(3): 765–768.

Carson, R. T.; Wright, J.; Carson, N.; Alberini A.; Flores, N. 1993. *A bibliography of contingent valuation studies and papers.* La Jolla, CA: Natural Resource Damage Assessment, Inc.

Caulkins, P.; Bishop, R. C.; Bouwes, N. W. 1986. The travel cost model for lake recreation: a comparison of two methods for incorporating site quality and substitution effects. *Amer. J. Agr. Econ.* 68(2):291–297.

Cropper, M. L.; Arriaga-Salinas, A. S. 1980. Inner-city wage differentials and the value of air quality. *J. Urban Econ.* 8(3):236–254.

Cummings, R. G.; Brookshire, D. S.; Schulze, W. D. 1986. *Valuing environmental goods: an assessment of the contingent valuation method.* Totowa, NJ: Rowan and Allanheld.

Desvousges, W. H.; Naughton, M. C.; Parsons, G. R. 1992. Benefit transfer: conceptual problems in estimating water quality benefits using existing studies. *Water Resour. Res.* 28(3):675–683.

Desvousges, W. H.; Johnson, F. R.; Banzhaf, H. S. 1999. *Environmental policy analysis with limited information: principles and applications of the transfer method.* Cheltenham, UK: Edward Elgar.

Diamond, P. A. 1996. Testing the internal consistency of contingent valuation surveys. *J. Environ. Econ. Manage.* 30:337–347.

Downing, M.; Ozuma, T. 1996. Testing the reliability of the benefit transfer approach. *J. Environ. Econ. Manage.* 30(5):316–322.

Englin, J.; Mendelsohn, R. 1991. A hedonic travel cost analysis for valuation of multiple components of site quality: the recreational value of forest management. *J. Environ. Econ. Manage.* 21(3):275–290.

Environment Canada. 1998. *Environmental valuation reference inventory. A Web-based tool* (http:www.evri.ec.gc.ca/EVRI/).

Freeman, A. M. 1993. *The measurement of environmental and resource values: theory and methods.* Washington, DC: Resources for the Future.

Hagler Bailly Consulting, Inc. 1995. *Economic evaluation of environmental impacts: a workbook.* Prepared for the Asian Development Bank. Boulder, CO: Hagler Bailly Consulting, Inc.

Hardin, G.; Baden, J., editors. 1977. *Managing the commons.* San Francisco: W. H. Freeman.

Hausman, J. A., editor. 1993. *Contingent valuation: a critical assessment.* Amsterdam: Elsevier/North-Holland.

Heimlich, R. E.; Wiebe, K. D.; Claassen, R.; Gadsby, D.; House, R. M. 1998. *Wetlands and agriculture: private interests and public benefits.* Agriculture Economic Report No. 765. Washington, DC: U.S. Dept. Agric., Resour. Econ. Div., Econ. Res. Serv.

Hoehn, J. P.; Berger, M. C.; Blomquist, G. C. 1987. A hedonic model of interregional wages, rents and amenity values. *J. Reg. Sci.* 27(4):605–620.

Johnson R. L.; Johnson, G. V., editors. 1990. *Economic valuation of natural resources: issues, theory and applications.* Boulder, CO: Westview Press.

Kask, S. B.; Shogren, J. F. 1994. Benefit transfer protocol for long-term health risk valuation: a case of surface water contamination. *Water Resour. Res.* 30(10).

Kirchoff, S.; Colby, B.; LaFrance, J. 1997. Evaluating the performance of benefit transfer: an empirical inquiry. *J. Environ. Econ. Manage.* 33(5):75–93.

Kneese, A. V.; Schulze, W. D. 1985. Ethics and environmental economics. In: Kneese, A.; Sweeney, J. L., eds. *Handbook of natural resource and energy economics*, vol. 1. Amsterdam: Elsevier/North Holland.

Loomis, J. B. 1992. The evolution of a more rigorous approach to benefit transfer: benefit function transfer. *Water Resour. Res.* 28(3):701–705.

Loomis, J. B.; Roach, B.; Ward, F.; Ready, R. 1995. Testing transferability of recreation demand models across regions: a study of Corps of Engineer reservoirs. *Water Resour. Res.* 33(3).

McConnel, K. E. 1992. Model building and judgement: implications for benefit transfers with travel cost models. *Water Resour. Res.* 28(3):695–700.

Mitchell, Carson, R. C. 1989. *Using surveys to value public goods: the contingent valuation method.* Washington, DC: Resources for the Future.

Norton, B. 1987. *Why preserve natural variety?* Princeton, NJ: Princeton University Press.

Palmquist, R. B. 1984. Estimating the demand for characteristics of housing. *Rev. Econ. Statist.* 64(3):224–232.

Palmquist, R. B. 1991. Hedonic methods. In: Braden, J. B.; Kolstad, C., eds. *Measuring the demand for environmental quality*. Amsterdam: Elsevier/North-Holland.

Pearce, D. W. 1993. *Economic values and the natural world.* London: Earthscan Publications.

Pearce, D.; Whittington, D.; Georgiou, S.; James, D. 1994. *Project and policy appraisal: integrating economics and the environment.* Chapter 10, prepared for the Organization for Economic Co-operation and Development.

Randall, A. 1987. *Resource economics: an economic approach to natural resource and environmental policy.* New York: John Wiley & Sons.

Randall, A.; Peterson, G. 1984. The value of wildlife benefits: an overview. In: Peterson, G. L.; Randall, A., eds. *Valuation of wildland resource benefits.* Boulder, CO: Westview Press.

Randall, A.; Stoll, J. 1983. Existence value in a total valuation framework. In: Rowe, R. D.; Chestnut, L. G., eds. *Managing air quality and scenic resources at national parks and wilderness areas.* Boulder, CO: Westview Press.

Rescher, N. 1980. *Unpopular essays on technical progress.* Pittsburgh, PA: University of Pittsburgh Press.

Rollins, K.; Ivy, M. 1997. *The use of the environmental reference inventory (EVRI) in the environmental assessment process: the case of Alpha Creek.* Report to the Environmental Assessment Branch and the Environmental Economics Branch of Environment Canada: 105 p.

Rollins, K. S.; Lyke, A. 1998. The case for diminishing marginal existence values. *J. Environ. Econ. Manage.* 36(3):324–344.

Rollins, K. S.; Wistowsky, W. 1997. Benefits of back-country canoeing in Ontario Wilderness Parks. *J. Appl. Recreation Res.* 22(1):9–31.

Rollins, K. S.; Wistowsky, W.; Jay, M. 1997a. Wilderness canoeing in Ontario: using cumulative results to update dichotomous choice contingent valuation offer amounts. *Can. J. Agr. Econ.* 45:1–16.

Rollins, K. S.; Frehs, J.; Tate, D.; Zachariah, O. 1997b. Resource valuation and public policy: the case of water pricing. *Can. Water Resour. J.* 22(2):185–195.

Rosen, S. 1974. Hedonic prices and implicit markets: product differentiation in perfect competition. *J. Polit. Econ.* 82(1):34–55.

Sagoff, M. 1988. *The economy of the earth: philosophy, law, and the environment.* Cambridge, UK: Cambridge University Press.

Sarkar, R.; McKenney, D. 1992. *Measuring un-priced values in Ontario's forests: an economic perspective and annotated bibliography.* Sault Ste. Marie, ON: Forest Resource Economics Section, Forestry Canada.

Smith, V. K. 1992. On separating defensible benefits transfers from "smoke and mirrors." *Water Resour. Res.* 28(3):685–694.

Smith, V. K. 1993. Nonmarket valuation of environmental resources: an interpretive appraisal. *Land Econ.* 69:1–26.

Smith, V. K. 1997. Pricing what is priceless: a status report on non-market valuation of environmental resources. In: Folmer, H.; Tietenburg, T., eds. *International yearbook of environmental and resource economics, 1997/1998: a survey of current issues.* Cheltenham, UK: Edward Elgar: 156–204.

Smith, V. K.; Gilbert, C. S. 1984. The valuation of environmental risks using hedonic wage models. In: David, M.; Smeeding, T., eds. *Horizontal equity, uncertainty, and economic wellbeing.* Chicago: University of Chicago Press.

Smith, V. K.; Kopp, R. J., editors. 1993. *Valuing natural assets: the economics of natural resource damage assessment.* Washington, DC: Resources for the Future.

Tietenberg, T. 1996. *Environmental and natural resource economics*, 4th ed. New York: Harper Collins.

VandenBerg, T. P.; Poe, G. L.; Powell, J. R. 1995. *Assessing the accuracy of benefits transfers: evidence from a multi-site contingent valuation study of groundwater quality.* WP95-01. Ithaca, NY: Department of Agriculture, Resource and Managerial Economics, Cornell University.

Van De Veer, D.; Pierce, C. 1986. *People, penguins, and plastic trees: basic issues in environmental ethics.* Belmont, CA: Wardsworth.

Walsh, R. G.; Johnson, D. M.; McKean, J. R. 1992. Benefit transfer of outdoor recreation demand studies, 1968–1988. *Water Resour. Res.* 28(3):664–673.

17 Economic Linkages to Natural Resources

Daniel E. Chappelle

17.1 Introduction

17.1.1 Linkages of Environment to and from the Economic System

Directly or indirectly, everything is connected to everything else. We hear this trivial comment continually. It is true, of course, but linkages are so complex that not much is known and the available information is inadequate. Ecology focuses on linkages in nature. What often is not recognized is that economics focuses on linkages as well, many of which extend back to nature. Ecological linkages and economic linkages are connected in important ways. Boulding (1970) even explored economics as an ecological science. Also, since about 1970, economists have become more concerned about environmental quality. There are many recent inquiries regarding externalities in the form of pollution and schemes to use market forces to improve environmental quality (e.g., marketable pollution rights). During the 1970s the subfield of ecological economics began. This subfield has focused on nonmarket goods and services and pollution abatement, in contrast to resource economics, which has traditionally focused on production of market goods and services (for an early example concerned with forestry, see Fernow, 1902). It is evident, therefore, that both ecology and economics are concerned with natural resource linkages.

When we engage in ecological assessment, it is clear that we must be able to quantify both ecological linkages and economic linkages. A traditional view of natural resource linkages to the economic system is as raw materials in primary industrial production. These linkages are quantified as dollar flows in national economic accounting systems (i.e., income and product accounts and input–output accounts). Data in economic accounts are most detailed, but, from the standpoint of resource management, they provide little information on ecological impacts because they do not quantify in physical units or are provided at small geographical scales. From the standpoint of ecological assessment, therefore, it is a matter of taking fragmentary data from economic accounts and connecting them with even more fragmented resource and ecosystem data. The potential for errors in such an approach is substantial and must be recognized.

However, in spite of any measurement difficulties, it is necessary to include quantification of economic impacts in an overall ecological assessment. Without assessment of economic considerations, it is impossible to evaluate impacts on humans. The objective of this chapter is to suggest approaches to accomplish this task.

In recent years linkages from the economic system to the ecological system have been highlighted using various approaches. Flows of residuals back to the environment constitute in the aggregate what we term *pollution*. To the extent that the assimilative capacity of the environment is adequate to process pollutants, damage to the environment is negligible, although some species may be affected either in a positive or negative way. When this capacity is exceeded, serious long-term damage may result. When human population is sparse, the environmental assimilative capacity is generally more than adequate to process flows of residuals. As population increases, pollution tends to increase unless technology is brought to bear to reduce residual flows from production and consumption processes and steps are taken to increase the assimilative capacity of the environment. Unfortunately, environmental assimilative capacity is very difficult to measure. In the long run, healthy environments are generally associated with healthy economies.

Ideally, all linkages of the economic system from and to the environment should be quantified in models (see Chapter 21). The reality is that "less than ideal" is likely to be the case for some time, although theoretical models can account for all linkages. We will consider both ideal and realistic analytical approaches.

17.1.2 Structure of the Regional Economy

Regional economic structure is of major concern when linking the economy with natural resources. The structures of regional economies in the United States vary a great deal, which means that appropriate natural resource policies (as well as those relating to industry and government) will also vary from region to region. Some regions are well endowed with natural resources, whereas many are not. Degree of industrialization varies much from region to region, as does the amount of publicly owned land. Even where natural resources are important contributors to the economic base, the types, quantities, and qualities of natural resources in the mix vary a great deal.

Measurement of regional economic structure is neither straightforward nor standardized. However, a simple way to look at it is to measure percentages of total income, employment, or output accounted by sectors consisting of establishments having similar mixes of outputs of goods or services. We can then contrast these percentage distributions from one region to another. In regional economics the economic base often is delineated and includes those sectors that export most or all of their outputs outside the region. Economic growth is thought to depend on growth of these sectors, and the rest of the regional economy (i.e., consisting of the nonbasic sectors) grows as basic sectors grow. Often resource sectors have been and continue to be numbered among the basic sectors of the economy. As these sectors expand, there is a ripple effect, meaning that a more than proportionate growth occurs in the total regional economy. Of course, the ripple effect works in both directions, as a basic sector is contracted, there is a greater than proportionate decrease in economic activity. This effect is one reason people in a region become very concerned when policy changes reduce outputs in basic sectors (e.g., the Pacific Northwest reductions in National Forest timber harvests).

17.1.3 Derived Demand for Natural Resources and Environmental Services

Most natural resources and environmental services are cases of derived demand. That is, final consumers do not purchase them, but rather they are producer goods and services (an exception would be wilderness recreation). Demands for timber, for example, depend on demands for houses (for lumber and building board, etc.), for various types of paper, and for recreation in developed areas. This makes measurement of supply and demand equilibrium very difficult, because linkages between various stages of processing must be quantified correctly. Also, the analysis can be seriously compromised if possible substitute inputs (e.g., different technology or sources) are not considered. This has been an embarrassing problem among resource planners from the beginning of the professions—witness the prophets of doom regarding the long-run economic availability of timber, coal and petroleum at various times in the nation's history.

Natural resources are linked to the economy through intermediate (i.e., primary and secondary) economic sectors. Establishments in these sectors process natural resource commodities (e.g., timber, forage, water) or, in a few unique cases, sell natural resource services directly to final consumers. To comprehensively quantify contributions of natural resources to the economy, it is necessary to measure transactions between natural resource sectors and intermediate sectors of the regional economy.

The linkages of the economy to natural resources are very complex and difficult to quantify. For example, the greatest demand for timber products comes from the sectors of lumber and wood products, wood furniture and fixtures, pulp and paper products, and Christmas trees and greens. Other sectors that also use wood in significant amounts include (McKeever and Hatfield, 1984) ship building and repair; boat building and repair; travel trailers and campers; musical instruments; games, toys, and children's vehicles, except dolls and bicycles; sporting and athletic goods, not elsewhere classified; brooms and brushes; signs and advertising displays; and burial caskets. Similar detailed linkages exist for other types of natural resource areas (e.g., rangelands, wetlands, recreational areas).

17.1.4 Role of Economic Impact Assessment in the Ecological Assessment

Economic analysis is needed in ecological assessments to provide a measure of economic performance in the past and a measure of potential economic impact of changes in both ecological and economic systems. Valuation of the worth of resource and ecological assets in dollar terms is necessary to provide a measure common to the rest of the economic system. Economic analysis provides a measure of economic feasibility of various alternatives that are viable in ecological terms. Decisions in natural and environmental resources, both in public and private sectors, are made on the basis of economic feasibility, among many other considerations (e.g., ecological, social, political, psychological). Financial feasibility is usually the first to be considered, because proposed management activities require a budget that must be justified on economic grounds.

Ideally, economic impact assessments should show how current and future uses of natural resources affect the economic system and how workings of the economic system have and will in the future affect the environment. Unfortunately, this ideal condition has not been achieved and is unlikely to be achieved in the near future.

An ecological assessment is concerned with all linkages in a region. Broadly interpreted, it includes social and economic linkages, although these concerns are usually considered separate fields.

17.1.5 Valuation Problem: The Benefit Side and the Cost Side. Incidence of Benefits and Costs

A serious problem in environment and natural resources is that organized markets do not value many goods and services. This situation is found both on the benefit and cost sides. For example, on the benefit side, amenities are not valued directly by the market. On the cost side, damages to natural resources by water and air pollution are not valued directly by markets.

Economic analysis assumes that both inputs to the analysis in physical terms and financial terms are correct. For example, consider timber, a product valued by markets. Total market value of a timber stand consists of a measurement of merchantable volume and market price. Economic analysis assumes that both variables are correctly measured. Note, however, that significant costs (e.g., increased erosion, decreased visual quality) generally are not included in setting market prices. Similar cases can be cited for other resources.

Economic impact analysis requires the use of market prices. Use value, not utility value, is measured. Of course, utility values (e.g., from contingent valuation studies) could be used if all linkages were valued in that way (Chappelle and Webster, 1993). Generally, such an approach would be extremely costly, since primary survey data would be required for every good and service in every sector of the economy.

The question then is how to value nonpriced goods and services. A practical approach is to derive impacts on the basis of expenditures made for the array of products involved. In such cases, impacts of nonpriced goods and services have to be derived indirectly, because there is no direct expenditure. It is possible to trace spending by users of natural resources. For example, even if recreationists are not charged entry to a recreational area or for use of the resources, other costs are incurred. This method quantifies any spending involved in making use of the recreational area (e.g., spending for groceries, gasoline, restaurants, motels) (e.g., see Pedersen et al., 1989).

17.1.6 Need for Both Stock and Flow Information

To conduct ecological assessments, it is necessary to have both stock and flow information on the regional economy. Flow information is provided by input–output accounts, but there is a severe lack of stock information. Data on stocks are not easily collected since the federal government in this country does not regularly conduct stock accounts. Wealth accounts, the major type of stock accounts, have been developed only periodically and none have been completed recently.

In the natural resource and environment area, however, there are information sources that, although fragmentary, can provide a useful picture of the wealth represented by certain natural resources and environmental resources. For example, the Forest Service, USDA, conducts resource inventories that are essentially wealth accounts in physical units. For those goods and services priced by organized markets, these physical inventories can be multiplied by market prices to develop wealth in economic terms. Unfortunately, these data are not comprehensive, and there are always missing data that only can be overcome with great effort and cost. It stands to reason, therefore, that economic impact assessments, and hence ecological assessments, provide valuable information, but almost always are incomplete and likely to miss part of the resource and ecosystem picture.

17.2 Microeconomic Approach versus Regional Economic Approach

There are many techniques for evaluating effects on environmental quality of a proposed project in a site-specific context (e.g., production efficiency models and models of market processes). Methods for environmental quality valuation from both the benefit and cost sides are available and described in detail in Hufschmidt et al. (1983).

Microeconomic approaches have traditionally been desirable from the perspective of most professional economists. These approaches center on consumers, individual firms, and industries and focus on supply–demand analyses and benefit–cost analysis, for which determination of product prices and factor shares play essential roles. In contrast, macroeconomic approaches focus on the national economy and use national aggregates. These types of analyses tend to emphasize aggregate productivity, money supply, and interest rates. However, neither the microeconomic or macroeconomic approaches are most appropriate to broad-scale ecological assessments.

The ultimate type of microeconomic analysis would be full economic equilibrium analysis, by which quantity supplied and demanded, input and output prices, and location of all firms and inputs and outputs are determined. This type of analysis exceeds what can be achieved with either existing and foreseeable data resources. The regional analysis approach uses aggregation of space, time, and economic activities to simplify the analysis to the point that is practicable. However, when working on a regional scale, rather than in a site-specific context, general equilibrium approaches are needed. In the remainder of this chapter we will restrict coverage to the regional approach, which is discussed in greater detail in Hufschmidt et al. (1983).

17.3 Aggregation for Regional Economic Analysis

Regionalization involves determining boundaries of the subject region. In the case of ecological assessment, usually this is defined on ecological grounds and is a given for the entire assessment. However, the assessment team still must contend with how to aggregate space within the defined assessment region (i.e., boundaries of subregions). Usually, the assessment region will logically contain numerous ecological regions and perhaps even many economic regions. Given that an economic region usually will contain a number of ecological regions, the assessment team must decide on appropriate boundaries of the spatial entities considered in the analysis. Normally, boundaries are based on the concept of homogeneity in terms of appropriate variables (i.e., ecological and economic variables). Ecological regions have often been defined as watersheds and, more recently, as ecosystems.

Periodization is the aggregation of time in analysis. Most likely, this is also specified in the charter for the ecological assessment. Note, however, that the economic analysis methods likely will be of the static type (i.e., relate to a specific year). Generally, this scheme will conflict with some parts of the assessment, for which perhaps seasons within a year must be recognized. This can be readily handled, but requires planning. Of course, all data of every type must be for the same time period.

Sectorization is the process of aggregating activity or objects into a group that will be recognized as an entity in the analysis. In economic analysis it generally refers to aggregating establishments producing similar output mixes, as in the Standard Industrial Classification (SIC) of the federal government. There is no reason that the same idea cannot be applied to biological organisms, pollutants, or other issues in different sections of the assessment (e.g., see Isard et al., 1972; Roberts and Rettig, 1975). Ideally, ecological processes would be integrated into one model with the economic and social sectors, as will be shown below. Although this has not been accomplished in past assessments, it is a worthwhile goal.

17.4 Regional Economic Impact Analysis

Economic analysis involves carrying out studies to quantify economic impacts in a region. Economic assessment takes results of economic impact analysis and applies them to a specific context. The approach generally used in quantifying prospective economic impacts in ecological assessment is input–output (I–O) modeling. The function of the I–O model is to provide an estimate of direct, indirect, and induced effects. Direct effects are values of inputs used in the production of the next dollar of output. Actually, direct effects are data in the I–O model. Indirect effects are those changes in input demand (in dollar terms) caused by changes in direct effects (hence the ripple effect). Induced effects are changes generated as household incomes change. Indirect and induced effects are forecast by the I–O model.

Numerous I–O models may be used, ranging from simply applying I–O multipliers developed in applied studies to secondary data models often involving national coefficients modified by data reduction techniques to apply to the region or, more rarely, to models based on primary data collection or a combination of primary and secondary data. Approaches requiring primary data collection usually require so much time and money that they are not feasible for assessments. The rest of this chapter will focus on information systems for economic impact analysis requiring only secondary data.

17.5 Economic Impact Assessment as One Type of Socioeconomic Impact Assessment

Socioeconomic impacts have been categorized by Sell et al. (1998) as including economic impacts, demographic impacts, public service impacts, fiscal impacts, and social impacts. These impacts are very interconnected. Economic impact assessment is concerned with measurement of past and potential impacts of development projects in terms of (1) production of goods and services or total regional output, usually in dollar terms, (2) salaries and wages and broader, regional income, and (3) employment in terms of number of jobs, either full-time equivalents or aggregated full-time and part-time jobs. Demographic impacts quantify changes in the size, composition (in terms of cohorts), and distribution of the population. Public service impacts quantify public services, facilities, and infrastructure needed. Fiscal impacts quantify changes and costs and revenues to governmental jurisdictions. Social impacts, in contrast, usually are considered to be broader in scope, considering impacts in terms of well-being of people, including physical and psychological health in the context of their institutional framework and sociological relationships. Sell et al. (1998, p. 3) indicate this category quantifies "changes in the patterns of interaction, the formal and informal relationships resulting from such interactions, and the perceptions of such relationships among various groups in a social setting." An essential aspect of measuring social impacts is to obtain inputs from the appropriate publics. There are a number of methods and techniques needed in this type of assessment that are not characteristic of other aspects of the ecological assessment. As noted by Haque (1996, p. 345), detailed surveys of the population groups likely to be affected are essential in social impact assessment, in contrast to the other types of socioeconomic impacts, which may be analyzed by the use of secondary data:

> Although it offers a weak form of participation from a PI [public involvement] perspective, questionnaire surveys such as community surveys, community leader studies, and synchronized policy issue studies are considered a useful PI technique. PI inputs are needed in various stages of the SIA [Social Impact Assessment] process, especially in the formulation of alternatives, assessment, evaluation, and monitoring stages (Smith, 1984). Formulations of attitudes toward the project, perceptions of public welfare, and interest group activities can be considered in compiling a social profile that can be used in the decision-making and planning processes.

Also, as was stated above in terms of the economic and ecological modeling, Haque (1996, p. 345) notes a potential problem in incompatible regionalizations between economic impact analysis and social impact analysis:

> REIA [regional economic impact assessment] and SIA have apparent incompatibility because of their application of different spatial scales: REIA tends to emphasize macro-regional economic growth effects and SIA stresses local-level changes in social variables, such as health, education, social relations, and amenities. SIA underscores the importance of PI in the project decision-making process whereas the scope for PI in regional economic analysis is minimal.

Of course, REIA and SIA overlap in several important areas. As pointed out by Haque (1996, pp. 345–346):

> On the one hand, many social variables are relevant to REIA-based planning. Predicted changes in employment are influenced by the derived demand for social services, future demographic changes, future demand for infrastructure, and the likely regional social and community mix—all social variables. On the other hand, SIA overlaps with regional development objectives, which have as their main elements regional employment and regional infrastructure.

To provide accurate estimates of economic impact, it is necessary to provide funding and efforts at the same level of intensity as those allocated to the other phases of the assessment. An appropriate balance between measures of ecological, economic, and social performance and impacts should be achieved in the assessment.

17.6 Regional Accounts and Their Uses in Economic Impact Assessment

Economic accounting systems provide data for economic models. The most important sources of such data are the U.S. Department of Commerce and U.S. Department of Labor. The most important economic accounts are of the flow type. Two flow accounting systems are maintained on a regular basis: (1) income and product accounts and (2) input–output accounts. Income and product accounts are updated most frequently and are the basis of estimates of gross domestic product (GDP). Most important for economic impact assessment are the input–output accounts (I–O accounts). These are updated infrequently because they are very detailed—they quantify economic transactions between all sectors of the economy. The I–O accounting system is the basis of the economic impact models that rely on secondary data. Income and product accounts and the I–O accounts are compatible in that the identical gross national product (and GDP) can be calculated from either system.

To correctly quantify derived demand relationships, it is necessary to measure the structure of the regional economy. Input–output accounts quantify the structure of the regional economy in a sense. They essentially show flows of outputs between sectors. A sector is an aggregation of establishments producing a similar product (i.e., good or service) or mix of products.

> Input–output accounts provide a quantitative representation of a regional economy's structure. This accounting system is a flow system that includes all monetary transactions occurring in the region over a certain period of time, usually a year. In this accounting system, the economy is divided into three types of sectors: processing, final demand, and payments or value added. Processing sectors constitute intermediate industries made up of firms that purchase outputs of other sectors, which are in turn used in combination with labor and other inputs in manufacturing products. Final demand sectors are various categories of economic activities that are final consumers of goods and services produced in the regional economy. Payments sectors provide inputs other than those provided by processing sectors (e.g., households provide labor). (Chappelle et al., 1986, p. 3)

Input–output accounts quantify both final product and intermediate product flows to and from every sector of the regional economy. The basic design of I–O accounts is shown in Figure 17.1. In the case of manufacturing, intermediate products

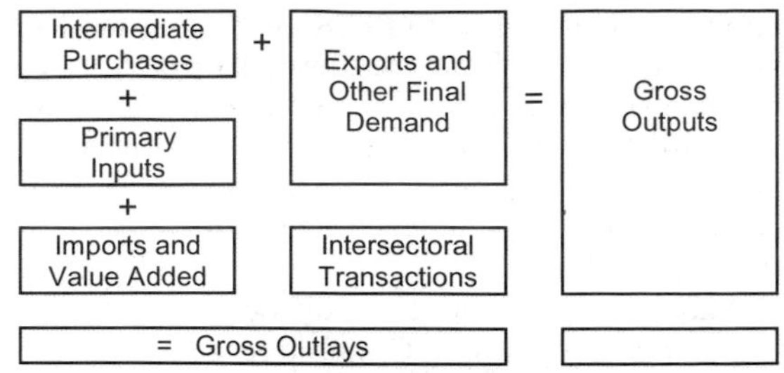

FIGURE 17.1. Basic design of input–output accounts.

are very important, and it is impossible to even estimate the economic importance of a sector without providing their estimation. Unfortunately, it is not possible to accurately express contributions of land management when using input–output accounts. Contributions of land management do not appear explicitly in the economic accounts, unless they are especially formulated for that purpose and data collected using surveys especially designed for that purpose (for an example, see the analysis for the Douglas County, Oregon, economy by Darr and Fight, 1974).

Input–output accounting results in a transactions table that summarizes all economic linkages in the regional economy for a specific time (usually a year). A simple hypothetical transactions table is shown in Table 17.1. This table shows all economic linkages in the region priced at the producer level. The transactions table accounts for not only the final consumer markets, but also all intermediate markets. Hence, within the aggregations selected, the table describes all regional economic linkages. It includes all market transactions from the final consumer back to the transactions for natural resources. As you read across the row of a transactions table, the distribution of outputs of that sector are shown. In contrast, when you read down a column, the distribution of inputs for that sector is shown. For example, if we had a residential construction sector, the table would describe the distribution of outputs, primarily to households. The distribution of costs would be for all inputs (e.g., building materials, fuel, equipment) and for labor, land, and capital.

This table is easily read and is a description of the regional economic structure at a given time. A row is read as describing how the sales of a sector are distributed. For example, the agriculture sector has a gross output of \$100 million, which is distributed as follows: it sells \$25 million to itself (i.e., transactions between firms in the sector itself), \$30 million to industry, \$10 million to trade (within re-

TABLE 17.1. Transactions table for the regional economy (in millions).

	Endogenous sectors			Exogenous sectors		
	(1)	(2)	(3)	(4)	(5)	
	Agriculture	Industry	Trade and services	Households	Exports and other final demand	Gross output
Agriculture	25	30	10	5	30	100
Industry	25	130	170	45	110	480
Trade and services	10	90	135	195	70	500
Households	30	115	135	10	10	300
Imports and value added	10	115	50	45	0	220
Gross outlay	100	480	500	300	220	1600

gion) and services, $5 million directly to households, and $30 million to exports (outside region) and other final demand. In an actual model, a vastly larger number of sectors would be required (a bare minimum would be about 40 for a developed industrial economy, and most examples include hundreds of sectors, some as many as 500, with elaboration particularly of industrial sectors). Also, capital accounts would be included to trace changes in inventory, depletion, and depreciation.

The columns describe how costs are distributed. For example, in our simple table, industry buys $30 million of inputs from agriculture, $130 million from itself, $90 million from trade and services, $115 million from households (i.e., primarily labor), $115 million from imports (from outside the region) and value added (inputs other than labor, including government). The total of the column, $480 million, is the gross outlay. If every dollar circulating in the economy is accounted for, then gross outlay will be exactly equal to gross output. In any actual application, there will be some statistical discrepancy.

17.7 Ecological and Demographic Accounts Integrated with Economic Accounts

A major problem in applying I–O to natural resources is that few models trace economic linkages adequately to show relationships of what is being done on the land to what is being done in the factories and stores of the regions and nation. During the last three decades in the literature of economics and regional science, numerous models proposed using input–output analysis to bring together production, consumption, and the environment into a single comprehensive model. Key papers were by Cumberland (1966), Isard (1969), Ayres and Kneese (1969), Leontief (1970), Kneese et al. (1970), Isard et al. (1972), and Kneese and Bower (1979).

Models presented in these papers differ from one another in various aspects, but they all consider dimensions that had been virtually ignored in the past, namely: the impacts of our diverse economic activities on environmental quality and the impacts of environmental quality on the economy. An essential feature of such extended input–output analysis is that it focuses not only on linkages between types of economic activity within the area being studied, but also on each economic activity to its pollutants, one pollutant to another pollutant, and each pollutant to the assimilative capacity of the environment. Just as we must recognize interdependencies between producing and consuming sectors in the regional economy, so must we recognize that residual problems (e.g., disposal of gas, liquid, and solid wastes) are interrelated. Of course, not all residuals produce negative impacts on the economy and society in general.

To the extent that recycling is done at market prices, economic accounts already include its effects. If we were able to completely recycle our wastes and pollutants, they would become intermediate products and would enter the usual processing sectors of the I–O model. We would then have achieved a closed materials economy. Of course, this has not been achieved, but this is the direction in a modern economy, at least in the industrial sectors. It appears more difficult to achieve this state in the final consumption sector, however, particularly because of the lack of concentration of wastes and high collection costs.

As pointed out by Isard and VanZele (1975, p. 8), the assumptions of I–O modeling are severely strained when applied to environmental analysis, particularly those of "fixed production and emission coefficients, lack of substitutability, fixed supply channels, fixed tastes and mechanistic behavior. . . . The linearity assumption, which is a strong one regarding economic commodities and processes, is considerably more severe for the ecologic commodities. Interactions between different pollutants and between pollutants and the environment at different concentrations and at different levels of economic activity are typically nonlinear." They further point out that these problems can be overcome, but only with side calculations and modifications of coefficients used in the I–O model. A rather notable hurdle in applying such a model is to analytically describe ambient environmental quality. This requires the use of diffusion processes and effectively modifies the scheme of ecological regions, generally requiring that they be very small.

There are a number of approaches to including environmental sectors with the economic sectors in a single model, including the (1) Cumberland model, (2) Isard–Daly models, (3) Leontief model, and (4) Victor (1972) model. Space does not permit describing these models here, but the reader may consult Richardson (1972) or Miller and Blair (1985) for details.

Development of data sets to input these models as part of an ecological assessment is almost totally out of the question for both time and funding considerations. To my knowledge, not one of these approaches has been attempted by assessment teams, and for good reason. Implementation of any of these models would constitute a large-scale research project in itself. However, these models are important in that they provide clarification as to how the joint ecological–economic system operates. The stumbling block in all cases is the lack of an integrated set of ecological accounts showing flows into and out of the environment. When an analyst wishes to include ecological sectors in a model, it is necessary to distill a mass of case study results into a coherent, compatible data set. For a very interesting application of this approach, see Isard et al. (1972).

17.8 Integration of Economic Impact Estimates with Other Social Impact Estimates and Ecological Impact Estimates

There are some variables that link economic impacts with social impacts, for example, employment and, by implication, population. Population (including immigration) and per capita consumption forecasts are needed to calculate future demands for all goods and services. These exogenous variables are used in I–O analysis to forecast future output of all goods and services, both final and intermediate sales. Also, a number of social indicators are quantified in economic accounts in monetary terms, including unemployment compensation, retirement benefits, medical payments, and welfare payments. Social impacts of changes in these economic indicators should be explored in the social impact section of the assessment. Impacts in monetary terms have to be translated into values that measure human suffering, well-being, and promise. Also, a cumulation of values that provides a measure of community well-being is necessary in the social impact analysis.

Linkages between economic and ecological impacts are less clearly defined. The difficulty, as noted previously, is that physical units of resources, or even resource constraints, do not appear anywhere in the typical I–O analysis. Rather, resources have to be considered in side calculations. For example, if we calculate output forecasts of resources, using sectors in an I–O analysis, it is a rather simple process to check if these production levels are feasible, given expected regional resource stocks. However, often it is very difficult to determine the amounts of each resource truly available at prices implicit in the I–O model. There are ways to modify price assumptions, but these are difficult, as well as somewhat arbitrary (see Miller and Blair, 1985, pp. 351–357). Also, outflows of pollutants implied by estimated production levels forecasted by the I–O model may be checked against available environmental assimilative capacity data, although such data are generally both limited and fragmentary.

Feedback of environmental quantity and quality likely to constrain future economic production is particularly difficult to measure. Ideally, resource stocks at every year in the forecast period should be measured in terms of ecological impacts of the previous year's production, but this degree of finetuning is not feasible given available data resources. About the best that can be done now is to temper future production schedules with expected changes in natural resources.

It is important that economic impact analysis be viewed from the standpoint of political economics and social analysis, rather than from a narrow technical viewpoint. The point of the analysis is to measure the extent to which natural resources influence human welfare, in both the short and long term. Measurement scales noted above (i.e., output, income, and employment) have broad social implications. More output, income, and employment are considered to be better for society. However, ecological assessment is concerned with links back to the environment from the economy and society. If more output, income, and employment lead to a degraded environment, this implies a decline even in these economic variables in the long run and hence a decline in human welfare. Also, some declines in quality of life aspects are not fully captured by economic variables. To the extent that models can be made dynamic (or at least comparative static) and feedbacks to the environment are included in the models, this trend could be measured quantitatively. This should be a goal of analysts involved in this field.

17.9 Steps in Model Building to Link the Ecological and Natural Resources to the Economic System

The following steps are involved in linking natural resources to the economic system.

1. Work within the temporal and spatial aggregations specified for the ecological assessment. For reasons explained previously, economic regions may be larger than ecological regions. In such a case, ecological regions should nest within economic regions.
2. Delineate sectors of the economy that will be recognized in the economic impact analysis (sectorization).
3. Develop an inventory of assets in the natural resource and environmental sectors (usually this is done in physical terms).
4. Collect market prices on the natural resource goods and services traded in organized markets.
5. Calculate market values of the priced goods and services related to natural resources.
6. For nonmarket goods and services (including amenities), develop value estimates based on expenditures (by sector) required to secure the product.
7. Develop input–output accounts appropriate to the region.
8. Develop an input–output modeling system.
9. Run the model for the region, time period, and sectorization scheme specified in the assessment. If the model does not contain all data needed, some primary data collection may be required.
10. Develop multipliers in the economic dimensions (e.g., output, income, and employment) desired.
11. Apply multipliers to quantify expected economic effects in the future.
12. Check for ecological feasibility of future expected economic changes. If economic changes are not ecologically feasible, then it is necessary to scale back the multiplicands applied to the multipliers.

Steps 7 to 12 are discussed in Chapter 21.

17.10 Conclusions

Three conclusions appear evident from the preceding exploration.

1. Economic impact assessment is a necessary and required part of ecological assessment. This type of assessment provides measures of economic contributions of natural resources. Economic impacts are necessary to link ecosystems with the human population.
2. Economic accounts constitute the information system used in economic models to generate broad-scale ecological assessment guidelines. Of the various economic accounting systems, the most useful for impact analysis is the input–output type. Producer prices are generally used in this accounting system, although partial analysis can be completed with physical data.
3. Lack of regional ecological accounts severely hinders economic analysis of environmental resources.

17.11 Recommendations for Future Work to Advance Integrated Ecosystem Assessments

1. We need to progress to the point of estimating changes in wealth and other stocks (including quantity and quality of natural resources). We have not yet arrived to this point and are not likely to in the foreseeable future, given the amount of resources devoted to data collection. Lack of stock accounts seriously limits economic impact assessments and hence limits policy implications that can be drawn from any ecological assessment.
2. Ecologists need to develop regional accounts of ecological linkages. An input–output framework could be useful in guiding this effort.
3. Efforts need to be accelerated in valuing environmental assets and flows. In particular, more

work is needed to delineate the linkages that are already included in regional input–output accounts.

4. Research efforts need to be devoted to introducing landownership sectors into input–output accounts. If this were done, it would be possible to more definitively state the impacts of changes in land ownership or changes in resource mix from ownerships.
5. More emphasis should be placed on the economic and social importance of environmental resources. Otherwise, it is likely that they will be mismanaged, because they will not be able to compete with other parts of the economic system for investment.

17.12 References

Ayres, R. U.; Kneese, A. V. 1969. Production, consumption and externalities. *Amer. Econ. Rev.* 59(3): 282–297.

Boulding, K. E. 1970. *Economics as a science*. New York: McGraw-Hill.

Chappelle, D. E.; Webster, H. H. 1993. Consistent valuation of natural resource outputs to advance both economic development and environmental protection. *Renewable Resour. J.* 11(4):14–17.

Chappelle, D. E.; Heinen, S. E.; James, L. M.; Kittleson, K. M.; Olson, D. D. 1986. *Economic impacts of Michigan forest industries: a partially survey-based input–output study*. Res. Rep. 472. East Lansing, MI: Natural Resour., Agric. Exp. Sta., Michigan State Univ.

Cumberland, J. H. 1966. A regional interindustry model for analysis of development objectives. *Reg. Sci. Assoc. Papers* 17:65–94.

Darr, D. R.; Fight, R. D. 1974. *Douglas County, Oregon: Potential economic impacts of a changing timber resource base*. PNW-RP-179. Portland, OR: U.S. Dept. Agric., For. Serv., Pacific Northw. Res. Sta.

Fernow, B. E. 1902. *Economics of forestry*, 3rd ed. New York: Thomas Y. Crowell.

Haque, C. E. 1996. Integration of regional economic impact assessment (REIA) with social impact assessment (SIA): the case of water improvement service projects in rural Manitoba, Canada. *Impact Assess.* 14(4):343–370.

Hufschmidt, M. M.; James, D. E.; Meister, A. D.; Bower, B. T.; Dixon, J. A. 1983. *Environment, natural systems, and development: an economic valuation guide*. Baltimore, MD: Johns Hopkins University Press.

Isard, W. 1969. Some notes on the linkage of the ecologic and economic systems. *Reg. Sci. Assoc. Papers* 22:85–96.

Isard, W.; VanZele, R. 1975. Practical regional science analysis for environmental management. *Int. Reg. Sci. Rev.* 1(1):1–25.

Isard, W.; Choguill, C. L.; Kissin, J.; Seyfarth, R. H.; Tatlock, R. 1972. *Ecologic–economic analysis for regional development*. New York: The Free Press.

Kneese, A. V.; Bower, B. T. 1979. *Environmental quality and residuals management*. Baltimore, MD: Johns Hopkins University Press.

Kneese, A. V.; Ayres, R. V.; D'Arge, R. C. 1970. *Economics and the environment: a material balance approach*. Baltimore, MD: Resources for the Future.

Leontief, W. 1970. Environmental repercussions and the economic structure: an input–output approach. *Rev. Econ. and Statis.* 52(3):262–271.

McKeever, D.; Hatfield, C. A. 1984. *Trends in the production and consumption of major forest products in the United States*. FPL-RB-14. Madison, WI: U.S. Dept. Agric., For. Serv., For. Products Lab.

Miller, R. E.; Blair, P. D. 1985. *Input–output analysis: foundations and extensions*. Upper Saddle River, NJ: Prentice Hall.

Pedersen, L.; Chappelle, D. E.; Lothner, D. C. 1989. *The economic impacts of Lake States forestry: an input–output study*. NC-GTR-136. St. Paul, MN: U.S. Dept. Agric., For. Serv., North Central For. Exp. Sta.

Richardson, H. W. 1972. *Input–output and regional economics*. New York: Halsted Press, John Wiley & Sons.

Roberts, K. J.; Rettig, R. B. 1975. *Linkages between the economy and the environment: an analysis of economic growth in Clatsop Co., Oregon*. Agric. Exp. Sta. Bull. 618. Corvallis, OR: Oregon State Univ.

Sell, R. S.; Leistritz, F. L.; Murdock, S. H.; Spies, S.; White, S.; Krannich, R. S.; Wrigley, K.; Wulfhorst, J. D. 1998. Economic and fiscal impacts of waste and non-waste development in rural United States. *Impact Assess. Project Appraisal* 16(1):3–13.

Smith, L. G. 1984. Public participation in policy making. *Geoforum* 15:253–259.

Victor, P. 1972. *Pollution: economy and environment*. London: George Allen & Unwin.

Part 5
Analytical Methods: Applications

18
Ecosystem Structure and Function Modeling

Hope C. Humphries and Jill Baron

18.1 Introduction

An important component of ecological assessments is the ability to predict and display changes in ecosystem structure and function over a variety of spatial and temporal scales. These changes can occur over short (less than 1 year) or long time frames (over 100 years). Models may emphasize structural responses (changes in species composition, growth forms, canopy height, amount of old growth, etc.) or functional responses (cycling of carbon, nutrients, and water). Both are needed to display changes in ecosystem components for use in robust ecological assessments. Structure and function models vary in the ecosystem components included, algorithms employed, level of detail, and spatial and temporal scales incorporated. They range from models that track individual organisms to models of broad-scale landscape changes. This chapter describes models appropriate for ecological assessments. The models selected for inclusion can be implemented in a spatial framework and for the most part have been run in more than one system. Model assumptions, advantages, limitations, and applications are discussed, and model features are summarized in Table 18.1.

Models are needed in ecological assessments for several reasons. The relatively short time periods and limited resources often allotted for assessments may preclude extensive empirical studies of ecosystem responses to changes. Experimental manipulation may not be feasible at the scale of a landscape or region due to prohibitive costs, and it may be difficult to control for environmental changes that have both direct and indirect effects on ecosystem components (Baker, 1989a). Adequate replication of large-scale experiments may not be possible, but many replications can be performed for stochastic model simulations, providing an estimate of the range of potential effects (Turner et al., 1995). Models can generate output data that are fine grained as well as spatially and temporally extensive (Smith and Urban, 1988).

Ecosystem processes such as succession may operate over very long periods of time due to the longevity of many plant species. Modeling can be used to extend the temporal and spatial scales considered in an assessment and can allow the user to explore ecosystem dynamics following different disturbance regimes with specified initial conditions (Dale et al., 1986). Environmental variables can be changed while holding organism attributes constant, and vice versa (Dunning et al., 1995). Responses to various management scenarios can be evaluated as a series of what-if questions. In addition, a simulation model represents a set of hypotheses about the functioning of a system and can be used to summarize understanding of ecosystem relationships in a quantitative and explicit manner and suggest areas where further research is needed.

18.2 Models of Ecosystem Structure

18.2.1 Transition Matrix Models

Transition matrix models, characterized by a matrix representing transitions in states of an ecological system over a specified period of time, include

Hope C. Humphries wishes to acknowledge partial funding provided by a Science to Achieve Results grant from the U.S. Environmental Protection Agency ("Multi-Scaled Assessment Methods: Prototype Development Within the Interior Columbia Basin"). Jill Baron wishes to acknowledge the support of the U.S. Geological Survey Global Change Program.

TABLE 18.1. Summary of model features.

Model	Spatial extent	Spatial resolution (cell size)	Temporal extent (years)	Entity simulated
Ecosystem structure				
Transition matrix models	Local to global	Any	Any	Successional and age classes
Individual-based plant models	Local to regional	<.01 to 0.2 ha	Decades to thousands	Species composition, size and age distributions
Individual-based animal models	Local to regional	1 to >100 ha	One to hundreds	Individual and population distributions
Gradient models	Local to regional	0.01 to 1 ha	<One to decades	Fire spread, fuels, vegetation class
LANDIS	Local to regional	Any appropriate for trees	Decades to thousands	Species composition, age class distributions
CRBSUM	Regional	1 km^2	Decades to hundreds	Successional classes
MAPSS	Regional to global	>10 km^2	One to decades	Potential vegetation type, runoff
Ecosystem function				
FOREST-BGC/ BIOME-BGC	Local to continental	<1 to >10 km^2	One to decades	Photosynthesis, evapotranspiration
RHESSys	Local to regional	<1 to >10 km^2	One to decades	Photosynthesis, evapotranspiration, watershed hydrology, nitrogen flux
CENTURY	Local to global	<1 to >10 km^2	Decades to tens of thousands	Primary productivity, soil organic matter, evapotranspiration, soil moisture
CELSS	Regional	1 km^2	One to decades	Ecosystem type

Markov chain models, Leslie matrix models, and spatial automata. In a transition matrix model, an initial set of states is projected forward in time to an output set of states; for example, changes in vegetation succession as a result of disturbance can be projected. Markov chain models, consisting of a matrix of probabilities of transition from one state to another, are stochastic because their output is probability-based (Baker, 1989a). Leslie matrix models are deterministic transition matrix models that have been widely used to model changes in plant, animal, and human populations (Baker, 1989a). When populations are modeled using this method, they are divided into discrete age classes or life stages (Turner and Dale, 1991). Leslie matrix models differ from Markov models in that their matrix values are rates of change, rather than probabilities, and include birth functions as well as transition rates among population classes (Hunsaker et al., 1993). However, comparable results have been obtained with both Markov and Leslie matrix models (Baker, 1989a).

Like other transition matrix models, the set of models known as spatial automata track changes in state over time as a function of rules expressed in a transition matrix (Childress et al., 1996). In addition, spatial automata are grid-based models in which each cell in the grid is an independently behaving entity. Changes in the state of a cell depend not only on its current state, but also on the states of neighboring cells (Shugart, 1998). The components of the transition matrix for a spatial automata model may reflect deterministic rules for changes in state, or they may include probabilities of change reflecting stochastic rules (Childress et al., 1996). Markov models could be considered special cases of spatial automata in which the neighborhood effect on a cell is absent.

The system to be simulated with a transition matrix model must be classified into a discrete set of states. Data are required to determine the components of the transition matrix and provide initial conditions for the model. Time series of remotely sensed data provide one source of such estimates for Markov models and spatial automata focusing on changes in vegetation and land-use class (Hunsaker et al., 1993; Childress et al., 1996). The data requirements for Leslie matrix models include detailed information about rates of change among age classes or life stages (Turner and Dale, 1991). In spatial automata models, the size and shape of the neighborhood surrounding a cell must be selected and the rules for tallying neighborhood states determined and parameterized. Given an appropriate data source, the models are relatively easy to construct and implement, and the mathematics are well understood (Sklar and Costanza, 1991). They have been applied at a variety of spatial scales from less than one hectare to thousands of hectares (Turner and Dale, 1991; Childress et al., 1996).

Markov models, in their traditional form, are limited by the assumption of stationarity (i.e., probabilities do not change with time), as well as their lack of mechanism for transitions, fixed time step, and lack of spatial and historical effects. Modifications to the models have been implemented to overcome these limitations. However, model extensions can significantly increase data requirements and processing time. Nonstationarity of transition can be incorporated by constructing more than one stationary matrix, with a process in the model for switching among matrices (Baker, 1989a). The influence of environmental variables has been incorporated by making the transition probabilities functions of these variables, including nonlinear functions. Examples of environmental variables that have been included in Markov models are climate attributes, fire frequency, and probability of insect attack (Henderson and Wilkins, 1975; Marsden, 1983; Woolhouse and Harmsen, 1987). Semi-Markov models can be used to simulate transition probabilities that vary with time (Ginsberg, 1971, 1972). Functions are developed that determine the duration of stay or sojourn time in each state (Baker, 1989a). The effects of landscape heterogeneity on changes in state, in which transitions may depend on location in the landscape, can be simulated by developing individual transition matrices for homogeneous subareas of the landscape (Baker, 1989a). The effect of previous history can be modeled by defining matrices that include both current and preceding states. This, however, results in an exponential increase in the number of transition probabilities that must be estimated (Baker, 1989a).

Spatial automata offer the advantage of explicit inclusion of spatial effects within a neighborhood (Shugart, 1998). However, some choices of modeling structure can lead to intractable complexity. Model complexity is determined by the number of possible neighborhood configurations, a function of the number of states, neighborhood size, and method of tallying neighborhood states. Childress et al. (1996) present formulas for calculating the number of neighborhood configurations. The method of tallying neighborhood states can include identifying each cell uniquely by its spatial position, the *unique neighbor* method, or tallying the number of cells without considering their spatial position, the *voting* method (Childress et al., 1996). Use of the voting method can reduce the number of neighborhood configurations by several orders of magnitude and is likely to meet the needs of many ecological applications, with the exception of those in which a gradient is included in the model grid (Childress et al., 1996).

Pastor et al. (1993) developed a Markov model to evaluate historical landscape changes caused by beaver pond construction in Voyageurs National Park, Minnesota. Aerial photographs taken over a 50-year period were the basis for delineating four hydrologic regimes in maps of floodable areas. Baker (1989b) used a set of Markov models to determine whether a stable mosaic of fire-induced vegetation patches was present in the Boundary Waters Canoe Area, Minnesota. Acevedo et al. (1996) implemented a semi-Markov model for forest dynamics at a landscape scale, estimating transition probabilities and holding times within states for tree functional types from a series of runs of an individual-based plant model (ZELIG, discussed later). This enabled integration of both landscape and local scales in the modeling effort. Hall et al. (1988) believed that landscape transition models could be coupled to general atmospheric circulation models to project changes in vegetation as a result of global climate change; they proposed substituting observed vegetation transition probabilities over a spatial climatic gradient for probabilities of transition as a result of climate change over time. Often, insects or disease-causing organisms operate on very different spatial and temporal scales than their hosts (Turner and Dale, 1991). Dale et al. (1991) developed two separate Leslie matrix models to couple the dynamics of a tree species, the Fraser fir, with an insect predator, the balsam woolly aphid, in the southern Appalachians. Trees were modeled with an annual time step, whereas insects were modeled using a two-day time step. Green (1989) used spatial automata to simulate changes in tree locations in response to the effects of fire, seed dispersal, and environmental gradients on a rectangular grid. Thiéry et al. (1995) developed a spatial automata model to examine change in tiger-bush vegetation in Niger, which has a distinctive striped spatial pattern. The model was able to reproduce the vegetation pattern when initialized with a random pattern (Shugart, 1998).

18.2.2 Individual-based Plant Models

The group of stochastic individual-based plant models derived from the JABOWA model (Botkin et al., 1972) has been used to model vegetation distributions throughout the world (Hunsaker et al., 1993). Models have been developed for a wide range of forests (Urban and Shugart, 1992), in addition to a grassland (Coffin and Lauenroth, 1990, 1996) and an alpine meadow (Humphries et al., 1996). These models track the establishment, growth, and death of individual plants on a small plot using an annual time step. As a consequence of differences in regeneration, growth, and mortality among species and individuals of different sizes, the models are useful in simulating mixed species and mixed age vegetation dynamics. Shugart (1984) describes the early formulation and application of these models. Urban and Shugart (1992) and Shugart and Smith (1996) provide comprehensive recent reviews for forest models.

The more than two dozen model versions vary somewhat in emphasis and the manner in which processes are simulated, but all share certain assumptions and model structures. Individuals are tracked by species, size, and vigor (Urban and Shugart, 1992). The size of the plot is scaled to the zone of influence of a mature individual of a canopy-dominant species. The death of such an individual creates a gap in resource space that can be filled by other individuals (gap-phase regeneration) (Watt, 1947); hence the models are often called *gap models*.

For most gap models, horizontal placement of individuals within a plot is not specified (Dale and Shugart, 1985). Competition for light occurs through shading of shorter individuals by taller individuals. In modeling the growth of an individual, a maximum potential response is constrained by environmental conditions such as available light, growing-degree-days temperature, soil moisture, and soil fertility (Shugart and Smith, 1996). The diversity of formulations of tree growth responses in the models is described in Bugmann et al. (1996). Plant establishment is a stochastic function based on species-specific responses to the plot environment, such as light, temperature, and moisture conditions. Mortality occurs as a result of plant age, loss of vigor, and, in some models, disturbances such as fire, insect attack, windthrow, flooding, hurricanes, and timber harvesting for forests (Urban and Shugart, 1992). Disturbances modeled for a grassland include ant and small mammal dirt mounds and cattle fecal pats (Coffin and Lauenroth, 1990).

The models have incorporated a variety of extensions, including nutrient cycling (e.g., Aber and Melillo, 1982; Pastor and Post, 1986; Bonan, 1990a, 1990b; Botkin, 1992) and detailed representation of soil moisture and soil thermal conditions (Bonan, 1989a, 1989b). Plots in the forest gap model ZELIG (Urban, 1990) can be linked in a grid or transect to represent spatial interactions among plots. ZELIG has been used to investigate spatial phenomena such as shading effects (Urban et al., 1991), seed dispersal (Urban and Smith, 1989), and spatial pattern in comparison with remotely sensed data (Weishampel et al., 1992).

Most gap models use simple functions that are relatively easy to parameterize (Urban and Shugart, 1992). For trees, species parameter values can often be obtained from the silviculture and forestry literature. The models are relatively computationally efficient (Keane et al., 1996b). They have been employed successfully in a wide variety of systems, from tropical to boreal forests and from floodplains to alpine vegetation. Locations that have been modeled include many areas in North America, as well as in Europe, South Africa, Australia, and New Zealand.

The simplified nature of the functions in a gap model, an advantage for rapid model parameterization and implementation, may be a limitation for some applications. In particular, the growth functions are descriptive rather than mechanistic; as a consequence, the models may be inadequate for predicting some kinds of responses to global change, such as how tree growth might differ under changed climate and increased CO_2 (Friend et al., 1993). In addition, processes occurring on very short time scales are not considered. Friend et al. (1993) developed the model HYBRID by modifying a gap model (ZELIG) to include physiological growth processes for individual trees using a daily time step. However, the incorporation of functionally realistic processes for carbon fixation and partitioning required many species-specific physiological parameters and greatly increased model processing time. A similar approach was taken in the model FORCLIM (Bugmann and Fischlin, 1996) to explicitly model the environmental constraint on photosynthesis and respiration.

The lack of plot-to-plot spatial interactions, a limitation of many gap models, has been overcome in ZELIG, which provides a framework for modeling spatial interactions for forests (Urban, 1990). The size of a modeled plot (0.01 to 0.2 ha for forests) may be a limitation for some applications, because it is very small compared to many areas of interest in ecological assessments. Continuous coverage of large areas using gap models could require simulating many thousands of plots. This problem has been addressed by running models in representative subareas and generalizing results to larger areas (Shugart, 1998).

Gap models are useful for considering variability within and between forest and grassland systems, including examination of age structure, species diversity, and disturbance effects (Dale and Shugart, 1985; Coffin and Lauenroth, 1990). Output from gap models has also been coupled to other models to predict outcomes of management activities. For example, ZELIG was used to model the effects of alternative silvicultural regimes on forest structure and composition, bird species densities, and the economic value of wood products in the western Cascades, Oregon (Hansen et al., 1995).

18.2.3 Individual-based Animal Models

Individual-based animal simulation models that incorporate habitat complexity in a spatially explicit framework can be used to evaluate species responses to changes in heterogeneous landscapes, including changes in management strategies (Dunning et al., 1995). Such models require habitat-specific information about demography, dispersal behavior, and habitat selection for the animal species modeled, as well as a map describing the spatial arrangement of habitat patches in the landscape. Individuals are tracked by location and acquire fitness characteristics based on the landscape cells and hence the habitat they occupy. The models can explore the consequences for animal populations of habitat fragmentation, isolation, shape, and patch size resulting from various management alternatives, including species recovery plans (Dunning et al., 1995, Turner et al., 1995). Questions can be addressed regarding animal responses to resource dynamics, effect of landscape composition on extinction probability, and efficiency of reserve design (Dunning et al., 1995).

The models assume that habitat can be adequately described by a set of mapped categories such as vegetation types or suitable–unsuitable habitat. Detailed demographic and behavioral data derived from field studies are required to parameterize the models (Dunning et al., 1995). Lack of such data for an area or at a particular scale may limit implementation of the models (Turner et al., 1995). In addition, highly accurate quantitative predictions of population responses to specific environmental changes are generally not possible at present (Dunning et al., 1995). Most spatially explicit, individual-based animal models consider only one or a few animal species at a time, which may further limit their utility for ecological assessments.

Plant simulation models can be linked to animal models to provide a more detailed characterization of the vegetation structure and composition comprising habitat than is generally included in most animal models (Holt et al., 1995). Individual-based plant models such as those considered previously can be utilized to predict landscape dynamics over long time periods and under conditions of climate change. For example, output from a spatially explicit plant model could be used to generate transition rules for changes in landscape cells from one habitat type to another (Holt et al., 1995).

The model BACHMAP simulates Bachman's sparrow populations in pine forests of the southeast United States (Pulliam et al., 1992). Individuals are tracked in grid cells scaled to the size of sparrow territories. Each cell is classified into one of 22 vegetation types. Sparrow birth, death, and movement rates are functions of vegetation type, which may change annually in response to age-related management practices. Another Bachman's sparrow model, ECOLECON, incorporates the economics of timber production so that managers can balance sparrow population size with timber outputs (Liu, 1992, 1993; Turner et al., 1995).

MOSAIC is an individual-based model designed to project owl numbers and distribution across large landscapes (McKelvey et al., 1992; Bart, 1995). Habitat in grid cells is classified as either suitable or unsuitable. Each cell is scored according to the amount of suitable habitat, number of owls using the cell, and the amount of habitat in surrounding cells. Territories and home ranges have benefits to owls based on the sum of cell scores and costs based on the size of the territory or home range. A spatially explicit routine is invoked to allow dispersing owls to search for suitable territories.

NOYELP, a model used to explore mortality of wintering elk and bison in response to the scale and pattern of fires in northern Yellowstone National Park, includes simulation of forage availability, snow conditions, ungulate movement and foraging, and ungulate energetics (Turner et al., 1993, 1994, 1995). The model is initialized with forage distributed among six vegetation types according to their actual spatial distribution. Resource levels in occupied patches are depleted by foraging animals such that the distribution of resources changes with time. The model provides managers with a tool to evaluate how species are affected by management activities.

18.2.4 Gradient Models

Gradient models, developed by Kessell and colleagues, predict the effects of fire on vegetation and fuels based on biotic–environmental gradient relationships (Kessell, 1976, 1977, 1979, 1990; Kessell and Cattelino, 1978; Kessell and Good, 1982; Kessell et al., 1984). The models are implemented on a grid of cells, with cell sizes of 0.01 to 1 ha. For each cell, values of environmental attributes such as elevation, topography, moisture, time since last burn, primary succession, drainage system, and nonfire disturbance categories determine species composition and fuel levels using gradient submodels. Daily spread of a fire front is predicted based on vegetation, fuels, a weather submodel, and a fire behavior submodel. Simulated fire behavior includes rate of fire spread and height of flames. The heat flux from fire in a cell raises the temperature of fuel in an adjacent cell, allowing for the prediction of fire spread rate (Sklar and Costanza, 1991). A submodel also has been developed to estimate the response of small and large mammals to fire disturbance. The effect of thinning or clear-cutting on fire spread is included in the gradient model FORPLAN (Kessell, 1976; Potter et al., 1979).

Models have been constructed for various coniferous forest locations in Montana, the Blue Mountains of Oregon, southern California chaparral, and Australian parks and reserves. Operation of the models requires a comprehensive resource inventory to populate grid cells with the necessary environmental attributes. For a model implemented in Glacier National Park, Montana, the necessary data were derived from aerial photographs, topographic maps, and fire history maps (Kessell, 1976).

The simplicity of the submodels allows large numbers of grid cells to be processed (Baker, 1989a). The models have been designed to be easily operated by resource managers in the management of small to medium fires. Limitations include simulation of succession as a deterministic process and the lack of submodels for plant dispersal and disturbances other than fire. In addition, natural or artificial fire breaks cannot be accurately simulated if the cell size (e.g., 1 ha) is too coarse to represent them (Sklar and Costanza, 1991).

18.2.5 LANDIS

LANDIS is a stochastic, spatially explicit grid-based model of forest succession that incorporates the effects of fire and windthrow disturbances as well as harvesting (He and Mladenoff, 1999; Mladenoff et al., 1996). A stochastic approach is taken to simulate landscape dynamics over broad spatial scales and long time periods, rather than predicting individual events, such as are simulated using the gradient models described previously. As in gap models, interactions among species life history characteristics, site conditions, and disturbance regimes determine the course of succession (Mladenoff et al., 1996). However, LANDIS can simulate larger areas than gap models with the use of a ten-year time step and aggregation of tree species into ten-year age classes. An object-oriented approach is taken in which each cell in the simulated grid is a spatial object (He and Mladenoff, 1999). The cell size can be modified for operation of the model at different scales of resolution.

Species life history parameters, including dispersal and establishment characteristics, longevity, age at sexual maturity, and response to disturbance, are obtained from the literature (Mladenoff et al., 1996). Seed dispersal is a spatially explicit process based on dispersal distance curves. Establishment of trees is a function of site characteristics, which are provided as input to the model in the form of a land-type data layer. Land types are intended to correspond to soil moisture and nutrient conditions. Site characteristics as expressed by land types also affect fuel accumulation and decomposition (Mladenoff et al., 1996). Disturbances are generated stochastically by selecting from disturbance-sized distributions. The actual size of a disturbance also depends on local susceptibility conditions. Fire disturbance is assigned to a fire severity class based on time since last fire. Fire severity and species fire tolerance determine which trees are killed, beginning with younger age classes (He and Mladenoff, 1999). Windthrow susceptibility increases with tree age, leading to removal of older age classes first. Great flexibility in model output is provided, including maps of cover types or age classes and an interface with a spatial analysis package for production of summaries and landscape indices such as fractal dimension, connectivity, and patch statistics.

The model has been parameterized for a northern Wisconsin landscape and verification and sensitivity analysis have been performed (He and Mladenoff, 1999; Mladenoff and He, 1999). A predecessor model, LANDSIM (Roberts, 1996a, 1996b), was parameterized for the southwest United States. A potential limitation is the representation of environmental characteristics by a single data layer.

18.2.6 CRBSUM

CRBSUM is a vegetation succession simulation model that was developed for predicting landscape characteristics as part of the interior Columbia River basin (ICRB) ecological assessment (Keane et al., 1996a). Successional development is simulated as a change in structural stage and cover type keyed to successional age, using an annual time step over a set of grid cells. The ICRB assessment included two spatial scales: the coarse scale, covering the entire basin at a grid cell size of 1 km^2, and the mid-scale, with a cell size of 0.01 km^2, covering a set of representative watersheds. CRBSUM was developed for the coarse-scale assessment, but also was used for evaluating management alternatives as part of the mid-scale effort.

The basis for vegetation succession in CRBSUM is a conceptual model of post-fire succession along multiple pathways that converge to a 'climax' community in the absence of disturbance (Kessell and Fischer, 1981). This approach was expanded for CRBSUM to include disturbances other than fire (such as insect and disease attack, grazing, thinning, and cutting) and to incorporate nonforest vegetation types. The endpoint of a particular successional pathway is a potential vegetation type (PVT) with a corresponding biophysical setting. Fifty-five PVTs were defined for CRBSUM runs using expert workshops. For each PVT, a set of successional classes, each described by a structural stage and cover type, are linked along pathways converging to the PVT. Rule-based functions determine changes in successional class as a result of the occurrence of disturbances and successional development in the absence of disturbance. Disturbances are implemented through management scenarios comprised of sets of probabilities that a management action or disturbance will occur on a cell in any given year. Probabilities are specified by geographic region, PVT, and successional class.

Maps of management regions, PVT, cover type, initial structural stage, and initial succession age are needed to run CRBSUM. The latter two maps can be input to the model or can be generated stochastically by the model. A limitation of CRBSUM is the simulation of disturbances as independent events in each cell, which precludes propagation of fire or insect attack across large areas or synchronization of disturbance occurrences. As a consequence, maps of disturbances generated by CRBSUM are applicable only for projection of general geographic trends, rather than for describing specific disturbance locations or year to year variability in disturbance occurrence. Limitations of the successional pathway approach include a deterministic specification of the endpoint of succession and lack of detail in defining some successional trajectories (Keane et al., 1996a). Many parameters were required to implement succession pathway models for each PVT and disturbance probabilities. Parameters were quantified using teams of experts; problems were noted with a lack of consistency and level of expertise across vegetation types (Keane et al., 1996a).

CRBSUM was used to predict coarse-scale vegetation changes for four management strategies termed consumptive demand, historical, passive management, and active management. Each management strategy was implemented by adjusting disturbance probabilities to reflect long-term management goals. The objective of the simulations was to contrast trends in landscape characteristics (e.g., changes in cover type) among alternative land management policies over a period of 100 years.

18.2.7 MAPSS

MAPSS is a biogeographic model constructed to simulate potential vegetation type, vegetation leaf area index (LAI), and site water balance and runoff (Neilson, 1995). The model incorporates thermal and site water balance constraints on vegetation. Thermal constraints are simulated with simple physiologically based rules. A process-based site water balance model integrates vegetation leaf area and stomatal conductance in canopy transpiration and soil hydrology. Because the model simulates potential climax vegetation, environmental constraints are implemented to determine vegetation carrying capacity and potential type. Therefore, it is assumed that in any given location vegetation leaf area production will be maximized to just utilize available soil water.

MAPSS includes woody vegetation (trees and shrubs) and grasses that compete for light and water. The model operates on a monthly time step. Fire, the only disturbance included in the model, is a rule-based function of available grass or shrub fuel and high summer rainfall as an index of convective activity. Vegetation is classified by rules incorporating thermal, humidity, and productivity constraints, which translate modeled life form and LAI into a vegetation type. The classification, which includes biome physiognomy (trees, shrubs, and grasses), leaf form, closed or open vegetation based on LAI, and thermal zone (tundra, taiga–boreal, boreal, temperate, subtropical, and tropical), results in 35 possible vegetation types.

The model was run over the conterminous United States and was found to simulate the distribution of forest, grassland, and desert areas with reasonable accuracy, as well as reproducing monthly runoff. The largest biogeographic error was in the prediction of the occurrence of forest in most of the prairie peninsula region of the central United States. Simulation of episodic summer droughts followed by fire may correct this problem.

MAPSS can be used across a broad range of spatial scales to predict changes in potential vegetation and water regimes under altered climates. An advantage of the model is its incorporation of the linkage between vegetation and water balance processes along with the ability to simulate changes in vegetation type. Neilson (1995) proposes that, after minor enhancements, the model be used for simulating the effects of alternative land-use patterns on vegetation and water resources.

As described by Neilson (1995), MAPSS is calibrated for overall accuracy and would be expected to perform less well for specific sites. Because the model operates on mean monthly climate, effects of year to year climate variability such as extreme events cannot be incorporated. In addition, the fire disturbance rules are inadequate for altered climate and fire in closed forests. The effect of fire is also too severe, removing all woody vegetation (Neilson, 1995).

The model has been used to compare worldwide changes in LAI and terrestrial water balance among five different general circulation models under predicted climate conditions where current CO_2 levels are doubled ($2 \times CO_2$) to determine if there are regionally consistent sensitivities to climate change (Neilson and Marks, 1994). The model was one of six models (including the BIOME-BGC and CENTURY models discussed next) implemented to compare simulations over the conterminous United States under current climate and under a range of climate change scenarios provided by three general circulation models (VEMAP Members, 1995).

18.3 Models of Ecosystem Function

18.3.1 FOREST-BGC and BIOME-BGC

FOREST-BGC was originally developed to simulate forest stand development through careful calculation of a daily plant water budget (Running and Coughlan, 1988; Running and Gower, 1991). It has been generalized to other biomes in the model BIOME-BGC (Running and Hunt, 1993; Hunt et al., 1996). Both FOREST-BGC and BIOME-BGC simulate hydrologic, carbon, and nitrogen cycles for generalized ecosystems (Hunt and Running, 1992; Running and Hunt, 1993). The models have been used to map photosynthesis, respiration, evapotranspiration, decomposition, nitrogen mineralization, and biospheric carbon exchange at point, regional, and global scales (Running and Hunt, 1993; Hunt et al., 1996).

BIOME-BGC runs on both daily and annual time steps concurrently. The daily model simulates photosynthesis, autotrophic and heterotrophic respiration, and a hydrologic budget. The annual time step allocates carbon and nitrogen among leaves, coarse and fine roots, litter, and soil. Inputs to the models include daily temperature and precipitation data from weather stations, biome type and LAI derived from remote sensing imagery, topography (slope, elevation, aspect), and soils data (texture, depth, water holding capacity) (Nemani et al., 1991; Hunt et al., 1996).

The use of LAI in the models to define the canopy makes acquisition of an extensive database of tree heights and diameters unnecessary (Running et al., 1989). The strong control of model processes by climate is an advantage, because daily climate data are among the most readily available data globally (Running and Hunt, 1993). Use of the models is limited for some applications by the entities simulated (e.g., changes in stand carbon over time), which may not address some ecological assessment needs. In addition, the models are computationally intensive and cannot be used to predict changes in species distribution or internal changes in LAI (Friend et al., 1993).

FOREST-BGC has been used to evaluate forest processes in response to site quality (McLeod and Running, 1988; Korol et al., 1991), as well as air pollution and $2 \times CO_2$ effects on forests (Kremer, 1991; Running and Nemani, 1991). More recently, BIOME-BGC has been used to estimate global net terrestrial carbon exchange and atmospheric CO_2 concentrations and global-scale responses of vegetation to $2 \times CO_2$ (Hunt et al., 1996). Attempts have been made to integrate some aspects of individual-based models with the process-based components of FOREST-BGC. One such model, HYBRID (Friend et al., 1993), is discussed in the section on individual-based models. Another model, FIRE-BGC, combines properties of an individual-based simulation model, FIRESUM (Keane et al., 1989, 1990a, 1990b), and FOREST-BGC (Keane et al., 1996b). FIRE-BGC is a mechanistic, individual tree succession model in which tree growth, organic matter decomposition, litterfall, and other processes are simulated using detailed physical relationships. Mechanistic components of FOREST-BGC provide the framework for FIRE-BGC, with FIRESUM routines added and modified to utilize FOREST-BGC information. The model includes fire and its effects on ecosystem components in a spatial context, as well as effects of insects and disease.

18.3.2 RHESSys

RHESSys, a combination of mapped data and land surface characteristics with integrated hydrologic and ecological models, allows exploration of landscape processes on a watershed scale (Band et al.,

1993). RHESSys addresses the dependency of spatial patterns of forest processes and hydrological fluxes on topography through the variation topographic relief causes to solar radiation distribution, precipitation, temperature, and soil water drainage (Band et al., 1993). By including hillslope hydrology with representations of biogeochemical processes, RHESSys represents seasonal plant productivity and evapotranspiration trends very differently than previous ecosystem models, which were restricted to homogeneous biogeochemical fluxes and vertical soil water transport equations.

RHESSys integrates FOREST-BGC, described previously, with TOPMODEL (Beven and Kirkby, 1979) to investigate the distributed feedbacks between ecological and hydrological processes at the watershed scale. TOPMODEL provides for lateral subsurface drainage of soil water from hillslopes and production of saturation runoff from partial contributing areas of the watershed. It also distributes soil moisture on the basis of hillslope position, introducing important topographic controls on canopy processes and forest growth through the zonation of soil moisture deficits. Daily weather is similarly distributed across the landscape with MT-CLIM (Running et al., 1987). This subroutine extrapolates from a base meteorological station to adjust and distribute temperature, relative humidity, precipitation, and solar radiation across the landscape by incorporation of elevational lapse rates, atmospheric optical depth, and illumination angles using site elevation, gradient, and exposure.

The spatial resolution of processes and surface representation is given two ways: hydrologically, where hillslopes and stream reaches are explicitly located within a watershed (Band, 1989), and by surface properties, derived from digital terrain data and remotely sensed imagery, that are statistically summarized by hillslope. Input data include vegetation type and LAI; digital elevation models for calculation of surface slope, aspect, elevation, and the contributing drainage area for each pixel (Band and Wood, 1988; Band, 1989); and soil maps for calculation of soil depth, texture, and water transmissivity. Meteorological data include daily maximum and minimum temperature, precipitation, relative humidity or dewpoint, and solar radiation, if available (Running et al., 1987).

There are several advantages to coupling hydrological and ecological models, in addition to that of more realistic heterogeneous representation of forest or ecosystem processes: the model can be employed to examine spatial fluxes of materials and to integrate terrestrial and aquatic processes. In addition, the digital terrain data allow the user to determine the desired level of spatial heterogeneity, thus allowing for expansion from local to regional scale simulations (Band, 1989, 1991). Use of the system may be limited by the availability of high-quality soils data (Nemani et al., 1993). Topographically based corrections to coarse-resolution soils maps can be implemented in RHESSys, but the effectiveness of such corrections depends on the accuracy of the digital elevation model used.

RHESSys has been used to address the effects of spatial scale on representation of ecological and hydrologic processes (Lammers et al., 1997), to determine whether the flushing of nitrogen from forested catchments into streams is controlled by topographic properties (Creed et al., 1996), to examine the effects of land-use change on ecological and hydrologic processes (Baron et al., 1998a), and to explore watershed responses to climate change (Baron et al., 1998b).

18.3.3 CENTURY

The CENTURY terrestrial ecosystem model simulates the major pathways for carbon and nitrogen exchange for all major vegetation types of the world (VEMAP Members, 1995; Schimel et al., 1996). Initially developed to simulate fluxes of carbon, nitrogen, and sulfur for grasslands, it has been expanded to address management questions related to agricultural cropping and grazing practices (e.g., Paustian et al., 1990; Holland et al., 1992; Cole et al., 1993; Rasmussen and Parton, 1994; Lyon et al., 1995; Metherell et al., 1995), forest responses to disturbance (Sanford et al., 1991), ecosystem responses to nitrogen deposition (Baron et al., 1994; Wedin and Tilman, 1996), and land-use and climate change scenarios (Burke et al., 1991; Schimel et al., 1991, 1994; VEMAP Members, 1995). CENTURY simulates primary productivity, soil nutrient dynamics, soil water, trace gas emissions, and hydrologic losses of nitrogen over long periods of time (50 to 2000 years) (Parton et al., 1987, 1988, 1994; Schimel et al., 1990, 1991, 1994, 1996; Ojima et al., 1994).

The key feature of CENTURY is a soil organic matter model that integrates changes in production and decomposition over time. Organic matter is central to the cycling of plant nutrients, influences soil water relations and erosion, and is important to soil structure. In CENTURY, soil organic matter is made up of three fractions: an active fraction of soil carbon and nitrogen consisting of live microbes and microbial products (1- to 5-year turnover time), a slow fraction that is more resistant to decomposition (20- to 40-year turnover time), and a passive fraction that is highly resistant to decomposition (200- to 1500-year turnover time) (Parton et al., 1987). Different vegetation submodels are used in conjunction with the soil organic matter submodel.

The model runs on a monthly time step, although a hydrologic version under development will run daily (Parton et al., 1994). Input variables include monthly mean maximum and minimum air temperature, precipitation, lignin content of plant material, plant nitrogen, phosphorus, and sulfur content, soil texture, atmospheric and soil nitrogen inputs, and initial soil nutrient levels (Metherell et al., 1993). A gridded version of the model allows spatial representation (Schimel et al., 1996). Disturbances such as grazing, fire, and storms have been incorporated (Sanford et al., 1991; Holland et al., 1992). The model is limited for some applications because vegetation type is considered only indirectly and changes in species composition cannot be predicted.

CENTURY can be joined with a number of other models to address more comprehensive ecosystem properties. It has been linked with a mesoscale atmospheric model, RAMS (Pielke et al., 1992), to better understand the feedbacks between land surface and atmospheric responses to climate. The simulation model MC1 incorporates biogeochemical processes from CENTURY with vegetation distribution functions from MAPSS (see Section 18.2.7) to represent vegetation and ecosystem process responses to changing climate (Lenihan et al., 1998; Daly et al., 2000).

18.3.4 CELSS

The CELSS model is a process-based wetland simulation model implemented for a marsh–estuarine system in southern Louisiana (Sklar et al., 1985; Sklar and Costanza, 1986, 1991). Each of nearly 2500 interconnected 1-km^2 cells contains a simulation model with eight state variables. Cells are connected through exchange of water and materials. Each cell is assigned to one of seven possible habitat types. If environmental conditions change sufficiently such that the environment is inappropriate for the assigned habitat type, succession occurs by switching habitat-type parameters to a new, more representative set of parameters. Interactions among water storage and connectivity (a function of habitat type, drainage density, waterway orientation, and levee height) at cell boundaries influence sediment deposition and erosion, which are believed to be important in habitat succession and productivity of the area (Sklar and Costanza, 1991).

The model simulated habitat changes over the period 1956 to 1983, correctly predicting 83% of 1983 verification data. A range of past and future climate scenarios and management scenarios was also modeled (Costanza et al., 1990). The advantages of using a model of this type are the explicit incorporation of spatial interactions and the linkage of cause and effect by mechanistic processes. The relatively coarse spatial and temporal scales and computational requirements may limit the use of the model for some applications (Sklar and Costanza, 1991).

18.4 Validation of Models

Model performance must be evaluated in relation to its purpose to assess the applicability of the results to the problem at hand (Grant, 1988; Hunsaker et al., 1993). This is particularly important when model output is used to help solve management problems (Bart, 1995). Models should be validated by comparing their outputs with relevant data not used in model construction or parameterization. However, such data may not be available, particularly for models simulating broad spatial and temporal scales. The models described in this chapter differ in the extent to which they have been validated.

The variety of validation tests to determine the reliability of individual-based plant models has included prediction of independent tree diameter increment data, forestry yield tables, change in species composition along environmental gradients, and reconstruction of vegetation under past climates (see Shugart and Smith, 1996, Table 18.1, for references). A recent version of ZELIG implemented in the western Cascades, Oregon, was tested by comparing modeled tree species basal area over a 500-year period with extensive field data (Hansen et al., 1995). However, no independent data were available to validate model functions simulating bird species densities in this model. Validation of the gradient model for Glacier National Park was conducted for overstory species composition, in which the percent similarity between observed and predicted stand composition was greater than 70% for 91% of test stands (Kessell, 1979). In addition, very accurate predictions of fire spread rate and fire perimeters were obtained in tests of fire behavior for three fires. Behavior of CRBSUM was validated using "known" conditions in which a historical vegetation-cover-type map was used as input to run the model for 100 years under a consumptive demand management scenario (Keane et al., 1996a). The output cover-type map was compared with a current vegetation-cover-type map. As described, the test produced promising but unsatisfactory results due in part to errors in the historical and current vegetation maps and consumptive demand disturbance probabilities. Predicted runoff from the MAPSS model was validated over the conterminous United States and was compared to two runoff maps constructed for the globe and Asia (Neilson and Marks, 1994; Neilson, 1995). A bias toward underprediction of runoff by the model, detected in all tests, was ascribed to underestimation of precipitation in mountainous terrain and increased observed runoff resulting from land cover conversions (Neilson and Marks, 1994). Vegetation distribution predicted by MAPSS was partially validated over the entire globe using a global existing vegetation data set (Olson et al., 1983) and output from another biogeographic model, BIOME (Prentice et al., 1992).

Regional evapotranspiration and net photosynthesis estimates produced by FOREST-BGC were difficult to validate directly (Running et al., 1989). Observed and predicted evapotranspiration were similar in two watersheds in the western Montana study area, and comparable water balance simulations were successfully validated against stand-level field data (e.g., timing and magnitude of seasonal leaf water potential and seasonal soil moisture depletion) (Donner and Running, 1986; Nemani and Running, 1989; Running et al., 1989). Comparison of predicted net photosynthesis with observed stem volume growth in six sites near the study area yielded $R^2 = 0.94$ (McLeod and Running, 1988). The CENTURY model was validated by simulating steady-state soil carbon and nitrogen levels and aboveground plant production for 24 sites in the Great Plains and then comparing simulation values with mapped plant production and soil carbon and nitrogen levels at these sites (Parton et al., 1987). The model adequately represented the effects of soil texture and climate on soil carbon and nitrogen in the Great Plains and was judged to have done an excellent job of simulating aboveground plant production. It was not possible to validate the CELSS model because of its highly aggregated and generalized implementation (Sklar et al., 1985).

18.5 Conclusion

The models considered in this paper differ in spatial and temporal extent and resolution as well as in the primary entities simulated (Table 18.1). Transition matrix models have been constructed for a wide variety of spatial and temporal scales. Most other ecosystem structure models have been implemented at finer spatial extents and resolutions than most of the ecosystem function models considered; the ecosystem structure model MAPSS is a notable exception. Determination of the appropriate variables and hierarchical level of organization to be modeled is critical for successful use of a simulation model in an ecological assessment (Allen and Starr, 1982; Sklar and Costanza, 1991). It may not be straightforward or even possible to implement a model at a different resolution than that for which it was developed (Turner et al., 1995). Therefore, the level of resolution of the model should closely match the resolution required to address the objectives of the assessment, and modeling results should be explicitly interpreted in view of the model scale (Baker, 1989a; Hunsaker et al., 1993). Models should be selected not only for appropriate scale, but also for appropriate level of complexity. The models discussed differ in the complexity with which processes are represented and in the incorporation of spatial feedbacks among model cells (Sklar and Costanza, 1991). For example, successional dynamics can be simulated with cell to cell interactions using LANDIS or the individual-based plant model ZELIG (Urban, 1990) or without such spatial interactions using CRBSUM (Keane et al., 1996a). Models should be selected for use in an ecological assessment that contain just enough complexity to adequately meet assessment objectives (Holt et al., 1995).

Lack of necessary data needed to parameterize a model or provide initial conditions may be an important factor limiting applicability of a model to a particular problem. When existing data are not available, implementation of a model may depend on whether users can afford to conduct studies or make measurements required to test and run the model (Turner et al., 1995). In addition, consideration should be given to the problem of propagating uncertainty in existing data through their use as model inputs (Hunsaker et al., 1993). Mapped data used in spatially explicit models are often generated by interpolation methods based on a finite number of observations, potentially leading to increased uncertainty in model output (Hunsaker et al., 1993).

In conclusion, simulation models may be the only means for investigating some ecological phenomena (Baker, 1989a). In such cases, careful selection of appropriately scaled, robust (verified and validated) models is needed to meet the objective of an ecological assessment and to allow the consequences of alternative management or conservation strategies to be identified.

18.6 Suggested Reading

Baker (1989a), Hunsaker et al. (1993), Shugart (1998), Sklar and Costanza (1991), and Turner and Dale (1991) review various approaches to simulation modeling with an emphasis on spatial models of landscape change.

18.7 References

Aber, J. D.; Melillo, J. M. 1982. *FORTNITE: a computer model of organic matter and nitrogen dynamics in forest ecosystems*. Madison: Univ. of Wisconsin Research Bulletin #R3130.

Acevedo, M. F.; Urban, D. L.; Ablan, M. 1996. Landscape scale forest dynamics: GIS, gap, and transition models. In: Goodchild, M. F.; Steyaert, L. T.; Parks, B. O., eds. *GIS and environmental modeling: progress and research issues*. Fort Collins, CO: GIS World Books: 181–185.

Allen, T. F. H.; Starr, T. B. 1982. *Hierarchy: perspectives for ecological complexity*. Chicago: University of Chicago Press.

Baker, W. L. 1989a. A review of models of landscape change. *Landscape Ecol.* 2:111–133.

Baker, W. L. 1989b. Landscape ecology and nature reserve design in the Boundary Waters Canoe Area, Minnesota. *Ecology* 70:23–35.

Band, L. E. 1989. A terrain based watershed information system. *Hydrol. Process.* 3:151–162.

Band, L. E. 1991. Distributed parameterization of complex terrain. *Surv. Geophys.* 12:249–270.

Band, L. E.; Wood, E. F. 1988. Strategies for large-scale, distributed hydrologic simulation. *Appl. Math. Comput.* 27:23–37.

Band, L. E.; Patterson, P.; Nemani, R.; Running, S. W. 1993. Forest ecosystem processes at the watershed scale: incorporating hillslope hydrology. *Agr. Forest Meteorol.* 63:93–126.

Baron, J.; Ojima, D. S.; Holland, E. A.; Parton, W. J. 1994. Analysis of nitrogen saturation potential in Rocky Mountain tundra and forest: implications for aquatic systems. *Biogeochemistry* 27:61–82.

Baron, J. S.; Hartman, M. D.; Band, L. E.; Lammers, R. L. 1998a. *Sensitivity of high elevation Rocky Mountain watersheds to climate change. Proceedings of the fifth national watershed coalition annual meeting*, May 1997. Reno, NV: 269–273.

Baron, J. S.; Hartman, M. D.; Kittel, T. G. F.; Band, L. E.; Ojima, D. S.; Lammers, R. B. 1998b. Effects of land cover, water redistribution, and temperature on ecosystem processes in the South Platte Basin. *Ecol. Appl.* 8:1037–1051.

Bart, J. 1995. Acceptance criteria for using individual-based models to make management decisions. *Ecol. Appl.* 5:411–420.

Beven, K. J.; Kirkby, M. J. 1979. A physically-based, variable contributing area model of basin hydrology. *Hydrolog. Soc. Bull.* 24:43–69.

Bonan, G. B. 1989a. A computer model of the solar radiation, soil moisture, and soil thermal regimes in boreal forests. *Ecol. Model.* 45:275–306.

Bonan, G. B. 1989b. Environmental factors and ecological processes controlling vegetation patterns in boreal forests. *Landscape Ecol.* 3:111–130.

Bonan, G. B. 1990a. Carbon and nitrogen cycling in North American boreal forests. I. litter quality and soil thermal effects in interior Alaska. *Biogeochemistry* 10:1–28.

Bonan, G. B. 1990b. Carbon and nitrogen cycling in North American boreal forests. II. biogeographic patterns. *Can. J. For. Res.* 20:1077–1088.

Botkin, D. B. 1992. *The ecology of forests: theory and evidence*. Oxford, UK: Oxford University Press.

Botkin, D. B.; Janak, J. F.; Wallis, J. R. 1972. Some ecological consequences of a computer model of forest growth. *J. Ecol.* 60:849–872.

Bugmann, H. K. M.; Fischlin, A. 1996. Simulating forest dynamics in a complex topography using gridded climatic data. *Climatic Change* 34:201–211.

Bugmann, H. K. M.; Yan, X.; Sykes, M. T.; Martin, P.; Lindner, M.; Desanker, P. V.; Cumming, S. G. 1996. A comparison of forest gap models: model structure and behavior. *Climatic Change* 34:289–313.

Burke, I. C.; Kittel, T. G. F.; Lauenroth, W. K.; Snook, P.; Yonker, C. M.; Parton, W. J. 1991. Regional analysis of the Central Great Plains: sensitivity to climate variability. *BioScience* 41:685–692.

Childress, W. M.; Rykiel, Jr., E. J.; Forsythe, W.; Li, B. L.; Wu, H. 1996. Transition rule complexity in grid-based automata models. *Landscape Ecol.* 11:257–266.

Coffin, D. P.; Lauenroth, W. K. 1990. A gap dynamics simulation model of succession in a semiarid grassland. *Ecol. Model.* 49:229–266.

Coffin, D. P.; Lauenroth, W. K. 1996. Transient responses of North-American grasslands to changes in climate. *Climatic Change* 34:269–278.

Cole, C. V.; Paustian, K.; Elliott, E. T.; Metherell, A. L.; Ojima, D. S.; Parton, W. J. 1993. Analysis of agroecosystem carbon pools. *Water Air Soil Pollut.* 70: 357–371.

Costanza, R.; Sklar, F. H.; White, M. L. 1990. Modeling coastal landscape dynamics. *BioScience* 40:91–107.

Creed, I. F.; Band, L. E.; Foster, N. W.; Morrison, I. K.; Nicolson, J. A.; Semkin, R. S.; Jeffries, D. S. 1996. Regulation of nitrate-N release from temperate forests: a test of the N flushing hypothesis. *Water Resour. Res.* 32:3337–3354.

Dale, V. H.; Shugart, H. H. 1985. A comparison of tree growth models. *Ecol. Model.* 29:145–169.

Dale, V. H.; Hemstrom, M.; Franklin, J. 1986. Modeling the long-term effects of disturbances on forest succession, Olympic Peninsula, Washington. *Can. J. For. Res.* 16:56–67.

Dale, V. H.; Gardner, R. H.; DeAngelis, D. L.; Eagar, C. C.; Webb, J. W. 1991. Elevation-mediated effects of balsam woolly adelgid on southern Appalachian spruce–fir forests. *Can. J. For. Res.* 21:1639–1648.

Daly, C.; Bachelet, D.; Lenihan, J.; Neilson, R. P.; Parton, W. J.; Ojima, D. 2000. Dynamic simulation of tree-grass interactions for global change studies. *Ecol. Appl.* 10:449–469.

Donner, B. L.; Running, S. W. 1986. Water stress response after thinning *Pinus contorta* stands in Montana. *For. Sci.* 32:614–625.

Dunning, J. B., Jr.; Stewart, D. J.; Danielson, B. J.; Noon, B. R.; Root, T. L.; Lamberson, R. H.; Stevens, E. E. 1995. Spatially explicit population models: current forms and future uses. *Ecol. Appl.* 5:3–11.

Friend, A. D.; Shugart, H. H.; Running, S. W. 1993. A physiology-based gap model of forest dynamics. *Ecology* 74:792–797.

Ginsberg, R. B. 1971. Semi-Markov processes and mobility. *J. Math. Soc.* 1:233–262.

Ginsberg, R. B. 1972. Critique of probabilistic models: application of the semi-Markov model to migration. *J. Math. Soc.* 2:83–103.

Grant, W. E. 1988. Models for conservation and wildlife management. *Ecol. Model.* 41:325–326.

Green, D. G. 1989. Simulated effects of fire, dispersal, and spatial pattern on competition within forest mosaics. *Vegetatio* 82:139–153.

Hall, F. G.; Strebel, D. E.; Sellers, P. J. 1988. Linking knowledge among spatial and temporal scales: vegetation, atmosphere, climate and remote sensing. *Landscape Ecol.* 2:3–22.

Hansen, A. J.; Garman, S. L.; Weigand, J. F.; Urban, D. L.; McComb, W. C.; Raphael, M. G. 1995. Alternative silvicultural regimes in the Pacific Northwest: simulations of ecological and economic effects. *Ecol. Appl.* 5:535–554.

He, H. S.; Mladenoff, D. J. 1999. Dynamics of fire disturbance and succession on a heterogeneous forest landscape: a spatially explicit and stochastic simulation approach. *Ecology* 80:81–99.

Henderson, W.; Wilkins, C. W. 1975. The interaction of bushfires and vegetation. *Search* 6:130–133.

Holland, E. A.; Parton, W. J.; Detling, J. K.; Coppock, D. L. 1992. Physiological responses of plant populations to herbivory and their consequences for ecosystem nutrient flow. *Amer. Naturalist* 140:685–706.

Holt, R. D.; Pacala, S. W.; Smith, T. W.; Liu, J. 1995. Linking contemporary vegetation models with spatially explicit animal population models. *Ecol. Appl.* 5:20–27.

Humphries, H. C.; Coffin, D. P.; Lauenroth, W. K. 1996. An individual-based model of alpine plant distributions. *Ecol. Model.* 84:99–126.

Hunsaker, C. T.; Nisbet, R. A.; Lam, D. C. L.; Browder, J. A.; Baker, W. L.; Turner, M. G.; Botkin, D. B. 1993. Spatial models of ecological systems and processes: the role of GIS. In: Goodchild, M. F.; Parks, B. O.; Steyaert, L. T., eds. *Environmental modeling with GIS*. Oxford, UK: Oxford University Press: 248–264.

Hunt, E. R., Jr.; Running, S. W. 1992. Simulated dry matter yields for aspen and spruce stands in North American boreal forest. *Can. J. Remote Sens.* 18:126–133.

Hunt, E. R., Jr.; Piper, S. C.; Nemani, R.; Keeling, C. D.; Otto, R. D.; Running, S. W. 1996. Global net carbon exchange and intra-annual atmospheric CO_2 concentrations predicted by an ecosystem process model and three-dimensional atmospheric transport model. *Global Biogeochem. Cycle.* 10:431–456.

Keane, R. E.; Arno, S. F.; Brown, J. K. 1989. *FIRESUM—an ecological process model for fire succession in western conifer forests*. Gen. Tech. Rep. INT-266. Ogden, UT: U.S. Dept. Agric., For. Serv., Intermountain Res. Sta.

Keane, R. E.; Arno, S. F.; Brown, J. K. 1990a. Simulating cumulative fire effects in ponderosa pine/Douglas-fir forests. *Ecology* 71:189.

Keane, R. E.; Arno, S. F.; Brown, J. K.; Tomback, D. F. 1990b. Modelling stand dynamics in whitebark pine (*Pinus albicaulis*) forests. *Ecol. Model.* 51:73–95.

Keane, R. E.; Long, D. G.; Menakis, J. P.; Hann, W. J.; Bevins, C. D. 1996a. *Simulating coarse-scale vegetation dynamics using the Columbia River basin succession model—CRBSUM*. INT-GTR-340. Ogden, UT: U.S. Dept. Agric., For. Serv., Intermountain For. and Range Exp. Sta.

Keane, R. E.; Morgan, P.; Running, S. W. 1996b. *FIRE-BGC—a mechanistic ecological process model for simulating fire succession on coniferous forest landscapes of the Northern Rocky Mountains*. INT-RP-484. Ogden, UT: U.S. Dept. Agric., For. Serv., Intermountain Res. Sta.

Kessell, S. R. 1976. Gradient modeling: a new approach to fire modeling and wilderness resource management. *Environ. Manage.* 1:39–48.

Kessell, S. R. 1977. Gradient modeling: a new approach to fire modeling and resource management. In: Hall, C. A. S.; Day, Jr., D. W., eds. *Ecosystem modeling in theory and practice: an introduction with case histories*. New York: John Wiley & Sons: 575–605.

Kessell, S.R. 1979. Phytosociological inference and resource management. *Environ. Manage.* 3:29–40.

Kessell, S. R. 1990. An Australian geographical information and modeling system for natural area management. *Int. J. Geogr. Inf. Syst.* 4:333–362.

Kessell, S. R.; Cattelino, P. J. 1978. Evaluation of a fire behavior information integration system for southern California chaparral wildlands. *Environ. Manage.* 2:135–159.

Kessell, S. R.; Fischer, W. C. 1981. *Predicting postfire plant succession for fire management planning*. INT-GTR-94. Ogden, UT: U.S. Dept. Agric., For. Serv., Intermountain For. Range Exp. Sta.

Kessell, S. R.; Good, R. B. 1982. *PREPLAN (Pristine Environment Planning Language and Simulator) user's guide for Kosciusko National Park*. Sydney: National Parks and Wildlife Service of New South Wales, Special Publication.

Kessell, S. R., Good, R. B.; Hopkins, A. J. M. 1984. Implementation of two new resource management information systems in Australia. *Environ. Manage.* 8: 251–270.

Korol, R. L.; Running, S. W.; Milner, K. S.; Hunt, Jr.,

E. R. 1991. Testing a mechanistic carbon balance model against observed tree growth. *Can. J. For. Res.* 21:1098–1105.

Kremer, R. G. 1991. Simulating forest response to air pollution: integrating physiological responses to sulphur dioxide with climate-dependent growth processes. *Ecol. Model.* 54:111–126.

Lammers, R. B.; Band, L. E.; Tague, C. L. 1997. Scaling behaviour of watershed processes. In: van Gardingen, P.; Foody, G.; Curran, P., eds. *Scaling up from cell to landscape.* Cambridge, UK: Cambridge University Press: 295–317.

Lenihan, J. M.; Daly, C.; Bachelet, D.; Neilson, R. P. 1998. Simulating broad scale fire severity in a dynamic global vegetation model. *Northwest Sci.* 72:91–103.

Liu, J. 1992. *ECOLECON: a spatially-explicit model for ecological economics of species conservation in complex forest landscapes.* Dissertation. Athens: University of Georgia.

Liu, J. 1993. ECOLECON: an ECOLogical–ECONomic model for species conservation in complex forest landscapes. *Ecol. Model.* 70:63–87.

Lyon, D.; Monz, C. A.; Brown, R.; Metherell, A. K. 1995. Soil organic matter changes over two decades of winter wheat–fallow cropping in western Nebraska. In: Paul, E. A.; Cole, C. V., eds. *Soil organic matter in temperate agricultural ecosystems: a site network approach.* Chelsea, UK: Lewis Publishers.

Marsden, M. A. 1983. Modeling the effect of wildfire frequency on forest structure and succession in the northern Rocky Mountains. *J. Environ. Manage.* 16: 45–62.

McKelvey; K. B.; Noon, B. R.; Lamberson, R. H. 1992. Conservation planning for species occupying fragmented landscapes: the case of the northern spotted owl. In: Kareiva, P. M.; Kingsolver, J. G.; Huey, R. B., eds. *Biotic interactions and global change.* Boston: Sinauer: 424–450.

McLeod, S. D.; Running, S. W. 1988. Comparing site quality indices and productivity in ponderosa pine stands of western Montana. *Can. J. For. Res.* 18:346–352.

Metherell, A. K.; Harding, L. A.; Cole, C. V.; Parton, W. J. 1993. *CENTURY soil organic matter model environment, technical documentation, agroecosystem version 4.0.* Unit Technical Report No. 4. Fort Collins, CO: U.S. Dept. Agric., Agric. Res. Serv.; Great Plains System Res.

Metherell, A. K.; Gambardella, C. A.; Parton, W. J.; Peterson, G. A.; Harding, L. A.; Cole, C. V. 1995. Simulation of soil organic matter dynamics in dryland wheat–fallow cropping systems. In: Lal, R., Kimball, J.; Levine, E.; Stewart, B. A., eds. *Soil management and greenhouse effect.* Boca Raton: CRC Press: 259–270.

Mladenoff, D. J.; He, H. S. 1999. Design, behavior and application of LANDIS, an object-oriented model of forest landscape disturbance and succession. In: Mladenoff, D. J.; Baker, W. L., eds. *Advances in spatial modeling of forest landscape change: approaches and applications.* Cambridge, UK: Cambridge University Press: 125–162.

Mladenoff, D. J.; Host, G. E.; Boeder, J.; Crow, T. R. 1996. LANDIS: a spatial model of forest landscape disturbance, succession, and management. In: Goodchild, M. F.; Steyaert, L. T.; Parks, B. O., eds. *GIS and environmental modeling: progress and research issues.* Fort Collins, CO: GIS World Books: 175–179.

Neilson, R. P. 1995. A model for predicting continental-scale vegetation distribution and water balance. *Ecol. Appl.* 5:362–385.

Neilson, R. P.; Marks, D. 1994. A global perspective of regional vegetation and hydrologic sensitivities and risks from climatic change. *J. Veg. Sci.* 5:715–730.

Nemani, R.; Running, S. W. 1989. Testing a theoretical climate–soil–leaf area hydrologic equilibrium of forests using satellite data and ecosystem simulation. *Agr. Forest Meteorol.* 44:245–294.

Nemani R.; Pierce, R.; Band, L. E.; Running, S. W. 1991. Forest ecosystem processes at the watershed scale: sensitivity to remotely sensed leaf area index observations. *Int. J. Remote Sens.* 14:2519–2534.

Nemani, R.; Running, S. W.; Band, L. E.; Peterson, D. L. 1993. Regional hydroecological simulation system: an illustration of the integration of ecosystem models in a GIS. In: Goodchild, M. F.; Parks, B. O.; Steyaart, L. T., eds. *Environmental modeling with GIS.* Oxford, UK: Oxford University Press: 296–304.

Ojima, D. S.; Schimel, D. S.; Parton, W. J.; Owensby, C. E. 1994. Long- and short-term effects of fire on nitrogen cycling in tallgrass prairie. *Biogeochemistry* 24:67–84.

Olson, J. S.; Watts, J. A.; Allison, L. J. 1983. *Carbon in live vegetation of major world ecosystems.* ORNL-5862. Oak Ridge, TN: Oak Ridge Natl. Lab.

Parton, W. J.; Schimel, D. S.; Cole, C. V.; Ojima, D. S. 1987. Analysis of factors controlling soil organic matter levels in Great Plains grasslands. *Soil Sci. Soc. Amer. J.* 51:1173–1179.

Parton, W. J.; Stewart, J. W. B.; Cole, C. V. 1988. Dynamics of C, N, P, and S in grassland soils: a model. *Biogeochemistry* 5:109–131.

Parton, W. J.; Schimel, D. S.; Ojima, D. S. 1994. Environmental change in grasslands: assessment using models. *Climate Change* 28:111–141.

Pastor, J.; Post, W. M. 1986. Influences of climate, soil moisture, and succession on forest carbon and nitrogen cycles. *Biogeochemistry* 2:3–27.

Pastor, J.; Bonde, J.; Johnston, C.; Naiman, R. J. 1993. Markovian analysis of the spatially dependent dynamics of beaver ponds. *Lectures Math. Life Sci.* 23:5–27.

Paustian, K.; Andren, O.; Clarholm, M.; Hansson, A. C.; Johansson, G.; Lagerlof, J.; Lindberg, T.; Pettersson, R.; Sohlenius, B. 1990. Carbon and nitrogen budgets of four agro-ecosystems with annual and perennial crops, with and without N fertilization. *J. Appl. Ecol.* 27:60–84.

Pielke, R. A.; Cotton, W. R.; Walko, R. L.; Tremback, C. J.; Lyons, W. A.; Grasso, L. D.; Nichols, M. E.;

Moran, M. D.; Wesley, D. A.; Lee, T. J.; Copeland, J. H. 1992. A comprehensive meteorological modeling system—RAMS. *Meteorol. Atmos. Phys.* 49:69–91.

Potter; M. W.; Kessell, S. R.; Cattelino, P. J. 1979. FORPLAN: a forest planning language and simulator. *Environ. Manage.* 3:59–72.

Prentice, I. C.; Cramer, W.; Harrison, S. P.; Leemans, R.; Monserud, R. A.; Solomon, A. M. 1992. A global biome model based on plant physiology and dominance, soil properties and climate. *J. Biogeography* 19:117–134.

Pulliam, H. R.; Dunning, Jr., J. B.; Liu, J. 1992. Population dynamics in complex landscapes: a case study. *Ecol. Appl.* 2:165–177.

Rasmussen, P. E.; Parton, W. J. 1994. Long-term effects of residue management in wheat/fallow: I. inputs, yield, and soil organic matter. *Soil Sci. Soc. Amer. J.* 58:523–530.

Roberts, D. W. 1996a. Landscape vegetation modelling with vital attributes and fuzzy systems theory. *Ecol. Model.* 90:175–184.

Roberts, D. W. 1996b. Modelling forest dynamics with vital attributes and fuzzy systems theory. *Ecol. Model.* 90:161–173.

Running, S .W.; Coughlan, J. C. 1988. A general model of forest ecosystem processes for regional application. *Ecol. Model.* 42:125–154.

Running, S. W.; Gower, S. T. 1991. FOREST-BGC, a general model of forest ecosystem processes for regional applications. II. dynamic carbon allocation and nitrogen budgets. *Tree Physiol.* 9:147–160.

Running, S. W.; Hunt, Jr., E. R.1993. Generalization of a forest ecosystem process model for other biomes, BIOME-BGC, and an application for global-scale models. In: Ehleringer, J. R.; Field, C., eds. *Scaling physiological processes: leaf to globe*. San Diego, CA: Academic Press: 141–158.

Running, S. W.; Nemani, R. R. 1991. Regional hydrologic and carbon balance responses of forests resulting from potential climate change. *Climatic Change* 19:349–368.

Running, S. W.; Nemani, R. R.; Hungerford, R. D. 1987. Extrapolation of synoptic meteorological data in mountainous terrain, and its use for simulating forest evapotranspiration and photosynthesis. *Can. J. For. Res.* 17:472–483.

Running, S. W.; Nemani, R. R.; Peterson, D. L.; Band, L. E.; Potts, D. F.; Pierce, L. L.; Spanner, M. A. 1989. Mapping regional forest evapotranspiration and photosynthesis by coupling satellite data with ecosystem simulation. *Ecology* 70:1090–1101.

Sanford, R. L., Jr.; Parton, W. J.; Ojima, D. S.; Lodge, D. J. 1991. Hurricane effects on soil organic matter dynamics and forest production in the Luquillo Experimental Forest, Puerto Rico: results of simulation modeling. *Biotropica* 23:364–372.

Schimel, D. S.; Parton, W. J.; Kittel, T. G. F.; Ojima, D. S.; Cole, C. V. 1990. Grassland biogeochemistry: links to atmospheric processes. *Climate Change* 17:13–25.

Schimel, D. S.; Kittel, T. G. F.; Parton, W. J. 1991. Terrestrial biogeochemical cycles: global interactions with the atmosphere and hydrology. *Tellus* 43AB: 188–203.

Schimel, D. S.; Braswell, B. H.; Holland, E. A.; McKeown, R.; Ojima, D. S.; Painter, T. H.; Parton, W. J.; Townsend, A. R. 1994. Climatic, edaphic, and biotic controls over storage and turnover of carbon in soils. *Global Biogeochem. Cycle.* 8:279–293.

Schimel, D. S.; Braswell, B. H.; McKeown, R.; Ojima, D. S.; Parton, W. J.; Pulliam, W. 1996. Climate and nitrogen controls on geography and timescales of terrestrial biogeochemical cycling. *Global Biogeochem. Cycle.* 10:677–692.

Shugart, H. H. 1984. *A theory of forest dynamics: the ecological implications of forest succession models.* New York: Springer-Verlag.

Shugart, H. H. 1998. *Terrestrial ecosystems in changing environments*. Cambridge, UK: Cambridge University Press.

Shugart, H. H.; Smith, T. M. 1996. A review of forest patch models and their application to global change research. *Climatic Change* 34:131–153.

Sklar, F. H.; Costanza, R. 1986. A spatial simulation of ecosystem succession in a Louisiana Coastal landscape. In: Crosbie, R.; Luker, P., eds. *Proceedings of the 1986 Summer Computer Simulation Conference, Society for Computer Simulation*. San Diego, CA: 467–472.

Sklar, F. H.; Costanza, R. 1991. The development of dynamic spatial models for landscape ecology: a review and prognosis. In: Turner, M. G.; Gardner, R. H., eds. *Quantitative methods in landscape ecology: the analysis and interpretation of environmental heterogeneity.* New York: Springer-Verlag: 239–288.

Sklar, F. H.; Costanza, R.; Day, J. W. 1985. Dynamic spatial simulation modelling of coastal wetland habitat succession. *Ecol. Model.* 29:261–281.

Smith, T. M.; Urban, D. L. 1988. Scale and resolution of forest structural pattern. *Vegetatio* 74:143–150.

Thiéry, J. M.; D'Herbès, J. M.; Valentin, C. 1995. A model simulating the genesis of banded vegetation patterns in Niger. *J. Ecol.* 83:497–507.

Turner, M. G.; Dale, V. H. 1991. Modeling landscape disturbance. In: Turner, M. G.; Gardner, R. H., eds. *Quantitative methods in landscape ecology: the analysis and interpretation of environmental heterogeneity.* New York: Springer-Verlag: 323–351.

Turner, M. G.; Wu, Y.; Romme, W. H.; Wallace, L. L. 1993. A landscape simulation model of winter foraging by large ungulates. *Ecol. Model.* 69:163–184.

Turner, M. G.; Wu, Y.; Wallace, L. L.; Romme, W. H.; Brenkert, A. 1994. Simulating winter interactions among ungulates, vegetation, and fire in northern Yellowstone National Park. *Ecol. Appl.* 4:472–496.

Turner, M. G.; Arthaud, G. J.; Engstrom, R. T.; Hejl, S. J.; Liu, J.; Loeb, S.; McKelvey, K. 1995. Usefulness of spatially explicit population models in land management. *Ecol. Appl.* 5:12–16.

Urban, D. L. 1990. *A versatile model to simulate forest pattern: a user's guide to ZELIG version 1.0.*

Urban, D. L.; Shugart, H. H. 1992. Individual-based models of forest succession. In: Glenn-Lewin, D. C.; Peet, R. K.; Veblen, T. T. eds. *Plant succession: theory and prediction*. London: Chapman and Hall: 249–292.

Urban, D. L.; Smith, T. M. 1989. Extending individual-based forest models to simulate large-scale environmental patterns. *Bull. Ecol. Soc. Amer*. 70:284.

Urban, D. L.; Bonan, G. B.; Smith, T. M.; Shugart, H. H. 1991. Spatial applications of gap models. *For. Ecol. Manage*. 42:95–110.

VEMAP Members.1995. Vegetation/ecosystem modeling and analysis project: comparing biogeography and biogeochemistry models in a continental-scale study of terrestrial ecosystem responses to climate change and CO_2 doubling. *Global Biogeochem. Cycle*. 9: 407–437.

Watt, A. S. 1947. Pattern and process in the plant community. *J. Ecol*. 35:1–22.

Wedin, D. A.; Tilman, D. 1996. Influence of nitrogen loading and species composition on the carbon balance of grasslands. *Science* 274:1720–1723.

Weishampel, J. F.; Urban, D. L.; Smith, J. B.; Shugart, H. H. 1992. A comparison of semivariograms from a forest transect model and remotely sensed data. *J. Veg. Sci*. 3:521–526.

Woolhouse, M. E. J.; Harmsen, R. 1987. A transition matrix model of seasonal changes in mite populations. *Ecol. Model*. 37:167–189.

19 Methods for Determining Historical Range of Variability

Hope C. Humphries and Patrick S. Bourgeron

19.1 Introduction

Previous chapters have emphasized the dynamic nature of ecosystems, including the occurrence of periodic disturbances. Consequently, current ecosystem composition, structure, and function are likely to operate within ranges of variability that arise from climatic variability, disturbance, and the effects of human activities (Bourgeron and Jensen, 1994; Kaufmann et al., 1994; Morgan et al., 1994; Cissel et al., 1998). Understanding the magnitude and direction of anthropogenic impacts requires knowledge of the range of fluctuations historically experienced by ecosystems as a result of variability in climatic conditions, disturbance regimes, and their interactions (Swetnam and Betancourt, 1998). Therefore, the determination of the historical range of variability (HRV) in key ecosystem patterns and processes is an important part of ecological assessments and results in the characterization of the range of variability in conditions to which ecosystem components (e.g., species) are adapted (Bourgeron and Jensen, 1994; Morgan et al., 1994; Swanson et al., 1994). HRV provides a baseline for evaluating anthropogenic changes and a means for identifying the potential for surprise events to occur (Holling, 1986). Historical conditions serve as a model of the functioning of ecosystems under unmodified disturbance regimes and alternative land-use scenarios. Ecosystem patterns and processes operate at multiple hierarchically structured spatial and temporal scales, and therefore the determination of HRV should be conducted at scales that both meet the objectives of the assessment and are appropriate for the patterns and processes of interest (Bourgeron et al., 1994).

In North America, HRV has been reconstructed for the period prior to significant impact from Euro-American settlement (Fulé et al., 1997; Kaufmann et al., 1998). The characterization of HRV includes species composition and structure, as well as the disturbance processes that are important determinants of biotic patterns. Such patterns occur at a variety of spatial and temporal scales. Methods for determining HRV usually have associated scales of resolution and spatial and temporal extents. The appropriateness of these method-specific scales for meeting the objectives of the assessment should be taken into account in selecting HRV methods. No single widely applicable period of time for assessing HRV exists for North America or elsewhere (Landres et al., 1999). In addition, HRV methods are limited in their ability to reconstruct many ecosystem patterns and processes (e.g., understory vegetation, animal populations, nutrient cycling); these limitations should be clearly understood by those conducting ecological assessments. This chapter describes methods for determining HRV in biotic patterns and disturbance processes (Table 19.1), their characteristic temporal and spatial scales, their advantages and limitations, and selected examples of their use, including studies that combine more than one HRV method. The methods primarily address HRV in terrestrial ecosystems, but selected methods for aquatic systems are also included. The examples span a wide range of geographic and ecological conditions.

Hope C. Humphries and Patrick S. Bourgeron wish to acknowledge partial funding provided by a Science to Achieve Results grant from the U.S. Environmental Protection Agency ("Multi-Scaled Assessment Methods: Prototype Development within the Interior Columbia Basin").

TABLE 19.1. Characteristics of historical range of variability (HRV) methods.

Method	HRV target	Temporal resolution	Spatial resolution	Techniques often used in concert	Limitations
Dendroecological: fire scar	V, D[a]	Seasonal to annual	Points, aggregated to local to regional	Charcoal analysis, repeat photography	Tree-based, lack of spatial precision
Dendroecological: stand age	V, D	Decades	Stand	Repeat photography	Tree-based
Pollen analysis	V	Annual to hundreds of years	Local to regional, depends on lake size	Plant macrofossils, charcoal analysis	Coarse taxonomic resolution, lack of spatial precision
Plant macrofossils from sediments	V	Points, associated with pollen chronologies	Local watershed	Pollen analysis, charcoal analysis	Heterogeneous distribution in sediments
Plant macrofossils from middens	V	Points	~1 ha	Pollen analysis	Restricted locations and numbers of samples
Charcoal analysis	D	Annual to hundreds of years	Local to regional, depends on lake size	Dendroecological, pollen analysis, plant macrofossils	Lack of spatial precision, heterogeneity in charcoal deposition
Land survey records	V, D	Points	Points, aggregated to local to regional	Repeat photography, historical maps	Availability of records, surveyor bias, limited temporal extent
Repeat photography: ground	V, D	Points	Points	Dendroecological methods	Photographer bias, limited temporal and spatial extent
Repeat photography: aerial	V, D	Points, may constitute time series	Local to regional	Dendroecological, land survey, historical maps	Limited temporal extent
Maps from historical data	V, D	Points	Local to regional	Land survey, repeat photography	Limited temporal extent
Simulation modeling	V, D	Annual to hundreds of years	Local to regional	Dendroecological, pollen analysis, repeat photography, historical maps	Predicted HRV
Biophysical environment characterization	Inferred V,D	Not applicable	Local to regional	Dendroecological, repeat photography	Predicted HRV

[a]V, vegetation composition; D, disturbance regimes.

19.2 Dendroecological Methods

Dendroecological methods are a major source of information about environmental variability in areas where trees grow. Tree responses to changes in climate, disturbance, insect and disease effects, and other environmental factors can be examined through tree-ring analysis. Fritts and Swetnam (1989) provide a comprehensive review of dendroecological methods, from which the following description is derived. Tree-ring sequences obtained from sections or increment cores are analyzed to date specific events, characterize disturbance regimes, and reconstruct climatic or hydrological conditions. Accurate dendrochronological dating is accomplished by cross-dating tree rings from different trees within an area based on synchrony in variations of ring characteristics among trees, resulting in an annual or seasonal temporal resolution for time sequences that can extend to several centuries (Swetnam and Baisan, 1996). Cross-dating methods can establish the exact year a ring was formed, which may not be possible with simple ring-counting techniques because of absent or doubled tree rings. Madany et al. (1982) compared simple ring counting with cross-dating and found that dates established by ring counting alone were accurate only 26% of the time.

Two applications of dendrochronological methods are the dating of fire scars and the reconstruction of stand ages. Tree-ring dated fire scars are often used to reconstruct fire histories (Fritts and Swetnam, 1989). Fire-scarred trees occur in forests in which surface fires may repeatedly scar but not kill trees; they also occur less frequently with fewer scars in forests with moderate- to high-severity fire regimes. Records from fire-scarred trees constitute minimum estimates of fire occurrence, because some fires may not have been recorded by the trees sampled (Swetnam et al., 1999). The length of the fire scar record can be extended by cross-dating records from snags and logs. Fire scar records from individual trees are points in space, and careful selection of trees for sampling may produce a detailed record of past fires if the sampled trees are well distributed spatially and have multiple well-preserved fire scars (Swetnam and Baisan, 1996). For example, in the southwestern United States, approximately 10 to 30 trees per site were sampled over a number of sites that generally had spatial extents of 10 to 100 ha (Swetnam and Baisan, 1996). Reconstruction of the spatial extent of low-intensity fires by means of fire-scar records is difficult because the fires may not affect some trees. Spatial extents can be estimated, however, by examining patterns of synchrony and asynchrony in fire dates among trees within and among sites, under the assumption that synchronous fire dates represent larger areas burned than do asynchronous fire dates (Swetnam and Baisan, 1996). Where trees have been sampled nonrandomly for fire scars, though, caution is warranted in extrapolating reconstructed fire histories outside the areas sampled (T. Veblen, personal communication).

For some ecosystems, such as subalpine forests, fire histories usually cannot be constructed from fire-scarred trees because the dominant tree species are highly fire sensitive and rarely survive fires (Fall, 1997). In these systems, dendroecological methods may be used to establish stand age the time since the last stand-replacing fire (Romme and Knight, 1981). This method has a less precise temporal resolution than does fire-scar dating because of uncertainty introduced by the time taken for trees to establish in a burned area and grow to a height detectable by dendrochronological methods (Millspaugh and Whitlock, 1995). The spatial extent of a stand-replacing event can be estimated by delineating the area occupied by the stand itself, a more exact determination of spatial scale than is possible using fire-scar methods. However, uncertainty is introduced by obliteration of the spatial pattern of older disturbances by more recent fires (Millspaugh and Whitlock, 1995).

Compilations of fire-scar records include the proceedings of a fire history workshop held in 1980 (Stokes and Dieterich, 1980), which contains fire history summaries for many areas in North America and in Sweden. Swetnam and Baisan (1996) compiled fire-scar chronologies from 63 sites in the southwestern United States and analyzed regional variability in fire regimes in a range of vegetation types to provide baseline information and show associations with climatic conditions. Barrett et al. (1997) acquired long-term records of fire histories in the interior Columbia River basin and mapped fire locations as fire episodes (defined as five-year periods having abundant and widespread fire evidence) in the region from 1540 to 1940. They also related fire episodes to regional drought and calculated mean annual acreage burned. Heyerdahl et al. (1995) assembled a fire history database for the Pacific Northwest. Maps of twentieth-century fires (1900–1993) on National Forests in the Sierra Nevada, California, were digitized (USDA Forest Service, 1996), and characteristics of mapped fire patterns were analyzed by McKelvey and Busse (1996).

Dendroecological methods have been used to reconstruct disturbances other than fire. Flood events can be detected by examining flood-scarred streambank trees (McCord, 1996; Poff et al., 1997). Stand ages can be reconstructed in riparian woodlands to investigate the effects of flooding (Baker, 1988). Tree-ring growth patterns associated with known insect or disease outbreaks can be the basis for reconstructing disturbance histories in forest stands (Baker and Veblen, 1990; Swetnam and Betancourt, 1998).

In addition to their limitation to wooded areas, dendroecological methods may not be suitable if trees have indistinct rings, rings that are not the result of an annual cycle, too little variability among rings, or too many missing rings (Fritts and Swetnam, 1989). Trees must also attain a sufficient age to provide useful information. Other important limitations of tree-ring methods are their restriction to the time periods set by maximum tree lifespans and the loss of detail in the record with time due to tree mortality (Clark, 1990). Clark (1990) obtained relatively few fire scars on red pine in northwestern Minnesota prior to 1800 because of tree mortality, but concurrent charcoal analysis (see Section 19.5) indicated that fire was equally frequent prior to 1800 in the area.

Examples

1. Fulé et al. (1997) used fire-scar and tree-ring dating to quantify the presettlement fire disturbance regime and forest structure of a ponderosa pine forest in Arizona, including determination of the mean fire interval, to establish reference presettlement conditions for comparison with current conditions and to serve as a goal for restoration.

2. In the Boundary Waters Canoe Area, Minnesota, Heinselman (1973) used fire-scarred trees and historical records such as General Land Office notes (see Section 19.6) to determine stand origin

dates, which in conjunction with aerial photographs and forest-type maps were the basis for production of stand origin and fire year maps for the period 1595 to 1972.

3. Kitzberger et al. (1997) derived fire chronologies from fire-scarred trees in northern Patagonia, Argentina, for the period 1820–1974 and detected a strong influence of annual climatic variation on fire frequency and extent, although human activities were also found to have affected fire occurrence.

4. Arno et al. (1993) developed a method for characterizing disturbance patterns in lodgepole pine forests of the northern Rocky Mountains by sampling plots in a grid system within a landscape, including fire-scar sampling and determination of age classes of trees. Synthesis of disturbance histories for grid points enabled estimation of the spatial extent and severity of past fires in the landscape.

5. Comparisons of reconstructed stand ages following stand-replacing fires were used to discern differences in fire frequency and rate of secondary succession on uplands versus valley bottoms in a subalpine watershed in the Medicine Bow Mountains, Wyoming, where fire-scarred trees are rare (Romme and Knight, 1981).

6. Veblen et al. (1991) incorporated the analysis of growth patterns in tree rings to assess the relative effects of dated fire and windthrow events on subalpine old-growth stand dynamics in the Front Range of Colorado.

7. The long-term preservation of woody stems in peatland enabled Arsenault and Payette (1997) to document a fire-induced shift from conifer to tundra vegetation in Quebec in the 16th century. Tree-ring chronologies were constructed for remains of spruce trees living prior to and following a fire in 1567–1568. A well-drained site responded immediately to fire-induced deforestation and consequent modification of snow accumulation, but a delayed response to landscape changes was observed for a protected site.

8. McCord (1996) studied stream-bank trees scarred by flood events to reconstruct the flood history of Frijoles Canyon, New Mexico, which was examined in conjunction with fire history data to detect possible correlation of floods with large-scale fires.

9. Baker (1988) determined stand ages using tree-ring analysis in riparian woodland stands on a reach of the Animas River in Colorado. Stand structure was found to be strongly influenced by the timing and character of large-scale flooding disturbances.

19.3 Pollen Content Analysis

Pollen and other plant materials such as macrofossils and charcoal may accumulate over time and be preserved in lake sediments and wetlands (Delcourt and Delcourt, 1988). When extracted from sediment cores, pollen data can be used to reconstruct the vegetation of an area over periods of up to thousands of years (Whitlock, 1992). The methodology involves collection and extrusion of sediment cores that are subsampled for pollen (as well as macrofossils and charcoal; see Sections 19.4 and 19.5) at measured stratigraphic intervals along the core (Foster and Zebryk, 1993). Berglund (1986) provides a comprehensive summary of paleoecologic techniques. A book edited by Bryant and Holloway (1985) contains summaries of pollen analyses across North America.

The number and spacing of samples taken from the sediment core determines the temporal resolution of the pollen analysis (Grimm, 1988; Whitlock, 1992). Paleoecological studies that examine changes over periods of thousands of years usually have a relatively coarse temporal resolution of 300- to 1000-year intervals (Whitlock, 1992). The finest temporal resolution is provided by sites with annually laminated (varved) sediments, but varved lakes are relatively uncommon (Millspaugh and Whitlock, 1995). Methods for determining chronological frameworks for sediment cores include radiocarbon and lead 210 dating of bulk sediment or organic fossils (Whitlock, 1992). In addition, varves, if present, can be counted, and volcanic ash layers can be associated with eruptions of known age (Clark, 1990; Whitlock, 1992). The pollen source area and hence the spatial extent sampled by pollen analysis is a function of the size of the lake or wetland; for example, small- to medium-sized lakes (1 to 50 ha) collect pollen from an area of 100 to 1000 km^2 (Jacobson and Bradshaw, 1981; Whitlock, 1992). The smaller the lake or wetland, the stronger the local pollen signal will be compared to regional pollen (Foster and Zebryk, 1993).

Tracer pollen is added to samples in known quantities to calculate pollen concentrations and pollen accumulation rates (Whitlock, 1993). Pollen grains are identified by comparison with published atlases and reference collections (Whitlock, 1993). Identification of pollen to species is often not possible, but greater taxonomic resolution may be achieved by combining pollen analysis with identification of associated plant macrofossils (see Section 19.4; Whitlock, 1993; Fall, 1997). Plant species differ in the amount of pollen produced and pollen susceptibility to destruction (Fall, 1997).

Greater accuracy in interpreting the relative abundance of taxa in the fossil pollen record can be obtained by calibrating pollen accumulation values based on influx into modern pollen traps (Whitlock, 1993; Fall, 1997). Although pollen records provide information about changes in relative plant species composition over time, patch or age class structure of vegetation cannot be detected.

EXAMPLES

1. Postglacial vegetation history was examined in the Harvard Forest, Massachusetts, at two sites that differed in size and hence in the spatial resolution of the vegetation reconstructed from each site (Foster and Zebryk, 1993). A 10-ha swamp provided information about regional changes in vegetation, whereas a 0.006-ha hollow provided a local vegetation history within a *Tsuga canadensis* forest. Postsettlement land-use activities were found to produce the greatest changes observed in regional and local vegetation in the postglacial period.

2. Pollen analysis was an important component of the reconstruction of change in Indiana oak–prairie vegetation over multiple temporal scales (Cole and Taylor, 1995). The reconstruction confirmed a gradual successional change from pine forest to prairie–oak savanna during the period A.D. 200 to 1000 and provided a historical context for rapid changes now taking place.

3. Whitlock (1993) examined pollen and plant macrofossils from lake sediments in Grand Teton and Yellowstone National parks to determine postglacial vegetation changes in response to climate, geology, and elevation. Modern pollen samples were also collected from 58 sites in the region to understand the relationship of the modern pollen rain to present-day vegetation. Pollen spectra were found to compare well with current vegetation zones, although *Pinus* and *Artemisia* pollen types were consistently overrepresented in samples.

19.4 Plant Macrofossil Assemblages

Two sources of plant macrofossils, or plant material visible to the naked eye, are available for reconstructing vegetation through time: those present in sediments from lakes and wetlands (Birks and Birks, 1980) and those deposited in middens by packrats (*Neotoma* species) and other small animals (Betancourt et al., 1990).

19.4.1 Plant Macrofossils in Sediments

Plant macrofossils are often analyzed in conjunction with pollen from sediment or peat cores. Segments of cores are washed through screens of various sizes and the plant material retained, such as seeds, fruits, and leaves, is identified through comparison with herbarium material and other reference collections (Whitlock, 1993). Because of the heterogeneous nature of macrofossil deposition in sediments, macrofossil analysis alone cannot be used to reconstruct the kinds of time sequences of vegetation change that are possible using pollen analysis (Birks and Birks, 1980). However, plant macrofossils can be associated with chronological frameworks established for pollen records taken from the same sediment sample. The relatively large size of macrofossils results in limited transport, producing a sample with a spatial resolution within the local watershed (Birks and Birks, 1980). Plant macrofossils and pollen provide complementary information for vegetation reconstruction. The consideration of plant macrofossils along with pollen records is useful because macrofossils, in contrast to pollen, can often be identified to species and can also confirm the local presence of a species (Birks and Birks, 1980; Mayle and Cwynar, 1995). In addition, some species that are infrequently represented as pollen may occur as macrofossils (e.g., *Dryas octopetala*) (Birks and Birks, 1980). Limitations of the use of plant macrofossils include their relatively small quantities and heterogeneous distribution within sediments (Birks and Birks, 1980).

EXAMPLES

1. Mayle and Cwynar (1995) investigated the pollen and plant macrofossil content of sediment from six lakes in New Brunswick and Nova Scotia, Canada, to determine vegetation response to an 800-year cooling event with rapid onset and rapid subsequent warming. Macrofossil evidence enabled Mayle and Cwynar (1995) to identify particular arctic–alpine species in the vegetation. The detection of very rapid vegetation changes (50 to 100 years), as well as lags in response for some species, has implications for understanding vegetation response to future climate change.

2. Changes in species composition in a subalpine forest were studied over the past 8000 years using pollen and plant macrofossil analysis of samples derived from a small bog in western Colorado (Fall, 1997). The combination of the two methods, along with calibration of values based on modern pollen influx, resulted in a detailed reconstruction

of relative tree species dominance over time in conjunction with changes in climate and fire regime. Plant macrofossil analysis provided crucial species-level identification for the vegetation reconstruction. Forests dominated by *Abies lasiocarpa* and *Picea engelmannii* occupied the site from 8000 to 2600 years ago, but a stable *Pinus contorta* forest has been present during the past 2600 years, perhaps reflecting a shift to drier climate conditions (Fall, 1997).

19.4.2 Plant Macrofossils in Middens

Members of the genus *Neotoma*, commonly known as packrats, share the habit of accumulating plant material in middens within caves and rock crevices or within stick houses (Wells, 1976). This habit has been extensively exploited by researchers in North America to reconstruct vegetation over the past 40,000 years (Betancourt et al., 1990). The contribution of fossil packrat middens has been especially important in arid regions of western North America, which may have relatively few other sources of fossil plant records (Van Devender and Spaulding, 1979). When deposited in a dry, protected environment and encased in crystalline urine, middens contain extremely well preserved plant fragments that are suitable for radiocarbon dating (Betancourt et al., 1986). Each midden or distinguishable layer in a midden is deposited over one to several years and therefore represents a point in time. Packrats collect plant material within foraging ranges that are usually less than 1 ha in size, resulting in a sample of vegetation whose spatial extent is very local in nature (Betancourt et al., 1986). Because middens that are preserved over time are located in caves and rocky shelters, they generally contain vegetation samples that are restricted to rocky habitats (Spaulding, 1990). No information about patch or age class structure of the vegetation is provided.

The contents of a midden result from the habitat preferences, den-building behavior, and diet of packrats (Vaughan, 1990). Some packrat species are dietary generalists and widely sample the plant species present in an area, while other species are dietary specialists (Vaughan, 1990). For this reason, packrat middens may provide a biased representation of the vegetation in an area, and it is desirable to determine which animal species is the agent of deposition whenever possible. It may be difficult to distinguish modern from ancient midden deposits in the field (Finley, 1990). Some deposits appear to have been produced within a very short time, and multiple radiocarbon dates from such deposits give a uniform age (Wells, 1976). However, in other cases, layers in a midden may have been deposited at different times, requiring that care be taken in collecting samples to avoid mixing material of different ages (Wells, 1976).

In addition to plant macrofossils, middens also contain pollen, which is derived from several sources and has been studied in conjunction with macrofossils (King and Van Devender, 1977; O'Rourke, 1991). Pollen present in packrat fecal pellets reflects the animal's dietary habits (King and Van Devender, 1977). In nonfecal portions of the midden, pollen may have been airborne, settling on the midden during its construction, or may have adhered to plants brought to the midden by packrats (King and Van Devender, 1977; O'Rourke, 1991). Therefore, nonfecal pollen is potentially contributed by both local and regional sources (Anderson and Van Devender, 1995). Midden pollen is analyzed using methods developed for pollen from sediments (King and Van Devender, 1977). The pollen analysis can be used to confirm the macrofossil data and, conversely, identification of macrofossils can narrow taxonomic assignments to pollen found in middens (King and Van Devender, 1977). However, such pollen records have been found to be highly variable between middens of similar age and location, and the spatial extent represented by the sample is unknown (i.e., can vary from local to regional; King and Van Devender, 1977).

Examples

1. A number of chapters in Betancourt et al. (1990) describe studies of postglacial vegetation history from packrat middens in various parts of the western United States. Other animals besides packrats produce middens that can be analyzed to reconstruct plant assemblages. They include porcupines (*Erethizon dorsatum*) in North America (Van Devender et al., 1984; Betancourt et al., 1986) and various rodents in South America (Markgraf et al., 1997). Chapters in Betancourt et al. (1990) describe middens from hyraxes (*Procaviidae*) and dassie rats (*Petromuriidae*) in Africa (Scott, 1990), hyrax middens in the Middle East (Fall et al., 1990), and stick-nest rats (*Leporillus*) in Australia (Nelson et al., 1990).

2. Spaulding (1990) compared separate pollen analysis and packrat midden studies conducted in localities in western North America as a means of testing pollen-based vegetation reconstructions. Comparisons enabled exploration of the strengths

and weaknesses of each method. Agreement in vegetation reconstructions between methods appeared to decrease with distance between pollen and midden sites. In addition, disparity in results between the two methods may arise from the restriction of midden data to rocky habitats and its relatively sparse geographic coverage.

3. Anderson and Van Devender (1995) examined fossil pollen from several middens and compared the results to an existing macrofossil analysis in northwest Sonora, Mexico, an area with few suitable sources of pollen from sediments. Pollen analysis from middens added both local and regional taxa to the suite of pollen types identified from macrofossils.

19.5 Charcoal Analysis in Sediments

Changes in the abundance of microscopic and macroscopic charcoal particles in lake and wetland sediments can be analyzed to determine fire frequency (Patterson et al., 1987). The temporal resolution of this method, like pollen analysis, depends on the sediment interval sampled and ranges from annual records in varved lakes to intervals of hundreds of years for very long sediment records. The spatial extent of fires inferred from charcoal analysis is a function of lake or wetland size, in which small lakes collect charcoal from a local area while large lakes receive charcoal from a region (Whitlock and Millspaugh, 1996). In addition, larger charcoal particles are assumed to be transported shorter distances in the atmosphere than smaller particles and thus indicate occurrence of fire in a more local area (Whitlock and Millspaugh, 1996). Quantification of charcoal particle abundance is frequently conducted in conjunction with pollen analysis from sediment cores, with chronologies established by varve counting or radiocarbon or lead 210 dating (Whitlock and Millspaugh, 1996). Intervals of abundant charcoal are interpreted as evidence of fire events (Millspaugh and Whitlock, 1995). Examination of pollen records in conjunction with charcoal analyses provides information about vegetation response to fire.

Reconstruction of fire histories using charcoal analysis is often combined with dendroecological methods, in which comparisons of data from increment cores and cut sections of nearby fire-scarred trees are used to date local fires (Clark, 1990). Both methods have advantages and limitations. Fire-scar chronologies are spatially more precise than charcoal analysis, but records become more sparse with time and do not extend beyond tree life-spans, although records from snags and logs can lengthen chronologies (Clark, 1990). Charcoal analyses can provide temporal sequences of thousands of years, although generally with less precise resolution than fire-scar chronologies (Fall, 1997); they can also be directly correlated with pollen records (Whitlock and Millspaugh, 1996). Comparison of fire-scar dates with charcoal profiles can be used to calibrate thresholds below which charcoal amounts are considered background levels (Millspaugh and Whitlock, 1995). Charcoal peaks above the threshold are interpreted as fire events. Charcoal analysis can also provide an important source of information about fire frequencies in systems in which trees are killed by fire and thus do not provide fire scars for dating fire events (Fall, 1997).

A number of studies have demonstrated a close relationship between peaks in various charcoal particle size classes and occurrence of local fires (Whitlock and Millspaugh, 1996). However, no measure of charcoal abundance has provided complete accuracy in detecting known local fires. No charcoal size examined by MacDonald et al. (1991) consistently identified all known local fires at Wood Buffalo National Park, Alberta, although the abundance of microscopic charcoal was found to be influenced by regional fires. Charcoal deposition patterns are a function of a number of factors, including the size, intensity, fuel type, and meteorological conditions of fires, as well as lake and watershed characteristics (Patterson et al., 1987; MacDonald et al., 1991; Whitlock and Millspaugh, 1996). In addition, charcoal studies have employed a variety of analytical techniques. Given this variability in techniques, deposition patterns, and results, Whitlock and Millspaugh (1996) recommend that different charcoal size classes and analytical techniques be compared in a fire history study to determine the best method for detecting local fires in an area.

Examples

1. Millspaugh and Whitlock (1995) constructed a 750-year fire history for the Central Plateau of Yellowstone National Park. Regional and local fires were distinguished by comparing the charcoal record of a large lake with those of nearby small lakes. In addition, local fires (within the watershed) were discriminated from regional fires in charcoal records by determining magnetic susceptibility, which is high in burned soils eroded into lake sed-

iments following fire. Close agreement of a 450-year dendrochronologic fire history with charcoal peaks was used as justification to extend the charcoal chronology to 750 years. The results indicated that the fire regime was characterized by punctuated episodes of severe fires.

2. The same 750-year period was analyzed using varved lake sediments as the source of annual charcoal records for small lakes in northwestern Minnesota (Clark, 1990). Charcoal peaks were compared to dated fire-scar records. Fire was found to be most frequent in the relatively warm and dry 15th and 16th centuries, and longer fire intervals were observed after A.D. 1600 during the Little Ice Age.

3. In Indiana, charcoal analysis over the past 4000 years showed an increase in abundance of charcoal as vegetation shifted from pine forest to oak savanna and prairie, as well as extremely high amounts of charcoal during the period 1850 to 1900 associated with intense fires that accompanied European settlement (Cole and Taylor, 1995).

4. Larsen and MacDonald (1998) combined pollen and charcoal records from varved lake sediments to determine boreal forest fire and vegetation history over an 840-year period in Wood Buffalo National Park, Alberta. A general sequence of herb, shrub, deciduous tree, and conifer pollen was observed following fires, but peak values for taxa differed among fires, suggesting that a site may not consistently undergo the same vegetation sequence following fires. This may be due to climate change, fires with varying characteristics, or rearrangement of patches of forest dominants around the lake (Larsen and MacDonald, 1998).

19.6 Land Survey Records

Land survey records can provide information about the biotic and environmental conditions that existed at the time the survey was conducted. Such information may include vegetation composition, occurrence of disturbance, and land-use patterns. Survey records generally represent points in time and a set of sample points in space whose spatial extent is determined by the extent of the survey. In North America, several sources of information describe the vegetation of an area prior to European settlement, including colonial surveys, warrant maps, and public land surveys. Key components of each of these sources are witness trees, or trees that mark boundary corners and points along survey lines (Loeb, 1987). Surveys of unsettled lands in the United States often included the identification of witness trees in areas where trees were present. Old survey records are available for most of the continental United States, and when surveys were conducted prior to settlement, witness tree records provide a source of information about presettlement tree species distributions and associations. However, care must be taken in the use of witness tree records due to possible surveyor bias in the selection of the species and sizes of witness trees (discussed further in Section 19.6.3). In addition, unreliable survey records have resulted in some cases from careless work or outright fraud (Bourdo, 1956).

19.6.1 Colonial Surveys

Land surveys conducted during the colonial era of the United States, as well as most surveys conducted east of Ohio, did not follow standardized procedures and described properties that were irregular in size and shape, unlike the rectangular public land surveys described later (Galatowitsch, 1990). However, such surveys can be a source of information about presettlement vegetation using witness tree records from surveys of unsettled lands. Loeb (1987) compiled witness tree records for unappropriated lands prior to 1790 for eastern New Jersey and southern New York and compared the resulting coarse-scale presettlement species distributions with a current forest-type map and with pollen records.

19.6.2 Warrant Maps

County surveyors produced warrant maps at the time of first settlement of an area to describe a tract of land. Witness trees were used along with posts and stone monuments to mark property corners prior to issuing a warrant conveying the property to an owner (Abrams and Ruffner, 1995). Witness tree records obtained from warrant maps were used to determine presettlement tree species distributions over physiographic units and landforms in Pennsylvania and West Virginia and to assess differences in overall percent species composition compared to present-day forest composition (Abrams and Ruffner, 1995; Abrams and McCay, 1996).

19.6.3 Public Land Surveys

The U.S. General Land Office conducted rectangular surveys of public land to delineate townships (36 square miles in size) and sections (1 square mile) within townships for states west of Pennsylvania (Galatowitsch, 1990). The U.S. Congress authorized public land surveys using the rectangular system in 1785, but only a small fraction of public land was surveyed prior to 1815 (Bourdo, 1956). Each township, section, and quarter-section corner was marked by up to four trees, whose common name, diameter, compass bearing, and distance to the corner were recorded in surveyors' field notes (Grimm, 1984). Such trees are also known as bearing trees (Bourdo, 1956). Characteristics of trees intersecting section lines were also recorded. Public land survey records represent a systematically collected estimate of tree abundance, distribution, and association (Schwartz, 1994) and can provide a relatively reliable method for reconstructing presettlement vegetation (Leitner et al., 1991), given the limitations described in the following. Most studies utilizing public land survey data have been conducted in upland forests, but survey notes have been used to establish baseline conditions of riparian habitat in Colorado and Oregon and grassland in Colorado and New Mexico (Galatowitsch, 1990). Galatowitsch (1990) lists dates of survey and location of survey notes (state government or Bureau of Land Management office) for states covered by General Land Office surveys.

Limitations in the use of survey data include the fact that public land surveys were not fully standardized prior to 1855 and instructions to surveyors changed from time to time (Bourdo, 1956). In addition, the time period over which survey data were recorded was only a few years in any given area. Surveyors may have selected witness trees that were not the closest trees to corners because they were intended to perpetuate the corner. Criteria for selection included size, age, species longevity, distance from the corner, and conspicuousness in the stand (Grimm, 1984). Consequently, witness tree records underestimate tree density and do not permit reconstruction of absolute tree density or size, but can provide relative density or size information for comparisons within the data (Grimm, 1984).

Surveyors were shown to select a disproportionate number of medium-sized trees (10 to 14 inches diameter at breast height) to provide sufficient room for the inscription of the township, range, and section on the tree, leading to bias against species that do not attain this size (Bourdo, 1956). Bias may also have existed in favor of selection of species that were easy to inscribe, such as beech, or that attained a large size, such as red pine (Whitney, 1986). However, tests of differences in post-to-tree distances between beech and sugar maple in lower Michigan did not reveal any significant preference for beech, indicating a lack of bias in species selection, at least in that location (Whitney, 1986). A check on possible biases in the selection of witness trees is provided by the descriptions written by surveyors for each 1-mile section line; such descriptions include a ranking of tree species by abundance (Leitner et al., 1991).

Public land survey information has been employed to reconstruct disturbance regimes using section line descriptions that record disturbance events, such as fire or windthrow, encountered along the survey line. Whitney (1986) reconstructed presettlement disturbance regimes in Lower Michigan using section line descriptions to estimate return times for windfall and fire. Limitations to such an approach include the small time interval over which the survey was conducted, lack of consideration of the effects of multiple events, and documentation of only the more severe events (e.g., destructive crown fires) (Whitney, 1986).

Examples

Witness tree data from public land surveys have been used to reconstruct presettlement vegetation in a number of locations in the United States, especially in the upper Midwest.

1. Grimm (1984) reconstructed vegetation of the Bigwoods and adjacent areas along the prairie–woodland border in Minnesota and related vegetation distributions to environmental factors.
2. Delcourt and Delcourt (1996) mapped presettlement vegetation cover types and conducted spatial landscape analyses in West Mackinac County, Michigan. The results included determination of the most appropriate cell size for characterizing landscape heterogeneity using survey data.
3. In Michigan's Upper Peninsula, comparison of presettlement with present-day forest composition revealed that in deer-yarding areas plant species palatable to deer or intolerant of browsing have decreased in abundance and browsing-tolerant species have increased (Van Deelen et al., 1996).
4. The relative structure and composition of presettlement vegetation were estimated from survey data along the 5th Principal Meridian in Mississippi and Arkansas (Nelson, 1997). The evidence suggested that an open woodland was present in 1815 in areas where mesophytic tree species abundance is currently increasing.
5. Survey records for 127,653 trees used to map species and community type distributions across northern Florida supported the overwhelming dominance of pine in presettlement vegetation (Schwartz, 1994).

6. Habeck (1994) used survey data and present-day sampling to quantify successional changes in ponderosa pine–Douglas fir forests in western Montana following altered fire regimes.

19.7 Repeat Photography

Repeat photography involves relocating the site of an early photograph and taking a new photograph of the same scene from the same camera position (G. F. Rogers et al., 1984). Both ground and aerial photographs have been repeated to evaluate changes in landscapes following settlement (Bahre, 1991). The photographs are points in time with spatial extents ranging from a few hectares (ground photographs) to kilometers (aerial photographs). Historical photographs, if taken prior to settlement-induced changes, can serve as benchmark records of landscape condition, and their depiction of landscapes may have an element of impartiality because they were made for purposes other than studying landscape change (G. F. Rogers et al., 1984). However, repeat photography has a number of limitations. Ground photographs are unlikely to be spatially or temporally representative samples of a region because they are fairly recent points in time that cover only a small portion of the total surface area in a region (Bahre and Shelton, 1993). They are also the result of the photographer's bias in selecting sites to photograph (G. F. Rogers et al., 1984). Some historical photographs used for repeat photography studies were taken after settlement activities had already affected an area (e.g., Hastings and Turner, 1965; Gruell et al., 1982). In addition, land-use histories are not usually available for the sites rephotographed, leading to difficulties in ascribing landscape changes to particular causes (Bahre, 1991). Repeated aerial photography has the advantage of enabling measures of areal cover to be made and providing continuous coverage over large areas. However, the time spans between repeated aerial photographs are short, because the first vertical aerial photographs were not taken in many areas until the 1930s (Bahre and Shelton, 1993).

EXAMPLES

1. Johnson (1987) repeated photographs taken in presettlement Wyoming grassland as part of the Hayden expedition of 1870 and concluded that, although sagebrush steppe, wooded areas, and river-bottoms had experienced changes in the intervening years since 1870, grassland-dominated areas had remained fairly stable.

2. Progulske (1974) repeated photographs taken during the Custer Expedition in 1874 to the Black Hills, South Dakota. The expedition photographs documented landscape conditions prior to settlement.

3. Bahre and Bradbury (1978), Gehlbach (1981), and Humphrey (1987) repeated photographs taken in the early 1890s of the international boundary markers between the United States and Mexico to determine vegetation changes and erosion effects. The photographs provide a more systematic sample of the vegetation than is present in most historical photographs because the markers are evenly spaced along the border (Bahre, 1991).

4. Rogers (1982) matched photographs taken in the Bonneville area of Utah and Idaho between 1868 and 1916. Because all the historical photographs were taken following settlement, changes prior to domestic livestock grazing could not be documented, but the comparisons were used to understand changes under continuous grazing since settlement. Trends included increasing prevalence of weedy annuals and increases in woody species such as juniper and oak.

5. Historical photographs taken between 1880 and 1915 in the Colorado Front Range were rephotographed and the comparisons supplemented with plot data collected on 24 forest stands (Veblen and Lorenz, 1991). Increasing conifer density was noted, especially in montane stands, and young ponderosa pine trees invaded former grasslands. Changes were attributed to a more mesic climate, overgrazing, and fire suppression.

6. Analyses of matches to photographs taken between 1873 and 1915 of subalpine forests in western Colorado were combined with tree-ring chronologies to distinguish patterns of mortality due to spruce beetle attack, fire, and wind (Baker and Veblen, 1990). Complex interactions among the disturbance agents produced a spatially heterogeneous mosaic of disturbance-created patches in the forest.

7. Veblen and Markgraf (1988) considered evidence from comparisons of 35 historical photographs taken between 1883 and 1913 of the forest–steppe ecotone in northern Patagonia, Argentina, along with pollen, tree-ring, and forest-stand structure analyses to evaluate vegetation changes at the ecotone over the past 100 years.

8. Lehmkuhl et al. (1994) used historical (1932 to 1959) and current (1985 to 1992) aerial photographs to determine HRV in vegetation composition and structure in six basins in eastern Oregon and Washington as part of the Eastside Forest Ecosystem Health Assessment (Everett et al., 1994). Changes in landscape pattern, wildlife habi-

tat, and insect and disease hazard were described by comparing maps derived from the aerial photographs. Mapped vegetation patches were also assigned potential fire behavior attributes to evaluate the effects of vegetation change on fire behavior and smoke production (Huff et al., 1995).

9. Bahre (1991) compared Soil Conservation Service aerial photographs taken between 1935 and 1937 with National High Altitude Photography photographs from 1983 and 1984 in conjunction with other sources of information to evaluate changes in vegetation distribution in southeastern Arizona. He concluded that anthropogenic impacts have played a substantial role in these changes.

10. Cole and Taylor (1995) examined aerial photographs covering the period 1959 to 1987 to document encroachment of forest into formerly open prairie in Indiana.

11. Miller et al. (1995) examined changes in a riparian landscape in southeastern Wyoming using aerial photographs taken in 1973 and 1990. Quantification of changes included a 75% decline in wetted area and indicated a shift from young dense cottonwood stands to older, more open stands following changes in the frequency and intensity of flooding in the North Platte River.

12. Allen and Breshears (1998) analyzed a sequence of aerial photographs taken between 1935 and 1975 to quantify an extremely rapid shift in an ecotone between ponderosa pine forest and pinyon–juniper woodland in northern New Mexico. The ecotonal shift, occurring over 2 km or more in less than five years, resulted from mortality of ponderosa pine in response to a severe drought in the 1950s. Other factors, such as fire suppression, increased pinyon–juniper density, and drought-triggered bark beetle infestation, amplified the effect of the drought on ponderosa pine. The shift has persisted for more than 40 years, resulting in greater fragmentation of forest patches and changes in ecosystem properties due to soil erosion (Allen and Breshears, 1998).

19.8 Maps from Historical Data

Some sources of historical data on vegetation pattern are either provided in the form of maps or contain sufficient detail that maps can be prepared from them.

Examples

1. Goldberg and Turner (1986) studied a set of permanent vegetation plots in the Sonoran Desert near Tucson, Arizona, in which locations of individual woody and succulent plants had been mapped as early as 1906 for some plots and subsequently remapped at irregular intervals over the next 72 years. Time series analyses revealed no consistent directional changes in vegetation composition during this period.

2. White and Mladenoff (1994) examined forest landscape transitions in north-central Wisconsin across three points in time. A presettlement forest cover map (1860) was constructed from General Land Office surveys using associations of tree species with major forest types to contour witness tree samples by forest cover type. The presettlement map was compared with forest cover maps produced in 1930 (based on inventory data and aerial photography) and 1989 (based on color infrared photography).

3. Foster (1992) constructed a vegetation map for a county in central Massachusetts based on a historical description of the landscape and vegetation within each township in the county in 1793. This map of forest vegetation prior to extensive impact by settlement was compared with data and maps describing postsettlement forest clearing followed by subsequent reforestation.

4. Chou et al. (1993) used digitized maps of major fires occurring between 1911 and 1984 in the San Jacinto Mountains, California, as a basis for constructing statistical models of probability of fire occurrence.

19.9 Simulation Modeling

Simulation models provide information about ranges of variability in patterns and processes over time. They can be used to explore the implications and limitations of reconstructed presettlement biotic distributions and disturbance regimes derived from the other methods described in this chapter. Chapter 18 describes a number of simulation models, including some of the models discussed here (transition matrix models, CRBSUM, individual-based models, and LANDIS). When selecting a simulation model, consideration should be given to operation at appropriate spatial and temporal scales, parameter and input data needs, and robustness of output as assessed by verification and validation procedures.

Examples

1. Simulation modeling was incorporated with dendroecological reconstruction of fire regimes and forest structure in an Arizona ponderosa pine forest (Covington and Moore, 1994a; 1994b). Plot-derived data on presettlement trees was entered into

ECOSIM, a multiresource forest growth-and-yield model (J.J. Rogers et al., 1984). Model output provided an estimate of changes with time in tree density, fuel loading, herbage production, streamflow, and scenic beauty index under postsettlement conditions.

2. Baker (1989) implemented transition matrix simulation models to determine whether a stable mosaic of forest patches existed during the period 1727–1868 in the Boundary Waters Canoe Area, Minnesota, based on fire history data reconstructed by Heinselman (1973). No stable patch mosaic was found at any of the spatial scales examined (units 25,000- to 400,000-ha in size). Such a result implies that even a very large conservation reserve would not experience steady-state environments, leading to potential conflict between the conservation goals of perpetuating fluctuations in landscape structure versus maintaining populations of species that depend on the presence of particular landscape structures.

3. CRBSUM is a vegetation succession model that was developed for use in the interior Columbia River basin ecological assessment (Keane et al., 1995). The model simulates changes in vegetation structural stages and cover types over time in response to stochastic disturbance. Coarse-scale vegetation changes were predicted for a set of management strategies as part of the assessment.

4. Individual-based plant models (gap models; Shugart, 1998) contain features that make them suitable for reconstructing the range of variability in species composition and structure under presettlement conditions, as well as predicting consequences of postsettlement activities. These features include incorporation of stochastically implemented disturbance processes (such as fire, flooding, and storms), climate and other environmental effects, and simulations over long time periods (Urban and Shugart, 1992). Gap models have been tested for their ability to reproduce reconstructed vegetation from pollen analysis over the postglacial period in the eastern United States, producing results that are consistent with long-term variations in forests, including temporal sequences and spatial patterns (Solomon and Shugart, 1984; Solomon and Webb, 1985; Shugart, 1998).

5. LANDIS is a simulation model that is related to gap models, but it is able to simulate larger areas than gap models by means of aggregated tree age-class distributions and longer time steps, which reduce processing constraints (Mladenoff et al., 1996; He and Mladenoff, 1999). The model is suitable for representing spatially explicit forest landscape dynamics in response to fire and windthrow disturbance.

19.10 Characterization of Biophysical Environments

In the absence of direct sources of information about the range of pattern or process variability or as a supplement to such sources (e.g., to extrapolate from sampled to unsampled areas), aspects of the biophysical environments of landscapes can be characterized to describe their expected biotic components and associated disturbance regimes (see Chapters 3 and 22). Biophysical attributes (e.g., topographic, climatic, hydrologic, or geologic variables) are selected based on their hypothesized control of biotic distributions and low temporal variability at a given scale of interest (Bourgeron et al., submitted). The resulting biophysical environments are land units in which each biophysical environment type is expected to contain a particular suite of biotic and disturbance responses (Bailey et al., 1994).

Testing of the use of biophysical environments for extrapolating HRV between spatial or temporal points has been conducted informally for some time. Stratification by categorical variables (e.g., geologic substrate) or biophysical environments has been performed for a variety of analyses, such as construction of species–environment models (e.g., Austin et al. 1990); development of fire models, both conceptual (e.g., Fischer and Clayton, 1983; Fischer and Bradley, 1987) and quantitative (e.g., Kessell, 1976; Kessell and Fischer, 1981); and characterization of forest insect and disease regimes (e.g., Harvey, 1994; Hessburg et al., 1994; Lehmkuhl et al., 1994; Filip et al., 1996). Formal assessment of biophysical environment characterization as a guide to stratifying sampling effort and extrapolating results is discussed in Chapter 7.

Examples

In areas with topographic complexity, changes to the disturbance regime may be strongly correlated with topography, enabling disturbance characteristics to be mapped based on topographic features.

1. Romme and Knight (1981) developed a graphic model of the relationships among topographic position, fire-free interval, and development of mature spruce–fir forests. The model accounts for vegetation composition differences in upland areas compared to sheltered ravines and valley bottoms in the subalpine zone of the Medicine Bow Mountains, Wyoming.

2. Camp et al. (1997) produced statistical models to predict historical fire *refugia* (forest patches minimally affected by recurrent fires) based on topographic variables in the Wenatchee Mountains,

Washington. Such models can be used to identify presettlement refugia in unsampled areas from their topographic settings.

3. Fire group models were developed using biophysical environments to characterize the fire regimes that drive vegetation succession in forested landscapes of Montana (Fischer and Clayton, 1983; Fischer and Bradley, 1987).

19.11 Methods for Aquatic Ecosystems

Some of the previously described methods can be used to characterize HRV for riparian or aquatic ecosystems, such as detection of flood events using dendroecological methods or characterization of aquatic plant species composition using pollen analysis and plant macrofossils from sediments. Methods such as repeat photography, maps from historical data, simulation modeling, and characterization of biophysical environments can be applied to aquatic as well as terrestrial ecosystems. The streamflow regime, a driving force in river ecosystems, controls key habitat parameters such as flow depth, velocity, and habitat volume (Richter et al., 1998). Long-term streamflow gauge observations provide information about the quantity, timing, and variability of a river's flow (Kondolf and Larson, 1995; Poff et al., 1997). Sediment cores extracted from lakes and wetlands characterize temporal fluctuations in physical properties, such as sediment size and chemical composition, in addition to presence of fish populations and aquatic plant species composition and abundance (Uutala, 1990; Sullivan et al., 1992; Steedman et al., 1996). Long-term fish surveys and commercial harvest data describe fish species composition and abundance over time (McIntosh et al., 1994; Kelso et al., 1996; Patton et al., 1998). In reconstructing HRV, consideration should be given to differences in the response of aquatic and terrestrial ecosystems to particular types of disturbance, such as the greater effect of flooding and erosion and the less pronounced effect of fire on aquatic than terrestrial ecosystems (Frissell and Bayles, 1996).

EXAMPLE

1. Richter et al. (1997) described a new method for setting streamflow-based ecosystem management targets on rivers where biodiversity conservation and protection of ecosystem function are management objectives. The method involves determining the range of variability in 32 hydrological parameters derived from daily streamflow values measured during a period with negligible human perturbations to the hydrological regime. River management targets are identified based on a statistical characterization of variability in the suite of hydrological parameters. Predam daily streamflow data were used to characterize the natural range of streamflow variability for the Roanoke River, North Carolina, as a basis for recommending modifications in reservoir operations rules for dams on the river. The method was also used to evaluate the impacts of dam construction on hydrologic variability in two rivers in the upper Colorado River basin in Colorado and Utah (Richter et al., 1998).

19.12 Summary

The methods for characterizing HRV described in this chapter produce information about the range of variability in vegetation patterns, disturbance regimes, or both (Table 19.1). Pollen and plant macrofossil analyses describe changes in plant species abundance with time and include both canopy and understory terrestrial species; pollen from lake sediments also contains records of aquatic plant species. Charcoal analysis identifies fire events. The other methods can potentially contribute information about both vegetation and disturbance; canopy species records are predominantly represented. Disturbance regime reconstructions for these methods can include flood events and outbreaks of insect and disease in addition to fire. Simulation modeling and characterization of biophysical environments produce predicted vegetation and disturbance HRV, rather than measuring the range of actual landscape conditions.

Methods differ in temporal and spatial extent. Some methods, such as plant macrofossils from middens, land surveys, repeat ground photography, and maps from historical data, reconstruct historical conditions at one or a few time periods. Such snapshots in time cannot characterize temporal fluctuations. Wherever possible, these methods should be used in conjunction with others that have the potential to provide sequences of change over long time periods, such as dendroecological methods, pollen analysis, macrofossils from sediment, charcoal analyses, or simulation modeling. Reconstruction of long-term temporal variation is limited by loss of records with time, the "fading record" problem (Swetnam et al., 1999). This problem is prominent in dendroecological methods, in which records are lost through tree death and decay, but records derived from most other methods are also subject to loss or degradation with time (Landres et al., 1999; Swetnam et al., 1999). Reconstructions at local scales are obtained using plant macrofossil analysis and repeat photography. Results from the other methods discussed can range from local to regional in spatial scale. When historical data are not available for a particular site, extrapolation of data from another site may be unavoidable, but the validity of such an extrapolation may depend on the distance between the sites, the "distance decay" problem (White and Walker, 1997; Swetnam et al., 1999).

All the methods described are limited in the patterns and processes they reconstruct, their spatial and temporal resolutions, and their availability in particular ecosystems. Limitations to many methods include the degree to which they describe vegetation structural characteristics, understory vegetation, spatial heterogeneity, and interactions among disturbances. Other patterns and processes potentially of interest in an assessment, such as animal populations or terrestrial biogeochemical processes, are not characterized by any of the HRV methods described. For some ecosystems, such as the grasslands of the Great Plains and Columbia Plateau, with sparse or absent historical records and extensive human alteration, reconstructing HRV may be difficult or impossible (Frissell and Bayles, 1996).

The limitations of HRV methods do not lessen the value of historical information for ecological assessments, but emphasize the need to consider as many sources of information for determining HRV as possible (Swetnam et al., 1999). As a result of methodological limitations, combinations of methods provide the most complete information about HRV and can broaden the suite of entities and spatial and temporal extents considered in an assessment (Table 19.1). Multiple HRV sources provide increased confidence in reconstructions when different lines of evidence converge (Swetnam et al., 1999). Many of the studies described in this chapter contain more than one HRV data source, such as the combinations of pollen with plant macrofossil analysis and charcoal with dendroecological analysis. Examples of other combinations of methods include one study in which information was derived from repeat photography, vegetation plots, and analyses of tree rings and pollen (Veblen and Markgraf, 1988) and another combining repeat photography, fire-scar analysis, and simulation modeling (Covington and Moore, 1994a). Shinneman and Baker (1997) evaluated a number of sources of information, such as dendroecological studies, historical photographs, maps, and written accounts, to assess HRV in ponderosa pine forests in the Black Hills of South Dakota and Wyoming. By combining multiple sources, they were able to suggest that presettlement fire disturbance regimes differed by location in the Black Hills. Kaufmann et al. (1998) reconstructed presettlement biotic and disturbance variability as part of an ecological assessment of the Sacramento Mountains, New Mexico, using General Land Office survey data, packrat and porcupine midden records, tree-ring analysis, and repeat photography, along with old survey records, historic accounts, and archeological records. Cole and Taylor's (1995) examination of change in dune prairie–oak savanna contained extensive use of multiple quantitative sources of information, including vegetation plots, aerial photography, fire-scar and multistemmed tree analysis, land survey data, and pollen and charcoal records. In this case, the use of a range of methods covering a number of temporal scales enabled current rapid rates of vegetation change to be placed in the context of 4000 years of vegetation and fire history. Information that is integrated across several scales is most likely to produce a comprehensive description of HRV (Cole and Taylor, 1995).

Determination of HRV in critical ecosystem components can provide important information for the planning process following an assessment. Cissel et al. (1998) explicitly incorporated HRV into a landscape plan developed to guide management activities in a 7600-ha planning area in western Oregon, in which landscape and watershed management objectives were based on the inferred HRV in landscape conditions and disturbance processes. The determination of historical disturbance regimes included (1) use of dendroecological methods and plot-level data to compile a fire history and construct fire-event maps, (2) interpretation of landslide and debris-flow occurrences from aerial photographs, maps of existing condition, and field surveys, and (3) assessment of riparian vegetation dynamics and disturbance history by means of a 40-year time series of aerial photographs. This case study illustrates how information on landscape conditions and disturbance regimes can be integrated for developing objectives and management prescriptions based on HRV to guide management activities.

The objectives of an ecological assessment should be the primary guide in selecting methods for obtaining HRV, including determining the appropriate spatial and temporal scales for conducting analyses. However, the characteristics of a particular area included in an assessment will determine in part which methods can be used (i.e., presence of fire-scarred trees, packrats, General Land Office survey records, historical photographs, etc.). The advantages, limitations, and spatial and temporal resolutions of methods should be clearly understood and, where possible, complementary methods should be used together to compensate for limitations in each method and paint a more complete picture of an area's HRV.

19.13 References

Abrams, M. D.; McCay, D. M. 1996. Vegetation-site relationships of witness trees (1780–1856) in the presettlement forests of eastern West Virginia. *Can. J. For. Res.* 26:217–224.

Abrams, M. D.; Ruffner, C. M. 1995. Physiographic analysis of witness-tree distribution (1765–1798) and present forest cover through north central Pennsylvania. *Can. J. For. Res.* 25:659–668.

Allen, C. D.; Breshears, D. D. 1998. Drought-induced shift of a forest–woodland ecotone: rapid landscape response to climate variation. *Proc. Nat. Acad. Sci.* 95:14839–14842.

Anderson, R. S.; Van Devender, T. R. 1995. Vegetation history and paleoclimates of the coastal lowlands of Sonora, Mexico—pollen records from packrat middens. *J. Arid Environ.* 30:295–306.

Arno, S. F.; Reinhardt, E. D.; Scott, J. H. 1993. *Forest structure and landscape patterns in the subalpine lodgepole pine type: a procedure for quantifying past and present conditions.* INT–GTR-294.Ogden, UT: U.S. Dept. Agric., For. Serv., Intermountain Res. Sta.

Arsenault, D.; Payette, S. 1997. Landscape change following deforestation at the arctic tree line in Quebec, Canada. *Ecology* 78:693–706.

Austin, M. P.; Nicholls, A. O.; Margules, C. R. 1990. Measurement of the realized qualitative niche: environmental niches of five *Eucalyptus* species. *Ecol. Monogr.* 60:161–177.

Bahre, C. J. 1991. *A legacy of change*. Tucson, AZ: University of Arizona Press.

Bahre, C. J.; Bradbury, D. E. 1978. Vegetation change along the Arizona–Sonora boundary. *Ann. Assoc. Amer. Geogr.* 68:145–165.

Bahre, C. J.; Shelton, M. L. 1993. Historic vegetation change, mesquite increases, and climate in southeastern Arizona. *J. Biogeogr.* 20:489–504.

Bailey, R. G.; Jensen, M. E.; Cleland, D. T.; Bourgeron, P. S. 1994. Design and use of ecological mapping units. In Jensen, M. E.; Bourgeron, P. S., tech. eds. *Volume II: ecosystem management: principles and applications*. PNW-GTR-318. Portland, OR: U.S. Dept. Agric., For. Serv., Pacific Northw. Res. Sta.: 95–106.

Baker, W.L. 1988. Size-class structure of contiguous riparian woodlands along a Rocky Mountain river. *Phys. Geogr.* 9:1–14.

Baker, W. L. 1989. Landscape ecology and nature reserve design in the Boundary Waters Canoe Area, Minnesota. *Ecology* 70:23–35.

Baker, W. L.; Veblen, T. T. 1990. Spruce beetles and fires in the nineteenth-century subalpine forest of western Colorado, U.S.A. *Arctic Alp. Res.* 22:65–80.

Barrett, S. W.; Arno, S. F.; Menakis, J. P. 1997. *Fire episodes in the inland Northwest (1540–1940) based on fire history data.* INT-GTR-370. Ogden, UT: U.S. Dept. Agric., For. Serv., Intermountain Res. Sta.

Berglund, B. E., editor. 1986. *Handbook of holocene paleoecology and palaeohydrology*. New York: John Wiley & Sons.

Betancourt, J. L.; Van Devender, T. R.; Rose, M. 1986. Comparison of plant macrofossils in woodrat (*Neotoma* sp.) and porcupine (*Erethizon dorsatum*) middens from the western United States. *J. Mammal.* 67:266–273.

Betancourt, J. L.; Van Devender, T. R.; Martin, P. S., editors. 1990. *Packrat middens: the last 40,000 years of biotic change*. Tucson, AZ: University of Arizona Press.

Birks, H. J. B.; Birks, H. H. 1980. *Quaternary palaeoecology*. London: Edward Arnold Ltd.

Bourdo, E. A., Jr. 1956. A review of the General Land Office survey and of its use in quantitative studies of former forests. *Ecology* 37:754–768.

Bourgeron, P. S.; Jensen, M. E. 1994. An overview of ecological principles for ecosystem management. In: Jensen, M. E.; Bourgeron, P. S., eds. *Volume II: ecosystem management: principles and applications.* PNW-GTR-318. Portland, OR: U.S. Dept. Agric., For. Serv., Pacific Northw. Res.: 45–67.

Bourgeron, P. S.; Humphries, H. C.; DeVelice, R. L.; Jensen, M. E. 1994. Ecological theory in relation to landscape evaluation and ecosystem characterization. In: Jensen, M. E.; Bourgeron, P. S., eds. *Volume II: ecosystem management: principles and applications.* PNW-GTR-318. Portland, OR: U.S. Dept. Agric., For. Serv., Pacific Northw. Res. Sta.: 58–72.

Bourgeron, P. S.; Humphries, H. C.; Jensen, M. E.; Auerbach, N. A.; Barber, J. A. Development of hierarchical land classifications based on biophysical environment criteria: a test of the predictive power of three systems. Submitted, *Landscape Ecol.*

Bryant, V. M., Jr.; Holloway, R. G., editors. 1985. *Pollen records of Late–Quaternary North American sediments*. Dallas, TX: Amer. Assoc. Stratigraphic Palynologists Foundation.

Camp, A.; Oliver, C.; Hessburg, P.; Everett, R. 1997. Predicting late-successional fire refugia pre-dating European settlement in the Wenatchee Mountains. *For. Ecol. Manage.* 95:63–77.

Chou, Y. H.; Minnich, R. A.; Chase, R. A. 1993. Mapping probability of fire occurrence in San Jacinto Mountains, California, USA. *Environ. Manage.* 17: 129–140.

Cissel, J. H.; Swanson, F. J.; Grant, G. E.; Olson, D. H.; Gregory, S. V.; Garman, S. L.; Ashkenas, L. R.; Hunter, M. G.; Kertis, J. A.; Mayo, J. H.; McSwain, M. D.; Swetland, S. G.; Swindle, K. A.; Wallin, D. O.

1998. *A landscape plan based on historical fire regimes for a managed forest ecosystem: the August Creek study*. PNW-GTR-422. Portland, OR: U.S. Dept. Agric., For. Serv., Pacific Northw. Res. Sta.

Clark, J. S. 1990. Fire and climate change during the last 750 yr in northwestern Minnesota. *Ecol. Monogr.* 60:135–159.

Cole, K. L.; Taylor, R. S. 1995. Past and current trends of change in a dune prairie/oak savanna reconstructed through a multiple-scale history. *J. Veg. Sci.* 6:399–410.

Covington, W. W.; Moore, M. M. 1994a. Southwestern ponderosa forest structure: changes since Euro-American settlement. *J. For.* 92:39–47.

Covington, W. W.; Moore, M. M. 1994b. Postsettlement changes in natural fire regimes and forest structure: Ecological restoration of old-growth ponderosa pine forests. *J. Sustainable For.* 2:153–181.

Delcourt, H. R.; Delcourt, P. A. 1988. Quaternary landscape ecology: relevant scales in space and time. *Landscape Ecol.* 2:23–44.

Delcourt, H. R.; Delcourt, P. A. 1996. Presettlement landscape heterogeneity: evaluating grain of resolution using General Land Office Survey data. *Landscape Ecol.* 11:363–381.

Everett, R.; Hessburg, P.; Jensen, M.; Bormann, B. 1994. *Volume I: executive summary*. PNW-GTR-317. Portland, OR: U.S. Dept. Agric., For. Serv., Pacific Northw. Res. Sta.

Fall, P. L. 1997. Fire history and composition of the subalpine forest of western Colorado during the Holocene. *J. Biogeogr.* 24:309–325.

Fire, P. L. 1997. Fire history and composition of the subalpine forest of western Colorado during the Holocene. *J. Biogeogr.* 24:309–325.

Fall, P. L.; Lindquist, C. A.; Falconer, S. E. 1990. Fossil hyrax middens from the Middle East: a record of paleovegetation and human disturbance. In: Betancourt, J. L.; Van Devender, T. R.; Martin, P. S., eds. *Packrat middens: the last 40,000 years of biotic change*. Tucson, AZ: University of Arizona Press: 408–427.

Filip, G. M.; Torgersen, T. R.; Parks, C. A. 1996. Insect and disease factors in the Blue Mountains. In: Jaindl, R. G.; Quigley, T. M., eds. *Search for a solution: sustaining the land, people, and economy of the Blue Mountains*. Washington, DC: American Forests: 169–202.

Finley, R. B., Jr. 1990. Woodrat ecology and behavior and the interpretation of paleomiddens. In: Betancourt, J. L.; Van Devender, T. R.; Martin, P. S., eds. *Packrat middens: the last 40,000 years of biotic change*. Tucson, AZ: University of Arizona Press: 28–42.

Fischer, W. C.; Bradley, A. F. 1987. *Fire ecology of western Montana forest habitat types*. INT–GTR-223. Ogden, UT: U.S. Dept. Agric., For. Serv., Intermountain Res. Sta.

Fischer, W. C.; Clayton, B. D. 1983. *Fire ecology of Montana forest habitat types east of the Continental Divide*. INT-GTR-141. Ogden, UT: U.S. Dept. Agric., For. Serv., Intermountain For. Range Exp. Sta.

Foster, D. R. 1992. Land-use history (1730–1990) and vegetation dynamics in central New England, USA. *J. Ecol.* 80:753–772.

Foster, D. R.; Zebryk, T. M. 1993. Long-term vegetation dynamics and disturbance history of a *Tsuga*-dominated forest in New England. *Ecology* 74:982–998.

Frissell, C. A.; Bayles, D. 1996. Ecosystem management and the conservation of aquatic biodiversity and ecological integrity. *Water Resour. Bull.* 32:229–240.

Fritts, H. C.; Swetnam, T. W. 1989. Dendroecology: a tool for evaluating variations in past and present forest environments. *Adv. Ecol. Res.* 19:111–188.

Fulé, P. Z.; Covington, W. W.; Moore, M. M. 1997. Determining reference conditions for ecosystem management of southwestern ponderosa pine forests. *Ecol. Appl.* 7:895–908.

Galatowitsch, S. M. 1990. Using the original land survey notes to reconstruct presettlement landscapes in the American West. *Great Basin Natur.* 50:181–191.

Gehlbach, F. R. 1981. *Mountain islands and desert seas*. College Station, TX: Texas A&M University Press.

Goldberg, D. E.; Turner, R. M. 1986. Vegetation change and plant demography in permanent plots in the Sonoran Desert. *Ecology* 67:695–712.

Grimm, E. C. 1984. Fire and other factors controlling the Big Woods vegetation of Minnesota in the mid-nineteenth century. *Ecol. Monogr.* 54:291–311.

Grimm, E. C. 1988. Data analysis and display. In: Huntley, B.; Webb, T., III, eds. *Vegetation history*. Dordrecht: Kluwer Academic: 43–76.

Gruell, G. E.; Schmidt, W. C.; Arno, S. F.; Reich, W. J. 1982. *Seventy years of vegetative change in a managed ponderosa pine forest in Western Montana—implications for resource management*. INT-GTR-130. Ogden, UT: U.S. Dept. Agric., For. Serv., Intermountain For. Range Exp. Sta.

Habeck, J. R. 1994. Using general land office records to assess forest succession in ponderosa pine/Douglas-fir forests in western Montana. *Northw. Sci.* 68:69–78.

Harvey, A. E. 1994. Integrated roles for insects, diseases and decomposers in fire dominated forests of the inland western United States: past, present, and future forest health. *J. Sustainable For.* 2:211–220.

Hastings, J. R.; Turner, R. M. 1965. *The changing mile: an ecological study of vegetation change with time in the lower mile of an arid and semiarid region*. Tucson, AZ: University of Arizona Press.

He, H. S.; Mladenoff, D. J. 1999. Spatially explicit and stochastic simulation of forest-landscape fire disturbance and succession. *Ecology* 80:81–99.

Heinselman, M. L. 1973. Fire in the virgin forests of the Boundary Waters Canoe Area, Minnesota. *Quaternary Res.* 3:329–382.

Hessburg, P. F.; Mitchell, R. G.; Filip, G. M. 1994. *Historical and current roles of insects and pathogens in eastern Oregon and Washington forested landscapes*. PNW-GTR-327. Portland, OR: U.S. Dept. Agric., For. Serv., Pacific Northw. Res. Sta.

Heyerdahl, E. K.; Berry, D.; Agee, J. K. 1995. *Fire history database of the western United States*. EPA/600/R-96/081. Washington, DC: Environ. Protect. Agency, Office of Res. and Develop.

Holling, C. S. 1986. The resilience of terrestrial ecosystems: local surprise and global change. In: Clark, W. M.; Munn, R. E., eds. *Sustainable development in the biosphere*. Oxford, UK: Oxford University Press: 292–320.

Huff, M. H.; Ottmar, R. D.; Alvarado, E.; Vihnanek, R. E.; Lehmkuhl, J. F.; Hessburg, P. F.; Everett, R. L. 1995. *Historical and current forest landscapes in eastern Oregon and Washington*. PNW-GTR-355. Portland, OR: U.S. Dept. Agric., For. Serv., Pacific Northw. Res. Sta.

Humphrey, R. R. 1987. *90 years and 535 miles: vegetation changes along the Mexican border*. Albuquerque, NM: University of New Mexico Press.

Jacobson, G. L., Jr.; Bradshaw, R. H. W. 1981. The selection of sites for paleovegetational studies. *Quaternary Res.* 16:80–96.

Johnson, K. L. 1987. *Rangeland through time*. Miscellaneous publication 50. Laramie, WY: Agric. Exp. Sta., University of Wyoming.

Kaufmann, M. R.; Graham, R. T.; Boyce, D. A., Jr.; Moir, W. H.; Perry, L.; Reynolds, R. T.; Bassett, R. L.; Mehlhop, P.; Edminster, C. B.; Block, W. M.; Corn, P. S. 1994. *An ecological basis for ecosystem management*. RM-GTR-246. Fort Collins, CO: U.S. Dept. Agric., For. Serv., Rocky Mountain For. Range Exp. Sta.

Kaufmann, M. R.; Huckaby, L. S.; Regan, C. S.; Popp, J. 1998. *Forest reference conditions for ecosystem management in the Sacramento Mountains, New Mexico*. RM-GTR-19. Fort Collins, CO: U.S. Dept. Agric., For. Serv., Rocky Mountain For. Range Exp. Sta.

Keane, R. E.; Long, D. G.; Menakis, J. P.; Hann, W. J.; Bevins, C. D. 1995. *Simulating coarse-scale vegetation dynamics using the Columbia River basin succession model—CRBSUM*. INT-GTR-340. Ogden, UT: U.S. Dept. Agric., For. Serv., Intermountain Res. Sta.

Kelso, J. R. M.; Steedman, R. J.; Stoddart, S. 1996. Historical causes of change in Great Lakes fish stocks and the implications for ecosystem rehabilitation. *Can. J. Fisheries Aquat. Sci.* 53(Suppl. 1):10–19.

Kessell, S. R. 1976. Gradient modeling: a new approach to fire modeling and wilderness resource management. *Environ. Manage.* 1:39–48.

Kessell, S. R.; Fischer, W. C. 1981. *Predicting postfire plant succession for fire management planning*. INT-GTR-94. Ogden, UT: U.S. Dept. Agric., For. Serv., Intermountain For. Range Exp. Sta.

King, J. E.; Van Devender, T. R. 1977. Pollen analysis of fossil packrat middens from the Sonoran Desert. *Quaternary Res.* 8:191–204.

Kitzberger, T.; Veblen, T. T.; Villalba, R. 1997. Climatic influences on fire regimes along a rain forest-to-xeric woodland gradient in northern Patagonia, Argentina. *J. Biogeogr.* 24:35–47.

Kondolf, G. M.; Larson, M. 1995. Historical channel analysis and its application to riparian and aquatic habitat restoration. *Aquat. Conserv.* 5:109–126.

Landres, P. B.; Morgan, P.; Swanson, F. J. 1999. Evaluating the utility of natural variability concepts in managing ecological systems. *Ecol. Appl.* 9:1179–1188.

Larsen, C. P. S.; MacDonald, G. M. 1998. An 840-year record of fire and vegetation in a boreal white spruce forest. *Ecology* 79:106–118.

Lehmkuhl, J. F.; Hessburg, P. F.; Everett, R. L.; Huff, M. H.; Ottmar, R. D. 1994. *Historical and current forest landscapes of eastern Oregon and Washington*. PNW-GTR-328. Portland, OR: U.S. Dept. Agric., For. Serv., Pacific Northw. Res. Sta.

Leitner, L. A.; Dunn, C. P.; Guntenspergen, G. R.; Stearns, F.; Sharpe, D. M. 1991. Effects of site, landscape features, and fire regime on vegetation patterns in presettlement southern Wisconsin. *Landscape Ecol.* 5: 203–217.

Loeb, R. E. 1987. Pre-European settlement forest composition in east New Jersey and southeastern New York. *Amer. Midland Naturalist* 118:414–423.

MacDonald, G. M.; Larsen, C. P. S.; Szeicz, J. M.; Moser, K. A. 1991. The reconstruction of boreal forest fire history from lake sediments: a comparison of charcoal, pollen, sedimentology, and geochemical indices. *Quaternary Sci. Rev.* 10:53–71.

Madany, M. H.; Swetnam, T. W.; West, N. E. 1982. Comparison of two approaches for determining fire dates from tree scars. *For. Sci.* 28:856–861.

Markgraf, V.; Betancourt, J.; Rylander, K. A. 1997. Late–Holocene rodent middens from Rio Limay, Neuquen Province, Argentina. *Holocene* 7:323–327.

Mayle, F. E.; Cwynar, L. C. 1995. Impact of the Younger Dryas cooling event upon lowland vegetation of Maritime Canada. *Ecol. Monogr.* 65:129–154.

McCord, V. A. S. 1996. Flood history reconstruction in Frijoles Canyon using flood-scarred trees. In: Allen, C.D., tech. ed. *Proceedings of the second La Mesa fire symposium*, March 29–31, 1994, Los Alamos, NM; RM-GTR-286. Fort Collins, CO: U.S. Dept. Agric., For. Serv., Rocky Mountain For. Range Exp. Sta.: 114–122.

McIntosh, B. A.; Sedell, J. R.; Smith, J. E.; Wissmar, R. C.; Clarke, S. E.; Reeves, G. H.; Brown, L. A. 1994. Historical changes in fish habitat for select river basins of eastern Oregon and Washington. *Northw. Sci.* 68:36–53.

McKelvey, K. S.; Busse, K. K. 1996. Twentieth-century fire patterns on Forest Service lands. In: *Sierra Nevada Ecosystem Project: final report to Congress, volume II, assessments and scientific basis for management options*. Davis, CA: Centers for Water and Wildland Resour., Univ. of California. 41:1119–1138.

Miller, J. R.; Schulz, T. T.; Hobbs, N. T.; Wilson, K. R.; Schrupp, D. L.; Baker, W. L. 1995. Changes in the landscape structure of a southeastern Wyoming riparian zone following shifts in stream dynamics. *Biol. Conserv.* 72:371–379.

Millspaugh, S. H.; Whitlock, C. 1995. A 750-year fire history based on lake sediment records in central Yellowstone National Park, USA. *Holocene* 5:283–292.

Mladenoff, D. J.; Host, G. E.; Boeder, J.; Crow, T. R. 1996. LANDIS: a spatial model of forest landscape disturbance, succession, and management. In: Goodchild, M. F.; Steyaert, L. T.; Parks, B. O.; Johnston, C.; Vaidment, D.; Crane, M.; Glendinning, S., eds. *GIS and environmental modeling: progress and research issues*. Fort Collins, CO: GIS World Books: 175–179.

Morgan, P.; Aplet, G. H.; Haufler, J. B.; Humphries, H. C.; Moore, M. M.; Wilson, W. D. 1994. Historical range of variability: a useful tool for evaluating ecosystem change. *J. Sustainable For.* 2:87–111.

Nelson, D. J.; Webb, R. H.; Long, A. 1990. Analysis of stick-nest rat (*Leporillus*: Muridae) middens from central Australia. In: Betancourt, J. L.; Van Devender, T. R.; Martin, P. S., eds. *Packrat middens: the last 40,000 years of biotic change*. Tucson, AZ: University of Arizona Press: 428–434.

Nelson, J. C. 1997. Presettlement vegetation patterns along the 5th Principal Meridian, Missouri Territory, 1815. *Amer. Midland Naturalist* 137:79–94.

O'Rourke, M. K. 1991. Pollen in packrat middens. *Grana* 30:337–341.

Patterson, W. A., III; Edwards, K. J.; Maguire, D. J. 1987. Microscopic charcoal as a fossil indicator of fire. *Quaternary Sci. Rev.* 6:3–23.

Patton, T. M.; Rahel, F. J.; Hubert, W. A. 1998. Using historical data to assess changes in Wyoming's fish fauna. *Conserv. Biol.* 12:1120–1128.

Poff, N.; Allan, J. D.; Bain, M. B.; Karr, J. R.; Prestegaard, K. L.; Richter, B. D.; Sparks, R. E.; Stromberg, J. C. 1997. The natural flow regime. *BioScience* 47:769–784.

Progulske, D. R. 1974. *Yellow ore, yellow hair, yellow pine: a photographic study of a century of forest ecology*. Bulletin 616. Brookings, SD: South Dakota Agric. Exp. Sta.

Richter, B. D.; Baumgartner, J. V.; Wigington, R.; Braun, D. P. 1997. How much water does a river need? *Freshwater Biol.* 37:231–249.

Richter, B. D.; Baumgartner, J. V.; Braun, D. P.; Powell, J. 1998. A spatial assessment of hydrologic alteration within a river network. *Regul. River.* 14:329–340.

Rogers, G. F. 1982. *Then and now*. Salt Lake City, UT: University of Utah Press.

Rogers, G. F.; Malde, H. E.; Turner, R. M. 1984. *Bibliography of repeat photography for evaluating landscape change*. Salt Lake City, UT: University of Utah Press.

Rogers, J. J.; Prosser, J. M.; Garrett, L. D.; Ryan, M. G. 1984. *ECOSIM: a system for projecting multiresource outputs under alternative forest management regimes*. Fort Collins, CO: U.S. Dept. Agric., For. Serv., Rocky Mountain For. Range Exp. Sta.

Romme, W. H.; Knight, D. H. 1981. Fire frequency and subalpine forest succession along a topographic gradient in Wyoming. *Ecology* 62:319–326.

Schwartz, M. W. 1994. Natural distribution and abundance of forest species and communities in northern Florida. *Ecology* 75:687–705.

Scott, L. 1990. Hyrax (Procaviidae) and dassie rat (Petromuridae) middens in paleoenvironmental studies in Africa. In: Betancourt, J. L.; Van Devender, T. R.; Martin, P. S., eds. *Packrat middens: the last 40,000 years of biotic change*. Tucson, AZ: University of Arizona Press: 398–407.

Shinneman, D. J.; Baker, W. L. 1997. Nonequilibrium dynamics between catastrophic disturbances and old-growth forests in ponderosa pine landscapes of the Black Hills. *Conserv. Biol.* 11:1276–1288.

Shugart, H. H. 1998. *Plant and ecosystem functional types: terrestrial ecosystems in changing environments*. Cambridge, MA: Cambridge University Press.

Solomon, A. M.; Shugart, H. H. 1984. Integrating forest-stand simulations with paleoecological records to examine long-term forest dynamics. In: Agren, G. I., ed. *State and change of forest ecosystems: indicators in current research*. Report Number 13. Uppsala: Swedish University of Agricultural Science: 333–357.

Solomon, A. M.; Webb, T., III. 1985. Computer-aided reconstruction of late Quaternary landscape dynamics. *Ann. Rev. Ecol. Syst.* 16:63–84.

Spaulding, W. G. 1990. Comparison of pollen and macrofossil based reconstructions of Late Quaternary vegetation in western North America. *Rev. Palaeobot. Palynol.* 64:359–366.

Steedman, R. J.; Whillans, T. H.; Behm, A. P.; Bray, K. E.; Cullis, K. I.; Holland, M. M.; Stoddart, S. J., White, R. J. 1996. Use of historical information for conservation and restoration of Great Lakes aquatic habitat. *Can. J. Fisheries Aquat. Sci.* 53(Suppl. 1): 415–423.

Stokes, M. A.; Dieterich, J. H., technical coordinators. 1980. *Proceedings of the fire history workshop*: October 20–24, Tucson, AZ; RM-81. Fort Collins, CO: U.S. Dept. Agric., For. Serv., Rocky Mountain For. Range Exp. Sta.

Sullivan, T. J.; Turner, R. S.; Charles, D. F.; Cummings, B. F.; Smol, J. P.; Schofield, C. L.; Driscoll, C. T.; Cosby, B. J.; Birks, B. J. B.; Uutala, A. J.; Kingston, J. C.; Dixit, S. S.; Bernert, J. A.; Ryan, P. F.; Marmorek, D. R. 1992. Use of historical assessment for evaluation of process-based model projections of future environmental change—lake acidification in the Adirondack Mountains, New York, USA. *Environ. Pollut.* 77:253–262.

Swanson, F. J.; Jones, J. A.; Wallin, D. O.; Cissel, J. H. 1994. Natural variability—implications for ecosystem management. In Jensen, M.E.; Bourgeron, P.S., technical editors. *Volume II: ecosystem management: principles and applications*. PNW-GTR-318. Portland, OR: U.S. Dept. Agric., For. Serv., Pacific Northw. Res. Sta.: 80–94.

Swetnam, T. W.; Baisan, C. H. 1996. Fire effects in southwestern forests. In: Allen, C.D., tech. ed. *Proceedings of the second La Mesa fire symposium*, March 29–31, 1994, Los Alamos, NM. RM-GTR-286.

Fort Collins, CO: U.S. Dept. Agric., For. Serv. Rocky Mountain For. Range Exp. Sta.: 11–32.

Swetnam, T. W.; Betancourt, J. L. 1998. Mesoscale disturbance and ecological response to decadal climatic variability in the American Southwest. *J. Climate* 11:3128–3147.

Swetnam, T. W.; Allen, C. D.; Betancourt, J. L. 1999. Applied historical ecology: using the past to manage the future. *Ecol. Appl.* 9:1189–1206.

Urban, D. L.; Shugart, H. H. 1992. Individual-based models of forest succession. In: Glenn-Lewin, D. C.; Peet, R. K.; Veblen, T. T., eds. *Plant succession: theory and prediction*. London: Chapman & Hall: 249–293.

USDA Forest Service (USFS). 1996. *Final environmental impact statement: Managing California spotted owl habitat in the Sierra Nevada National Forests of California—An ecosystem approach (CALOWL EIS)*. 2 vols. Berkeley, CA: U.S. For. Serv., Pacific Southw. Res. Sta.

Uutala, A. J. 1990. *Chaoborus* (Diptera: Chaoboridae) mandibles: paleolimnological indicators of the historical status of fish populations in acid-sensitive lakes. *J. Paleolimnol.* 4:139–152.

Van Deelen, T. R.; Pregitzer, K. S.; Haufler, J. B. 1996. A comparison of presettlement and present-day forests in two northern Michigan deer yards. *Amer. Midland Naturalist* 135:181–194.

Van Devender, T. R.; Spaulding, W. G. 1979. Development of vegetation and climate in the southwestern United States. *Science* 204:701–710.

Van Devender, T. R.; Betancourt, J. L.; Wimberly, M. 1984. Biogeographic implications of a packrat midden sequence from the Sacramento Mountains, south-central New Mexico. *Quaternary Res.* 22:344–360.

Vaughan, T. A. 1990. Ecology of living packrats. In: Betancourt, J. L.; Van Devender, T. R.; Martin, P. S., eds. *Packrat middens: the last 40,000 years of biotic change*. Tucson, AZ: University of Arizona Press: 14–27.

Veblen, T. T.; Lorenz, D. C. 1991. *The Colorado Front Range: a century of ecological change*. Salt Lake City, UT: University of Utah Press.

Veblen, T. T.; Markgraf, V. 1988. Steppe expansion in Patagonia. *Quaternary Res.* 30:331–338.

Veblen, T. T.; Hadley, K. S.; Reid, M. S. 1991. Disturbance and stand development of a Colorado subalpine forest. *J. Biogeogr.* 18:707–716.

Wells, P. V. 1976. Macrofossil analysis of wood rat (*Neotoma*) middens as a key to the Quaternary vegetational history of arid America. *Quaternary Res.* 6:223–248.

White, M. A.; Mladenoff, D. J. 1994. Old-growth forest landscape transitions from pre-European settlement to present. *Landscape Ecol.* 9:191–205.

White, P. S.; Walker, J. L. 1997. Approximating nature's variation: selecting and using reference information in restoration ecology. *Restor. Ecol.* 5:338–349.

Whitlock, C. 1992. Vegetational and climatic history of the Pacific Northwest during the last 20,000 years: implications for understanding present-day biodiversity. *Northw. Environ. J.* 8:5–28.

Whitlock, C. 1993. Postglacial vegetation and climate of Grand Teton and southern Yellowstone National Parks. *Ecol. Monogr.* 63:173–198.

Whitlock, C.; Millspaugh, S. H. 1996. Testing the assumption of fire-history studies: an examination of modern charcoal accumulation in Yellowstone National Park, USA. *Holocene* 6:7–15.

Whitney, G. G. 1986. Relation of Michigan's presettlement pine forests to substrate and disturbance history. *Ecology* 67:1548–1559.

20 Representativeness Assessments

Patrick S. Bourgeron, Hope C. Humphries, and Mark E. Jensen

20.1 Introduction

Conservation scientists, planners, and managers generally agree that the ultimate goal of conservation planning is the comprehensive protection of all aspects of biodiversity (Noss and Cooperrider, 1994); consequently maintaining all ecosystem components over time and over large areas is a primary objective (Noss, 1983). To achieve its goal, conservation planning requires knowledge of the linkages among the various components of ecosystems (e.g., species, biogeochemical processes), including human factors (see Chapters 27 and 29), across gradients of spatial and temporal variability (see Chapters 2, 3, and 26), and integrated within the socioeconomic context (see Chapter 9). Therefore, basic characterization of ecological patterns and processes of interest at all scales relevant to ecosystems (see Chapters 2, 3, 22, and 23) is needed. This knowledge should be used to formulate realistic conservation goals and strategies and to place networks of conservation areas into the proper ecological and socioeconomic context for implementation at relevant ecological, social, and economic scales (see Chapter 35).

Conservation planning has four major components: identification of conservation targets and selection, design, and management of conservation areas. Biodiversity patterns of interest are identified and mapped. All such patterns or a subset of the patterns becomes conservation targets. Selected areas that support the targets are assigned a conservation status (e.g., wilderness, nature reserve). Finally, the designated conservation areas are managed to ensure the persistence of the patterns of interest. Current broad-scale conservation assessments have been designed largely to ensure that networks of conservation areas contain all conservation targets (i.e., representation of species or communities) by identifying areas where these targets are found (e.g., USGS GAP Analysis Program, or GAP, Scott and Jennings, 1998; World Wildlife Fund ecoregional conservation assessment, Olson and Dinerstein, 1998; Ricketts et al., 1999a). Determination of the processes maintaining the targets is usually conducted during the site conservation planning phase (e.g., Poiani et al., 1998), which is concerned with the delineation of conservation area boundaries, identification and involvement of stakeholders, and development of management strategies at the landscape level. A landscape approach to conservation is used increasingly often (e.g., Kim and Weaver, 1994; Naveh, 1994; Steinitz et al., 1996; Leser and Nagel, 1998).

Representativeness assessment (RA) is a relatively recently developed process first described by Austin and Margules (1986), which originated in Australia as an aid to conservation planning (see also Mackey et al., 1988). As formulated by Austin and Margules (1986), it differs from other conservation strategies in the inclusion of biotic–environmental relationships as part of the definition of conservation targets and also in the inclusion of the concept of scale (e.g., ecologically meaningful regions). RA has been widely used in the southern hemisphere (Lombard et al., 1995; Haila and Margules, 1996; DeVantier et al., 1998; Freitag et al., 1998; Preen, 1998), but RA principles have only recently been widely applied in North America

Patrick S. Bourgeron and Hope C. Humphries wish to acknowledge partial funding provided by a Science to Achieve Results grant from the U.S. Environmental Protection Agency ("Multi-scaled Assessment Methods: Prototype Development within the Interior Columbia Basin").

(e.g., Davis et al., 1996; Kiester et al., 1996; White et al., 1999), although conservation of representative examples of major U.S. ecosystems was proposed for the continent early in the 20th century by the Committee on the Preservation of Natural Conditions (Shelford, 1926).

RA includes three components of conservation planning: identification, selection, and design of conservation areas. Routine implementation of RA has been infrequent because it requires not only the distribution of conservation targets, but also knowledge of the environmental space they occupy. While this knowledge exists for some elements of biodiversity (Noss et al., 1997; Soulé and Terborgh, 1999a), it is rarely available for the entire suite of species and ecosystems found in an area (but see Cox et al., 1994, for an unusual example of the analysis of the environmental relationships of a large number of conservation targets in Florida). The scope (Chapters 1 and 35), methods (Chapters 3, 13 through 18, 20, and 21), and data requirements (Chapters 5 and 7) of integrated regional ecological assessments (IREAs) provide the means to conduct RAs at large scales.

The purpose of this chapter is to present a framework for using the principles and tools of IREA, as presented in this book, to perform RAs to determine the status of biodiversity, as well as the status of ecosystem conditions relevant to conservation planning at national, regional, state, and local levels. The following aspects of RAs are discussed: (1) definition and purpose (Section 20.2), (2) basic components and requirements (Sections 20.3 and 20.4), (3) principles and process of RA (Section 20.5), and (4) practical implementation strategies (Sections 20.6 and 20.7).

20.2 Defining Representativeness Assessment

20.2.1 What Is Representativeness Assessment?

RA is a process for (1) representing all natural features of interest found in a region within a network of conservation areas (Austin and Margules, 1986; Pressey et al., 1993) and (2) characterizing the ecological context of the biota found in any area of a region (Austin and Margules, 1986; Mackey et al., 1988; Engelking et al., 1994; Bourgeron et al., 1995a). The primary objective in conducting RA is to design a network of conservation areas in a region that includes all aspects of regional biotic variability, all aspects of regional environmental variability driving biotic responses, and the maintenance over time of patterns and processes within the network. Although this agenda is very ambitious, it defines the initial conditions under which specific and realistic objectives can be determined given particular biological, economic, and social constraints.

20.2.2 What Are the Requirements for Representativeness Assessment?

RA includes three main components of conservation planning: (1) clear definition of conservation goals and hence of conservation targets and their geographic distributions (generally known as identification), (2) selection of conservation areas to achieve the conservation goals, and (3) preliminary evaluation of the design of the network as a whole and of individual conservation areas. RA requires the definition of an area for conducting the assessment and analyses, multiscale descriptions and mapping of biotic and environmental variability, and quantification of the environmental relations of the biota (Austin and Margules, 1986; Margules et al., 1987; Mackey et al., 1988, 1989; Margules and Stein, 1989; McKenzie et al., 1989). RA can be conducted at any spatial scale necessary to meet the objectives of a project, from assessing the rangewide distribution of a single feature of biodiversity (e.g., distribution of tree, grizzly bear, or endemic snail species), to assessing the distributions of all biotic or biophysical features within an area.

RA has four distinct steps: (1) delineation of areas in which to conduct the assessment and analyses, (2) delineation of land units to serve as the basis for selection of conservation areas, (3) determination of the conservation suitability of all land units, and (4) a process for the selection of land units for inclusion in a regional network of conservation areas.

20.2.3 How Does Representativeness Assessment Differ from and Integrate with Other Conservation Criteria and Strategies?

Common approaches to conservation evaluation and the selection of conservation areas historically have included multicriteria scoring procedures (Pressey and Nicholls, 1989a). Areas are ranked in order of significance or priority according to scores derived from various criteria (e.g., rarity, diversity, naturalness; see review in Usher, 1986). Sites are chosen independently of each other on the basis of the selected scoring criteria (e.g., Master, 1991). In contrast, at each step of the site selection process, RA takes into account the attributes of both candidate and previously selected sites (Pressey and Nicholls, 1989a). RA allows for consideration of other major criteria, such as rarity and diversity during the selection process. A wide array of selection procedures has been developed recently using RA principles (see Section 20.5).

RA has been reported as being more efficient than scoring procedures (Pressey and Nicholls, 1989a, 1989b). Efficiency is defined as the ability of a criterion to capture the maximum number of attributes under consideration in the smallest number of sites. For example, Pressey and Nicholls (1989b) calculated that the most efficient scoring procedure required 45% of the total land area in New South Wales, Australia, to represent all the land systems present in the region at least once. In contrast, an iterative procedure for representativeness included one example of each land system in 5.7% of the land area (Pressey and Nicholls, 1989a). However, efficiency of RA varies with the degree of implementation of the principles of flexibility and irreplaceability (see Section 20.5.1) and with the region under consideration (Noss, 1996).

Over the last two decades, a number of large-scale assessments have been initiated that emphasize the protection of geographic areas ranking high in species richness, levels of endemism, number of rare and threatened species, and intensity of threat. Three such assessments include GAP (Scott and Jennings, 1998), The Nature Conservancy's ecoregional planning (The Nature Conservancy, 1996), and the World Wildlife Fund's ecoregional conservation assessment (Ricketts et al., 1999a) and Global 200 (Olson and Dinerstein, 1998) initiatives. Although they differ in specific objectives and scope, these efforts share an emphasis on representation of conservation targets (see also discussion in Reid, 1998) as the initial step of conservation planning. In contrast, RA aims at representing processes as well as patterns. The four strategies have in common some data requirements and methods.

In practice, conservation scientists have emphasized different aspects of the design of conservation area networks, depending on the objectives. In some cases, large, well-connected, unmanaged areas are thought to provide the best solution for the long-term maintenance of biodiversity (e.g., Noss, 1992, 1993; Noss and Cooperrider, 1994; DellaSala et al., 1995; Noss et al., 1999; Soulé and Terborgh, 1999a). In other cases, emphasis has been placed on the importance of embedding conservation areas within a matrix of land managed according to ecological principles to successfully sustain biodiversity (e.g., Alverson et al., 1994). The latter emphasis is based on the implementation of the principles of ecosystem management at landscape and regional levels (Christensen et al., 1996; Jensen et al., 1996). Three conservation approaches use the concept that well-managed ecosystems will maintain biodiversity: biodiversity management areas (Davis et al., 1996), diversity maintenance areas (Alverson et al., 1994), and emphasis–use (Everett and Lehmkuhl, 1997). The general framework presented in this chapter provides the means to evaluate the impact of various aspects of network design on a case by case basis by including assessment of biotic–environmental relationships and suitability of land units at multiple scales, in addition to representation of conservation targets (Noss and Cooperrider, 1994; Christensen et al., 1996).

20.2.4 Is Representativeness Assessment Relevant to Site Conservation and Management?

RA is relevant to site conservation and management in two ways. First, it is used to identify networks of sites capable of sustaining all conservation targets. Second, it provides information on biotic and environmental variation at multiple spatial and ecological scales to guide the design of each individual conservation area. RA does not produce conservation area designs in the traditional sense. The land areas selected during RA, as well as their surroundings, need to be analyzed in a site conservation planning process such as the framework developed by Poiani et al. (1998). In contrast to current practices, RA provides site conservation planners with information that places the conservation and management activities of individual conservation areas in the context of the sustainability of conservation targets within the entire network. For example, landscapes within each conservation area in a regional network may be allowed to fluctuate within the bounds of natural disturbance regimes and other processes, and therefore conservation targets may be lost locally within some areas, but maintained or acquired within others.

20.3 Delineation of the Assessment Area and Land Units

20.3.1 Delineation of the Assessment Area

The assessment area is the region in which data are gathered and analyzed to characterize biodiversity (see Chapters 1 and 7) and to recommend ecologically sensible conservation actions that meet the conservation objectives. This requires a first approximation of broad-scale correlations among regional driving variables (e.g., climate, landforms such as mountain ranges), broad biotic patterns (e.g., biomes), and planning units. A number of ecoregional frameworks (e.g., see Chapters 22 and 23; also Grossman et al., 1999) can be used as is or easily modified. Alternatively, de novo frameworks can be created for specific objectives. A candidate system should be evaluated for its usefulness for a given conservation objective before its implementation (see Chapter 22), because ecoregions have been reported as both appropriate (Host et al., 1996; Bourgeron et al., 1999a) and not appropriate (Wright et al., 1998) as a level of analysis for ecological patterns.

A second level, the analysis area, may also need to be delineated. The analysis area is the geographic area in which conservation and management actions directly influence patterns in the area's component land units. Often, the assessment area is several times larger than the analysis area and may need to be subdivided. In this case, RA is conducted within each analysis area. Such a subdivision of the assessment area was performed during the Sierra Nevada Ecosystem Project (SNEP, 1996) for the selection of biodiversity management areas (Davis et al., 1996) and during the Interior Columbia Basin Ecosystem Management Project (ICBEMP) (Quigley and Arbelbide, 1997) for conducting land-

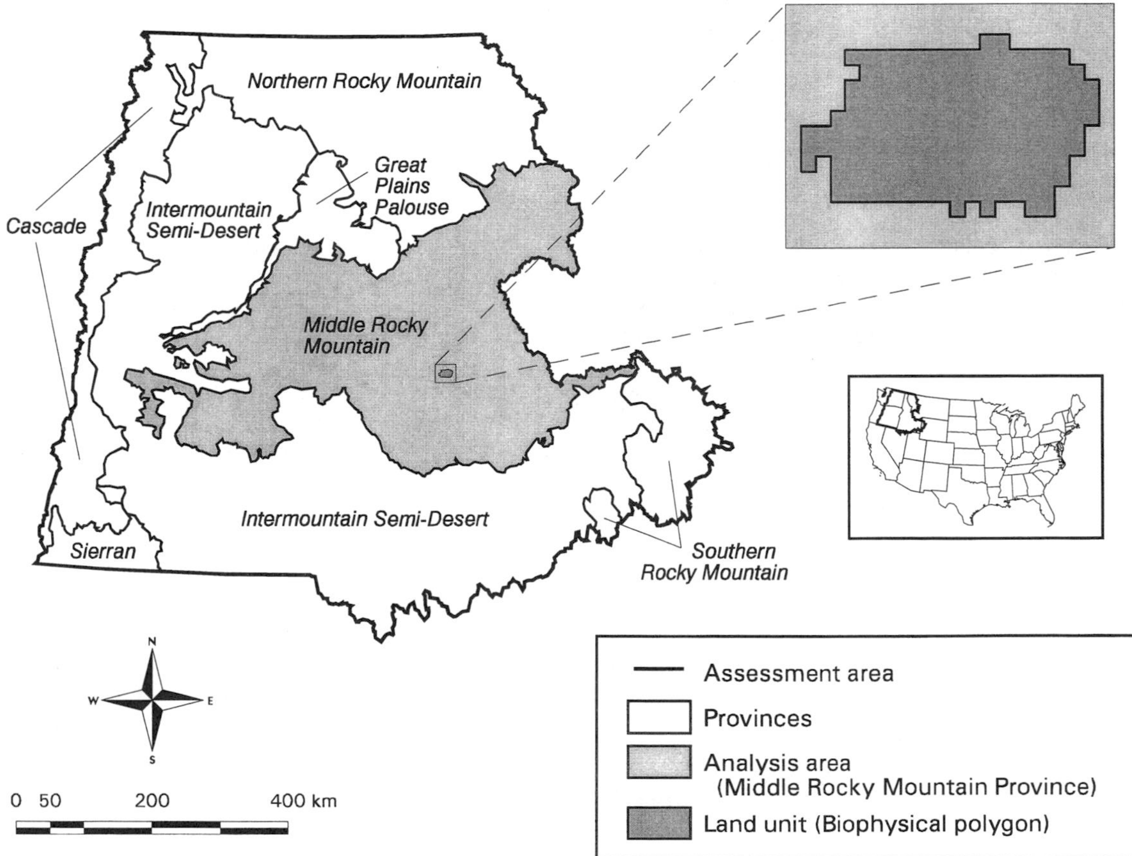

FIGURE 20.1. Map of interior Columbia River basin showing relationships among assessment area, analysis area (province), and land unit (basic unit of selection for conservation areas).

scape analyses and the assessment of ecological conditions (Hann et al., 1997; Jensen et al., 1997). For example, in the ICBEMP, the assessment area was defined by an analysis of biophysical, ecological, and administrative maps in light of political directives to develop an ecosystem management strategy for forests east of the Cascade crest in the U.S. Pacific Northwest (Figure 20.1) (Quigley and Arbelbide, 1997). Provinces (sensu Bailey, 1995) were used as the units for many biophysical and vegetation analyses (Figure 20.1) (Jensen et al., 1997).

Delineation of assessment and analysis areas is necessary because there is often a mismatch between traditional planning units and the boundaries required by the objectives of an assessment (Knight and Landres, 1998; Groom et al., 1999). Traditional planning units include such administrative boundaries as National Forests, National Parks, and counties, which historically were delineated without explicit consideration of ecosystem distribution, functioning, and persistence.

20.3.2 Delineation of Land Units

Within assessment and analysis areas, a land unit for RA is defined as the basic unit of selection for inclusion in a network of conservation areas. Such a land unit may contain more than one conservation target of interest (e.g., species, vegetation type, ecosystem). Delineation of land units for RA follows the procedures of ecosystem characterization (see Chapters 3, 22, and 23) and requires (1) characterization of biotic ecosystem components, (2) characterization of environmental ecosystem components relevant to the biota, (3) characterization of biotic–environmental interactions to define the environmental space occupied by each biotic element, (4) characterization of ecosystems as a whole and of ecosystem properties, and (5) a method of mapping these components, interactions, and properties. Hierarchical databases (see Chapters 5 and 7) and classifications and maps of ecological and environmental units (Table 20.1) (Austin and Margules, 1986) are used to derive an appropriate hierarchical system of ecological mapping units for a conservation objective (see Chapters 22 and 23; also see Grossman et al., 1999). Maps of biophysical environments (see Chapters 22 and 23; also see Zonneveld, 1989; Bailey et al., 1994; Jensen et al., 1997) form the basis for defining these units. Maps of biotic elements (species, vegetation) are overlaid on a biophysical environment map to develop integrated ecological units at the scales appropriate for a specific conservation goal. A particular level of the hierarchy must be selected for determining the land units used in the selection process. Higher levels in the hierarchy provide the foundation for assessing the context of these selection units. Existing frameworks (see Chapters 22 and 23; also see Grossman et al., 1999) should be evaluated carefully for possible use. For example, over 17,000 land units were delineated in the 58 million hectares of the interior Columbia River basin using a combination of biophysical (climate and biogeochemical) variables and potential vegetation (Figure 20.1) (Bourgeron et al., 1999b).

Multiscaled analyses determine the ecological properties of the land units needed to characterize biodiversity (Urban et al., 1987; Bourgeron and Jensen, 1994; Bourgeron et al., 1995b; Jensen et al., 1996). These analyses include ordering biotic components along environmental gradients (e.g., Austin and Margules, 1986; Faith, 1992) and modeling biotic distributions using key environmental factors and processes (e.g., Austin et al., 1990, 1996; Yee and Mitchell, 1991; Franklin, 1998; Leathwick, 1998).

The resulting map of land units is used to interpret the temporal variability, that is, landscape dynamics (see Chapter 23), which is a major criterion in defining suitability (see Section 20.4). The map also provides a template for interpreting the spatial distribution and variability of the elements of biodiversity in an area. Of particular interest for RA is incorporation of the spatial pattern of gradual change in landscape properties (Bourgeron et al., 1994) into the definition of land units to ensure that all portions of relevant environmental variability are incorporated in the network. For example, occurrences of a regional vegetation type that is distributed along an environmental gradient will show changes in species composition across the gradient (e.g., Bourgeron et al., 1994). These differences are

TABLE 20.1. Regional classifications and maps for representativeness assessment.

Attribute	Assessment Unit	Product
Climate	Climatic regions	Map description
Bioenvironments, ecoregions	Land systems, terrain patterns	Maps of bioenvironments or ecoregions
Floristics, fauna	Species distribution maps	Maps of biogeographic regions
Vegetation	Communities, vegetation types	Vegetation distribution maps

important for two reasons: (1) a representative network should incorporate all relevant aspects of biotic and environmental variability, and (2) a conservation target may respond differently to conservation or management practices depending on location-specific biotic–environmental interactions.

Two types of surrogate data are used to delineate or attribute land units if biotic data are limited. An environmental stratification may serve as a surrogate for biotic data when they are not available regionally (Mackey et al., 1988, 1989; Belbin, 1993; Kirkpatrick and Brown, 1994; Bourgeron et al., 1995a; White et al., 1999), or the distribution of one or more group of indicator taxa may be used as a surrogate for total biodiversity (Pressey et al., 1993; Margules et al., 1994; Scott and Jennings, 1998). Ideally, any surrogate data should be tested for their actual surrogacy value (e.g., Bedward et al., 1992; Kirkpatrick and Brown, 1994), but it is often not possible to do so before conservation decisions are required (Faith and Walker, 1996). Therefore, regardless of the type of surrogate, any surrogacy methods used to make conservation decisions should be robust (Ricketts et al., 1999b) and should not make unwarranted assumptions (Faith and Walker, 1996).

20.4 Suitability of Land Areas

Once land units have been delineated and their ecological properties determined, their suitability for conservation must be assessed. The presence or absence of key targets, defensibility, viability, land-use practices, road density, and management status are common criteria used to determine conservation suitability value (McHarg, 1969; McKendry and Machlis, 1993; Davis et al., 1996) in light of the specific objectives of a conservation assessment. In this section, we discuss some of the factors that constrain the suitability of sites for the long-term protection of regional diversity by affecting ecological conditions and persistence of patterns and processes.

20.4.1 Assessment of Ecological Conditions

The effects of management practices and natural processes on biotic patterns should be assessed at multiple scales by contrasting the current condition of an area with other managed or unmanaged areas that occur in the same biophysical environment. It is often useful to conduct an analysis of the historic range of variability (HRV) (see Chapter 19) to characterize past fluctuations in pattern and process, resulting in determination of the upper and lower limits or range of variability within which an ecosystem historically operated. Knowledge of how far an ecosystem departs from its historical range is gained by comparison of its current state with HRV (Morgan et al., 1994; Landres et al., 1999; Swetnam et al., 1999). Analyses can focus on the processes or disturbance regimes operating on an ecosystem, such as the frequency and severity of disturbance, or they can focus on the states of the system resulting from processes, such as descriptions of the size, shape, and other characteristics of patches (Swanson et al., 1994; Cissel et al., 1998; Kaufmann et al., 1998). There are known limitations to the power of HRV analyses, but careful application of the method can provide important information (see Chapter 19; also see Landres et al., 1999).

20.4.2 Maintenance of Patterns and Processes

Estimates of the persistence of patterns and processes are required for each land unit as well as at larger scales. Assessment of persistence should take into consideration information about biotic patterns, biophysical environments, processes, and existing conditions at different hierarchical levels (Levin, 1992; Bourgeron and Jensen, 1994). For example, both coarse-scale and site-level conditions are important for assessing the population viability of many species (Glenn and Collins, 1994). Across the landscape, reproducing populations in high-quality habitats may be sources for populations in other areas (sinks; see Pulliam, 1988). In areas with less suitable habitats, where mortality exceeds reproduction, populations are dependent on immigration from source populations to maintain existing numbers (Ritchie, 1997).

Within a landscape, the quality of the local habitat depends on many factors, including ecosystem complexity, biodiversity, vegetation composition and structure, land-use patterns, road density, and fragmentation (see Chapters 17 and 27 through 29 for discussions of economic and sociopolitical factors). The potential for manipulating ecosystem components (e.g., vegetation) to improve habitat should also be included in the determination of suitability. Analyses should be developed that integrate relevant factors into a single index or a suite of indices of suitability (e.g., Davis et al., 1996) that meet the specific objectives of an assessment. For example, Bourgeron et al. (1999b) determined the conservation suitability of over 17,000 land units in the interior Columbia River basin using a knowledge-based system operating within a geographic information system-integrated application framework. The knowledge-based model used fuzzy logic networks to describe relationships and assign suitability ratings to land units based on information about biotic (historic and current vegetation, species distributions) and abiotic landscape elements (biophysical environments, disturbances), ecological conditions (landscape connectivity and diversity, departure from historic conditions), management criteria, and road density. The resulting suitability rankings were used to assess the representativeness of conservation areas in the interior Columbia River basin.

20.5 Principles and Methods for the Selection of Conservation Areas

20.5.1 Principles for Selecting Regional Conservation Areas

Three principles are central to conducting RA (Pressey et al., 1993):

1. Given limited resources and land available for designation as conservation areas, new sites should be added to a network that are most complementary in the targets that they contain to the sites already included (Pressey et al., 1993). Complementary is closely related to efficiency, which is highly dependent on the size of the land units and the scale of the conservation targets (Pressey et al., 1993). This dependency should be considered when selecting an appropriate hierarchical level for the land units.
2. Flexibility refers to alternative ways of combining land units to form representative networks and has implications for selection procedures. Land units are usually not unique in their representativeness, and several may potentially fill the same role in a network. Therefore, specific components of a given network can be substituted by others of similar conservation value to fit planning needs. It may also be possible to chose among alternative networks of comparable representativeness. Both types of substitution may lower the efficiency of representation. This reduction in efficiency and its impact on the overall conservation strategy should be carefully evaluated when the flexibility principle is applied.
3. Irreplaceability may be defined in two different ways: (1) as the potential contribution of a site to a conservation goal and (2) as the extent to which options for conservation are lost if the site is forfeited (Pressey et al. 1993, 1994). The notion of a gradient of irreplaceability is compatible with the traditional concept of rarity, in which sites have outstanding value because of the presence of a rare or threatened element, and is also compatible with the concept of endemism. The impact of assigning irreplaceability values to land units on efficiency and overall conservation strategy should be evaluated.

20.5.2 Methods for the Selection Process

A substantial number of formal methods exist for selecting conservation areas from regional conservation databases (see reviews in Csuti et al., 1997, and Pressey et al., 1997). They fall into two general categories, heuristic and optimizing algorithms; all share the common goal of making the selection process explicit, repeatable, and efficient. Heuristic procedures are usually iterative and include the principles discussed previously (e.g., Pressey et al., 1993, 1994). They generally proceed in a stepwise manner, adding sites at each step that are the most complementary to those already in the network. They may be initiated with sites having the greatest number of conservation targets (i.e., richness), or they may start with sites containing rare targets (irreplaceability) (Pressey et al., 1997). Optimizing algorithms emphasize efficiency (i.e., select the network that will include all targets within a minimum area). Such techniques seek an optimal solution, that is, the most efficient set of sites that meet all user-defined constraints (e.g., Church et al., 1996; Kiester et al., 1996; Csuti et al., 1997). The drawback of many of these algorithms is that their implementation might result in ignoring population viability considerations and maintenance of ecological processes.

The relative merits of the two categories and the methods within each category have been discussed in terms of their contribution to conservation planning. In selecting a method, the following points should be carefully considered (after Pressey et al., 1997). The results of the application of any method should be considered as indicative, not prescriptive. Final choices are made through the planning process, which involves many partners and stakeholders (see Section 20.7). Therefore, methods that clearly define what is negotiable and what is not, that offer alternatives, and that are sufficiently flexible and interactive to permit real-time changes can be used in negotiations with parties of different interests and can strengthen the case for a conservation goal.

20.6 Representativeness Assessment in Practice

The RA process can be summarized in 11 generic steps (Figure 20.2). Like all IREAs, RAs start with societal and managerial input, which drives the scale of the assessment (see Chapter 1). The manner in which RA should be integrated in regional planning is discussed in the following (see also Chapters 9 and 35).

Step 1. Define the specific conservation targets of interest (e.g., plant or animal species or communities), according to the assessment objectives (see Chapter 1). The process of setting conservation goals and conservation targets is a social, economic, and political activity and is included in the planning process (see Chapters 1, 3, and 6; also Section 20.7 and Figure 20.3).

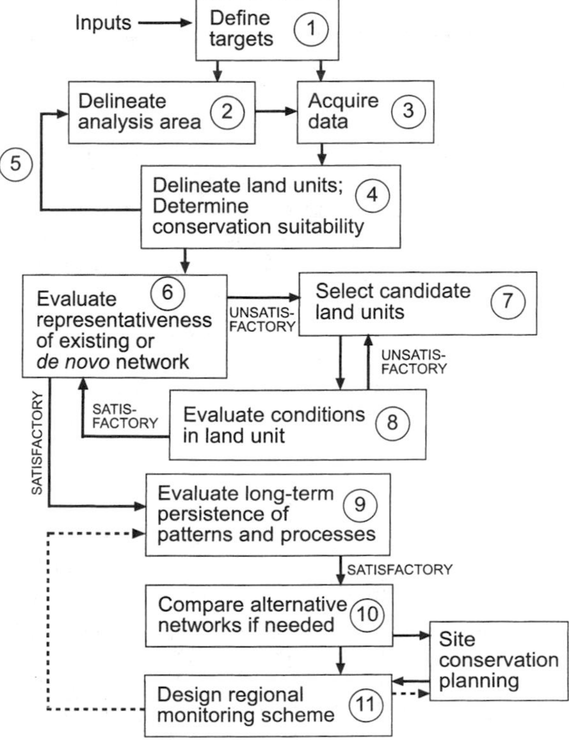

FIGURE 20.2. Representativeness assessment in practice. Circled numbers correspond to steps in Section 20.6 of the text.

Step 2. Delineate the assessment area and analysis areas (see Chapter 7), as appropriate.

Step 3. Gather relevant information and data concerning the natural history, ecology, and spatial extent of the targets and associated processes (see Chapters 5 through 7).

Step 4. Delineate land units for selection, and determine their suitability for conservation as defined by the objectives. Perform ecosystem and biodiversity characterization (see 3, 22, and 23), including relevant information on human activities (see Chapters 16, 17, and 27 through 29), and determine ecosystem properties and dynamics (see Chapter 19). Most analytical and modeling work, including spatial analysis, takes place at this stage (see Chapters 5 and 13 through 15).

Step 5. Assess the adequacy of the assessment and analysis areas and correct their boundaries if necessary (step 2). Make changes only to provide results that more closely approximate the conservation goal.

Step 6. If conservation areas already exist, evaluate them to determine whether they are representative of all conservation targets and associated processes. If not, the process continues in step 7. If the network is judged representative, the process moves to step 9. If no conservation areas exist, the process goes initially to step 7 and then iterates through steps 6, 7, or 8 as needed.

Step 7. Select candidate land units for inclusion in the network based on the representativeness of their conservation targets. The principles of complementary and irreplaceability are applied. Initial conditions for networks may vary from random initialization to network seeds made up of combinations of the most irreplaceable sites or of existing conservation areas (Pressey et al., 1993).

Step 8. Use the suitability of the candidate land units and of their surroundings at multiple scales (e.g., immediate neighboring units, analysis area) to assess whether to include them in the network. Questions that reflect the assessment objectives should be formulated. Table 20.2 provides examples of questions pertaining to suitability at multiple scales. Answers to the questions result in a decision about whether a land unit should be included in the network. If so, the process returns to step 6. If the decision is not to keep a land unit, its irreplaceability level must be evaluated, and if the irreplaceability of a land unit warrants its retention, the process moves to step 6. If the land unit is not irreplaceable, it is

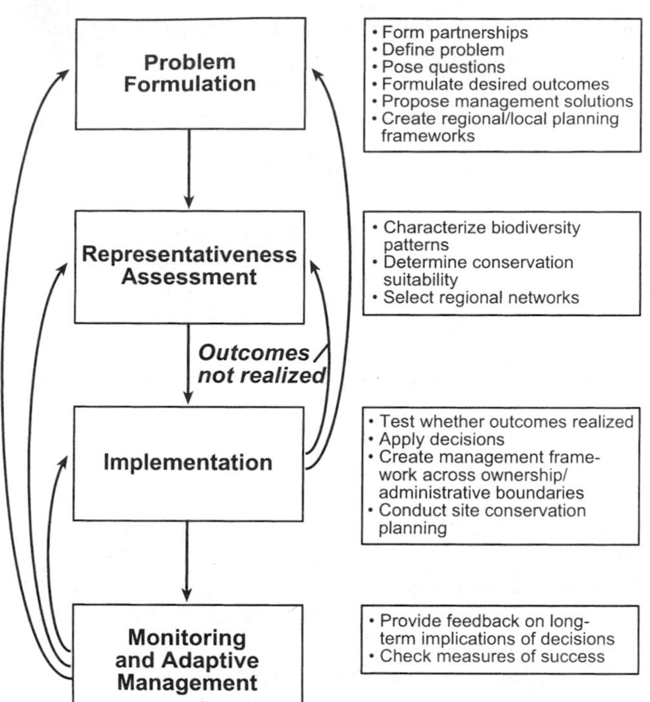

FIGURE 20.3. Representativeness assessment as part of the planning process.

discarded, and another unit with a similar representativeness value is selected (proceeding to step 7).

Step 9. A preliminary, broad-brush evaluation is conducted to assess whether all the patterns and processes included in the network will persist over time. The following are sample questions that should be answered at this stage. Are the network and the sum of the conservation actions consistent with maintaining biodiversity and associated processes within HRV, given the economic development, population growth, and social values expected in the assessment area? Do the network and associated conservation actions within the assessment area contribute to the maintenance of biodiversity at higher hierarchical levels, especially for targets that are not confined to the assessment area? For example, a conservation network representative of the high-elevation forests of the Northern Rocky Mountains should be evaluated in terms of its contribution to the conservation of these forests over the entire Rocky Mountain cordillera. If the answers to the questions are not satisfactory, the flexibility principle (modification of the network) should be applied to attempt to improve the results (returning to step 7). If modification of the network does not produce satisfactory answers, an alternative network should be considered. If the answers to the questions are satisfactory, the process moves to step 10. A conceptually similar approach was suggested by Lambeck (1997), who used umbrella species for defining the attributes required to meet the needs of the biota in a landscape and the management regimes that should be applied.

Step 10. The network is complete. Alternative networks may be compared. When a final choice of a network is made, site conservation planning can begin on individual conservation areas.

TABLE 20.2. Screening questions to determine suitability of land units as candidates for inclusion in a regional conservation network.

1	Are vegetation patterns (physiognomic–floristic and structural) currently outside HRV within the land unit, analysis area, assessment area, or any other scale of interest?
2	Are ecological processes currently outside HRV within the land unit, analysis area, assessment area, or any other scale of interest?
3	Are ecological processes temporally synchronized within the land unit, analysis area, assessment area, or any other scale of interest?
4	Is landscape connectivity a concern for biodiversity within the land unit, analysis area, assessment area, or any other scale of interest?
5	Are fragmentation and patch size a concern within the land unit, analysis area, assessment area, or any other scale of interest?

Step 11. A regional monitoring scheme should be designed using RA information to ensure the long-term maintenance of all the targets and associated processes. Ecological indicators should be developed to indicate changes within each conservation area that are significant at the network level.

The process is flexible enough to allow for emphasizing some steps and deleting others, depending on specific conservation objectives, available data and resources, the sociopolitical context, and the selection methods used. The impact of such decisions on the conservation value of the planning process (see Section 20.7) and on the properties of the network should be evaluated carefully. For example, omitting assessment of ecological conditions (e.g., fire and hydrologic conditions) at multiple scales may appear to be convenient and time and cost effective when databases are assembled during the assessment. However, site conservation planning (sensu Poiani et al., 1998) may reveal that (1) extensive and long-term management actions are required immediately to alleviate the stresses of fire suppression and altered hydrologic regimes and maintain conservation targets in the conservation area, and (2) the management costs far exceed the additional expenditures that would have been incurred during the assessment. The process described has not explicitly considered the spatial distribution of land units. A large body of work documents increased protection of biodiversity generally afforded by large, well-connected conservation areas (e.g., Noss et al., 1997; Soulé and Terborgh, 1999a, 1999b). However, there is also evidence that populations scattered over several sites might be less vulnerable to environmental disturbance than populations in a single, large site (Harrison and Quinn, 1989). Spatial considerations should be incorporated into the RA process to provide solutions that best fit the targets of conservation under consideration. The spatial connectivity of the selected sites should be evaluated at different spatial scales.

20.7 Representativeness Assessment, Partnerships, and Conservation Planning

Like all IREAs, RAs involve large areas of multiple ownership, and therefore many management options should be evaluated. Assessments that cover large geographic domains can provide a common scientific foundation for coordinated planning and management among all concerned parties (see Chapter 35; also see Slocombe 1993a, 1993b, 1998). Effective conservation planning and management within the bounds of sustainable local and regional economies requires coordination among landowners and managers (see Chapter 35; also see Moore and Lee, 1999). The problems involved in conducting conservation across multiple ownerships are substantial (Landres et al., 1998), but known general solutions can be applied (Knight and Clark, 1998; Landres, 1998; Groom et al., 1999).

A major obstacle to the integration of knowledge (see Chapter 9) is that management and planning units do not relate to ecosystems, ecosystem connections to economic and social processes, or the cultural and political identities of local people (Slocombe, 1993b; Knight and Landres, 1998). A solution is the creation of regional and local planning and management frameworks that can be used to decide on a common strategy that will maintain a representative network of conservation areas and provide for a stable economy (e.g., Slocombe, 1993b; Yaffee, 1998; Moore and Lee, 1999).

To be successful and effective, RAs should be embedded in a flexible, adaptive, management-oriented planning process (see Chapter 35 and Figure 20.3). First, the problem to be managed is formulated and its boundaries are either set or clarified. Stakeholders are identified and their involvement is sought. Partnerships are formed and planning frameworks are created. Specific actions are proposed to solve problems and answer questions. Scientists work within the planning framework to clearly formulate the problems, questions, and alternatives as hypotheses with explicit predicted outcomes that will be tested by the RA (Underwood, 1995).

The next step involves conducting RAs that are designed specifically to test the hypotheses. If the predicted outcomes are not realized, an assessment is made of how the process led to failure. In particular, an evaluation is required of how the outcomes were determined. The criteria for testing hypotheses should be examined as well. If the predicted outcomes are realized, RA results are implemented. Activities shift to scientific and management activities (e.g., site conservation planning). Monitoring schemes are designed at the site, landscape, and regional levels to determine whether the desired outcomes are maintained over time; the results of monitoring are incorporated into the planning process (Hockings, 1998).

20.8 Conclusions

The goal of representing and sustaining regional biodiversity in an efficient network of conservation areas requires a set of procedures for determining the conservation value of the areas using explicit methods. Ad hoc designation of conservation areas has been demonstrated to increase the final costs of achieving representation of biodiversity in conservation areas (Margules, 1989; Pressey and Nicholls, 1989a; Pressey 1990, 1994; Pressey et al., 1990; Rebelo and Siegfried, 1992), without the guarantee of long-term persistence over large areas.

This chapter has presented a set of procedures for conducting RAs for conservation planning at regional and continental scales. RA is the process by which networks of conservation areas are designed to capture the ecological patterns that constitute biodiversity and their associated processes so that biodiversity is sustainable at all relevant scales. RA should be incorporated into the planning process, and methods should be chosen that strengthen the on the ground implementation of conservation goals. Many of the steps described are the focus of active research and development. RAs have been conducted increasingly often and have been incorporated into the planning system, especially in Australia. Few conservation assessments have had the data, time, budget and human resources available to follow the entire process. However, as IREAs at the scale of the ICBEMP and SNEP (see also Chapters 30 through 33) get underway or are completed, there is an opportunity to implement the full RA framework.

Data, analyses, and interpretations are required to incorporate information about the dynamic nature of ecosystems into conservation planning at a regional scale. Human activities should be considered as integral components of ecosystems. The best applicable science and technology should be identified so that the required knowledge about ecosystems can be arranged hierarchically from very broad continental or global scales down to the smallest local levels. Several assessment scales are important in deciding what actions should be taken and in evaluating the implications and effectiveness of such actions. The most important consideration is incorporation of the RA process into a flexible planning process that includes all interested parties from the outset and that follows the principles of adaptive management. Such a planning process provides the best way to implement the recommended actions. For example, coordination across all boundaries is necessary if regional networks of conservation areas are to be efficient, effective, and sustainable.

RAs have been widely conducted in Australia, but their implementation in North America has been less frequent. RAs and other large-scale conservation assessments conducted in the past have omitted important steps of the process as a result of data, budgetary, and time constraints. Well-funded current and future IREAs provide the opportunity to conduct effective RAs. Establishing a system of conservation areas to represent selected conservation targets and the processes that maintain them is only the first step in the long-term protection of biodiversity. The application of RA may ensure the sustainability of conservation targets in a network, but it is not a substitute for the exploration of management practices conducive to the conservation of biodiversity in other lands.

20.9 References

Alverson, W.S.; Kuhlmann, W.; Waller, D.A. 1994. *Wild forests: Conservation Biology and Public Policy*. Washington, DC: Island Press.

Austin, M.P.; Margules, C.R. 1986. Assessing representativeness. In: Usher, M.B., ed. *Wildlife conservation evaluation*. London: Chapman and Hall: 45–67.

Austin, M.P.; Nicholls, A.O.; Margules, C.R. 1990. Measurement of the realized qualitative niche: environmental niches of five eucalyptus species. *Ecol. Monogr*. 60:161–177.

Austin, M.P.; Pausas, J.G.; Nicholls, A.O. 1996. Patterns of tree species richness in relation to environment in southeastern New South Wales, Australia. *Australian J. Ecol*. 21:154–164.

Bailey, R.G. 1995. *Description of the ecoregions of the United States*. Washington, DC: U.S. Dept. Agric., For. Serv. Misc. Publ. 1391.

Bailey, R.G.; Jensen, M.E.; Cleland, D.T.; Bourgeron, P.S. 1994. Design and use of ecological mapping units. In: Jensen, M.E.; Bourgeron, P.S., eds. *Volume II: Ecosystem management: principles and applications*. PNW-GTR-318. Portland, OR: U.S. Dept. Agric., For. Serv., Pacific Northw. Res. Sta.: 95–106.

Bedward, M.; Pressey, R.L.; Keith, D.A. 1992. A new approach for selecting fully representative reserve networks: addressing efficiency, reserve design and land suitability with an iterative analysis. *Biol. Conserv*. 62:115–125.

Belbin, L. 1993. Environmental representativeness: regional partitioning and reserve selection. *Biol. Conserv*. 66:223–230.

Bourgeron, P. S.; Jensen, M. E. 1994. An overview of ecological principles for ecosystem management. In: Jensen, M. E.; Bourgeron, P. S., eds. *Volume II: Ecosystem management: principles and applications*.

PNW-GTR-318. Portland, OR: U.S. Dept. Agric., For. Serv., Pacific Northw. Res. Sta.: 45–67.

Bourgeron, P. S.; Humphries, H. C.; DeVelice, R. L.; Jensen, M. E. 1994. Ecological theory in relation to landscape and ecosystem characterization. In: Jensen, M. E.; Bourgeron, P. S., eds. *Volume II: Ecosystem management: principles and applications*. PNW-GTR-318. Portland, OR: U.S. Dept. Agric., For. Serv., Pacific Northw. Res. Sta.: 58–72.

Bourgeron, P. S.; Engelking, L. D.; Humphries, H. C.; Muldavin, E.; Moir, W. H. 1995a. Assessing the conservation value of the Gray Ranch: Rarity, diversity and representativeness. *Desert Plants* 11:1–68.

Bourgeron, P. S.; Jensen, M. E.; Engelking, L. D.; Everett, R. L.; Humphries, H. C. 1995b. Landscape ecology, conservation biology and principles of ecosystem management. In: Everett, R. L.; Baumgartner, D. M., eds. *Proceedings from the ecosystem management in Western Interior Forests: symposium proceedings*; May 3–5, 1994. Pullman, WA: Dept. of Natural Resource Sciences, Washington State University: 41–47.

Bourgeron, P. S.; Humphries, H. C.; Barber, J. A.; Turner, S. J.; Jensen, M. E.; Goodman, I. A. 1999a. Impact of broad- and fine-scale patterns on regional landscape characterization using AVHRR-derived land cover data. *Ecosystem Health* 5:234–258.

Bourgeron, P. S.; Humphries, H.C.; Reynolds, K. 1999b. *Selecting a regional conservation network using fuzzy logic and optimization procedures*. Amer. Assoc. Geogr. 95th Annual Meeting; March 23–27, 1999, Honolulu, HI.

Christensen, N. L.; Bartuska, A. M.; Brown, J. H.; Carpenter, S.; D'Antonio, C.; Francis, R.; Franklin, J. F.; MacMahon, J. A.; Noss, R. F.; Parsons, D. J.; Peterson, C. H.; Turner, M. G.; Woodmansee, R. G. 1996. The report of the Ecological Society of America Committee on the scientific basis for ecosystem management. *Ecol. Appl.* 6(3):665–691.

Church, R. L.; Stoms, D. M.; Davis, F. W. 1996. Reserve selection as a maximal covering location problem. *Biol. Conserv.* 76:105–112.

Cissel, J. H.; Swanson, F. J.; Grant, G. E.; Olson, D. H.; Gregory, S. V.; Garman, S. L.; Ashkenas, L. R.; Hunter, M. G.; Kertis, J. A.; Mayo, J. H.; McSwain, M. D.; Swetland, S. G.; Swindle, K. A.; Wallin, D. O. 1998. *A landscape plan based on historical fire regimes for a managed forest ecosystem: the Augusta Creek study*. PNW-GTR-422. Portland, OR: U.S. Dept. of Agric., For. Serv., Pacific Northw. Res. Sta.

Cox, J.; Kautz, R.; MacLaughlin, M.; Gilbert, T. 1994. *Closing the gaps in Florida's wildlife habitat conservation system*. Tallahassee: Office of Environmental Services, Florida Game and Fresh Water Fish Commission.

Csuti, B.; Polansky, S.; Williams, P. H.; Pressey, R. L.; Camm, J. D.; Kershaw, M.; Kiester, A. R.; Downs, B.; Hamilton, R.; Huso, M.; Sahr, K. 1997. A comparison of reserve selection algorithms using data on terrestrial vertebrates in Oregon. *Biol. Conserv.* 80:83–97.

Davis, F. W.; Stoms, D. M.; Church, R. L.; Okin, W. J.; Johnson, K. N. 1996. Selecting biodiversity management areas. In: *Sierra Nevada ecosystem project: final report to Congress, Volume II: Assessments and scientific basis for management options*. Davis, CA: University of California, Centers for Water and Wildland Resources: 1503–1528.

DellaSala, D. A.; Olson, D. M.; Crane, S. L. 1995. Ecosystem management and biodiversity conservation: applications to inland Pacific Northwest forests. In: Baumgartner, D. M.; Everett, R. L., eds. *Proceedings of ecosystem management in Western Interior Forests*. Pullman, WA: Washington State University: 139–160.

DeVantier, L.M.; De'ath, G.; Done, T. J.; Turak, E. 1998. Ecological assessment of a complex natural system: A case study from the Great Barrier Reef. *Ecol. Appl.* 8:480–496.

Engelking, L.D.; Humphries, H. C.; Reid, M. S.; DeVelice, R. L.; Muldavin, E. H.; Bourgeron, P. S. 1994. Regional conservation strategies: assessing the value of conservation areas at regional scales. In: Jensen, M. E.; Bourgeron, P. S., eds. *Volume II: Ecosystem management: principles and applications*. PNW-GTR-318. Portland, OR: U.S. Dept. Agric., For. Serv., Pacific Northw. Res. Sta.: 208–224.

Everett, R. L.; Lehmkuhl, J. F. 1997. A forum for presenting alternative viewpoints on the role of reserves in conservation biology? A reply to Noss (1996). *Wildlife Soc. Bull.* 97:575–577.

Faith, D. P. 1992. Conservation evaluation and phylogenetic diversity. *Biol. Conserv.* 61:1–10.

Faith, D. P.; Walker, P. A. 1996. Environmental diversity: on the best-possible use of surrogate data for assessing the relative biodiversity of sets of areas. *Biodiversity Conserv.* 5:399–415.

Franklin, J. 1998. Predicting the distribution of shrub species in southern California from climate and terrain-derived variables. *J. Veg. Sci.* 9:733–748.

Freitag, S.; Nicholls, A. O.; van Jaarsveld, A. S. 1998. Dealing with established reserve networks and incomplete distribution data sets in conservation planning. *South African J. Sci.* 94:79–86.

Glenn, S. M.; Collins, S. L. 1994. Species richness and scale: the application of distribution and abundance models. In: Jensen, M. E.; Bourgeron, P. S., eds. *Volume II: Ecosystem management: principles and applications*. PNW-GTR-318. Portland, OR: U.S. Dept. Agric., For. Serv., Pacific Northw. Res. Sta.: 135–142.

Groom, M.; Jensen, D. B.; Knight, R. L.; Gatewood, L. M.; Boyd-Heger, D.; Mills, L. S.; Soulé. 1999. Buffer zones: benefits and dangers of compatible stewardship. In: Soulé, M. E.; Terborgh, J., eds. *Continental conservation: scientific foundations of regional reserve networks*. Washington, DC: Island Press: 171–197.

Grossman, D. H.; Bourgeron, P. S.; Busch, W.-D. N.; Cleland, D.; Platts, W.; Ray, G. C.; Robins, C. R.; Roloff, G. 1999. Principles of ecological classification. In: Johnson, N. C.; Malk, A. J.; Sexton, W. T.;

Szaro, R. C. *Ecological stewardship: a common reference for ecosystem management*. Amsterdam: Elsevier Science: 353–393.

Haila, Y.; Margules, C. R. 1996. Survey research in conservation biology. *Ecography* 19:323–331.

Hann, W. J.; Jones, J. L.; Karl, M. G.; Hessburg, P. F.; Keane, R. E.; Long, D. G.; Menakis, J. P.; McNicol, C. H.; Leonard, S. G.; Gravenmier, R. A.; Smith, B. G. 1997. Landscape dynamics of the basin. In: Quigley, T. M.; Arbelbide, S. J., eds. *An assessment of ecosystem components in the Interior Columbia Basin and portions of the Klamath and Great Basins: Volume II*. PNW-GTR-405. Portland, OR: U.S. Dept. Agric., For. Serv., Pacific Northw. Res. Sta.: 337–1055.

Harrison, S.; Quinn, J. F. 1989. Correlated environments and the persistence of metapopulations. *Oikos* 56:293–98.

Hockings, M. 1998. Evaluating management of protected areas: integrating planning and evaluation. *Environ. Manage*. 22:337–345.

Host, G. E.; Polzer, P. L.; Mladenoff, M. A.; White, M. A.; Crow, S. J. 1996. A quantitative approach to developing regional ecosystem classifications. *Ecol. Appl.* 6:608–618.

Jensen, M. E.; Bourgeron, P. S.; Everett, R.; Goodman, I. 1996. Ecosystem management: A landscape ecology perspective. *J. Amer. Water Resour. Assoc.* 32: 208–216.

Jensen, M.; Goodman, I.; Brewer, K.; Frost, T.; Ford, G.; Nesser, J. 1997. Biophysical environments of the basin. In: Quigley, T. M.; Arbelbide, S. J., eds. *An assessment of ecosystem components in the Interior Columbia Basin and portions of the Klamath and Great Basins: Volume II*. PNW-GTR-405. Portland, OR: U.S. Dept. Agric., For. Serv., Pacific Northw. Res. Sta.: 99–314.

Kaufmann, M. R.; Huckaby, L. S.; Regan, C. S.; Popp, J. 1998. *Forest reference conditions for ecosystem management in the Sacramento Mountains, New Mexico*. RMRS-GTR-19. Fort Collins, CO: U.S. Dept. Agric., For. Serv., Rocky Mountain Res. Sta.

Kiester, A. R.; Scott, J. M.; Csuti, B.; Noss, R.; Butterfield, B.; Sahr, K.; White, D. 1996. Conservation prioritization using GAP data. *Conserv. Biol.* 10:1332–1342.

Kim, K. C.; Weaver, R. D., editors. 1994. *Biodiversity and landscapes*. Cambridge, UK: Cambridge University Press.

Kirkpatrick, J. B.; Brown, M. J. 1994. A comparison of direct and environmental domain approaches to planning reservation of forest higher plant communities and species in Tasmania. *Conserv. Biol.* 8:217–224.

Knight, R. L.; Clark, T. W. 1998. Boundaries between public and private lands: defining obstacles, finding solutions. In: Knight, R. L.; Landres, P. B., eds. *Stewardship across boundaries*. Washington, DC: Island Press: 175–191.

Knight, R. L.; Landres, P. B., editors. 1998. *Stewardship across boundaries*. Washington, DC: Island Press.

Lambeck, R. J. 1997. Focal species: a multi-species umbrella for nature conservation. *Conserv. Biol.* 11:849–856.

Landres, P. B. 1998. Integration: A beginning for landscape-scale stewardship. In: Knight, R. L.; Landres, P. B., eds. *Stewardship across boundaries*. Washington, DC: Island Press: 337–345.

Landres, P. B.; Knight, R. L.; Pickett, S. T. A.; Cadenasso, M. L. 1998. Ecological effects of administrative boundaries. In: Knight, R. L.; Landres, P. B., eds. *Stewardship across boundaries*. Washington, DC: Island Press: 39–64.

Landres, P. B.; Morgan, P.; Swanson, F. J. 1999. Evaluating the utility of natural variability concepts in managing ecological systems. *Ecol. Appl.* 9:1179–1188.

Leathwick, J. R. 1998. Are New Zealand's Nothofagus species in equilibrium with their environment? *J. Veg. Sci.* 9:719–732.

Leser, H.; Nagel, P. 1998. Landscape diversity—a holistic approach. In: Barthlott, W.; Winiger, M., eds. *Biodiversity: a challenge for development research and policy*. New York: Springer-Verlag: 129–143.

Levin, S. A. 1992. The problem of pattern and scale in ecology. *Ecology* 73:1942–1968.

Lombard, A. T.; Nicholls, A. O.; August, P. V. 1995. Where should nature reserves be located in South Africa? a snake's perspective. *Conserv. Biol.* 9:363–372.

Mackey, B. G.; Nix, H. A.; Hutchinson, M. F.; MacMahon, J. P.; Fleming, P. M. 1988. Assessing representativeness of places for conservation reservation and heritage listing. *Environ. Manage*. 12(4):501–514.

Mackey, B. G.; Nix, H. A.; Stein, J. A.; Cork, S. E.; Bullen, F. T. 1989. Assessing the representativeness of the wet tropics of Queensland World Heritage Property. *Biol. Conserv*. 50:279–303.

Margules, C. R. 1989. Introduction to some Australian developments in conservation evaluation. *Biol. Conserv*. 50:1–11.

Margules, C. R.; Stein, J. L. 1989. Patterns in the distributions of species and the selection of nature reserves: an example from Eucalyptus forests in south-eastern New South Wales. *Biol. Conserv.* 50:219–238.

Margules, C. R.; Nicholls, A. O.; Austin, M. P. 1987. Diversity of Eucalyptus species predicted by a multivariable environmental gradient. *Oecologia (Berl.)* 71:229–232.

Margules, C. R.; Creswell, I. D.; Nicholls, A. O. 1994. A scientific basis for establishing networks of protected areas. In: Forey, P. L.; Humphries, C. J.; Vane-Wright, R. I., eds. *Systematics and conservation evaluation*. Oxford, UK: Clarendon Press: 327–350.

Master, L. L. 1991. Assessing threats and setting priorities for conservation. *Conserv. Biol.* 5:559–563.

McHarg, I. L. 1969. *Design with nature*. Garden City, NY: Natural History Press.

McKendry, J. E.; Machlis, G. E. 1993. The role of geography in extending biodiversity gap analysis. *Appl. Geog.* 13:135–152.

McKenzie, N. L.; Belbin, L.; Margules, C. R.; Keighery, G. J. 1989. Selecting representative reserve systems in remote areas: a case study in the Nullarbor Region, *Australia. Biol. Conserv.* 50:239–261.

Moore, S. A.; Lee, R. G. 1999. Understanding dispute resolution processes for American and Australian public wildlands: towards a conceptual framework for managers. *Environ. Manage.* 23:453–465.

Morgan, P.; Aplet, G. H.; Haufler, J. B.; Humphries, H. C.; Moore, M. M.; Wilson, W. D. 1994. Historical range of variability: a useful tool for evaluating ecosystem change. *J. Sustain. For.* 2:87–111.

Naveh, Z. 1994. Biodiversity and landscape management. In: Kim, K. C.; Weaver, R. D., eds. *Biodiversity and landscapes*. Cambridge, UK: Cambridge University Press: 187–207.

Noss, R. F. 1983. A regional landscape approach to maintain biodiversity. *BioScience* 33:700–706.

Noss, R. F. 1992. The wildlands project: land conservation strategy. *Wild earth* (Special Issue: The Wildlands Project): 10–25.

Noss, R. F. 1993. Wildlife corridors. In: Smith, D. S.; Hellmund, P. C., eds. *Ecology of greenways*. Minneapolis, MN: University of Minnesota Press: 43–68.

Noss, R. F. 1996. How much is enough? In Wright, R. G., ed. *National Parks and protected areas*. Oxford, UK: Blackwell Science.

Noss, R. F.; Cooperrider, A. 1994. *Saving nature's legacy: protecting and restoring biodiversity*. Washington, DC: Defenders of Wildlife and Island Press.

Noss, R. F.; O'Connell, M. A.; Murphy, D. D. 1997. *The science of conservation planning: habitat conservation under the Endangered Species Act*. Washington, DC: Island Press.

Noss, R. F.; Dinerstein, E., Gilbert, B.; Gilpin, M.; Miller, B. J.; Terborgh, J.; Trombulak, S. 1999. Core areas: where nature reigns. In: Soulé, M. E.; Terborgh, J., eds. *Continental conservation: scientific foundations of regional reserve networks*. Washington, DC: Island Press: 99–128.

Olson, D. M.; Dinerstein, E. 1998. The global 200: a representation approach to conserving the earth's most biologically valuable ecoregions. *Conserv. Biol.* 3:502–515.

Poiani, K. A.; Baumgartner, J. V.; Buttrick, S. C.; Green, S. L.; Hopkins, E.; Ivey, G. D.; Seaton, K. P.; Sutter, R. D. 1998. A scale-independent, site conservation planning framework in The Nature Conservancy. *Landscape Urban Planning* 43:143–156.

Preen, A. 1998. Marine protected areas and dugong conservation along Australia's Indian Ocean coast. *Environ. Manage.* 22:173–181.

Pressey, R. L. 1990. Preserve selection in New South Wales: where to from here? *Austral. Zoologist* 26: 70–75.

Pressey, R. L. 1994. Ad hoc reservations: forward or backward steps in developing representative reserve systems? *Conserv. Biol.* 8:662–668.

Pressey, R. L.; Nicholls, A. O. 1989a. Efficiency in conservation evaluation: scoring vs. iterative approaches. *Biol. Conserv.* 50:199–218.

Pressey, R. L.; Nicholls, A. O. 1989b. Application of a numerical algorithm to the selection of reserves in semi-arid New South Wales. *Biol. Conserv.* 50:263–278.

Pressey, R. L.; Bedward, M.; Nicholls, A. O. 1990. Reserve selection in mallee lands. In: Noble, J. C.; Joss, P. J.; Jones, G. K., eds. *The Mallee lands: a conservation perspective*. Melbourne: Commonwealth Scientific and Industrial Research Organization (CSIRO): 167–178.

Pressey, R. L.; Humphries, C. J.; Margules, C. R.; Vane-Wright, R. I.; Williams, P. H. 1993. Beyond opportunism: key principles for systematic reserve selection. *Trends Ecol. Evol.* 8:124–128.

Pressey, R. L.; Johnson, I. R.; Wilson, P. D. 1994. Shades of irreplaceability: towards a measure of the contribution of sites to a reservation goal. *Biodiversity Conserv.* 3:242–262.

Pressey, R. L.; Possingham, H. P.; Day, J. R. 1997. Effectiveness of alternative heuristic algorithms for identifying indicative minimum requirements for conservation reserves. *Biol. Conserv.* 80:207–219.

Pulliam, H. R. 1988. Sources, sinks, and population regulation. *Amer. Naturalist* 132:652–661.

Quigley, T. M.; Arbelbide, S. J., editors. 1997. *An assessment of ecosystem components in the interior Columbia Basin and portions of the Klamath and Great Basins: Volume I*. PNW-GTR-405. Portland, OR: U.S. Dept. Agric., For. Serv., Pacific Northw. Res. Sta.

Rebelo, A. G.; Siegfried, W. R. 1992. Where should nature reserves be located in the Cape Floristic Region, South Africa? Models for the spatial configuration of a reserve network aimed at the protection of floral diversity. *Conserv. Biol.* 6:243–252.

Reid, W. V. 1998. Biodiversity hotspots. *Trends Ecol. Evol.* 13:275–280.

Ricketts, T. H.; Dinerstein, E.; Olson, D. M.; Loucks, C. J.; Eichbaum, W.; DellaSala, D.; Kavanagh, K.; Hedao, P.; Hurley, P. T.; Carney, K. M.; Abell, R.; Walters, S. 1999a. *Terrestrial ecoregions of North America: a conservation assessment*. Washington, DC: Island Press.

Ricketts, T. H.; Dinerstein, E.; Olson, D. M.; Loucks, C. J. 1999b. Who's where in North America: patterns of species richness and the utility of indicator taxa for conservation. *BioScience* 49:381–396.

Ritchie, M. E. 1997. Population in a landscape context: Sources, sinks, and metapopulations. In: Bissonette, J. A., ed. *Wildlife and landscape ecology*. New York: Springer-Verlag: 160–184.

Scott, J. M.; Jennings, M. D. 1998. Large-area mapping of biodiversity. *Annals Missouri Botanical Garden* 85:34–47.

Shelford, V. E. 1926. *Naturalist's guide to the Americas*. Baltimore, MD: Williams and Wilkins.

Slocombe, D. S. 1993a. Implementing ecosystem-based management. *BioScience* 43(9):612–622.

Slocombe, D. S. 1993b. Environmental planning, ecosystem science, and ecosystem approaches for integrating environment and development. *Environ. Manage.* 17:289–303.

Slocombe, D. S. 1998. Defining goals and criteria for ecosystem-based management. *Environ. Manage.* 22: 483–493.

SNEP. 1996. *Sierra Nevada ecosystem project: final report to Congress, assessments and scientific basis for management options*. Davis, CA: Center for Water and Wildlands Resources, University of California.

Soulé, M. E.; Terborgh, J., editors. 1999a. *Continental conservation: scientific foundations of regional reserve networks*. Washington, DC: Island Press.

Soulé, M. E.; Terborgh, J. 1999b. Conserving nature at regional and continental scales—a scientific program for North America. *BioScience* 49:809–817.

Steinitz, C.; Binford, M.; Cote, P.; Edwards, Jr., T.; Ervin, S.; Forman, R. T. T.; Johnson, C.; Kiester, R.; Mouat, D.; Olson, D.; Shearer, A.; Toth, R.: Wills, R. 1996. *Biodiversity and landscape planning: alternative futures for the region of Camp Pendleton, California*. Cambridge, MA: Harvard University School of Design.

Swanson, F. J.; Jones, J. A.; Wallin, D. O.; Cissel, J. H. 1994. Natural variability—implications for ecosystem management. In: Jensen, M. E.; Bourgeron, P. S., tech. eds. *Volume II: ecosystem management: principles and applications*. PNW-GTR-318. Portland, OR: U.S. Dept. Agric., For. Serv., Pacific Northw. Res. Sta.: 80–94.

Swetnam, T. W.; Allen, C. D.; Betancourt, J. L. 1999. Applied historical ecology: using the past to manage the future. *Ecol. Appl.* 9:1189–1206.

The Nature Conservancy. 1996. *Conservation by design: a framework for mission success*. Arlington, VA: The Nature Conservancy.

Underwood, A. J. 1995. Ecological research and (and research into) environmental management. *Ecol. Appl.* 5:232–247.

Urban, D. L.; O'Neill, R. V.; Shugart, H. H., Jr. 1987. Landscape ecology: a hierarchical perspective can help scientists understand spatial patterns. *BioScience* 37:119–127.

Usher, M. B. 1986. Wildlife conservation evaluation: Attributes, criteria and values. In Usher, M. B., ed. *Wildlife conservation evaluation*. London: Chapman & Hall: 3–44.

White, D.; Preston, E. M.; Freemark, K. E.; Kiester, R. A. 1999. A hierarchical framework for conserving biodiversity. In: Klopatek, J. M.; Gardner, R. H., eds. *Landscape ecological analysis*. New York: Springer-Verlag: 127–153.

Wright, R. G.; Murray, M. P.; Merrill, T. 1998. Ecoregions as a level of ecological analysis. *Biol. Conserv.* 86:207–213.

Yaffee, S. L. 1998. Cooperation: A strategy for achieving. In: Knight, R. L.; Landres, P. B., eds. *Stewardship across boundaries*. Washington, DC: Island Press: 299–324.

Yee, T. W.; Mitchell, N. D. 1991. Generalised additive models in plant ecology. *J. Veg. Sci.* 2:587–602.

Zonneveld, I. S. 1989. The land unit—a fundamental concept in landscape ecology, and its applications. *Landscape Ecol.* 3(2):67–89.

21 Methods of Economic Impact Analysis

Daniel E. Chappelle

21.1 Introduction

Chapter 17 described the steps involved in examining economic linkages for a particular region at a specific time. This chapter discusses how this description can be used to analytically determine economic impacts that may result if a particular action is carried out in the region. These economic impacts, when combined with other types of impacts (e.g., ecological, social, sociological, and political), provide a comprehensive assessment of the region, which can be of enormous assistance in the decision-making process for the management of the region's natural resources.

As indicated in the earlier chapter, input–output accounts are the most useful type in economic impact analysis. Various types of models, appropriate to the given situation, can be developed using these accounts as an information system. Alternative model structures include (1) input–output models (I–O), (2) linear programming models, (3) econometric models, (4) simulation models, and (5) hybrid models of various types. In this chapter, only applications that use input–output modeling are discussed, because this type of modeling strategy is necessary to fully express linkages in the regional economy.

Textbooks providing a thorough description of regional input–output modeling are Richardson (1972) and Miller and Blair (1985). Progress reports can be found in Richardson (1985) and Jensen (1990). The special double issue of the *International Regional Science Review* (volume 13, numbers 1 and 2, 1990) contains a series of articles on the construction and use of regional input–output models. See Otto and Johnson (1993) for a series of articles on input–output modeling on microcomputers.

21.2 Input–Output Modeling

Input–output models of various types are commonly used in economic impact assessments. It is important to recognize the major assumptions of input–output modeling.

1. Each product (or group of products) is supplied by a single sector of production.
2. No two products are produced jointly. Each sector produces only one homogeneous output or a mix of products in fixed proportions.
3. "The basic design of . . . interindustry accounts is derived from the division of uses into two categories—intermediate and final—and the corresponding division of inputs into 'produced' and 'primary' . . . , there is some choice as to the uses which will be considered autonomous (or 'final' . . .) which must be determined from both theoretical and empirical considerations" (Chenery and Clark, 1959, p. 33). These choices must be specified in the model.

A major task in the assessment is to delineate boundaries of the system being studied, that is, what is internal and what is external. Blair (1991, p. 179) is explicit about boundaries for the economic system:

> Many applications of input–output analysis require that a distinction be made between exogenous (outside system) and endogenous (inside system) activity. Exogenous activities are determined by forces outside the region. Exports are determined by forces outside the region. Exports for each of the sectors within the economy are normally the major exogenous component. . . . The assumption that exports are the only exogenous factor is consistent with the export-base theory.

4. In demand-driven models, final demands for each sector are exogenous variables and intermediate transactions are endogenous variables. In supply-driven models, payments for inputs for each sector are exogenous variables and intermediate transactions are endogenous variables.
5. Quantities of inputs purchased by each sector are a function only of that sector's level of output. No substitution of factors (e.g., capital for labor) is permitted (i.e., the relative proportion of factors when forecasting with the model remains identical to that mix represented in the input–outputs accounts).
6. "Supply and demand in each market are equated, not through changes in price and resulting movements along supply and demand curves, but through a horizontal shift in the demand function of each industry resulting from changes in production levels in other sectors. . . . The assumption of maximizing behavior, which is central to partial equilibrium analysis [the primary tool of analytical economics], plays no explicit role in the Leontief system [i.e., the I–O system of equations]. . . . [Rather] it is assumed that producers have little or no choice as to factor proportions in the short run and react to demand changes by changing output rather than price . . . (Chenery and Clark, 1959, p. 4).

   However, there is an implicit objective function built into the structure of an input–output model. Basically there are two options: (1) a supply-driven model in which the implicit objective function is to find the production schedule for all sectors that will exactly exhaust the exogenous supplies of resources (as defined in the payments sectors of the transactions table), or (2) a demand-driven model in which the implicit objective function is to find the production schedule for all sectors that will exactly meet the exogenous schedule of final demands.

   Most input–output models, even those that focus on natural resources, are of the demand-driven type. There are a number of reasons for this, primarily because such seems more logical given the economic system and the fact that inputs become more specialized as they proceed through the production system (e.g., a tree has more alternative uses than a 1- by 4-inch piece of lumber). Also, for the most part, resources are actually valued by their expected contributions to the value of final products. For a comparison of supply- and demand-driven models, see Hoover and Giarratani (1984, p. 330) and Schallau and Maki (1983).
7. Any resource constraints (e.g., natural resource stocks, industrial capacity, labor supply) must be handled by side calculations that are exogenous to the input–output model. Expanding the input–output model to a linear programming model permits building in such constraints, but only at great costs in terms of model development and generalization of results (see Section 21.5).
8. "The total effect of carrying on several types of production is the sum of the several effects. This is known as the additivity assumption, which rules out external economies and diseconomies" (Chenery and Clark, 1959, p. 33).

21.2.1 Brief Description of an Input–Output Model*

Input–output accounts form the basis for the transactions table, which indicates the dollar value of each sector's transactions with all others in the regional economy. Input–output accounting results in a transactions table summarizing all economic linkages in the regional economy for a specific time (usually a year). A simple hypothetical transactions table is shown in Table 21.1.

This table shows all economic linkages in the region priced at the producer level. It is easily read and is a description of the regional economic structure at a given time. A row is read as describing how the sales of a sector are distributed. For example, the agriculture sector has a gross output of \$100 million, which is distributed as follows: it sells \$25 million to itself (i.e., transactions between firms in the sector itself), \$30 million to industry, \$10 million to trade (within region) and services, \$5 million directly to households, and \$30 million to exports (outside region) and other final demand. In a realistic model, a vastly larger number of sectors would be required (a bare minimum would be about 40 for a developed industrial economy, and most examples include hundreds of sectors, some as many as 500, with elaboration particularly of industrial sectors). Also, capital accounts would be included to trace changes in inventory, depletion, and depreciation.

The columns describe how costs are distributed. For example, in our simple table, industry buys \$30 million of inputs from agriculture, \$130 million from itself, \$90 million from trade and services, \$115 million from households (i.e., primarily labor), and \$115 million from imports (from outside

*Quoted sections from Chappelle et al., 1986.

TABLE 21.1. Transactions table for the regional economy.

	Endogenous sectors			Exogenous sectors		
	(1)	(2)	(3)	(4)	(5)	
	Agriculture	*Industry*	*Trade and services*	*Households*	*Exports and other final demand*	*Gross output*
			Million dollars			
Agriculture	25	30	10	5	30	100
Industry	25	130	170	45	110	480
Trade and services	10	90	135	195	70	500
Households	30	115	135	10	10	300
Imports and value added	10	115	50	45	0	220
Gross outlay	100	480	500	300	220	1600

Rows in this table can be represented by the following equation:

$$X_i = x_{i1} + x_{i2} + \cdots + x_{in} + Y_i = \sum_j^n x_{ij} + Y_i,$$

where X_i = total output of sector i,
x_{ij} = sector i output that is delivered as input to sector j (sales of producers' goods or intermediate sales)
Y_i = sales to final demand for sector i,
n = number of sectors in the transactions table.
In contrast, columns can be represented by

$$X_j = x_{1j} + x_{2j} + \cdots + x_{nj} + V_j = \sum_i^n x_{ij} + V_j,$$

where X_j = total outlay of sector j,
V_j = charges against final demand or payments to a factor of production in sector j.

the region) and value added (inputs other than labor, including government). The total of the column, \$480 million, is the gross outlay. If every dollar circulating in the economy is accounted for, then gross outlay will be exactly equal to gross output. In any actual application, there will be some statistical discrepancy.

A simple production function can be calculated by dividing each transaction in a column by the total outlay for that particular endogenous sector, that is, the percentage of total outlay expended on each sector. Reading down the columns of Table 21.2, we can think of these coefficients as indicating the ingredients (i.e., the recipe) of the sector's product mix. The coefficients, termed direct coefficients, indicate " . . . the percentages of total expenditures that are transacted to each sector of the regional economy on the first transaction. Another way of looking at [the table of direct coefficients] is that the figures represent the number of cents out of each purchase dollar that must be paid by the column sector to intermediate industries . . . indicated on the rows to produce one dollar's worth of product or product mix. Therefore, each column . . . represents the expenditure distribution for the average firm in the sector indicated by the column heading. (p. 4)

TABLE 21.2. Direct coefficients for the hypothetical regional economy.

	1	2	3
1	0.25	0.06	0.02
2	0.25	0.27	0.34
3	0.10	0.19	0.27

This operation may be expressed as follows:

$$[A] = a_{ij} = \frac{x_{ij}}{X_j},$$

where a_{ij} is the direct coefficient that quantifies input requirements to be purchased from sector i by sector j, and $[A]$ is the matrix of a_{ij} (see Chapelle et al., 1986, pp. 3–4).

The direct coefficients table expresses only initial transactions by producing sectors as they purchase inputs from other sectors of the regional economy. Normally we are interested in quantifying the stream of transactions as the successive rounds of respending occur that are initiated by these first transactions. The view generally taken in input–output analysis is that purchases of products by final demand sectors set off rounds of respending that generate a level of economic activity within the regional economy that will be some multiple of the amount initially spent by the final demand sectors. (p. 4)

The inverted Leontief matrix $[(I - A)^{-1})]$ provides a precise calculation of the stream of respendings that occur in the regional economy. This matrix is a multiplier matrix, in that it provides information on the amount of sales generated by all sectors of the regional economy when final demand is increased by one dollar for each sector. It is possible to use this matrix to forecast production required to satisfy expected final demand because the following relation holds:

$$X = (I - A)^{-1}\, Y,$$

where X is a vector of output levels and Y is a vector of final demands. (p. 4)

Considering the household sector exogenous, the inverted Leontief matrix for a hypothetical regional economy is shown in Table 21.3.

This matrix indicates both direct and indirect linkages in the regional economy in terms of output (or sales). Note that the diagonal of this matrix has numbers all higher than unity. The reason for this is that the one dollar of final demand that initiated the respending process (or ripple effect) for each sector is included. To this is added the direct effect (see Table 21.2). Subtracting the final demand and direct effects from the elements in Table 21.3 leaves the indirect effects (Table 21.4). The numbers express the respending that occurs after the initial final demand of one dollar and direct expenditures required to meet that final demand. Therefore, the numbers quantify all the respending that occurs in the economy to meet the exogenous final demand.

This important principle of input–output analysis might be best summed up by considering the following power series, which approximates the equation for the Leontief inverse (the accuracy of which increases as the power in the series is increased). This power series is useful in illuminating the process of spending and respending that reverberates through complex linkages of the economy.

$$X = I + A + A^2 + A^3 + A^4 + \cdots + A^n,$$

TABLE 21.3. Direct and indirect coefficients for the hypothetical regional economy.

	1	2	3
1	1.40	0.15	0.11
2	0.64	1.63	0.78
3	0.36	0.44	1.58

TABLE 21.4. Indirect coefficients for the hypothetical regional economy.

	1	2	3
1	0.15	0.09	0.09
2	0.39	0.36	0.44
3	0.26	0.25	0.31

where I equals the one dollar of initial spending (final demand) and the power of A denotes the round of respending through intersectoral linkages. Because the A matrix consists of numbers less than 1, the contribution of each subsequent round decreases until no significant addition can be measured.

21.2.2 The Multiplier Process: Types and Measurement Scales of Multipliers

Normally, in economic development analysis we are interested in calculating various types of multipliers that will indicate magnitudes of impacts likely to occur in the regional economy if a certain strategy is pursued instead of some other strategy. These multipliers are of various types, depending on the dimensions of economic life in which we are interested, for example, sales [or output], income, employment, and pollution. For each dimension, a multiplier may be calculated for each sector to measure economic impacts on the regional economy resulting from an increase of one dollar in final demand for the product mix of the particular processing sector.

All multipliers are based on the inverted Leontief matrix, which expresses interactions of sectors in the economy. Other types of multipliers (e.g., income and employment) may be calculated by transforming from the output measure (which is the measure used in the transactions table) to the measure of interest. Most commonly, income and employment multipliers are developed. Output multipliers are calculated directly from this matrix, which is quantified in terms of output or sales. All other multipliers are calculated by transforming from output to other measures.

All measures may be expressed as either Type I or Type II multipliers. Type I expresses the direct plus indirect effects divided by direct effects. Type II multipliers are larger in that they include direct, indirect, and induced effects in the numerator. The induced effects quantify the increased or decreased consumption by households as the economy expands or contracts.

Therefore, the simplest way to view multipliers is that

$$\text{Type I multiplier} = \frac{\text{direct effect} + \text{indirect effect}}{\text{direct effect}}$$

and

$$\text{Type II multiplier} = \frac{\text{direct effect} + \text{indirect effect} + \text{induced effect}}{\text{direct effect}}.$$

Explanations of the different multipliers are rather straightforward. "An output multiplier for sector j is defined as the total value of production in all sectors of the economy that is necessary in order to satisfy a dollar's worth of final demand for sector j's output" (Miller and Blair, 1985, p. 102). In contrast, income multipliers, "translate an initial \$1.00 output estimate (which comes from an initial \$1.00 final-demand change) into an expanded . . . estimate of the value of resulting employment (household income) (Miller and Blair, 1985, p. 105). "The direct income change for each sector is given by the household row entry of the regional I–O table when expressed in input coefficient form (i.e., the direct coefficients table)" (Richardson, 1972, p. 32). As before, "The Type II multiplier takes into account the repercussionary effects of secondary rounds of consumer spending in addition to the direct and indirect interindustry effects" (Richardson, 1972, p. 33).

Following the same argument as was presented for Types I and II income multipliers, we may wish to relate the total employment effect to an initial change in employment, not final demand (and output) in monetary terms. That is, a dollar's worth of new output by sector j means additional jobs in sector j in the amount of the physical labor coefficient (Miller and Blair, 1985, p. 112). The employment multiplier is simply the expanded employment effect divided by the direct employment change, as before. Equations for these multipliers can be found in Miller and Blair (1985, pp. 103–112).

Note that the Type II multiplier is larger than the Type I by the addition of the induced effect to the numerator. The induced effect is the increased consumption of all goods and services, directly and indirectly, as households have increased income, due to increased production (i.e., considering the labor share of increased income). The household sector is taken as an exogenous sector for the Type I multiplier, but as endogenous for the Type II. Blair (1991, p. 180) discusses the question as to whether households should be considered endogenous:

> Should households be included with the endogenous sectors? In calculating . . . [the augmented Leontief inverted matrix], household consumption was considered to be determined by the amount of spending within the system. Thus, household spending was endogenous. Such a treatment was consistent with the export-base theory of growth, which claims that all local economic activity is supported by exports. However, some input–output analyses have treated household consumption as independent of the level of exports; in other words, household spending has been treated as exogenous. Clearly, some consumption would occur even if households had no income. The consumption could be financed from past savings. Thus, to some extent, household consumption is exogenous.
>
> When household consumption is treated as independent of the level of exports, the size of the multiplier is smaller than when household consumption is considered to be induced by the level of exports. When consumption is induced by exports, an increase in exports will not only stimulate industry trade, but also local consumption; hence, the multiplier will be larger when household consumption is considered to be dependent upon exports.
>
> [T]he sales multiplier for each sector can be calculated directly from the inverted Leontief matrix. . . . This calculation is accomplished by simply adding coefficients in a given column of this matrix for the processing sector. This multiplier indicates the amount of economic activity generated in the economy by an additional dollar of final demand for the products of the specific sector. Higher output multipliers indicate a higher degree of interdependence among sectors of the economy. (p. 4)

For our hypothetical regional economy, the output multipliers are 2.40, 2.22, and 2.47 for sectors 1, 2, and 3, respectively (see Table 21.3).

In general, prevailing opinion among analysts is that Type I multipliers underestimate impacts, whereas Type II multipliers overestimate impacts. Some analysts maintain that Type III multipliers are the most realistic, because they recognize the fact that the household consumption function changes as income level changes. These are the same in concept as Type II (i.e., the numerator is the direct effect + indirect effect + induced effect), but the induced effect is modified to reflect changes in average per capita consumption as income changes. Essentially the consumption function is changed as income changes. This is in contrast to the fixed consumption function used to derive Type II multipliers. Unfortunately, there are a number of different formulations of Type III, so it is not possible to generalize this type. For an example, see the IMPLAN models (Section 21.3.1). Techniques for deriving other multipliers may be found in Richardson (1972, Chapter 3) and, at a more advanced level, in Miller and Blair (1985, Chapter 4).

Multipliers for goods and services not delineated as sectors in the transactions table (e.g., tourism and recreation) cannot be derived directly. However, it is possible to derive multipliers indirectly by including expected changes in final demand for the various sectors involved, forecasting the total output, and then dividing total output by the direct effect (e.g., see Pedersen et al., 1989). For example, sectors that would be included for tourism are hotels, motels, and resorts; eating and drinking places; and automobile service stations.

It should be recognized that multipliers change as the spatial aggregation changes. The larger the geographic area is, the larger the multiplier effect, because additional interaction effects are quantified in the multiplier. This emphasizes the importance of correctly specifying regional boundaries.

21.2.3 Limitations of Multipliers

Although multipliers are very useful in evaluating economic alternatives, they have some definite limitations.

> The Type I and Type III multipliers are useful as measures of the additional activity which can be expected to occur throughout the economy from changes in a particular economic sector if additional inputs are available. Because they are ratios of total impacts to the direct impact, use of the Type I and III multipliers can be misleading if used by themselves. . . . the size of impacts differs depending on the economic measure considered and whether the induced component is included. (Pedersen et al., 1989, p. 15)

Although multipliers are multiplied by direct impacts to estimate the ripple effect from initial changes in economic activity, magnitudes of multipliers by themselves say little directly regarding total economic impact. For example, if a sector has a very high employment multiplier, but the number of direct jobs per one dollar of production is very low, then total employment impact will be low.

21.2.4 Distribution of Output, Income, and Employment

Input–output models contain a tremendous amount of information that most people never use. Unfortunately, multiplier analysis has tended to dominate any impact analysis. Often distributional impacts are not explored at all. This is unfortunate, because an input–output model can provide estimates of sectoral distributions of output, income, and employment. The direct coefficients table shows the distribution of direct effects of outputs and input. The inverted Leontief matrix shows the distribution of direct and indirect effects in output terms, which can be transformed to income and employment terms if desired. It is likely that distributional impacts are more important in the public policy arena than total regional impacts. Discussion of income distribution impacts is found in Rose et al. (1988).

21.3 Selected Available Input–Output Modeling Systems: Ready-made Models

Because of the extreme costs and time required to conduct a survey-based input–output analysis, much less an economic impact assessment, usually the most practicable way to proceed is to use one of the available secondary data models. Many such models have been reported in the literature. Comparisons of many of the input–output modeling systems are found in Brucker et al. (1987, 1990). For practitioners, it is important to select one that is kept up to date (i.e., the information system used by the model is continually updated). The most commonly available supported secondary-data-based modeling systems are (1) IMPLAN (developed by Forest Service, USDA, currently sold and maintained by MIG, Inc.); (2) REMI (Regional Economic Models, Inc.); and (3) RIMS (Bureau of Economic Analysis, U.S. Commerce Department). A brief comparison of the popular ready-made regional economic impact models is shown in Table 21.5.

21.3.1 IMPLAN

A brief description of IMPLAN follows (Shields et al., 1996, p. 19):

> IMPLAN, an acronym for Impact Analysis for Planning, is the I–O model developed by the USDA Forest Service. It is a nonsurvey-based modeling tool that provides front- and back-end user interface software plus a matrix-inverter that creates the Leontief inverse matrix. Data sets can be manipulated using the software, and scenarios can be developed using the modified information. A standard run of IMPLAN generates the Leontief inverse, various multipliers, and a wide variety of reports.
>
> IMPLAN is a demand-driven I–O model, useful for examining backward linkages. . . . IMPLAN data sets represent a single year and utilize make and use matrices, so analyses can be conducted from an industry or a commodity perspective. . . .

TABLE 21.5. Comparison of popular ready-made regional economic impact models.[a]

Criterion	RIMSII	IMPLAN	REMI
Basic structure	Input–output	Input–output	Conjoined input–output and econometric forecasting model
User inputs	Desired aggregations	Desired aggregations; changes in final demands	Desired aggregations; changes in employment or income
Outputs	Types I and II multipliers; total requirements; final demands; direct earnings coefficients	Types I and III multipliers; transactions table; direct and total requirements	Types I, II, and III multipliers; final demands; direct and total requirements
Minimum spatial unit	County	County	County
Maximum number of sectors	531	538	53
Relative cost	Low	Medium	High
Relative complexity	Low	Medium	High

[a]All three models have unique features not covered by this simple comparison. See Brucker et al. (1987, 1990) and Crihfield and Campbell (1991) for more details.

IMPLAN is unique among I–O models in that its technology matrices are fully developed for each county in the United States. These accounts are developed using published county statistics and the national input–output accounts published by the Bureau of Economic Analysis (BEA), . . . Analysis can be conducted at the level of one or a group of counties, up to the national level.

IMPLAN Type III multipliers are formulated as follows (Pedersen et al., 1989, p. 5, fn. 3):

The induced impact calculated by IMPLAN is different from traditional estimates of induced spending associated with a Type II multiplier. The IMPLAN induced effect is based on direct and indirect impacts causing changes in employment. These changes are used with ratios of employment to population to estimate population changes which are, in turn, multiplied by estimates of average per capita consumption to generate induced estimates. These induced estimates are then fed back through IMPLAN as final demands.

21.3.2 REMI

REMI models and databases have been used extensively in the United States for planning activities and economic forecasting. The REMI approach is described in detail by Treyz (1993). Crihfield and Campbell (1991, pp. 3–4) characterize REMI as follows:

The REMI model is a conjoined input–output and econometric forecasting model. It resembles the standard input–output approach in that the direct and the first round of indirect changes are generated by the input–output model. Changes in employment, output, and income are then passed to an econometric forecasting module. Although simulation results from the forecasting module are presented at either the 14-sector or 53-sector level of aggregation, the initial change can be specified for any of the 490 sectors in the input–output table. Additionally, users have access to nearly 800 "policy" variables which can be activated at any stage of the forecast to simulate future changes in corporate tax rates, sales tax rates, unemployment compensation, etc. Detailed occupational demands are estimated through a link with an occupational requirements matrix. The use of a conjoined model attempts to overcome the rigidities of a static input–output model while maintaining its detail for impact definition.

The mechanics of implementing the REMI model differ from IMPLAN. In IMPLAN the exogenous change in economic activity is typically modeled as a change in final demand in terms of output for the sector(s) of interest. In REMI final demand changes are typically entered in the model as changes in employment or income. These changes are then converted to their corresponding output values for the base year of simulation. The input–output component of REMI calculates the direct effect and the first round of indirect interindustry spending. Changes in output, employment, and income are passed to the forecasting module which simulates the remaining indirect effects and a variety of induced effects to the end of the forecast period. . . .

21.3.3 RIMS

Chappelle (1995) provides a detailed review of the RIMS model.

RIMS is a non-survey shortcut method (sometimes termed a "ready-made" model) that provides economic impact multipliers of various types and measured in various scales without producing a full input–output matrix.

Multipliers are based on BEA's national input–output accounts of 1982 and BEA's four-digit Standard Industrial Classification (SIC) county wage and salary data (generally no more than three years old), which are used to reduce national coefficients to represent structure and trading patterns of the regional economy specified by the user. Regional multipliers are developed using a modified simple location quotient approach. Regions may consist of one or more counties. (Chappelle, 1995, p. 3)

Documentation of the RIMS II model is found in Bureau of Economic Analysis (1992).

21.4 Including Ecological, Demographic, and Social Variables in Input–Output Analysis

Laurent and Hite (1972) provided a simplified model that shows some promise for ecological assessment work. They summarize their model as follows (Laurent and Hite, 1971, p. 9):

An economic input–output table was constructed for the Charleston, South Carolina SMSA on the basis of a field survey. Considering this Leontief system as a general theory of production, an environmental matrix showing the inflow from the environment and outflow to the environment associated with one dollar of gross sales was developed to fit within this general production model. The linking of the economic model to the environmental matrix completed the general model.

The model can be represented most simply as

$$R = (I - A)^{-1}(E),$$

where E = a matrix of inflows to and outflows from the economy to the environment

$(I - A)^{-1}$ = inverse matrix of an input–output model of the region

R = matrix of the direct and indirect environmental impact of each economic sector

Laurent and Hite (1972) recommend developing resource-income or environmental-income multipliers by dividing the usual income multipliers into the R matrix. These multipliers provide a measurement of the trade-offs between pollution and income. Roberts and Rettig (1975) used the Laurent and Hite approach in an analysis of the Clatsop (Oregon) County economy.

The use of input–output to forecast demographic change is discussed in Stone (1966). Population accounting is necessary to develop a social accounting matrix (SAM), which, when linked to labor market information and household formation information, can provide an element of dynamics to the input–output model. For a discussion of a social accounting matrix, see Holland and Wyeth (1993). A model designed to trace linkages between the regional economy, labor force, and the population is IPASS (Interactive Policy Analysis Simulation System), a dynamic simulation model that includes an input–output module to calculate economic impacts. Olson et al. (1984, abstract p. i) describes this model as follows:

The IPASS model is a dynamic analytical tool that forecasts growth and development of an economy. It allows the user to introduce changes in selected parameters based on different assumptions about socioeconomic variables. The user can analyze the impact those changes would have on the economy by comparing results of the alternative situation with the original forecast. The program is interactive, and the user needs no previous experience in programming or model building.

This model has features worth considering when conducting an ecological assessment because it links the regional economy in an integral way to the human population. For an application of this model, see Maki et al. (1985).

21.5 Advantages and Disadvantages of Transforming Input–Output Models to a Linear Programming Format

There are a number of advantages to transforming a regional input–output model into a linear programming (LP) model (for an early example of this transformation, see Gass, 1958, pp. 167–170). This is possible because an input–output model is essentially the simplest linear programming model with the objective function to find that program of production that exactly meets the exogenous demand schedule (demand-driven model) or the production program that exactly exhausts resource supplies (supply-driven model). The most notable advantages of transforming to an LP model are (1) to incorporate of multiple policy objectives, (2) to incorporate resource and other types of constraints, (3) to have the ability to develop shadow pricing of environmental quality variables, and (4) to isolate the best technological mix. The greatest disadvantages of such a modeling strategy are (1) with resource and other types of constraints built into the model, it becomes very specific to the situation being modeled, that is, it loses generality; (2) considerable cost and effort is involved in determining the appropriate constraints; (3) it is difficult to determine appropriate objective function(s) since they apply to a region rather than an individual decision-making unit (i.e., a firm); and (4) if the best technological mix is a major study objective, much effort, time, and cost are required to construct appropriate cost functions for each alternative technology. For an application of this type of model, see Chappelle et al. (1982).

An approach suggested by Chappelle (1993, p. 16) may be useful in quantifying nonpriced goods and services in a comprehensive model:

> [E]xcept by measuring transactions required to experience nonpriced goods and services, those [nonpriced] products would be ignored in the analysis. Therefore, many amenity and other nonpriced products will continue to be ignored if accounting systems are not expanded to include them as entities. However, these products can be included in the analysis by entering appropriate constraints into a demand-driven mathematical programming model based on input–output accounts. The model could be made dynamic by embedding the mathematical programming model within a simulation modeling system that could continually update exogenous variables along their forecasted trends to form a simoptimization model. This type of model could generate scenarios exhibiting consequences of policy changes in the face of changing conditions.

21.6 Forecasting Economic Impacts for Purposes of the Assessment

The following steps must be considered when forecasting comprehensive economic impacts in assessments:

1. Decide whether changes need to be made in the input–output model to reflect expected changes over the forecast period (e.g., in technology, in the household consumption function). If changes are needed, strategies will have to be devised for this.
2. Develop forecasts of exogenous variables (e.g., final demands of all goods and services) for each forecast year. These forecasts should be based on authoritative reports if at all possible.
3. Calculate forecasts of endogenous variables by running the model. Using an input–output model, the forecasted endogenous variables consist of production levels for each sector of the economy.
4. If at all possible, provide some information on the likely precision of forecasts. This will be difficult because of the lack of statistical information on many inputs. From the standpoint of evaluating alternative scenarios, the client should be able to surmise whether there are any operationally significant differences between the scenarios. A major conclusion of the assessment may be that data are so poor (in quantity or quality) that significant differences between alternative scenarios cannot be adequately appraised.

After a decision has been made regarding sources of economic impact information, it can be applied to the assessment. This generally involves use of input–output multipliers. The objective in this phase is to quantify all economic effects—direct (which are data to the input–output analysis), indirect, and induced. In this way, economic impacts likely to occur with changes to the regional environment can be quantified, which will then permit planners to weigh these impacts against other types of impacts (e.g., ecological, social, and political). Most impact assessments involve making forecasts—both of current conditions and likely future conditions if current direction continues.

Forecasting is carried out in three stages: (1) using multipliers to forecast total impacts from direct impacts, (2) forecasting future production schedules, given expected final demands for all goods and services, and (3) forecasting future distributions of transactions, given forecasted production levels.

21.6.1 Use of Multipliers in the Assessment

Applications of the various multipliers, whether derived in the assessment or simply taken from some other study, are fairly straightforward. Since all multipliers are in the form of total effect divided by the direct effect, for forecasting purposes all we need is the direct effect, which we must remember is input data for the input–output model (or implied by the transactions data). Of course, to use multipliers or any other formulation, it is necessary that the assumptions associated with input–output modeling can safely be incurred. If the assumptions can be made, direct effects of changes in the region can be used with the multipliers to derive estimates of total effects. For example, if it is envisioned that an additional 145 jobs in the lumber industry will be created because of the increase of regional timber resources, justifying entry of a new mill, and the employment multiplier is 2.60, then we would expect 145 × 2.60 = 377 total jobs in the economy because of the utilization of this additional timber.

21.6.2 Forecasting Future Production Based on Environmental Conditions

Another important type of forecasting is that of production levels of all goods and services, given expected environmental conditions. This is based on the relationship, introduced earlier,

$$X = (I - A)^{-1}(Y).$$

If we accept the inverted Leontief inverse matrix $[(I - A)^{-1}]$ as representative of economic linkages over the forecast period and are able to specify a schedule of expected final demands (Y), then a future production schedule (X) may be calculated. However, note that the ecological feasibility of meeting such a production schedule must be appraised using side calculations, because supply constraints are not included in the input–output model. This can be done readily if we can forecast resource use per unit of product and if resource inventories can be forecasted. These calculations can be completed for all years in the forecast period. Note, however, that we should be wary of forecasting resource use per unit of product. The natural inclination is to assume no change in the future. History indicates that it is difficult to accurately predict resource use per unit of product in the long run. Technological change has resulted in fuller use of natural resources (i.e., less waste or residuals) and lower resource use per unit of product (i.e., less material is needed to fulfill the function of the good), and the introduction of new materials.

21.6.3 Forecasting Future Distributions Based on Environmental Conditions

If the input–output relations as expressed in the direct coefficients table $[A]$ can be expected to be maintained over the forecast period, the model can be used to allocate future production to the various sectors of the regional economy. This can be shown by remembering that

$$[A] = a_{ij} = \frac{x_{ij}}{X_j}$$

Therefore,

$$x_{ij} = a_{ij}X_j$$

Assuming stability in the production coefficients (a_{ij}) and the production schedule for the forecast year (X_j), it is possible to allocate the intersectoral flows (x_{ij}).

However, note that it is implicit that the sum of the payments sectors, which are exogenous, also will be assumed to remain stable for each sector. Therefore, it is important to check for resource availabilities, even those included in the economic accounts (e.g., labor) that may become scarce and hence higher priced in the future.

21.7 Major Problems in Impact Analysis Modeling

Specific problems include the following:

1. Lack of theory or existence of widely divergent theories regarding operation of the real-world system, both in terms of economic and ecological linkages.
2. Difficulty in achieving appropriate aggregation (temporal, spatial, and activities). When secondary data from various sources are used, it is difficult to maintain consistency in aggregations. A major problem of estimating economic impacts is that the smaller the region considered when using secondary data, the less reliable are the estimates. Also, there are major problems in collecting primary data for the region (particularly related to disclosure rules). Also, less than ideal sampling may be required by funding and time limitations. Because of these problems, aggregation error may be large.
3. "Input–output models focus primarily on flows of dollars between sectors of the economy, and stock relationships are generally handled by variables that measure changes in the stocks during the time period covered by the model. In [biological] productive enterprises, especially those that are long term such as forestry, there are a number of stock relationships that need to be considered over a longer period than is covered by the short period covered by the input–output model (generally a single year)" (Chappelle, 1988, p. 18). Normally, it is safe to assume that multipliers and percentage distributions are stable over a fairly long period (e.g., ten years),

but the possibility of structural change should be seriously considered.

4. Poor-quality data (i.e., lack of precision). Lack of balance in precision of the many variables involved in the model.
5. Lack of agreement on appropriate measurement scales and techniques for some variables is likely to be important. In such cases, there is a tendency to ignore these variables. Probably the best example of this is the treatment of nonpriced goods and services.
6. Generally, there are difficulties in measuring interrelationships falling between academic disciplines.
7. There may not be enough adequate information about many linkages to develop accurate forecasts, particularly if magnitudes of statistical and nonstatistical errors are considered. A major limitation of all approaches is data scarcity, particularly ecological and social data. There is a definite tendency in this type of work to assume that we can quantify impacts more scientifically than is really possible, given the state of the art.
8. Even if we can measure a variable, it is difficult to collect data on responses of species in reaction to a treatment (or a pollutant or a catastrophe), except under experimental conditions. Because of this, there is a tendency to retreat to a crude "standards approach," based largely on laboratory tests or guesses, because of a lack of knowledge. There is always a question as to the validity of the standard.
9. It is difficult to recognize trade-offs between environmental media (e.g., between air pollution and water pollution), and comparable valuation is very difficult to achieve.
10. The institutional framework (e.g., ownership patterns and goals of owners) is generally ignored in regional economic impact analyses. This aspect is usually left to the implementation phase of environmental planning, but ideally such institutional variables would be recognized in the economic impact assessment. This is possible, but requires primary data collection by survey.
11. It is difficult to model diffusion processes and impacts on receptors, even when these can be identified. For a discussion of these problems, see Kneese and Bower (1979).
12. Since we are interested in comprehensive assessment of various types of impacts (e.g., ecological, social, and economic), models tend to be very large and complex. Since it is generally not possible to have a single integrated model, usually a number of interrelated models must be developed and linked in some way. However, it should be noted that highly complex models do not seem to work very well because of the propagation of error in using data of varying precision (for a classic discussion of this problem in the context of planning, see Lee, 1973).

21.8 Conclusions

A number of conclusions appear evident from the preceding exploration:

1. Input–output modeling is generally the best modeling strategy in ecological assessments. Also, because of high monetary and time costs, usually it will be necessary to use one of the input–output modeling systems that includes secondary data on economic linkages.
2. Although input–output multipliers are important, they are not sufficient to quantify economic impacts in the assessment. It is necessary to apply them to expected changes in the regional economy, which must be based on ecological impacts, among others.
3. Forecasts of production levels and distributions of intermediate and final outputs are highly desirable.
4. Given that the input–output model does not include resource constraints, it is necessary to carry out side calculations to be sure that forecasted production levels, income, and employment are feasible in terms of the environment as a source of raw materials.
5. Results of the economic impact analysis may provide guidelines as to which ecological variables need to be quantified more precisely.

21.9 References

Blair, J. P. 1991. *Urban and regional economics*. Homeward, IL: Irwin.

Brucker, S.; Hastings, S. E.; Latham, W. R., III. 1987. Regional input–output analysis: a comparison of five "ready-made" model systems. *Rev. Reg. Studies* 17(2): 1–16.

Brucker, S.; Hastings, S. E.; Latham, W. R., III. 1990. The variation of estimated impacts from five regional input–output models. *Int. Reg. Sci. Rev.* 13(1&2):119–139.

Bureau of Economic Analysis. 1992. *Regional multipliers: a user handbook for the regional input–output modeling system (RIMS II)*, 2nd ed. Washington, DC:

Economics and Statistics Admin., U.S. Dept. of Commerce.

Chappelle, D. E. 1988. Regional economic impacts using input–output analysis. In: Lothner, D. C.; Bradley, D. P.; Gambles, R. L., eds. *Proceedings, IEA/BA workshop. Economic evaluations of short-rotation biomass energy systems; Information Report 88:2*. Toronto, ON: International Energy Agency, Bioenergy Agreement: 177–189.

Chappelle, D. E. 1993. *Consistent valuation of natural resource outputs to advance both economic development and environmental protection*. Presented at the Annual Southern Forest Economics Workshop, Duke University, Durham, NC (1993 April 22), at the Western Forest Economists Conference at Wemme, OR (1993 May 4) and at the Midwest Forest Economists (1993, August 18).

Chappelle, D. E. 1995. *An evaluation of RIMS (regional input–output modeling system) II: a white paper*. Sacramento, CA: Unpublished report submitted to the California Energy Commission.

Chappelle, D. E.; Miley, R. C.; Gustafson, R. 1982. Land use models. In: Countryman, D. W.; Sofranko, D. M., eds. *Guiding land use decisions: planning and management for forests and recreation*. Baltimore, MD: Johns Hopkins University Press: 11–49.

Chappelle, D. E.; Heinen, S. E.; James, L. M.; Kittleson, K. M.; Olson, D. D. 1986. *Economic impacts of Michigan forest industries: a partially survey-based input–output study*. Research Report 472. East Lansing, MI: Natural Resources, Agricultural Experiment Station, Michigan State University.

Chenery, H. B.; Clark, P. G. 1959. *Interindustry economics*. New York: John Wiley & Sons.

Crihfield, J. H.; Campbell, H. S., Jr. 1991. Evaluating alternative regional planning models. *Growth and Change* (Spring):1–16.

Gass, S. I. 1958. *Linear programming: methods and applications*. New York: McGraw-Hill Book Co.

Holland, D.; Wyeth, P. 1993. SAM multipliers: their interpretation and relationship to input–output multipliers. In: Otto, D. M.; Johnson, T. G., eds. *Microcomputer-based input–output modeling: applications to economic development*. Boulder, CO: Westview Press: 181–197.

Hoover, E. M.; Giarratani, F. 1984. *An introduction to regional economics*, 3rd ed. New York: Knopf.

Jensen, R. C. 1990. Construction and use of regional input–output models: progress and prospects. *Int. Reg. Sci. Rev.* 13(1&2):9–25.

Kneese, A. V.; Bower, B. T. 1979. *Environmental quality and residuals management*. Baltimore, MD: Johns Hopkins University Press.

Laurent, E. A.; Hite, J. C. 1972. Economic–ecologic linkages and regional growth: a case study. *Land Econ.* 47(1):70–72.

Lee, D. B., Jr. 1973. Requiem for large-scale models. *J. Amer. Inst. Planners* 39(3):163–178.

Maki, W. R.; Olson, D.; Schallau, C. H. 1985. *A dynamic simulation model for analyzing the importance of forest resources in Alaska*. PNW-RN-432. Portland, OR: U.S. Dept. Agric., For. Serv., Pacific Northw. For. Range Exp. Sta.

Miller, R. E.; Blair, P. D. 1985. *Input–output analysis: foundations and extensions*. Upper Saddle River, NJ: Prentice Hall.

Olson, D.; Schallau, C.; Maki, W. 1984. *IPASS: an interactive policy analysis simulation system*. PNW-GTR-170. Portland, OR: U.S. Dept. Agric., For. Serv., Pacific Northw. For. Range Exp. Sta.

Otto, D. M.; Johnson, T. G., ed. 1993. *Microcomputer-based input–output modeling: applications to economic development*. Boulder, CO: Westview Press.

Pedersen, L.; Chappelle, D. E.; Lothner, D. C. 1989. *The economic impacts of Lake States forestry: an input–output study*. NC-GTR-136. St. Paul, MN: U.S. Dept. Agric., For. Serv., North Central For. Exp. Sta.

Richardson, H. W. 1972. *Input–output and regional economics*. New York: Halsted Press, John Wiley & Sons.

Richardson, H. W. 1985. Input–output and economic base multipliers: looking backward and forward. *J. Reg. Sci.* 25(4):607–661.

Roberts, K. J.; Rettig, R. B. 1975. *Linkages between the economy and the environment: an analysis of economic growth in Clatsop Co., Oregon*. Agric. Exp. Sta. Bull. 618. Corvallis, OR: Oregon State University.

Rose, A.; Stevens, B.; Davis, G. 1988. *Natural resource policy and income distribution*. Baltimore, MD: Johns Hopkins University Press.

Schallau, C. H.; Maki, W. R. 1983. Interindustry model for analyzing the regional impacts of forest resource and related supply constraints. *For. Sci.* 29(2):384–394.

Shields, D. J.; Winter, S. A.; Alward, G. S.; Hartung, K. L. 1996. *Energy and minerals industries in national, regional, and state economies*. FPL-GTR-95. Madison, WI: U.S. Dept. Agric., For. Serv., For. Products Lab.

Stone, R. 1966. Input–output demographic accounting: a tool for educational planning. *Minerva* 4(2):365–380.

Treyz, G. I. 1993. *Regional economic modeling: a systematic approach to economic forecasting and policy analysis*. Hingham, MA: Kluwer Academic Publishers.

Part 6
Ecosystem Characterization

22 Elements of Ecological Land Classifications for Ecological Assessments

Patrick S. Bourgeron, Hope C. Humphries, and Mark E. Jensen

22.1 Introduction

An ecological land classification (ELC) is the product of the formal definition of land-based ecosystems and ecosystem complexes (Rowe and Sheard, 1981; Sims et al., 1996), based on the ecological and mapping principles of ecosystem characterization (see Chapters 2 and 3). In addition to the specific requirements of ecosystem characterization (see Chapter 3), constructing an ELC requires making decisions about the classification concepts to follow and the specific uses of the ELC (Grossman et al., 1999). In practice, the classification process is a balance between science and art (Sims et al., 1996).

In the broad fields of environmental management and planning, ELCs have been used for a variety of purposes, such as management of natural resources, land-use planning, and delineation of ecological regions, among others (Zonneveld, 1988, 1989; Klijn, 1994; Sims et al., 1996). As a result of recent developments in ecosystem management, sustainable land development, the need to detect ecosystem responses to global and land-use changes, and advances in technology (e.g., remote sensing data, geographic information systems, and databases; see Chapters 8, 10, and 11), there has been renewed interest in environmental stratification and therefore in ELCs. Accordingly, there has been a proliferation of ELC systems constructed to meet a variety of objectives, using an array of methods and technologies (Sims et al., 1996). Furthermore, there has been increasing recognition that traditional planning and management units (e.g., countries, states, National Forests) generally do not match the ecological boundaries that bound patterns and processes of interest (Uhlig and Jordan, 1996). ELCs have been used to delineate these ecological boundaries and hence provide a flexible means to analyze specific issues addressed during ecological assessments (EAs). Most importantly, ELCs have been used to provide the spatially explicit framework needed during EAs (see Chapter 13) to address multiple issues and to monitor trends.

The main objective of this chapter is to provide an overview of the approaches taken in the development and implementation of ELCs in EAs for various objectives, at a variety of spatial scales. This overview is intended to allow EA practitioners to decided what type of system(s) may work for their particular goals. The chapter makes use of and complements the material covered in Chapters 2, 3, and 19. It is organized into four main sections: principles and methods of ELCs, general areas of application of ELCs, overview of the major ELCs used worldwide, and examples of specific applications and evaluation of ELCs. The chapter is not intended to be an exhaustive review of all ELC concepts, systems, and implementation. More extensive reviews can be found in Küchler and Zonneveld (1988), Klijn (1994), Sims et al. (1996), and Grossman et al. (1999).

Patrick S. Bourgeron and Hope C. Humphries wish to acknowledge partial funding provided by a Science to Achieve Results grant from the U.S. Environmental Protection Agency ("Multi-scaled Assessment Methods: Prototype Development within the Interior Columbia Basin").

22.2 Overview of the Principles and Methods of Ecological Land Classifications in Ecological Assessments

ELCs have the general properties of any classification (Sokal, 1974); that is, in their simplest form, they group objects into sets that share particular attributes. However, ELCs are always used to generate maps of ecological land units (ELUs), whereas many other types of classification (e.g., taxonomic) do not include map generation. The basic characteristic that all ELCs share is their use in the environmental stratification of geographic space (Bunce et al., 1996a, 1996b; also see Chapter 3): environmental space is stratified by ELC classes derived from the variables used in their construction. The purpose of environmental stratification is to (1) quantify and minimize variation in physical, chemical, biological, and ecological patterns and processes, or any other measure of interest (Hughes et al., 1990; Bunce et al., 1996a) and (2) represent the distributions of these patterns and processes. Furthermore, it is hypothesized that this reduction in variation allows better quantification of the responses of patterns and processes to global changes and land-use practices. ELCs are the product of a formal process in which the results of local ecosystem characterization (Küchler, 1973; also see Chapter 3) are regionalized within a spatial framework. This chapter does not discuss mapping and stratification per se or the relationship between classification and mapping. Hereafter, we refer to an ELC class defined at any scale as an ELU. Reviews of the topics of ecosystem characterization, ecological classification and mapping, and the various terms employed can be found in Küchler and Zonneveld (1988), Sims et al. (1996), and Grossman et al. (1999); also see Chapter 3.

Although they may differ in nature and scope, ELCs share the goal of representing some or all ecosystem components and their interactions. To achieve this goal, some ELCs emphasize abiotic (physical) ecosystem components, whereas others emphasize biotic components or attempt to represent both. Figure 22.1a represents a simplified general model of the ecosystem patterns and processes and their hypothesized relationships that ELCs are intended to represent. The model has four major compartments: regional climate (A), the physical environment (B), disturbances and land use (C), and existing biotic characteristics (D). In this simplified ecosystem model, potential vegetation (PV) (compartment B) is defined as the vegetation that an area can potentially support in the absence of disturbances and human activities (Tüxen, 1959; Pfister and Arno, 1980). Therefore, according to this definition, PV is an expression of the physical attributes of the environment and is rarely actually observed. As defined, PV is useful for the characterization of ELUs (Küchler, 1964; Jensen et al., 1997; Zerbe, 1998); in some ELCs, ELUs are defined solely on the basis of PV (e.g., Daubenmire, 1966; Pfister and Arno, 1980). In this chapter, we refer to PV as an attribute of the physical environment. The attributes of compartment C are rarely incorporated in the definition of ELUs; they are often analyzed using ELUs as templates for deriving other ecosystem properties (e.g., the historic range of variability, HRV; see Section 22.3 and Chapter 19). Attributes of compartments A, B, and D have been used alone or in combination to define ELUs. Only some of the factors identified in the model in Figure 22.1 may be relevant to the definition of an ELC, based on specific goals and spatial scales.

Three main approaches to ELCs have been followed. The first approach employs the recognition of biotic communities (e.g., assemblages of species or species attributes, such as physiognomy) as a surrogate for the environment (Figure 22.1b). The most widely used method delineates vegetation types, based on the assumption that vegetation is a faithful expression of site characteristics (Küchler, 1988). This method has a venerable history (McIntosh, 1985). A recent example of this approach in the United States is the national Gap Analysis Project (Scott et al., 1993). However, when testing the value of an ecological relationship, arguments about assessing the correlations between biotic units and environmental factors can become circular, with the focus being switched between biotic and abiotic factors (Mackey et al., 1988; Short and Hestbeck, 1995).

The second approach utilizes broad environmental patterns alone (Figure 22.1c) or broad correlations between the biota and indirect environmental variables (Figure 22.1d) to describe and delineate ELUs. Indirect variables (Figure 22.1c) are defined as those factors that do not have a direct influence on the ecosystem components of interest (definition adapted from Austin et al., 1984; Austin, 1985; Austin and Smith, 1989). For example, elevation is an indirect variable in determining the distribution of the vegetation (Austin et al., 1984). The definition of an indirect variable is specific to the pattern of interest. A variety of systems based on combinations of soils, lithology, and landform have been used alone or combined with biotic data to produce classifications of ecological regions (ecoregions) or natural landscape units

→

FIGURE 22.1. Simplified general model of relationships among ecosystem components used in ecological land classifications. (a) all patterns and processes. See text for explanation of compartments A, B, C, and D; (b) existing biotic characteristics alone; (c) indirect variables alone (see text for definition); (d) existing biotic characteristics and indirect variables; (e) direct variables alone (see text for definition).

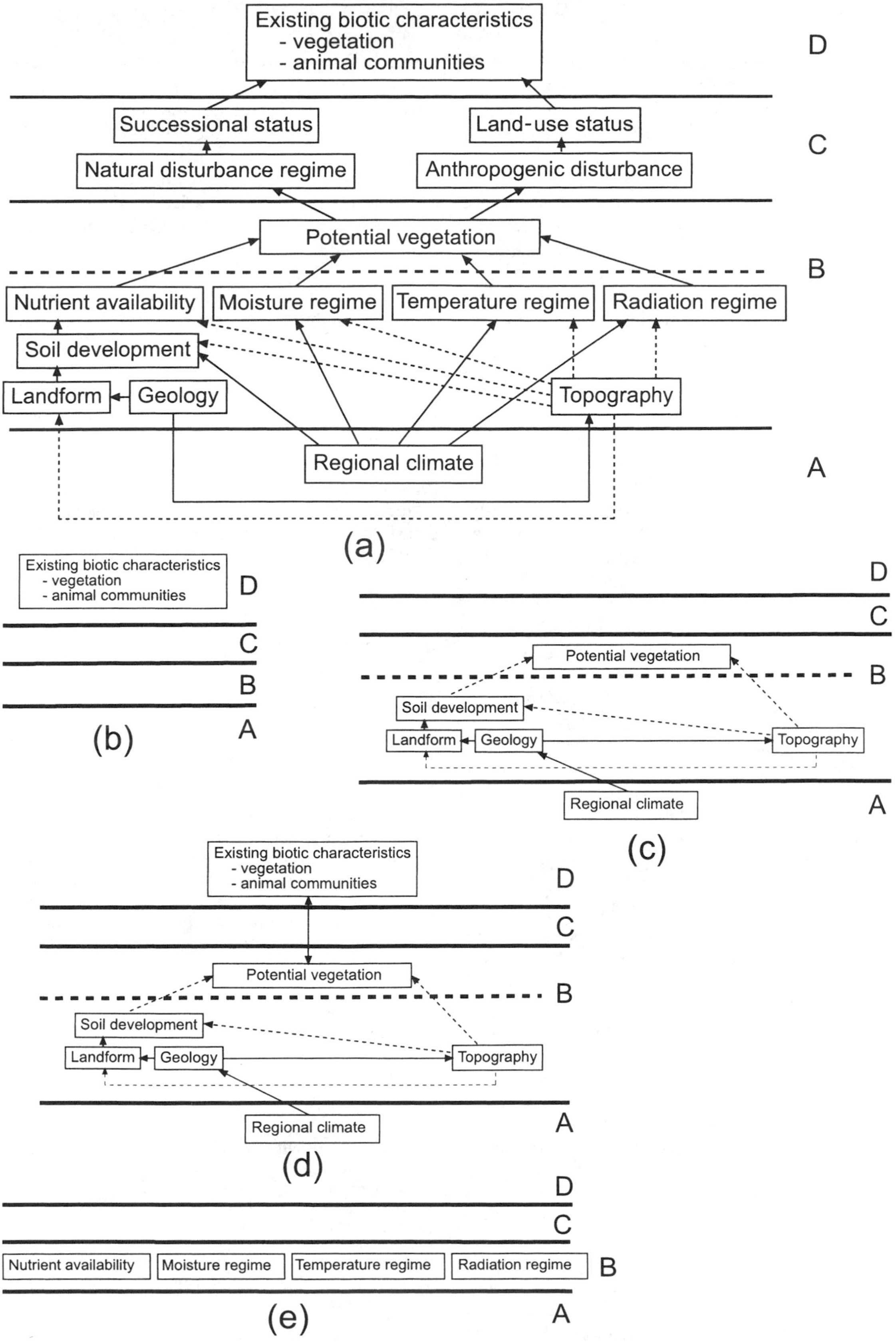

Existing biotic characteristics
- vegetation
- animal communities
D
Successional status
Land-use status
Natural disturbance regime
Anthropogenic disturbance
C
Potential vegetation
B
Nutrient availability
Moisture regime
Temperature regime
Radiation regime
Soil development
Landform
Geology
Topography
Regional climate
A
(a)
Existing biotic characteristics
- vegetation
- animal communities
D
C
B
A
(b)
D
C
Potential vegetation
B
Soil development
Landform
Geology
Topography
Regional climate
A
(c)
Existing biotic characteristics
- vegetation
- animal communities
D
C
Potential vegetation
B
Soil development
Landform
Geology
Topography
Regional climate
A
(d)
D
C
Nutrient availability
Moisture regime
Temperature regime
Radiation regime
B
A
(e)

(Bailey, 1976, 1995; Austin and Yapp, 1978; Walter, 1979; Rowe and Sheard, 1981; Omernik, 1987; Zonneveld, 1989; ECOMAP, 1993; Bailey et al., 1994; Maxwell et al., 1995; Jensen et al., 1997).

The main weakness of this approach is that the hypothesized ecological relationships between the biotic components and the indirect variables are not always explicitly stated and tested. For example, landform patterns have often been used to stratify areas into natural landscape units to represent natural levels of ecosystem integration with respect to environmental regimes and key ecosystem processes, and indeed there is compelling evidence for such relationships (Swanson et al., 1988; Ligon et al., 1995). Geomorphic pattern, through erosion–sedimentation processes, has been shown to control carbon, nitrogen, and phosphorus cycles in soils of riparian forests in southern France (Pinay et al., 1992). However, Kovalchik and Chitwood (1990) used geomorphology in addition to a floristic classification of the vegetation of riparian zones in central Oregon, but did not explicitly test the purported vegetation–geomorphological process relationship.

On the other hand, the strength of this approach is that it allows a direct analysis of the spatial and temporal scales of landscape features (Urban et al., 1987; Delcourt and Delcourt, 1988; Zonneveld, 1989), which is necessary to match patterns and processes (Levin, 1992; see also Chapter 3). A terrestrial example is provided in a mountainous landscape in northwestern Montana by Lathrop and Peterson (1992). They established that watershed morphological characteristics exhibited the same basic properties at various spatial scales (self-similarity).

A third approach is based on the argument that, to be meaningful, ELCs should be based on pattern–process relations (Hutchinson, 1959; Whittaker, 1972; Nix, 1982; Brown, 1984). Therefore, the need is to estimate a pattern or group of patterns' responses to a limited set of dominant direct variables (Nix, 1982). Direct variables are defined as factors that have a direct influence on the distribution of the pattern(s) (definition after Austin et al., 1984; Austin, 1985; Austin and Smith, 1989). For example, direct variables for species comprise their primary niche dimensions, such as temperature, radiation, moisture, and nutrient regimes for plants (Nix, 1982). Direct variables (Figure 22.1e) have been used to generate ELUs (Mackey et al., 1989; DeVelice et al., 1994). The definition of a direct variable depends on the pattern of interest.

Regardless of the approach taken, the accuracy and utility of describing and mapping ELUs are functions of the variables selected, the strength of the purported relationship between the biota and these variables, the relationship between indirect and direct variables, and the estimation procedures and mapping scales used. Ideally, the interactions between landscape features, climatic factors, and ecologically meaningful variables should be obtained using a combination of geographical information systems (GIS) and simple process models. In practice during an EA, testing biotic–abiotic relationships and the interactions between direct and indirect variables is generally difficult, time consuming, and costly. Consequently, until recently, indirect variables have served, largely untested, as surrogates for direct variables in the delineation of ELUs. An example of an ELC framework that incorporates the elements described in this section, including testing, is presented in Perera et al. (1996) for Ontario, Canada. A close examination of ELCs, including evaluation of their performance, is necessary for validating their accuracy and usefulness (see Section 22.4).

22.3 General Applications of Ecological Land Classifications in Ecological Assessments

Table 22.1 summarizes some of the general areas of application of ELCs in EAs. The specific application implemented depends on the EA objectives, the spatiotemporal scale of the data (see Chapters

TABLE 22.1. Example of potential areas of application of ecological land classifications in ecological assessments.

Delineating geographic areas
Assessment area
Characterization area
Analysis area
Cumulative impact area
Reporting unit
Basic characterization unit
Integrated response unit
Determining geographic–spatial context for defining ecosystem properties
Characterizing historic range of variability
Assessing ecological and socioeconomic conditions and pattern and process persistence
Assessing potential for restoration
Assessing suitability for specific purposes
Providing environmental stratification for strategic surveys
Setting priorities for monitoring
Selecting monitoring sites

7 and 13), and the spatial scale of the area under consideration.

ELCs are often used to define the spatial scales and boundaries of seven types of areas that are associated with different activities in an EA: assessment, characterization, analysis, and cumulative impact areas; and reporting, basic characterization, and integrated response units.

1. The assessment area is the area under consideration for a specific EA. Although the assessment area is usually defined by sociopolitical concerns and issues, all or part of its boundary may be adjusted to reflect ecological boundaries.
2. The characterization area includes the ecological boundaries of the major ecosystem and socioeconomic components of interest in an EA. The spatial analysis of these components (see Chapter 13) is conducted in the characterization area, which is often much larger than the assessment area.
3. The analysis area defines the area in which analyses are conducted and management recommendations directly influence the patterns and processes of interest.
4. In the cumulative impact area, the effects of management decisions are assessed.
5. Reporting units are constructed to facilitate the reporting of results from different components of an assessment (e.g., terrestrial, aquatic, and socioeconomic). Reporting units are important in very large assessment areas, because they are the means by which the results of analyses conducted at different spatial scales are communicated to the general public, stakeholders, and decisionmakers.
6. The basic characterization unit defines the grain of the assessment.
7. Integrated response units are generated using basic ELU maps at all relevant scales, in conjunction with similarly scaled maps of the complementary biophysical, biotic, and socioeconomic themes that are appropriate for the type of resource interpretation under consideration, but were not included in the ELC.

The first five types of areas are the basis for creating regional and local planning and management frameworks (Slocombe, 1993a, 1993b). In a few cases, some scales of conservation, political concern, management actions, and ecological responses may be congruent and comparable, and therefore some of the types of areas may have the same boundaries. For example, for the Interior Columbia Basin Ecosystem Management Project (ICBEMP) (Figure 22.2), the assessment area was defined by an analysis of biophysical, ecological, and administrative maps in light of political directives to develop an ecosystem management strategy for forests east of the Cascade crest in the U.S. Pacific Northwest (Quigley and Arbelbide, 1997). Based on specific objectives, a larger characterization area was delineated that encompassed the distributions of the patterns and processes of interest (Quigley and Arbelbide, 1997). Subbasins (4th-field hydrologic unit codes) were used as the units for many biophysical and aquatic analyses (Jensen et al., 1997). Thirteen reporting units (Jensen et al., 1997) were defined for reporting the effects of land management activities by the landscape ecology, terrestrial, aquatic, and socioeconomic science groups. Subwatersheds (6th-field hydrologic unit codes) (Jensen et al., 1997) comprised the basic characterization units for many ICBEMP studies (soil erosion, stream recovery potential, fine-scale vegetation patterns, and fish distributions) (Hann et al., 1997; Jensen et al., 1997; Lee et al., 1997). For a conservation assessment in the interior Columbia River basin (ICRB), over 17,000 land units, representing 689 classes of integrated response units, were delineated using ELUs defined with direct (climate, biogeochemical) variables stratified by PV types (see Chapter 20).

The ecosystem properties that need to be defined may differ among EAs, but generally include those that are required to predict the effects of land management, economic development, global or land-use change (O'Callaghan, 1996), or any other factor of interest in a specific EA and region. They may include hydrological, biogeochemical, ecological, and socioeconomic characteristics (see Chapters 9 and 16). The properties are usually derived from the multiscaled data analyses that produced the ELUs or ELU-derived units (Urban et al., 1987; Bourgeron and Jensen, 1994; Bourgeron et al., 1994; Jensen et al., 1997; also see Chapters 2 and 3). ELU or ELU-derived maps provide a basic template for interpreting the spatial and the temporal variability of a landscape, that is, its dynamics (see Chapters 13, 23, 24, 26, 28, and 29). For example, predicting the response of a vegetation type to disturbance is facilitated by identifying the ELUs in which it occurs (e.g., Arno et al., 1986; Host and Pastor, 1998).

ELUs are also used as environmental strata to characterize the range of variability in which an ecosystem historically operated. Knowledge of how far an ecosystem departs from HRV at multiple scales is useful in EAs (Morgan et al., 1994; Landres et al., 1999; Swetnam et al., 1999; also see

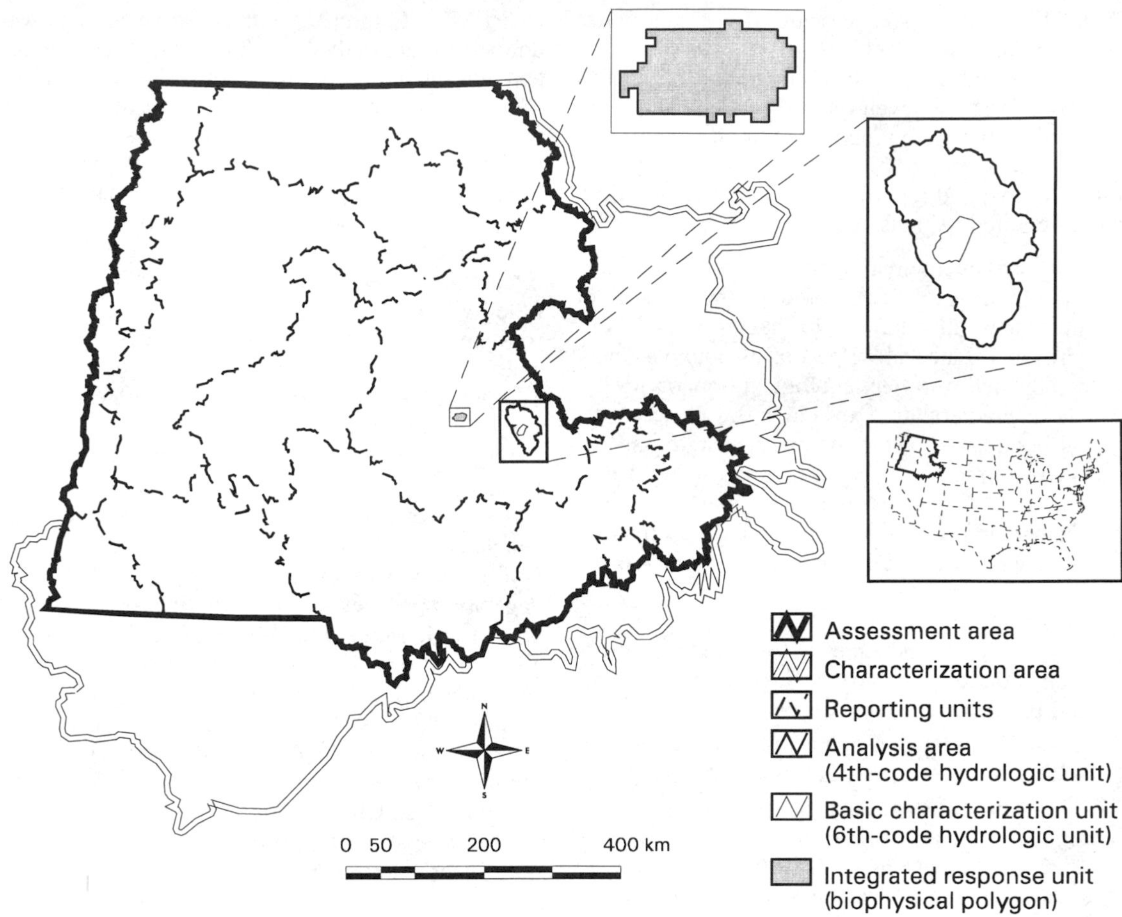

FIGURE 22.2. Types of areas delineated in interior Columbia River basin for ecological assessment activities.

Chapter 19). The assessment of ecological conditions and pattern and process persistence requires the interpretation and integration of ecological properties, spatiotemporal variability, and HRV. This assessment may be conducted in light of the potential response expected at any appropriate scale, using ELUs or ELU-derived units as a basic template for the interpretation of change in biotic patterns (e.g., existing vegetation, plant or animal abundance) and processes that commonly display changes after management or the impact of natural disturbances (e.g., fire, flood, insect outbreak). Effects of land-use practices and natural processes can be assessed by contrasting the current ecological conditions of an area with conditions in other managed or unmanaged areas that occur in the same ELU or ELU-derived units. Finally, ELU or ELU-derived maps may be coupled with models (see Chapter 18) to combine information on physical environments, biotic and socioeconomic patterns, past and present conditions, and processes at different hierarchical scales to provide a template for estimating spatial heterogeneity and its effects on landscape and ecosystem dynamics (Pickett and Cadenasso, 1995; also see Chapter 13).

Figure 22.3 shows a hypothetical EA scenario that could be followed as part of the process for defining ELUs and ELU-derived units and for characterizing various attributes useful in EAs (Table 22.1). In the example in Figure 22.1, basic ELUs are defined with selected climate and other physical variables, which also determine PV units. Both basic ELUs and PV units define potential response units (PRUs). The term *potential* is used because, at this stage, no other property has been attributed to the PRUs. PRUs then become the basis for characterizing (1) disturbance regimes and land-use patterns, (2) HRV, and (3) biotic distributions and responses derived from models. These three groups of characteristics, as well as existing vegetation, are attributed to the PRUs to define integrated response units. The latter serve as the basis for modeling

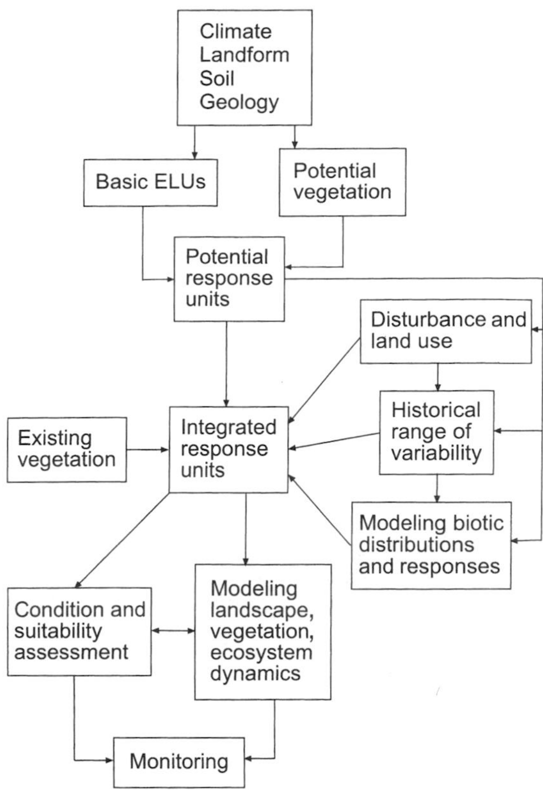

FIGURE 22.3. Hypothetical steps followed to delineate ecological land units and other attributes needed for ecological assessment.

ecosystem dynamics and for assessing ecological conditions and suitability for different purposes. The final step is to interpret the results of all previous steps to formulate monitoring schemes.

Table 22.2 summarizes the major spatial scales at which ELUs can be defined, the general ranges of scales used for ecological mapping, the types of data that are usually available at a given spatial scale, the corresponding political stratification, and examples of the applications possible at each scale. In some cases, ranges of map scales overlap spatial scale categories. Examples of ELC use in EAs at different scales are presented in Section 22.5.

22.4 Overview of Major Ecological Land Classifications

The large number of ELCs developed worldwide reflects the variety of applications for which they were designed. Differences in objectives lead to ELC-specific nomenclature for ELUs. This inconsistency in nomenclature is often confusing, because similar terms may have different meanings or apply to different scales, and different terms may have the same meaning. In accordance with the principles discussed in Section 22.2, two main approaches have been taken in the construction of ELCs and the subsequent delineation of ELUs. In the first approach, floristic provinces or elements are defined by recurrent patterns of plant distributions that reflect similarities among species in their evolutionary histories and ecological tolerances (e.g., McLaughlin, 1986, 1989). The assemblages of various floristic elements in a given area guide the interpretation of past environments and ecological events (e.g., Whittaker and Niering, 1964, 1965; Bourgeron et al., 1995). Floristic provinces have had limited use in EAs. In the second approach, biotic provinces (e.g., Dice, 1943) have been delineated as areas with assumed characteristic physiography, climate, vegetation, flora, and fauna. This category includes ELCs that incorporate biotic, climatic, and physical landscape characteristics, alone or in combination (e.g., Bailey, 1976; Walter, 1979; Omernik, 1987; see discussions in Zonneveld, 1989; Bailey, 1995). ELCs that have combined biotic and floristic provinces include a biotic community-based system (Brown et al., 1979, 1998) and the U.S. vegetation scheme of Küchler (1967). These two ELCs have been used by government agencies in the United States for landscape assessments of regional terrestrial ecosystems.

Table 22.3 (adapted and modified from Grossman et al., 1999) summarizes the following aspects of 30 ELCs currently used by various institutions: whether or not they are hierarchical, the types of data used in their development, their intended geographic coverage, and the broad categories of purposes for which they were initially designed. Many ELCs are hierarchical, allowing the definition of ELUs at multiple scales according to specific classification rules. Hierarchical ELCs vary in number of levels, reflecting specific objectives and concepts. Qualitative data (Table 22.3) are defined as categorical data; quantitative data result from measurements on a variable (including nominal data). ELCs are multifactored when multiple variables were used in their construction. Although the geographic range of an ELC has sometimes been expanded after its initial development, the coverage in the table refers only to the area indicated in the cited reference. ELCs are grouped in Table 22.3 according to the dominant category of variable on which the classification is based: biotic, abiotic, and combination of biotic and abiotic. ELCs range

TABLE 22.2. Major spatial scales of application for ecological land classifications, data required and/or generally available, political stratification, and possible applications.

Spatial scale	Map scale	Remote sensing data	Demographic and economic data	Environmental factors	Species distributions	Existing vegetation[a]	Political stratification	Possible applications
Continental to global	<1 : 10,000,000	AVHRR	Land use and administrative boundaries	Potential vegetation class or formation Climate type Physiography	Range maps	Class or formation	Country State Province	International planning Strategic national planning Coarse ecological assessments (e.g., land-use change)
Regional	1 : 15,000,000 to 1 : 250,000	Thematic mapper SPOT	Land use or capability Ownership Managed areas Road density	Potential vegetation series or groups of series Climatic province or region Landform Geology	Population centers Range maps	Alliances or groups of alliances	State Province County	Strategic, multiinstitution planning Biodiversity assessments Reporting results of ecological assessment analyses
Landscape	1 : 250,000 to 1 : 10,000	Aerial photography	Zoning Roads	Potential vegetation association Local climate Topography Soils	Endemic species populations Other species populations	Associations or groups of associations	County Managed area District Population center	Area-wide planning Forest, range management
Site or plot	>1 : 24,000	Aerial photography	Site use Management prescriptions	Potential vegetation association or phase Topography Elevation Aspect	Species population	Association or subassociation	Specific ownership	Site planning and management Site monitoring Habitat management

[a]Sensu Grossman et al., 1998.

TABLE 22.3. Characteristics of 30 ecological land classifications.[a]

		Development methodology						Geographic coverage				Purpose					
Basis	Classification effort	Hierarchical (multiscaled)	Direct Variables	Indirect Variables	Qualitative	Quantitative	Multifactored	Global	Continental	Regional	Landscape	Management Planning	Conservation Planning	Monitoring	Modeling and Sampling	Assessments	Inventory
Biotic	Braun-Blanquet (1965)	√	√	√	√			√				√	√	√		√	√
	Cherrill and Lane (1995)		√		√						√					√	√
	Cherrill et al. (1995)		√			√	√				√	√			√	√	
	Grossman et al. (1998)	√	√	√	√					√	√		√				
	UNESCO (1973)	√		√	√		√	√	√	√						√	√
Abiotic	Bailey (1995)	√	√	√	√		√	√	√	√		√				√	√
	Beauchesne et al. (1996)	√	√	√	√	√	√			√	√	√	√			√	
	Christian and Stewart (1968)	√		√	√				√	√	√	√					
	ECOMAP (1993)	√	√	√			√			√	√	√		√		√	√
	Holdridge (1947)		√	√	√	√	√			√	√	√					√
	Pfister et al. (1977)	√	√	√	√	√	√			√	√	√				√	√
	Smith and Carpenter (1996)	√	√	√	√		√				√	√		√		√	
Combination	Albert (1995)	√	√	√	√	√	√			√		√	√	√			√
	Anderson et al. (1976)	√		√	√		√			√		√				√	√
	Banner et al. (1996)	√	√	√	√		√			√	√	√		√			√
	Bunce et al. (1996a)		√	√		√	√			√		√		√	√	√	
	Cherrill et al. (1994)			√		√	√				√					√	√
	Driscoll et al. (1984)	√		√	√		√			√	√	√				√	√
	Klijn and Udo de Haes (1994)	√	√	√	√		√			√	√	√			√	√	
	Krajina (1965)	√	√	√	√		√			√	√	√				√	√
	Lenz (1994)	√	√			√	√			√	√	√		√		√	
	Omernik (1995)	√	√	√	√		√		√	√		√		√		√	
	Perera et al. (1996)	√	√	√		√	√			√	√	√	√	√		√	
	Pregitzer and Barnes (1982)		√	√		√	√				√	√				√	√
	Reiners and Thurston (1997)	√	√	√	√	√	√				√	√			√	√	
	Rowe (1984)		√	√	√		√			√	√	√				√	√
	Spies and Barnes (1985)			√		√	√				√	√		√			
	Wiken (1979)	√	√	√	√	√	√			√	√	√	√	√		√	√
	Witmer (1978)	√		√	√	√	√			√	√	√	√	√		√	√
	Zonneveld (1979)	√	√	√	√	√	√			√	√	√	√	√	√	√	√

[a]See text for further explanation.

from systems based solely on vegetation (e.g., Braun-Blanquet, 1965) to systems based solely on physical attributes (e.g., Christian and Stewart, 1968). Land cover classifications differ in the degree to which biotic and abiotic data are incorporated [e.g., use of predominantly biotic data in Cherrill et al.'s (1995) classification and use of both biotic and abiotic data in Bunce et al.'s (1996) classification].

Table 22.4 compares scales and nomenclature for nine hierarchical ELCs applied to assessments and planning in their country or province of origin. These ELCs vary in the specific data that define the ELUs at different levels. Some incorporate biotic data in conjunction with abiotic data (e.g., the British Columbia and Netherlands systems), whereas others are based only on abiotic data (e.g., the Quebec and U.S. systems). The scales at which levels are recognized differ among systems; these differences range from slight to considerable. The names of levels in each ELC have been placed in the table at their approximate locations in the spectrum of spatial and map scales; some overlap exists between ranges of map scales describing ELC spatial scales.

TABLE 22.4. Scales and names of levels of nine ecological land classifications applied to assessment and planning in country or province of origin.

Spatial scale	Mapping scale	Minimum mapping unit	Australia[a]	Canada: British Columbia[b]: Regional	Canada: British Columbia[b]: Zonal	Canada: British Columbia[b]: Site	Canada: National[c]	Canada: Quebec[c]	Netherlands[d]	Russia[e]	USA[f]
Global	<1 : 30 × 10^6	>62,500 km^2		Ecodomain					Ecozone	Zone	Domain
Continental	1 : 30 × 10^6 to 1 : 10 × 10^6	2500–62,500 km^2		Ecodivision			Ecozone	Natural province	Ecoprovince		Division
	1 : 15 × 10^6 to 1 : 5 × 10^6	100–2500 km^2					Ecoregion	Natural region	Ecoregion	Province	Province
Regional	1 : 7.5 × 10^6 to 1 : 2.5 × 10^6	625–10,000 ha						Physiographic unit			
				Ecoprovince	Biogeoclimatic zone		Ecodistrict	Ecological district	Ecodistrict		Section
										Landscape	
	1 : 3.5 × 10^6 to 1 : 250,000	25–625 ha									Subsection
			Land system	Ecoregion	Biogeoclimatic subzone	Site association	Ecosection	Topographic complex	Ecosection		
Landscape	1 : 250,000 to 1 : 50,000	1.5–25 ha							Ecoseries	Urochishcha	Land-type association
			Land unit								
	1 : 60,000 to 1 : 10,000	0.25–1.5 ha	Land type	Ecosection	Biogeoclimatic variant	Site series	Ecosite	Topographic entity	Ecotope		Land type
Site or plot	>24,000	<0.25 ha	Site			Site type	Ecoelement		Ecoelement		Land-type phase

[a]Christian and Stewart, 1968.
[b]Banner et al., 1996.
[c]Wiken, 1979.
[d]Beauchesne et al., 1996.
[e]Isachenko et al., 1973.
[f]ECOMAP, 1993.

Table 22.4 compares scales and nomenclature for nine hierarchical ELCs applied to assessments and planning in their country or province of origin. These ELCs vary in the specific data that define the ELUs at different levels. Some incorporate biotic data in conjunction with abiotic data (e.g., the British Columbia and Netherlands systems), whereas others are based only on abiotic data (e.g., the Quebec and U.S. systems). The scales at which levels are recognized differ among systems; these differences range from slight to considerable. The names of levels in each ELC have been placed in the table at their approximate locations in the spectrum of spatial and map scales; some overlap exists between ranges of map scales describing ELC spatial scales.

22.5 Examples of the Use of Ecological Land Classifications in Ecological Assessments and Performance Evaluation

In this section, several examples of uses of ELCs are presented to illustrate the wide variety of specific purposes and spatial scales for which ELCs have been applied. Section 22.3 included a discussion of the use of ELCs in the ICBEMP for defining the areas needed to implement this ecoregion-based assessment. In addition, selected results from the ICBEMP assessment are described in Chapter 34. The examples presented here are derived from four areas of application of ELCs, which cover the range of spatial scales defined in Table 22.2. ELCs have often been used to address particular management practices (e.g., silviculture) at fine scales. For example, the ecosystem classification developed in British Columbia (Banner et al., 1996) was designed to guide the management of small units within a forested landscape (MacKinnon et al., 1992), among other goals. Such uses are not discussed further in this section; examples of ELC-based management practices are described in Sims et al. (1996). Finally, examples of ELC performance evaluation are discussed.

The first example concerns the use of higher-level ELUs to stratify areas on a continental scale for conservation. As part of a recent trend in conducting biodiversity assessments at continental and/or regional scales with ELC frameworks (TNC, 1996; Noss, 1999; Soulè and Terborgh, 1999), a conservation assessment of the status and threats to biodiversity in North America used 110 terrestrial ecoregions covering the continental United States and Canada (Ricketts et al., 1999). The ecoregions are based on three established ecoregional frameworks (ESWG, 1995; Gallant et al., 1995; Omernik, 1995). Information on various aspects of biodiversity (distribution of major habitat types, number of species, endemic species, and rare ecological elements) and on conservation status (habitat loss, remaining habitats, landscape fragmentation, and protection) was considered to set broad-scale conservation priorities. Other aspects of the use of ELCs for conservation are discussed in Chapter 20.

A second group of examples comprises the use of ELCs to identify restoration targets and strategies for restoration of terrestrial and wetland ecosystems. Palik et al. (2000) used the ELUs of an ELC at the site–landscape scale to predict PV of disturbed terrestrial ecosystems in a 11,400-ha landscape in southwestern Georgia, USA. Higher-level ELUs were combined with physical (geomorphology, soil characteristics) and land-use data to predict PV distribution within lower-level ELUs and to assess their disturbance levels. Disturbance level was combined with information on the conservation status of each ELU to prioritize restoration efforts. At the regional scale, Detenbeck et al. (1999) applied an ecoregional framework to evaluate perturbations in wetland ecosystems and to develop restoration strategies for these wetlands in the Great Lakes region, USA. Wetland distribution, land-use pattern, and disturbance-level variation were determined; ecoregions exhibited characteristic patterns of variation in these attributes. General strategies and goals for wetland restoration at the ecoregional scale were derived from information on historic wetland distribution patterns and the distributions of species of special concern, which depend on specific wetland types or mosaics of habitat types.

EAs often require precise information on the distribution of species for many analyses. However, although general species range maps may exist, precise species distribution maps are seldom available. In the third set of examples, ELCs were used in innovative ways to predict the distributions of species at the regional scale. In the United Kingdom, Cherrill et al. (1995) combined a landscape classification with land cover and plant community maps to predict the probability of occurrence of 579 species. The model performed well for common native species, but relatively poorly for rare and introduced species. He et al. (1998) used a combination of abiotic ELC stratification, a satellite-derived land cover map, and plot-specific information to predict the distribution of selected tree species and age classes in northern Wisconsin, USA. Within each ecoregion defined by the abiotic ELC, field-inventory data were aggregated to provide information on secondary and subcanopy tree species occurrence and age distributions. A probabilistic algorithm was derived to assign information from a point coverage (forest inventory sampling points) and a polygon coverage (ecoregion boundaries) to a raster map (satellite land cover classification). This method was the basis for assessment of forest patterns across regional landscapes and the potential for restoration at the ecoregion scale.

The fourth set of application examples incorporates the use of ELCs for nonbiological purposes. Bhat et al. (1998) employed an ecoregional approach to estimate the economic value of land- and water-based recreation in the United States. The value of outdoor activities was modeled in ten coarse-level ecoregions with an individual travel cost model, a widely used nonmarket valuation technique (Bowker et al., 1996). Bhat et al. (1998) concluded that outdoor recreation values exhibit spatial differences across ecoregions in the United States. In another example, an ELC was incorporated into an impact assessment for the rapid (TGV) rail link between Rotterdam and Amsterdam in the Netherlands (Canters et al., 1991).

Although ELCs receive frequent use, their performance is often not validated, even for the purpose for which they were designed. Notable exceptions include the performance evaluation of the ecoregional framework used by the US EPA for establishing recovery criteria in aquatic ecosystems (Hughes et al., 1990); testing the correspondence between existing vegetation cover and the ELUs of an ELC developed with abiotic data in northwestern Wisconsin, USA (Host et al., 1996) and in the Pacific Northwest (Wright et al., 1998); evaluation of the use of an ELC at a fine scale for forest management with a forest succession model (LINKAGES; Pastor and Post, 1986) in northern Minnesota, USA (Host and Pastor, 1998); and evaluation of the impact of changing landscape pattern and land cover variability on ecoregion discrimination in Minnesota (Mladenoff et al., 1997) and the ICRB (Bourgeron et al., 1999).

ELCs are often used for purposes for which they were not initially designed. For example, terrestrial ELCs designed for the management of vegetation are often used for conservation, although biodiversity maintenance generally was not an initial concern in their development. Two existing ELCs (the US-EPA ELC: Omernik, 1995; the USFS ELC: ECOMAP, 1993; Bailey, 1995) and two de novo ELCs (Bourgeron et al., unpublished) were compared to determine their ability to recover patterns not used in their construction in the ICRB (Bourgeron et al., unpublished). The two de novo ELCs were constructed using direct and indirect variables (see Section 22.2), respectively. ELCs were evaluated under the hypothesis that, to be useful for a specific objective, the ELUs of an ELC at a given scale should differ in the composition of the patterns of interest (Host et al., 1996). Three levels were addressed: provinces, sections (ECOMAP, 1993; Bailey, 1995), and subsection groups. The latter groupings of subsections (ECOMAP, 1993; Bailey, 1995) were performed for analyses in the ICBEMP (Jensen et al., 1997). The US-EPA ELC was tested solely at the province level, because it was the only level available for this system that covered the entire ICRB. An electivity index (Jacobs, 1974; Jenkins, 1979; Pastor and Broschart, 1990) indicated a positive or negative association of selected patterns with ELUs.

Table 22.5 summarizes electivity index results for two cover-type classifications (coarse: 10 classes; fine: 37 classes; Hann et al., 1997), cover-type diversity (number of fine-cover-type classes), hydrologic interpretations (Jensen et al., 1997), soil

TABLE 22.5. Electivity index association of interior Columbia River basin patterns with ecological land classifications (ELCs) at three levels.

Pattern	Hierarchical level	Number of ELUs	ELC			
			US-EPA	USFS	Indirect	Direct
37 Fine	Province	7	Fair	Fair	Good	Best
existing	Section	20	NA	Fair	Best	Good
cover types	Subsection group	37	NA	Fair	Best	Good
10 Coarse	Province	7	Good	Fair	Good	Best
existing	Section	20	NA	Fair	Good	Best
cover types	Subsection group	37	NA	Good	Good	Best
Cover-type	Province	7	Fair	Fair	Good	Best
diversity	Section	20	NA	Best	Good	Good
(37 types)	Subsection group	37	NA	Best	Good	Good
Hydrologic	Province	7	Fair	Fair	Best	Good
interpretations	Section	20	NA	Good	Best	Good
(8 coefficients)	Subsection group	37	NA	Fair	Best	Good
Soil taxonomy	Province	7	Good	NA	Best	Good
(orders and	Section	20	NA	NA	Best	Good
suborders)	Subsection group	37	NA	NA	Best	Good
Fish species	Province	7	Good	Fair	Fair	Best
distributions	Section	20	NA	Fair	Fair	Best
(38 species)	Subsection group	37	NA	Fair	Fair	Best

Degree of association of electivity index values with ELCs subjectively assessed as fair, good, and best. NA, not applicable.

taxonomy (USDA, 1993), and distributions of 38 fish species (Lee et al., 1997). Soil taxonomy, although derived from an independent database, was not assessed for the USFS ELC, because soil was used as a criterion in defining the levels of this ELC. The performance of the ELCs varied with the pattern of interest and also in some cases with level. For example, coarse-cover-type classes were most strongly associated with the direct-variable ELC at the province level; however, at the section and subsection group levels, the strongest association was with the indirect-variable ELC. The other two ELCs did not perform as well at any level. For the fine-cover-type classification, the direct-variable ELC performed best at all three levels. For cover-type diversity, the direct-variable ELC had the strongest association at the province level, but the USFS ELC performed best at the section and subsection group levels. This analysis suggests caution when using an ELC for a purpose other than that for which it was designed.

22.6 Conclusion

ELCs are essential to EAs at all assessment scales. They are used for many different components of an EA, including stratifying areas for different purposes, serving as input to models, and evaluating ecological conditions. Four important recommendations in the definition or use of ELCs are the following:

1. Construct ELCs with appropriate variables that address the objectives of an EA.
2. Test hypothesized relationships between variables used to construct ELCs and the patterns and processes of interest.
3. Evaluate the performance of ELCs with respect to the objectives.
4. Evaluate the performance of ELCs before using them for purposes other than those for which they were designed and tested.

In practice, it is likely that arbitrary decisions may be made about which variables to use in the design of an ELC or which existing ELC to use. Therefore, emphasis should be placed on the validation of the ELC (recommendations 3 and 4). Suggestions for future research in the design and implementation of ELCs for EAs include the following:

1. Quantifying the delineation of ELUs (e.g., Hargrove and Hoffman, 1999).
2. Developing additional techniques to evaluate ELCs for specific applications.
3. Developing methods to more effectively characterize variability within ELUs.

22.7 References

Albert, D. A. 1995. *Regional landscape ecosystems of Michigan, Minnesota, and Wisconsin: a working map and classification. Fourth revision.* GTR-NC-178. St. Paul, MN: U.S. Dept. Agric., For. Serv., North Central For. Exp. Sta.

Anderson, J. R.; Hardy, E. E.; Roach, J. T. 1976. *Land use and land cover classification system for use with remote sensing data.* Geological Survey Professional Pap. 964 (a revision of the land use classification system as presented in U.S. Geological Circular 671). Washington, DC: U.S. Govt. Printing Off.

Arno, S. F.; Simmermann, D. G.; Keane, R. E. 1986. *Characterizing succession within a forest habitat type—an approach designed for resource managers.* Research Note INT-357. Ogden, UT: U.S. Dept. Agric., For. Serv., Intermountain Res. Sta.

Austin, M. P. 1985. Continuum concept, ordination methods and niche theory. *Ann. Rev. Ecol. Syst.* 16:39–61.

Austin, M. P.; Smith, T. M.. 1989. A new model for the continuum concept. *Vegetatio* 83:35–47.

Austin, M. P.; Yapp, G. A. 1978. Definition of rainfall regions of southeastern Australia by numerical classification methods. *Archiv fur Meterologie, Geophysik und Bioklimatologie, Series B* 26:121–142.

Austin, M. P.; Cunningham, R. B.; Fleming, P. M. 1984. New approaches to direct gradient analysis using environmental scalars and statistical curve fitting procedures. *Vegetatio* 55:11–27.

Bailey, R. G. 1976. *Ecoregions of the United States.* Ogden, UT: U.S. Dept. Agric., For. Serv., Intermountain Region. Scale 1 : 7,500,000.

Bailey, R. G. 1995. *Description of the ecoregions of the United States.* Misc. Publ. 1391. Washington, DC: U.S. Dept. Agric., For. Serv.

Bailey, R. G.; Jensen, M. E.; Cleland, D. T.; Bourgeron, P. S. 1994. Design and use of ecological mapping units. In: Jensen, M. E.; Bourgeron, P. S., eds. *Volume II: Ecosystem management: principles and applications.* PNW-GTR-318. Portland, OR: U.S. Dept. Agric., For. Serv., Pacific Northw. Res. Sta.: 95–106.

Banner, A.; Meidinger, D. V.; Lea; E. C.; Maxwell, R. E.; Von Sacken, B. C. 1996. Ecosystem mapping methods for British Columbia. In: Sims, R. A.; Corns, I. G. W.; Klinka, K., eds. *Global to local: ecological land classification.* Dordrecht, The Netherlands: Kluwer Academic Publishers: 97–117.

Beauchesne, P.; Ducruc, J.-P.; Gerardin, V. 1996. Ecological mapping: a framework for delimiting forest management units. *Environ. Monit. Assess.* 39:173–186.

Bhat, G.; Bergstrom, J.; Teasley, R. J.; Bowker, J. M.; Cordell, H. K. 1998. An ecoregional approach to the economic valuation of land- and water-based recreation in the United States. *Environ. Manage.* 22:69–77.

Bourgeron, P. S.; Jensen, M. E. 1994. An overview of ecological principles for ecosystem management. In: Jensen, M. E.; Bourgeron, P. S., eds. *Volume II: Ecosystem management: principles and applications.* PNW-GTR-318. Portland, OR: U.S. Dept. Agric., For. Serv., Pacific Northw. Res. Sta.: 45–67.

Bourgeron, P. S.; Humphries, H. C.; Jensen, M. E. 1994. Landscape characterization: a framework for ecological assessment at regional and local scales. *J. Sustain. For.* 2(3–4):267–281.

Bourgeron, P. S.; Engelking, L. D.; Humphries, H. C.; Muldavin, E.; Moir, W. H. 1995. Assessing the conservation value of the Gray Ranch: rarity, diversity and representativeness. *Desert Plants* 11:1–68.

Bourgeron, P. S.; Humphries, H. C.; Barber, J. A.; Turner, S. J.; Jensen, M. E.; Goodman, I. A. 1999. Impact of broad- and fine-scale patterns on regional landscape characterization using AVHRR-derived land cover data. *Ecosystem Health* 5:234–258.

Bowker, J. M.; English, D. B. K.; Donovan, J. A. 1996. Toward a value for guided rafting on southern rivers. *J. Agric. Appl. Econ.* 28:423–432.

Braun-Blanquet, J. 1965. *Plant sociology: the study of plant communities (Pflanzensoziologie).* New York: Hafner Publishing Co.

Brown, D. E.; Lowe, C. H.; Pase, C. P. 1979. A digitized classification system for the biotic communities of North America, with community (series) and association examples for the Southwest. *J. Arizona–Nevada Acad. Sci.* 14(Suppl.1):1–16.

Brown, D. E.; Reichenbacher, F.; Franson, S. E. 1998. *A classification of North American biotic communities.* Salt Lake City, UT: University of Utah Press. 152 p.

Brown, J. H. 1984. On the relationship between the abundance and distribution of species. *Amer. Naturalist* 124:255–279.

Bunce, R. G. H.; Barr, C. J.; Clarke, R. T.; Howard, D. C.; Lane, A. M. J. 1996a. Land classification for strategic ecological survey. *J. Environ. Manage.* 47: 37–60.

Bunce, R. G. H.; Barr, C. J.; Gillespie, M. K.; Howard, D. C. 1996b. The ITE land classification: providing an environmental stratification of Great Britain. *Environ. Monit. Assess.* 39:39–46.

Canters, K. J.; den Herder, C. P.; de Veer, A. A.; Veelenturf, P. W. M.; de Waal, R. W. 1991. Landscape–ecological mapping of the Netherlands. *Landscape Ecol.* 5:145–162.

Cherrill, A.; Lane, M. 1995. The survey and prediction of land cover using an environmental land classification. *Appl. Geogr.* 15(1):69–85.

Cherrill, A. J.; Lane, A.; Fuller, R. M. 1994. The use of classified Landsat-5 thematic mapper imagery in the characterization of landscape composition: a case study in northern England. *J. Environ. Manage.* 40: 357–377.

Cherrill, A. J.; McClean, C.; Watson, P.; Rushton, S. P.; Sanderson, R. 1995. Predicting the distributions of plant species at the regional scale: a hierarchical matrix model. *Landscape Ecol.* 10:197–207.

Christian, C. S.; Stewart, G. A. 1968. Methodology of integrated surveys. In: *Aerial surveys and integrated studies.* Paris: Proc. Toulouse Conf. 1964, UNESCO: 233–280.

Daubenmire, R. 1966. Vegetation: identification of typal communities. *Science* 151:291–298.

Delcourt, H. R.; Delcourt, P. A. 1988. Quaternary landscape ecology: relevant scales in space and time. *Landscape Ecol.* 2:45–61.

Detenbeck, N. E.; Galatowitsch, S. M.; Atkinson, J.; Ball, H. 1999. Evaluating perturbations and developing restoration strategies for inland wetlands in the Great Lakes basin. *Wetlands* 19:789–820.

DeVelice, R. L.; Daumiller, G. J.; Bourgeron, P. S.; Jarvie, J. O. 1994. Bioenvironmental representativeness of nature preserves: assessment using a combination of a GIS and a rule-based model. In: Despain, D. G., ed. *Technical Report NPS/NRYELL/NRTR-93/XX: Plants and Their Environments: Proceedings of the First Biennial Scientific Conference on the Greater Yellowstone Ecosystem,* September 16–17, 1991, Yellowstone National Park, WY. Denver, CO: USDI National Park Service: 131–138.

Dice, L. R. 1943. *The biotic provinces of North America.* Ann Arbor, MI: University of Michigan Press.

Driscoll, R. S.; Merkel, D. L.; Radloff, D. L.; Snyder, D. E.; Hagihara, J. S. 1984. *An ecological land classification framework for the United States.* Misc. Publ. 1439. Washington, DC: U.S. Dept. Agric., For. Serv.

ECOMAP. 1993. *National hierarchical framework of ecological units.* Washington, DC: U.S. Dept. Agric., For. Serv.

ESWG (Ecological Stratification Working Group). 1995. *A national ecological framework for Canada.* Ottawa: Agriculture and Agri food Canada, Research Branch, Centre for Land and Biological Resources Research; and Environment Canada, State of the Environment Directorate, Ecozone Analysis Branch. (map)

Gallant, A. L.; Binnian, E. F.; Omernik, J. M.; Shasby, M. B. 1995. *Ecoregions of Alaska.* U.S. Geological Survey Professional Pap. 1567. Washington, DC: U.S. Govt. Printing Off.

Grossman, D. H.; Faber-Langendoen, D.; Weakley, A. S.; Anderson, M.; Bourgeron, P.; Crawford, R.; Gooding, K.; Landaal, S.; Metzler, K.; Patterson, K. D.; Pyne, M.; Reid, M.; Sneddon, L. 1998. *International classification of ecological communities: terrestrial vegetation of the United States. Volume I. The national vegetation classification standard.* Arlington, VA: The Nature Conservancy.

Grossman, D. H.; Bourgeron, P. S.; Busch, W.-D. N.; Cleland, D.; Platts, W.; Ray, G. C.; Robins, C. R.; Roloff, G. 1999. Principles for ecological classification. In: Johnson, N. C.; Malk, A. J.; Sexton, W. T.; Szaro, R. C., eds. *Volume II. Ecological stewardship: a common reference for ecosystem management.* Amsterdam: Elsevier Science: 353–393.

Hann, W. J.; Jones, J. L.; Karl, M. G.; Hessburg, P. F.; Keane, R. E.; Long, D. G.; Menakis, J. P.; McNicol, C. H.; Leonard, S. G.; Gravenmier, R. A.; Smith, B. G. 1997. Landscape dynamics of the basin. In: Quigley, T. M.; Arbelbide, S. J., eds. *An assessment of ecosystem components in the Interior Columbia Basin and portions of the Klamath and Great Basins: Volume II*. PNW-GTR-405. Portland, OR: U.S. Dept. Agric., For. Serv., Pacific Northw. Res. Sta.: 337–1055.

Hargrove, W. W.; Hoffman, F. M. 1999. Using multivariate clustering to characterize ecoregion borders. *Comput. Sci. Eng.* 1:18–25.

He, H. S.; Mladenoff, D. J.; Radeloff, V. C.; Crow, T. R. 1998. Integration of GIS data and classified satellite imagery for regional forest assessment. *Ecol. Appl.* 8:1072–1083.

Holdridge, L. R. 1947. Determination of world plant formations from simple climatic data. *Science* 105:367–368.

Host, G.; Pastor, J. 1998. Modeling forest succession among ecological land units in northern Minnesota. *Conserv. Ecol.* 2(2):15 [online] URL: http://www.consecol.org/vol2/iss2/art15.

Host, G. E.; Polzer, P. L.; Mladenoff, M. A.; White, M. A.; Crow, S. J. 1996. A quantitative approach to developing regional ecosystem classifications. *Ecol. Appl.* 6:608–618.

Hughes, R. M.; Whittier, T. R.; Rohm, C. M. 1990. A regional framework for establishing recovery criteria. *Environ. Manage.* 14:673–683.

Hutchinson, G. E. 1959. Homage to Santa Rosalia, or why are there so many kinds of animals. *Amer. Naturalist* 93:145–159.

Isachenko, A. G. 1973. *Principles of landscape science and physical geographic regionalization*. Melbourne: J. S. Massey.

Jacobs, J. 1974. Quantitative measurement of food selection: a modification of the forage ratio and Ivlev's electivity index. *Oecologia* 14:413–417.

Jenkins, S. H. 1979. Seasonal and year-to-year differences in food selection by beavers. *Oecologia* 44:112–116.

Jensen, M.; Goodman, I.; Brewer, K.; Frost, T.; Ford, G.; Nesser, J. 1997. Biophysical environments of the basin. In: Quigley, T. M.; Arbelbide, S. J., eds. *An assessment of ecosystem components in the Interior Columbia Basin and portions of the Klamath and Great Basins: Volume II*. PNW-GTR-405. Portland, OR: U.S. Dept. Agric., For. Serv., Pacific Northw. Res. Sta.: 99–314.

Klijn, F., editor. 1994. *Ecosystem classification for environmental management. Ecology & environment.* Dordrecht, The Netherlands: Kluwer Academic Publishers.

Klijn, F.; Udo de Haes, H. A. 1994. A hierarchical approach to ecosystems and its implications for ecological land classification. *Landscape Ecol.* 9(2):89–104.

Kovalchik, B. L.; Chitwood, L. A. 1990. Use of geomorphology in the classification of riparian plant associations in mountainous landscapes of central Oregon, U.S.A. *For. Ecol. Manage.* 33/34:405–418.

Krajina, V. J. 1965. Biogeoclimatic zones and classification of British Columbia. In: Krajina, V. J., ed. *Ecology of Western North America*. Vancouver, BC: University of British Columbia: 1–17.

Küchler, A. W. 1964. Potential natural vegetation of the conterminous United States. *Am. Geogr. Soc. Spec. Publ.* 36:1–54.

Küchler, A. W. 1967. *Vegetation mapping*. New York: Ronald Press.

Küchler, A. W. 1973. Problems in classifying and mapping vegetation for ecological regionalization. *Ecology* 54:512–523.

Küchler, A. W. 1988. The nature of vegetation. In: Küchler, A. W.; Zonneveld, I. S., eds. *Vegetation mapping*. Dordrecht, The Netherlands: Kluwer Academic Publishers: 321–329.

Küchler, A. W.; Zonneveld, I. S., editors. 1988. *Vegetation mapping*. Dordrecht, The Netherlands: Kluwer Academic Publishers.

Landres, P. B.; Morgan, P.; Swanson, F. J. 1999. Evaluating the utility of natural variability concepts in managing ecological systems. *Ecol. Appl.* 9:1179–1188.

Lathrop, R. G.; Peterson, D. L. 1992. Identifying structural self-similarity in mountainous landscapes. *Landscape Ecol.* 6:233–238.

Lee, D. C.; Sedell, J. R.; Rieman, B. E.; Thurow, R. F.; Williams, J. E.; Burns, D.; Clayton, J.; Decker, L.; Gresswell, R.; House, R.; Howell, P.; Lee, K. M.; MacDonald, K.; McIntyre, J.; McKinney, S.; Noel, T.; O'Connor, J. E.; Overton, C. K.; Perkinson, D.; Tu, K.; Van Eimeren, P. 1997. Broadscale assessment of aquatic species and habitats. In: Quigley, T. M.; Arbelbide, S. J., eds. *An assessment of ecosystem components in the Interior Columbia Basin and portions of the Klamath and Great Basins: Volume III*. PNW-GTR-405. Portland, OR: U.S. Dept. Agric., For. Serv., Pacific Northw. Res. Sta.: 1057–1496.

Lenz, R. J. M. 1994. Ecosystem classification by budgets of material; the example of forest ecosystems classified as proton budget types. In: Klijn, F., ed. *Ecosystem classification for environmental management*. Dordrecht, The Netherlands: Kluwer Academic Publishers: 117–137.

Levin, S. A. 1992. The problem of pattern and scale in ecology. *Ecology* 73:1942–1968.

Ligon, F. K.; Dietrich, W. E.; Trush, W. J. 1995. Downstream ecological effects of dams: a geomorphic perspective. *BioScience* 45:183–192.

Mackey, B. G.; Nix, H. A.; Hutchinson, M. F.; MacMahon, J. P.; Fleming, P. M. 1988. Assessing representativeness of places for conservation reservation and heritage listing. *Environ. Manage.* 12(4):501–514.

Mackey, B. G.; Nix, H. A.; Stein, J. A.; Cork, S. E.; Bullen, F. T. 1989. Assessing the representativeness of the wet tropics of Queensland World Heritage Property. *Biol. Conserv.* 50:279–303.

MacKinnon, A.; Meidinger, D.; Klinka, K. 1992. Use of

the biogeoclimatic ecosystem classification system in British Columbia. *For. Chron.* 68:100–120.

Maxwell, J. R.; Edwards, C. J.; Jensen, M. E.; Paustian, S. J.; Parrott, H.; Hill, D. M. 1995. *A hierarchical framework of aquatic ecological units in North America.* NC-176. St. Paul, MN: U.S. Dept. Agric., For. Serv., North Central For. Exp. Sta.

McIntosh, R. P. 1985. *The background of ecology: concept and theory.* Cambridge, UK: Cambridge University Press.

McLaughlin, S. P. 1986. Floristic analysis of the southwestern United States. *Great Basin Naturalist* 46: 45–65.

McLaughlin, S. P. 1989. Natural floristic areas of the western United States. *J. Biogeogr.* 16:239–248.

Mladenoff, D. J.; Niemi, G. J.; White, M. A. 1997. Effects of changing landscape pattern and U.S.G.S. land cover data variability on ecoregion discrimination across a forest–agriculture gradient. *Landscape Ecol.* 12:379–396.

Morgan, P.; Aplet, G. H.; Haufler, J. B.; Humphries, H. C.; Moore, M. M.; Wilson, W. D. 1994. Historical range of variability: a useful tool for evaluating ecosystem change. *J. Sustain. For.* 2(3–4):57–72.

Nix, H. A. 1982. Environmental determinants and evolution in Terra Australia. In: Barker, W. M.; Greenslade, P. J. M., ed. *Evolution of the flora and fauna of Arid Australia.* South Australia: Peacock: 47–66.

Noss, R. F. 1999. Conservation assessments: a synthesis. In: Ricketts, T. H.; Dinerstein, E.; Olson, D. M.; Loucks, C. J.; Eichbaum, W.; DellaSala, D.; Kavanaugh, K.; Hedao, P.; Hurley, P. T.; Carney, K. M.; Abell, R.; Walters, S. *Terrestrial ecoregions of North America: a conservation assessment.* Washington, DC: Island Press: 89–92.

O'Callaghan, J. R. 1996. *Land use: The interaction of economics, ecology and hydrology.* London: Chapman & Hall.

Omernik, J. M. 1987. Ecoregions of the conterminous United States. *Ann. Assoc. Amer. Geogr.* 77:118–125.

Omernik, J. M. 1995. Ecoregions: A framework for environmental management. In: Davis, W.; Simon, T., eds. *Biological assessment and criteria: tools for water resource planning and decision making.* Boca Raton, FL: Lewis Publishers: 49–62.

Palik, B. J.; Goebel, P. C.; Kirkman, L. K.; West, L. 2000. Using landscape hierarchies to guide restoration of disturbed ecosystems. *Ecol. Appl.* 10:189–202.

Pastor, J.; Broschart, M. 1990. The spatial pattern of a northern conifer–hardwood landscape. *Landscape Ecol.* 4:55–68.

Pastor, J.; Post, W. M. 1986. Influence of climate, soil moisture and succession on forest carbon and nitrogen cycles. *Biogeochemistry* 2:3–27.

Perera, A. H.; Baker, J. A.; Band, L. E.; Baldwin, D. J. B. 1996. A strategic framework to ecoregionalize Ontario. In: Sims, R. A.; Corns, I. G. W.; Klinka, K., eds. *Global to local ecological land classification.* Dordrecht, The Netherlands: Kluwer Academic Publishers: 85–96.

Pfister, R. D.; Arno, S. F. 1980. Classifying forest habitat types based on potential climax vegetation. *For. Sci.* 26:52–70.

Pfister, R. D.; Kovalchik, B. L.; Arno, S. F.; Presby, R. C. 1977. *Forest habitat types of Montana.* INT-34. Ogden, UT: U.S. Dept. Agric., For. Serv., Intermountain For. Range Exp. Sta.

Pickett, S. T. A.; Cadenasso, M. L. 1995. Landscape ecology: spatial heterogeneity in ecological systems. *Science* 269(5222):331–334.

Pinay, G.; Fabre, A.; Vervier, P.; Gazelle, F. 1992. Control of C, N, P distribution in soils of riparian forest. *Landscape Ecol.* 6(3):121–132.

Pregitzer, K. S.; Barnes, B. V. 1982. The use of ground flora to indicate edaphic factors in upland ecosystems of the McCormick Experimental Forest, Upper Michigan. *Can. J. For. Res.* 12:661–672.

Quigley, T. M.; Arbelbide, S. J., editors. 1997. *An assessment of ecosystem components in the interior Columbia Basin and portions of the Klamath and Great Basins: Volume I.* PNW-GTR-405. Portland, OR: U.S. Dept. Agric., For. Serv., Pacific Northw. Res. Sta.

Reiners, W. A.; Thurston, R. C. 1997. *Delineations of landtype associations for southeast Wyoming.* Bureau of Land Management/University of Wyoming, Final Report, Contract K-910-P960124. Laramie, WY: Dept. of Botany, University of Wyoming.

Ricketts, T. H.; Dinerstein, E.; Olson, D. M.; Loucks, C. J.; Eichbaum, W.; DellaSala, D.; Kavanaugh, K.; Hedao, P.; Hurley, P. T.; Carney, K. M.; Abell, R.; Walters, S. 1999. *Terrestrial ecoregions of North America: a conservation assessment.* Washington, DC: Island Press.

Rowe, J. S. 1984. Forestland classification: limitations of the use of vegetation. In: Bockheim, J. G., ed. *Proceedings of the symposium on forest land classification: experiences, problems, perspectives,* March 18–20, 1984. Madison, WI: University of Wisconsin: 132–147.

Rowe, J. S.; Sheard, J. W. 1981. Ecological land classification: a survey approach. *Environ. Manage.* 5:451–464.

Scott, J. M.; Davis, F.; Csuti, B.; Noss, R.; Butterfield, B.; Groves, C.; Anderson, H.; Caicco, S.; D'Erchia, F.; Edwards, T. C.; Ulliman, J.; Wright, R. G. 1993. Gap analysis: a geographic approach to protection of biological diversity. *Wildlife Monogr.* 123:1–41.

Short, H. L.; Hestbeck, J. B. 1995. National biotic resource inventories and GAP analysis. *BioScience* 45: 535–539.

Sims, R. A.; Corns, I. G. W.; Klinka, K., editors. 1996. *Global to local ecological land classification.* Dordrecht, The Netherlands: Kluwer Academic Publishers.

Slocombe, D. S. 1993a. Implementing ecosystem-based management. *BioScience* 43(9):612–622.

Slocombe, D. S. 1993b. Environmental planning, ecosystem science, and ecosystem approaches for integrating environment and development. *Environ. Manage.* 17:289–303.

Smith, M.-L.; Carpenter, C. 1996. Application of the USDA Forest Service national hierarchical framework of ecological units at the sub-regional level: The New England–New York example. In: Sims, R. A.; Corns, I. G. W.; Klinka, K., eds. *Global to local: ecological land classification.* Dordrecht, The Netherlands: Kluwer Academic Publishers: 187–198.

Sokal, R. R. 1974. Classification: purposes, principles, progress, prospects. *Science* 185:1115–1123.

Soulé, M.E.; Terborgh, J., editors. 1999. *Continental conservation: scientific foundations of regional reserve networks.* Washington, DC: Island Press.

Spies, T. A.; Barnes, B. V. 1985. A multifactor ecological classification of the northern hardwood and conifer ecosystems of Sylvania Recreation Area, Upper Peninsula, Michigan. *Can. J. For. Res.* 15:949–960.

Swanson, F. J.; Kratz, T. K.; Caine, N.; Woodmansee, R. G. 1988. Landform effects on ecosystem patterns and processes. *BioScience* 38:92–98.

Swetnam, T. W.; Allen, C. D.; Betancourt, J. L. 1999. Applied historical ecology: using the past to manage the future. *Ecol. Appl.* 9:1189–1206.

TNC (The Nature Conservancy). 1996. *Conservation by design: a framework for mission success.* Arlington, VA: The Nature Conservancy.

Tüxen, R. 1959. Typen von Vegetationskarten und ihre Erarbeitung. In: Tüxen, R., ed. *Vegetationskartierung.* Weinheim, Germany: Cramer: 139–154.

Uhlig, P. W. C.; Jordan, J. K. 1996. A spatial hierarchical framework for the co-management of ecosystems in Canada and the United States for the upper Great Lakes region. *Environ. Monit. Assess.* 39:59–73.

UNESCO. 1973. *International classification and mapping of vegetation.* Paris: United Nations Educational, Scientific and Cultural Organization.

Urban, D. L.; O'Neill, R. V.; Shugart, H. H., Jr. 1987. Landscape ecology: a hierarchical perspective can help scientists understand spatial patterns. *BioScience* 37:119–127.

USDA (U.S. Department of Agriculture, Soil Conservation Service). 1993. *State soil geographic database (STATSGO)—data user's guide.* Misc. Pub. 1492. Washington, DC: U.S. Govt. Printing Off.: 88 p.

Walter, H. 1979. *Vegetation of the earth and ecological systems of the geobiosphere.* Heidelberg, Germany: Springer-Verlag.

Whittaker, R. H. 1972. Evolution and measurement of species diversity. *Taxon* 21:213–351.

Whittaker, R. H.; Niering, W. A. 1964. Vegetation of Santa Catalina Mountains, Arizona. *J. Arizona Acad. Sci.* 3:12–33.

Whittaker, R. H.; Niering, W. A. 1965. Vegetation of Santa Catalina Mountains: a gradient analysis of the south slope. *Ecology* 46:429–452.

Wiken, E. B. 1979. Rationale and methods of ecological land surveys: an overview of Canadian approaches. In: Taylor, D. G., ed. *Land/wildlife integration: proceedings of a technical workshop to discuss the incorporation of wildlife information into ecological land surveys,* May 1–2, 1979, Saskatoon, Saskatchewan. Ottawa: Ecological Land Classification Series No. 11, Lands Directorate, Environment Canada: 160 p.

Witmer, R. E. 1978. U.S. Geological Survey land-use and land-cover classification system. *J. For.* 76:661–668.

Wright, R. G.; Murray, M. P.; Merrill, T. 1998. Ecoregions as a level of ecological analysis. *Biol. Conserv.* 86:207–213.

Zerbe, S. 1998. Potential natural vegetation: validity and applicability in landscape planning and nature conservation. *Appl. Veg. Sci.* 1:165–172.

Zonneveld, I. S. 1979. *Landscape science and land evaluation.* ITC-textbook VII-4, 2nd ed. 134 p.

Zonneveld, I. S. 1988. Landscape (ecosystem) and vegetation maps: their relation and purpose. In: Küchler, A. W.; Zonneveld, I. S., ed. *Vegetation mapping.* Dordrecht, The Netherlands: Kluwer Academic Publishers: 481–486.

Zonneveld, I. S. 1989. The land unit—a fundamental concept in landscape ecology, and its applications. *Landscape Ecol.* 3:67–89.

23 Dynamic Terrestrial Ecosystem Patterns and Processes

Stephanie P. Wilds and Peter S. White

23.1 Introduction

Ecological assessments often begin with the inventory of a site's biological resources and culminate in the generation of a static, two-dimensional map or other fixed report of present site conditions. Using this approach, we may fail to appreciate the short- and long-term dynamics of the populations and ecosystems that we are assessing. Spatial and temporal dynamics of patterns and processes are important to consider for at least three reasons. First, the very resources we assess may be changing, with consequences for the conclusions of the assessment itself. Second, the dynamics have implications for whether and under what conditions the populations and ecosystems are sustainable on the site of interest. Finally, key processes may occur at broader spatial scales or on longer temporal scales than we would otherwise consider in the assessment; that is, the populations and communities at one site may be attributed in part to the spatial context surrounding that site or the occurrence of rare past events. In essence, the processes responsible for the biological resources on the site may themselves not be contained within the site being assessed, and the resources may not be in balance with current site processes. Overlooking the effects of dynamic patterns and processes can therefore lead to misinterpretations of observations and erroneous conclusions concerning a given site's condition and functional nature.

Ecosystem dynamics can encompass a broad spectrum of phenomena, including global climate change, succession, exotic invasions, and episodic disturbances. These phenomena can be classified into three general categories. *Gradual changes* in the environment, such as climate change or long-term geomorphic and soil development, occur at broad scales and over very long time periods. *Disturbances* are relatively discrete, disruptive events and may include fires, floods, storms, wind, ice, droughts, freezes, and disease. *Natural periodicities*, such as seasonal variations and semiperiodic environmental fluctuations like those caused by the Southern Oscillation, include annual hydrologic cycles and regular temperature- and solar radiation-dependent fluctuations (DeAngelis and White, 1994) (Figure 23.1). These three classes of ecosystem dynamics readily interact with one another; global warming, for example, may alter the hydrologic cycle, and seasonal wet or dry periods influence the probability of droughts, floods, or fire events. Often, the immediate effects of both gradual environmental change and natural periodicities (e.g., changes in precipitation caused by the Southern Oscillation) are disturbances to the existing ecosystem. In this chapter, we focus on disturbances as the prime agents of change in dynamic pattern; natural periodicities and gradual changes occur on such a broad scale that they are beyond the scope of most assessment efforts.

Disturbance regimes are part of a more general group of agents of pattern formation that also includes the *biophysical template* (e.g., physical landform constraints, precipitation patterns, and substrate conditions) and *biotic processes* (e.g., dispersal, colonization, extinction and other demographic processes) (Urban et al., 1987; Bourgeron and Jensen, 1994). By altering the availability of space and resources and creating patchy environments, disturbances play a particularly critical role in the determination of landscape pattern and in the development of ecosystem composition, structure, and function.

Disturbances are also important in maintaining biological diversity. A site's ecological attributes result from the way disturbances interact with environmental gradients, substrates, and topography

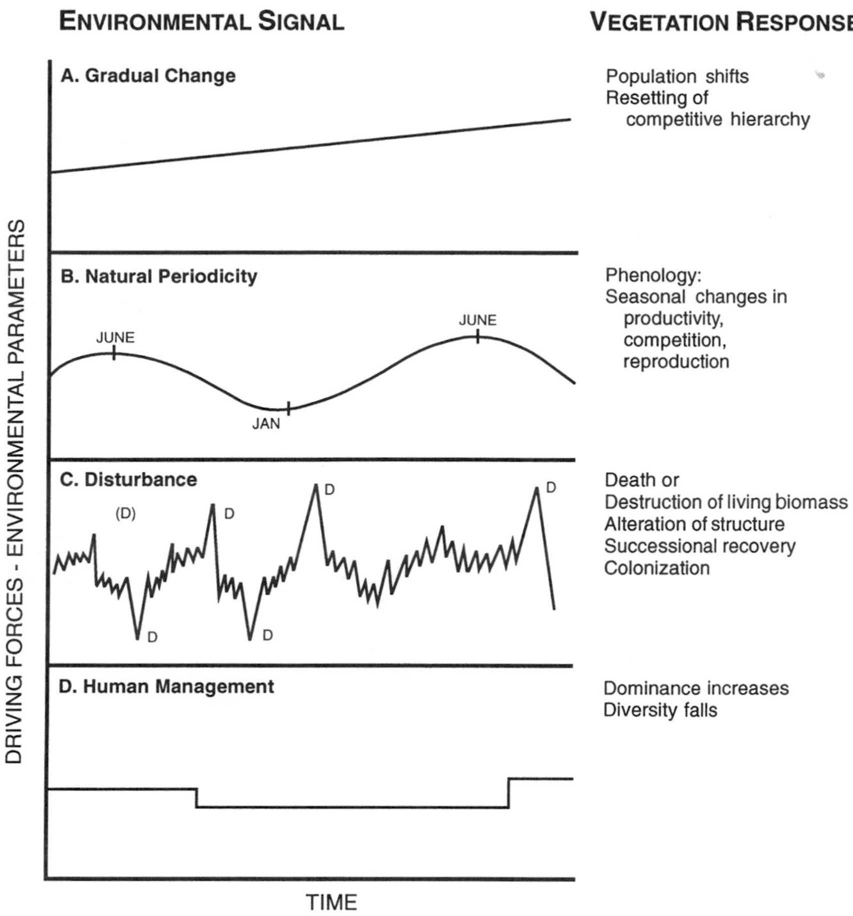

FIGURE 23.1. Classification of ecosystem dynamics and types of vegetation response (DeAngelis and White, 1994).

(Baker, 1992a) and, more fundamental, underlying biophysical processes. Disturbances influence and are frequently influenced by biotic fluctuations, such as pathogenic invasion or extinction of keystone species. The resultant spatial structure of the landscape (i.e., the range of sizes, arrangement, and distance between patches) is important to the ecology of many species, both floral and faunal. Species differ greatly in their specificity for patch types; some are associated with or even restricted to older patches, others reproduce in or are restricted to younger patches, and still others must have a mix of successional ages for survival and reproduction.

23.2 Definition of Terrestrial Dynamic Pattern

Terrestrial dynamic pattern refers to the arrangement and distribution of biotic and abiotic phenomena. The two major components of landscape pattern include *composition* (the variety of patch types) and *configuration* (position and orientation of patches). The myriad pattern descriptors include *size*, *shape*, *orientation*, and *edge-to-area ratio* (or *patch shape index*) of individual patches; *distance* between patches; and *patch density*, *patch diversity*, *patch connectivity*, and *patch contiguity* (Forman, 1995). Pattern in the landscape is generated by various biotic and abiotic processes operating at various scales (Urban et al., 1987). The driving forces behind pattern are complex. They include not only the formative processes themselves, but also interactions among processes and a feedback effect of pattern on process, wherein pattern itself may act as an environmental constraint to disturbance events and biotic processes. It is this dynamic nature of pattern that we are attempting to characterize and understand in assessing landscape dynamics.

Terrestrial patterns with which we are familiar include the distribution of vegetation communities, such as the distinct transition from mixed hardwood

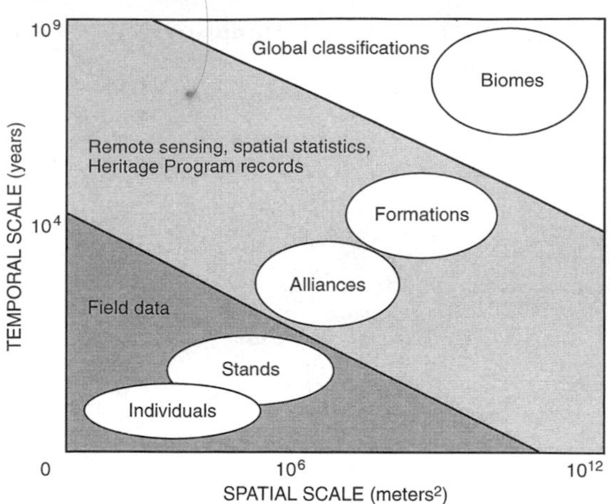

FIGURE 23.2. Hierarchical spatial–temporal arrangement of patterns in vegetation and corresponding conventional documentation.

forest to spruce–fir forest at high elevations in the Southern Appalachians or the almost regular spacing of individual trees in the longleaf pine–wiregrass savannas of the southeastern coastal plain. Animal species also display dynamic patterns in their distributions: avian species guilds are known to exhibit habitat affinity for certain landscape structure types, and southern pine bark beetle exhibits distinct and predictable patterns of infestation in southern pine forests. Such dynamic patterns can be arranged into hierarchies that increase in predictability and ease of documentation with broader time and space scales (Figure 23.2). Vegetation patterns, for example, can be observed on a range of levels, beginning with individual plants, grouped then into stands, alliances, formations, physiognomic classes, and biomes. Animal populations can also be viewed as grouped by individuals, species, populations, metapopulations, guilds, assemblages, and communities.

Although patterns may be determined by diverse factors, when arranged in this hierarchical structure they bear one trait in common: patterns that are unpredictable at fine scales or at lower levels of the hierarchy become much more predictable at broader scales or at higher levels of the hierarchy (Levin, 1992). The implication that stems from this is that nonequilibrium dynamics evident at a fine scale can be translated into more stable and predictable patterns at a higher level (Urban et al., 1987; Levin, 1992) and that a significant initial step in performing ecological assessments is to match the appropriate level of classification of phenomena to the type of assessment (Bourgeron and Jensen, 1994). As will become increasingly evident, the correct selection of scale of observation is critical to adequately describing dynamic pattern (Levin, 1992).

Sources of information used to characterize dynamic pattern may also follow this same hierarchical structure (Figure 23.2). Information for fine-scale phenomena tends to be sparse and is often developed from field observations done especially for the assessment. Vegetation pattern at the alliance, formation, and community levels is frequently assessed using a combination of remotely sensed data and spatial statistics, much of which is available from agencies charged with managing natural resources. Characterizations of animal distribution patterns of approximately the same scale can be developed from state Heritage Program data (Carr, 1994), such as element occurrence observations, or through modeling, as in gap analysis modeling (Cogan, 1994; see also http://www.gap.uidaho.edu/gap/Projects/Index.htm).

23.3 Definition of Terrestrial Disturbance Regime

Disturbances are relatively discrete events in time that disrupt ecosystem, community, or population structure and change resources, substrate availability, or the physical environment (White and Pickett, 1985). Disturbance descriptors include *kind* of disturbance, *spatial characteristics* (e.g., area, shape, spatial pattern), *temporal characteristics* (e.g., frequency, regularity, return interval), *specificity* (e.g., to species, size, or age classes), *magnitude*, and the *synergisms* among disturbances (Harmon et al., 1983; White et al., in press). Each kind of disturbance can be characterized by a set of frequency distributions for each of these descriptors (Baker, 1992a). A basic understanding of the spatial and temporal characteristics of disturbance is critical to assessing the importance of disturbance in a landscape, and understanding the synergisms among disturbances is critical to assessing a landscape's overall disturbance regime.

23.3.1 Components of Spatial and Temporal Scale: Grain and Extent

Any assessment of dynamic pattern in a landscape requires explicit attention to both spatial and temporal scales of observation. Understanding the components of scale and their impact on our perception of dynamic patterns is a critical element in the assessment process. Scale has two components: *grain*, the degree of resolution or smallest recognizable element, and *extent*, the total area or time span over which observations are made (Turner et al., 1989; Wiens, 1989). Varying either the grain or extent of a sample alters our observations. Increasing grain size tends to reduce observed variance among samples because each sample is larger and encompasses more variation internally; increasing extent tends to increase observed variance because, in larger areas, more kinds of environmental conditions and events are observed (Wiens, 1989).

When conducting ecological assessments, it is preferable to avoid a single arbitrary choice of scale based purely on expediency (Wiens, 1989). Different kinds of disturbance create patterns at different scales. Even for a single disturbance event within one landscape, considerable variation occurs in the scale of resulting pattern. For both reasons, a multiscale approach is essential. The scale at which patterns are identified influences the assessment of these patterns, and patterns that are discerned at different scales may not be comparable. For this reason, the grain and extent of any observation should be stated explicitly (Turner et al., 1989; Reed et al., 1993).

23.3.2 Spatial Scale

Disturbances differ in a variety of spatial characteristics, including *size* (e.g., patch area, total disturbance area, area per time period), *distribution* (location with respect to various physical or biotic gradients), and *landscape pattern* (e.g., patch shape, complexity, area-to-perimeter ratio, orientation and dispersion, degree of connectivity or contiguity, and relationship to the surrounding matrix). These parameters affect the rate, degree, and nature of colonization, succession, and recovery from disturbance in both terrestrial and marine systems (Runkle, 1985; Sousa, 1985; White et al., 1999).

Smith and Urban (1988) explored the effect of different scales of observation in developing an individual tree-based forest succession model. They concluded that at a broad scale a forest stand might appear to be in a near-equilibrium state, with succession progressing toward a specific overall mean species composition. An individual gap within the stand, however, constantly undergoes structural changes driven by the life-history traits of a few dominant individuals. Stand dynamics, therefore, tend to be driven by the demographics of individuals at a small spatial scale and by overall stand biomass and composition at a broader scale (Smith and Urban, 1988). This work illustrates the potential to overlook relevant mechanisms if we limit ourselves to one spatial scale.

23.3.3 Temporal Scale

Temporal characteristics of disturbances include *frequency* (the number of disturbance events per unit time), *rotation period* (the time needed to disturb an area equal in size to the study area), *return interval* (the time between disturbances), *regularity* (or *periodicity*), and *contagion* (rate and probability of spread) (White et al., 1999). Many factors influence temporal variation, especially the periodicity of a particular disturbance type. These may include endogenous feedback mechanisms (e.g., increased fuel loading and subsequent flammability with long fire return intervals) or exogenous factors (e.g., long-term climatic cycles, the recurrence of *El Niño*). Deducing a characteristic or representative temporal periodicity may prove elusive, however, because disturbance regimes may not be temporally stable, especially in light of recent and ongoing human alterations to many systems.

Ecological assessments often seek to separate recent human influences from naturally occurring dynamics. However, human effects often may extend backward for longer time periods than we have generally addressed. In many locations in the United States, there is growing evidence of widespread manipulation of the fire regime by native Americans and European settlers, who used fire to drive game and prepare spring forage. The degree to which these anthropogenic activities may have affected the development and structure of the most fire-prone ecosystems is unknown, but it is speculated that their influence might have been profound in areas that were otherwise naturally fire protected (Baker, 1992b; Ware et al., 1993).

Different assessment objectives will require different temporal scales. Assessing the historical range of variability for a forested site will involve a time scale on the order of centuries in order to encompass several generations of the dominant trees (Morgan et al., 1994). An assessment of shorter-term successional trends may be based on data collected during a single decade.

As with spatial scale, the amount of observable variation depends on the resolution of the observations, in this case the time span of individual observations. At small temporal scales, an assessment of dynamic pattern may seem straightforward, but may also be shortsighted. Consider the example of assessing dynamic patterns in riparian environments. Stream meandering results in continuous

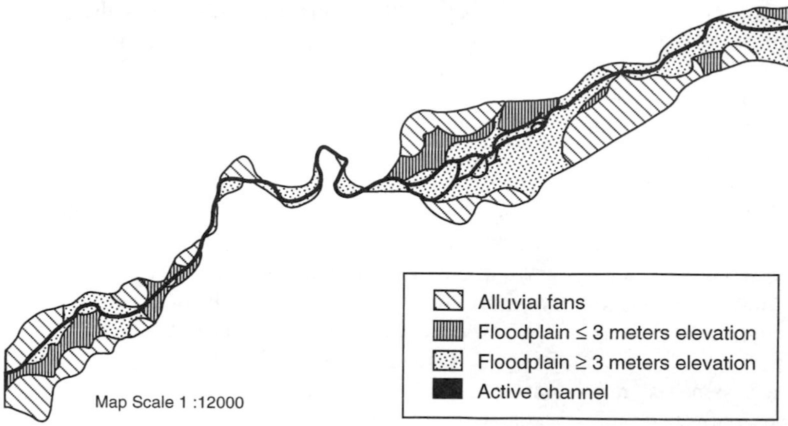

FIGURE 23.3. Variation in geomorphological characteristics, including substrate and hydroperiod, as a function of stream dynamics. Such variation affects the compositional diversity and dispersal characteristics of associated vegetation communities (Gregory et al., 1991).

aggradation and degradation throughout the floodplain (Barnes, 1997). These dynamic fluvial processes create constantly changing mosaics of landforms, with a concomitant biotically diverse and highly variable suite of biological communities. Everitt (1968) determined that during a 100-year period a meandering reach of the Little Missouri River in North Dakota traveled an average distance of 5.9 channel widths across the valley floor by continuously eroding and redepositing its floodplain banks (Swanson et al., 1988).

Riparian zones are not easily delineated and must be considered in the context of the river valley as a whole. In broad flat reaches, the riparian zone will support a complex mosaic of plant communities in various developmental stages. Narrower reaches exhibit less diversity, but are important corridors for dispersal between larger floodplain sites (Figure 23.3). Whether a reach develops as a broad flat floodplain or a narrow valley depends on local edaphic and topographic characteristics, as well as disturbance history. The geomorphic processes that shape and control riparian zones operate at many temporal scales, from minor but intense flooding episodes that scour stream banks in a matter of days, to chronic erosion that ultimately can relocate entire river channels (Gregory et al., 1991). How these processes are perceived and assessed depends on both the area and the time frame considered. How biotic communities found in these areas are identified and classified similarly depends on the spatial and temporal scale of observation.

23.3.4 Dynamic Interactions

Many divergent types of dynamics interact with one another, complicating their analysis and assessment. These interactions may be synergistic, with one type of disturbance promoting another. For example, in many pine-dominated areas of the southeastern United States, droughts stress trees, which then promote outbreaks of southern pine bark beetle. Such insect outbreaks cause tree mortality, increasing available fuels, which in turn increases the severity of fires (Knight, 1987; White, 1987). Synergistic interactions may be negative, wherein one disturbance type decreases the chance of another. Postfire, early successional stands, for example, are less affected by windstorms that may produce extensive blowdowns in neighboring mature forest stands. Disturbances may also exhibit feedback effects in which a disturbance influences subsequent similar disturbances, as in the case of fire-return intervals. These types of interactions may all be present in the same landscape.

In Florida's Everglades National Park, fire incidence and frequency are closely related to local topography and water table position. When fire occurs, it removes organic matter and thus lowers relative topographic position. However, the position of the water table may be altered by droughts, by artificial impoundments or drainage projects, or by the evapotranspiration of an introduced tree species, *Melaleuca*, which is increasingly common in this normally open, grass-dominated landscape. Higher decomposition rates occur in areas where the water table has been lowered, reducing the amount of organic matter and thus altering fire behavior (White et al., 1999). Such complex processes are likely to produce heterogeneous, patchy landscapes that defy straightforward analysis.

Disturbances also vary along environmental gradients in much the same way as vegetation may vary along such gradients. Lightning-ignited fire, for example, has a higher probability of occurrence and a much greater impact on exposed upper slopes, where limited soil moisture and increased solar radiation already result in more xerophytic, flammable vegetation (Harmon et al., 1983). Where disturbance gradients parallel environmental gradients, distinguishing one from another may prove challenging. Human-caused disturbances are less likely to originate in response to environmental gradients. For example, fires are more likely ignited by humans in low-lying settled areas, and exotic pests and diseases frequently are introduced near commercial ports of entry. Once initiated, however, these disturbances may then be perpetuated by natural processes, thus obscuring their anthropogenic origins.

These types of interactions, despite their importance, are poorly documented and understood components of disturbance regimes. Although common in many ecosystems, they have been studied in only a few. Their presence should be considered in any detailed assessment of dynamic pattern.

23.4 Ecosystem Parameters Affected by Ecosystem Dynamics

Disturbances not only occur at a variety of intensities and frequencies, but they also affect ecosystem composition and structure in different ways. Many disturbances alter *ecosystem composition* because species have different susceptibilities to disturbance and different responses to the conditions that disturbances create. Biotic disturbances, such as disease or insect outbreaks, often have very specific effects on composition because they affect selected species. For example, the introduction of chestnut blight in the early 20th century eliminated the American chestnut as a canopy species throughout the hardwood forests of the Eastern United States (Woods and Shanks, 1959). The dramatic impacts of chestnut blight have resulted in a major shift in forest composition from widespread dominance by chestnut to dominance by a variety of oak and other species, depending on geography, elevation, and topographic position (Whittaker, 1956). Disturbance susceptibility also varies with age and size class, successional state, and topographic position (White et al., 1999).

Ecosystem structure is also affected in a variety of ways by disturbance. Within one ecosystem type, different kinds of disturbance can cause a wide array of structural changes, and even for a single disturbance type within one ecosystem, a wide array of structural effects can result as a function of local disturbance intensity (Lang, 1985). For example, some less intense fires may consume only minor amounts of organic matter and may even create more fuels than they burn, because some trees are killed without being consumed by the fire. At the other extreme are intense fires that consume all the organic material on the forest floor, cause heavy mortality in the canopy, and consume most available fuels. Ecosystem structure may be the key consideration for maintaining reproducing populations of some species. For example, the red-cockaded woodpecker in southeastern longleaf pine ecosystems depends on large but isolated nest trees, a structure that requires frequent ground fire (to reduce understory woody plants) and rarer intense fires (to allow pine reproduction) (Noss, 1989; Ware et al., 1993).

Disturbance also has various effects on *ecosystem function*, specifically the flow of energy, materials, and species among ecosystem components. On a local scale, storm-related tree gaps experience greater solar radiation, lower relative humidity, and reduced leaf litter, resulting in a shift of available resources. Larger disturbances, such as avalanches, landslides, volcanic eruption, flooding, and erosion, all create or destroy substrate while altering soil chemistry, texture, and drainage characteristics, all of which will affect ecosystem function.

23.5 Characterization of Dynamics and Evidence Used in Assessment

Although practicality may require the analysis of dynamic pattern to be driven by the type, scale, and quality of the assessment data set itself, the preceding sections serve as a guide to determining a scale and level of analysis appropriate to the system being examined and the assessment objectives. A second important step before analysis is to determine the methods that will be used to characterize ecosystem dynamics. Although many different descriptors of disturbance have already been identified (e.g., type, shape, magnitude, spatial distribution, temporal distribution, duration), most disturbance regimes have been typified by in-depth analysis of just a few of these attributes (Baker, 1992a). These are often limited to *size distributions*, *temporal distributions* (particularly interval between disturbances and frequency per unit area), and *spatial distributions* (density per area and topographic position) (Figure 23.4). Mean values of

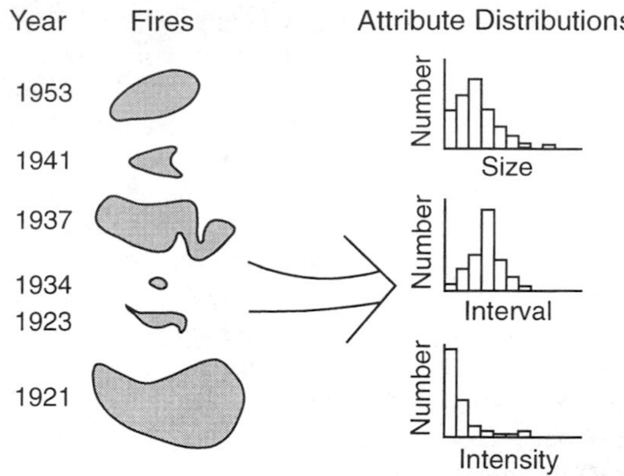

FIGURE 23.4. Typical attribute distributions for a fire disturbance regime (Baker, 1992a). Histograms reveal more pattern characteristics than simple statistics.

each of these are not nearly as informative as frequency distributions (histograms), which can supply much additional information. Charting attributes over time may reveal historical trends worth extrapolating into the future.

Determining sources of data for the analysis of dynamic patterns is not as straightforward as simply locating maps and historical records and performing field surveys. There are three general approaches that can be used to assess disturbance patterns, each of which has its own range of applications (White et al., 1999). Ecologists are most familiar with the *historical approach*, wherein past events and conditions are documented through extensive field survey, including fossil pollen, charcoal layer, fire scar, and regeneration pattern analysis, as well as through archival research of stand records, timber surveys, General Land Office survey notes, hydrographs or other hydrologic data, climate data, historical photographs, narrative reports, and previous site-specific research. The historical approach can be broadened to include the practice of using data from a similar site to draw conclusions concerning the site in question. This type of research is often augmented with the *observational approach*, which focuses on the present-day analysis of existing conditions, prevalent disturbance patterns, successional and reproductive trends, and general ecosystem responses. In the *simulation approach*, statistical and spatial models can be derived, parameterized, and employed to explore a variety of disturbance regime scenarios. These three approaches differ considerably in their respective spatial and temporal resolutions, applicability to a given situation, and practicality.

23.5.1 Using Historical Data

Historical records have been used in many applications to assess past disturbance events. In the Great Smoky Mountains National Park, Harmon (1981) derived fire history statistics and the distribution of fire events along environmental gradients from a combination of park records (date, probable location, and source of ignition) and field survey (fire-scar and tree-ring analysis). Working in the same area, Pyle (1986) studied anthropogenic disturbance, determining dates for logging activities and settlement through lumber company records, land title abstracts, unpublished written summaries, U.S. Census records, and cemetery burial records, among other documents. More recent disturbance histories can be derived from archival photography and satellite imagery.

These types of assessments, based on historical data, are beset with a number of difficulties. A common problem is the incompatibility of multiple maps in terms of map projection and scale and a resultant need to generalize detail to the map of largest cartographic scale. In this process, the transference of data from one scale or projection to another involves a certain amount of imprecision, and generalizing data to one scale involves a loss of detailed information. Maps may also be incompatible in terms of classification schemes and representation of data. Another problem is the time specificity of the data. Information from one time period is only indicative of conditions at that time period and cannot and should not be used to extrapolate conditions at another time period. Historical data may have been recorded for an insufficient period to capture the full range of variation, and they may be of inappropriate spatial resolution. Historical data may entirely lack a spatial component (narrative only) and can be greatly affected by human bias and misrepresentation. In the case of Pyle's work, the historical records she used contained few if any maps and little site-specific information, and those associated with condemnation proceedings leading to Park formation often were prone to bias and intentional misrepresentation.

Another drawback to historical data is the possible presence of unmeasured factors that confound interpretation. Unmeasured but correlated changes in the environment, complex responses to past environmental change, and response to past or current management practices may not be evident through analysis of some data. Characterizing landscape dynamics during periods of human influence can be particularly subject to synergisms of multiple processes occurring in the landscape.

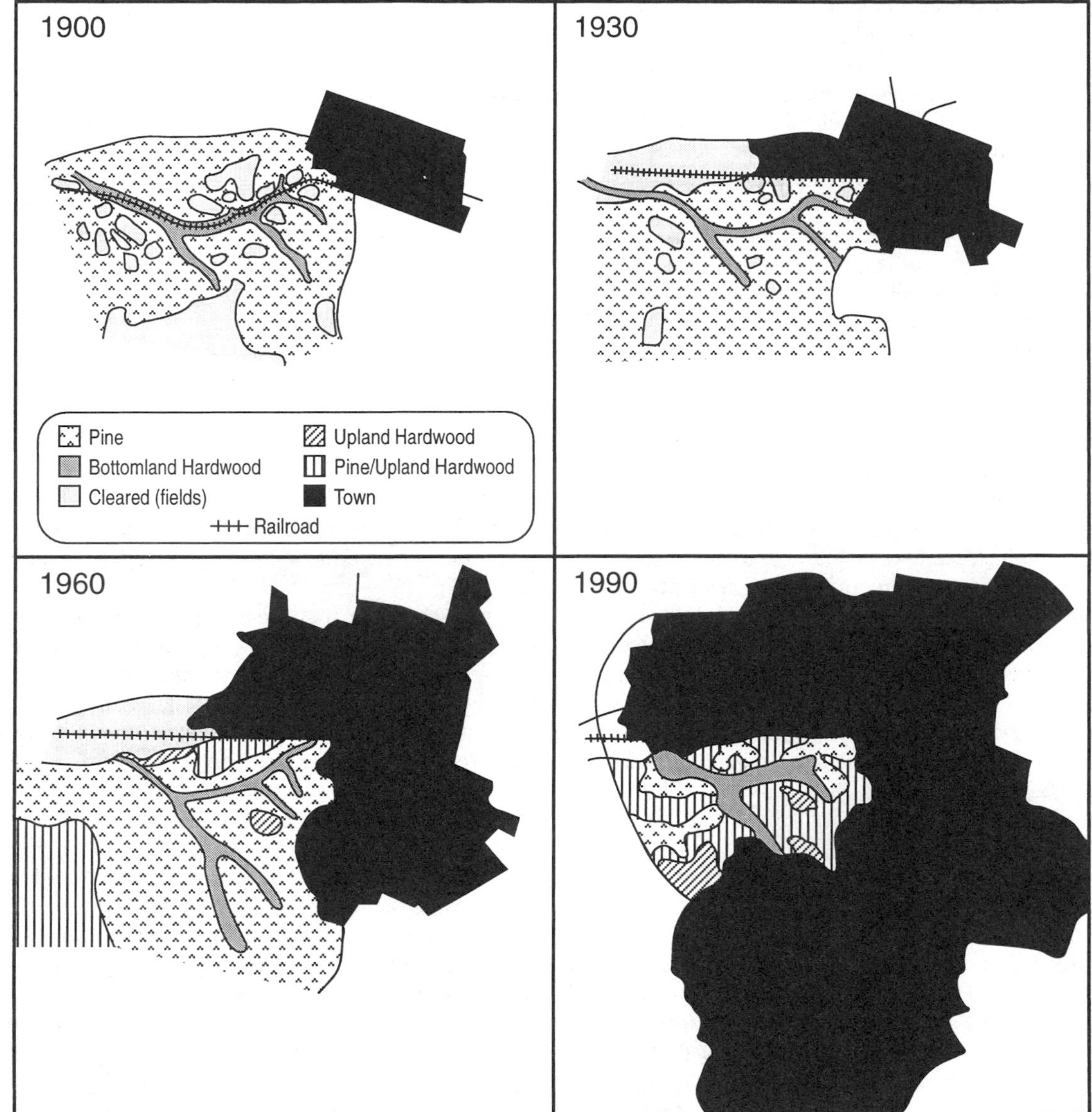

FIGURE 23.5. Growth in area of the City of Aiken, South Carolina, and the associated alteration in vegetation composition in Hitchcock Woods Nature Preserve. Transition of the site from primarily longleaf pine savanna to mixed pine–hardwoods has greatly accelerated in the past 40 years due to expansion of surrounding impervious surfaces and increased runoff.

For instance, the effects of fire suppression on vegetation composition and structure in a formerly fire-influenced landscape may be exacerbated by simultaneous alterations in the hydrologic regime. These patterns may not be evident until information from multiple sources is assembled. In assessing the effects of fire suppression in an 800-ha nature preserve in the Sandhills region of South Carolina (S.W. Wilds, unpublished data), the rapid conversion (ca. 70 years) from fire-adapted upland longleaf pine–wiregrass savanna to fire-sensitive mixed pine–hardwood forest with an extensive bottomland hardwood component cannot be explained by fire suppression alone. Examining the rate and extent of adjacent urban spread surrounding the site reveals that the increased influx of storm water has altered soil moisture characteristics and established extensive barriers to fire spread (Figure 23.5). Pyle, too, found that, when multiple maps derived from independent sources were used in conjunction with written records, a detailed analysis of anthropogenic disturbance history could be constructed. Identifying the sequence of historic disturbance events, including fire, logging, and conversion to farmland, makes it possible to separate these events from natural disturbances, such as windthrows, lightning-set fires, ice damage, and floods.

A final limit to using historical data pertains to the nature of the site and the disturbance regime itself. In areas subject to infrequent and catastrophic disturbance, not only may the historical record be too limited in duration to capture the disturbance pattern, but more recent disturbances may erase all trace of previous disturbances. In areas where human influence has been particularly high, evidence of a natural disturbance regime may be lacking. Ecosystems dominated by nonwoody vegetation, lacking tree rings, fire scars, and structural evidence of release events, are also difficult to assess from the historical record (Morgan et al., 1994). Sorting out anthropogenic from natural disturbance in some landscapes can be difficult, particularly when past human effects alter the occurrence of natural disturbances. A particular problem is assessing the impact of Native Americans on the natural fire regime, which, until recently, has been generally underestimated (White et al., 1999). Only the results of fires can be observed, and differentiating between fires ignited by lightning, Native Americans, or Euro-American settlers can be challenging.

Limitations in availability or quality of on-site historical data can be augmented by using more complete data from intact similar systems. Areas believed to have similar vegetation, topography, climate characteristics, and unaltered disturbance regimes can be used as models for systems for which data are unavailable. Large wilderness areas and public lands are often used in this manner. Intact sites are rare, however, because the effects of fire suppression, climate alteration, air pollution, and exotic invasion are widespread. In many parts of the western United States, for example, the extirpations of native herbivores (e.g., bison) or the introduction of nonnative herbivores (cattle and sheep) have permanently altered vegetation structure on a regional scale (Billings, 1990). An additional impediment to substituting one site for another is that protected areas are often limited in their range of ecological conditions (Morgan et al., 1994). Most large wilderness areas in the United States, for example, are limited to mountainous regions.

23.5.2 Interpreting Observational Data

Observational data typically consist of descriptive stand data and an assessment of obvious ecosystem trends, including overall stand structure and composition and an approximation of successional stage. These data are usually augmented with objectively collected quantitative data, derived using techniques such as the relevé method, random quadrats, or various plotless methods. Ecosystem attributes that should be noted include any changes in ecosystem components, such as the absence of species or the presence of exotic species, as well as changes in ecosystem processes, such as relict fire-dependent species and the absence of species reproduction (White and Walker, 1997). These types of evidence are critical to understanding the recent and long-term disturbance history of a site, especially the interval since last disturbance and the magnitude and frequency of regular and/or past disturbance.

An assessment of successional stages and trends is usually an important objective in the collection of observational data. Successional analysis may be based on any number of factors, depending on the ecosystem involved, but in forested stands it typically focuses on a comparison of the composition and structure of the overstory stratum to the composition and structure of the seedling and sapling strata. Successional stage analysis provides a great deal of information about the intensity of and interval since disturbance. The size, shape, and dispersal pattern of different successional patches across the landscape convey a wealth of information about current disturbance dynamics. Although succession is best quantified by repeated measures over time in the same location or by analyzing historical records for the location in question, an extensive array of sampling points across the landscape can be used to both characterize typical successional stages in areas of similar topographic and edaphic conditions (especially when the date of disturbance is known) and to identify aberrations from these typical stages, which may indicate distinctive disturbance events.

Although observational data clearly offer the most effective way to understand a particular system's present and recently past dynamic status, it must be restated that single-time-frame observations can be misleading (White and Walker, 1997); it is only when they are coupled with long-term data that a realistic perception of the disturbance regime can be constructed. Many ecosystem processes function at nonstationary rates depending on interactions with other processes, and data confined to one time period cannot be used to extrapolate conditions over other time periods.

23.5.3 Simulating Dynamic Patterns

Ecological modeling, or the explicit mathematical simulation of physical and biological relationships, has been used in numerous applications to estimate and predict ecosystem processes. Modeling has been used on a global scale to estimate the impacts of climate change (Claussen and Esch, 1994; Dale and Rauscher, 1994), to simulate biogeochemical cycles (Parton et al., 1988; Bonan, 1995), and to estimate net primary production (Lieth, 1975; Running and Coughlan, 1988). It has also been used extensively to model hydrologic processes (Beven and Kirkby, 1979; Moore et al., 1991), usually through the use of raster, or cell-based, geographical information systems (GIS) and digital elevation models (Davis and Goetz, 1990; Drayton et al., 1992). Application to terrestrial dynamic patterns has been limited largely to fire behavior (Kessell, 1979; Baker, 1994; Davis and Burrows, 1994; Vasconcelos et al., 1994), insect outbreak (Gribko et al., 1995), and the dynamics of vegetation change (forest succession, or gap-phase, models) (Urban and Shugart, 1992). This is likely due to the fact that these disturbance types are largely endogenous (a product of life-history traits in the case of forest succession and insect outbreak) or controlled by characteristics of the system (i.e., landform and interval since last fire in the case of fire behavior). Exogenous disturbances, such as weather events, flooding, earthquakes, and exotic invasions, are nearly impossible to predict and therefore model effectively.

Models are useful for projecting future landscape structure and for investigating the impacts of a variety of management scenarios (Baker, 1992a). This is usually perceived as their primary function. However, models are also useful for testing the validity of assumptions made about a system. By using historic conditions as a starting point and comparing results to current conditions, spatial models can reveal the extent and limits of our understanding. Models can provide estimates of historical range of variability when historical data are incomplete or inadequate (Morgan et al., 1994).

Models can also be used to determine the circumstances under which a system might be considered to be in an equilibrium state, wherein the proportion of the landscape in various successional states remains fairly constant, although the location of patches constantly shift. This basic definition of equilibrium can be modified in highly dynamic landscapes to include systems in which all successional stages are constantly present, although highly variable in relative proportion (White and Walker, 1997). Being able to model a patch dynamic equilibrium state for a particular landscape is a significant route by which the natural, stable disturbance regime can be identified. Several researchers have concluded that dynamic equilibrium is a function of the relationship between the mean size of disturbances in a stable disturbance regime and the size of the area of interest (Shugart, 1984; White, 1987). Thus simulation modeling can aid in determining (1) the stable disturbance regime of a region, (2) whether a particular study area is large enough to capture this disturbance regime, and (3) the appropriate temporal scale for assessing the disturbance regime. For example, in applying a spatially explicit, GIS-based fire model (REFIRES) to a chaparral ecosystem in southern California, Davis and Burrows (1994) found that the variation between long model runs with different initial conditions but the same parameter settings was only slight. This suggests that, within regions with uniform climate, vegetation, and physiography and over long periods of time, a fire-controlled chaparral system may approach an equilibrium state. Such an assessment can aid in determining the important components in a system's natural disturbance regime, and the appropriate spatial and temporal scales for describing that disturbance regime.

Models range widely in their degree of specificity and spatial detail, from whole landscape models, which estimate the value of a variable over an entire landscape, to distributional models, which estimate the distribution of values of some variables(s) over time, and finally to spatial landscape models, in which the spatial location and configuration of a variable can be tied to any number of landscape measures and submodels (Baker, 1989). Models may be either continuous or discrete in their treatment of time. Although spatial models are the most detailed and informative, their computational demands are prohibitive. Distributional models, therefore, have been most widely developed and applied, whereas whole landscape models are usually integrated as submodels. Many recent models have been developed and implemented in a GIS environment (see Goodchild et al., 1993, and Haines-Young et al., 1993) and structured on a raster format, often using a hexagonal grid tessellation to facilitate contagious diffusion among grid cells. Such a format allows for easy integration with other raster-based data, especially satellite imagery; a drawback is the automatic limitation to a specific resolution of data (Goodchild, 1994).

Spatial models, although computationally demanding, offer the greatest flexibility for parameterization and realistic responses. They rely heavily on the concepts of gradient analysis, that is, the way in which vegetation structure and composition steadily change along an array of environmental conditions, such as altitude, moisture availability, sheltering, nutrient availability, or disturbance impact. In simulation modeling, gradient analysis comes into play when a multidimensional gradient analysis and a site-specific stand inventory are linked by computer modeling software (Kessell, 1979). Parameterizing a spatially explicit disturbance simulation model involves linking a specific set of conditions (physiographic, climatic, vegetative) with a set of rules, usually mathematical equations, that describe disturbance behavior under a range of conditions. Early models were usually based on a spatially heterogeneous but entirely simulated landscape. Recent advances in GIS-based data allow for the use of real landscapes, with precise georeferencing of environmental conditions, which can then be used to simulate any number of alternative management scenarios (Vasconcelos et al., 1994).

A prime example of the model-building process is provided by models of fire behavior. Many fire models are based on the fire spread equations of Rothermel (1972), which predict fire behavior (intensity and rate of spread) with homogeneous and continuous fuel availability as a function of detailed meteorological information and fuel characteristics (Davis and Burrows, 1994). These rules of fire behavior are then given spatial dimension by applying them to a digital elevation model. Topography controls direction and rate of spread by affecting the exposure of fuels to wind and the drying effects of solar radiation (through slope aspect), flame length and degree of flame contact (through slope steepness), and the presence and extent of contagion barriers and corridors, a function of the degree of topographic dissection. This landform-based analysis of disturbance propagation (fire spread) takes into account any existing natural directions of disturbance movement and the effects of natural barriers. The final element, land cover information in the form of either detailed stand inventories or satellite imagery, is incorporated into the spatial data to provide information as to the types and condition of available fuels. These data are augmented by information concerning meteorological conditions, postfire succession, fuel accumulation, the effects of disturbance on vegetation, and climatic variation. Fire ignition frequency and timing may be generated statistically or explicitly, depending on the model's intentions.

Simulation modeling of fire behavior is not without several drawbacks. An obvious obstacle is a need for access to sophisticated computing facilities and skilled programmers. A second obstacle is a need for detailed data for parameterization, including detailed meteorological data (temperature, humidity, wind speed, wind direction) and detailed fuel data (fuel per land area, fuel size distributions, height of the fuel bed, fuel heat content, particle density, and moisture content) (Davis and Burrows, 1994). For long-term modeling, as would be required to determine the historical range of variability or to assess the likelihood of a patch dynamic equilibrium, a model must also be coupled with successional models in order to represent variation in fuel conditions over time. Finally, models will likely have to be spatially explicit, that is, to include the effects of patch size, shape, and adjacency. Because of the lack of such detailed data, many models depend heavily on simplifying assumptions. In most modeling activities, it is not a lack of technology that prevents effective model building, but rather a lack of knowledge about the processes in the landscape and how to best represent these processes. A final drawback is the difficulty of testing models against observations to discern where assumptions have been made incorrectly. Nonetheless, simulation models are valuable tools that allow managers to predict the consequences of alternative management scenarios, particularly when there are many and interacting variables of interest.

23.5.4 Performing the Assessment

The process of assessing dynamic pattern in a particular landscape includes, first, the analysis of all available historical data, followed by the analysis of observational data from the site. Gaps in this information may be filled by both historical and observational data from similar sites and/or by simulation modeling when no other information is available. Assessing dynamic pattern is not merely a matter of describing the patterns of past and current disturbances, along with the successional stage and functionality of the study area; it uses these analyses to place the site in its broader spatial and temporal context. Table 23.1 outlines an appropriate approach, as well as a series of questions that should be addressed during the assessment process.

TABLE 23.1. Questions and issues for biological assessment of ecosystem dynamics.

A. Information sources for the assessment of terrestrial dynamics
  1. What historical data are available for the site?
     Aerial photographs — Hydrologic and climate records
     Land-use records — Historical photographs
     Timber surveys — Narrative reports
  2. What observational data are available or can be collected?
     Overall stand and individual strata structure and composition
     Presence or absence of significant species
     Presence or absence of stand regeneration
     Size, shape, and spatial pattern of individual successional stands
  3. What reference sites are available?
     Sites that are similar in topographic, hydrologic, edaphic, and climatic properties
     Sites where evidence is present of similar vegetation cover in presettlement era
     Sites where the disturbance regime is determined to be largely intact
  4. Can ecosystem dynamics be simulated?

B. Questions to guide the assessment of terrestrial dynamics
  1. Does the site contain all the dynamic elements that make it self-sustaining?
  2. Is the source of dynamics consistently outside the study area?
  3. Are there exotic species, missing species, or species failing to reproduce?
     If so, what processes are absent from the site or have been introduced to the site that may cause this?
  4. Is there evidence of synergistic effects occurring between two or more altered processes, which may not be evident in a stable reference site?
  5. Has the size, shape, and/or pattern of successional patches shifted over time?
  6. Are there patterns of zonation in the landscape, indicating a systematic trajectory, such as a rising or lowering water table?

23.6 Conclusion: The Value of a Multiscale Approach and Adaptive Management

Within the framework of an assessment's objectives, uniquely appropriate spatial and temporal scales can almost always be established for each ecosystem pattern observed. It is unlikely, however, that one spatial or temporal scale will be appropriate for all discernible processes occurring in any given landscape or that all scales will be equally useful. It is equally unlikely that a set of scales appropriate for one landscape will be applicable to another. Even a single process, such as fire, is best understood when examined at a variety of spatial and temporal scales, in order to best capture the interactions which may be taking place with other processes. Depending on an assessment's scope, objectives, and relation to other management activities, the selection of a single scale of observation may be appropriate and necessary, but this should only be done with an awareness of the possibility of unseen and unmeasured dynamic phenomena.

Many researchers have urged the adoption of a multiscale perspective in ecological research (Wiens, 1989; Levin, 1992; Morgan et al., 1994; White et al., 1999). The search for an appropriate scale, or suite of scales, need not be random and arbitrary, however. Studies in which the grain and extent of scale are varied systematically and independently provide an efficient means by which dynamic processes and dynamic interactions and the interrelationships among scales can be comprehensively investigated (Wiens, 1989). This systematic scaling process is particularly useful in capturing those disturbance events that are infrequent, irregular, and/or spatially patchy and thus may appear to occur at a variety of scales. Without a multiscale approach, a comprehensive assessment must acknowledge the potential limitations of conclusions based on a single scale of observation.

Defining and analyzing disturbance regimes and terrestrial dynamic patterns is challenging. Perhaps no other ecological process is more fraught with interactions, synergisms, and complex interrelationships than naturally occurring disturbance. Coupled with the effects of human activities operating in the short term and regional and global processes operating over a much longer period, the driving forces of disturbance and other cyclical processes can be very difficult to perceive and assess. Rare events may occur at such long intervals that they are beyond our temporal realm of observation, or events may be of such broad spatial scale that they are driven by mechanisms outside the area of investigation. Ecosystem components may exhibit a wide range of reactions to disturbance; some species may be highly disturbance sensitive, others hardly affected. And identical disturbances may have different impacts depending on the successional state of the ecosystem or on other, perhaps unmeasured, processes that may be occurring simultaneously.

Because of possibly immeasurable long-term ecosystem variability, climate change, system inertia or synergistic effects, and undocumented human influence, a comprehensive documentation of natural disturbance regimes may be an unattainable goal in many ecosystems. Also, even when documented, such disturbance regimes may not be stable. While such assessments must be incorporated into restoration and management programs, there must always be an allowance for unanticipated effects resulting from the application of incomplete knowledge, an approach that has been termed *adaptive management* (see Chapter 4). Collecting new observations and testing prior assumptions should be perceived as an opportunity to add to our body of knowledge about dynamic interactions.

23.7 References

Baker, W. L. 1989. A review of models of landscape change. *Landscape Ecol.* 2(2):111–133.

Baker, W. L. 1992a. The landscape ecology of large disturbances in the design and management of nature reserves. *Landscape Ecol.* 7(3):181–194.

Baker, W. L. 1992b. Effects of settlement and fire suppression on landscape structure. *Ecology* 73:1879–1887.

Baker, W. L. 1994. Restoration of landscape structure altered by fire suppression. *Conserv. Biol.* 8:763–769.

Barnes, W. J. 1997. Vegetation dynamics on the flood plain of the lower Chippewa River in Wisconsin. *J. Torrey Bot. Soc.* 124(2):189–197.

Beven, K. J.; Kirkby, M. J. 1979. A physically based, variable contributing area model of basin hydrology. *Hydrolog. Sci. Bull.* 24:43–69.

Billings, W. D. 1990. Bromus tectorum, a biotic cause of ecosystem impoverishment in the Great Basin. In: Woodwell, G. M., ed. *The Earth in transition*. Cambridge, UK: Cambridge University Press: 301–322.

Bonan, G. B. 1995. Land–atmosphere interactions for climate system models: coupling biophysical, biogeochemical, and ecosystem dynamical processes. *Remote Sens. Environ.* 51:57–73.

Bourgeron, P. S.; Jensen, M. E. 1994. An overview of ecological principles for ecosystem management. In: Jensen, M. E.; Bourgeron, P. S., tech eds. *Vol. II: Ecosystem management: principles and applications*. Gen. Tech. Rep. PNW-GTR-318. Portland, OR: U.S. Dept. Agric., For. Serv., Pacific Northwest Res. Sta.: 45–57.

Carr, K. 1994. *The biological and conservation data (BCD) system and the natural heritage network*. Prepared for the 1994 Computerworld Smithsonian Awards Program. The Nature Conservancy; available on the Internet (http://www.heritage.tnc.org/nhp/award.html#REF).

Claussen, M.; Esch, M. 1994. Biomes computed from simulated climatologies. *Clim. Dynam.* 9:235–243.

Cogan, C. 1994. Terrestrial and aquatic Gap Analysis summary of national data resources and identification of cooperators. In: Scott, J. M.; Jennings, M. D., eds. *A handbook for gap analysis*. Moscow, ID: Idaho Coop. Fish and Wildlife Res. Unit, University of Idaho: 4.1–4.13.

Dale, B. H.; Rauscher, H. M. 1994. Assessing impacts of climate change on forests: the state of biological modeling. *Climatic Change* 28:65–90.

Davis, F. W.; Burrows, D. A. 1994. Spatial simulation of fire regime in Mediterranean-climate landscapes. In: Moreno, J. M.; Oechel, W. C., eds. *The role of fire in Mediterranean-type communities*. New York: Springer-Verlag: 117–139.

Davis, F. W.; Goetz, S. 1990. Modeling vegetation pattern using digital terrain data. *Landscape Ecol.* 4: 69–80.

DeAngelis, D. L.; White, P. S. 1994. Ecosystems as products of spatially and temporally varying driving forces, ecological processes, and landscapes—a theoretical perspective. In: Davis, S. M.; Ogden, J. C., eds. *Everglades: the ecosystem and its restoration*. Delray Beach, FL: St. Lucia Press: 9–28.

Drayton, R. S.; Wilde, B. M.; Harris, J. H. K. 1992. Geographical information system approach to distributed modeling. *Hydrol. Process.* 6:361–368.

Everitt, B. L. 1968. Use of the cottonwood in an investigation of the recent history of a floodplain. *Amer. J. Sci.* 266:417–429.

Forman, R. T. 1995. *Land mosaics: the ecology of landscapes and regions*. Cambridge, UK: Cambridge University Press.

Goodchild, M. F. 1994. Integrating GIS and remote sensing for vegetation analysis and modeling: methodological issues. *J. Veg. Sci.* 5:615–626.

Goodchild, M. F.; Parks, B. O.; Steyaert, L. T. 1993. *Environmental modeling with GIS*. New York: Oxford University Press.

Gregory, S. V.; Swanson, F. J.; McKee, W. A.; Cummins, K. W. 1991. An ecosystem perspective of riparian zones. *BioScience* 41(8):540–551.

Gribko, L. S.; Liebhold, A. M.; Hohn, M. E. 1995. Model to predict gypsy moth (Lepidoptera: Lymantriidae) defoliation using kriging and logistic regression. *Environ. Entomol.* 24(3):529–537.

Haines-Young, R.; Green, D. R.; Cousins, S. 1993. *Landscape ecology and geographic information systems*. London and New York: Taylor and Francis.

Harmon, M. E. 1981. *Fire history of Great Smoky Mountains National Park, 1940–1979*. U.S. Dept. Interior, Natl. Park Serv., Southeast Region. Off., Uplands Field Res. Lab, Research/Resource Manag. Rep. No. 46.

Harmon, M. E.; Bratton, S. P.; White, P. S. 1983. Disturbance and vegetation response in relation to environmental gradients in the Great Smoky Mountains. *Vegetatio* 55:129–139.

Kessell, S. R. 1979. *Gradient modeling: resource and fire management*. New York: Springer-Verlag.

Knight, D. H. 1987. Parasites, lightning, and the vegetative mosaic in wilderness landscapes. In: Turner, M. G., ed. *Landscape heterogeneity and disturbance*. New York: Springer-Verlag: 59–83.

Lang, G. E. 1985. Forest turnover and the dynamics of bole wood litter in subalpine balsam fir forest. *Can. J. For. Res.* 15:262–268.

Levin, S. A. 1992. The problem of pattern and scale in ecology. *Ecology* 73(6):1943–1967.

Lieth, H. 1975. Modeling the primary productivity of the world. In: Lieth, H.; Whitaker, R.H., eds. *Primary productivity of the biosphere*. New York: Springer–Verlag: 237–263.

Moore, I. D.; Grayson, R. B.; Ladson, A. R. 1991. Digital terrain modeling: a review of hydrological, geomorphological, and biological applications. *Hydrol. Process.* 5:3–30.

Morgan, P.; Aplet, G. H.; Haufler, J. B.; Humphries, H. C.; Moore, M. M.; Wilson, W. D. 1994. Historical range of variability: a useful tool for evaluating ecosystem change. *J. Sustain. For.* 2(1/2):87–111.

Noss, R. F. 1989. Longleaf pine and wiregrass: keystone components of an endangered ecosystem. *Nat. Areas J.* 9(4):211–213.

Parton, W. J.; Stewart, J. W. B.; Cole, C. V. 1988. Dynamics of C, N, P, and S in grassland soils: a model. *Biogeochemistry* 5:109–131.

Pyle, C. 1986. The use of vegetation disturbance history source materials: three examples from Great Smoky Mountains National Park. In: Bratton, S., ed. *Vegetation change and historic landscape management: proceedings of the conference on science in the National Parks*; July 13–18, 1986, Colorado State University: 76–94.

Reed, R. A.; Peet, R. K.; Palmer, M. W.; White, P. S. 1993. Scale dependence of vegetation–environment correlations: a case study of a North Carolina piedmont woodland. *J. Veg. Sci.* 4:329–340.

Rothermel, R. C. 1972. *A mathematical model for predicting fire spread in wildland fuels*. Ogden, UT: U.S. Dept. Agric., For. Serv., Res. Paper INT-115. Intermountain For. and Range Exp. Sta.

Runkle, J. R. 1985. Disturbance regimes in temperate forests. In: Pickett, S. T. A.; White, P.S., eds. *The ecology of natural disturbance and patch dynamics*. New York: Academic Press: 17–34.

Running, S. W.; Coughlan, J. C. 1988. A general model of forest ecosystem processes for regional application. I. hydrologic balance, canopy gas exchange and primary production processes. *Ecol. Model.* 42:125–154.

Shugart, H. H. 1984. *A theory of forest dynamics*. New York: Springer-Verlag.

Smith, T.; Urban, D. L. 1988. Scale and resolution of forest structural pattern. *Vegetatio* 74:143–150.

Sousa, W. P. 1985. Disturbance and patch dynamics on rocky intertidal shores. In: Pickett, S. T. A.; White, P. S., eds. *The ecology of natural disturbance and patch dynamics*. New York: Academic Press: 101–124.

Swanson, F. J.; Kratz, T. K.; Caine, N.; Woodmansee, R. G. 1988. Landform effects on ecosystem patterns and processes. *BioScience* 38(2):92–98.

Turner, M. G.; O'Neill, R. V.; Gardner, R. H.; Milne, B. T. 1989. Effects of changing spatial scale on the analysis of landscape pattern. *Landscape Ecol.* 3(3/4): 153–162.

Urban, D. L.; Shugart, H. H. 1992. Individual-based models of forest succession. In: Glenn–Lewin, D. C.; Peet, R. K.; Veblen, T. T., eds. *Plant succession: theory and prediction*. London: Chapman & Hall: 249–290.

Urban, D. L.; O'Neill, R. V.; Shugart, H. H. 1987. Landscape ecology: a hierarchical perspective can help scientists understand spatial patterns. *BioScience* 37: 119–127.

Vasconcelos, M. J.; Pereira, J. M. C.; Zeigler, B. P. 1994. Simulation of fire growth in mountain environments. In: Price, M. F.; Heywood, D. I., eds. *Mountain environments and geographic information systems*. Bristol, PA: Taylor and Francis: 167–185.

Ware, S.; Frost, C.; Doerr, P. D. 1993. Southern mixed hardwood forest: the former longleaf pine forest. In: Martin, W. H.; Boyce, S. G.; Echternacht, A. C., eds. *Biodiversity of the southeastern United States: lowland terrestrial communities*. New York: John Wiley & Sons: 447–493.

White, P. S. 1987. Natural disturbance, patch dynamics, and landscape pattern in natural areas. *Nat. Areas J.* 7(1):14–22.

White. P. S.; Pickett, S. T. A. 1985. Natural disturbance and patch dynamics, an introduction. In: Pickett, S. T. A.; White, P. S., eds. *The ecology of natural disturbance and patch dynamics*. New York: Academic Press: 3–13.

White, P. S.; Walker, J. L. 1997. Approximating nature's variation: selecting and using reference information in restoration ecology. *Restor. Ecol.* 5:338–349.

White, P. S.; Harrod, J.; Romme, W.; Betancourt, J. 1999. The role of disturbance and temporal dynamics. In: Szaro, R. C.; Johnson, N. C.; Sexton, W. T.; Malk, A. J., eds. *Ecological Stewardship*. Oxford: Elsevier Science. 2:281–312.

Whittaker, R. H. 1956. Vegetation of the Great Smoky Mountains. *Ecol. Monogr.* 26:1–80.

Wiens, J. A. 1989. Spatial scaling in ecology. *Ecology* 3:385–397.

Woods, F. W.; Shanks, R. E. 1959. Natural replacement of chestnut by other species in the Great Smoky Mountains National Park. *Ecology* 40:349–361.

24 Ecological Classification and Mapping of Aquatic Systems

Mark E. Jensen, Iris A. Goodman, Christopher A. Frissell, C. Kenneth Brewer, and Patrick S. Bourgeron

24.1 Introduction

Ecological classifications group similar items to provide a framework for organizing our knowledge about ecosystems (Driscoll et al., 1984; Jensen et al., 1991). Examples of such classification schemes include soil taxonomy (USDA-SCS, 1975), potential vegetation types (Driscoll et al., 1984), channel units (Hawkins et al., 1993), stream types (Rosgen, 1994), valley bottom types (USDA, 1978), and watershed types (Jensen et al., 1997). These classifications can be given a spatial component by describing their composition within ecological mapping units (Cleland et al., 1997).

Ecological mapping units are commonly delineated based on different criteria than those used in ecological classification; that is, ecological unit maps are often based on coarser-scale differentia designed to predict repeatable patterns of the finer-scale classifications they are assumed to constrain. For example, ecological unit maps at the landtype scale are commonly delineated based on landform and geologic criteria (Cleland et al., 1997). The classification components they are assumed to constrain include soil families and plant associations; the former are usually identified based on soil texture, climate, and mineralogy and the latter on presence of upper- and lower-layer indicator plant species. This approach to mapping (moving up one scale and predicting the patterns of interest) is extremely cost effective because survey costs commonly increase exponentially with increased resolution. Furthermore, this approach is appropriate from an ecological, as well as an economic, standpoint because many ecosystem patterns are hierarchically organized.

In this chapter, we present a hierarchical framework for the ecological classification and mapping of freshwater aquatic systems (i.e., riverine, lacustrine, and groundwater systems). We begin our discussion by reviewing basic concepts important to the classification and mapping of aquatic systems and follow by describing selected classification and mapping systems commonly used in aquatic ecosystem characterization. This discussion focuses primarily on classification systems for streams, lakes, groundwater, hyporheic zones, and wetlands. Finally, we present general recommendations on the use of terrestrial ecological units in aquatic ecosystem characterization and the appropriateness of aquatic ecological units to multiscale descriptions of ecosystem patterns and processes.

24.2 Basic Concepts

24.2.1 Scaled Relations of Aquatic Systems

Ecological systems are groups of interacting, interdependent parts (e.g., species, habitats) linked to each other by the exchange of energy, matter, and information. These systems are considered complex because they are characterized by strong interactions among components, complex feedback loops, discontinuities, thresholds, and limits (Constanza et al., 1993). Such attributes make it difficult to separate cause and effect relations and to use knowledge of ecosystem behavior at fine scales to predict behavior at coarser scales. Accordingly, descriptions and classifications of ecosystems must acknowledge the nested characteristics of terrestrial and aquatic systems at multiple spatial scales (Bourgeron and Jensen, 1994).

Complex ecosystem patterns and the multitude of processes that form them exist within a hierarchical framework (Allen and Starr, 1982; Allen et al., 1984; O'Neill et al., 1986). Recognizing this, considerable recent attention has been directed at describing the organization of ecological systems for improved land management planning (Cleland et al., 1997; Maxwell et al., 1995). Hierarchy theory (Allen and Starr, 1982; O'Neill et al., 1986) suggests that multiscaled systems can be viewed as a system of constraints in which a higher level of organization provides to some extent the environment in which lower levels develop. A critical characteristic of hierarchical systems is the "whole/part" duality of their components, whereby every level is a discrete entity and at the same time part of a larger whole (Koestler, 1967; Allen and Starr, 1982; Allen et al., 1984).

Hierarchy theory provides a needed framework for ecological classification and mapping because it facilitates scaled definitions of ecosystem components and identification of the linkages that exist between different scales of ecological organizations (Bourgeron and Jensen, 1994; Jensen et al., 1996). Hierarchical approaches to ecosystem characterization also simplify the description and prediction of complex ecological pattern/process relations at all relevant scales of system organization (Forman and Godron, 1986; Urban et al., 1987; Levin, 1992). For example, the types of channel units found in a given environment are constrained by the stream types in which they are nested, which in turn are determined by the valley bottom types and geoclimatic watershed types in which they are nested (Maxwell et al., 1995). In a similar manner, the physical/environmental processes that create these aquatic patterns (e.g., channel scour, major floods, glaciation) are also constrained hierarchically by scaled geoclimatic settings (e.g., lithologic groups, landforms, climate zones) (Maxwell et al., 1995; Jensen et al., 1997). Recognition of these relations greatly simplifies the prediction of finer-scale patterns, such as channel units and stream types, in classification and mapping of aquatic systems (Maxwell et al., 1995).

24.2.2 Hierarchical Framework for Description of Aquatic Systems

Aquatic systems are commonly identified as riverine (streams and rivers), lacustrine (lakes and reservoirs), and groundwater (aquifer) systems (Cowardin et al., 1979). These systems are commonly mapped directly (at appropriate scales) in most ecological assessment efforts (Maxwell et al., 1995). Understanding of these systems, however, is facilitated through description of the terrestrial biophysical environment, including zoogeographic and geoclimatic properties, of the drainage basins within which they are nested. The ECOMAP Working Group of the USDA, Forest Service, has developed a hierarchical framework (Maxwell et al., 1995) useful for describing these aquatic ecosystems. A generalized summary of the ECOMAP approach follows.

Maps of the biophysical environment are used to identify landscapes or aquatic habitats that behave in a similar manner, given their potential ecosystem composition, structure, and function (Bailey et al., 1994). Such maps commonly delineate areas with similar response potential and resource production capabilities and are constructed based on landscape components that display low temporal variability at a given scale of mapping (e.g., regional climate, geology, and landform). In designing biophysical environment maps, differentiating criteria are selected to include those that exert primary control on the ecosystem patterns and processes of interest in an assessment area (e.g., vegetation, flooding, and fire).

Biophysical environment maps are commonly used to describe how the landscape could look or function under historical, current, or potential future ecosystem process regimes (e.g., fire and successional pathway relations), as well as different management scenarios. They provide a semipermanent map theme that can be used to extrapolate ecosystem pattern/process relations from sampled areas to unsampled locales and are useful to stratified sampling design strategies for environmental monitoring purposes (Bailey et al., 1994). Ecological units (McNab and Avers, 1994; Cleland et al., 1997), land units (Zonneveld, 1989), ecoregions (Omernik, 1987), biogeoclimatic ecosystems (Meidinger and Pojar, 1991), and land systems (Christian and Stewart, 1968) are examples of mapping systems that delineate ecologically homogeneous biophysical environments at different spatial scales based primarily on climatic, geomorphic, and biotic criteria. Hierarchical watershed/stream network maps are additional examples of biophysical environment maps that are increasingly being used in aquatic ecosystem assessment efforts (Maxwell et al., 1995).

The biophysical environments that delineate aquatic ecosystems include hydrologic units (drainage basins), geoclimatic settings (terrestrial ecological units), groundwater systems, riverine systems, and lake systems. All these environments are hierarchically organized and are useful for the identification of aquatic ecological units (Figure 24.1).

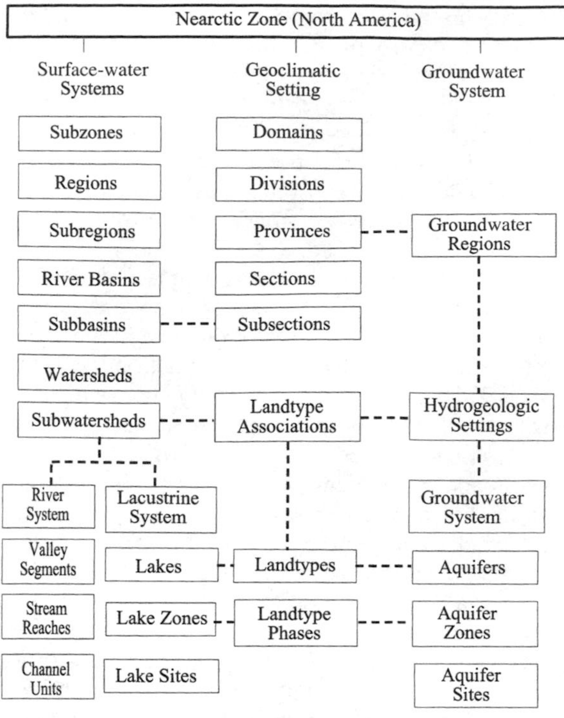

FIGURE 24.1. General framework of an aquatic ecological unit hierarchy (Maxwell et al., 1995). Sizes of units decrease from top to bottom, and primary functional linkages between aquatic systems and terrestrial (geoclimatic) systems are shown as dashed lines.

Hydrologic units are of particular importance because they provide a natural nested hierarchy for stratification of aquatic ecological units over a wide range of scales (Odum, 1971; Lotspeich, 1980; Hornbeck and Swank, 1992). Classification of hydrologic units based on zoogeographic and geoclimatic criteria is especially useful for identifying aquatic ecological units at global to landscape scales (Table 24.1) (Maxwell et al., 1995). Biophysical environments that support aquatic ecosystems are described below.

Geoclimatic Setting

Hydrologic units and aquatic systems with similar geoclimatic properties often have similar ecological patterns and processes (Hack, 1957; Strahler, 1957; Bailey, 1983). Geoclimatic settings (terrestrial ecological units) also influence zoogeographic distributions of aquatic biota and govern arrangements of aquatic habitats. Developed by the USDA, Forest Service, ECOMAP Working Group (Cleland et al., 1997), the National Hierarchical Framework of Ecological Units provides a multifactor (Spies and Barnes, 1985) hierarchical system useful for characterizing and describing both terrestrial and aquatic ecosystems. The ECOMAP framework suggests methods appropriate to the classification and description of land systems based on association of geoclimatic factors that integrate the effects of direct biophysical variables (e.g., energy, moisture, and nutrient gradients) on finer-scale ecosystem patterns (e.g., plant community distribution and stream reaches) and processes (e.g., fire and sedimentation).

The theoretical basis for the ECOMAP framework is that patterns of landforms, soils, potential plant communities, valley segments, and stream reaches are determined by the interaction of climatic forces with the geologic structure of Earth's surface (Bailey et al., 1994). This framework (Table 24.2) recognizes climatic ecoregions as de-

TABLE 24.1. Generalized scales used in ecological assessment efforts and the types of biophysical environments commonly described at each scale.

Assessment scale	Typical polygon size (km^2)	Biophysical environments	
		Terrestrial units	Aquatic units
Global	>1,000,000	Domain	Aquatic region (HUC-1)[a]
Continental	100,000–1,000,000	Division	Aquatic subregion (HUC-2)
Regional	10,000–100,000	Province	River basin (HUC-3)
Subregional	1000–10,000	Section	Subbasin (HUC-4)
	100–1000	Subsection	
Landscape	10–100	Landtype association	Watershed (HUC-5) Subwatershed (HUC-6)
Land unit	1–10	Landtype	Valley bottoms, lake types
	0.1–1	Landtype phase	Stream reach, lake zones
Site	<0.1	Ecological site	Channel unit, lake sites

Source: Jensen et al. 1997.
[a]Approximate relations between USGS hydrologic unit codes (HUCs) and aquatic ecological units are denoted by HUC number. For example, 4th code watersheds (HUC-4) are commonly used to delineate aquatic ecological units in ecological assessments at subregional scales.

TABLE 24.2. Overview of the USDA, Forest Service, national hierarchical framework for terrestrial ecological units.

Mapping level	Typical size (scale)	Primary design criteria	Associated characteristics
Domain	100,000 square miles (1 : 30,000,000)	Climatic zone or group	Repeatable patterns of vegetation classes or subclasses and soil orders or suborders
Division	50,000 square miles (1 : 15,000,000)	Climatic type (Koppen, 1931)	Repeatable patterns of vegetation subclasses or formations and soil suborders or great groups
Province	5000 square miles (1 : 3,000,000)	Hammond's (1954) land surface form, plant climax formation patterns	Repeatable patterns of vegetation formations and soil great groups
Section	1000 square miles (1 : 1,000,000)	Climax plant series patterns following Kuchler (1964)	Repeatable patterns of vegetation series and soil subgroups or great groups
Subsection	100 square miles (1 : 500,000)	Geologic (e.g., lithology structure), physiographic (e.g., glaciated mountain slopes), and statewide climatic zones	Repeatable patterns of vegetation series and soil subgroups
Landtype association	10 square miles (1 : 250,000)	Physiographic and geologic criteria (e.g., fluvial dissected, granitic, mountain breaklands)	Repeatable patterns of plant association groups and soil subgroups
Landtype	1 square mile (1 : 63,000)	Physiographic criteria (e.g., landform, shape, elevation, range, drainage, aspect, dissection characteristics)	Repeatable patterns of plant associations, soil families, and stream types
Landtype phase	0.1 square mile (1 : 24,000)	Topographic criteria (e.g., percent slope, position, aspect, plant association soil family, stream type)	Repeatable patterns of soil series, ecological sites, and fishery habitat components
Site	0.01 square mile (1 : 15,840)	Ecological site, phase of soil family, fishery habitat components (e.g., pools)	

Source: Cleland et al., 1997.

scribed by Bailey (1980, 1983, 1995) at coarser scales. Climatic and geologic properties of the land are emphasized at the province, section, and subsection level of biophysical stratification. At the lower levels of the system hierarchy (from sites to landtype associations), terrestrial ecological units are differentiated primarily by landforms, soils, and potential plant communities. Valley segments, stream reaches, and channel units are also used to delineate riverine systems at scales corresponding to the landtype, landtype phase, and site levels of mapping.

Hydrologic Units

Hydrologic units (Table 24.1) follow drainage divides and delineate multiscale drainage systems (Seaber et al., 1987) containing linked aquatic networks that significantly affect the biotic components of aquatic systems. For example, river basins affect speciation because their boundaries have isolated aquatic populations. These basins contain groups of aquatic populations that have been isolated by geomorphic and biotic coevolution, explaining historic patterns of native species distributions (Moyle and Cech, 1988). Zoogeographic, geoclimatic, and morphometric information can all be efficiently used in the classification of hydrologic units at multiple spatial scales dependent on specific needs for ecosystem characterization (Maxwell et al., 1995).

Riverine System

The riverine system consists of stream networks within subwatersheds. Stream networks are divided into valley segments based on geomorphic and hydroclimatic factors. Classifications of valley segments commonly use valley confinement, slope, and stream-size criteria in describing valley bottom types (Table 24.3). Valley segments are divided into stream reaches based on their geomorphic features. Stream reach classifications commonly use channel and morphometric criteria in the identification of stream types (Table 24.3). Stream reaches are divided into channel units based primarily on hydraulic and substrate features. Channel unit classifications include pool and riffle types, as described by Hawkins et al. (1993).

Lacustrine System

The lacustrine system consists of lakes, ponds, and reservoirs. Whole lake types are classified by their geology, hydrology, and morphometry and are

TABLE 24.3. Criteria commonly used in the classification of aquatic systems.

Aquatic system	Classification level	Classification criteria
Riverine	Valley segments	Confinement, slope, stream size, flow regime, water source
	Stream reaches	Channel pattern, entrenchment, width–depth ratio, gradient, bedform, substrate type
	Channel units	Flow hydraulics, bed roughness, substrate type, morphometry
Lacustrine	Lake types	Morphometry, riverine–groundwater linkage, geomorphology
	Lake zones	Depth classes
	Lake sites	Point features (e.g., springs, deltas)
Groundwater	Groundwater regions	Bedrock lithology, porosity, mineralogy
	Hydrogeologic settings	Landforms, drainage pattern
	Aquifers	Geology, hydrology, water quality
	Aquifer zones	Recharge, discharge zones
	Aquifer sites	Characteristics of springs and sinks

Source: (Maxwell et al., 1995).

commonly described by physical, chemical, and biological features (Table 24.3). Lake zones are based on depth classes. Lake sites represent specific lake habitats based on substrate, flora, and other features.

Groundwater System

Groundwater regions delineate patterns of aquifer systems with similar occurrence and availability of groundwater as determined by bedrock lithology, porosity, and mineralogy (Table 24.3). These regions are divided into hydrogeologic settings that define associations of aquifers whose hydrogeologic factors affect groundwater movement. Aquifer types within these settings are based on their geology, hydrology, and water quality. Aquifer zones distinguish recharge areas from discharge areas, and aquifer sites delineate springs and sinks where the water table intersects the land surface (Table 24.3).

24.3 Review of Classification Systems for Aquatic Biophysical Environments

24.3.1 Stream Classification

There is a long and distinguished history of effort in stream ecology to develop a universal system for classifying streams, stream habitats, or their biotic communities. However, it is widely recognized that none of the approaches proposed to date successfully achieves this goal, and in fact many ecologists have wondered whether a truly universal classification could ever be achieved or would be useful in a practical sense even if it could be. For comprehensive reviews of stream classification, we refer readers to Illes and Botosaneanu (1963), Hawkes (1975), Warren (1979), and particularly Naiman et al. (1992). The following brief summary highlights converging themes that are directly relevant to ecological assessment efforts.

The earliest efforts at stream classification focused on downstream zonation of riverine systems. Huet (1954), Illes (1961), and others in Europe described downstream zonation as evidenced by shifts in the composition of fish or macroinvertebrate assemblages. Within a region, such longitudinal transitions were consistently associated with physical gradients, such as variation in channel slope and thermal regime. The discreteness of such transitions within a particular riverine system was debated (reviewed by Naiman et al., 1992), but, especially in montane regions, transitions were seen to occur at relatively abrupt discontinuities at tributary confluences or lithologic, geomorphic, or bioclimatic junctures. Other critics pointed out that any approach that differentiated classes by the presence of particular indicator taxa or taxonomic profiles would not be directly transferable across biogeographic regions (although ecological analogs could often be discerned).

The River Continuum Concept (Vannote et al., 1980) was an important North American effort to synthesize a theory of longitudinal (downstream) variation in stream ecosystems and biotic communities. The concept provided a useful empirical template for a "typical" downstream sequence of biotic community characteristics and carbon pools, arrayed by stream order from headwater tributaries to large rivers. Numerous less well integrated hypotheses about the ecological mechanisms underlying the pattern has proved less theoretically and conceptually robust. Although numerous exceptions to the predictions of the River Continuum Concept have been observed (see Culp and Davies,

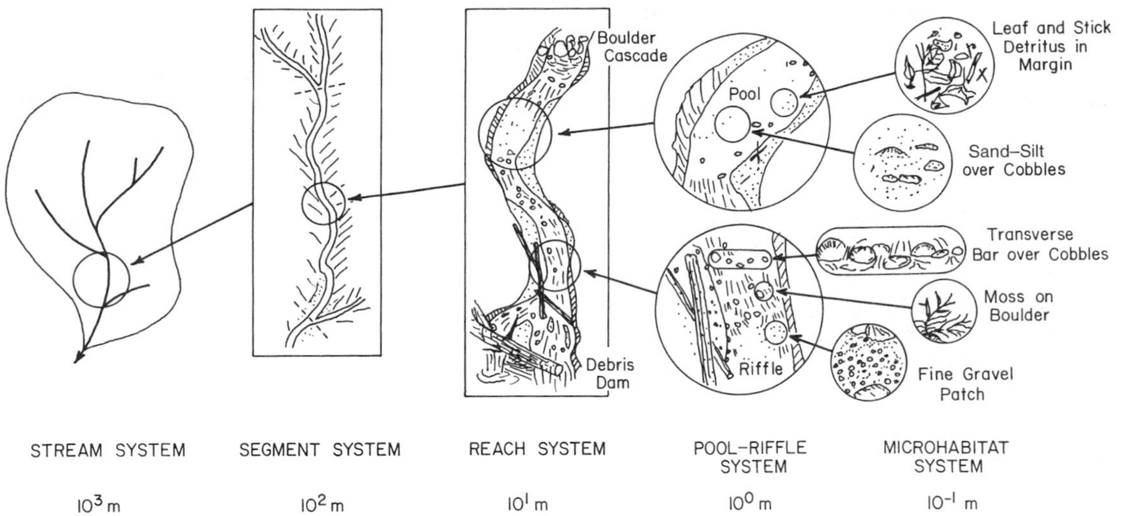

FIGURE 24.2. Spatial and temporal hierarchy of habitat units defined for streams and rivers by Frissell et al. (1986). Smaller-scale habitats are nested both spatially and temporally within larger-scale biophysical systems. Therefore, changes in the larger and more temporally persistent systems should be assumed to change biophysical conditions in the smaller-scale habitats that they subsume.

1982; Minshall, 1988; Naiman et al., 1992), the concept has served usefully as a simplified hypothetical construct against which these exceptions can be instructively compared and understood. The utility of the River Continuum Concept as a tool for direct prediction or management of particular streams or stream types, however, is questionable (Naiman et al., 1992).

Two presently popular classification schemes for stream habitats are categorical, but, as typically applied, are not hierarchical in the scaled sense of Frissell et al. (1986) or Maxwell et al. (1995). Classification of *channel units* (Bisson et al., 1982; Hawkins et al., 1993), essentially low-flow, wetted-channel features at the scale of individual pools and riffles (Figure 24.2), has been widely prescribed and applied for characterization of stream habitat, especially for purposes of fish habitat management. Poole et al. (1997) recently pointed out serious but commonly unrecognized limitations of this method when applied to monitoring and many assessment purposes, although it clearly can have utility as one level of description and stratification within a hierarchical classification system. Rosgen's (1996) scheme for classification of channel types (i.e., at the level of stream reaches) is widely used by federal agencies for inventory and assessment of streams. In practice, Rosgen's classification has serious limitations, including (1) difficulty in unambiguously applying it in regions other than those where the original categories were developed, (2) unresolved ambiguity about the time scale over which categories are presumed to persist and in anticipating what categorical transitions are possible or likely, and (3) resulting frequent disagreement among users about the actual delineation and identification of reaches.

Other scientists have developed classifications of channel types (e.g., Montgomery and Buffington, 1993) or valley types (e.g., Cupp, 1989; Frissell, 1992) that are at least as robust and clearly defined as those of Rosgen and whose geomorphic, ecological, and successional context is better articulated (Naiman et al., 1992). However, these have not yet come into widespread use and, just as with the Rosgen system, concerns such as repeatability and precision when applied by multiple users have yet to receive much systematic scrutiny (Poole et al., 1997). It appears likely that all such classifications are most useful when criteria for differentiation and mapping are tailored to a particular region and when such criteria are applied by highly qualified experts familiar with both methodology and region. Perhaps the most valid use of any of these categorical classifications in ecological assessments is to stratify stream systems into smaller-scale units (Frissell et al., 1986). Within these qualitative strata, quantitative surveys should focus on simple, unambiguous, and accurately repeatable measurements of aspects of habitat structure (Poole et al., 1997) that can be extracted from aerial photographs or made easily in the field. Measurements that can be taken from aerial photographs include sinuosity, channel-width variation, and channel pattern; those readily obtained in the field include maximum, mean, and residual depth of large pools; linear frequency of pools greater than some threshold depth, and density and physical orientation of coarse woody debris of a specified size range.

In recent years, the complex and diverse geomorphic and hydrologic fabric that shapes stream ecosystems across the landscape has been increasingly appreciated by ecologists. Hynes (1975) clearly anticipated and hailed a new focus on the valley or catchment as a direct and indirect determinant of stream biota. The groundbreaking work of Platts (1979) established that fish assemblages sometimes appeared to be more closely tied to the geomorphic class of the surrounding drainage catchment than to measured channel or in-stream habitat features, although in-stream physical conditions themselves are clearly at least partly correlated with landscapes (Nelson et al., 1992).

Stemming in part from this recognition of the potential importance of broader-level ecosystem features in shaping habitats and biotic communities in streams, hierarchical models of stream habitat organization and classification have emerged (Warren, 1979; Frissell et al., 1986; Gregory et al., 1991) (Figure 24.1). These constructs place stream habitats explicitly in the context of their linkages with the surrounding drainage catchment and terrestrial landscape. Although this thinking has clearly spawned important conceptual advances in stream ecology and more recently in environmental assessment (Imhoff et al., 1996; Jensen et al., 1996, 1997), it remains unclear whether the approach will eventually give rise to generic classifications in the traditional sense of the word. Rather, hierarchical approaches offer conceptual tools that allow more systematic analysis, comparison, and ordination of streams and stream habitats; they also provide an explicit way of viewing stream conditions in relationship to the landscape in which they are embedded (Frissell et al., 1986; Naiman et al., 1992; Imhoff et al., 1996; Jensen et al., 1996, 1997).

The aquatic ECOMAP approach of the USDA Forest Service (Maxwell et al., 1995; see also Jensen et al., 1996, 1997, and previous discussion under Section 24.2.2) exemplifies formal application of these hierarchical concepts, listing systematic criteria for defining aquatic mapping units for inventory and ecological assessment. This approach identifies a nested set of land regions of varying scale defining the geoclimatic and zoogeographic setting of watersheds (Figure 24.1) and then decomposes stream networks within watersheds into valley segments, stream reaches, and channel units, successively (similar to Frissell et al., 1986; see Figure 24.2). ECOMAP strives to develop the regional geographical context for local aquatic habitat features, but brings relatively little new clarity to the question of how specific habitat features in streams should be described and assessed in a generic or categorical sense.

In a similar manner, ecoregions have been developed for the United States by the Environmental Protection Agency (Omernik, 1987) for broad-level environmental monitoring and assessment efforts. These mapping units have proved useful in describing fish assemblage distributions across much of the United States; however, their utility is limited to broad-level assessments of large geographic areas (Lyons, 1989). Accordingly, prediction of fish species distribution at finer scales will require more refined classification systems that account for local features not addressed at the regional level (Lyons, 1989).

The Nature Conservancy (1997) has recently offered a prototype classification of aquatic habitats and biotic communities in the form of a dramatically truncated geographical hierarchy. Based on the Conservancy's previous efforts to establish a global plant community classification, this approach begins with extensive ecoregions defined at the province and section scales, stepping down to *macrohabitat types* corresponding with biotic *alliances* (akin to the valley segment level of Frissell et al., 1986, or *sections* of Gregory et al., 1991), and then to *habitat unit types* corresponding with biotic *associations* (equivalent to the riffle and pool or channel unit scale). The intent is that after comprehensive surveys eventually a discrete set of habitat and biotic community categories will be identified at each of these levels. This system skips the level of watersheds or drainage networks altogether; consequently, much information about local biogeographic controls, diversity, and natural and impact history is not preserved. This no doubt severely limits its potential utility for the vast majority of ecological assessments, which require characterization and understanding of dynamic linkages between drainage catchments and aquatic habitats. Whether this scheme will be judged adequate as a tool for the general cataloguing and protection of biological diversity in streams remains to be seen.

24.3.2 Lake Classification

In lakes, as for streams, the confounding influences of geographical and biophysical covariates (hydrologic position, lake size, groundwater influence, substrate type, human accessibility for species introductions and harvest, top-down influences of fishes on trophic structure, and others) complicate understanding of the mechanisms underlying lake biotic organization. For many management purposes, however, it may be sufficient to recognize and accurately map general patterns of covariation that exist across the landscape, using physical and geographical variables as correlative proxies of biotic community type. In this spirit, for Aquatic ECOMAP, Maxwell et al. (1995) propose classifying lakes based on several simple categorical or easily scaled *primary* criteria: geology (lake basin genesis and surrounding landscape physiography), hydrology (riverine linkage, groundwater linkage, and water-stage regime), and lake morphometry (surface area, mean depth, lake orientation and shoreline complexity, and ratio of lake area to catchment area).

In their lake classification scheme, Maxwell et al. (1995) identify a suite of *secondary attributes* that are highly correlated with the primary classification criteria, but also use additional information in some circumstances. These attributes are not so easily quantified and scaled, and most data are difficult or impossible to obtain without direct field observations from each lake. They include lake thermal regime, stratification pattern, hydrologic retention time, water column color and clarity, water chemistry, trophic status, and composition of the aquatic biota. In the Maxwell et al. approach, after lakes are classified into groups or ordinated according to these criteria, an individual lake can be decomposed into depth-based zones (littoral, pelagic, and profundal), and zones can be further subdivided into *lake sites*, including special features such as deltas, bays, inlets, outlets, and spring sources (Table 24.3).

Busch and Sly (1992) present a comprehensive examination of approaches to lake habitat classification. Presumably based on the work of Busch and Sly, The Nature Conservancy (1997) has proposed a simplified lake habitat classification hierarchy directly analogous to their stream habitat types described previously. This scheme includes a *macrohabitat type–alliance* level that applies to whole small lakes or basins within larger lakes and a lower *habitat unit type–association* level defined by smaller-scale features, such as substrate variation and depth or thermal strata. These categories are nested within larger ecoregional classifications. It is somewhat unclear how this approach accounts for other factors known to determine lake biota, such as location within the drainage network, the presence of keystone predators, or variation in lake size or depth spanning several orders of magnitude.

Although many studies have demonstrated the correspondence of lake chemical and physical parameters with the geological and environmental context of catchment basins (e.g., Gibson et al., 1994), relatively few published studies have yet demonstrated coherence of these relationships with lake biota across a region. Local assemblage structure of fishes has been shown to reflect the nonequilibrium, biogeographic differences among lakes in the suite of colonizing species, sifted through the local habitat suitability template or filter that is imposed by within-lake biophysical conditions (Tonn et al., 1990; Kelso and Minns, 1996).

24.3.3 Groundwater and Hyporheic Zone Classification

Aquatic ECOMAP (Maxwell et al., 1995) provides at least a minimal framework for characterizing regional groundwater resources, but none of the classification systems for streams and stream habitats published to date recognizes and incorporates the hyporheic zone, that portion of the valley fill where stream water and groundwater freely intermix (Stanford and Ward, 1992, 1993; Brunke and Gonser, 1997; Ward, 1997, and citations therein). The hyporheic zone itself constitutes a unique and important habitat for a specialized fauna, and it plays a poorly recognized but potentially critical role in mediating the exchange of nutrients and carbon between stream and riparian or floodplain ecosystems. At least in temperate, nondesert environments, upwelling hyporheic waters are thermally buffered compared to waters that have remained at the surface. As a result of such thermal and nutrient fluxes, virtually anywhere hyporheic water returns to the surface it substantially increases aquatic habitat diversity and supports localized hotspots of aquatic productivity.

Hyporheic zones and other near-channel groundwater bodies need urgent attention in future research. However, enough is already known that identification and protection of the ecosystem elements that create and maintain hyporheic exchange should be among the highest priorities of any ecological assessment that encompasses areas of sand- or coarser-grained alluvial or glacial valley fill materials. Within a region, the spatial extent and influences of hyporheic processes are likely to be correlated with valley and channel geomorphic descriptors that are already in common use. Harris (1988) inferred that subsurface hydrology and thus riparian vegetation corresponded with valley morphology in the Sierra Nevada Mountains of California. Recent studies have identified features of valley and channel morphology of montane streams that correlate with lateral flow exchange, upwelling of hyporheic waters, and the presence of summer thermal refugia (Ebersole, 1994) and spawning of bull trout in upwelling-influenced reaches (Baxter et al., in press).

24.3.4 Wetland and Estuarine Classification

Wetlands are important aquatic systems that experience seasonally high groundwater tables, which in turn create anaerobic soil conditions that favor the dominance of hydrophytic plant communities. Groundwater hydraulics, soil properties, and hydrophytic plant community composition are currently used in the identification of jurisdictional wetlands within the United States. However, hierarchical wetland classification systems that integrate these components are currently lacking in most locales. The most inclusive classification of wetlands for the United States is that of the USDI Fish and Wildlife Service; this system's goal is to create "boundaries on natural ecosystems for the purpose of inventory, evaluation and management" (Cowardin et al., 1979). This classification system represents a generalized hierarchical taxonomy that has proved useful in a variety of regional scale assessment efforts; however, mapping of finer-level taxa within this system is often difficult and expensive.

Estuaries occur along the continental shelf and represent important mixing zones of fresh and marine waters. Classification of estuarine systems has relied primarily on the integration of geomorphic, hydrologic, and climatic criteria. The most simplistic estuarine classification is the Venice System, which uses salinity regime as its sole criterion (Anon., 1959). This classification system has long been considered to be overly simplistic, prompting researchers to develop multivariate statistical classifications of salinity regimes in the description of biologically relevant estuarine types (Bulger et al., 1993). Geomorphic classifications of estuaries commonly emphasize the importance of ecological processes, such as climatic inputs and runoff. These factors, in turn, influence biological responses such as migration and spawning. The estuarine classification of New South Wales Australia by Roy (1984) is one useful example of such a geomorphic process-based classification system.

Aquatic classifications of coastal-marine systems have only recently been developed and are based in large part on hierarchical watershed classification methodologies (Lotspeich, 1980; Seaber et al., 1987). For example, Ray and Hayden (1993) adopted hierarchical watershed classification concepts in the development of a watershed–seashed classification that recognized five coastal-zone subdivisions: uplands, coastal plains, tidelands, continental shelf shoreface entrainment zone, and continental shelf offshore entrainment zone. The correspondence between such classification systems and the regional diversity patterns of reef fishes has been reviewed by Robins (1991).

24.4 Use of Terrestrial Ecological Units in Description of Aquatic Ecosystems

Terrestrial ecological units commonly delineate geoclimatic environments that serve two primary purposes in aquatic ecosystem description. They are used (1) to group (classify) hydrologic units (e.g., river basins and watersheds) into similar types based on their structure and function and (2) to predict finer-scale aquatic patterns (e.g., channel units and stream reaches) within hydrologic units. The following is a description of how terrestrial ecological units can be used to characterize aquatic ecosystems.

Hydrologic units can be effectively characterized based on the geoclimatic settings in which they are nested (Jensen et al., 1997). The transport of water, sediment, and solutes is governed by geoclimatic factors such as elevation, relief, slope, landforms, soils, and climate. Because terrestrial ecological units distinguish important hydrogeomorphic properties of a hydrologic unit, they are very useful in grouping hydrologic units into types with similar hydrologic responses. Understanding the relations that exist between land and aquatic systems is key to predicting their response to natural or human-induced disturbances. For example, predicting effects of mass erosion on sediment delivery to streams and sediment routing downstream to critical fish habitat commonly requires knowledge of the nature and interrelationship of the landtype associations, landtypes, stream networks, and valley segments that occur in an area.

Terrestrial ecological unit settings such as subsections and landtype associations (Table 24.2) are useful criteria for grouping hydrologic units at different scales. The mix of specific criteria used to stratify hydrologic units, however, must be governed by local climate, geology, topography, and potential plant cover (Minshall, 1994) and by specific ecosystem assessment needs. The geoclimatic setting that immediately encompasses the hydrologic unit of interest defines its context (e.g., similar geoclimatic settings identify similar watershed types). Geoclimatic patterns such as landtypes (Table 24.2) divide watersheds into meaningful hydrologic response units and help us to understand the functional components of the watershed. Infiltration and evapotranspiration, runoff and erosion, and surface and subsurface flow are examples of processes influenced by landtype-scale ecological units.

Because the ecoregion maps of Bailey (1983, 1995) and Omernik (1987) both utilize geoclimatic settings as defining criteria, there should be some correlation between these ecoregions and aquatic biotic patterns (e.g., fish distributions) at different assessment scales. Hughes et al. (1990) have shown a coarse-scale correlation between ecoregions and fish distributions over widely separated geographic areas of the United States. However, both Lyons (1989) and Poff and Allan (1995) have suggested that an understanding of fish assemblages and distributions within ecoregions is improved considerably by using habitat variables and hydrology, respectively. Moreover, Bayley and Li (1992) have explained that some of the inconsistency of ecoregions in predicting fish distributions can be attributed to the fact that the ecological potential of aquatic ecosystems may be dominated by geoclimatic conditions in the headwaters of their watersheds, and not by the ecoregion in which they occur. These results reinforce the need to match aquatic pattern predictions to appropriately scaled biophysical environment templates in ecological assessment efforts.

24.5 Use of Aquatic Ecological Units in Multiscale Description of Ecosystem Patterns and Processes

Implementation of hierarchy theory in the description of ecological systems is achieved by explicitly characterizing the scaled relations that exist between the patterns of interest and the ecological factors that determine such patterns, that is, the agents of pattern formation (Urban et al., 1987). This type of ecosystem characterization is commonly called *pattern analysis* (Bourgeron and Jensen, 1994) and can be illustrated using fish distributions as the ecological pattern of interest. Fish species may be visualized as exhibiting different patterns of organization (e.g., individuals to metapopulations) that follow different spatial and temporal scales (Figure 24.3a). The formal definition of the hierarchical arrangement of fish distribution patterns is important because it requires explicit statements about (1) the spatial and temporal bounds of each pattern and (2) the order in which these patterns are nested. Such an objective-specific exercise provides the basis for identification of pattern formation agents (Urban et al., 1987; Bourgeron et al., 1994).

The agents of pattern formation can be organized into different hierarchies of biotic processes (Figure 24.3b), disturbance processes (Figure 24.3c), and environmental constraints or biophysical environments (Figure 24.3d). Biotic processes important to an understanding of fish distribution patterns may include behavior or physiologic adjustment at the channel unit or stream reach level, dispersal or genetic exchange at the watershed level, and speciation or extinction at the river basin (or broader scales) in this example. The specific spatial and temporal relations that exist between fish distribution patterns and biotic processes are efficiently described through this type of characterization. In a similar manner, the relation between fish distribution patterns and disturbances and environmental constraints can be described if each agent is specified and its spatial and temporal bounds clearly identified. At the watershed level, for example, population or guild distributions may be viewed as responding to mass wasting or flooding disturbance processes, which in turn are a function of local climate, geology, and landform environmental constraints (Figure 24.3).

The aquatic biophysical variables (or ecological units) described previously are very important to the understanding of scaled relations between aquatic ecosystem patterns and processes. The aquatic processes that create habitat for species (e.g., flooding, sedimentation, and temperature loading) are most efficiently described and mapped in most ecological assessment efforts through an implicit relation to aquatic ecological units. In a similar manner, the potential distribution patterns of aquatic species are commonly described through an understanding of aquatic ecological units and the ecological processes that they constrain. The relations between aquatic ecological units, aquatic system processes, and species distribution patterns are described more fully in Chapters 25 and 26.

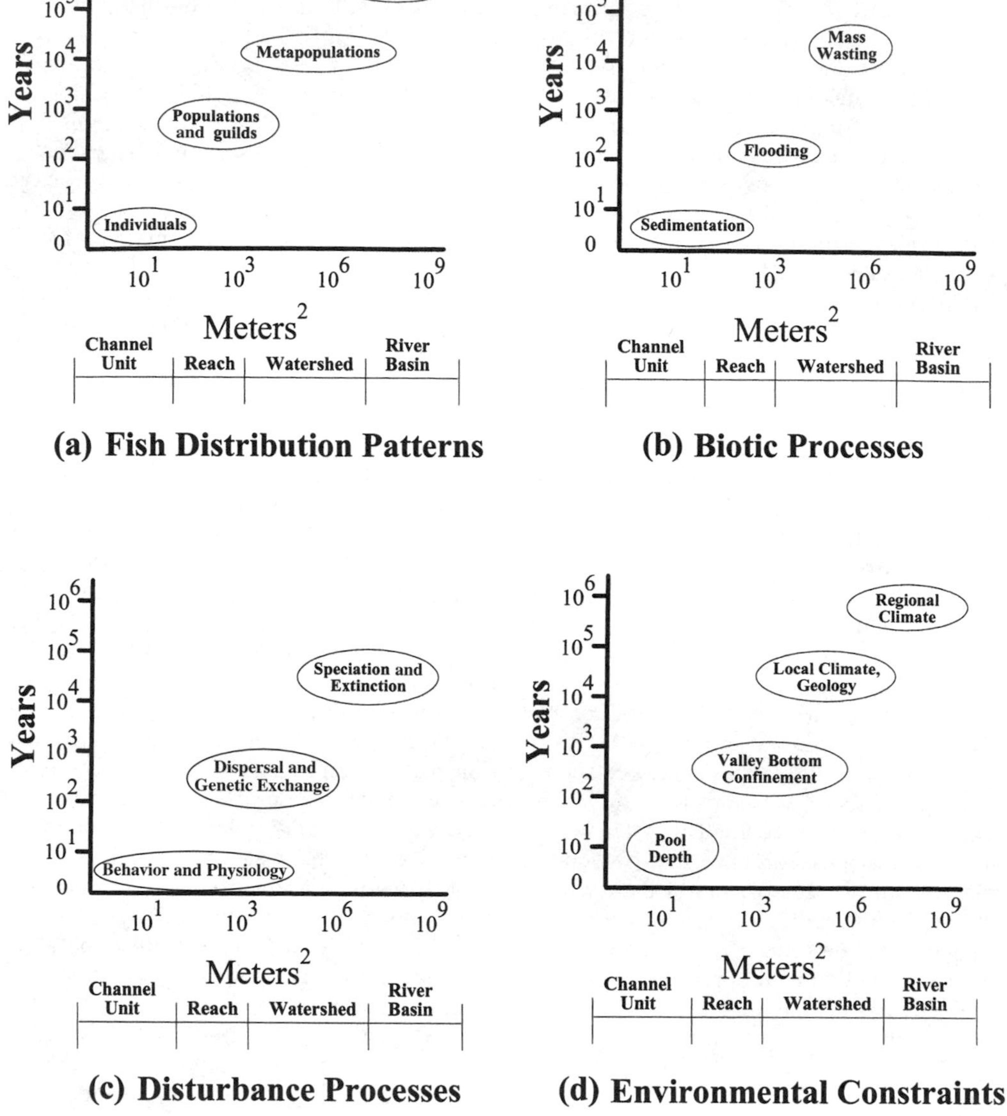

FIGURE 24.3. Agents of pattern formation (Jensen et al., 1996).

24.6 Conclusions

Freshwater aquatic systems (riverine, lacustrine, and groundwater) are nested within hydrologic units (e.g., river basins, watersheds) that may be used to define aquatic ecological units at regional, subregional, and landscape assessment scales. Hierarchical classifications of hydrologic units and aquatic systems are critical components of ecological assessments that are used in land-use planning to implement ecosystem management. This chapter presented an overview of the theory, design, and use of ecological classifications and mapping units in aquatic ecosystem characterization. The hierarchical schemes presented for aquatic system description emphasize the importance of using geoclimatic and zoogeographic criteria in the identification of aquatic biophysical environments at multiple spatial scales.

From the review of aquatic system classifications

presented, it is apparent that there is no one universally accepted method for identifying aquatic biophysical environments. This is not surprising given the fact that ecosystem classifications are human constructs; they only have value if they assist us in understanding complex biophysical environments and aid us in conserving our natural resource base while meeting societal expectations (Jensen et al., 1996). Accordingly, we should expect many schemes for aquatic system classification, given the large variety of issues and spatial scales of assessment such classifications must address. To assist such efforts, the following key points should be considered when designing aquatic ecological classifications or maps for multiscale ecological assessments.

1. The major issues that must be addressed in an ecological assessment determine the appropriate level of classification, data resolution, interpretations, and mapping scale of an ecological unit.
2. Depending on the issues being addressed and resultant interpretation needs, a given assessment may require more than one classification or map of biophysical environments.
3. Definition of an ecological map unit is driven by the finer-scale ecological patterns and processes it is supposed to predict; these patterns and processes must be clearly articulated to facilitate subsequent map unit design (identification of mapping unit differentia) and testing of mapping unit accuracy.

24.7 Recommended Reading

Frissell, C. A.; Liss, W. J.; Warren, C. E.; Hurley, M. D. 1986. A hierarchical framework for stream habitat classification: viewing streams in a watershed context. *Environ. Manage.* 10(2):199–214.

Gregory, S. V.; Swanson, F. J.; McKee, W. A.; Cummins, K. W. 1991. An ecosystem perspective of riparian zones. *BioScience* 41:540–551.

Jensen, M. E.; Bourgeron, P.; Everett, R.; Goodman, I. 1996. Ecosystem management: a landscape ecology perspective. *J. Amer. Water Resources Assoc.* 32: 203–216.

Lotspeich, F. B. 1980. Watersheds as the basic ecosystem: this conceptual framework provides a basis for a natural classification system. *Water Resources Bull.* 16(4):581–586.

Maxwell, J. R.; Edwards, C. J.; Jensen, M. E.; Paustian, S. J.; Parrot, H.; Hill, D. M. 1995. *A hierarchical framework of aquatic ecological units in North America.* GTR-NC-176. St. Paul, MN: U.S. Dept. Agric., For. Serv., North Central For. Exp. Sta.

Naiman, R. J.; Lozarich, D. G.; Beechie, T. J.; Ralph, S. C. 1992. General principles of classification and the assessment of conservation potential. In: Boon, P. J.; Caloa, P.; Petts, G. E., eds. *River conservation and management.* New York: John Wiley & Sons: 93–123.

24.8 References

Allen, T. F. H.; Starr, T. B. 1982. *Hierarchy: perspectives for ecological complexity.* Chicago: University of Chicago Press.

Allen, T. F. H.; Hoekstra, T. W.; O'Neill, R. V. 1984. *Interlevel relations in ecological research and management: some working principles from hierarchy theory.* GTR-RM-110. Fort Collins, CO: U.S. Dept. Agric., For. Serv., Rocky Mountain Res. Sta.

Anon. 1959. *Symposium on the classification of brackish waters.* Venice, April 8–14, 1958. Archivio di Oceanografia e Limnologia, Vol. 11 Supplemento.

Bailey, R. G. 1980. *Descriptions of the ecoregions of the United States.* Misc. Publ. 1391. Washington, DC: U.S. Dept. Agric., For. Serv.

Bailey. R. G. 1983. Delineation of ecosystem regions. *Environ. Manage.* 7(4):365–373.

Bailey, R. G. 1995. *Description of the ecoregions of the United States*, 2nd ed., rev. and expanded (1st ed. 1980). Misc. Publ. No. 1391 (rev.). Washington, DC: U.S. Dept. Agric., For. Serv.

Bailey, R. G.; Jensen, M. E.; Cleland, D. T.; Bourgeron, P. S. 1994. Design and use of ecological mapping units. In: Jensen, M. E.; Bourgeron, P. S., eds. *Ecosystem management: principles and applications.* GTR-PNW-318. Portland, OR: U.S. Dept. Agric., For. Serv., Pacific Northw. Res. Sta.: 101–112.

Baxter, C. V.; Frissell, C. A.; Hauer, F. R. (2001). Geomorphology, logging roads, and the distribution of bull trout (*Salvelinus confluentus*) spawning in a forested river basin: implications for management and conservation. *Trans. Amer. Fisheries Soc.*

Bayley, P. B.; Li, H. W. 1992. Riverine fishes. In: Calow, P.; Petts, G. E., eds. *The rivers handbook.* London: Blackwell Scientific Publications: 251–281.

Bisson, P. A.; Nielsen, J. L.; Palmason, R. A.; Grove, L. E. 1982. A system of naming habitat types in small streams, with examples of habitat utilization by salmonids during low streamflow. In: Armentrout, N. B., ed. *Acquisition and utilization of aquatic habitat inventory information: Proceedings*; October 28–30, 1981, Portland, OR. West. Div. Amer. Fisheries Soc.: 62–73.

Bourgeron, P. S.; Jensen, M. E. 1994. An overview of ecological principles for ecosystem management. In: Jensen, M. E.; Bourgeron, P. S., eds. *Ecosystem management: principles and applications.* GTR-PNW-318. Portland, OR: U.S. Dept. Agric., For. Serv., Pacific Northw. Res. Sta.: 49–60.

Bourgeron, P. S.; Humphries, H. C.; DeVelice, R. L.; Jensen, M. E. 1994. Ecological theory in relation to landscape evaluation and ecosystem characterization.

In: Jensen, M. E.; Bourgeron, P. S., eds. *Ecosystem management: principles and applications*. GTR-PNW-318. Portland, OR: U.S. Dept. Agric., For. Serv., Pacific Northw. Res. Sta.: 58–72.

Brunke, M.; Gonser, T. 1997. The ecological significance of exchange processes between rivers and groundwater. *Freshwater Biol*. 37:1–33.

Bulger, A. J.; Hayden, B. P.; Monaco, M. E.; Melson, D. M.; McCormick-Ray, M. G. 1993. Biologically-based estuarine salinity zones derived from a multivariate analysis. *Estuaries* 16(2):311–322.

Busch, W. D. N.; Sly, P. G. 1992. *The development of an aquatic habitat classification system for lakes*. Boca Raton, FL: CRC Press.

Christian, C. S.; Stewart, G. A. 1968. Methodology of integrated surveys. In: *Aerial surveys and integrated studies*. New York: UNESCO: 233–268.

Cleland, D. T.; Avers, P. E.; McNab, W. H.; Jensen, M. E.; Bailey, R. G.; King, T.; Walter, R. E. 1997. National hierarchical framework of ecological units. In: Boyce, M. S.; Haney, A., eds. *Ecosystem management: applications for sustainable forest and wildlife resources*. New Haven, CT: Yale University Press: 181–200.

Constanza, R.; Wainger, L.; Folke, C.; Maler, K. 1993. Modeling complex ecological economic systems. *BioScience* 43:545–555.

Cowardin, L. M.; Carter, V.; Golet, F. C.; LaRoe, E. T. 1979 (reprinted in 1992). *Classification of wetlands and deepwater habitats of the United States*. FWS/OBS-79/31. Corvallis, OR: U.S. Dept. Interior, Fish and Wildl. Serv.

Culp, J. M.; Davies, R. W. 1982. Analysis of longitudinal zonation and the river continuum concept in the Oldman–South Saskatchewan River system. *Can. J. Fisheries Aquat. Sci*. 39:1258–1265.

Cupp, C. E. 1989. *Stream corridor classification for forested lands of Washington*. Olympia, WA: Washington Forest Protection Assoc.

Driscoll, R. S.; Merkel, D. L.; Radloff, D. L.; Snyder, D. E.; Hagihara, J. S. 1984. *An ecological land classification framework for the United States*. Misc. Publ. 1439. Washington, DC: U.S. Dept. Agric., For. Serv.

Ebersole, J. L. 1994. *Stream habitat classification and restoration in the Blue Mountains of northeast Oregon*. M.S. thesis. Corvallis, OR: Oregon State University.

Forman, R. T. T.; Godron, M. 1986. *Landscape ecology*. New York: John Wiley & Sons.

Frissell, C. A. 1992. *Cumulative effects of land use on salmon habitat in southwest Oregon coastal streams*. Doctoral dissertation. Corvallis, OR: Oregon State University.

Frissell, C. A.; Liss, W. J.; Warren, C. E.; Hurley, M. D. 1986. A hierarchical framework for stream habitat classification: viewing streams in a watershed context. *Environ. Manage*. 10(2):199–214.

Gibson, C. E.; Wu, Y.; Smith, S. J.; Wolfe-Murphy, S. A. 1994. Synoptic limnology of a diverse geological region: catchment and water chemistry. *Hydrobiology* 306:213–227.

Gregory, S. V.; Swanson, F. J.; McKee, W. A.; Cummins, K. W. 1991. An ecosystem perspective of riparian zones. *BioScience* 41:540–551.

Hack, J. T. 1957. *Studies of longitudinal stream profiles in Virginia and Maryland*. Prof. Pap. 294. Corvallis, OR: U.S. Dept. Interior, Geological Survey. B:45–97.

Hammond 1954. Small-scale continental landform maps. *Ann. Assoc. Am. Geogr*. 44:33–42.

Harris, R. R. 1988. Associations between stream valley geomorphology and riparian vegetation as a basis for landscape analysis in eastern Sierra Nevada, California, USA. *Environ. Manage*. 12:219–228.

Hawkes, H. A. 1975. River zonation and classification. In: Whitton, B. A., ed. *River ecology*. London: Blackwell Scientific: 312–374.

Hawkins, C. P.; Kershner, J. L.; Bisson, P. A.; Mason, D. B.; Decker, L. M.; Gregory, S. V.; McCullough, D. A.; Overton, C. K.; Reeves, G. H.; Steedman, R. J.; Young, M. K. 1993. A hierarchical approach to classifying stream habitat features. *Fisheries* 18(6): 181–188.

Hornbeck, J. W.; Swank, W. T. 1992. Watershed ecosystem analysis as a basis for multiple-use management of eastern forests. *Ecol. Appl*. 2(3):238–247.

Huet, M. 1954. Biologie, profil en long et en travers des eaux courantes. *Bull. Francais Pisciculture* 175:41–53.

Hughes, R. M.; Whittier, T. R.; Rohm, C. M.; Larsen, D. P. 1990. A regional framework for establishing recovery criteria. *Environ. Manage*. 14:673–683.

Hynes, H. B. N. 1975. The stream and its valley. *Intern. Vereinigung Theoreoeetische Angewandte Limnologie* 19:1–15.

Illes, J. 1961. Versuch einer allgemein biozonotischen Gleiderung der Fliessgewasser Verhandlungen. *Intern. Vereinigung Theroetische Angewandte Limnologie* 13:834–844.

Illes, J.; Botosaneanu, L. 1963. Problems et methodes de la classification et de la zonation ecologique des eaux courantes, considerees surtout du point de vue faunistique. Milleilungen. *Intern. Vereinigung Theoretische Angwandte Limnologie* 12:1–57.

Imhoff, J. G.; Fitzgibbon, J.; Annable, W. K. 1996. A hierarchical evaluation system for characterizing watershed ecosystems for fish and habitat. *Can. J. Fisheries Aquat. Sci*. 53(Suppl. 1):312–326.

Jensen, M.; McNicoll, C.; Prather, M. 1991. Application of ecological classification to environmental effects analysis. *J. Environ. Qual*. 20:24–30.

Jensen, M. E.; Bourgeron, P.; Everett, R.; Goodman, I. 1996. Ecosystem management: a landscape ecology perspective. *J. Amer. Water Resour. Assoc*. 32:203–216.

Jensen, M. E.; Goodman, I.; Brewer, K.; Frost, T.; Ford, G.; Nesser, J. 1997. Chapter 2: biophysical environments of the basin. In: Quigley, T. M.; Arbelbide, S. J., tech. eds. *An assessment of ecosystem components in the interior Columbia Basin and portions of the Klamath and Great Basins: Volume 1*. GTR-PNW-405. Portland, OR: U.S. Dept. Agric., For. Serv., Pacific Northw. Res. Sta.: 99–335.

Kelso, J. R. M.; Minns, C. K. 1996. Is fish species richness at sites in the Canadian Great Lakes the result of local or regional factors? *Can. J. Fisheries Aquat. Sci.* 53(Supplement 1):175–193.

Koestler, A. 1967. *The ghost in the machine*. New York: Macmillan.

Köppen, W. 1931. *Grundriss der klimakunde*. Berlin: Walter de Gruyter: 388.

Kuchler, A.W. 1964. Potential natural vegetation of the conterminous United States. *American Geographic Society Special Publication* 36:116.

Levin, S. A. 1992. The problem of pattern and scale in ecology. *Ecology* 73:1942–1968.

Lotspeich, F. B. 1980. Watersheds as the basic ecosystem: this conceptual framework provides a basis for a natural classification system. *Water Resour. Bull.* 16(4): 581–586.

Lyons, J. 1989. Correspondence between the distribution of fish assemblages in Wisconsin streams and Omernik's ecoregions. *Amer. Midland Naturalist* 122: 163–182.

Maxwell, J. R.; Edwards, C. J.; Jensen, M. E.; Paustian, S. J.; Parrott, H.; Hill, D. M. 1995. *A hierarchical framework of aquatic ecological units in North America (Nearctic Zone)*. GTR-NC-176. St. Paul, MN: U.S. Dept. Agric., For. Serv., North Central For. Exp. Sta.

McNab, W. H.; Avers, P. E., compilers. 1994. *Ecological subregions of the United States: section descriptions*. Admin. Publ. WO-WSA-5. Washington, DC: U.S. Dept. Agric., For. Serv.

Meidinger, D.; Pojar, J. 1991. *Ecosystems of British Columbia*. Special Report. Series 6. Victoria, BC: Res. Branch, Ministry For.

Minshall, G. W. 1988. Stream ecosystem theory: a global perspective. *J. N. Amer. Benthol. Soc.* 7:263–288.

Minshall, G. W. 1994. Stream–riparian ecosystems: rationale and methods for basin-level assessments of management effects. In: Jensen, M. E.; Bourgeron, P. S., eds. *Ecosystem management: principles and applications*. GTR-PNW-318. Portland, OR: U.S. Dept. Agric., For. Serv., Pacific Northw. Res. Sta.: 149–173.

Montgomery, D. R.; Buffington, J. M. 1993. *Channel classification, prediction of channel response, and assessment of channel condition*. Report FW-SH10-93-002. Seattle, WA: University of Washington, Dept. Geological Sci. and Quaternary Res. Center.

Moyle, P. B.; Cech, J. J., Jr. 1988. *Fishes: an introduction to icthyology*. Upper Saddle River, NJ: Prentice Hall.

Naiman, R. J.; Lonzarich, D. G.; Beechie, T. J.; Ralph, S. C. 1992. General principles of classification and the assessment of conservation potential in rivers. In: Boon, P. J.; Calow, P.; Petts, G. E., eds. *River conservation and management*. New York: John Wiley & Sons: 93–123.

Nelson, R. L.; Platts, W. S.; Larsen, D. P.; Jensen, S. E. 1992. Trout distribution and habitat in relation to geology in the North Fork Humboldt River drainage, northeastern Nevada. *Trans. Amer. Fisheries Soc.* 121: 405–426.

Odum, E. P. 1971. *Fundamentals of ecology*. Philadelphia: W.B. Saunders.

Omernik, J. M. 1987. Ecoregions of the conterminous United States. *Ann. Assoc. Amer. Geogr.* 77:118–125.

O'Neill, R. V.; DeAngelis, D. L.; Waide, J. B.; Allen, T. F. H. 1986. A hierarchical concept of ecosystems. *Monogr. Pop. Biol.* 23:1–272.

Platts, W. S. 1979. *Including the fishery system in land planning*. GTR-INT-60. Ogden, UT: U.S. Dept. Agric., For. Serv., Intermountain Res. Sta.

Poff, N. L.; Allan, J. D. 1995. Functional organization of stream fish assemblages in relation to hydrological variability. *Ecology* 76:606–627.

Poole, G. C.; Frissell, C. A.; Ralph, S. C. 1997. In-stream habitat unit classification: inadequacies for monitoring and some consequences for management. *J. Amer. Water Resour. Assoc.* 33:879–896.

Ray, G. C.; Hayden, B. P. 1993. Marine biogeographic provinces of the Bering, Chukchi, and Beaufort Seas. In: Sherman, K.; Alexander, L.; Gold, B. D., eds. *Large marine ecosystems: stress, mitigation, and sustainability*. Washington, DC: AAAS Press: 175–184.

Robins, C. R. 1991. Regional diversity among Caribbean reef fishes. *BioScience* 41(7):458–459.

Rosgen, D. L. 1994. A classification of natural rivers. *Catena* 22:169–199.

Rosgen, D. R. 1996. *Applied river morphology*. Pagosa Springs, CO: Wildland Hydrology.

Roy, P. S. 1984. New South Wales estuaries: their origin and evolution. In: Thom, B. G., ed. *Coastal geomorphology in Australia*. Sydney: Academic Press: 99–121.

Seaber, P. R.; Kapinos, F. P.; Knapp, G. L. 1987. *Hydrologic unit maps*. Water-Supply Paper 2294. Corvallis, OR: U.S. Geological Survey.

Spies, T. A.; Barnes, B. V. 1985. A multifactor ecological classification of the northern hardwood and conifer ecosystems of Sylvania Recreation Area, Upper Peninsula, Michigan. *Can. J. For. Res.* 15:949–960.

Stanford, J. A.; Ward, J. V. 1992. Management of aquatic resources in large catchments: recognizing interactions between ecosystem connectivity and environmental disturbance. In: Naiman, R. J., ed. *Watershed management*. New York: Springer-Verlag.

Stanford, J. A.; Ward, J. V. 1993. An ecosystem perspective of alluvial rivers: connectivity and the hyporheic corridor. *J. N. Amer. Benthol. Soc.* 12:48–60.

Strahler, A. N. 1957. Quantitative analysis of watershed geomorphology. *Trans. Amer. Geophys. Union* 33: 913–920.

The Nature Conservancy. 1997. *A classification framework for freshwater communities. Proceedings of The Nature Conservancy's aquatic community classification workshop*; New Haven, MO; April 9–11, 1996. Chicago: Great Lakes Program Office.

Tonn, W. M.; Magnuson, J. J.; Rask, M.; Toivonen, J. 1990. Intercontinental comparison of small-lake fish assemblages: the balance between local and regional processes. *Amer. Naturalist* 136:345–375.

Urban, D. L.; O'Neill, R. V.; Shugart, H. N., Jr. 1987. Landscape ecology: a hierarchical perspective can help scientists understand spatial patterns. *BioScience* 37:119–127.

USDA (U.S. Department of Agriculture, Forest Service). 1978. *Valley bottomland inventories with management implications: a working draft*. Ogden, UT: U.S. Dept. of Agric., For. Serv., Intermountain Res. Sta.

USDA-SCS (U.S. Department of Agriculture, Soil Conservation Service). 1975. *Soil taxomony: a basic system of soil classification for making and interpreting soil surveys*. Agric. Handb. No. 436. Washington, DC: U.S. Dept. of Agric., For. Serv.

Vannote, R. L.; Minshall, G. W.; Cummins, J. R.; Sedell, J. R.; Cushing, C. E. 1980. The river continuum concept. *Can. J. Fisheries Aquat. Sci.* 37:130–137.

Ward, J. V. 1997. An expansive perspective of riverine landscapes: pattern and process across scales. *Gaia* 6(1):52–60.

Warren, C. E. 1979. *Toward classification and rationale for watershed management and stream protection*. Report No. EPA-600/3-79-059. Corvallis, OR: U.S. Environ. Protect. Agency.

Zonneveld, I. S. 1989. The land unit—a fundamental concept in landscape ecology, and its application. *Landscape Ecol.* 3:67–86.

25 Geomorphic Patterns, Processes, and Perspectives in Aquatic Assessment

Michael Hurley and Mark E. Jensen

25.1 Introduction

The variability that characterizes fluvial–aquatic systems has confounded both interpretation and development of conceptual approaches for these dynamic systems. Because aquatic systems are fluid, they are much more responsive to change and thus more dynamic than terrestrial systems. This inherent dynamism is compounded by the spatial variability of geoclimatic settings and stage of recovery from natural and cultural disturbances (Graf, 1982). However, it is this same dynamic quality or variability, often in the context of disturbance, that is becoming the key to understanding aquatic systems and the processes operating on them (Poff and Ward, 1989; Reeves et al., 1995). We can best order this variability in hierarchically structured classification frameworks that link aquatic patterns and processes with the controlling and driving variables of different geoclimatic settings (Frissell et al., 1986; Maxwell et al., 1995).

The increasing focus on more holistic watershed and ecosystem assessments illustrates the need to integrate disciplines and synthesize concepts. For example, we need to think of physical processes in the context of how they function in the ecosystem. In this chapter we present physical processes that create fluvial landscape heterogeneity and influence aquatic ecosystem composition and dynamics, with a primary focus on stream systems. To begin, we offer some key historical and contemporary conceptual views of processes and process-based investigations. The process categories that we address include geologic, climatic, hydrologic, erosional, and sediment transport. For each of these categories, we discuss the importance of processes in ecosystem function, various analytical procedures, and biophysical factors that influence aquatic processes and pattern. Finally, we consider the need for incorporating spatial and temporal variability in ecosystem assessments.

25.2 Geomorphic Perspectives

How we view processes determines the usefulness as well as limitations of a process-based approach to landscape analyses and ecosystem assessment. In this respect, a brief history of key geomorphic perspectives provides context for our application of current approaches. The current emphasis in geomorphology and landscape analysis on surface processes, disturbance, and life-history characteristics parallels the individualistic view of biotic communities. However, during the early 20th century, geomorphic and ecological investigations focused on historical interpretation, description, and classification. During this period, the geographic cycle (Davis, 1899) and the Clementsian organismic concept of community development (Clements, 1936) were closed-system models of progressive landscape and ecosystem change. In these views, a landscape progresses through continuous erosional stages to a climax, erosional steady state of the peneplain; likewise, species assemblages progress through successional states to a self-perpetuating, steady-state climax community. Despite deficiencies, these views were popular and persist in some form today, because they simplify ecosystem–landscape change as linear progression.

Later in the century, geologists stressed a more quantitative, dynamic approach to counter the qualitative cyclic perspective (Strahler, 1957). In an attempt to finally displace the cyclic theory, two landmark papers emphasize a process-based open-systems approach (Chorley, 1962) and resurrection and refinement of the dynamic equilibrium concept (Hack, 1960). In an attempt to synthesize the seemingly divergent concepts of cyclic and dynamic theories, Schumm (1974) proposes different temporal scales of equilibrium adjustment. This hierarchical structure of stream dynamics recognizes the scale dependency of variables in cause–effect relations (Schumm and Lichty, 1965).

Dynamic equilibrium describes the mutual adjustment of hillslope and fluvial form in the process–response adjustment of the stream to erosional processes, hydrodynamics, and sediment delivery and transport. Equilibrium and stability (grade or steady state) are attractive concepts, because they explain variability as a result of channel self-adjustment to complex interaction and process–response behavior of stream and watershed energetics at different scales (Mackin, 1948; Schumm, 1974), although differences exist on how to view equilibrium (Ahnert, 1994; Lane and Richards, 1997). To explain the complexity of geomorphic change, Chorley (1962) considers a system as always trying to track or attain a steady-state equilibrium condition due to changing conditions, but that it seldom attains it and is best viewed as disturbance or change at various spatial scales in a hierarchy (Frissell et al., 1986) in that "those parts of the system whose time scale is shorter than the disturbance react to produce a new steady state" (Hack and Goodlett, 1960).

As with most theoretical perspectives, there are limits to applying the equilibrium–steady-state concept, which has led to use of alternative approaches. The variable nature of channel response to climatic, hydrologic, and erosional processes, especially as expressed in mountainous and arid climates of the western United States, is often considered stochastic in nature and more amenable to probabilistic approaches in landscape analysis (Megahan, 1996; Benda and Dunne, 1997). This is reflected in Graf's (1988) concept of the "probabilistic river." Some investigators use the concept of critical power (stream power related to resisting power) in place of equilibrium. Critical power can be used to explain threshold stream response to varying sediment conditions at various temporal and spatial scales (Bull, 1979), as well as for stream response to climate change (Bull, 1991). Thresholds represent an important quantitative and qualitative concept for watershed management and assessment (Megahan, 1996) and for understanding geomorphic change (Bull, 1979) and the relative impact of processes on the landscape. Thresholds are characterized by sudden interruptions of progressive landscape change where process–response mechanisms are triggered that lead to significant landscape alteration (Schumm, 1974).

The concept of magnitude–frequency of processes is essential for understanding process–landform relations. Climatic and geomorphic processes differ in frequency of occurrence, magnitude (size), and ecosystem function, such as for habitat formation and maintenance. The relative magnitude of events, their relative effectiveness or impact, and cause–effect relations depend on the temporal and spatial scale at which a system is viewed and the general biophysical setting. The effectiveness of processes at altering the landscape and the relative rate of recovery following disturbance depend on the climatic variability, sensitivity of the geologic setting, and history of past events (Wolman and Gerson, 1978; Kochel, 1988). In general, the greater the climatic and hydrologic variability is, the larger the relative influence that large-magnitude events such as floods have on amount of sediment transported, channel morphology, and recovery time (Wolman and Gerson, 1978). For example, 100-year floods have a greater and longer lasting impact on channel conditions in arid regions than in humid temperate regions of the northeast United States (Graf, 1988). Magnitude–frequency relations of channel-forming processes also change with valley setting or basin size (Kochel, 1988; Grant and Swanson, 1995).

There also are limitations with the quantitative process-based approach that dominates contemporary geomorphology and landscape ecology. Knowledge of hillslope and fluvial processes is necessary to understand system behavior, cause–effect relations, and the impacts of management actions and application of recovery prescriptions. However, we need to consider processes and extrapolation of models in the spatial context of environmental controlling factors and driving variables, which is uncommon. Also, the emphasis on prediction modeling and quantifying processes has overshadowed the benefits of perspectives that emphasize interpretation and explanation (Baker, 1988). Climatic geomorphology (Derbyshire, 1976) and geologic evolution (Tricart, 1974) are examples of underutilized perspectives that provide an appropriate scale of inquiry for coarse-scale regional assessments.

Another concern is whether the emphasis on surface processes can adequately explain landscape evolution (Douglas, 1982; Sugden et al., 1997), which is a similar concern with landscape ecology in linking disturbance processes and evolution. A more evolutionary-based perspective is partially addressed by the habitat template concept, which can link evolutionary attributes of organisms, such as life-history adaptation, to processes and the dynamic nature of aquatic systems (Southwood, 1977; Poff and Ward, 1990). In this respect, persistence, stability, and predictability are important concepts that can link life-history and habitat attributes with processes of hydrodynamics and sediment transport. A major contribution in our understanding of coarse-scale ecosystem dynamics is how geomorphic perspectives of landscape evolution and behavior lend an essential time perspective to landscape patterns and processes (Swanson, 1979; Thornes, 1983). Although often overlooked in assessments, behavior is considered an important dynamic property of ecosystems (Allen and Starr, 1982). In aquatic systems, behavior is the dynamic result of the collective, interactive operation of climatic, hydrological, and erosional processes.

25.3 Geologic Processes

Analysis of aquatic landscape patterns must focus on more than surface processes and vegetation dynamics and is best considered in context of landscape origin and development. Large-scale climatic and geologic processes form the landscape patterns on which contemporary processes operate. To better understand landscape variability and the controlling factors on aquatic ecosystem composition and dynamics, we must recognize and formalize these geologic patterns as geomorphic or geoclimatic templates. Geologic processes, primarily tectonics and vulcanism, act as endogenous energy inputs, which are reflected in patterns of channel networks and landform relief (Tricart, 1974). In turn, landform relief influences potential energy of aquatic systems as expressed in watershed morphology and stream profile relief (Morisawa, 1985). Regional landform developmental processes and resultant properties impart a particular character to regions (Coates, 1974; Hack, 1982). Such processes and features include tectonic uplift and faulting, isostatic release and pluton development, lithology, stratigraphy and structural deformation, and Pleistocene glacial processes. Our knowledge of these processes provides predictive and explanatory power of the spatial distribution and compositional properties of watershed, stream, and biotic patterns. For example, Pleistocene glacial processes form different landscapes with variable hydrologic and habitat conditions, which explains corresponding distribution of salmonids (Benda et al., 1992) and bivalves (mussels) (Strayer, 1983). We also need to consider coarse-scale fish distribution in the evolutionary context of geologic processes such as tectonics and watershed evolution (Minckley et al., 1986).

We cannot adequately interpret aquatic patterns without considering that watersheds have evolved with and adjusted to different geologic processes and erosional regimes. In this respect, both basin and channel patterns reflect climate and landform evolution since the Quaternary period (Petts and Foster, 1985). Differences in bedrock and surficial lithology, in the context of basin and channel evolution, help frame explanation of within-region differences in watershed properties (Cunha et al., 1975; Kowall, 1976; Miller and Ritter, 1990) and contemporary variation in channel characteristics (Hack, 1965; Benda et al., 1992). For example, processes such as tectonic uplift and landform tilting influence spatial variation in valley entrenchment and stream profile characteristics in mountainous areas (Merritts and Vincent, 1989; Rhea, 1993). In lowland alluvial rivers, areas of slow tectonic activity affect channel patterns, channel direction, and sediment transport (Ouichi, 1985; Marple and Talwani, 1993). Because bedrock composition and structural arrangement are static representations of geologic processes, they are fundamental in the analysis and interpretation of basin and channel patterns (Morisawa, 1985). Because the longitudinal stream profile reflects historical geologic processes and properties of watersheds (Hack, 1957; Wheeler, 1979), we can use the profile for preliminary assessment and interpretation of basin and stream characteristics and even for stream segment stratification (Hack, 1973; Shepherd, 1985; Bisson and Montgomery, 1996). Also, patterns of basin morphology related to geologic properties and processes can be identified by multivariate analysis, an underutilized tool in aquatic assessments (Mather and Doornkap, 1970; Miller and Ritter, 1990).

25.4 Hydrological Processes

25.4.1 Climate and Hydrology

Climate controls exogenous energy inputs to ecosystems as precipitation and radiation and thus is the driving force behind geomorphic and hydrologic processes and aquatic ecosystem temporal variability. This section focuses on how climatic processes regulate the dynamic variability of regional hydrology. Climatic factors control the character of the hydrologic regime through the timing, duration, and intensity of rainfall and by how the seasonal temperature regime affects types of precipitation and water storage and release (snow, permafrost, frozen soils, ice cover, glaciers). Climate also strongly influences regional aquifer characteristics, such as whether stream relations with groundwater are either influent or effluent.

Climatic classifications can be either genetic, which reflect the actual climatic mechanisms or processes, or parametric (Thornthwaite, 1948), using metric proxies for climate potential. The climatic metrics used in parametric climatic classification, such as evapotranspiration, average annual precipitation, precipitation effectiveness, and intensity, influence biotic distribution and erosion as well as hydrological variability. These metrics can be derived from various map and information sources, such as regional atlases of the United States (e.g., Miller et al., 1973).

To effectively interpret hydrologic variability, we need to recognize the actual climatic processes or mechanisms that affect hydrology. Different types of atmospheric processes have different temporal–spatial scales of influence and can be scaled by size, duration, and aerial coverage of precipitation (Orlanski, 1975; Hirschboeck, 1988). Hayden (1988) presents a global atmospheric classification for flood potential based on primary factors of temperature and water stored in the atmosphere and secondarily as water stored in snow and ice. Other climatic–hydrologic classifications include seasonal flood-climate regions for the United States (Hirschboeck, 1991) and a flash-flood hazard map for the United States (Beard, 1975). Climatic factors that influence flash floods are also classified for the western United States (Maddox et al., 1980).

The seasonal sequencing of these various climatic processes determines the hydrologic variation within and between regions (Hirschboeck, 1987). For instance, the variability of peak flows in flood frequency curves between mountainous regions of the western United States reflects various climatic mechanisms, such as snow, thunderstorms and frontal systems (Pitlick, 1994). However, Hirschboeck (1988) warns that stochastic–probabilistic hydrological analyses may mask the actual climatic factors that cause variation and thus inhibit adequate hydrologic explanation and prediction. Thus hydrologic analysis needs to recognize that populations of peak flows of similar magnitude–frequency can result from different types of climatic mechanisms and their seasonal sequencing (Black, 1989; Hirschboeck, 1991). We also need to consider different spatial scales in hydrological analysis, because the hydrologic response to an atmospheric event is relative to the location within and size of the watershed (Black, 1989), as well as the size of the event.

25.4.2 Hydrological Characterization

Hydrologic processes perform a variety of ecosystem functions, such as formation of streambed and floodplain fluvial features. Dynamic hydrologic processes strongly influence riparian vegetation composition and distribution (Gregory, 1991; Gurnell, 1995). Riparian composition is in part controlled by the frequency of inundation of different fluvial surfaces across a valley (Hupp, 1990). In turn, variability in valley characteristics constrains hydrologic processes and consequent riparian composition (Hupp, 1990). Hydrologic dynamic character also affects distribution of aquatic organisms relative to their life history characteristics (Poff and Ward, 1989).

Most standard hydrological analyses are oriented toward engineering or economic ends. However, more ecologically oriented hydrologic metrics that are being developed need to be considered for ecological assessments (e.g., Richter et al., 1997). Stream hydrology can be characterized by a variety of methods that are well summarized in texts, especially those by Leopold (1994), Gustard (1992), and Rogers and Armbruster (1990). Two basic metrics in watershed hydrology are the unit hydrograph and hyetograph, which can be combined to compare streamflow relations to a single precipitation event. These metrics compare the relationship between the time lag of a flow event to a precipitation event relative to the magnitude of the flow event, which is important in the interpretation of basin conditions. Such conditions include antecedent soil moisture, the influence of basin characteristics on water delivery processes, and alteration of the watershed by various natural and land management processes (Leopold, 1991).

The flow record is a major tool in hydrological interpretation and analysis, primarily as flood frequency or flow duration curves. However, the duration of the flow record severely limits interpretation, because the addition of very high or very low flows changes the character of flood frequency and flow duration curves to varying degrees, depending on the length of the flow record. For intensive or long-term assessments, methods such as paleohydrological analyses can be used to extend the record of large peak flows (Jarret, 1991). Flood frequency curves depict the relative magnitude and frequency of peak flow events (the highest stage of a flow event) and can be used to interpret hydrologic variability. The mean and variance of peak flows are linked to the nature of climatic mechanisms (Hirschboeck, 1988); furthermore, the slope of flood frequency curves reflects the variability of flood-producing climactic mechanisms (Pitlick, 1994). Flow duration curves represent cumulative frequencies of all flows and are viewed as the percentage of time that certain flow events are exceeded. The shape, slope, and log plot of the curve are related to both the variability of flows and flow-generating processes. The flow duration curve is particularly useful for interpreting low-flow characteristics and for assessing the impact of factors such as groundwater, water diversions, and reservoirs. Base flows are also characterized from flow recession curves, low-flow frequency curves and from a Base Flow Index (Rogers and Armbruster, 1990; Gustard, 1992). Base flows in turn have been used to determine groundwater aquifer contribution to catchment hydrology (Johnston, 1971), as exemplified by a proposed geologic low-flow index (Brown, 1981). A simple metric of flow variability is the ratio of 100-year flood (or base flow) to the mean annual discharge. This ratio also scales flow events to relative geomorphic effectiveness (Wolman and Gerson, 1978). The annual flow regime reflects the influence of climatic processes on seasonal flow development and is useful for understanding seasonal behavior, distribution, and life histories of aquatic organisms. Flow regime types show particular promise for ecological assessments, because they reflect flow variability from factors such as snow melt, rain on snow, precipitation, and groundwater. Such flow regime types have been developed for the United States (Poff, 1996) and the Columbia River Basin (Quigley and Arbelbide, 1997).

Hydrological regions have been established at various scales, often oriented toward management interests such as culvert design and flood hazard. Geographic extrapolation of flood frequency relations (summarized by Jennings et al., 1994) is one of many methods used to establish hydrologic regions. Other classifications that are more oriented for ecosystem assessments include a hierarchical, hydrogeophysical watershed classification of the Columbia River Basin (Quigley and Arbelbide, 1997). Also of use in assessments are hydrogeomorphic regions of the United States that relate physiographic and climatic influences on groundwater properties (Heath, 1984; Coates, 1990a). Finer-scale physiographic subregions refine groundwater relations with geologic processes and bedrock and surface geology (Coates, 1974).

25.4.3 Biophysical Factors and the Hydrological Cycle

The hydrological cycle is an elementary but essential illumination of the connection between climate, hillslopes, and streams (Figure 25.1). A stream, lake, or wetland should be viewed as the surface expression of the hydrology of a watershed. Water in the channel is linked to the watershed by processes such as rainfall interception, overland and subsurface flow, floodplain inundation and water storage, groundwater and base flow, and subsurface channel hyporheic flow. Black (1997) identifies the three watershed hydrologic functions of water collection, storage, and discharge, which integrate the major watershed hydrological processes of precipitation, interception, infiltration, evapo-

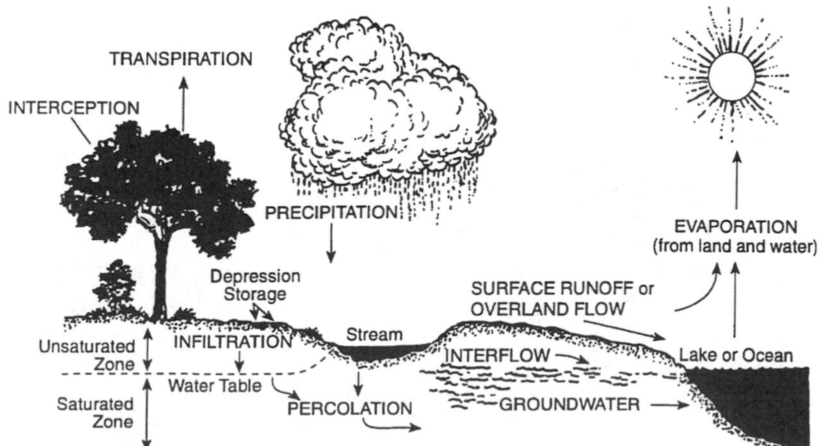

FIGURE 25.1. The hydrological cycle.

transpiration, percolation, and runoff. In assessments we need to understand how various biophysical factors and land use affect the processes and functions in the hydrologic cycle, all within a geographic spatial context.

Groundwater, an often ignored factor in aquatic assessments, deserves special attention because it is linked to surface-water conditions and has important geomorphic and biological functions. The increasing recognition of groundwater influence on the stream corridor, and particularly on aquatic ecology through the hyporheic zone, emphasizes the need to include the hydrological connection of the valley and floodplain in aquatic and riparian assessment and management (Stanford and Ward, 1993). Groundwater influences fish distribution in that fish may key in on geomorphic areas with beneficial groundwater influence (Baxter et al., 2001), such as local temperature refugia or areas with intragravel flow for egg incubation (Keller and Kondolf, 1990). The linkage between groundwater and watershed is also shown in patterns of upland vegetation types closely tied to subsurface hydrology (Winkler and Rothwell, 1983), as well as in differences in riparian composition among valley types (Harris, 1988). In both management and assessments, we should recognize that groundwater availability is a key element for riparian vigor and channel erosion (Kondolf et al., 1987; Beschta, 1991). Fluvial surfaces such as floodplains function as a temporary source of water storage from high flows (bank storage). Differences in floodplain sediment composition affect bank storage through water infiltration, storage, and release, with variable effects on channel flow (Whiting and Pomerants, 1997). The classifications and procedures that we use in assessments should explicitly incorporate these various geomorphic parameters that influence groundwater conditions (Coates, 1990a; Maxwell et al., 1995).

The complex interaction of biophysical factors controls the variable expression of both erosional and hydrologic processes; such factors include climate, bedrock or surficial lithology and soil characteristics, and basin, hillslope, and channel morphology. We can order and analyze the complex interplay of these factors in a geographic context to characterize and predict hydrologic relations at different scales. Base flows are influenced by the annual precipitation regime, temporary storage in soils and floodplains, and groundwater. Larger flood flows are controlled by intensity and duration of precipitation, whereas basin geology and morphology have greater influence on within-region differences in smaller peak flows and low flows. Hillslope delivery processes are important in flow response and peak flows in small watersheds; in larger watersheds, routing by the stream network controls peak flow expression. Water is delivered to the stream as either surface flow (overland flow or saturated overland flow), subsurface flow (throughflow), or groundwater, with losses from evapotranspiration and to groundwater, and is temporarily stored in soils and vegetation. Figure 25.2 depicts environmental variation in these flow delivery processes, which are regionally controlled by precipitation regime and vegetation density. In general, the relative amount of surface flow is low in forested watersheds and greater in semiarid and arid areas (Figure 25.2). Degree of overland flow depends on physical factors such as soil infiltration capacity, antecedent soil moisture, rainfall intensity, and slope shape, gradient and length; it is moderated by interception by vegetation.

Vegetation is a critical factor in management and assessments in that it influences erosion and the hydrological cycle through interception of precipitation and evapotranspiration, fog drip, interception of surface flow, and stabilization of the soil surface. Loss of vegetation through fire or management actions can increase water yield, especially in the growing season (Burton, 1997). Vegetation has a strong effect on snow accumulation, as well as timing and duration of snow melt; removal of vegetation can significantly alter timing and magnitude of peak flows.

Antecedent conditions such as frozen soils and soil moisture (variable source area) affect flow variability by increasing surface flow and hence the effect a precipitation event has on peak flow (Dunne et al., 1975). For example, rain on snow events can generate a disproportionately large flow response for a given amount of precipitation, which may trigger slope movements and floods (Harr, 1981). Transitional rain-on-snow zones occur at certain ranges of latitude and elevation and are an important management and assessment consideration. Summaries are available of processes and models of hydrology and hillslope erosion for the Pacific Northwest (Swanston, 1991) and for regional evaluation and modeling of forest practices on watershed hydrology and erosion processes in humid temperate regions (EPA, 1980).

Drainage density can be used to illustrate the complex interactions between biophysical setting and water and erosional delivery processes (Figure 25.3). High drainage density is correlated with faster delivery of water, higher peak flows and flow variability, and higher sediment delivery to the stream. Climatic parameters such as average pre-

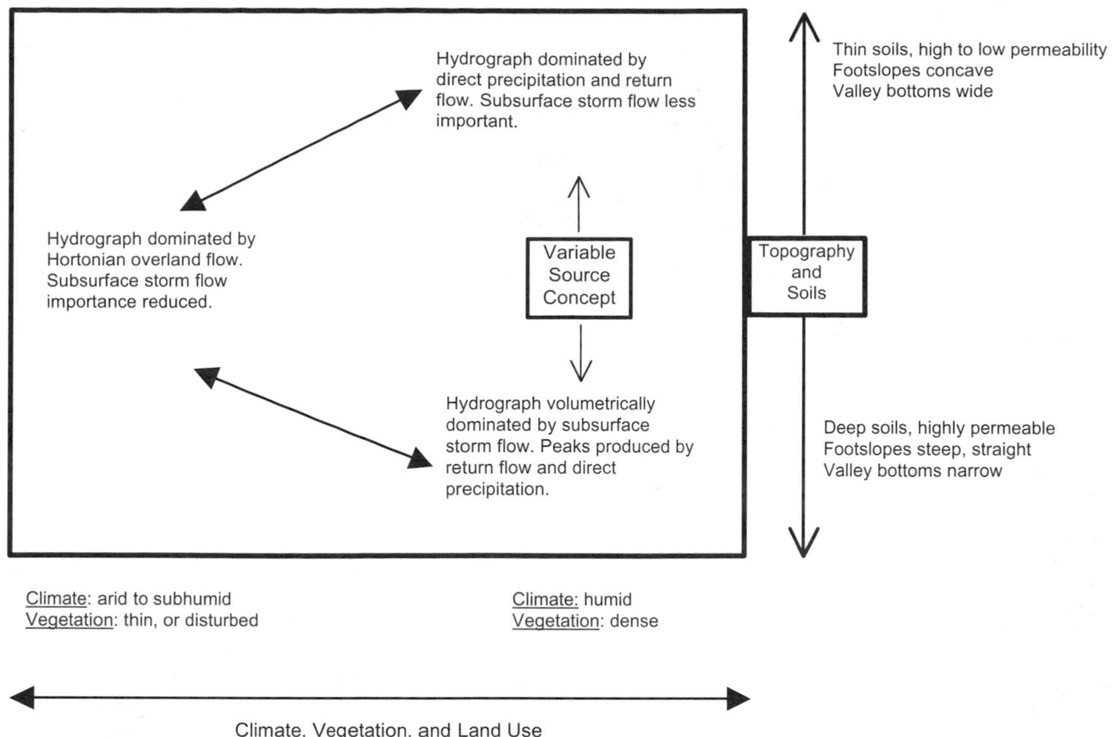

FIGURE 25.2. Watershed and climate controls on water delivery processes and their relative contribution to hydrographs, from Dunne (1982). Direct precipitation and return flow refer to surface flow processes related to saturated soils (e.g., variable source area).

cipitation and precipitation intensity control large-scale patterns of drainage density (Abrahams and Ponczynski, 1985; see Figure 25.3a), which is highest in semiarid climates. At smaller scales, permeability and transmissibility of the geologic substrate exert strong controls on drainage density, groundwater interaction with the stream, and thus flow variability (Carlston, 1963; Coates, 1990a; see Figure 25.3b). Thus we need to include bedrock composition and secondary structural deformation and fracturing in any hydrological analysis.

Bedrock or surficial lithology and soil characteristics influence many factors that contribute to streamflow variability and groundwater characteristics, such as rates of infiltration, storage, and transmission of subsurface water flow (Swanston, 1991). Thomas (1966) demonstrates that a systematic change in flow duration curves is due to variation in glacial processes that produce differences in properties of geomorphic surfaces. Variable glacial processes also explain drastic differences in compositional properties and discharge on north–south aspect slopes (Coates, 1990a). Variation in streamflow between basins can also be caused by differences in the depth and location of channel intersection with the groundwater aquifer. At the reach scale, groundwater recharge and streamflows can vary along a river due to differential input from bedding geology (Keller and Kondolf, 1990). Additionally, the amount and persistence of surface flows (e.g., intermittent or perennial) are affected by variation in valley fill and depth to impervious layers.

Human-induced change such as road construction can effectively extend or increase drainage networks and density, affecting both routing and rates of delivery of water (Jones and Grant, 1996). Also, increased surface impermeability caused by land uses such as urban and suburban development significantly change runoff characteristics, peak flow response, habitat, and fish distribution (Booth and Jackson, 1997; Moscrip and Montgomery, 1997).

Basin morphometry and the stream network are additional physical factors to consider in routing and delivery of water and thus for the character and relative magnitude of peak flows. Hillslope topog-

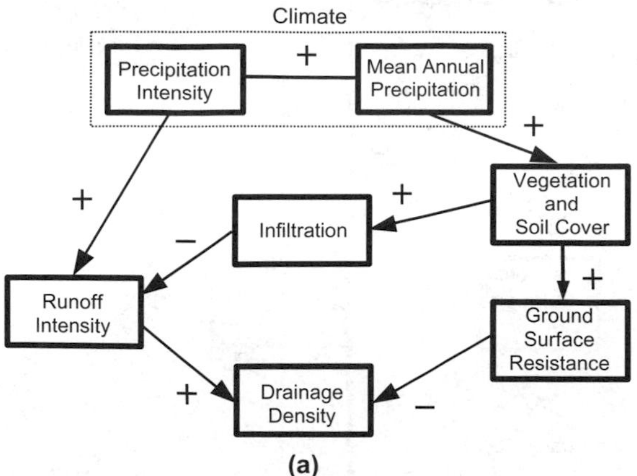

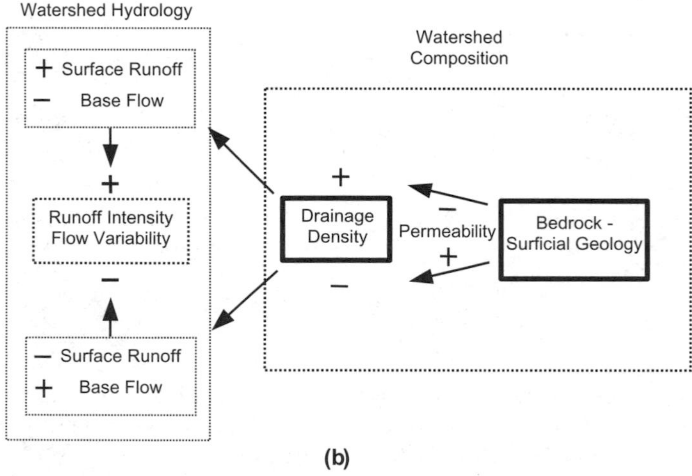

FIGURE 25.3. Drainage density as an example of scaled controls on watershed characteristics and hydrology: (a) Regional climate controls on drainage density, from Abrahams and Ponczynski (1985). (b) Influence of bedrock/surficial geology permeability on drainage density and flow variability, adapted from Coates (1990a).

raphy and relief affect water table slope, depth and rate of subsurface water movement (Coates, 1990a), and flow delivery processes (Figure 25.3). Higher-relief watersheds have lower retention time, faster water delivery, and higher peak flows. Basin shape influences water delivery and thus timing and magnitude of peak flows (Taylor and Schwartz, 1952). Circular basins deliver water more efficiently, which results in flashy hydrology and higher peak flows, whereas elongated basins spread water delivery over time, producing peak flows of lower magnitude but longer duration.

If not overemphasized, floods are of particular economic and ecological interest in hydrological analysis because they can dramatically reshape portions of the landscape. The factors that influence the relative impact or effectiveness of floods in terms of landscape change vary at different scales, ranging from climatic regions to basin physiography and specific location within a basin. Kochel (1988) identifies external basin factors that include climate and hydrological variability, basin morphology, sediment load, vegetation and soils; local factors are valley, channel, and floodplain morphology, channel gradient, and bank cohesion. Flood records for regions of the United States are available in USGS water supply papers 1671–1688 and on the USGS web page.

Multivariate approaches can be used to integrate many of the previously mentioned factors for interpreting and predicting flow characteristics. Factors such as geology, geomorphology, and basin morphometrics have been used to order hydrologic characteristics (e.g., White, 1975; Marston, 1978).

25.5 Erosional and Sediment Processes

25.5.1 Erosional Processes

The challenge before us in ecosystem assessment of erosion and sediment is twofold. First, we need to connect erosional and sediment processes to ecosystem functions and biota in some meaningful way. Second, we need to decipher habitat conditions and management impacts in the context of the complex natural historical–temporal and spatial variability inherent in these processes. It is essential that we recognize that the long-term function of the watershed is to transport sediment as the landscape is eroded. Erosional processes and sediment production are a natural part of this function, which can be altered or accelerated by changes in climate, fire, or land use. Sediment delivery to the channel has a duality in aquatic ecosystems in that it is necessary to create and maintain habitat, yet can negatively effect biota. Stream habitat is simply part of the sediment, water, and organic material delivered to and stored in streams (Lisle, 1983). In this view, pools are temporarily stored water in channel depressions that are scoured at high flows, and riffles flow over sediment deposits. The concept of the sediment budget is important in watershed assessments, because it links stream habitat composition and dynamics to watershed processes of sediment delivery, transport, and storage (Swanson et al., 1987; Swanston, 1991; Dunne, 1998). Hillslope erosional processes transfer materials such as boulders and large wood to the stream, adding structure to the stream, and in some instances greatly improve fish habitat in structurally deficient channels (Benda et al., 1998). However, in areas of intensive management, increased frequency or episodic mass wasting over large areas and consequent channel adjustment can be detrimental to fish habitat and rearing (Poulin, 1991).

For effective assessments, we need to recognize the types of processes involved in sediment production and transport and the triggering and transport mechanisms for these processes. There are essentially three sediment sources: hillslopes, floodplain features, and in-stream storage. The watershed functions of erosion–sediment are sediment production, delivery, storage, and transport. Sediment sources from hillslope processes include surface erosion and slope movement processes; the relative types and contribution of these processes vary with geoclimatic setting, location within a watershed, and watershed history or condition. Swanson et al. (1982) provide a good summary of sediment transfer processes applicable for steepland humid temperate climates. Sources of in-stream sediment include storage in bedforms and erosion of fluvial surfaces. The floodplains of wide alluvial valleys can be a significant in-channel source of sediment from accelerated lateral or vertical channel migration (e.g., as bank erosion; Beschta, 1991). Most erosional processes involve forces of water and gravity, which are countered by resistance by vegetated soil, soil cohesion, and regolith geotechtile properties. Surface erosion is the dominant process in regions with overland flow, and slope movement processes are more prevalent in areas predominated by subsurface flow (Coates, 1990b). Surface erosion requires detachment, transport, and deposition of soil particles and is primarily expressed as rill, gully or sheet erosion and secondarily as rainsplash and dry ravel. Soil mass movement types are classified by the rate of delivery, size and type of materials, and type of delivery (Varnes, 1979). Triggering mechanisms for mass movement include precipitation and soil properties that affect soil saturation, cohesion, and increased pore pressures (Coates, 1990b). We should also consider that groundwater and subsurface flow processes could be important agents of erosion and landform development in a variety of settings (Dunne, 1990).

Erosional processes, sediment delivery, and sensitivity to disturbance are controlled by the same multiscaled biophysical factors that affect hydrology. Climate controls erosion processes through degree of vegetation development and the precipitation and storm regime. Peltier (1975) has formalized coarse-scale controls on the effectiveness of climate and vegetation on erosional processes for regions of the United States. These regional climatic controls are reflected in sediment load and drainage density, both of which are highest in semiarid climates (Langbein and Schumm, 1958). Geomorphic sensitivity to change in vegetation depends on the effectiveness of vegetation on controlling geomorphic processes, which changes with geoclimatic setting and factors such as slope and soil cohesion (Swanson, 1981). Vegetation influences erosion by moderating precipitation impact, soil moisture, and water delivery processes and by stabilizing the soil surface. Changes in landscape vegetation from fire, grazing, tree removal, or climate change may lower threshold response and cause a precipitation event to have a much larger impact on the system than it would have had under previous vegetation conditions. In forested regions, for example, rooting strength is an important factor in slope stability on steep hillslopes, and the change in timing and degree of slope instability depends on vegetation and site conditions (Sidle, 1992). Fire strongly influences historic and contemporary sediment and channel conditions (Meyer et al., 1995) and is an essential consideration in assessments. Fire intensity and extent determine the amount of vegetation removed, exposure of soil, and soil hydrophobisity (water repellency), which in turn influence surface runoff and erosion (Beschta, 1990; DeBano et al., 1996). We must also recognize how regional constraints of precipitation regimes and physiographic setting affect geographic variation in fire cycles and resultant erosional regimes (Swanson, 1981). Another consideration in forested watersheds is road-related sediment production and transfer, which vary in different bedrock lithologies (Duncan and Ward, 1985).

Both physiographic and compositional properties of watersheds influence spatial variability in slope stability, erosional processes, and efficiency and rates of sediment production and delivery (Nolan and Marron, 1985). Of particular concern in watershed management is the efficiency of sediment delivery from the watershed to the channel. Watershed physiographic factors related to sediment delivery efficiency include watershed morphology and drainage density; local factors are hillslope shape and morphology (such as slope–length), valley confinement, and hillslope and riparian vegetation characteristics. Higher-relief watersheds have greater sediment yields in part due to greater mass wasting erosion, reduced storage of sediment in floodplains, and greater sediment transport potential (Swanson et al., 1987). Valley confinement is a good channel morphometric because it reflects the relative impingement of the hillslope on the channel and effective delivery of sediment from hillslope processes.

Physical factors alone do not adequately account for slope stability, as shown by variation in feedback initiation mechanisms among different bedrock types (Onda, 1992). Bedrock lithology is a primary determinant in slope stability and sediment production in that it influences geotechtile properties, basin and hillslope morphology, and soil characteristics, but unfortunately these factors are not incorporated in assessments in a consistent manner. The role of geology is exemplified in the Pacific Northwest, where spatial patterns, types, and frequency and magnitude of mass movement processes are related to geologic factors such as hillslope relief, lithology, geologic contact areas, and structural dip slopes (Swanston, 1978; Graham, 1985; Swanson and Leinkaemper, 1985; Poulin, 1991). Both watershed management and assessment need to consider spatial distribution of physiographic features that influence mass wasting, such as bedrock hollows, because they significantly affect frequency–magnitude relations of erosion and sediment dynamics (Dunne, 1998). Also, geologic processes such as earthquakes can generate episodic mass wasting and excessive sediment loading to channels (Pearce and Watson, 1986), which can be a significant regional consideration in sediment dynamics.

Because of the dynamic nature and variability of erosional processes and initiation mechanisms in different geoclimatic settings, modeling and measuring erosion is problematic, and this complexity is compounded in mountainous areas managed for multiple resources. Furthermore, sediment modeling has not adequately addressed the current desire to continue land-use activities in watersheds while maintaining the natural frequency and magnitude of erosional processes, especially in the context of ecological function. An assessment strategy that needs further refinement is modeling and analyzing sediment at multiple scales (Arnett, 1979; Ahnert, 1987). Benda and Dunne (1997) address the dynamic nature of sediment supply with stochastic models at different basin scales. These models emphasize probability occurrence of combinations of fire, precipitation events, and mass wasting and show promise for regional characterization of sediment dynamics. Another approach is hierarchical physiographic land classification, which provides a useful spatial context for the controlling factors that influence geologic processes and land management (Godfrey, 1978). This type of framework has particular promise for assessments, especially if coupled with GIS-based models for spatial analysis of erosion. Other more traditional models used to assess land management impacts, slope stability, and mass wasting erosional processes are outlined by Megahan (1996) and in the WREN document (EPA, 1980). The universal soil loss equation, originally developed for agricultural applications and surface erosion, has been applied to forested areas (e.g., EPA, 1980), but is not effective for mass wasting erosional processes. Sediment rating curves are useful for determining realized and expected sediment delivery for individual basins and lithologies or for regional comparisons (EPA, 1980).

25.5.2 Sediment Transport, Storage, and Stream Habitat

Sediment transport is a key but weak link in ecosystem assessments. Although the mechanics of sediment transport are well studied, the application to stream habitat is not adequately understood, especially in the context of variability at different spatial and temporal scales. Fluvial processes of sediment transport and storage are directly linked with ecosystem concepts of habitat formation, distribution, and dynamics. To better connect stream habitat with properties and processes of the watershed, recognized geomorphic features of channels, such as channel units, bedforms, and structural elements, can be perceived as habitat features (Frissell et al., 1986; Lisle, 1986; Grant et al., 1990). Sediment transport and bed composition are highly variable in space and time (VanSickle and Beschta, 1983; Benda et al., 1998), confounding the development of methods and regulatory standards for impacts on fish (Chapman and McLeod, 1987). Excess sediment from extreme natural events and land use causes adjustments in channel morphology, and the impacts on aquatic habitat can be quite variable spatially and temporally. Sediment input from large-scale events tends to travel through the basins as sediment slugs or waves. Valley confinement has a strong influence on rates and patterns of transport of sediment waves (Meade et al., 1990). Conversely, channel morphology and bedform size change over time as sediment is eroded and transported from landslide-derived in-channel sediment deposits (Perkins, 1989). Pool volume (Hilton and Lisle, 1993) is a sensitive index for monitoring the effect that transport of high sediment volumes has on habitat parameters of pool density and volume (Lisle, 1982).

Fish spawning success depends in part on the timing and extent of bed scour. Thus the distribution and interannual variability of salmonid populations are related to the geographic range and timing of geomorphic, land-use, and hydrologic factors that influence bed stability (Kondolf et al., 1991; Montgomery et al., 1996). Channel or bed stability indices can potentially assess both geographic variation in stability and the sensitivity of channels to disturbance (Kappesser, 1993; Olsen et al., 1997). Elevated fine-sediment content is considered detrimental to fish egg development and affects overwintering habitat for juvenile fish in ice-covered streams. Furthermore, increases in fine sediment can differentially increase sediment transport, bed stability, and channel adjustment (Buffington, 1995). Indices based on surface bed material size have been used to infer changes in sediment transport and monitor fine sediment, but have limitations in application and interpretation (Potyondy and Hardy, 1994; Kondolf, 1995a). Other methods for measurement and analysis of channel and sediment characteristics related to fisheries are outlined in various manuals (Platts et al., 1983; Chapman and McLeod, 1987; Schuett-Hames et al., 1993).

It is essential that assessments identify the dynamics of long-term sediment supply from fire regimes, climate fluctuations, and historical land use. It is only in this context of historical process dynamics that we can interpret contemporary conditions in fluvial surfaces, sediment storage, habitat characteristics, and fish distribution (Costa, 1975; Swanson, 1981; Meyer et al., 1995; Benda et al., 1998). For instance, streams may have inherited a sediment regime from the Pleistocene, reflected in variations of floodplain and terrace morphology (McDowell, 1983), that strongly influence contemporary processes (Heed, 1981).

We also need to consider that sediment transport processes and resultant fluvial landforms vary within a watershed, and their distribution is controlled by watershed and channel size, slope, and morphology (Ohmori and Shimazu, 1994). The travel behavior of hyperconcentrated flows (e.g., debris torrents) can demonstrate the influence of basin characteristics on sediment transport. Geoclimatic factors affect the initiation processes and composition (water and sediment composition) of hyperconcentrated flows, which in turn influence travel behavior, channel impact, and depositional characteristics (fluvial landforms) of these flows (Costa, 1988). For instance, flows with higher water content and finer sediment composition tend to flow farther and stop at lower gradients than those with lower water content. Travel and deposition of debris torrents are also controlled by basin morphometry, such as stream network configuration, tributary junction angles, and valley slope and confinement (Benda, 1990). Thus assessments must incorporate variation in geoclimatic factors and basin characteristics that affect both regional and within-basin behavior of sediment transport mechanisms, such as debris torrents (Swanson et al., 1987; Benda, 1990).

Large wood is an important geomorphic agent controlling sediment transport and channel morphology (Keller and Swanson, 1979). Wood steps in steeper streams function in dissipation of energy (Heed, 1981), which results in localized areas of lower gradient and scour for pool formation. Wood steps also function in sediment storage (Mosley, 1980; Bojonell et al., 1993) and act as in-channel storage buffers to sediment transferred from hillslopes. Watershed assessments and riparian management schemes need to consider input of large wood from both local sources and transport processes, such as debris torrents and landslides,

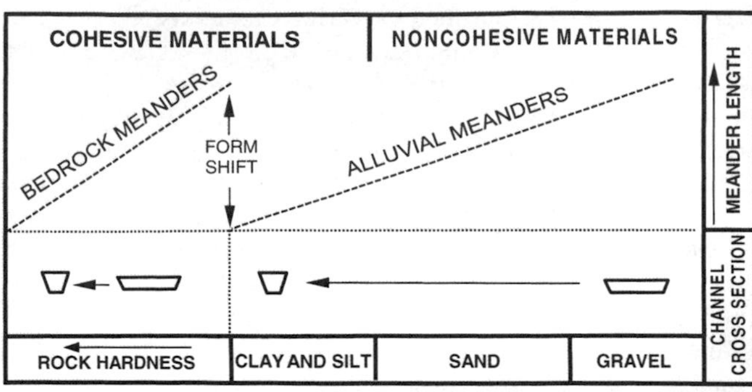

FIGURE 25.4. Influence of bank composition and resistance on channel form, from Braun (1985).

that deliver wood and materials to the channel from upland areas.

Channel morphology, both planform and cross-sectional shape, is an important factor in monitoring and assessment, because it represents how the channel transport capacity is adjusted to the prevailing sediment and discharge regime. This relationship between sediment supply and channel morphology is reflected in classic channel patterns for lowland streams: straight, meandering, and braided (Leopold and Wolman, 1957; Schumm, 1977). Channel form is also influenced by feedback mechanisms from local in-stream processes (Lane and Richards, 1997). Stream power and shear stress are indices of channel energetics used to measure sediment transport that are also useful for assessing channel conditions and change in channels over time. Stream power, determined by channel slope and discharge, reflects the energy available to mobilize and transport sediment. Shear stress is the energy required to initiate sediment movement and is related to the size and structural arrangement of sediment. The adjustment of the channel to change in the discharge and sediment regime causes channel adjustment of planform slope, width, and depth as summarized in general equations presented by Schumm (1977) and modified for upland streams by Montgomery and Buffington (1993). Hydraulic geometry as the channel cross section is a useful tool for monitoring and assessment in that it reflects local controlling factors on channel adjustment to the discharge and sediment regime (Leopold and Maddock, 1953; Williams, 1987; Wharton, 1995).

A somewhat universal consideration in fluvial geomorphology is that channel morphology, specifically bankful width, is formed by the *dominant discharge* (e.g., 1.6- to 2-year return flow). However, channel morphology often reflects the multiscaled geologic and climatic influence on upstream and local constraints, such as flow magnitude–frequency relations and floodplain resistance to erosion. For example, coarse-scale patterns of channel geometry reflect regional climatic controls (Park, 1977). At local scales, different flow events variably affect channel and terrace morphology through bank erosion and sediment transport. Pickup and Warner (1976) recognizes that variable channel morphology is due in part to how basin lithology conditions affect factors such as flow relations and bank cohesion. Channel shape, through-bank cohesion, and erosion depend on the timing and stage of flows relative to seasonal bank stabilization by vegetation (Pickup and Warner, 1976). Bank cohesion or resistance is also a function of bank materials (and vegetation) that influence channel shape and heterogeneity (Miller and Onesti, 1979; Braun, 1985; see Figure 25.4). Patterns of valley meandering, channel planform morphology, and channel cross-sectional shape are also related to bedrock lithology and bank resistance (Hack, 1965; Howard and Dolan, 1981; Braun, 1985). Bedrock lithology affects relative channel slope directly through hardness or resistance to erosion and indirectly through sediment size and bed transport relations (Hack, 1957, 1973).

25.5.3 Habitat Formation and Distribution

For effective assessment and in-channel restoration strategies, we need to understand the processes and factors that influence habitat formation and distribution patterns and recognize that these patterns and processes change with geomorphic setting. Classic concepts for pool–riffle formation and distribution were developed for lowland or low-gradient alluvial unconstrained channels in humid-temperate regions, where the discharge regime easily transports the predominately smaller sized sediment. In these settings the dominant discharge regime of high-frequency, low-magnitude events (e.g., 2-year return period) that forms the active channel width also drives formation of pools by scour through secondary flow processes (Thompson, 1986). The hydrological scour that forms patterns of meander formation and pool scour is considered equilibrium dissipation of energy in planform and vertical dimensions, respectively (Langbein and Leopold, 1966) and are regularly spaced. The optimal spacing of pools is considered 5 to 7 channel widths, and this distributional pattern is considered universal in a wide variety of bedrock and alluvial streams (Keller and Melhorn, 1978).

Factors that cause variability and deviation from a normal rhythmic distribution are important for interpretation of stream processes and features and development of restoration strategies. Channel planform morphology and pool spacing are altered by contact with high terraces or valley hillslopes, variable floodplain, and sediment sizes (Milne, 1982). Features such as bedrock outcrops, large wood, large boulders, and rootwad-protected banks are considered large roughness elements and cause local scour that forms variably sized stationary pools (Lisle, 1986). The size of pools formed by these structural features is related to the size of the feature scaled to the width of the stream (Lisle, 1986). Pool spacing and size relations can be skewed by large-scale paleohydrological events, which create large bedforms and pools that are spaced relative to the magnitude of the event (O'Conner et al., 1986). Size and density of pools vary in geomorphic river segments of different lithology and channel morphology, such as in the Colorado River, Grand Canyon section (Howard and Dolan, 1981). Also in this setting, large-pool distribution is indirectly influenced by coarse-scale patterns of structural properties such as jointing–faulting, because debris torrents and tributary formation follow fault patterns, and the larger pools are associated with debris torrent deposits.

The processes and patterns related to habitat formation and distribution differ in steeper, more confined upland streams (Grant et al., 1990; Montgomery and Buffington, 1997). This is partly because the predominant bankfull discharge may not be competent to effectively transport and sort the larger substrate sizes encountered in smaller upland streams, and thus pools may be poorly or irregularly formed (Miller, 1958; Milne, 1982). In steeper upland streams with large substrate sizes, bed entrainment that forms channel units occurs at higher flows of relatively large magnitude and low frequency (e.g., 25 year + return period; Grant et al., 1990). Also, hillslope processes and valley features often directly impinge on upland channels (Grant and Swanson, 1995), which affects habitat patterns. Channel bedforms and structural features such as boulders may have a clustered distribution and are associated with mass wasting processes that are more prevalent in certain valley reach types (Benda, 1990; Grant et al., 1990; Grant and Swanson, 1995). The coarse-scale distribution of structural pool types in the Oregon Coast Range reflect these upland controls and are related to stream power, essentially channel slope, and drainage area (Stack and Beschta, 1989). This relation with habitat formation and stream size is demonstrated in forested streams, where pool spacing is inversely related to wood loading, and wood influence on pool spacing decreases with increasing stream width (watershed size) (Montgomery et al., 1995).

25.6 Variability and Ecosystem Dynamics in Assessments

In ecological assessment and restoration, we need to recognize that aquatic systems are not static entities, but are naturally dynamic. This requires a framework that incorporates the properties and processes that both cause and constrain ecosystem dynamic variability (such as channel dynamics). Thus we must manage and restore aquatic systems in context of these dynamics, not necessarily for some stable or optimum state. Four general types of variability need to be accounted for in ecological assessments. The first two types are spatial variability due to change in geoclimatic or biophysical characteristics. Horizontal variability reflects changes across a single spatial scale (e.g., changes in lithology, climate). Vertical variability results from change in scale (e.g., from regional to local scale). A third type of scale change is change in size, which in lakes, and especially for streams, fundamentally affects physical processes. A fourth general type is temporal variability as a result of process dynamics. The first two types of spatial variability are well accounted for in classifications that emphasize process, form, and composition of aquatic systems at various spatial scales (Maxwell et al., 1995; also see Chapter 24). Although the spatial context of controlling and driving variables are incorporated in these classifications, the actual processes and dynamics are usually only implicitly implied. This process-driven dynamic variability is emphasized in this section, particularly channel dynamics and climatic fluctuations.

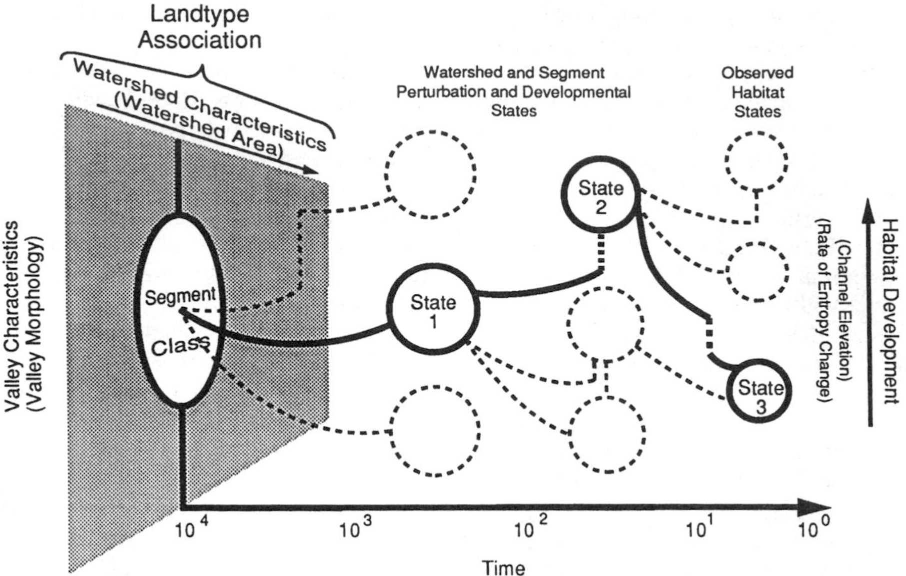

FIGURE 25.5. Conceptualization of potential stream habitat development within the context of watershed and stream segment classification. Bold lines represent actual habitat development under a particular set of historical environmental process regimes. Dashed lines represent possible developmental paths given different environmental conditions.

In any aquatic assessment we must first consider the geomorphic system of the watershed in its entirety (Sear, 1994), because channel dynamics and morphology can only be understood in context of local and upstream controlling factors. Many geomorphologists consider these controls on channel composition and dynamics as essentially the climate and geology of the watershed. Basin relief, geology, soils, vegetation, and land use influence the discharge regime and the size and quantity of sediment delivered (Schumm, 1977). Schumm also identifies local controls as the morphology and composition of valleys and floodplains, bank cohesion, and vegetation.

Within a basin, the main divisions of the stream network for aquatic classification can be considered stream or valley segment, channel reach, and stream habitat features (Frissell et al., 1986; Maxwell et al., 1995). The valley or stream segment is an often ignored component in aquatic ecosystem classifications, yet is an essential scale for integration and interpretation of the influence of terrestrial properties on the aquatic system. Many properties of the valley are time invariant relative to channel dynamics, and these properties exert strong controls on channel morphology and dynamics. Variability of channel characteristics is better understood in the context of the valley as stages of development or recovery after disturbance (Figure 25.5). For assessments, it is essential that we formalize the constraints of the valley on channel form and dynamics. In addition, a genetic floodplain classification as proposed by Nanson and Croke (1992) can potentially aid segment classification by integrating the valley and riparian vegetation with the composition and formative processes of floodplains.

Understanding processes of channel change is imperative in assessment and restoration. Although general equations have been developed for channel adjustment, channel dynamics and adjustment to disturbance are fairly complex and difficult to predict and may be better understood in the context of channel classification (Downs, 1995; Montgomery and Buffington, 1997). Channel classification is increasingly used for resource assessment, management, and restoration (Potyondy and Hardy, 1994; Kondolf, 1995b). Channel types and channel patterns are based on channel morphology and sediment composition (Rosgen, 1985) and on the processes and features that form and characterize channels (Frissell et al., 1986; Montgomery and Buffington, 1997). Regional channel types based on sediment type and channel dimensions have also been developed (Osterkamp and Hedman, 1982). These show promise for coarse-scale analysis if incorporated with multivariate spatial analysis of channel morphometry. We should realize that channel classifications based primarily on morphology are limited because the relations to processes are not clearly defined. This impedes our need to understand particular process relations as they pertain to channel change or evolution, particularly in the context of local and upstream constraints. More evolutionary, process-based models of channel change have been developed for specific regional conditions (Simon, 1994; Downs, 1995). These specific process-based types of classifications would be a natural extension for regional assessments and recovery plans. The concept of sensitivity to disturbance, recovery, and types and direction of channel change can be incorporated into classifications by including the factors discussed previously that constrain channel morphology and dynamics, such as climate, bedrock, bank materials, and valley–floodplain morphology (Downs and Gregory, 1993; see Figure 25.5).

Classic channel developmental models in alluvial lowland streams emphasize planform development from straight to more stable meandering forms (Keller, 1972). In steeper upland streams, energy dissipation and channel adjustment is often in a vertical dimension, which is well summarized by Montgomery and Buffington (1993). Beschta (1981) presents a model of channel response to increased sediment load for gravel streams of less than 5% slope, which results in increased width and reduction of pools as exemplified in streams in California (Lisle, 1982). Various land-use practices in the stream corridor affect processes and channel degradation (Galay, 1983). Multivariate techniques are useful to elucidate the complexity of these various land-use impacts on streams and threshold response of channel change. (Touysinhthiphonexaym and Gardner, 1984; Wood-Smith and Buffington, 1996).

The processes and features that influence aquatic systems and channel form and dynamics vary with position within the landscape and stream network and with watershed, lake, or channel size. Methods, models, and interpretation in assessments must account for these spatial changes. For example, progressive change in a downstream direction is a dimension of variability unique to stream systems that confounds within- and between-stream comparisons and the influence of terrestrial impacts on the stream. To account for this functional change with increasing size, stream assessments need to be scaled to watershed size or discharge (Hughes and Omernick, 1983) and by stratifying stream segments by position both within the basin and the landscape (geoclimatic setting) (Maxwell et al., 1995).

The generalized, coarse-level resolution of downstream change in channel size, morphology, discharge, sediment characteristics, longitudinal profile shape, and physical and biotic processes is reflected in various conceptual models. Schumm (1977) simplifies the function of watershed processes as three zones of sediment supply, transport, and deposition. This watershed zonation is also linked with channel processes and channel classification (Montgomery and Buffington, 1997) and the river continuum concept of downstream change in stream–riparian connection, retention, and biotic composition and processes (Vannote et al., 1980; Petts and Foster, 1985). We also need to scale the relative function and effect of landscape features and processes to watershed size. For example, the influence and function of wood boles (Bilby and Ward, 1989; Ralph et al., 1994) and the persistence and impact of hillslope processes (Swanson et al., 1985) are related to stream discharge and stream power. Channel features that affect habitat formation are often scaled to the size of the channel (Lisle, 1986), as are channel patterns such as pool spacing and meander wavelengths. Likewise, geomorphic "reach" dimensions used in assessments can be scaled by channel widths (length–channel width, valley width–channel width).

25.6.1 Climate and Ecosystem Dynamics

Because hydrodynamics are a major driving force in aquatic systems (Legendre and Demers, 1984), it is imperative that we place ecological assessments and restoration strategies in both geographic and temporal contexts of multiscaled climate change and ecosystem dynamics. Climatic fluctuations drive various aspects of ecosystem dynamics, such as semicyclic patterns of drought and variation of streamflow, fire and erosion cycles, and temperature regimes (Currie, 1984; Ebbesmeyer et al., 1991). Scientists have recently recognized interdecadal climate change (20- to 30-year cycles) that is a coarse-scale driver of ecosystem dynamics. This scale provides an illumination of how life-history strategies and regional fish productivity are related to long-term fluctuations in climate and ocean conditions (Hare et al., 1999). Various climatic fluctuations also affect fish production by controlling storm events and bed scour, streamflow timing, and volume, which in turn affect survival, rearing, and both upstream and downstream migration of fish (Pultwarty and Redmond, 1997).

El Niño–La Niña (ENSO, 2- to 7-year cycles)

are well recognized shorter-term climatic patterns that influence hydrodynamics through regional flood frequency relations (Webb and Betancourt, 1990). Extreme flood events are temporally clustered around ENSO events (Hirschboeck, 1987). Periodicity of fire and associated erosion and patterns of sediment production also are driven by ENSO events (Swetnam and Betancourt, 1990; Meyer et al., 1995). Such semicyclic variability in magnitude of precipitation, peak flow events, and fire regimes results in alternating periods of quiescent erosion and clustering of mass wasting events, creating a punctuated sediment delivery and transport regime (Benda et al., 1998). Climatic induced lake mixing also is responsible for temporal patterns of productivity and clarity in oligotrophic lakes (Goldman et al., 1989).

Assessments must also recognize that these climatic fluctuations and consequent processes, such as streamflow dynamics, vary regionally and can even produce opposite effects in different regions. For example, regional differences in drought conditions are identified in the western United States (Rogers and Armbruster, 1990), and in ENSO-related streamflow variability for the entire United States (Piechota and Dracup, 1996) and for the western states (Redmond and Koch, 1991). Recognition of these geographic and temporal differences is important for planning, interpretation, and comparison of interregional assessments. This is especially critical for assessments of large-scale watersheds that integrate regional differences of climatic and hydrologic variability (Pultwarty and Redmond, 1997; Quigley and Arbelbide, 1997).

Seasonal climate development drives annual variability in aquatic ecosystems. Lake dynamics are characterized by seasonal patterns of thermocline stratification, water column stability, and mixing; the latter is considered the "master variable" in seasonal development of biotic processes in deep lakes (Sommer, 1985). Lake mixing and hydrodynamic character are key factors in lake classifications, which display regional distributional patterns (Winter, 1977; Lewis, 1983). For streams, annual variability and stability are characterized by the annual flow regime: the timing, stability, and persistence of base flows and the timing, variability, and magnitude of high flows. This is illustrated by fish distribution and habitat use variability along with regional differences in flow development. For instance, in Pacific coastal streams, high water occurs in winter, and important overwintering areas are areas of flow diversity or slower velocity, such as off-channel habitats (backwaters and side channels). In colder climates with snow and ice cover, there is a seasonal winter low flow, and juvenile fish use ice-free areas and interstitial spaces in the channel bed as refugia.

25.7 Conclusions

Any aquatic ecosystem assessment and restoration strategy must be placed in the context of the multiscaled dynamics of the system and the geoclimatic templates that constrain process dynamics, habitat composition, and evolutionary life-history strategies. Coarse-scale geologic and climatic evolutionary processes enhance our understanding of both patterns in the landscape and the distribution and adaptation of biota relative to fluvial landscape variability. Basin-level processes consist of delivery and routing of water and sediment (and other materials) to the channel. Both sediment and hydrodynamics are interconnected and influenced by climate, vegetation, watershed morphology, soils, and geology. The geomorphic function of a watershed is to transport sediment and water delivered to it by climatic and hillslope erosional processes. In this sense, processes such as erosion and floods are a natural function of the watershed. In a natural setting, they create a balance between perturbation and creation of habitat and are an important component of the disturbance regime or process dynamics of a watershed (Reeves et al., 1995; Benda et al., 1998). Although the concept of disturbance is becoming well established as a key to understanding both terrestrial and aquatic ecosystems, disturbance has a negative connotation (Schroeder and Savonen, 1997). Thus *disturbance* may be best considered as the natural process dynamics of a system that organisms have adapted to in the context of evolutionary life-history strategies. We can improve management through understanding the inherent variability of populations and their ecosystems in the context of process dynamics and the strategies and diversity of life histories.

To understand ecosystem properties and management actions in the context of natural processes and variability, we need to formalize the climatic and geologic characteristics that influence thresholds, resiliency to disturbance, rates and pathways of recovery, and complex process–response systems of watersheds. Formalized geoclimatic or biophysical settings in a hierarchical, integrated terrestrial–watershed classification framework will greatly aid interpretation of landscape patterns and processes. By providing a common framework for communication, coordination, and information management, classifications would increase efficiency of assessments and management actions. For example, watershed assessments may follow common general procedures, but they are often conducted by different teams using different models, methods, and stratification schemes. Comparison of results on a common landscape across watershed boundaries thus becomes difficult. Watershed assessments can be much more efficient in terms of time and cost if a common biophysical hierarchical template is used to extrapolate knowledge, information, and generalizations from site-specific studies across watershed and political boundaries.

25.8 Suggested Readings

Callaham, R. Z. 1990. *Case studies and catalog of watershed projects in western provinces and states*. 22, 1–188. Berkeley, CA: Wildlife Resources Center, Division of Agriculture and Natural Resources, University of California.

Graf, W. L. 1998. *A guidance document for monitoring and assessing the physical integrity of Arizona streams*. 95-0137. Phoenix, AZ: Arizona Department of Environmental Quality.

Gurnell, A.; Petts, G., editors. 1995. *Changing river channels*. New York: John Wiley & Sons.

Knighton, D. 1984. *Fluvial forms and processes*. Baltimore, MD: Edward Arnold.

Montgomery, D. R.; Buffington, J. M. 1993. *Channel classification, prediction of channel response, and assessment of channel condition*. TWF-SH10-93-002. Olympia, WA: Washington State Dept. of Natural Resources: Timber, Fish, and Wildlife.

Morisawa, M. 1985. *Rivers: form and process*. Geomorphology texts. London and New York: Longman.

U.S. Department of Agriculture. 1999. *Stream corridor restoration: principles, processes, practices*. NTIS PB98-1583481NQ. Washington, DC: U.S. Dept. Agric., For. Serv.

25.9 References

Abrahams, A. D.; Ponczynski, J. J. 1985. Drainage density in relation to precipitation intensity in the USA. *J. Hydrol.* 75:383–388.

Ahnert, F. 1987. Process–response models of denudation at different spatial scales. *Catena Suppl.* 10:31–50.

Ahnert, F. 1994. Equilibrium, scale and inheritance in geomorphological perspectives. *Geomorphology* 11: 125–140.

Allen, T. F. H.; Starr, T. B. 1982. *Hierarchy: perspectives for ecological complexity*. Chicago and London: University of Chicago Press.

Arnett, R. R. 1979. The use of differing scales to identify factors controlling denudation rates. In: Arnett, R. R., ed. *Geographical approaches to fluvial processes*. Norwich: Geo Abstracts: 300.

Baker, V. R. 1988. Geological fluvial geomorphology. *Geol. Soc. Amer. Bull.* 100:1157–1167.

Baxter, C. V.; Frissell, C. A.; Hauer, F. R., 2001. Geomorphology, logging roads, and the distribution of bull trout (*Salvelinus confluentus*) spawning in a forested river basin: implications for management and conservation. *Trans. Amer. Fish. Soc.*

Beard, L. R. 1975. *Generalized evaluation of flash-flood potential*. CRWW-124. Austin, TX: Center for Research in Water Resources, University of Texas.

Benda, L. 1990. The influence of debris flows on channels and valley floors in the Oregon Coast Range, USA. *Earth Surf. Process. Landf.* 15:457–466.

Benda, L.; Dunne, T. 1997. Stochastic forcing of sediment supply to channel networks from landsliding and debris flow. *Water Resour. Res.* 33(12):2849–2863.

Benda, L.; Beechie, T. J.; Wissmar, T. C.; Johnson, A. 1992. Morphology and evolution of salmonid habitats in a recently deglaciated river basin, Washington State, USA. *Can. J. Fisheries Aquat. Sci.* 49(6):1246–1256.

Benda, L.; Miller, D. J.; Dunne, T.; Reeves, G. H.; Agee, J. K. 1998. Dynamic landscape systems. In: Naiman, R.; Bilby, R., eds. *Ecology and management of streams and rivers in the Pacific Northwest coastal ecoregion*. New York: Springer-Verlag: 261–288.

Beschta, R. L. 1981. *Management implications of sediment routing research*. Special Report. Portland, OR: National Council of the Paper Industry for Air and Stream Improvement.

Beschta, T. 1990. Effects of fire on water quantity and quality. In: Walstad, J. D.; Radosevich, S. R.; Sandberg, D. V., eds. *Natural and prescribed fire in Pacific Northwest forests*. Corvallis, OR: Oregon State University Press: 219–232.

Beschta, T. L. 1991. Stream habitat management for fish in the northwest United States: the role of riparian vegetation. *Amer. Fish. Soc. Symp.* 10:53–58.

Bilby, R. E.; Ward, J. W. 1989. Changes in characteristics and function of woody debris with increasing size of streams in western Washington. *Trans. Amer. Fish. Soc.* 118:368–378.

Bisson, P. A.; Montgomery, D. R. 1996. Valley segments, stream reaches, and channel units. In: Hauer, F. R.; Lamberti, G. A., eds. *Methods in stream ecology*. San Diego, CA: Academic Press: 23–52.

Black, P. E. 1989. Stratified flood frequency analysis. In: Woessner, W. W.; Potts, D. F., eds. *Proceedings on the symposium on headwaters hydrology*. Missoula, MT: Amer. Water Resour. Assoc.: 563–572.

Black, P. E. 1997. Watershed functions. *J. Amer. Water Resour. Assoc.* 33(1):1–11.

Bojonell, H. A.; Tecle, A.; King, J. G. 1993. Assessing factors contributing to sediment storage in headwater streams in northern Montana. In: *Emerging critical issues in water resources of Arizona and Southwest; sixth annual symposium of the Arizona Hydrol. Soc.*: 155–165.

Booth, D. B.; Jackson, C. R. 1997. Urbanization of aquatic systems: degradation thresholds, stormwater detection, and the limits of mitigation. *J. Amer. Water Resour. Assoc.* 33(5):1077–1090.

Braun, D. D. 1985. Lithologic control of bedrock meander dimensions in the Appalachian Valley and Ridge Province: a comment on a comment. *Earth Surf. Process. Landf.* 10(6):635–642.

Brown, G. H. 1981. Derivation of a geological index for low flow studies. *Catena* 8:265–280.

Buffington, J. M. 1995. *Effects of hydraulic roughness and sediment supply on surface textures of gravel-bedded rivers.* TFW-SH10-95-002. Olympia, WA: Washington State Dept. of Natural Resources; Timber, Fish, and Wildlife.

Bull, W. A. 1979. Threshold of critical power in streams. *Geol. Soc. Amer. Bull.* 90:453–464.

Bull, W. B. 1991. *Geomorphic responses to climatic change.* New York: Oxford University Press.

Burton, T. A. 1997. Effects of basin-scale timber harvest on water yield and peak streamflow. *J. Amer. Water Resour. Assoc.* 33(6):1187–1196.

Carlston, C. W. 1963. *Drainage density and streamflow.* 422-C. Washington, DC: U.S. Geological Survey.

Chapman, D. W.; McLeod, K. P. 1987. *Development of criteria for fine sediment in the Northern Rockies ecoregion.* EPA-910/9-87-162. Boise, ID: U.S. Environ. Protect. Agency, Water Div.

Chorley, R. J. 1962. *Geomorphology and general systems theory.* 500-B. Washington, DC: U.S. Geological Survey.

Clements, F. E. 1936. Nature and structure of the climax. *J. Ecol.* 24:252–284.

Coates, D. R. 1974. Reappraisal of the glaciated Appalachian Plateau. In: Coates, D. R., ed. *Glacial geomorphology.* Binghamton, NY: State University Publications in Geomorphology: 205–243.

Coates, D. R. 1990a. Geomorphic controls of groundwater hydrology. *Geol. Soc. Amer. Special Paper* 252: 341–356.

Coates, D. R. 1990b. The relation of subsurface water to downslope movement and failure. In: Higgins, C. G.; Coates, D. R., eds. *Groundwater geomorphology: the role of subsurface water in Earth-surface processes and landforms.* Boulder, CO: Geol. Soc. Amer. Special Paper.

Costa, J. E. 1975. Effects of agriculture on erosion and sedimentation in the Piedmont Province, Maryland. *Geol. Soc. Amer. Bull.* 86:1281–1286.

Costa, J. E. 1988. Rheologic, geomorphic, and sedimentologic differentiation of water floods, hyperconcentrated flows, and debris flows. In: Baker, V. R.; Kochel, R. C.; Patton, P. C., eds. *Flood geomorphology.* New York: John Wiley & Sons: 113–122.

Cunha, S. B.; Machado, M. B.; Mousinho de Meis, M. R. 1975. Drainage basin morphometry on deeply weathered bedrocks. *Z. Geomorphologie N.F.* 19(2): 125–139.

Currie, R. G. 1984. Periodic (18.6 year) and cyclic (11 year) induced drought and flood in Western North America. *J. Geophys. Res.* 89:7215–7230.

Davis, W. M. 1899. The geographic cycle. *Geogr. J.* 14: 481–504.

DeBano, L.; Ffolliott, P.; Baker, M. 1996. *Fire severity effects on water resources, effects of fire on Madrean Province ecosystem.* RM-GTR-289. Fort Collins, CO: U.S. Dept. Agric., For. Serv., Rocky Mountain For. Range Exp. Sta.: 77–85.

Derbyshire, E. 1976. *Geomorphology and climate.* London: John Wiley & Sons.

Douglas, I. 1982. The unfulfilled promise: Earth surface processes as a key to landform evolution. *Earth Surf. Process. Landf.* 7:101.

Downs, P. W. 1995. River channel classification for channel management purposes. In: Gurnell, A.; Petts, G., eds. *Changing river channels.* New York: John Wiley & Sons: 347–365.

Downs, P. W.; Gregory, K. J. 1993. The sensitivity of river channels in the landscape. In: Thomas, D. S. G.; Allison, R. J., eds. *Landscape sensitivity.* New York: John Wiley & Sons: 15–30.

Duncan, S. H.; Ward, J. W. 1985. The influence of watershed geology and forest roads on the composition of salmon spawning gravel. *Northw. Sci.* 59(3):204–212.

Dunne, T. 1982. Models of runoff processes and their significance. In: Committee, G. S., ed. *Scientific basis of water-resource management.* Washington, DC: National Academy Press: 17–30.

Dunne, T. 1990. Hydrology, mechanics, and geomorphic implications of erosion by subsurface flow. In: Higgins, C. G.; Coates, D. R., eds. *Groundwater geomorphology: the role of subsurface water in earth-surface processes and landforms.* Boulder, CO: Geol. Soc. Amer. Special Paper 252.

Dunne, T. 1998. Critical data requirements for prediction of erosion and sedimentation in mountain drainage basins. *J. Amer. Water Resour. Assoc.* 34(4): 795–808.

Dunne, T.; Moore, T. R.; Taylor, C. H. 1975. Recognition and prediction of runoff-producing zones in humid regions. *Hydrol. Sci. Bull.* 20:305–327.

Ebbesmeyer, C.; Cayan, D.; McLain, D.; Nichols, F.; Peterson, D.; Redmond, K. 1991. 1976 step in the Pacific climate: forty environmental changes between 1968–1975 and 1977–1984. In: Betancourt, J. L.; Tharp, V. L., eds. *Proceedings of the seventh annual Pacific climate (PACLIM) workshop.* Sacramento, CA: California Department of Water Resources.

EPA (Environmental Protection Agency). 1980. *An approach to water resources evaluation of non-point silvicultural sources (a procedural handbook).* EPA-IAG-D6-0660. Athens, GA: U.S. Environ. Protect. Agency.

Frissell, C. A.; Liss, W. J.; Warren, C. A.; Hurley, M. D. 1986. A hierarchical framework for stream habitat classification: viewing streams in a watershed context. *Environ. Manage.* 10:199–214.

Galay, V. J. 1983. Causes of river bed degradation. *Water Resour. Res.* 19(5):1057–1090.

Godfrey, A. E. 1978. Land surface instability on the Wasatch Plateau, Central Utah. *Utah Geol.* 5(2):131–141.

Goldman, C. R.; Jasby, A.; Powell, T. 1989. Interannual fluctuations in primary productivity: meteorological forcing at two subalpine lakes. *Limnol. Oceanogr.* 34:310–323.

Graf, W. L. 1982. Spatial variations of fluvial processes in semi arid lands. In: Thorne, C. E., ed. *Space and time in geomorphology*. London: Allen and Unwin: 193–217.

Graf, W. L. 1988. *Fluvial processes in dryland rivers*. New York: Springer-Verlag.

Graham, J. D. 1985. *Mass movement dynamics, geomorphology and their relationship to geology in the North Fork Siuslaw River drainage basin, Oregon Coast Range*. M.S. thesis. Corvallis, OR: Oregon State University.

Grant, G. E.; Swanson, F. J. 1995. Morphology and processes of valley floors in mountain streams, Western Cascades, Oregon. In: Costa, J. E.; Miller, A. J.; Potter, K. W.; Wilcock, P. R., eds. *Natural and anthropogenic influences in fluvial geomorphology*. Washington DC: Amer. Geophys. Union: 83–101.

Grant, G. E.; Swanson, F. J.; Wolman, M. G. 1990. Patterns and origin of stepped-bed morphology in high-gradient streams, western Cascades, *Oregon. Geol. Soc. Amer. Bull.* 102:340–352.

Gregory, K. J., editor. 1985. The impact of river channelization. *Geogr. J.* 151(1):53–74.

Gregory, S. V. 1991. An ecosystem perspective of riparian zones. *BioScience* 41(8):541–551.

Gurnell, A. 1995. Vegetation along river corridors: hydrogeomorphological interaction. In: Gurnell, A.; Petts, G., eds. *Changing river channels*. New York: John Wiley & Sons: 237–260.

Gustard, A. 1992. Analysis of river regimes. In: Calow, P.; Petts, G., eds. *The rivers handbook*. Cambridge, MA: Blackwell Scientific: 28–47.

Hack, J. T. 1957. *Studies of longitudinal stream profiles in Virginia and Maryland*. 294-B, 97. Washington, DC: U.S. Geological Survey Professional Paper.

Hack, J. T. 1960. *Interpretation of erosional topography in humid temperate regions*. 258-A. Washington, DC: U.S. Geological Survey.

Hack, J. T. 1965. *Postglacial drainage evolution and stream geometry in the Ontonagon area, Michigan*. 504-B. Washington, DC: U.S. Geological Survey.

Hack, J. T. 1973. Stream-profile analysis and stream gradient index. *J. Res. U.S. Geol. Survey* 1:421–429.

Hack, J. T. 1982. *Physiographic divisions and differential uplift in the Piedmont and Blue Ridge*. 1265. Washington, DC: U.S. Geological Professional Paper.

Hack, J. T.; Goodlett, J. C. 1960. *Geomorphology and forest ecology of a mountain region in the Central Appalachians*. Washington, DC: U.S. Geological Survey Professional Paper 347:1–64.

Hare, S. E.; Mantua, N. J.; Francis, R. C. 1999. Inverse production regimes: Alaska and west coast Pacific salmon. *Fisheries* 24(1):6–14.

Harr, R. D. 1981. Some characteristics and consequences of snowmelt during rainfall in western Oregon. *J. Hydrol.* 53:277–304.

Harris, R. R. 1988. Association between stream valley geomorphology and riparian vegetation as a basis for landscape analysis in the eastern Sierra Nevada, California, USA. *Environ. Manage.* 12:219–228.

Hayden, B. P. 1988. Flood climate. In: Baker, V. R.; Kochel, R. C.; Patton, P.C., eds. *Flood geomorphology*. New York: John Wiley & Sons: 13–26.

Heath, R. C. 1984. *Ground-water regions of the United States*. Washington, DC: U.S. Geological Survey Water-Supply Paper: 2242:78.

Heed, B. H. 1981. Dynamics of selected mountain streams in the western United States of America. *Z. Geomorphology N. F.* 25(1):17–32.

Hilton, S.; Lisle, T. E. 1993. *Measuring the fraction of pool volume filled with fine sediment*. PSW-RN-414. Berkeley, CA: U.S. Dept. Agric., For. Serv, Pacific Southwest Res. Sta.

Hirschboeck, K. K. 1987. Catastrophic flooding and atmospheric circulation anomalies. In: Mayer, L.; Nash, D., eds. *Catastrophic flooding (Brighamton symposium in geomorphology)*. Boston: Allen & Unwin.

Hirschboeck, K. K. 1988. Flood hydroclimatology. In: Baker, V. R.; Kochel, R. C.; Patton, P. C., eds. *Flood geomorphology*. New York: John Wiley & Sons: 27–47.

Hirschboeck, K. K. 1991. *Hydrology of floods and droughts: climate and floods*. Washington, DC: U.S. Geological Survey Water Supply Paper: 2375:67–88.

Howard, A.; Dolan, R. 1981. Geomorphology of the Colorado River in the Grand Canyon. *J. Geol.* 89(3): 269–298.

Hughes, R. M.; Omernick, J. M. 1983. An alternative for characterizing stream size. In: Fontaine, T. D., Bartell, S. M., eds. *Dynamics of lotic ecosystems*. Ann Arbor, MI: Ann Arbor Sciences.

Hupp, C. R. 1990. Vegetation patterns in relation to basin hydrogeomorphology. In: Thornes, J. B., ed. *Vegetation and erosion*. Chichester, UK: John Wiley & Sons: 217–237.

Jarret, R. D. 1991. *Paleohydrology and its value in analyzing floods and droughts*. Washington, DC: U.S. Geological Survey Water Supply Paper: 2375: 105–116.

Jennings, M. E.; Thomas, W. O., Jr.; Riggs, H. C. 1994. *Nationwide summary of U.S. Geological Survey regression equations for estimating magnitude and frequency of floods for ungaged sites*. 94-4002. Washington, DC: U.S. Geological Survey.

Johnston, R. H. 1971. *Base flow as an indicator of aquifer characteristics in the coastal plain of Delaware*. P 0750-D. Reston, VA: U.S. Geological Survey.

Jones, J. A.; Grant, G. E. 1996. Peak flow response to clear-cutting and roads in small and large basins, western Cascades, Oregon. *Water Resour. Res.* 32(4): 959–974.

Kappesser, G. B. 1993. *Riffle stability index*. Coeur d'Alene, ID: U.S. Dept. Agric, For. Serv., Idaho Panhandle National Forest.

Keller, E. A. 1972. Development of alluvial stream channels: a five stage model. *Geol. Soc. Amer. Bull.* 83: 1531–1536.

Keller, E. A.; Kondolf, G. M. 1990. Groundwater and fluvial processes: selected observations. *Geol. Soc. Amer. Special Paper* 252:319–340.

Keller, E. A.; Melhorn, M. N. 1978. Rhythmic spacing and origin of pools and riffles. *Geol. Soc. Amer. Bull.* 89:723–730.

Keller, E. A.; Swanson, F. J. 1979. Effects of large organic material on channel form and fluvial processes. *Earth Surf. Process. Landf.* 4:361–380.

Kochel, R. C. 1988. Geomorphic impact of large floods: review and new perspectives on magnitude and frequency. In: Baker, V. R.; Kochel, R. C.; Patterson, P. C., eds. *Flood geomorphology*. New York: John Wiley & Sons: 169–187.

Kondolf, G. M. 1995a. Comment: use of pebble counts to evaluate fine sediment increase in stream channels by John. P. Potyondy and Terry Hardy. *Water Resour. Bull.* 31(3):537–539.

Kondolf, G. M. 1995b. Geomorphological stream channel classification in aquatic habitat restoration: uses and limitations. *Aquat. Conserv. Marine Freshwater Ecosystems* 5:127–141.

Kondolf, G. M.; Maloney, L. M.; Williams, J. G. 1987. Effects of bank storage and well pumping on base flow, Carmel River, Monterey County, California. *J. Hydrol.* 91(3–4):351–369.

Kondolf, G. M.; Cada, G. F.; Sale, G. F. 1991. Distribution and stability of potential salmonid spawning gravels in Steep Sierra Nevada. *Trans. Amer. Fish. Soc.* 120:177–186.

Kowall, S. J. 1976. The hypsometric integral and low streamflow in two Pennsylvania provinces. *Water Resour. Res.* 12(3):497–502.

Lane, S. N.; Richards, K. S. 1997. Linking river channel form and process: time, space and causality revisited. *Earth Surf. Process. Landf.* 22:249–260.

Langbein, W. B.; Leopold, L. B. 1966. *River meanders–theory of minimum variance*. 422-H. Washington, DC: U.S. Geological Survey Professional Paper.

Langbein, W. B.; Schumm, S. A. 1958. Yield of sediment in relation to mean annual precipitation. *Trans. Amer. Geophys. Union* 39(6):1076–1084.

Legendre, L.; Demers, S. 1984. Towards dynamic biological oceanography and limnology. *Can. J. Fisheries Aquat. Sci.* 41:2–19.

Leopold, L. B. 1991. Lag times for small drainage basins. *Catena* 18:157–171.

Leopold, L. B. 1994. *A view of the river*. Cambridge, MA: Harvard University Press.

Leopold, L. B.; Maddock, T. J. 1953. *The hydraulic geometry of stream channels and some physiographic implications*. Washington, DC: U.S. Geological Survey Professional Paper 252:57.

Leopold, L. B.; Wolman, M. G. 1957. *River channel patterns: braided, meandering, and straight*. 282-B. Washington, DC: U.S. Geological Survey.

Lewis, W. M. 1983. A revised classification of lakes based on mixing. *Aquat. Sci.* 40:1779–1787.

Lisle, T. E. 1982. Effects of aggradation and degradation on riffle-pool morphology in natural gravel channels, northwestern California. *Water Resour. Res.* 18(6): 1643–1651.

Lisle, T. E. 1983. The role of structure in the physical habitat of salmonids. In: Toole, S., ed. *Reports of the first California salmon and steelhead restoration conference*. Bodega Bay, CA: California Sea Grant College Program.

Lisle, T. E. 1986. Stabilization of a gravel channel by large streamside obstructions and bedrock bends, Jacoby Creek, northwestern California. *Geol. Soc. Amer. Bull.* 97:999–1011.

Mackin, J. H. 1948. Concept of the graded river. *Bull. Geol. Soc. Amer.* 59:463–512.

Maddox, R. A.; Canova, F.; Hoxit, L. R. 1980. Meteorological characteristics of flash flood events over the western United States. *Mon. Weather Rev.* 108:1866–77.

Marple, R. T.; Talwani, P. 1993. Evidence of possible tectonic upwarping along the South Carolina coastal plain from an examination of river morphology and elevation data. *Geology* 21:651–654.

Marston, R. A. 1978. *Morphometric indices of streamflow and sediment yield from mountain watersheds in western Oregon*. 00-04t0-9-73. Corvallis, OR: U.S. Dept. Agric., For. Serv., Siuslaw National Forest.

Mather, P. M.; Doornkap, J. C. 1970. Multivariate analysis in geography. *Trans. Inst. Brit. Geogr.* 51:163–187.

Maxwell, J. R.; Edwards, C. J.; Jensen, M. E.; Paustian, S. J.; Parrott, H.; and Hill, D. M. 1995. *A hierarchical framework of aquatic ecology units in North America*. NC-GTR-176. St. Paul, MN: U.S. Dept. Agric., For. Serv., North Central For. Exp. Sta.

McDowell, P. F. 1983. Evidence of stream response to Holocene climatic change in a small Wisconsin watershed. *Quaternary Res.* 19:100–116.

Meade, R. H.; Yuzyk, T. R.; Day, T. J. 1990. Movement and storage of sediment in rivers of the United States and Canada. In: Wolman, M. G.; Riggs, H. C., eds. *Surface water hydrology*. Boulder, CO: Geol. Soc. Amer.: 255–280.

Megahan, W. F. 1996. A cross section of procedures used to assess cumulative watershed effects on forest lands in the U.S. In: Neary, D.; Ross, K. C.; Coleman, S. S., eds. *National hydrology workshop proceedings*. Phoenix, AZ: U.S. Dept. Agric., For. Serv.: 7–14.

Merritts, D.; Vincent, K. R. 1989. Geomorphic response of coastal streams to low, intermediate and high rates of uplift, Mendocino triple junction region, northern California. *Geol. Soc. Am. Bull.* 101:1373–1388.

Meyer, G. A.; Wells, S. G.; Jull, A. J. T. 1995. Fire and alluvial chronology in Yellowstone National Park: climatic and intrinsic control on Holocene geomorphic processes. *Geol. Soc. Amer. Bull.* 107:1211–1230.

Miller, J. P. 1958. *High mountain streams: effects of ge-*

ology on channel characteristics and bed material. Memoir 4. Albuquerque, NM: State Bureau of Mines and Mineral Resources, New Mexico Institute of Mining and Technology.

Miller, J. R.; Ritter, D. F. 1990. Morphometric assessment of lithologic controls on drainage basin evolution in the Crawford upland, south-central Indiana. *Amer. J. Sci.* 290:569–599.

Miller, J. R.; Frederick, R. H.; Tracey, R. J. 1973. *Precipitation frequency atlas of the western United States, 9.* Washington, DC: Natl. Oceanic and Atmospheric Admin.

Miller, T. K.; Onesti, L. J. 1979. The relationship between channel shape and sediment characteristics in the channel perimeter. *Bull. Geol. Soc. Amer.* I(90): 301–304.

Milne, J. A. 1982. Bed forms and bend-arc spacing of some coarse-bedload channels in upland Britain. *Earth Surf. Process. Landf.* 7:227–240.

Minckley, W. L.; Hendrickson, D. A.; Bond, C. E. 1986. Geography of western North American freshwater fishes: description and relationships to intracontinental tectonism. In: Hocutt, H. C.; Wiley, E. O., eds. *The zoogeography of North American freshwater fishes.* New York: John Wiley & Sons.

Montgomery, D. R.; Buffington, J. M. 1993. *Channel classification, prediction of channel response, and assessment of channel condition.* TWF-SH10-93-002. Olympia, WA: Washington State Department of Natural Resources—Timber Fish and Wildlife.

Montgomery, D. R.; Buffington, J. M. 1997. Channel-reach morphology in mountain drainage basins. *Geol. Soc. Amer. Bull.* 109(5):2–17.

Montgomery, D. R.; Buffington, J. M.; Smith, R. D.; Schmidt, K. M.; Press, G. 1995. Pool spacing in forest channels. *Water Resour. Res.* 31(4):1097–1104.

Montgomery, D. R.; Buffington, J. M.; Peterson, N. P.; Quinn, T. P.; Schuett-Hames, D. 1996. Stream-bed scour, egg burial depths, and the influence of salmonid spawning on bed surface mobility and embryo survival. *Can. J. Fisheries Aquat. Sci.* 53:1061–1070.

Morisawa, M. 1985. *Rivers: form and process.* Geomorphology texts. London and New York: Longman.

Moscrip, A. L.; Montgomery, D. R. 1997. Urbanization, flood frequency, and salmon abundance in Puget lowland streams. *J. Amer. Water Resour. Assoc.* 33(6): 1289–1297.

Mosley, M. P. 1980. The influence of organic debris on channel morphology and bedload transport in a New Zealand forest stream. *Earth Surf. Process. Landf.* 6: 571–579.

Nanson, G. C.; Croke, J. C. 1992. A genetic classification of floodplains. *Geomorphology* 4:459–486.

Nolan, K. M.; Marron, D. C. 1985. Stream-channel response to major storms in two mountainous areas of California. *Geology* 13:135–138.

O'Conner, J. E.; Webb, R. H.; Baker, V. R. 1986. Paleohydrology of pool and riffle development. Boulder Creek, UT: *Geol. Soc. Amer. Bull.* 97:410–420.

Ohmori, H.; Shimazu, H. 1994. Distribution of hazard types in a drainage basin and its relation to geomorphological setting. *Geomorphology* 10:95–106.

Olsen, D. S.; Whitaker, A. C.; Potts, D. F. 1997. Assessing stream channel stability thresholds using flow competence estimates at bankfull stage. *J. Amer. Water Resour. Assoc.* 33(6):1197–1207.

Onda, Y. 1992. Influence of water storage capacity in the regolith zone on hydrological characteristics, slope processes, and slope form. *Z. Geomorphologie* 36: 165–178.

Orlanski, I. 1975. A rational subdivision of scales for atmospheric processes. *Bull. Amer. Meteorol. Soc.* 56: 527–530.

Osterkamp, W. R.; Hedman, E. R. 1982. *Perennial-streamflow characteristics related to channel geometry and sediment in the Missouri River basin.* 1242. Washington, DC: Geological Survey Professional Paper.

Ouichi, S. 1985. Response of alluvial rivers to slow active tectonic movement. *Geol. Soc. Amer. Bull.* 96: 504–515.

Park, C. C. 1977. World-wide variations in hydraulic geometry exponents of stream channels: an analysis and some observations. *J. Hydrol.* 33:133–146.

Pearce, A. J.; Watson, A. J. 1986. Effects of earthquake-induced landslides on sediment budget and transport over a 50 yr. period. *Geology* 14:52–55.

Peltier, L. C. 1975. Concepts of climate geomorphology. In: Milhorn, W. N.; Flemal, R. C., eds. *Theories of landform development.* London: Allen and Unwin: 129–143.

Perkins, S. J. 1989. Landslide deposits in low-order streams—their erosion rates and effects on channel morphology. In: Woessner, W. W.; Potts, D. F., eds. *Symposium proceedings on headwater hydrology.* Missoula, MT: Amer. Water Resour. Assoc.: 173–182.

Petts, G.; Foster, I. 1985. *Rivers and landscape.* London: Edward Arnold.

Pickup, G.; Warner, R. F. 1976. Effects of hydrologic regime on magnitude and frequency of dominant discharge. *J. Hydrol.* 29:51–75.

Piechota, T. C.; Dracup, J. A. 1996. Drought and regional hydrologic variation in the United States: associations with El Nino—southern oscillation. *Water Resour. Res.* 32(5):1359–1373.

Pitlick, J. 1994. Relation between peak flows, precipitation, and physiography for five mountainous regions in the western USA. *J. Hydrol.* 158:219–240.

Platts, W. S.; Megahan, W. F.; Minshall, G. W. 1983. *Methods for evaluating stream, riparian, and biotic conditions.* INT-GTR-138. Ogden, UT: U.S. Dept. Agric., For. Serv., Intermountain For. Range Exp. Sta.

Poff, N. L. 1996. A hydrogeography of unregulated streams in the United States and an examination of scale dependence in some hydrological descriptors. *Freshwater Biol.* 36:71–91.

Poff, N. L.; Ward, J. V. 1989. Implication of streamflow variability and predictability for lotic community structure: a regional analysis of streamflow patterns. *Can. J. Fisheries Aquat. Sci.* 46:1805–1817.

Poff, N. L.; Ward, J. V. 1990. Physical habitat template and recovery. *Environ. Manage.* 14(5):629–645.

Potyondy, J. P.; Hardy, T. 1994. Use of pebble counts to evaluate fine sediment increase in stream channels. *Water Resour. Bull.* 30(3):509–520.

Poulin, V. A. 1991. *Fish–forestry interactions program: summary. Part 1: extent and severity of mass wasting on the Queen Charlotte Islands and impact on fish habitat and forest sites.* Victoria, BC: British Columbia Ministry of Forests and Lands.

Pultwarty, R.; Redmond, K. 1997. Climate and salmon restoration in the Columbia River Basin: the role and usability of seasonal forecasts. *Bull. Amer. Meteorol. Soc.* 7(3):381–397.

Quigley, T. M.; Arbelbide, S. J. 1997. *An assessment of ecosystem components in the interior Columbia basin and portions of the Klamath and Great Basins.* PNW-GTR-405. Portland, OR: U.S. Dept. Agric., For. Serv., Pacific Northw. Res. Sta.

Ralph, S. C.; Poole, G. C.; Conquest, L. L.; Naiman, R. J. 1994. Stream channel morphology and wood debris in logged and unlogged basins of western Washington. *Can. J. Fisheries Aquat. Sci.* 51:37–48.

Redmond, K. T.; Koch, R. W. 1991. Surface climate and streamflow variability in the western United States and their relationship to large-scale circulation indices. *Water Resour. Res.* 27(9):2381–2399.

Reeves, G. H.; Benda, L. E.; Burnett, K. M.; Bisson, P. A.; Sedell, J. R. 1995. A disturbance-based ecosystem approach to maintaining and restoring freshwater habitats of evolutionarily significant units of anadromous salmonids in the Pacific Northwest. *Amer. Fish. Soc. Symp.* 17:334–349.

Rhea, S. 1993. Geomorphic observations of rivers in the Oregon Coast Range from a regional reconnaissance perspective. *Geomorphology* 6:135–150.

Richter, B. D.; Baumgartner, J. V.; Wigington, R.; Braun, D. P. 1997. How much water does a river need? *Freshwater Biol.* 37:231–249.

Rogers, J. D.; Armbruster, J. T. 1990. Low flows and hydrologic droughts. In: Wolman, M. G.; Riggs, H. C., eds. *Surface water hydrology.* Boulder, CO: Geol. Soc. Amer.: 121–130.

Rosgen, D. L. 1985. A stream classification system. In: *Proceedings of the first North American riparian conference, riparian ecosystem and their management: reconciling conflicting uses.* RM-GTR-120; Tucson, AZ. Fort Collins, CO: U.S. Dept. Agric., For. Serv., Rocky Mountain For. Range Exp. Sta.

Schroeder, K.; Savonen, C. 1997. Lessons from floods. *Fisheries* 22(9):14–16.

Schuett-Hames, D.; Pleus, A.; Bullchild, L.; Hall, S. 1993. *Ambient monitoring program manual.* TFW-AM9-93-001. Olympia, WA: Washington State Dept. of Natural Resources; Timber, Fish, and Wildlife.

Schumm, S. A. 1974. Geomorphic thresholds and complex response of drainage systems. In: Morisawa, M. E., ed. *Fluvial geomorphology.* Binghampton: State University of New York.

Schumm, S. A. 1977. *The fluvial system.* New York: John Wiley & Sons: 2–16.

Schumm, S. A.; Lichty, R. W. 1965. Time, space, and causality in geomorphology. *Amer. J. Sci.* 263:110–119.

Sear, D. A. 1994. River restoration and geomorphology. *Aquat. Conserv. Marine Freshwater Ecosystems* 4: 169–177.

Shepherd, R. G. 1985. Regression analysis of river profiles. *J. Geol.* 93:377–384.

Sidle, R. C. 1992. A theoretical model of the effects of timber harvesting on slope stability. *Water Resour. Res.* 28:1897–1910.

Simon, A. 1994. *Gradation processes and channel evolution in modified west Tennessee streams: process, response, form.* 1470. Denver, CO: U.S. Government Printing Office.

Sommer, U. 1985. Seasonal succession of phytoplankton in Lake Constance. *BioScience* 35(6):351–357.

Southwood, T. R. E. 1977. Habitat, the template for ecological strategies? *J. Anim. Ecol.* 46:337–365.

Stack, W. R.; Beschta, R. L. 1989. Factors influencing pool morphology in Oregon coastal streams. In: Woessner, W. W.; Potts, D. F., eds. *Symposiums proceedings on headwater hydrology.* Missoula, MT: Amer. Water Resour. Assoc.: 401–411.

Stanford, J. A.; Ward, J. V. 1993. An ecosystem perspective of alluvial rivers: connectivity and the hyporheic corridor. *J. N. Amer. Benthol. Soc.* 12(1):48–60.

Strahler, A. N. 1957. Quantitative analysis of watershed geomorphology. *Amer. Geophys. Union Trans.* 38: 913–920.

Strayer, D. 1983. The effects of surface geology and stream size on freshwater mussel (*Bivalvia, Uniodiae*) distribution in southeastern Michigan, USA. *Freshwater Biol.* 13:253–264.

Sugden, D. E.; Summerfield, M. A.; Burt, T. P. 1997. Editorial: linking short-term geomorphic processes to landscape evolution. *Earth Surf. Process. Landf.* 22: 1193–1194.

Swanson, F. J. 1979. Geomorphology and ecosystems. In: Richard, W., ed. *Forests: fresh perspectives from ecosystem analysis, proceedings 40th annual biology colloquium.* Corvallis, OR: Oregon State University.

Swanson, F. J. 1981. *Fire and geomorphic processes.* WO-26, 401–420. Washington, DC: U.S. Dept. Agric., For. Serv.

Swanson, F. J.; Leinkaemper, G. W. 1985. Geological zoning of slope movements in Western Oregon, USA. In: *USA Proceedings of the IVth international conference and field workshop on landslides*, Tokyo, Japan. Portland, OR: U.S. Dept. Agric., For. Serv.: 41–46.

Swanson, F. J.; Fredricksen, R. L.; McCorison, F. M. 1982. Material transfer in a western Oregon forested watershed. In: Edmonds, R. L., ed. *Analysis of coniferous forest ecosystems in the western United States.* Stroudsburg, PA: Hutchingson Ross Publishing Co.

Swanson, F. J.; Graham, R. L.; Grant, G. E. 1985. Some effects of slope movements on river channels. In: *In-*

ternational symposium on erosion, debris flow, and disaster prevention, Tsukba, Japan: 273–278.

Swanson, F. J.; Benda, L. E.; Duncan, S. H.; Grant, G. E.; Megahan, W. F.; Reid, L. M.; Ziemer, R. R. 1987. Mass failures and other processes of sediment production in Pacific Northwest forest landscapes. In: Salo, E. O.; Cundy, T. W., eds. *Streamside management: forestry and fishery interactions*. Seattle, WA: Institute of Forest Resources, University of Washington: 9–38.

Swanston, D. N. 1978. The effect of geology on soil mass movement activity in the Pacific Northwest. In: Youngberg, C. T., ed. *Forest soils and land use; proceedings of the fifth North American forest soils conference*. Fort Collins, CO: Colorado State University: 89–115.

Swanston, D. N. 1991. Natural processes. In: Meehan, W. R., ed. *Influences of forest and rangeland management on salmonid fishes and their habitat*. Bethesda, MD: Amer. Fisheries Special Publication: 139–179.

Swetnam, T. W.; Betancourt, J. L. 1990. Fire–southern oscillation relations in the Southwestern United States. *Science* 249:1017–1020.

Taylor, A. B.; Schwartz, H. E. 1952. Unit hydrograph lag and peak flow related to basin characteristics. *Amer. Geophys. Union Trans.* 33:235–246.

Thomas, M. P. 1966. *Effects of glacial geology upon the time distribution of streamflow in eastern and southern Connecticut*. 550-B, 209–212. Washington, DC: U.S. Geological Survey.

Thompson, A. 1986. Secondary flows and the pool–riffle unit: a case study of the processes of meander development. *Earth Surf. Process. Landf.* 11:631–641.

Thornes, J. B. 1983. Evolutionary geomorphology. *Trans. Inst. Brit. Geogr.* 8:225–235.

Thornthwaite, C. W. 1948. An approach toward a rational classification of climate. *Geol. Rev.* 38:55–94.

Touysinhthiphonexaym, K. C. N.; Gardner, T. W. 1984. Threshold response of small streams to surface coal mining, bituminous coal fields, central Pennsylvania. *Earth Surf. Process. Landf.* 9:43–58.

Tricart, C. R. 1974. *Structural geomorphology*. New York: Longman.

Vannote, R. L.; Minshall, G. W.; Cummins, K. W.; Sedell, J. R.; Cushing, C. E. 1980. The river continuum concept. *Can. J. Fisheries Aquat. Sci.* 37:130–137.

VanSickle, J.; Beschta, R. L. 1983. Supply-based models of suspended sediment transport in streams. *Water Resour. Res.* 19(3):768–778.

Varnes, D. J., editor. 1979. *Slope movement types and processes. Landslides: analysis and control*, 176. Natl. Academy Sci. Special Report: 11–33.

Webb, R. H.; Betancourt, J. L. 1990. *Climate effects on flood frequency: an example from southern Arizona*. Sixth annual Pacific climate workshop, Tech. Rep. 23. Sacramento, CA: Interagency Ecology Study Program, California Department of Water Resources: 62–66.

Wharton, G. 1995. Information from channel geometry discharge relations. In: Gurnell, A.; Petts, G., eds. *Changing river channels*. New York: John Wiley & Sons: 325–340.

Wheeler, D. A. 1979. The overall shape of longitudinal profiles of streams. In: Pitty, A. F., ed. *Geographical approaches to fluvial processes*. Norwich: Geo Abstracts: 300.

White, E. L. 1975. Factor analysis of drainage basin properties: classification of flood behavior in terms of basin geomorphology. *Water Resour. Bull.* 11(4):676–687.

Whiting, P. J.; Pomerants, M. 1997. A numerical study of bank flow and its contribution to streamflow. *J. Hydrol.* 202:121–136.

Williams, G. P. 1987. *Unit hydraulic geometry—an indicator of channel changes*. Washington, DC: U.S. Geological Survey Water Supply Paper 2230:77–89.

Winkler, R. D.; Rothwell, R. L. 1983. Biogeoclimatic classification system for hydrological interpretation. *Can. J. For. Res.* 13:1043–1050.

Winter, T. C. 1977. Classification of the hydrologic settings of lakes in the north central United States. *Water Resour. Res.* 13:752–767.

Wolman, M. G.; Gerson, R. 1978. Relative scales of time and effectiveness of climate in watershed geomorphology. *Earth Surf. Process. Landf.* 3:189–208.

Wood-Smith, R. D.; Buffington, J. M. 1996. Multivariate geomorphic analysis of forest streams: Implications for assessment of land use impacts on channel conditions. *Earth Surf. Process. Landf.* 21:377–393.

26 Assessment of Biotic Patterns in Freshwater Ecosystems

Christopher A. Frissell, N. LeRoy Poff, and Mark E. Jensen

26.1 Introduction

Whether curse or blessing, scientists and managers have developed a multitude of approaches to the assessment of patterns in the biota of freshwater ecosystems. Several rather independent traditions in aquatic ecology have focused on the relationship of ecological pattern to physical pattern and process. Approaches that focus on broad-scale, regional patterns may completely miss aspects of the ecosystem that are critical to the survival of rare taxa in special habitats, just as approaches that focus on rare species or endangered populations may provide little information about the broader biotic community and its response to an ecological change (Table 26.1). Today's natural resource laws, societal expectations, limited fiscal resources, and declining biotic resources demand that these once separate strands converge in a more holistic approach to ecological assessment. This brief review cannot aspire to map a clear path toward meeting that major challenge, but is intended only to till the ground for future synthesis.

This chapter reviews selected examples from the literature that illustrate varied approaches to assessing and understanding the underlying causes of biological patterns in streams, rivers, and lakes. Some theoretical concepts are examined to help to explain the complementary capabilities and purposes of the various approaches employed in these studies. Most of our examples are drawn from the peer-reviewed scientific literature, as these are the best-documented cases.

Human actions, which affect terrestrial and atmospheric processes inexorably, affect the freshwater ecosystems that are nested in the same landscape. Thus assessment of ecosystems at the scale of whole watersheds can provide crucial information for understanding the dynamic responses of the diverse freshwater ecosystems residing within them.

26.2 Biotic Patterns in Streams

26.2.1 Temporal and Spatial Scales of Variation

To date, most efforts to assess responses of stream biota to environmental stresses rest on unexamined, implicit assumptions about the scales of ecological function and response. Scaling of response in freshwater ecosystems depends on two major and partly independent aspects of their biophysical organization. First, physical morphology—of the catchment, the stream network, and the lakes and other water bodies within that network—determines where different kinds of environmental stresses or disturbance events originate and how such events are translated or propagated across space and time through the network. In running-water ecosystems, strong physical connectivity downstream and laterally (and biotic connectivity upstream and laterally; Pringle, 1997) affords ample opportunity for rapid, direct transfer of disturbance effects. Second, the biological consequences of disturbances or stresses also depend on the life histories, interactions, and spatial distribution of the species present in the aquatic ecosystem (Southwood, 1977; Townsend and Hildrew, 1994). Both the physical and biotic aspects of scaling are crucial to understanding and predicting biotic response to environmental change (Statzner and Higler, 1986; Poff and Ward, 1989; Naiman et al., 1992; Townsend and Hildrew, 1994; Biggs, 1995; Ebersole et al., 1997; Poff, 1997). Seldom, however, have both as-

TABLE 26.1. Selected levels of biological organization and methods commonly used to assess biotic response to environmental change in aquatic ecosystems, descending from large-scale, low-resolution, or regionally focused data downward to fine-scale, high-resolution or locally focused data.

Level of resolution		*Assessment method*							
Biotic response unit	*Examples of data measures*	*Index of biotic integrity*	*Distribution mapping*	*Gap Analysis Program (GAP)*	*Change in taxon distribution*	*Qualitative status and trends*	*Quantitative trend or PVA*	*Habitat use, behavior*	*Genetic studies*
Assemblage, guild	Presence/absence	×	×	×	×	×			
	Relative abundance	×	×	×	×	×			
Species, subspecies, race	Absolute abundance and distribution	×	×	×	×	×	×	×	×
Metapopulation	Rate of change in abundance across multiple sites		×		×	×	×	×	×
Population or subpopulation	Rate of change in abundance or density at a site; habitat occupation		×		×	×	×	×	×
Family group or social group	Social behavior, mate selection, territoriality						×	×	×
Individual	Individual identity, health or well-being; foraging behavior; movement and migration	×						×	×

Horizontal relationships are meant to indicate loose association, not strict relationships, and × indicates the method is commonly applied to the level of resolution or biotic response units indicated. PVA is population viability analysis. Empty cells imply domains not commonly covered by conventional applications, where future assessments might benefit from novel conceptual and methodological applications.

pects been explicitly considered in research or comprehensively addressed in environmental assessments. Some scientists have begun to examine scale issues directly in recent research (e.g., Amoros et al., 1987b; Roth et al., 1996; Allan et al., 1997; Wiley et al., 1997); as a result, previous assumptions about appropriate scales for assessment and management intervention or regulation are coming into question.

In one such study, Wiley et al. (1997) examined time series data for fish and stream insects from multiple streams in Michigan to determine the proportional contribution of spatial and temporal variation to overall variance in local population density. They found that the spatial and temporal patterns in density appeared at different scales for different taxa. Trout populations and *Baetis* mayflies tended to vary somewhat synchronously among streams across the study region, suggesting that populations responded to common regional factors, such as the influence of climatic variation on groundwater levels, affecting reproduction or survival similarly among streams. By contrast, populations of two caddisflies, *Glossosoma nigrior* and *Goera stylata*, varied asynchronously among streams. Wiley et al. (1997) suggest that the *Glossosoma* densities were regulated primarily by localized outbreaks of sporozoan pathogens; *Goera* populations expanded in response to competitive release in years and at sites where *Glossosoma* declined. The most important point to emerge from the work of Wiley et al. (1997) is that, in the absence of an extended time series of biological data across the study sites (at least 20 years for trout populations and 5 years for insect populations), much of the variation observed among stream biota could have been ascribed mistakenly to geographic structuring of populations, not interannual population fluctuations at particular sites. This same point has been made for lakes (Magnuson et al., 1990). The need for appropriate temporal dimensionality is often overlooked in spatial studies based on one-time samples, when all observed variation among sites is commonly assumed to be associated with spatial processes.

For many years, stream ecologists have focused on local conditions within the channel and its adjacent riparian zone as the factors primarily controlling biotic responses to land use. Even the River Continuum Concept of Vannote et al. (1980), a notable attempt to synthesize many aspects of stream ecosystem structure and function in a land- or riverscape perspective, appeared to presume that land use and other habitat-altering impacts had significance only as site-specific deviations from the larger-scale norm dictated by stream order. Regulatory mechanisms (e.g., Best Management Practices) have similarly focused on site-specific application of procedures or technologies to ameliorate local consequences of human activity (Frissell and Bayles, 1996; Allan et al., 1997). However, Warren (1979) and Frissell et al. (1986) suggested that watershed- or catchment-scale processes and conditions interact hierarchically with local processes and conditions to regulate stream habitat and biotic conditions. Aspects of this approach were later folded into revisions of the River Continuum Concept (Minshall, 1988), elaborated in the form of patch-mosaic theory by Pringle et al. (1988) and integrated into comprehensive theoretical views of stream community organization by Townsend and Hildrew (1994) and Poff (1997) and of riverine floodplain landscapes by Ward (1997).

Underlying these theoretical developments is the generalization that the physical dynamics of stream ecosystems can be viewed as a hierarchy of nested processes and habitats that vary from large-scale systems that vary slowly and are disrupted infrequently down to smaller habitat patches that change more rapidly and are more frequently disrupted (see Figure 24.2). A second tenet is that different kinds of organisms, with different life histories, biological requirements, interactions with other biota, and spatial distributions, will vary differently in response to different scales and processes of habitat change. Thus biotic responses to any environmental event will not be uniform and synchronous; rather, they will be complex and will often appear taxon- or guild-specific (Townsend and Hildrew, 1994; Lewis et al., 1996; Frissell et al., 1997; Poff, 1997). Large-scale and long-term physical dynamics may constrain the species pool of an entire region or drainage basin, while prevailing local physical processes and conditions constrain the actual biotic community that develops from that species pool at the scale of a particular microhabitat patch (Figure 26.1). Until recently, however, there have been few attempts to empirically test the underlying assumptions of these multiscaled conceptual models at appropriate spatial scales.

In a compelling set of studies, Roth et al. (1996) and Allan et al. (1997) explored the response of streams throughout a Michigan river basin to various land-use activities, attempting to identify the

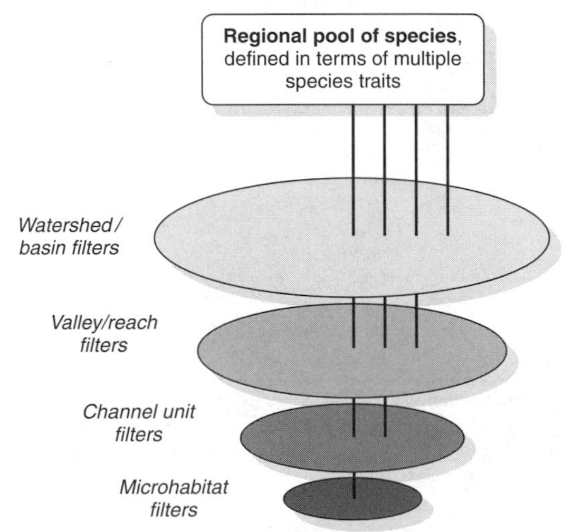

FIGURE 26.1. Conceptual model of stream community organization structured by nested habitat filters operating across a range of spatial scales (from Poff, 1997; similar to Tonn et al., 1990, for lakes). In this view, biotic patterns are determined not only by local biophysical conditions within local microhabitat patches, but also by the larger-scale mosaic of habitat types and regional physical processes that confer long-term constraints on the pool of species available to colonize a given site. Biotic assessments must consider not only effects at the spatial and temporal scale of the target project or program (typically a single site or restricted set of sites), but also larger-scale effects that have occurred or could foreseeably arise from aggregate or cumulative local changes in habitat templates and/or the species pool.

scales at which land-use impacts on stream ecosystems should be evaluated. Their work demonstrated that catchment-wide land use (measured as percent area in agricultural land use) explained a vastly larger share of site to site variation in stream habitat conditions and biotic integrity of fish populations than did indices of local riparian condition. The strong implication of similar research (e.g., Steedman, 1988; Corkum 1989, 1990; Quinn and Hickey, 1990; Fore et al., 1996; Richards et al., 1996) is that site-specific environmental assessments, which fail to account for the larger-scale, cumulative, basin-wide consequences of catchment alteration (by anthropogenic or natural processes), will not accurately characterize biotic response to environmental change.

Environmental assessments do not make a credible and comprehensive effort to address incremental or local effects of activities or projects in the context of catchment-wide processes and conditions (Schlosser, 1990; Stanford and Ward, 1992; Frissell and Bayles, 1996). Indeed, there is a recognized scarcity of decision-making mechanisms to avert the consequences of catchment-scale effects, which often encompass and implicate many landowners and jurisdictions. Widespread recognition that the effects of nonpoint sources of water pollution are hampering or overriding efforts to achieve water quality goals under the U.S. Clean Water Act is but one manifestation of this problem (Karr, 1991; Hunsaker and Levine, 1995; Naiman et al., 1995; Allan et al., 1997).

26.2.2 Indexes of Biotic Integrity

The index of biotic integrity (IBI) approach is becoming widely accepted as a standard method for ecological assessment. This approach, based on biological indicators of ecosystem state, is not intrinsically hierarchical in a geographical sense (although nothing would preclude it from being so adapted). As commonly applied, the IBI approach rests on the identification and sampling of representative *reference sites* within a larger ecoregion (Hughes et al., 1986; Omernik, 1987; Fore et al., 1996). Within the ecoregion, all other streams of roughly the same basic physical context (stream order, drainage area, elevation, etc.) are assumed to have the potential to exhibit the same biotic structure observed at reference sites. Following comprehensive biotic sampling of streams with a range of known conditions and environmental stresses, an index of biotic integrity (Steedman, 1988; Karr, 1991) is developed for a particular segment of the biota (typically, for fish or macroinvertebrates, sometimes algae). These *multimetric* indexes (sensu Karr and Chu 1997) incorporate data from several independent methods and biological levels of organization (Table 26.1). Fish IBIs, for example, often combine data on species richness, proportional abundance of certain taxa or life stages known to be sensitive to prevalent environmental stresses, and the incidence of external disease symptoms on individual fish as an indicator of sublethal stress. Sampling and data are typically organized at the level of a stream reach (sensu Frissell et al., 1986) or site. These data are combined in the form of a numerical index whose overall rating provides a relative ranking of how far the observed biotic community diverges from the community that we would expect to observe at an unimpacted reference site. Furthermore, the particular aspects of the index that contribute to a low rating can provide some insight into the nature of the environmental stress (e.g., absence of thermally tolerant species or depletion or proliferation of certain trophic specialists such as algivores) (Muntkittrick and Dixon, 1989). Most variations of the IBI have been developed for streams, but similar applications to lakes and other ecosystems are feasible (Minns et al., 1996; Karr and Chu, 1997).

Where they have been developed, tested, and widely applied, IBIs and similar multimetric bioindicator approaches can serve as convenient and effective monitoring tools. However, IBIs cannot be widely applied without careful regional calibration to accommodate geographical variation in species distribution and ecological tolerances (Fausch et al., 1984; Steedman, 1988). Tailoring and testing IBIs for each ecoregion can be an intensive and time-consuming process. Many nuances and criticisms of the use of multimetric indices in IBI-type applications are discussed in detail by Karr and Chu (1997).

In application to assessments of regional scope, certain limitations of the IBI approach sometimes are overlooked. First, some ecoregions are so pervasively altered by human activity that true reference sites are impossible to locate. The nature of unimpaired communities may be unknown, and the magnitude of biodiversity loss from past impacts will go unmeasured. Second, some ecoregions, particularly those in montane areas, where lithology is diverse or where karst groundwaters occur, have highly heterogeneous stream habitats. In such a region, it is not clear that even a large set of reference sites is sufficient to provide a reliable indication of a potential biotic community that could be expected to occur across the region. The more detailed stratification and interpretation of biotic data that may be necessary to cope with such intraregional variability may be so great as to preclude simple monitoring and assessment. Third, partly because the U.S. Environmental Protection Agency draws ecoregion boundaries without regard to watershed drainage divides (Omernik, 1987; Omernik and Bailey, 1997), many streams cross ecoregion boundaries, and their biota may manifest (often in idiosyncratic ways) the ecological influences of all the ecoregions that they traverse (Corkum, 1989, 1990; Culp and Davies, 1982). The latter problem becomes more acute if, to cope with unacceptable intraregional variability, mappers decide to further subdivide ecoregions (sometimes termed *subecoregions*) (Bryce and Clarke, 1997). The smaller the regional units mapped, the larger the proportion of streams that will be confounded by boundary effects. These difficulties are not necessarily insurmountable, and to the extent that they are based on patterns of biotic response that are demonstrated to be general and repeatable, the IBI and related multimetric, bioindicator approaches hold much promise for field assessment.

26.2.3 Guild-based Approach

The composition of local assemblages can be viewed nontaxonomically in terms of the traits possessed by the assemblage members. Such a trait- or guild-based approach allows biological communities to be compared across biogeographic boundaries, because the traits, not the species, are the unit of measurement. This coarse-grained approach has utility insofar as traits can be identified that are sensitive or responsive to measurable habitat variables that change naturally across, or in response to human modification of, the landscape. For example, Poff and Allan (1995) characterized fish species in the upper Midwest in terms of five life history and other biological traits hypothesized to be sensitive to hydrologic variability (a surrogate for habitat stability). In an analysis of the fish assemblages in 34 streams, they found that these ecological traits did correlate in a meaningful way with several measures of hydrologic variability. Interestingly, when assemblages were defined by strictly taxonomic criteria without regard to ecological information, there was no clear correspondence of the resulting assemblages with hydrologic variability. Establishing such relationships can help in predicting how assemblages may change in response to changes in important environmental or habitat variables. Increasingly, species-level data are being assembled for a number of ecological traits, including trophic requirements, reproductive behavior and habitat, life-history patterns, pollution tolerance, and response to disturbance (see Poff, 1997, for a review). This approach, while insensitive to individual species identity, has promise for assessing assemblage response to habitat change and comparing assemblages across broad scales.

26.2.4 Assessment of Hydrologic Variability in Rivers

Most larger rivers in the temperate world (Dynesius and Nilsson, 1994) and many smaller ones are altered by flow regulation, dams, and diversions. Flow regulation systematically affects many aspects of stream ecosystems and their biota. The life histories of the species indigenous to any river system are adapted to the naturally prevailing hydrologic regime of the stream (Amoros et al., 1987b; Poff and Ward, 1989, 1990; Poff et al., 1997), and alteration of flow patterns directly affects biota even when other aspects of the stream system (e.g., channel morphology, riparian vegetation) remain largely unchanged. Over time, flow alterations often cause substantial, deep-seated changes in stream ecosystems, including changes in geomorphic structure and dynamics, riparian and floodplain vegetation, and the exchange of surface and groundwater, all of which have additional impacts on biota (Bravard et al., 1986; Amoros et al., 1987a, b; Decamps et al., 1988; Johnson, 1994; Ligon et al., 1995; Stanford et al., 1996; Poff et al., 1997; Ward, 1997).

Recent studies have demonstrated that hydrologic regimes of streams and rivers can and should be characterized not just in terms of mean states of water flow across seasons, but also in terms of variability, constancy, and predictability of flow variation at several time scales, from hours to decades (Amoros et al., 1987a, b; Poff and Ward, 1989, 1990; Stanford and Ward, 1992; Poff, 1996; Richter et al., 1996, 1997). Virtually any of these aspects of flow regime may be altered by river regulation or other hydrologic changes in the catchment. Hydrologic variability and alterations can best be evaluated where detailed hydrographic records exist or can be accurately synthesized; but if these are not available, similar information may be developed through intraregional comparisons of river flow pattern (Poff, 1996). Restoration or rehabilitation of streams and rivers must include returning flow regimes to some semblance of their natural–historical flow regime (Amoros et al., 1987a, b; Stanford et al., 1996; Poff et al., 1997; Waru, 1997), making such assessments central to conservation and restoration efforts.

26.3 Biotic Patterns in Lakes

26.3.1 Historical Development

Historically, most scientific assessments of biotic differences among lakes addressed the ability of lakes to produce biomass of fishes or other harvestable resources (e.g., Matusek, 1978; Wetzel, 1983). More recently, limnologists and aquatic ecologists have begun to examine the life-history requirements and interactions among lake-dwelling species and the specific mechanisms by which differences among lakes and changes within lakes over time shape the organization and dynamics of their biotic communities. For example, Murphy et al. (1990) compared the life history and ecological strategies of submergent macrophyte species in four physically contrasting lakes, and Evans et al. (1996) reconstructed the limnological changes that could explain changes in fish populations in Lake Simcoe, Ontario, over more than a century. The groundbreaking work of Carpenter et al. (1985) on trophic cascades has transformed lake ecologists' views of trophic dynamics and revealed the disproportionate functional role that certain kinds of animals or even particular species may play in structuring in lake food webs. This change in perspective emphasizes that biotic communities in lakes are not simply passively structured by the chemical and physical template of the ecosystem, but that species themselves can exert strong control on the diversity, productivity, and chemistry of lakes.

26.3.2 Lake Biota in a Landscape Context

Until recently, most comparative limnological studies examined lakes as self-contained ecosystems, independent of their catchment settings and hydrologically connected habitats. Increasingly, scientists are examining the ecological organization of lakes in the context of the landscape in which they reside. For example, comparing headwater lakes with small catchments to downstream lakes with large catchments in northern Wisconsin, Kratz et al. (1997) found that hydrologic position is correlated with lake water retention, concentration of silica and dissolved organic carbon, physical and chemical response to drought, and biological factors, such as the vertical distribution of primary production and fish species richness. Kratz et al. (1997) suggest that silica and calcium concentrations, which tend to be lower in headwater lakes with small groundwater inflow rates, may influence the abundance and ecological roles of freshwater sponges and crayfish. The vertical distribution of primary production in lake water columns is a function of water clarity and light penetration; thus downstream lakes with higher nutrient and organic carbon loads may have shallower productive zones than do headwater lakes with deeper photic zones.

Lower fish species richness in Wisconsin headwater lakes compared with downstream lakes might stem from the tendency for downstream lakes to be larger, implying that in larger lakes more area of each habitat type is available to support rare species (Kratz et al., 1997). However, because connectivity with the drainage network and other lakes is clearly lower for headwater lakes, lower richness in headwater lakes could reflect the relative biogeographical isolation and lower probability of fish species colonization of these lakes (Tonn et al., 1990; Kelso and Minns, 1996). This is consistent with the relative ease of establishment of many fish or other aquatic predators newly introduced in such lakes (Tonn et al., 1983; Moyle, 1986) and the major ecological rearrangements that often result from such introductions (e.g., Spencer et al., 1991).

26.3.3 Ecosystem Influences and Alterations of Lake Biotas

Managers and scientists engaged in environmental assessment should recognize and address the limitations and potential pitfalls of any approach that relies exclusively on inferring general or potential patterns in the biota of lakes based on geographical or physical indices. Globally, few but the very smallest lakes have escaped major ecological disruption from human introduction of nonnative fishes, macrophytes, crayfishes, or other organisms. Establishment of a single predatory fish species can cause massive direct and indirect alterations of ecological structure and function that reverberate through several ecological levels both up and down the trophic web (Zaret, 1982; Carpenter et al., 1985; Moyle, 1986). The effects of introduced species that play key roles in the food web can spin off to alter terrestrial, riverine, and lake communities, as in the case of the introduction of the mysis shrimp to lakes outside their native range (Spencer et al., 1991). Furthermore, despite longstanding and familiar theories of pond to meadow succession, little is yet known about the long-term trophic evolution of larger lakes and the extent to which trophic succession is sensitive or reversible in response to changing environmental factors. This is an area of prolific research activity; recent studies suggest not only that human influences on lake eutrophication have been more pervasive and persistent than previously assumed (e.g., Smol et al., 1983; Reavie et al., 1995), but that even small shifts in trophic status might have substantial but complex effects on fish assemblage organization (e.g., Plante and Downing, 1993; Evans et al., 1996; Kelso et al., 1996; Steedman et al., 1996). Over periods of decades, such changes in fish assemblages could subsequently feed back to either hasten or slow eutrophication processes initially driven by changes in nutrient or carbon loading.

Because these influences have not been widely recognized, the long-term vulnerability of lakes to human activities (and other causes of landscape change or species dispersal) has likely been seriously underestimated in the past. The potential regional, as well as local, influences of human alteration of landscapes should be carefully considered in any assessment of lake ecosystems. Across many regions where recent human influences are pervasive, most or all lake communities may be altered from their historical condition due to both externally driven eutrophication and changes in fish assemblages and trophic structure. In such circumstances, location of unaltered reference systems may be difficult or impossible; historical analysis from lake bed cores may be the only remaining evidence of such pervasive changes. In core records, temporal changes in fish and many invertebate taxa usually must be inferred from evidence of general trophic status provided by the few diatom and invertebrate taxa that are consistently preserved as fossils. From a conservation point of view, except in regions that have not seen extensive human activity (such as much of arctic North America), any lakes that have escaped the ecological changes commonly wrought by introduced taxa should be recognized as potentially having exceedingly high value as reservoirs of biological diversity and as examples of naturally functioning ecosystems. Unfortunately, we know of no published, comprehensive regional conservation assessment for lake ecosystems that can be recommended as an example or prototype.

26.4 Assessing Conservation Status of Freshwater Biota

Recently, scientists have begun to document the advanced global extent of biotic deterioration and impoverishment of freshwater ecosystems. Allan and Flecker (1993) review evidence of the declining status of aquatic biota worldwide and discuss the varied causes of such declines, which include the loss, deterioration, and fragmentation of habitat; introduction and invasion of nonnative species; overexploitation; and effects of artificial propagation. Regional assessments of fish (Williams et al., 1989; Nehlsen et al., 1991), river mussels (Williams et al., 1993), and other taxa (Master, 1990) have drawn similar conclusions about the multifarious causes of decline and the apparent overriding importance of habitat alteration. These regional status assessments have consisted largely of qualitative assessments of species status based on synthesis of expert judgment of the available, typically fragmentary data on distribution, population trends, and threats. Typical data sources include distribution and occurrence data from state Natural Heritage Program databases, other state and federal surveys, population trend data gained from agency monitoring efforts or inferred indirectly from sources such as catch in fisheries, site-specific information compiled from research studies conducted at scattered sites, and qualitative information about the distribution of threats from many sources. These assessments are at best conservative in documenting biotic declines, because taxa, regions, and sites for which no information is available are usually not listed. Some assessments have attempted to correct for missing biases through mapping protocols based on explicit rules for interpolation from known sites (e.g., Frissell, 1993). However, although it is useful to map regional patterns of biotic decline, in regional assessments the absence of information for a given site should not automatically be assumed to mean that its conservation values are low.

To display conservation status data in a spatially explicit way, Frissell (1993) translated information from several relevant reviews and recent studies of fish conservation status along the Pacific Coast into a regional map of fish endangerment. When the numbers of taxa known to be declining, depressed, or already extinct in each watershed were mapped, zones of high endangerment were associated with urban areas and concentrations of dams and diversions. However, a background pattern was also apparent across most of the region, in which two to three or more fish taxa were imperiled or extinct, indicating that fish endangerment is a problem even in basins lacking dams, diversions, or concentrated human populations. Fish populations in these areas may be affected by the habitat-altering effects of pervasive land-use practices (e.g., logging, grazing, mining, and cropland agriculture). Frissell (1993) also presented examples of range-wide status maps for selected species, showing that several species with formerly extensive distribution across the region are already extinct or declining across vast portions of their native range. By conveying endangerment and extinction data in a spatially explicit way and by placing simple lists of populations and streams into a landscape context, the mapping approach better reveals the expansive nature of biotic loss in aquatic ecosystems. Recent federal government decisions to confer regional protection for certain threatened salmon species under the Endangered Species Act have considered regional patterns of endangerment, even where stream-specific population data are lacking. Allendorf et al. (1997) recommend that priorities for salmon protection include such large-scale spatial considerations as the regional extinction patterns among co-occurring fishes and other biota and the possible regional ecological and genetic consequences of local population extinction.

Fragmentation of species ranges at the regional scale is one consequence of local extinction and decline that is widely evident in species status maps such as those in Frissell (1993). As individual populations decline, potential metapopulation functions that provide demographic and genetic support to nearby populations are also compromised. At the smaller scale of populations and metapopulations within watersheds and river reaches, fragmentation is an even more widespread and imminent threat to biotic diversity. Physical barriers (e.g., dams, dewatered or polluted reaches) not only isolate populations and accelerate local extinction by preventing dispersal (Sheldon, 1988; Zwick, 1992; Dynesius and Nilsson, 1994; Rieman and McIntyre, 1995; Pringle, 1997), but they also disrupt the continuity of flow of water, sediment, nutrients, and heat loads, thus altering many ecosystem processes and favoring introduced over native species (Stanford et al., 1996; Poff et al., 1997; Ward, 1997).

Assessments of an aquatic population's robustness or probability of persistence should consider the actual and potential spatial distribution of both the population and its habitat within a catchment or region (Figure 26.2). *Connectivity* among populations or subpopulations refers in this case to the ability of individuals to occupy and disperse freely among a well-distributed or continuous array of habitats within the drainage. High biotic connectivity allows for global population persistence by promoting rapid recolonization should extinction occur within local habitat patches (e.g., Rieman and McIntyre, 1993, 1995). *Redundancy* refers to the distribution of the population or populations among hydrologically independent and geographically separated tributaries within a catchment. With high

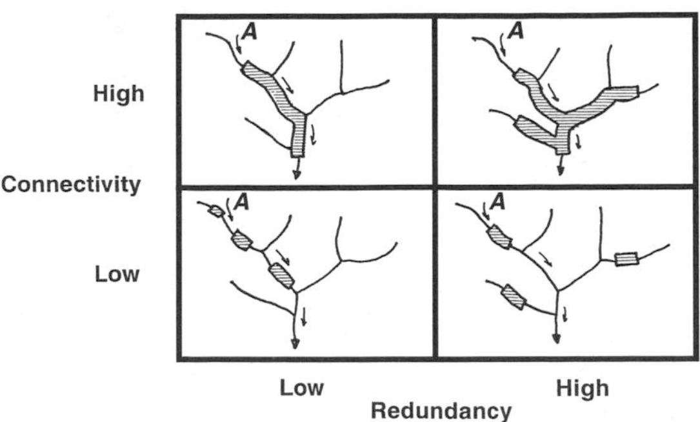

FIGURE 26.2. Two dimensions of ecological change associated with population simplification and fragmentation of a stream-dwelling species. Spatial continuity or *connectivity* among subpopulations declines as habitat fragmentation increases. As a result, the potential for persistence or recolonization after catastrophic disturbances can decline. Declining *redundancy* is a measure of spatial simplification that reduces the potential for population persistence within tributary or other hydrologically independent refugia. As spatial range shrinks and subpopulations are confined to fewer segments of the drainage network, remaining population fragments are more vulnerable to global extinction from a single hydrologic or geomorphic perturbation (mapped in the figure as arrows flowing downstream from source *A*). Such spatial or regional effects can threaten the persistence of wide-ranging species over areas of many hundreds or thousands of square kilometers or of locally distributed species at the scale of a small headwater stream or stream and lake network.

redundancy (i.e., many tributaries occupied), the loss of populations from a catastrophic event transmitted downstream through the mainstem and/or a subset of tributaries can be at least partly mitigated by the persistence of other populations or population fragments within unimpacted tributary refugia, which serve as centers of recolonization. The presence of such refugia is a critical factor in the persistence and recovery of fish populations (Sedell et al., 1990; Schlosser, 1991). Zoogeographic evidence suggests that dispersal and recolonization are also important for maintaining fish diversity in small lakes (e.g., Tonn et al., 1990; Kelso and Minns, 1996); hence these concepts may be generally relevant to other freshwater ecosystems as well. Endangered populations typically have experienced dramatic loss of both connectivity and redundancy relative to their natural condition.

Past ecosystem disruptions tend to concentrate in downstream portions of the stream network, and tributaries have inherently high biotic connectivity with the other stream segments that they join. As a result, headwater tributaries that harbor a high proportion of their native biota may be of exceptionally high value for biodiversity conservation and restoration (Sedell et al., 1990; Moyle and Sato, 1991; Doppelt et al., 1993; Frissell and Bayles, 1996). In ecological assessments for which protection of biological diversity is a goal, locations of sensitive species must be explicitly identified in a drainage network context to formulate plans that ensure their continued function as biodiversity refugia (Doppelt et al., 1993; Frissell and Bayles, 1996; Stanford et al., 1996). Moyle and Randall (1998) provide an example of such a spatially explicit assessment for watersheds of the Sierra Nevada area of California. Their multimetric approach incorporates data on distribution and status of native fish and amphibians, disruption of ecosystem integrity by introduced fishes (e.g., those that prey on native amphibians), and alteration of aquatic environments by dams, roads, and other human developments.

The global and regional assessments of the status of aquatic animals cited previously universally point to habitat alteration and loss as the most widespread cause of aquatic biodiversity decline. Yet, for both streams and lakes, there are few (if any) universally accepted protocols for general assessment of habitat condition or change in habitat quality over time. Thus few monitoring networks or comprehensive surveys provide comparable data for either spatial or temporal assessment of habitat trends. Many kinds of physical data can be useful in attempting to reconstruct such trends (see extensive discussions and citations earlier in this chapter), but in no case can this be expected to be a cookbook exercise. Any method that pretends to offer a simple or certain answer to this problem should not be considered reliable and most certainly has not met with widespread professional or scientific acceptance. For example, measures of channel unit composition of streams are often adopted as the survey and monitoring method of choice for federal land management agency assessments, yet this widespread approach has serious limitations for this purpose (Poole et al., 1997). Structural and dynamic changes in physical habitat that frequently impair biota are often not predictably associated with changes in water chemistry or violation of chemical water quality standards (Karr, 1991; Schlosser, 1991). Therefore, routine monitoring of chemical water quality conditions to meet water quality standards is not sufficient for ecological assessments.

Nevertheless, several studies offer useful models for comparative analysis of habitat data and selected biotic information at the scale of tributaries within a region or large river basin. For example, Roth et al. (1996), Minns et al. (1996), Richards et al. (1996), and Allan et al. (1997) provide evidence through such comparative studies indicating correspondence of various habitat measurements to metrics of biotic integrity and correlation of both habitat and biotic integrity to land use or land cover. In Europe, using somewhat more explicit but in many ways similar methodological framework, Bravard et al. (1986) and Amoros et al. (1987a, b) also developed a synthetic understanding of landscape-level patterns in stream habitat changes and biotic responses to them. On the other hand, Hughes (1985) found that, although invertebrate assemblages corresponded well to the principal gradient of environmental stress (mining wastes) among study streams in Montana, trout biomass did not show correspondence (perhaps partly because the trout populations included a mix of native and introduced species whose varying tolerances and interactions were not taken into account). Working at a similar scale with a comparative design across a montane landscape, Baxter et al. (1999) demonstrated strong correlations of bull trout spawning populations with land-use pattern (road density) and with valley geomorphology (reflecting groundwater efflux and other channel conditions). Following Frissell et al. (1986), Frissell and Bayles (1996) suggest that comparative studies of this type, incorporating a selected mosaic of catchments with similar geological and biophysical templates (including shared biogeographical history), but which have been differently altered by human activities and recent natural events, are likely to provide the most useful information for ecological assessments. Such a design plows the middle ground in scale between ecoregionally comprehensive assessments and local or site-focused assessments. Frissell and Bayles (1996) argue that this approach has already proved more fruitful than single-watershed analyses, which are presently fashionable among federal and state agencies and industrial landowners and some scientists (e.g., Montgomery et al., 1995). In a typical single-watershed analysis, the ability to address critical questions about physical and biological linkages rests on the end to end coupling of simplified, mechanistic simulation models. Individually and collectively, these models lack empirical validation, precision, and any means of appropriate scaling (among other problems).

Some conservation assessments of aquatic species have employed population viability analyses (PVA) to understand the long-term probability of persistence or extinction of a populations under alternative scenarios (e.g., Ratner et al., 1997). These analyses require much demographic information that is seldom available except for closely monitored and carefully enumerated populations (Rieman and McIntyre, 1993). Where sufficient data are available, it is prudent to incorporate PVAs for target species that may be particularly endangered or vulnerable. If sufficient data are not available, considering the data needs for PVA (especially through sensitivity analysis using plausible ranges of parameter estimates) can point out high priorities for monitoring and assessment, as well as domains of risk relative to prospective decisions (Rieman and McIntyre, 1993; Ruggiero et al., 1994; Kokko et al., 1997). For fishes, for which considerable data are sometimes available, the results of quantitative PVA are often in agreement with credible qualitative assessments performed by independent scientists (Nehlsen et al., 1991; Ratner et al., 1997). However, wherever listed or candidate species are at issue, managers should consider the lack of sufficient data to perform PVA as a red flag indicating that they might be unwittingly jeopardizing these populations through management action or inaction. Wishful thinking and pressing deadlines notwithstanding, sometimes analysis simply cannot substitute for field data, and the necessary data must be acquired for a reliable assessment to be completed.

26.5 Conclusions

It is an awkward challenge indeed to find a common organizing principle among the many approaches used to assess patterns in aquatic biota and linkages between aquatic biota and habitat or landscape conditions. In keeping with the recent literature, this chapter focuses on issues of scale, stressing the need to explicitly address the consequences of scale in determining which of the many available protocols and methods to apply in a given situation. Table 26.1 contrasts a number of ecological assessment methods and metrics commonly employed for aquatic biota in terms of the biological scale or level of organization that they address. Clearly, there is tension between objectives or requirements for assessing patterns in community or assemblage structure or integrity and assessments necessary to examine the viability of particular species, subspecies, metapopulations, and populations. The variation in spatial and temporal scale of biological systems of concern in Table 26.1 itself spans many orders of magnitude, and the full consideration of key biophysical linkages and the ecological context of an ecosystem can create additional levels of complexity and scale. Population and species viability are not just concerns at the level of a single lake or stream reach; they also must be addressed (especially for endangered species) at the cumulative level of the entire historic and present-day range of the taxon.

Deciding on a standard approach to this multidimensional complexity of scale is daunting, especially when we consider the diversity of management issues and challenges that ecological assessments are called on to address. Some key questions, if resolved, could potentially simplify the task of determining appropriate scales for assessment. For example, debate continues over the need to incorporate a discrete, catchment-level unit in the hierarchy of aquatic habitat. Some practitioners argue that, from the biological point of view, there is not sufficient need for such a unit (e.g., The Nature Conservancy, 1997), while others focus almost exclusively on the catchment as the organizing unit for analysis and synthesis (e.g., Montgomery et al., 1995). Still others (e.g., Frissell and Bayles, 1996; Omernik and Bailey, 1997) see it as a necessary and useful unit if viewed in a regional, comparative context.

In addition to resolving the question of appropriate scale and geographic scope, the ecological scope of an environmental assessment must also be defined. This second overarching question asks if it is necessary to treat all species or if it will be sufficient to conduct explicit analyses only for selected target species for which data or a large body of knowledge already exists. Some species are known to perform special roles as dominant interactors in aquatic ecosystems, either as *keystones* (e.g., predacious fishes) or as *ecosystem engineers* (e.g., beavers). It seems clear that ecological assessments must pay special attention to these species (Power et al., 1996). But can we also identify for aquatic ecosystems a subset of *focal* or *umbrella species* (Lambeck, 1997) that can be considered to have requirements that, if met, will result in the conservation of numerous other species with similar habitat requirements? This seems the only logical way to simplify the task of coping with the large diversity of taxa with which we need to be concerned. Yet little progress has been made in identifying a basis for aggregating species by habitat requirements or other ecological criteria (but see Townsend and Hildrew, 1994; Poff, 1997). Again, explicit consideration of scale of life-history traits, habitat templates, and ecological responses will be necessary to advance this front.

The concepts of spatial refugia and biodiversity

hotspots hold promise for coping with multiple taxa in ecological assessments in a spatially explicit way; however, criteria and procedures for identifying these phenomena remain poorly developed. Given the diverse scales of life history and ecological response exhibited across aquatic taxa from microbes to fishes, birds, and mammals (Townsend and Hildrew, 1994), it remains unclear how coherent these phenomena might be among various taxonomic groups. Clearly, however, refugia, hotspots, and hostile ecological barriers often possess distinct physical attributes that set them boldly apart from the remaining matrix of aquatic habitat. Physical scientists should help to address the ways these physical attributes are manifested across the range of biotically relevant spatial and temporal scales. The developing, general paradigm in stream ecology of deviation from a posited, locally normative dynamic state or regime (Richter et al., 1996; Stanford et al., 1996; Poff et al., 1997) holds promise in facilitating the physical description of these attributes in a new and holistic way.

Many opportunities for accurate and meaningful biological assessment and effective decision making are missed because scientists and managers have devoted insufficient attention to comparing the inherent strengths and weakness of the many approaches that have evolved over time. Assessment methods are often chosen based on habit, local tradition, available expertise, or apparent convenience and cost of application. Little critical evaluation is directed toward the empirical, analytic, and conceptual consequences of the choice and potential alternatives. In fact, because assessment methods vary in conceptual foundations, emphases, and assumptions, they can lead to markedly different management implications, even using the same set of empirical data. In many cases, the methods used do not adequately address the specific management issue at stake; administrative appeals, litigation, or unrealized expectations can then vastly compound the difficulty and time necessary to reach a technically defensible and socially acceptable resolution to a management problem. More deliberate consideration of methodological criteria, including taxonomic resolution, data requirements, and spatial and temporal scale of analyses, would contribute to more accurate, concise, and cost-effective scientific assessments of aquatic biota.

The past decade has seen major conceptual advances in the understanding of aquatic ecosystem organization and response to environmental change. Simultaneously, the globally precarious status of aquatic biological resources is becoming widely recognized. Past approaches and methods of aquatic ecological assessment are beginning to be reexamined in light of both these developments. The next decade promises a revolution in the theory and application of ecological science in the management and conservation of aquatic ecosystems and their biodiversity.

26.6 References

Allan, J. D.; Flecker, A. S. 1993. Biodiversity conservation in running waters. *BioScience* 43:32–43.

Allan, J. D.; Erickson, D. L.; Fay, J. 1997. The influence of catchment land use on stream integrity across multiple spatial scales. *Freshwater Biol.* 37:149–162.

Allendorf, F. W.; Bayles, D.; Bottom, D. L.; Currens, K. P.; Frissell, C. A.; Hankin, D.; Lichatowich, J. A.; Nehlsen, W.; Trotter, P. C.; Williams, T. H. 1997. Prioritizing Pacific salmon stocks for conservation. *Conserv. Biol.* 11:140–152.

Amoros, C.; Rostan, J. C.; Pautou, G.; Bravard, J. P. 1987a. The reversible process concept applied to the environmental management of fluvial hydrosystems. *Environ. Manage.* 11:607–617.

Amoros, C.; Roux, L.; Reygrobellet, J. L.; Bravard, J. P.; Pautou, G. 1987b. A method for applied ecological studies of fluvial hydrosystems. *Regul. River.* 1:17–36.

Baxter, C. V.; Frissell, C. A.; Hauer, F. R. 1999. Geomorphology, logging roads, and the distribution of bull trout (*Salvelinus confluentus*) spawning in a forested river basin: implications for management and conservation. *Trans. Amer. Fisheries Soc.* 128(5):854–867.

Biggs, B. J. F. 1995. The contribution of flood disturbance, catchment geology, and land use to the habitat template of periphyton in stream ecosystems. *Freshwater Biol.* 33:419–438.

Bravard, J. P.; Amoros, C.; Pautou, G. 1986. Impact of civil engineering works on the successions of communities in a fluvial system. *Oikis* 47:92–111.

Bryce, S. A.; Clarke, S. E. 1997. Landscape-level ecological regions: linking state-level ecological frameworks with stream habitat. *Environ. Manage.* 20:297–311.

Carpenter, S. R.; Kitchell, J. F.; Hodgson, J. R. 1985. Cascading trophic interactions and lake productivity. *BioScience* 35:634–639.

Corkum, L. D. 1989. Patterns of benthic invertebrate assemblages in rivers of northwestern North America. *Freshwater Biol.* 21:191–205.

Corkum, L. D. 1990. Intrabiome distributional patterns of lotic macroinvertebrate assemblages. *Can. J. Fisheries Aquat. Sci.* 47:2147–2157.

Culp, J. M.; Davies, R. W. 1982. Analysis of longitudinal zonation and the River Continuum Concept in the Oldman–South Saskatchewan River system. *Can. J. Fisheries Aquat. Sci.* 39:1258–1265.

Decamps, H.; Fortuné, M.; Gazelle, F.; Pautou, G. 1988.

Historical influence of man on the riparian dynamics of a fluvial landscape. *Landscape Ecol.* 1:163–173.

Doppelt, B.; Scurlock, M.; Frissell, C.; Karr, J. 1993. *Entering the watershed: a new approach to save America's river ecosystems.* Covelo, CA: Island Press.

Dynesius, M.; Nilsson, C. 1994. Fragmentation and flow regulation of river systems in the northern third of the world. *Science* 266:753–762.

Ebersole, J. L.; Liss, W. J.; Frissell, C. A. 1997. Restoration of stream habitats in the western United States: restoration as re-expression of habitat capacity. *Environ. Manage.* 21(1):1–14.

Evans, D. O.; Nicholls, K. H.; Allen, Y. C.; McMurtry, M. J. 1996. Historical land use, phosphorus loading, and loss of fish habitat in Lake Simcoe, Canada. *Can. J. Fisheries Aquat. Sci.* 53(Supplement 1):194–218.

Fausch, K. D.; Karr, J. R.; Yant, P. R. 1984. Regional application of an index of biotic integrity based on stream fish communities. *Trans. Amer. Fisheries Soc.* 113:39–55.

Fore, S.; Karr, J. R.; Wisseman, R. W. 1996. Assessing invertebrate responses to human activities: evaluating alternative approaches. *J. N. Amer. Benthol. Soc.* 15: 212–233.

Frissell, C. A. 1993. Topology of extinction and endangerment of fishes in the Pacific Northwest and California. *Conserv. Biol.* 7:342–354.

Frissell, C. A.; Bayles, D. 1996. Ecosystem management and the conservation of aquatic biodiversity and ecological integrity. *Water Resour. Bull.* 32:229–240.

Frissell, C. A.; Liss, W. J.; Warren, C. E.; Hurley, M. D. 1986. A hierarchical framework for stream habitat classification: viewing streams in a watershed context. *Environ. Manage.* 10:199–214.

Frissell, C. A.; Liss, W. J.; Gresswell, R. E.; Nawa, R. K.; Ebersole, J. L. 1997. A resource in crisis: changing the measure of salmon management. In: Stouder, D. J.; Bisson, P. A.; Naiman, R. J., eds. *Pacific salmon and their ecosystems: status and future options.* New York: Chapman and Hall: 411–444.

Hughes, R. M. 1985. Use of watershed characteristics to select control streams for estimating effects of metal mining wastes on extensively disturbed streams. *Environ. Manage.* 9:253–262.

Hughes, R. M.; Larsen, D. P.; Omernik, J. M. 1986. Regional reference sites: a method for assessing stream potential. *Environ. Manage.* 10:629–635.

Hunsaker, C. T.; Levine, D. A. 1995. Hierarchical approaches to the study of water quality in rivers. *BioScience* 45:193–203.

Johnson, W. C. 1994. River expansion in the Platte River, Nebraska: patterns and causes. *Ecol. Monogr.* 64:45–84.

Karr, J. R. 1991. Biological integrity: a long-neglected aspect of water resource management. *Ecol. Appl.* 1:61–68.

Karr, J. R.; Chu, E. W. 1997. *Biological monitoring and assessment: using multimetric indices effectively.* EPA 235-R97-001. Seattle, WA: University of Washington.

Kelso, J. R. M.; Minns, C. K. 1996. Is fish species richness at sites in the Canadian Great Lakes the result of local or regional factors? *Can. J. Fisheries Aquat. Sci.* 53(Supplement 1):175–193.

Kelso, J. R. M.; Steedman, R. J.; Stoddart, S. 1996. Historical causes of change in Great Lakes fish stocks and the implications for ecosystem rehabilitation. *Can. J. Fisheries Aquat. Sci.* 53(Supplement 1):10–19.

Kokko, H.; Lindström, J.; Ranta, E. 1997. Risk analysis of hunting of seal populations in the Baltic. *Conserv. Biol.* 11:917–927.

Kratz, T. K.; Webster, K. E.; Bowser, C. J.; Magnuson, J. J.; Benson, B. J. 1997. The influence of landscape position on lakes in northern Wisconsin. *Freshwater Biol.* 37:209–217.

Lambeck, R. J. 1997. Focal species: a multi-species umbrella for nature conservation. *Conserv. Biol.* 11:849–856.

Lewis, C. A.; Lester, N. P.; Bradshaw, A. D.; Fitzgibbon J.; Fuller, K.; Hakanson, L.; Richards, C. 1996. Considerations of scale in habitat conservation and restoration. *Can. J. Fisheries Aquat. Sci.* 53(Supplement 1):440–445.

Ligon, F. K.; Dietrich, W. E.; Trush, W. J. 1995. Downstream ecological effects of dams: a geomorphic perspective. *BioScience* 45:183–192.

Magnuson, J. J.; Benson, B.; Kratz, T. 1990. Temporal coherence in the limnology of a suite of lakes in Wisconsin, U.S.A. *Freshwater Biol.* 23:145–159.

Master, L. 1990. The imperiled status of North American aquatic animals. *The Nature Conservancy: Biodiversity Network News* 3:1–8.

Matusek, J. E. 1978. Empirical predictions of fish yields of large North American lakes. *Trans. Amer. Fisheries Soc.* 107:385–394.

Minns, C. K.; Kelso, J. R. M.; Randall, R. G. 1996. Detecting the response of fish to habitat alterations in freshwater ecosystems. *Can. J. Fisheries Aquat. Sci.* 53(Supplement 1):403–414.

Minshall, G. W. 1988. Stream ecosystem theory: a global perspective. *J. N. Amer. Benthol. Soc.* 7:263–288.

Montgomery, D. R.; Grant, G. E.; Sullivan, K. 1995. Watershed analysis as a framework for implementing ecosystem management. *Water Resour. Bull.* 31:369–386.

Moyle, P. B. 1986. Fish introductions into North America: patterns and ecological impact. In: Mooney, H.; Drake, J. A., eds. *Biological invasions in North America.* New York: Springer-Verlag: 155–170.

Moyle, P. B.; Randall, P. J. 1998. Evaluating biotic integrity of watersheds in the Sierra Nevada, California. *Conserv. Biol.* 12:1318–1326.

Moyle, P. B.; Sato, G. M. 1991. On the design of preserves to protect native fishes. In: Minckley, W. L.; Deacon, J. E., eds. *Battle against extinction: native fish management in the American West.* Tucson, AZ: University of Arizona Press: 155–170.

Muntkittrick, K. R.; Dixon, D. G. 1989. A holistic approach to ecosystem health assessment using fish population characteristics. *Hydrobiologia* 188/189:123–135.

Murphy, K. J.; Rorslett, B.; Springuel, I. 1990. Strategy analysis of submerged lake macrophyte communities: an international example. *Aquat. Bot.* 36:303–323.

Naiman, R. J.; Lozarich, D. G.; Beechie, T. J.; Ralph, S. C. 1992. General principles of classification and the assessment of conservation potential. In: Boon, P. J.; Calow, P.; Petts, G. E., eds. *River conservation and management.* New York: John Wiley & Sons: 93–123.

Naiman, R. J.; Magnuson, J. J.; McKnight, D. M.; Stanford, J. A., editors. 1995. *Freshwater imperative: a research agenda.* Washington, DC: Island Press.

Nehlsen, W.; Lichatowich, J. A.; Williams, J. E. 1991. Pacific salmon at the crossroads: stocks at risk from California, Oregon, Idaho, and Washington. *Fisheries* 16(2):4–21.

Omernik, J. M. 1987. Ecoregions of the conterminous United States. *Ann. Assoc. Amer. Geogr.* 77:118–125.

Omernik, J. M.; Bailey, R. G. 1997. Distinguishing between watersheds and ecoregions. *J. Amer. Water Resour. Assoc.* 33:935–949.

Plante, C.; Downing, J. A. 1993. Relationship of salmonine production to lake trophic status and temperature. *Can. J. Fisheries Aquat. Sci.* 50:1324–1328.

Poff, N. L. 1996. A hydrogeography of unregulated streams in the United States and an examination of scale-dependence in some hydrological descriptors. *Freshwater Biol.* 36:606–627.

Poff, N. L. 1997. Landscape filters and species traits: towards mechanistic understanding and prediction in stream ecology. *J. N. Amer. Benthol. Soc.* 16:391–409.

Poff, N. L.; Allan, J. D. 1995. Functional organization of stream fish assemblages in relation to hydrologic variability. *Ecology* 76:606–627.

Poff, N. L.; Ward, J. V. 1989. Implications of streamflow variability and predictability for lotic community structure: a regional analysis of streamflow patterns. *Can. J. Fish. Aquat. Sci.* 46:1805–1818.

Poff, N. L.; Ward, J. V. 1990. The physical habitat template of lotic ecosystems: recovery in the context of historical pattern of spatio-temporal heterogeneity. *Environ. Manage.* 14:629–646.

Poff, N. L.; Allan, J. D.; Bain, M. B.; Karr, J. R.; Prestegaard, K. L.; Richter, B. D.; Sparks, R. E.; Stromberg, J. C. 1997. The natural flow regime: a paradigm for river conservation and restoration. *BioScience* 47: 769–784.

Poole, G. C.; Frissell, C. A.; Ralph, S. C. 1997. In-stream habitat unit classification: inadequacies for monitoring and some consequences for management. *J. Amer. Water Resour. Assoc.* 33:879–896.

Power, M. E.; Tilman, D.; Estes, J. A.; Menge, B. A.; Bond, W. J.; Mills, L. S.; Daily, G.; Castilla, U. C.; Lubchenco, J.; Paine, R. T. 1996. Challenges in the quest for keystones. *BioScience* 46:609–620.

Pringle, C. M. 1997. Exploring how disturbance is transmitted upstream: going against the flow. *J. N. Amer. Benthol. Soc.* 16:425–438.

Pringle, C. M.; Naiman, R. J.; Bretschko, G.; Karr, J. R.; Oswood, M. W.; Webster, J. R.; Welcomme, R. L.; Winterbourne, M. J. 1988. Patch dynamics in lotic ecosystems: the stream as mosaic. *J. N. Amer. Benthol. Soc.* 7:411–428.

Quinn, J. M.; Hickey, C. W. 1990. Magnitude of effects of substrate particle size, recent flooding, and watershed development on benthic invertebrates in 88 New Zealand rivers. *N. Z. J. Freshwater Mar. Res.* 24:411–428.

Ratner, S.; Lande, R.; Roper, B. B. 1997. Population viability analysis of spring Chinook salmon in the South Umpqua River, Oregon. *Conserv. Biol.* 11:879–889.

Reavie, E. D.; Smol, J. P.; Carmichael, N. B. 1995. Post-settlement eutrophication of six British Columbia (Canada) lakes. *Can. J. Fisheries Aquat. Sci.* 52:2388–2401.

Richards, C.; Johnson, L. B.; Host, G. E. 1996. Landscape-scale influences on stream habitats. *Can. J. Fisheries Aquat. Sci.* 53(Supplement 1):295–311.

Richter, B. D.; Baumgartner, J. V.; Powell, J.; Braun, D. P. 1996. A method for assessing hydrologic alteration within ecosystems. *Conserv. Biol.* 10:1163–1174.

Richter, B. D.; Baumgartner, J. V.; Wigington, R.; Braun, D. P. 1997. How much water does a river need? *Freshwater Biol.* 37:231–249.

Rieman, B. E.; McIntyre, J. D. 1993. *Demographic and habitat requirements for conservation of bull trout.* INT-GTR-302. Ogden, UT: U.S. Dept. Agric., For. Serv., Intermountain For. Range Exp. Sta.

Rieman, B. E.; McIntyre, J. D. 1995. Occurrence of bull trout in naturally fragmented habitat patches of various size. *Trans. Amer. Fisheries Soc.* 124:285–296.

Roth, N. E.; Allan, J. D.; Erickson, D. L. 1996. Landscape influences on stream biotic integrity assessed at multiple spatial scales. *Landscape Ecol.* 11:141–156.

Ruggiero, L. F.; Hayward, G. D.; Squires, J. R. 1994. Viability analysis in biological evaluations: concepts of population viability analysis, biological population, and ecological scale. *Conserv. Biol.* 8:364–372.

Schlosser, I. J. 1990. Environmental variation, life history attributes, and community structure in stream fishes: implications for environmental management assessment. *Environ. Manage.* 14:621–628.

Schlosser, I. J. 1991. Stream fish ecology: a landscape perspective. *BioScience* 41:704–712.

Sedell, J. R.; Reeves, G. H.; Hauer, F. R.; Stanford, J. A.; Hawkins, C. P. 1990. Role of refugia in recovery from disturbances: modern fragmented and disconnected river systems. *Environ. Manage.* 14:711–724.

Sheldon, A. L. 1988. Conservation of stream fishes: patterns of diversity, rarity and risk. *Conserv. Biol.* 2: 149–156.

Smol, J. P.; Brown, S. R.; McNeely, R. 1983. Cultural disturbances and trophic history of a small meromictic lake from central Canada. *Hydrobiologia* 103:125–130.

Southwood, T. R. E. 1977. Habitata, the templet for ecological strategies. *J. Anim. Ecol.* 46:337–365.

Spencer, C. N.; McClelland, B. R.; Stanford, J. A. 1991. Shrimp stocking, salmon collapse, and eagle displacement. *BioScience* 41:14–21.

Stanford, J. A.; Ward, J. V. 1992. Management of aquatic resources in large catchments: recognizing interactions between ecosystem connectivity and environmental disturbance. In: Naiman, R. J., ed. *Watershed management*. New York: Springer-Verlag: 91–124.

Stanford, J. A.; Ward, J. V.; Liss, W. J.; Frissell, C. A.; Williams, R. N.; Lichatowich, J. A.; Coutant, C. C. 1996. A general protocol for restoration of regulated rivers. *Regul. River.* 12:391–413.

Statzner, B.; Higler, B. 1986. Stream hydraulics as a major determinant of benthic invertebrate zonation patterns. *Freshwater Biol.* 16:127–139.

Steedman, R. J. 1988. Modification and assessment of an index of biotic integrity to quantify stream quality in southern Ontario. *Can. J. Fisheries Aquat. Sci.* 45:492–501.

Steedman, R. J.; Whillans, T. H.; Behm, A. P.; Bray, K. E.; Cullis, K. I.; Holland, M. M.; Stoddart, S. J.; White, R. J. 1996. Use of historical information for conservation and restoration of Great Lakes aquatic habitat. *Can. J. Fisheries Aquat. Sci.* 53(Supplement 1):415–423.

The Nature Conservancy. 1997. A classification framework for freshwater communities. In: *Proceedings of The Nature Conservancy's aquatic community classification workshop*; April 9–11, 1996, New Haven, MO. Chicago: The Nature Conservancy, Great Lakes Program Office.

Tonn, W. M.; Magnuson, J. J.; Forbes, A. M. 1983. Community analysis of fishery management: an application with northern Wisconsin lakes. *Trans. Amer. Fisheries Soc.* 112:368–377.

Tonn, W. M.; Magnuson, J. J.; Rask, M.; Toivonen, J. 1990. Intercontinental comparison of small-lake fish assemblages: the balance between local and regional processes. *Amer. Naturalist* 136:345–375.

Townsend, C. R.; Hildrew, A. 1994. Species traits in relation to a habitat templet for river systems. *Freshwater Biol.* 31:265–275.

Vannote, R. L.; Minshall, G. W.; Cummins K. W.; Sedell, J. R.; Cushing, C. E. 1980. The river continuum concept. *Can. J. Fisheries Aquat. Sci.* 37:130–137.

Ward, J. V. 1997. An expansive perspective of riverine landscapes: pattern and process across scales. *Gaia* 6(1):52–60.

Warren, C. E. 1979. *Toward classification and rationale for watershed management and stream protection*. Report No. EPA-600/3-79-059. Corvallis, OR: U.S. Environ. Protect. Agency.

Wetzel, R. G. 1983. *Limnology*, 2nd ed. New York: CBS College Publishing.

Wiley, M. J.; Kohler, S. L.; Seelbach, P. 1997. Reconciling landscape and local views of aquatic communities: lessons from Michigan trout streams. *Freshwater Biol.* 37:133–148.

Williams, J. D.; Warren, M. L., Jr.; Cummings, K. S.; Harris, J. L.; Neves, R. J. 1993. Conservation status of freshwater mussels of the United States and Canada. *Fisheries* 18(9):6–22.

Williams, J. E.; Johnson, J. E.; Hendrickson, D. A.; Contreras Balderas, S.; Williams, J. D.; Navarro-Mendoza, M.; McAllister, D. E.; Deacon, J. E. 1989. Fishes of North America, endangered, threatened, or of special concern. *Fisheries* 14(6):2–38.

Zaret, T. M. 1982. The stability/diversity controversy: a test of hypotheses. *Ecology* 63:721–731.

Zwick, P. 1992. Stream habitat fragmentation—a threat to biodiversity. *Biodivers. Conserv.* 1:80–97.

27 Characterizing the Human Imprint on Landscapes for Ecological Assessment

Daniel G. Brown

27.1 Introduction

Human activity has altered natural landscape patterns and ecosystem functioning to varying degrees over the course of recorded history (Turner, 1990). Even the act of "preserving" a wilderness landscape results from a societal decision based on the value placed on that landscape. Such acts, as with many ecosystem uses, rarely happen by accident. Therefore, to assess ecosystem function and integrity, some understanding of the human dimension is necessary. It is useful to identify two critical aspects of the human dimension: factors that drive and can be used to predict human activities, and factors that characterize the impacts of these activities on ecosystems. By and large, the drivers of human activity are intangible (e.g., profit motive or concern for beauty) and the impacts result from tangible structures or activities on the landscape. At the same time, it is important to view human activity as a critical component of, and not separate from, ecosystem functioning. Ecosystem structure and function can influence society's valuation of resources, driving human activity. For example, the presence of a pristine or high-quality landscape might drive development nearby, degrading the quality of this landscape.

TABLE 27.1. Elements of the human imprint on landscapes that might be included in an ecological assessment.

Tangible elements	Intangible elements
Transportation corridors and junctions	Political and census boundaries
Utilities	Ownership boundaries
Land cover	Land use
Sites of cultural importance	
Key commercial and industrial concerns	

The human imprint on landscapes is variable in time and space and varies at a multitude of different time and space scales. Temporal variability implies that (1) data about human activities must be updated on a regular basis and (2) multitemporal or time series data may be necessary to characterize changes in the human dimension, especially when these changes drive significant ecological changes. Spatial variability demands that these elements be represented spatially, usually using a geographic information system (GIS). A GIS facilitates the management of multithematic spatial data and can be used to manage multiple scale representations of a given theme to deal with the variability in scale.

This chapter presents approaches for characterizing tangible and intangible elements of the landscape that are societal in nature (Table 27.1), discusses their ecological importance, and presents several practical and conceptual issues in dealing with these elements. The focus is on identifying the spatial distributions of each element. For each element, the discussion involves its ecological–environmental importance, potential data sources, and other issues related to spatial and temporal scale, spatial data structures, and related research needs. The spatial location and pattern of these elements are important determinants of their influence within an ecosystem, as are certain descriptive attributes.

27.2 Tangible Elements

Tangible elements often have a direct influence on ecosystem structure and function by actually changing the physical character of the landscape (e.g., pavement for roads and parking lots or the addition of structures). Sometimes they can have indirect influences on the way in which the ecosys-

TABLE 27.2. Classification of selected surface transportation features used by the U.S. Geological Survey for digital line graphs.

Feature type	Classes	Subclasses	Other attributes
Point	Bridge abutment		
	Tunnel portal		
	Gate		
	Cul-de-sac		
	Dead end		
	Drawbridge		
Line	Class 1 road (primary)	Undivided, divided by centerline, divided with lanes separated, one-way	Number of lanes Interstate route number U.S. route number State route number
	Class 2 road (secondary)	Undivided, divided by centerline, divided with lanes separated, one-way	Reservation, park, or military route number County route number Road width
	Class 3 road	Undivided, divided by centerline, divided with lanes separated, one-way	
	Class 4 road	Undivided, one-way	
	Class 5 trail	Four-wheel drive, other than four-wheel-drive vehicle	
	Footbridge		
	Road ferry crossing		
	Railroad	In tunnel, on bridge,	Number of tracks
	Railroad in street	Elevated, rapid transit,	Angle of clockwise rotation
	Carline	Private, U.S. government	
	Cog railroad		

Source: USGS, 1990.

tem is perceived and used by people. For example, the addition of a cell phone tower will have a minimal direct impact on an ecosystem, but it will improve the accessibility of people in this ecosystem to information, thereby increasing the range of human activities that can be conducted in this ecosystem.

27.2.1 Transportation Corridors and Junctions

Transportation systems are composed of corridors, which provide topological connection, and junctions or stations on the network. In terms of spatial data representation, the corridors are commonly represented as linear objects and the junctions as point objects or nodes on the network. The linear objects can represent roads of various types, railroads, navigable rivers and canals, transit networks, runways, and hiking or biking trails. The point objects are destinations, junctions, or facilities and can include airports, highway or rail transfer terminals, and water ports.

The direct effects of the transportation systems can often be discerned with simple locational information. For example, a forest that is bisected by a road will tend to provide lower-quality habitat for large mammals than the same forest without the road (i.e., few deer–car accidents occur in the forest). Also, roads can act as firebreaks and increase light penetration into a forest canopy. Attributes about the transportation network can improve our understanding of the direct and, especially, indirect effects of the transportation network. Attributes describing links and junctions in a transportation network include jurisdiction, capacity, actual volume (which is highly variable in time), pavement quality, and width. Indirect effects involve the relations between the quality of transportation links and the tendency for land around these links to be used for specific purposes (e.g., residential and commercial development). All other things being equal, a site with good transportation connections will tend to be more susceptible to development or "improvement" than a site without such connections.

Spatial and attribute data on transportation networks are usually available in some form because they provide an essential base layer for many management activities. In the United States, the U.S. Geological Survey (USGS) distributes digital line graphs (DLGs) that contain roads and other features. Table 27.2 lists some of the classes and attributes used in the DLGs to describe transportation features. However, the DLGs are not updated consistently. For up to date transportation-related information, transportation departments in various states are good places to start. The U.S. Department of Transportation (DOT), through its Bureau of Transportation Statistics (BTS), produces a digital data product called the National Transportation Atlas Databases (USBTS, 1996), with much of the transportation-related data mapped at a base scale of 1 : 100,000. Many private companies provide relatively up to date transportation-related information because of the growing demand for real-time vehicle navigation systems.

Research and development needs in this area include expanding the use of GIS network analysis capabilities in ecosystem assessments. For example, distances, which are often used to describe the influence that societal structures or features have on an ecosystem, are often more accurately depicted when based on the transportation network than using straight-line or Euclidean distance. For example, the potential of a place to be developed for residential use is a function of, among other things, nearness to urban employment centers and services. The distance perceived by a landowner, however, has more to do with travel time and distance on the transportation network than with straight-line distance (i.e., how far from the city as the crow flies). Distances, therefore, should be calculated along the road network, rather in a straight line.

27.2.2 Utilities

Utility networks have a topological structure that is similar to transportation networks, with a primary exception being that flow on the networks is usually one-way. Utility networks that may be of interest in an ecological assessment can be placed into two categories: pipelines (including gas, oil, water, and sewer) and cables (including telephone, cable television, and electricity). The direct effects of the former are more obvious than those of the latter. Installation of pipelines tends to be more disruptive to an ecosystem, and any break in the network can result in the release of a substantial quantity of polluting substances. In the case of both types of networks, there is usually some land clearing and maintenance involved along the route of the network, resulting in some habitat fragmentation. The indirect effects of both types of utility networks are similar to those of transportation networks. Although development can occur in their absence, it tends to be more rapid in their presence. As communication links are improved and information flows more efficiently (e.g., with the introduction of fiber-optic cable), rural areas are increasingly seen as alternative places for development of residential property and remote service industries.

Attributes of utility networks describe capacity and actual transmission rates and might be useful in understanding the potential for indirect influences on ecosystems. These attributes are more commonly used in automated mapping and facilities management (AM/FM) systems. Such systems, maintained by utility companies, are the best source for network data. For this reason there is relatively little in the way of standardized data over large areas, except for the largest, national-scale pipeline networks (e.g., the Alaska pipeline).

27.2.3 Land Cover

Although *land cover* is a general term for that which covers the soil surface (including vegetation, litter, water, structures), some specific types of land cover (e.g., crops, built-up areas) are more obviously anthropogenic than others. Vegetation is often managed for human use in some manner, and its pattern might also be viewed as anthropogenic to some extent. Among the more radical changes to land cover by humans is the paving of the soil surface, which interrupts hydrologic and nutrient cycles and affects microclimate (e.g., the urban heat island effect). Buildings and other structures have a similar effect. Urbanized ecosystems, therefore, tend to function very differently from nonurbanized ecosystems; they tend to have warmer and more contaminated atmospheres with more rapid surface runoff. Replacement of natural vegetation with agricultural crops, orchards, and feed for animals (i.e., pasture) has a range of implications for these cycles as well. Conversion of wetlands for agricultural or urban uses can also have drastic effects on ecosystem function, eliminating the ecosystem services provided by wetlands while altering nutrient cycles for crop production.

Land cover is distinct from land use, which refers to the way in which the land is managed for human goals. The critical land cover issues address what is covering the surface at any given point in time and how that cover affects hydrological, biogeochemical, and biological functions. Different land uses may lead to different spatial and temporal patterns of land cover, but land use does not predict land cover perfectly. Regardless, land use and cover are often mapped together. Anderson et al. (1976) present a system for use with remotely sensed data that has become a standard in the field for land-use–cover classification. Because the Anderson classification is not strictly a land cover or land-use classification and was developed to accommodate the method of remote sensing rather than analytical needs, it has limited usefulness for both modeling the human system and assessing its impact on the ecosytem. The NOAA Coastal Change and Analysis Program (C-CAP) developed a good land cover classification system (Dobson et al., 1995) for use with remotely sensed satellite data. Because this program focused on coastal areas, there was more detail in the wetland and water classes than is common for terrestrial investigations. Table 27.3 lists the classes identified as part of the C-CAP classification, with simplified wetland and water categories.

In addition to the type of land cover, spatial patterns of land cover can have very dramatic effects on ecosystem function. A growing literature in landscape ecology is focused on describing landscape pattern and its implications for landscape function (Forman, 1997). Especially important are the effects of relative landscape fragmentation and the diversity of landscape types, which have been shown to affect habitat quality and biodiversity (Iida and Nadashizuka, 1995). Descriptions of landscape pattern tend to be scale sensitive, producing very different descriptions depending on the scale of observation (Turner et al., 1989). Any ecological assessment should include a description (perhaps a multiscale analysis) of a landscape's spatial pattern, the dynamics in spatial pattern (O'Neill et al., 1988), and the socioeconomic drivers of changes in the pattern (Medley et al., 1995). Land cover patterns in much of North America and other portions of the developed world are the legacy of human use of the land.

TABLE 27.3. Classification of upland land cover types developed for NOAA C-CAP.

Class
Upland
Developed
High intensity
Low intensity
Cultivated
Orchards, groves, nurseries
Vines, bushes
Cropland
Grassland
Unmanaged
Managed
Woody
Deciduous forest
Deciduous scrub, shrub
Evergreen forest
Evergreen scrub, shrub
Mixed deciduous, evergreen
Bare
Tundra
Snow, ice
Perennial snow, ice
Glaciers
Wetland
Rocky shore
Unconsolidated shore
Emergent wetland
Woody wetland
Water and submerged land
Water
Reef
Aquatic bed

Source: Dobson et al., 1995.

Data on the amount of urbanized area or land cover in general can often be obtained through a national mapping agency (like the USGS) or a census-taking agency (like the U.S. Census Bureau). However, these data sources are rarely updated frequently enough to satisfy the demands of an ecological assessment, especially in areas that are undergoing rapid urbanization or other land cover changes. Also, these data attempt to measure land use more than land cover.

The most common approach to obtaining temporally sensitive information about land cover is through the use of remote sensing (Jensen, 1996). Remote sensing can take the form of aerial or satellite imaging and can involve the use of photography, multispectral imaging, radar and laser sensing, and thermal sensing. (Technical issues on the use of remote sensing are presented in Chapter 10.) Because all remote sensing systems record information about the electromagnetic or structural properties of the land surface, they can only provide adequate information about land cover (as opposed to land use) when used without supplemental information. The most commonly used approach to characterizing land cover is through a multivariate classification process based on information recorded in several spectral channels and some sort of ground truthing information. The result is a map in which each location (pixel, or picture element) is assigned to one of a number of predefined categories. New approaches, like spectral unmixing (Adams et al., 1995), attempt to obtain a more realistic portrayal of the landscape by determining the proportional area of each pixel in each land cover class. This process accounts for the mixed nature of many satellite-based image pixels, which have spatial resolutions of 20 to 1000 m.

We are currently examining changes in forest fragmentation in the upper Midwest to understand the link between landscape pattern and socioeconomic variability. We are using georeferenced triplicates of Landsat Multispectral Scanner (MSS) data, called North American Landscape Characterization (NALC) data and representing conditions in the early 1970s, mid-1980s, and early 1990s (Lunetta et al., 1998). Metrics of spatial pattern discussed in the landscape ecology literature (McGarigal and Marks, 1993) are being applied to characterize forest landscape patterns (Brown et al., 2000a). These patterns are being linked to information on land use, land ownership, demographics, economics, and institutional factors through empirical and modeling approaches (Brown et al., 2000b).

27.2.4 Cultural, Esthetic, Spiritual, and Scientific Resources

Although several features within an ecosystem may have the same apparent ecological significance, the value placed on these features may vary due to factors entirely unrelated to ecosystem function. Sites can have a relatively higher value because of their scientific value, historical or prehistorical significance, or esthetic value or some spiritual value. These resources affect the way in which ecosystems have been managed, are being managed, and will be managed in the future. These sites are often given special use designations, which legally limit the range of permitted activities. Although the value placed on the resource is intangible, in many cases the object of value is, in fact, quite tangible and it may have been managed or treated quite differently. For example, the Black Hills of South Dakota, and particular segments, have had a high spiritual importance to the Lakota (Sioux) people of North America (Geores, 1996), and this fact has led to conflicts with European descendants regarding the management of this ecosystem. The spiritual importance of features or ecosystems can only be understood through qualitative assessment, such as through surveys or interviews.

Alternatively, segments of an ecosystem may have higher value simply because they are rare or unique. For example, rare oak openings in southern Michigan are home to the endangered Karner blue butterfly (*Lycaeides melissa samuelis*). Ecosystem features may have high scientific value because of the uninvestigated properties of the species present. Their relative value in terms of ecosystem function may be unknown, and they may need to be preserved for scientific scrutiny. Also of scientific value is any archeological evidence that may be in place. An effort should be made, and in some cases is legislated, to preserve or catalog such cultural resources.

Information about resources with high cultural value is, by its nature, difficult to obtain and quantify. Known sites of archeological interest are often recorded and mapped. In the United States, state historical offices can be contacted to acquire some of this information. GIS-based models have been used by archeologists in an attempt to predict the locations of culturally important sites, based on a variety of site and situation factors (Dolanski, 1997). Sites of particular spiritual or esthetic value may or may not be designated as parks or reserves. Boundaries of most federal, state and local parks, reserves, and other areas designated for special use are available with the USGS DLGs (USGS, 1990), and the designations used are listed in Table 27.4.

The values placed on sites designated for special use are intangible elements of ecosystems. A variety of methods are available for characterizing these values. To understand the potential recreational value of an area, the U.S. Forest Service and the Bureau of Land Management have employed a tool called the recreational opportunity spectrum, or ROS. The ROS uses the following five criteria to determine the recreational potential of an area: remoteness, size, evidence of human use, social setting, and managerial setting. It can be applied to a particular site, or the criteria can be mapped to portray a spatial representation of recreational value. A project by the U.S. Bureau of Land Management and U.S. Forest Service (n.d.) illustrates such application in Coos Bay, Oregon.

Where maintenance of scenic quality is an important management objective, characterization of scenic quality is imperative. The usefulness of viewshed analysis, a common GIS application, to quantify the views from key vantage points or corridors has been demonstrated elsewhere (e.g., Krist and Brown, 1994; Lake et al., 1998).

TABLE 27.4. Forest, park, and reserve types as classified and mapped in digital line graphs.

National park, monument, lakeshore, seashore, parkway, battlefield, or recreation area
National forest or grassland
National wildlife refuge, game preserve, or fish hatchery
National scenic waterway, riverway, wild and scenic river, or wilderness
State park, recreation area, arboretum, or lake
State wildlife refuge, game preserve, or fish hatchery
State forest or grassland
County game preserve
Large park (city, county, or private)
Small park (city, county, or private)

Source: USGS, 1990.

27.2.5 Key Commercial and Industrial Concerns

To understand the functioning of an ecosystem and changes over time, it is important to understand the commercial or industrial activities underway in the ecosystem. For example, extractive industries like logging, oil and gas extraction, mineral mining, and peat mining all occur because of the economic value placed on the resources. An ecosystem can be dramatically altered when some of its elements are altered or removed, interrupting normal hydrological and biogeochemical functioning. Industries often metabolize raw natural resources and generate waste products in producing goods and services. Raw resource requirements might be met from biological, geological, hydrological, or atmospheric sources. Wastes can take the form of gaseous, liquid, and/or solid waste, which enters the atmosphere, hydrological system, or landfills. Although residences metabolize raw resources to create waste, the types of industries in an area will tend to have a greater influence on what resources are extracted and what wastes are produced.

Data on industrial activities are often collected through the use of agricultural and industrial censuses. Data may be in the form of materials ingested or wastes produced, but more often this information must be inferred from information on the dollar value or quantity produced or people employed (U.S. Bureau of the Census, 1992b). Although this information can be misleading if not used with care, it can provide a representation of the industrial activities in a region and the potential impact on the ecosystem. This information is typically, but not always, only available at an aggregate level (e.g., by county) and not for individual producers.

27.3 Intangible Elements

A number of intangible elements, although they do not necessarily directly affect ecosystem functioning, can have very obvious indirect influences. Intangible elements provide the framework for ecosystem functioning under human management. The geographies of land tenure, administration, use, and regulation can all have noticeable impacts on the structure and functioning of an ecosystem. The social, cultural, economic, historical, and institutional setting of an ecosystem tends to drive the human activities in the ecosystem (see Chapter 28). To be complete, an ecological assessment should include a characterization of these jurisdictional patterns.

The political, administrative, ownership, and other boundaries can be viewed as analogous to the ecological boundaries dividing various subdivisions within an ecosystem. Like ecological boundaries, the intangible social boundaries are porous to the movement of some materials and information and act as effective barriers to others. Just as a watershed boundary defines an area that is distinct hydrologically, but not necessarily meteorologically or biogeochemically distinct, administrative boundaries define areas with homogeneous legal characteristics that may have little to do with spatial variation in demographic or economic characteristics. Just as an animal may travel farther to access a higher-quality food source, people often travel greater distances to access employment opportunities where choices, pay, or benefits are better. When we view humans as a critical part of an ecosystem, information on these intangible elements is critical to understanding the societal components of ecosystems.

27.3.1 Political and Census Boundaries

It is helpful to have the boundaries for all the administrative units that have some regulatory control over the ecosystem. Political boundaries affecting ecosystem management, and therefore ecosystem function, may be present at a number of scales (e.g., national, subnational). In the United States it may be useful to know the boundaries of states, counties, townships or minor civil divisions, legislative districts, school districts, census districts, and other agency-specific management districts. Figure 27.1 displays three levels in the administrative hierarchy within the state of Michigan. Minor civil divisions (MCDs) in Michigan correspond to townships and incorporated cities, towns, and villages. The necessity of obtaining such boundary information arises from both the policy and management relevance of these boundaries on the ground and the availability of socioeconomic and environmental data summarized and available by these spatial units. For example, the data on the U.S. censuses of population, manufactures, business, and agriculture are all available by these units.

Many examples exist of very real landscape expressions of imaginary lines separating intangible entities (e.g., national or other jurisdictional boundaries). The landscape differences result from different policies, culture, history, institutions, or economic situations on either side of the boundaries. For example, usage of a forest or grassland may be very intense on one side of a national boundary because of economic pressure or fewer regulations,

FIGURE 27.1. Three levels of administration boundaries in Michigan: state, county, and minor civil division (MCD).

whereas on the other side it may be less intense. Specific factors affecting differential management include differential taxation, land-use regulations, enforcement of environmental regulations, and incentives.

To characterize the human dimension of ecosystem function, data on human activities are required. To allow for sampling of human activities and to protect the privacy of individuals, most socioeconomic data are collected in such a way that they are aggregated to some spatial unit. For example, the U.S. Census uses the census block as the smallest unit for which many data are available. Figure 27.2 displays four levels in the census geographical hierarchy for Grand Traverse County, Michigan. All units at one level are nested within the next larger units in the hierarchy. Census tracts, or block numbering areas (BNAs) in nonurban areas, are delineated to be relatively homogeneous with respect to population and economic characteristics, yet to remain relatively unmodified from census to census. In 1990, block groups contained about 400 housing units and tracts about 1700.

Administrative boundaries are typically easy to obtain from the agency that has authority over the policies within each unit. The U.S. Census Bureau produces a data set containing all the nested census units and many of the important administrative units (Figure 27.1). These topological integrated geographic encoding and referencing (TIGER) line files are produced for each census year and updated in some intervening years (U.S. Bureau of the Census, 1990).

The spatial aggregation of socioeconomic data has implications for ecological assessment. Any degree of aggregation masks detail in the data. For this reason, it seems reasonable to acquire data at the finest scale possible; but for many socioeco-

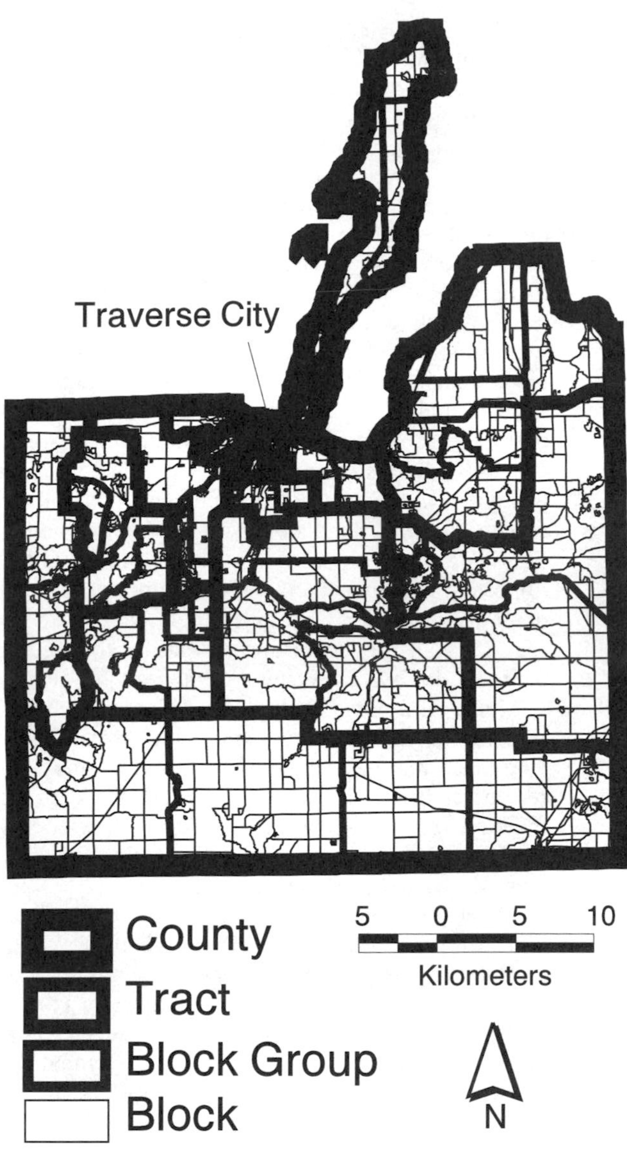

FIGURE 27.2. Four levels of census geography in Grand Traverse County, Michigan: county, tract or block numbering area, block group, and block.

nomic attributes, counties or MCDs are the finest scale available. Some (e.g., population counts) are available down to the block level. When aggregate data are used, it is preferable that the aggregation attempt to minimize heterogeneity within units. So census tracts and BNAs are more useful aggregation units than MCDs because they attempt to minimize heterogeneity. Other considerations in selecting the scale of data should include the scale of the process of interest and the scale of ecological data being used. Just as watersheds can be defined at a number of different scales (e.g., to correspond to first- versus 5-order streams) and the scale should be selected at the scale of the process of interest, be it local or regional in scope, socioeconomic data can be represented at a number of different scales.

Overlay and regional characterization techniques common to many GIS systems can be used to facilitate integration of the socioeconomic aggregate data with other data. Conversion between areal aggregation units or the creation of new areal units that represent the intersection of two sets of units provides mechanisms for comparing disparate data. For example, it may be necessary to summarize population data by watershed. Data collected and aggregated to some spatial aggregation unit (e.g., counties) can be converted from these units to other spatial units (e.g., watersheds) by making assumptions or adding information about the spatial distributions of a quantity (e.g., population density) within a set of enumeration units (e.g., counties or tracts). This process, called *areal interpolation*, can be carried out using techniques outlined by Tobler (1979), Flowerdew and Green (1989), Langford et al. (1991), Martin and Bracken (1991), and Goodchild et al. (1993). A common assumption for many of these methods is that the value of the variable (say, population density) is constant throughout each unit (Langford et al., 1991). Tobler's (1979) method relaxes this assumption by assuming smooth variation of a population density surface. The method of Goodchild et al. (1993) relaxes this assumption by employing control units that are more likely to satisfy the homogeneity assumption. None of these methods is without error, and it is important to acquire data measured directly for geographic areas of interest in an assessment (e.g., watersheds) if at all possible.

27.3.2 Ownership Boundaries

The goals and rules of land management vary depending on who owns the land. Land might be held in private hands (e.g., corporate or personal) or by public agencies (e.g., federal, state, or local). Each class of owner has a different set of constraints on use and management defined by the economy, law, culture, tax systems, and regulations.

The system of publicly held lands in the United States is a legacy of initial settlement policy, various preservation-oriented initiatives, and tax delinquencies that caused private land to revert back to public ownership. Governmental agencies that own lands can provide information about the locations and management of these lands. This information is often not up to date or in a form readily accessible for analysis, but data quality is improving. The U.S. Geological Survey maintains boundary files for land owned by a number of federal, state, county, and local agencies, stored as DLGs (Table 27.4). These files, however, are rarely complete or recent. To address the needs by the Gap Analysis Program (GAP) for information on the locations of public lands, as well as attribute information on the level of management for wildlife and habitat preservation, McGhie et al. (1996) compiled the managed areas database (MAD) at a scale of 1 : 2,000,000, for the conterminous United States. The database includes attributes on area name, designation, protection status, level of management, and data source. A variety of sources was used for this database, most of which ultimately came from the U.S. Geological Survey and the Bureau of Land Management.

Private lands are much more difficult to characterize, because many more different individuals and entities manage them and because they change hands more frequently than public lands. In some areas of the United States, plat map books are produced on a regular basis, which show the boundaries of ownership parcels and list the names of owners (Rockford Map Publishers, 1998). One trend that should improve the availability of information about private lands in the United States is the increasing tendency for county and local tax offices to maintain digital versions of their maps of parcels. The need for this information for ecological assessment is such that there should be no need to invade the privacy of individuals. A useful attribute of private land ownership is simply whether the owner is a corporation or an individual. Other important attributes include the size of parcels and diversity of owners in a given area, which can affect the land use–cover and the ability to manage or intervene in a ecosystem's function.

Many rural areas in the United States are experiencing declines in ownership parcel sizes on private lands that have a variety of implications. In a recent study of parcel sizes in rural areas of the forested upper Midwest, initially reported by Brown and Vasievich (1996), average parcel sizes were shown to have declined an estimated 1.2% per year between 1960 and 1990. Higher rates of decline were observed in the 1970s, when rural population growth rates were highest. Trends such as these have implications for how land is or can be used, the spatial patterning of land use and land cover, and appropriate landscape management approaches.

27.3.3 Land Use

Land use is the intangible relative of land cover and refers to the way in which the land is managed for human goals. The management of land is a very important, but difficult, factor to characterize. The way the land is managed will have indirect, and sometimes direct, effects on what cover results and on the aspects of this cover (e.g., intensity) that affect ecosystem function. At the same time, land use is reflective of the socioeconomic forces driving landscape changes. Changes in demographic, economic, political, and institutional characteristics often lead to changes in land use (e.g., Vesterby and Heimlich, 1991). Land management practices are included as a factor (P) in the universal soil loss equation (USLE) for characterizing average annual

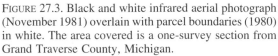

FIGURE 27.3. Black and white infrared aerial photograph (November 1981) overlain with parcel boundaries (1980) in white. The area covered is a one-survey section from Grand Traverse County, Michigan.

soil losses from agricultural fields (Wischmeier and Smith, 1978). How to quantify management is not always clear. The incentives for and information about various management approaches are ever changing.

Market forces drive land use to a large extent. Over time, economic individuals use the land to maximize the value that they can extract from the land. A variety of federal, state, and local government initiatives has been implemented to affect land management toward a particular goal (e.g., conservation). Although government agencies have some record of the numbers of people taking advantage of various incentive programs, they rarely keep spatially referenced (i.e., GIS) databases on such programs. Egbert et al. (1997) demonstrated the use of remote sensing to identify the locations and amount of land that was enrolled in the conservation reserve program (CRP).

Whereas land cover information may be obtained by satellite remote sensing, information about land use usually requires much more direct observation. Aerial photography is an invaluable tool for characterizing land use, but it can rarely be effective in the absence of other information. Figure 27.3 is an aerial photograph of a single square mile survey section (259 ha) in Grand Traverse County, Michigan. The aerial photograph can give clues to the land use. By overlaying the property boundaries on the photo, however, land-use interpretation is dramatically improved. In fact, the size of the parcel may be the characteristic that distinguishes a forested lot used for silviculture versus one used primarily for residential purposes. Table 27.5 lists the land-use classification scheme designed for an ongoing study of the socioeconomic drivers of land-use change in the upper Midwest. The classification scheme will aid in interpreting the land use of ownership parcels from aerial photographs.

Another method of land-use mapping involves "windshield" survey, in which properties are observed from a road and land use is recorded on a base map of ownership parcels. Also, the U.S. Bureau of the Census (1992a) reports proportions of various land-use types in each county as part of its Census of Agriculture. Some aspects of land use, for example, the intention of the owner or the actual day to day use of the land, can only be determined through surveys or interviews with individuals involved in use of the land.

The section displayed in Figure 27.3 is instructive for a number of reasons. The differences between land use and land cover are evident in the northern one-half of this 1-square-mile survey sec-

TABLE 27.5. Classification of land uses for a study of land use and cover change in the upper Midwest.

Developed
- Residential
  - High-density residential
  - Low-density residential
- Retail, office
- Industrial, warehouse
- Infrastructure, transportation
  - Airport
  - Transport corridor or terminal
  - Utility corridor or station
  - Institutional
- Site-based outdoor recreation
  - Campground
  - Golf course
  - Ski area
  - Marina
  - Park and outdoor assembly
- Mining, extractive
- Other developed

Agriculture
- Row crop
- Nonrow crop
- Pasture, grazing
- Other agriculture

Undeveloped
- Open, grass
- Old field, young forest
- Mature forest
- Tree plantation
- Open water
- Wetland
- Forested wetland
- Other undeveloped

Mapped parcels were classified through overlay with and interpretation of aerial photographs.

tion. The land cover would be classified as some sort of grassland, whereas the land use, at least in the majority of this area, is for pasture and hay production. The former has direct effects on ecosystem functioning (e.g., hydrologic and biogeochemical cycling functions differently on grasslands than forests), and the latter is driven by numerous economic, cultural, demographic, and political factors. The rows of planted trees in the southeastern portion of the section suggest something of the history of the site. This area, like many areas in the upper Midwest, was logged initially during the late 19th or early 20th century; then farming was attempted. Now, because of the difficulty of farming the glacial soils with a short growing season, many sites have been planted to forest or are reverting to forest naturally.

The impacts of the intangible boundaries on the landscape are also evident in Figure 27.3. The ownership boundaries retain their rectilinear shape imposed during the initial survey and disposal of the land. The parcel in the extreme southwestern corner of the section is a 16-ha (40-acre) parcel, or one-quarter of a quarter-section. In an attempt to reduce the amount of land subdivision in Michigan, the state legislature passed the Subdivision Control Act (or SCA) in 1967. To avoid paying high fees for subdividing their land, land owners could not divide and sell more than four parcels of 4 ha (10 acres) or less in a span of 10 years. This has led to proliferation of parcels that are just over 4 ha throughout rural Michigan (Norgaard, 1994), like the four narrow parcels found along the northern edge of this section. The importance of boundary shapes for ecosystem structure is illustrated by the rectilinear boundaries that follow property boundaries of the plantation in the southeast corner of the section. Also, the tonal variation between the two grassland parcels in the northwest corner of the section suggests that, although the land cover class is the same, the degree and type of management have been different. Finally, a utility corridor right-of-way, which is not explicitly owned by the utility company, is evident as a swath through the plantation forest on the southern edge of the section.

27.4 Conclusions

This chapter has outlined a number of tangible and intangible human elements that are present in most ecosystems and that should be considered in a complete ecological assessment. Not all these elements will be of importance in every situation, but taken as a whole they can dramatically affect ecosystem history, function, and potential. Although it is tempting to treat human activity as external to the functioning of natural ecosystems, the interrelationships among the value humans place on ecosystems, the goods and services that they derive from them, and natural processes dictate that the human dimension be understood as evolving with the ecological system.

A variety of anthropogenic issues have been presented that should be considered in an ecological assessment. Two important common issues to the characterization of many of these elements are spatial scale and the temporal nature of these elements. The issue of spatial scale, although often determined by the availability of data, affects our ability to characterize various processes. Although of more importance to some of the elements than others, the dynamic nature of the human dimension requires vigilance in the obtaining and maintaining of data.

Integrating the human dimension within an ecological assessment poses challenges to both our ability to characterize the essential elements and to understand their significance. Given the importance of human activities in driving, constraining, and redirecting ecological function, it is imperative that these challenges be met.

27.5 References

Adams, J. B.; Sabol, D. E.; Kapos, V.; Filho, R. A.; Roberts, D. A.; Smith, M. O.; Gillespie, A. R. 1995. Classification of multispectral images based on fractions of endmembers: application to land-cover change in the Brazilian Amazon. *Remote Sens. Environ.* 52: 137–154.

Anderson, J. R.; Hardy, E. E.; Roach, J. T.; Witmer, R. E. 1976. *A land use and land cover classification system for use with remote sensor data.* Geological Survey Professional Paper 964. Washington, DC: U.S. Govt. Printing Office.

Brown, D. G.; Duh, J. D.; Drzyzga, S. 2000a. Estimating error in an analysis of forest fragmentation change using North American Landscape Characterization Data. *Remote Sens Environ.* 71:106–117.

Brown, D. G.; Pijanowski, B. C.; Duh, J. D. 2000b. Modeling the relationships between land-use and land-cover on private lands in the Upper Midwest, USA. *J Environ Manage.* 59:247–263.

Brown, D. G.; Vasievich, J. M. 1996. A study of land ownership fragmentation in the Upper Midwest. In: *Proceedings, GIS/LIS '96 Conference*, November 19–21, 1996, Denver, CO. Bethesda, MD: Amer. Soc. Photogramm. Remote Sens.: 1199–1209.

Dobson, J. E.; Bright, E. A.; Ferguson, R. L.; Field, D. W.;

Wood, L. L.; Haddad, K. D.; Iredale, H.; Jensen, J. R.; Klemas, V. V.; Orth, R. J.; Thomas, J. P. 1995. *NOAA coastal change analysis program (C-CAP): guidance for regional implementation*. NOAA Technical Report NMFS 123. Washington, DC: Natl. Oceanic and Atmospheric Admin.

Dolanski, J. J. 1997. *Development, application, and evaluation of the archaeological site prediction system*. M.A. thesis. East Lansing, MI: Department of Geography, Michigan State University.

Egbert, S. L.; Lee, R.; Price, K. P.; Boyce, R. 1997. *Mapping conservation reserve program lands using multitemporal Thematic Mapper imagery*. Abstracts, Association of American Geographers 93rd Annual Meeting, April 1–5, 1997, Fort Worth, TX. Washington, DC: Association of American Geographers.

Flowerdew, R.; Green, M. 1989. Statistical methods for inference between incompatible zonal systems. In: Goodchild, M. F.; Gopal, S., eds. *Accuracy of spatial databases*. New York: Taylor and Francis: 239–247.

Forman, R. T. T. 1997. *Land mosaics: the ecology of landscapes and regions*. New York: Cambridge University Press.

Geores, M. 1996. *Common ground: struggle for ownership*. New York: Rowan and Littlefield.

Goodchild, M. F.; Anselin, L.; Deichmann, U. 1993. A framework for the areal interpolation of socioeconomic data. *Environ. Plan. A* 25:383–397.

Iida, S.; Nadashizuka, T. 1995. Forest fragmentation and its effect on species diversity in suburban coppice forest in Japan. *For. Ecol. Manage.* 73(1/3):197–207.

Jensen, J. R. 1996. *Introductory digital image processing: a remote sensing perspective*, 2nd ed. Upper Saddle River, NJ: Prentice Hall.

Krist, F. J.; Brown, D.G. 1994. GIS modeling of Paleo-Indian period caribou migrations and viewsheds in Northeastern Lower Michigan. *Photogramm. Eng. Remote Sensing* 60(9):1129–1137.

Lake, M. W.; Woodman, P. E.; Mithen, S. J. 1998. Tailoring GIS software for archaeological applications: an example concerning viewshed analysis. *J. Archaeol. Sci.* 25(1):27.

Langford, M.; Maguire, D. J.; Unwin, D. J. 1991. The areal interpolation problem: estimating population using remote sensing in a GIS framework. In: Masser, I.; Blakemore, M. B., eds. *Handling geographic information*. Essex, UK: Longman Scientific and Technical: 55–77.

Lunetta, R. S.; Lyon, J. G.; Guindon, B.; Elvidge, C. D. 1998. North American landscape characterization dataset development and data fusion issues. *Photogramm. Eng. Remote Sensing* 64(8):821–830.

Martin, D.; Bracken, I. 1991. Techniques for modelling population-related raster databases. *Environ. Plan. A* 23:1069–1075.

McGarigal, K.; Marks, B. J. 1993. *FRAGSTATS: spatial pattern analysis program for quantifying landscape structure*. Corvallis, OR: Forest Science Department, Oregon State University.

McGhie, R. G.; Scepan, J.; Estes, J. E. 1996. A comprehensive managed areas spatial database for the conterminous United States. *Photogramm. Eng. Remote Sensing* 62(11):1303–1306.

Medley, K. E.; Okey, B. W.; Renwick, W. H. 1995. Landscape change with agricultural intensification in a rural watershed, southwestern Ohio, U.S.A. *Landscape Ecol.* 10(3):161–176.

Norgaard, K. J. 1994. *Impacts of the Subdivision Control Act of 1967 on land fragmentation in Michigan's townships*. Ph.D. dissertation. East Lansing, MI: Department of Agricultural Economics, Michigan State University.

O'Neill, R. V.; Krummel, J. T.; Gardner, R. H.; Sugihara, G.; Jackson, B.; DeAngelis, D. L.; Milne, B. T.; Turner, M. G.; Zygmunt, B.; Christensen, S. W.; Dale, V. H.; Graham, R. L. 1988. Indices of landscape pattern. *Landscape Ecol.* 1(3):153–162.

Rockford Map Publishers, Inc. 1998. *County land atlas and plat books*. Rockford, IL: Rockford Map Publishers.

Tobler, W. 1979. Smooth pycnophylactic interpolation for geographical regions. *J. Amer. Statist. Assoc.* 74(367):519–536.

Turner, B. L., editor. 1990. *The Earth as transformed by human action: global and regional changes in the biosphere over the past 300 years*. New York: Cambridge University Press.

Turner, M. G.; O'Neill, R. V.; Gardner, R. H.; Milne, B. T. 1989. Effects of changing spatial scale on the analysis of landscape pattern. *Landscape Ecol.* 3:153–162.

USBTS (U.S. Bureau of Transportation Statistics). 1996. *National transportation atlas databases, for Unix*. BTS CD-14. Washington, DC: U.S. Dept. Transp., Bureau Transp. Statist.

U.S. Bureau of Land Management; U.S. Forest Service (n.d.). *Recreation opportunity spectrum (ROS)*, Coos Bay ROS demonstration page: (http://www.forestry.umt.edu/blm/), last accessed August 29, 1998.

U.S. Bureau of the Census. 1990. *TIGER: The coast-to-coast digital map data base*. Washington, DC: U.S. Dept. Commerce, Bureau of the Census.

U.S. Bureau of the Census. 1992a. *1992 Census of Agriculture—geographic area series—state and county data* (AC92-A). Washington, DC: Govt. Printing Office.

U.S. Bureau of the Census. 1992b. *1992 Census of Manufactures—geographic area series* (MC92-A). Washington, DC: Govt. Printing Office.

USGS (U.S. Geological Survey). 1990. *Digital line graphs from 1 : 24,000-scale maps. Data user's guide 1)*. Reston, VA: U.S. Geological Survey.

Vesterby, M.; Heimlich, R. E. 1991. Land use and demographic change: results from fast-growth counties. *Land Econ.* 67(3):279–291.

Wischmeier, W. H.; Smith, D. D. 1978. *Predicting rainfall erosion losses: a guide to conservation planning*. Agriculture Handbook No. 537.Washington, DC: U.S. Dept. Agric.

28 Assessing Human Processes in Society: Environment Interactions

David J. Campbell

28.1 Introduction

Human activity has transformed Earth's surface, is changing atmospheric conditions, and altering aquatic ecosystems. Human impact is pervasive but, depending on the activity and the aspects of the ecosystem affected, the effects may be direct or indirect, and the consequences range from localized and short term to extensive over both space and time.

Assessment of interactions between societies and natural ecosystems has to carefully consider the society involved—its values and aspirations, its science and technology, its economy and political institutions. These are not static, but dynamic. Yet the dynamics are not unidirectional or predictable. The complexity of interactions between ecological and societal processes over time and across space continues to challenge the scientific community. Efforts to understand climate change, land cover change, and land–atmosphere relationships struggle with both theoretical–conceptual issues and with the complexities of modeling and related data needs.

This chapter focuses on the importance of understanding the nature of interaction between societal and environmental processes that arise at and between the local, national, and global scales. It begins with an overview of approaches that social scientists have adopted to examine society–environment interactions, linkages to the biophysical sciences, and concepts that are critical to an appreciation of social processes. A presentation of land-use systems, selected models, and economic and noneconomic characteristics of human activity follows. The chapter ends with a discussion of the implications for methods of analyzing the interaction of human activity and ecosystems and concludes that many of the key analytical concepts identified by social scientists, such as scale, time, and uncertainty, are shared by ecologists. This common interest may encourage the emergence of new approaches to the study of interactions between societal and biophysical systems based on common goals, shared concepts, and respect for concepts and methods that are not shared, but which may contribute more effectively to understanding the complexity of these interactions.

28.2 Interaction between Society and Environment: Conceptual Frameworks in the Social Sciences

Much of the concern with ecological sustainability arises from divergent views of appropriate resource management strategies (Colby, 1990). Although accounts of the impact of society on the environment can be traced back to Aristotle and Hypocrites, the present century has witnessed the presentation of a series of conceptual frameworks in the social sciences. These frameworks include the following:

1. Those that see society as being an integral part of nature, as illustrated by Buddhist ideology, deep ecology, and the notion of Gaia (Lovelock, 1979, 1991). Policies stemming from these proposals include the exclusion of people from areas to re-create natural systems and less radical ideas such as conservation and sustainable development.
2. Environmental determinism that sees biophysical conditions as determinants of the attributes

of society. A temperate climate, for example, was said to stimulate cultural development. This discredited view emerged as an early attempt to explain different social and economic circumstances around the globe. It was used to justify colonialism as a God-given basis for the "superior" nations of Europe and North America to seize hegemony over "backward" areas of Africa, Asia, and Latin America (Peet, 1985; Peet and Thrift, 1989).
3. Cultural and human ecology emerged as popular frameworks in the 1950s and continue to be developed. These express the interaction of economic, demographic, and cultural aspects of society with the environment as one in which people both adapt and adapt to the environment to create their land-use systems. Two volumes capture the progression of these ideas in the past 50 years, Thomas (1956) and Turner et al. (1993a).
4. Approaches that assess the proposition that increasing human numbers are the primary cause of environmental problems. Neo-Malthusians following the writings of Malthus (1973), including (Hardin, 1968, 1974, 1993; Hardin and Baden, 1977) and Ehrlich and Ehrlich (1987, 1990), argue that there are finite limits to the ability of Earth's resources to support the demands of a large population. Disasters such as famine, disease, and war may represent nature's way of bringing the number of people back into balance with their resource base. Proponents often apply the concept of carrying capacity from ecology to human production systems and argue that policy should recognize the "tragedy of the commons" (Hardin, 1968).

Opposing views, such as those of Boserup (1965) and Simon (1986, 1992), present humans' search for means of overcoming shortages as the source of innovation that allows society to meet its resource needs. Boserup, for example, describes a common process of agricultural intensification that occurs as increasing population densities stimulate adoption of new technologies. The application of carrying capacity to humans has been criticized for its failure to take account of the complexity of human society. For example, food consumption in an area is affected by agricultural technology, trade, and the population's ability to purchase food, as determined by its wealth. Each of these factors reflects the interaction of the local population with the broader community, the local and national economic and political system, and the distribution of wealth within the local population. The "tragedy of the commons" perspective has been assessed by those who argue that "common property regimes" are far more complicated in their application than portrayed in the "tragedy" literature. They do include mechanisms to protect against mismanagement (Fratkin, 1997), and the heterogeneity and dynamics of society determine their effectiveness (Bromley, 1992).

These opposing arguments have given rise to a prolific and contentious literature. Other contributors have argued that none of the preceding frameworks sufficiently considered that societies are economically and otherwise heterogeneous and that political and economic power affects resource allocation and use. A more radical perspective places the society–resource discussion in the context of the wider political economy (Harvey, 1974, 1996; *Ecological Economics*, 1995).

Political ecology is one framework that examines societal and biophysical processes (Blaikie, 1985, 1994; Blaikie and Brookfield, 1987). This approach encompasses "interactive effects, the contribution of different geographic scales and hierarchies of socioeconomic organizations (e.g. person, household, village, region, state, world) and the contradictions between social and environmental changes through time" (Blaikie and Brookfield, 1987, p. 17). Campbell and Olson (1991) provide a heuristic approach to applying political ecology.

These latter ideas can be linked to those that recognize that environmental problems are often associated with social injustice (Harvey, 1996). This perspective promotes the move from sterile intellectual debate to contribute directly to political action. For example, Peet and Watts (1996, p. 37) argue that the intention of "liberation ecology" is "not simply to *add* politics to political ecology, but to raise the emancipatory potential of environmental ideas and engage directly with the larger landscape of debates over modernity, its institutions, and its knowledge." This activist perspective also posits the importance of recognizing the diversity of perspectives that exist within society, including those of women (Rocheleau et al., 1996). An effective representation of this diversity is found in works that examine situations through the experiences of different actors, using approaches such as narratives and analysis of the discourse, the vocabulary used by different actors (Brink, 1982; Rocheleau et al., 1994).

Social science perspectives on the relationship between human activity and ecosystems remain varied and are the subject of productive ongoing discussions. While the discussions continue, there is broad agreement that certain concepts are essential to understanding the interaction between societal and biophysical systems. These include the following:

1. The role of *power* in determining outcomes of interaction.
2. Awareness that *social heterogeneity*, in terms of categories such as social and economic class, gender, and age, is reflected in access to and in the distribution and use of resources. Processes that contribute to changing landscape patterns may be perceived differently by various groups and may affect them in unique ways. Understanding such variability contributes to an assessment of landscape change.
3. Understanding that interactions take place over *time*. Different processes have different dynamics and thus interactions over time are often fraught with uncertainty, rather than being predictable. Furthermore, some processes have immediate consequences, such as land clearance, while others, such as reforestation, may take longer and some impacts are temporary while others will be long lasting.
4. Awareness that interactions between *scales*, from the village to the nation and the world, determine patterns of resource use. While each level is linked, it is also to some extent independent. Imagination is needed to interpret the processes acting both within and between levels because "the interactions between areas or across scales are not limited to one sector; environmental change can influence economic outcomes and political decisions can alter social conditions. Interactions are therefore not unilinear, but rather form a *spiral* of interconnected impulses" (Campbell and Olson, 1991, p. 17).
5. Recognition that the complexity and levels of *uncertainty and unpredictability* in interactions between societal and biophysical systems call for a methodological pluralism (Norgaard, 1989) to the collection, analysis, and interpretation of information. Posturing over the relative scientific status of qualitative and quantitative approaches has become detrimental to the search for inclusive approaches that respect the scientific accomplishments of both.
6. When *policy implications* are drawn, this uncertainty should be considered. Clarity in terms of objectives, projected gains and losses, ecosystem impact and options for remediation, and social impact differentiated by societal groups will contribute to policies that promote flexibility in future options.

28.3 Modeling Social and Biophysical Processes

Analytical models of the complex interactions between human and biophysical systems have been developed (NASA, 1988; CIESIN, 1992). These indicate the wide range of variables and processes that characterize interactions between the human and the physical systems. However, the processes that link them are less clearly articulated than the components of the systems themselves. Yet it is this juncture between the societal and the environmental systems that is crucial to an understanding of human interactions with ecological systems. Where the farmer breaks the ground with her hoe in rural settings or where the smokestack reaches into the atmosphere in industrial areas are critical points of confluence between societal decisions and environmental systems.

How can the detailed models of environmental and social scientists be linked in ways that both promote scientific understanding of the interaction

TABLE 28.1. From urban smog to acid rain—a search for solutions.

Time	Source	Impact	Solution
T_1	Local industry	Local urban system: Weathering Smog	Local urban system: Acceptance Absorption
T_2	Local industry	Local urban system: Weathering Smog	Local urban system: Adaptation: smokestack
T_3	Local industry	Regional, national, international: Acid rain Effects: Atmospheric system: acid rain Drainage systems: water quality, lakes Vegetation system: tree and plant damage Economic system: costs to lumbering, fishing, tourism industries	Political, economic, and legal systems: Regional: East Coast states sue Midwest states US federal: EPA reports International: USA–Canada discussions; W. Europe and N. Europe discussions

and inform policy makers in ways that promote more effective understanding? Do conceptual structures exist that can enable a discourse and guide effective policy and research?

Much activity in this arena suffers from a lack of a clear conceptual framework that integrates societal and biophysical processes and from a single-issue focus that fails to portray the complexity of issues and their spatial and temporal characteristics. A simple illustration of the origins and remediation of the problem of urban smog can illustrate this complexity (Table 28.1).

Urban smog develops under high-pressure atmospheric conditions in which a stable air mass fails to disperse air pollutants from factories, vehicle emissions, and other sources. In the 1950s and 1960s, smog formed under high-pressure conditions that persisted over a number of days in London, UK. The incidence of bronchial and cardiac illnesses increased, and in one four-day smog in 1952 over 4000 people died. This situation was not uncommon in industrial centers in many countries.

The search for a solution to the problem of human health was predicated on reducing the incidence of smog. The issue was seen as a local one of weather: stable air, preventing dispersal of particulate matter from the urban area, resulted in air quality decline dangerous to health. Solutions were found in altering the industrial architecture by building tall smokestacks that emit factory wastes into the atmosphere above the observed height of the inversion layer. Once above the inversion layer, the emissions were distributed by the winds present at those levels. An architectural and economic solution effectively solved the local problem of urban smog.

This solution created a different problem, that of acid rain in areas downwind from the areas emitting acid particles higher in the atmosphere. What was initially a local problem of smog became a more widespread problem of acid rain. The systems involved were altered from the localized urban economic, atmospheric, and human health systems to atmospheric, terrestrial, aquatic, and economic systems in areas experiencing the acid rain. Frequently, these systems were distant from and were in states and even nations different from those originating the particle emissions. The search for solutions has necessitated involvement of political, legal, and economic systems at regional to international levels. For example Canada and the United States, western and northern European countries, and individual states in the United States have come into conflict over the impact of acid rain on their economies and the expectation of remedial action.

This brief example, presented at a most general level, illustrates the essential elements of a conceptual framework that would address the interaction of human and biophysical systems. The elements include those that are common to both human and biophysical systems, such as space, scale, and time, together with uncertainty and unpredictability. Both social and ecological scientists recognize the need for a comprehensive approach. Two related but separate conceptual frameworks, landscape ecology and political ecology, illustrate contemporary approaches that may provide a basis for developing frameworks that more effectively integrate the societal and the biophysical.

Although common elements are evident in the conceptual approaches of biophysical and social scientists, others are seldom shared. From the social science perspective, the issue of power is an essential component, but one that receives insufficient attention in ecological approaches. Power reflected in legal, political, and economic systems determines the outcome of conflicts in goals and policies that are inherent in the management of dynamic interactions between human and biophysical systems.

Critical categories for analyzing and explaining human impacts on ecosystems through natural resources management (NRM) are political, economic, and social–cultural (Table 28.2), although interactions occur between categories and some processes may fit in more than one category.

28.4 Human Activity and the Environment: Land-use Systems

Patterns of land use are the spatial imprint of interactions between society and the physical environment. Embedded in them are economic objectives, cultural traditions, esthetic values, political regulations, and the inertia of past decisions. Identical ecological settings can be interpreted and used in distinct ways by different societies and even by diverse groups within a society. Where such differences intersect, competition over land use results, and the outcome is a function of power expressed through negotiation or fiat.

28.4.1 Human Processes That Affect Ecosystems

Attention has long been paid to the economic and demographic processes that result in the transformation of the land surface and in the use and

TABLE 28.2. Critical components for analysis of the impact of human systems on ecosystems.

Economic production systems: *Industries*: mining, manufacturing, agriculture, forestry, fisheries *Labor force characteristics*: age, wages, hours worked, unionization, benefits, seasonality *Demands on natural resources*: local and imported **Economic policy:** *Macroeconomic policy*: structural adjustment, pricing policy, domestic and international trade, tax policy **Commercial institutions:** Commodity boards, markets, banking **Spatial distribution:** of activity, wealth/poverty **Social distribution:** wealth/poverty, employment, income (wage, unearned entitlements/retirement/ capital gains), dependency ratios, housing **Spatial interaction:** regional/national/global input–output characteristics **Infrastructure:** investment, recurrent costs **Private/public management** **Ownership of resources:** private individual, institutions, public	**Political:** **Individuals:** powerful individuals, of both national and local status **Institutions:** political parties, community organizations, local, national and international bureaucracies; multi- and bilateral organizations, NGOs, multinational corporations **Power:** power is vested in people and institutions with different political, economic, and social status and different influence over civil authorities and public policy. NRM issues are defined by the exercise of power between institutions, between and within countries, between gender, ethnic and socioeconomic groups, by fiat and/or negotiation. **Power determines:** • access to private and public means of decision making • access to public and private resources (land, labor, and capital) • access to means of enforcement (legal instruments, public safety, military) • impact on economic and social processes, and NRM	**Social/cultural values and mores:** reflected in institutions and behaviors that influence NRM: *cultural*: ethnicity, history *religious*: numbers with different affiliations *views* re natural resources (e.g., deep ecology vs. ecodevelopment) *social*: gender roles, respect for age, etc. **Demography** *population dynamics*: vital rates, migration *life cycle dynamics* *distribution of population*: by age, gender, economic status **Education and science** *education policy*: availability, curriculum *knowledge systems*: indigenous knowledge, western knowledge **Social organization** *gender roles* *individual versus group* *family, lineage, and social groups* **Social structure** *distribution of resources*: land tenure *distribution of power*: authority

quality of aquatic ecosystems. Economic and demographic processes are intimately linked. Industrialization and trade are tied to urbanization. Urbanization includes redistribution of population, migration from rural areas, and the commercialization of rural agricultural systems. Industrialization and urbanization are often associated with improved health care and education, sanitation, and water supply. These stimulate reductions in death rates and contribute to demographic transition in the form of urban and rural population growth.

Different types of industrial activity have distinctive impacts on ecosystems. Mining may have only a localized impact, whereas processing of minerals can contribute to urbanization with its attendant demands on ecosystems and, through waste production, can affect atmospheric and hydrological systems. Urban development entails conversion of land use to urban forms. Rural population growth is associated with extensification of agriculture, involving replacement of forest or other land cover by crop or animal production, or with intensification of land use.

Care is required in the assessment of the ecological impact of such land cover or use changes. The impact is most intense at the time of initial transformation from one use or cover to another, for example, from forest to agriculture, agriculture to suburbia, riparian forest to irrigated crop production, or lakeshore to recreational activities. As the new land uses mature, they may, over time, replace some ecosystem functions that were lost in the initial transformation. Others may be permanently damaged or lost.

There is a temptation to assume that the losses incurred at the time of initial transformation are permanent and irreparable. This is not necessarily the case. Whether impacts are irreparable depends on criteria for evaluation. For example, where an area of forest is replaced by cultivation, there may be initial disruptions or destruction of habitat, loss of biodiversity, diminished carbon cycling, and disturbance of drainage. Cultivators seldom seek to destroy the resource base; rather, actions are taken to enhance their livelihoods. For example, they may plant trees to maintain productivity, to achieve esthetic objectives, and to establish land rights. These trees may also restore drainage regimes and revive carbon cycling, and, while they may create a different biodiversity, they may not re-establish the same habitats. There may be gains, losses, or equivalence in outcomes depending on the objective criteria. Where the outcomes pit societal objectives, such as promotion of livelihoods, against ecological ones, then the objective criteria are complex, and the power of interest groups usually determines the political outcome.

Categories of Economic Activity

Land use can be classified into four categories of economic activity: primary, secondary, tertiary and quaternary. Each of these has different demands of and impacts on ecosystems. These may change in nature from the time they are initiated to those related to their long-term functioning.

Primary activities are those that derive their economic utility directly from aquatic or terrestrial resources. These include agriculture, fishing, mining, forestry, and wilderness or wildlife-based tourism. Some of these, such as agriculture, forestry, and often wilderness and wildlife, are extensive land uses. They therefore have a greater impact on overall land cover than other activities.

The density of population that primary activities support depends on a variety of factors. In industrial societies such as those of North America and Europe, rural population densities are low and the pattern of associated urban development is one of relatively small towns that act as markets and service centers. The processing of products often takes place far from the point of production. In other parts of the world, rural population densities are similar to those of suburban areas in the United States. In Rwanda, for example, densities of over 800 people per square kilometer were reported by the 1991 census.

The distribution of population associated with other primary activities, such as fishing and mining, is more localized. Although fishing makes extensive use of aquatic ecosystems, the fleets are based in ports, giving the industry a localized terrestrial pattern. The distribution and exploitation of mineral resources is also localized. Furthermore, the characteristics of the primary product, the perishability of fish and the impurity of most ores, attract industries that use these products as raw materials to the point of production, such as the ports and mines.

The processing of primary products creates *secondary economic activity*. Forestry is the basis for furniture making; mining of iron, coal, and limestone is the foundation of the steel industry; and agriculture supports the food-processing industry. These secondary activities have locational requirements that are internal to the production unit and others that relate to other associated industrial activities. The weighing of the relative importance of these determines the location of production.

Costs of land, labor, capital, and transport are the

major economic considerations. Some are local and firm specific. For example, production involving flammable processes may seek to locate near a body of water. Others are more general and apply to all industries in an area; for example, a tradition of unionization may influence labor costs, and an established transportation network reduces the costs of moving raw materials and the final product. For industries in which transport costs account for a significant proportion of production costs, the relationship between the nature of raw materials or components and the finished product has a strong role in determining the location of the firm. Where there is significant weight loss in the production process, activities are oriented to locations near raw materials, whereas those that gain weight in the manufacturing process are market oriented. These economic factors are modified by noneconomic factors such as government policy on taxation, environmental quality, and land zoning.

The location of secondary economic activity has a significant impact on the distribution of population. It provides employment directly to those involved in these activities and is also the basis for employment in *tertiary economic activities* that provide services to the industrial population. These include the marketing of secondary products and the organization of society, such as retailing, government services, and health services. These activities are demand oriented and thus have a strong tendency toward agglomeration. They both respond to and contribute to the process of urbanization.

Quaternary economic activities refer to those aspects of the economy that provide innovation. They include research and development in specific industries, information technology, and the emergence of "green" technology in response to concern about environmental sustainability. Many of these activities do not have specific locational requirements associated with land, labor, and capital. They can be widely distributed, and, while strongly linked to other sectors of the economy, these links are often not bounded spatially.

In all sectors, economies of scale, both internal and external, contribute to the spatial pattern of economic activity. Firms achieve *internal economies* through increasing the size of the production unit. Greater size raises demand for resources, including raw materials, land, electricity, water, waste disposal, and labor. The impact on ecosystems and infrastructure is thus related to the unit of production of individual firms. Firms also achieve cost savings by locating close to related activities. These are *external economies*. They are important where industrial processes are complex and require interaction between a number of activities. For example, the auto industry depends on the availability of component industries, including sheet metal, upholstery, and tires, together with advertising, banking, and insurance industries that specialize in the needs of the auto industry. The proximate location of these activities improves efficiency and innovation and thus reduces costs of production.

Distribution of Population

These distinct economic activities are associated with different distributions of population and demands on ecosystems. The variations occur between economic activities and also among economic activities under different social, economic, and political circumstances.

Industrialization has been a major force in influencing the growth and distribution of population. It began with the Industrial Revolution in Europe and is the preferred approach to economic and social development in most countries today. The transition from agrarian-based societies to industrial ones is accompanied by a shift in emphasis from primary to secondary, tertiary, and quaternary activities and by urbanization and its related redistribution of population from rural areas to towns and cities.

The unevenness of economic opportunity both within and between countries has resulted in complex patterns of migration. In the established industrial countries of Europe, North America, and Japan, the process of rural to urban migration has resulted in an urbanization rate of over 75%. The opportunities for industrialization elsewhere have been constrained by global economic patterns. Some, such as the newly industrialized economies of Asia, have developed complex industrial structures, and associated patterns of population redistribution to cities has occurred rapidly.

Where urbanization has been rapid, the ecological impacts on land, water, and air have been enormous and environmental quality has diminished. Examples include air pollution in cities such as Bangkok and Mexico City, urban sprawl in cities of Latin America and Africa, and overloading of waste disposal facilities in cities throughout the world.

In other countries the opportunities for economic gain in industrial cities are restricted by the limited growth of the industrial economy. While perceived economic opportunity in urban areas acts as a magnet to people in rural areas, the experience of many migrants is one of failed ambitions and expectations. In these circumstances the exodus to the towns may be retarded, resulting in increased in situ rural densities, and rural to rural migration and urban to rural migration expands. These processes extend the agricultural frontier and raise rural population densities.

In industrialized countries, extensive primary activities, such as agriculture, are generally associated with low population densities. Although immigration to cities has slowed, urban populations continue to increase and suburbanization is occurring rapidly, resulting in conversion of agricultural land. There is also a pattern of land conversion to residential uses for retirement and second homes characterized by areas with desirable environmental conditions such as forests and water bodies. An increase in recreational facilities, road and public utility infrastructure, and service activities augment the residential land-use pattern, and forest and land cover fragmentation and altered edge features often ensue with implications for ecosystem functions and processes, including wildlife habitats, biodiversity conservation, albedo, and carbon cycling.

28.5 Spatial Models of Human Impacts on Ecosystems

Many ecologists as well as social scientists recognize that prediction is limited by uncertainties about the nature of social and ecological systems. Increasing numbers argue that there is a need to move beyond the search for regularity and predictability as a basis for explanation. They recognize that these systems and the processes that determine society–environment interaction often include nonlinear relationships, and that there are temporal, spatial, and scalar characteristics of the interactions that defy determinism, or even determinism modified by stochastic processes. Also, critical variables, such as time, scale, and power, and the influences of values, beliefs, preferences and mores are too often omitted from the analysis.

These criticisms are significant, but they should not be used to dismiss the utility of data-based models. They do, however, demonstrate that models should not be used for policy making without explicit attention to the assumptions and caveats of the model builders and to the insights of indigenous knowledge systems (Warren, 1991; Seeland, 1997). This calls for a more robust approach, which can include both quantitative and qualitative information from a variety of sources. Such approaches would value not only data-based models, but also modeling strategies such as decision-making models and role-playing simulations.

28.5.1 Land-use Models

A variety of models have been developed to model land-use changes. The majority assumes that market forces dominate determination of the optimum utility of land and the location of economic activity. Industrial location theory assesses the relative importance of economic factors of land labor and capital and external and internal economies of the firm. Other models that generalize from the level of the firm to that of land-use patterns and argue that a bid rent process induces patterns of land use include von Thunen's model of agricultural land use and the urban land-use models associated with central place theory. Such models are illustrated in Haggett (1983) and in Stutz and de Souza (1998).

These models are predicated on the concept that land uses such as agriculture, industry, and commerce have specific requirements that make some locations more desirable than others. These include towns as markets for agricultural products, city centers as foci for commerce, and, for industry, places with available land and access to transport infrastructure. These desirable locations may be identical for different activities, in which case they will bid against each other. The outcome will reflect the will and ability of each potential land use to outbid others.

Although economic processes dominate these models, there is recognition that political, cultural, and historical factors may alter the outcome (Harvey, 1973; Massey, 1984). Furthermore, there exists an inertia of the created environment that constrains replacement of one land use with another that achieves optimum spatial economic utility. Such inertia may result from the existence of buildings that represent a significant investment, regulations reflecting historical or cultural considerations, and externalities such as the presence of toxic wastes.

28.5.2 Demographic Projections

Population dynamics reflect the relationship between births and deaths and the redistribution of population through migration. For any particular region the growth or decline of population can be calculated as

$$\text{population growth} = (\text{births} - \text{deaths}) + (\text{immigrants} - \text{emigrants})$$

Historical assessments of this relation have resulted in the categorical demographic transition model (Figure 28.1), which in its original formulation identified four stages that reflect changes in the bal-

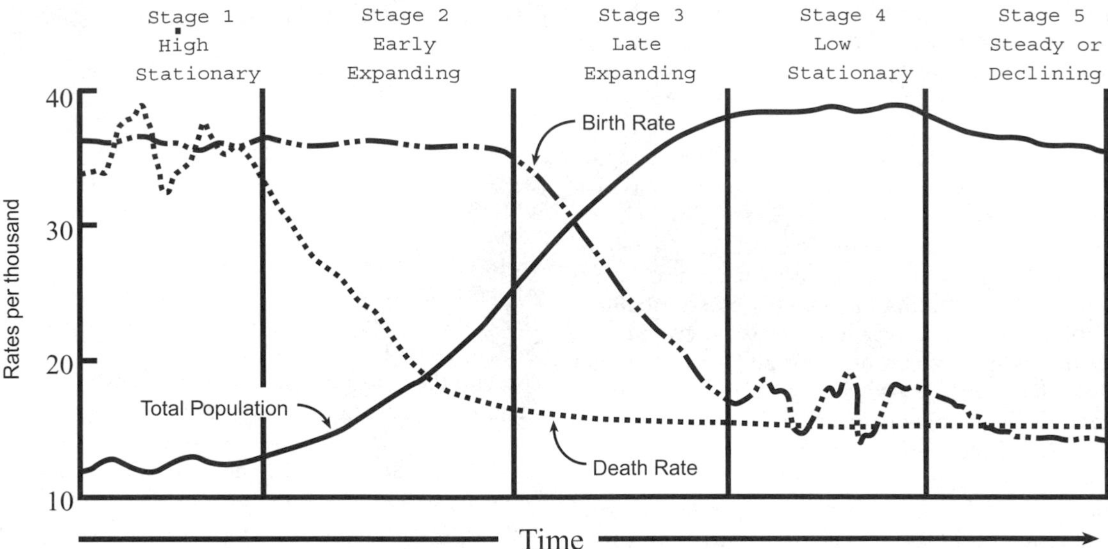

FIGURE 28.1. The demographic transition model.

ance between birth rates and death rates. It postulates that societies pass through these four stages in a sequence that is associated with economic and social conditions.

The first stage reflects preindustrial societies in which birth and death rates are high and cancel each other out, resulting in a relatively small population with a stable growth rate. In the second stage, death rates decline, but birth rates remain high. This initiates a period of population growth that extends into the third stage, in which death rates have stabilized at a low level and birth rates decline, but remain higher than the death rate. In the final, fourth stage, both death and birth rates are stable and low, the population growth rate is low, but the population itself has increased substantially during the second and third stages. Recent demographic trends in some European countries have led to an identification of a fifth stage in which death rates exceed birth rates and the population is declining (Rees, 1997).

The original demographic transition model and the recent addition of the fifth stage were based on the experience of northern Europe. The model has proved less suitable in other settings. Critics have focused on its inability to specify the conditions that trigger the changes that instigate the transition from one stage to another. Declining death rates have been associated with better nutrition and sanitation and improved medical care, including the control of diseases such as smallpox, malaria, and those related to diarrhea. Subsequent concern with global population growth led to programs designed to reduce the birth rate. A variety of strategies has been attempted, including the promotion of contraception, availability of abortion, and the encouragement of one-child families.

Although these have had some success, there also are failures. There is now widespread recognition that declining birth rates depend on changes in family circumstances that reflect improved economic conditions and a rise in the status of women. Such economic and cultural conditions have often proved more difficult to effect than those that reduce death rates. In addition, there is considerable contention as to economic implications and the religious and cultural values associated with family planning.

These issues have been debated in the policy and research literature and were controversially debated at the United Nations International Conference on Population and Development held in Cairo in 1994 (http://www.mbnet.mb.ca/linkages/cairo.html). This meeting

> emphasized the need to improve the status of women and act to ensure gender equity and equality. . . . Adopting this approach, the conference recognized, would lead to smaller, healthier families and slower global population growth. . . . The Cairo conference called for universal access to comprehensive reproductive and sexual health care including family planning by 2015. Other goals were universal primary education and to close the gender gap in education; and sharp reductions in infant, child and maternal mortality. (http://www.unfpa.org/news/pressroom/1999/pre-hague.htm)

The 1999 follow-up conference in The Hague has

an official objective of assessing the outcomes of the policies formulated in 1994, and reducing the debate over contentious issues (http://www.unfpa.org/).

The implications of population growth on ecosystems can be assessed through the application of migration models that examine redistribution of population. The impacts of the distribution of a growing population can be used as inputs to the economic models discussed previously and in rural settings have been examined through approaches that employ concepts of carrying capacity and intensification of land use.

28.5.3 Migration Models

Three classes of migration models can be identified: behavioral, economic, and spatial. Behavioral models are fundamentally descriptive. They examine the decision-making processes that an individual engages to decide whether to migrate. Essentially, the decision to migrate is based on an individual's assessment of satisfaction in the existing location when compared to alternatives. If alternatives offer sufficient incentive, then migration will occur. If not, the person remains in the present location, and issues that led a person to consider moving will have to be managed.

A wide range of variables contributes to the decision to move. While the weighing of these will be unique to each individual, the push–pull model of migration categorizes the most common factors that attract or repulse people to and from origins and destinations. The model was elaborated by Caldwell (1975) and has been applied most frequently in the context of rural to urban migration in countries of the south.

Studies have shown that the people involved in migration and the relative importance of rural–rural and rural–urban migration changes over time. Initially, rural to rural migration is the most frequent response to pressure on resources. However, with declining rural opportunities and an expansion of perceived opportunities in towns and cities, rural–urban migration increases. Age, gender, and level of education of migrants also change over time. The first rural–urban migrants tend to be young men, between 15 and 35 years old, with some education. These are the people most likely to succeed in finding urban employment. Later, as unskilled jobs are more available, the level of education of the migrant is less important. Relatively few women are involved at the early phases of rural–urban migration. However, recent evidence from South America suggests that as urbanization proceeds women come to predominate.

Empirical testing of these behavioral constructs has shown that, among the variety of variables that influence migrants, economic circumstances are fundamental (Todaro, 1969). Once economic concerns are specified, then other factors modify the process, including relatives or friends at origin and destination, environmental circumstances, and cultural values. Todaro's research shows that what is critical is not the reality of economic opportunity, but rather the opportunities that people in rural areas *perceive* to exist in the cities. He developed a numerical model that focuses on perceived employment opportunities and includes other factors that migrants consider, including many of those mentioned previously.

The push–pull and perceived employment opportunity models were developed in the countries of the south. In the north, these models would have had greater relevance in the past. Important contemporary redistributions of population are the continued pattern of suburbanization; in metropolitan areas, the movement of workers' residences to formerly rural communities with good transportation links to the metropoles; and migration of retirees to areas with attractive environments. For example, the migration trends in Michigan include the growth of towns located near Detroit and migration to recreational areas in the northern portions of the state (Manson and Groop, 1996). Contemporary patterns of international migration, for example, those of people from Mexico to the United States and from southern Europe and Africa to the cities of the European Union countries, are driven by economic circumstances and by social and political circumstances, such as war and political unrest. Recent studies indicate that the latter will become dominant instigators of international migration in future decades.

While migration models aid in understanding the redistribution of population from rural to urban areas, the growth of population within rural areas, particularly in countries of the south, has received considerable attention in studies of carrying capacity and of intensification of land use.

Carrying capacity studies of the relationship between societies and their land resources borrow from the ecological concept that a particular land area can support a finite population. The concept has been elaborated by scientists such as Hardin (1968, 1992, 1993) and Ehrlich and Ehrlich (1990) and is commonly discussed in journals such as *Population and Development Review*, *Population and Environment*, and *Population Bulletin*. This perspective has been challenged by a number of economists, including Boserup (1965, 1981) and Simon (1986, 1992), by anthropologists concerned with the concept of communal land tenure (Bromley, 1992; Bruce and Migot-Adholla, 1994), and by geographers (Turner et al., 1993b). The basis of the challenge is that people differ from animals and insects by their ability to make individual and institutional decisions about how to manage resources and to exercise some control over population numbers and their application of technology in the interests of sustained production.

Both carrying capacity models and the intensification literature imply a unidirectional process toward either collapse or continued expansion of output in response to innovations induced by population pressure. Olson's (1998) discussion of the driving forces of land-use changes in Africa links the discussion on the impact of increasing population densities with political ecology in rural Africa. She demonstrates that the common process of agricultural intensification associated with population increase is not inevitably unidirectional and progressive or environmentally neutral. Rather, societies may progress at one time and regress at another in their patterns of land use, land management, and production as political forces emanating at the global and national levels change local power structures and the local economy.

28.5.4 Noneconomic Factors

The preceding discussion has shown that, while many see economic factors as a determinant of people's interaction with the resource base, these models also recognize a wide range of noneconomic factors. These include political factors such as zoning and cultural ones such as the preservation of holy ground or forests (Nyamweru, 1998).

Much modeling and planning make the assumption that, by altering the spatial distribution of particular phenomena, desired outcomes in terms of land use can be facilitated. For example, in planning for improved social and economic development among a population, many have assumed that an even spatial distribution in the provision of infrastructural services, such as roads, markets, schools, and hospitals, will inevitably result in a more equable use of such facilities by the population. In reality, this is not inevitable, because political, cultural, and economic conditions often intervene to prevent the desired and predicted outcome of more equitable social and economic conditions.

For example, improved transportation infrastructure will have a greater impact on those whose livelihoods can benefit from reduced transport costs, such as commercial farmers, than on those whose economic focus is the local market. Social services such as schools and hospitals may be built to a plan of spatial equity. However, the ability of different members of the society to access the services will depend on their economic standing: Can they afford school fees, or the cost of medicine, or can children be spared from agricultural labor to attend school? Cultural values also influence outcomes. Will people accept the type of medical treatment provided or will they continue to approach traditional healers? In many societies the division of labor between men and women allows men more freedom than women, whose daily responsibilities in economic and domestic activities permit little time to attend clinics.

The recognition that spatial equity does not necessarily imply social equity is captured in the concept of *spatial separatism* (Sack, 1980; Gore, 1984). It is an important concept for modeling the relation between ecological and societal systems. It highlights the fact that societies are heterogeneous and that models that use highly generalized variables, such as population numbers, growth, and density, to represent society are inevitably restricted. They are also flawed in their application in circumstances where the inherent heterogeneity of the target population is not explicitly considered in either the design of the model or in the interpretation of its output.

The heterogeneity that needs to be considered in such circumstances includes variations in terms of the components listed in Table 28.2. Political power exercised by individuals, groups, and institutions may initiate a change in resource use that may become established, altered through experience, or rejected by all or part of a population in the short or longer term. The outcome will reflect economic and cultural factors and their expression through power held by affected individuals, groups, and institutions.

This dialectic among those with different access to power has been clearly developed in the work of Giddens (1982) under the term *structure and agency*. Giddens argues that the way society arranges the distribution of resources reflects both *structure*, that is, the ability of those with power to determine and enact structures that enable their particular objectives through laws, economic policy, investment strategies and the like, and also *agency*, that is, the power held by those affected to accept, reject, and modify these to represent their values, objectives, and mores. Resource use reflects an ongoing dialogue over objectives between the dominant classes, often a minority, and the less powerful majority.

28.6 Methodological Implications

Analytical methods for examining people–environment interactions should facilitate the inclusion of the components listed in Table 28.2. This implies that analysts should make them explicit at every stage of study, in conceptualizing research questions, defining data needs, and sampling frames, in analysis and interpretation of information, and in drawing conclusions.

The majority of these issues is common to approaches from the social science perspective, such as political ecology, and that of ecological scientists, such as landscape ecology. These approaches share many common elements, and the differences between them represent clear opportunities for exploring more effective ways to examine the interactions between societal and biophysical processes (Campbell, 1998). However, a comprehensive understanding of the complex interactions between biophysical and socioeconomic processes, particularly of the issues of scale and time, remains incomplete. In fact, some scientists are calling for a completely new approach, arguing that the integrating of biophysical and socioeconomic approaches "implies that the politico-social-ecological system be treated as a complex system, that an understanding of ecosocio relations is truly a new field that requires new methodologies, and those methodologies need not have anything to do with the standard methodologies of either ecology or the social sciences" (Vandermeer, 1994, p. 105).

28.6.1 Data, Measurement, and Verification[1]

In modeling patterns of human activity and their interactions with ecological systems, we are seldom in a position to specify all the variables and interactions in a particular setting. Model building calls for selection of some and omission of other variables and linkages. This selection should be based on a clear theoretical–conceptual framework or on empirical evidence that the variables and interactions omitted would not alter or add to the likely interpretation. Because the model is selective in the components that are included and excluded, some linkages will be measured empirically, while others will be assessed qualitatively.

The inability to measure or completely specify the system and the requirements of cost efficiency and ease of representation often result in indirect measurement and simplified representation of complex processes and situations. Pragmatic concerns such as cost and time may constrain the ability of an analyst to collect the specific information needed to reflect a particular process or pattern. Recourse is often made to available data, often spoken of as "the best available information." This is particularly true of macro-level data whose

> measurement is for the most part opportunistic and atheoretical. Lacking the necessary resources to collect their own data, investigators have usually searched for available indicators without adequately spelling out the assumed connections with the variables they are intended to measure. Very frequently, macro-level researchers are forced to rely on derived measurements that have been based on decisions made by persons whose purposes are very different from those of the investigators. Therefore, it becomes essential to examine very carefully just how these derived measures have been constructed, what homogeneity assumptions have been made, and how they are to be linked to the important constructs in macrotheories. (Blalock, 1982, p. 237)

Blalock's concern with theory has to be complemented by empirical and experimental activity to try to identify causal processes because

> the demonstration that a specific mechanism can in theory give rise to a range of observed patterns is not proof that that mechanism is indeed responsible for those patterns . . . in general, there will be many conceivable mechanisms that could give rise to any set of patterns. All that theory alone can do is to create a catalogue of possible mechanisms; experiments are then needed to distinguish among the candidate mechanisms. (Levin, 1992, p. 1948)

The patterns observed over broad areas represent the outcomes of complex processes acting over subareas. In many cases, information on variables gathered over large spatial units, for example, a province or nation, reflects a summary of complex events and processes that occur within that unit. These variables are therefore surrogates of this complexity. Their meaning and interpretation depends on a clear understanding of the complexity and what elements of it are reflected in such surrogate variables.

28.6.2 The Ecological Fallacy

Since the publication of Robinson's (1950) article, we have been on notice of the *ecological fallacy* whereby characteristics of individuals cannot be inferred from group characteristics. Aggregation can yield summary measures with a descriptive utility, but relationships between aggregated variables cannot be implied to individuals that compose the aggregate. There are interactions between the micro-

[1]This section draws on Campbell (1996, 1998).

and macrolevels, but they cannot be inferred from macrolevel conditions. Rather an explicit theoretical–conceptual framework is required to explicate them.

A recent study of policy making for natural resource management over time in Rwanda illustrates the ecological fallacy (Brown et al., 1996). In the early 1980s, national-level information indicated that the Rwandan population was growing rapidly and that deforestation and land degradation were accelerating. The association between them was interpreted as causal—the driving force of environmental deterioration was rapid population growth in the context of limited land resources. The government established policies on land reclamation, resettlement, and family planning. These continued into the 1990s, when policies to diversify rural livelihoods through the off-farm–nonfarm sector were added.

However, higher-resolution studies using air photography and household surveys reached a different conclusion. The analysis of air photography showed that government-sponsored tree cutting to establish a buffer zone around national forests and tea plantations accounted for a great deal of the deforestation. In areas under cultivation, the information from farm households demonstrated that, while the contiguous tree cover seen on satellite imagery and air photographs might have decreased over time, the number of trees had actually increased as a result of planting by poor farmers (Olson, 1994; Olson et al., 1995).[2]

Government policy was based on national-level information and assumed relationships among variables that had been imputed to the constituents making up the national-level information. Failure to respect the ecological fallacy resulted in inaccurate representation of rural conditions and in inappropriate policy. Understanding the ecological fallacy is critical to effective analysis of people–environment interaction. Without it, facile interpretations of environmental issues are possible, based on statistical association among generalized variables, resulting in remedial policies based on incomplete understanding of the circumstances. In these situations, both science and policy are poorly served.

28.6.3 Dynamics of Scale and Time

The dynamics of society–environment interactions and the scalar structure of information and its implications for interpretation call for explicit attention to be paid to the interacting concepts of scale and time in the analysis and modeling of such interactions. These interactions fall within the domain of a number of different disciplines, and the issues of scale and time are complex in these interdisciplinary settings.

Defining the appropriate spatial resolution for analysis of environmental issues is one of the most difficult problems facing environmental analysts (Robinson, 1950; Harvey, 1968; Blaikie and Brookfield, 1987; Meentemeyer and Box, 1987; Hall et al., 1988; Turner et al., 1989; Turner, 1990; Campbell and Olson, 1991; Izac and Swift, 1994; Klijn and de Haes, 1994). Different processes have different dominant scales of activity. For example, government economic policy is national in its scope, cultural characteristics and rainfall patterns are regional, and soil conditions and agricultural practices are at the scale of the community, household, or field. These interact in determining land-use options, but the available information may or may not permit the interactions to be depicted. It is important therefore that the gaps in information and knowledge that exist in linking across scales be examined and identified using methods such as description, statistical relationships, and simulation models.

The issues have been concisely summarized for land-use/cover change:

> It is well documented that as questions about land-use/cover change are addressed at different space–time scales, so the answers vary significantly. More importantly, these answers cannot be adequately linked. For example, the "cause" factors of land-use/cover change identified at the global scale cannot be integrated well with those found at the local level. As a result, coupling local, regional and global models is extremely difficult. Moreover, resolving or making significant advancement on the scales issue will go a long way towards resolving ideological and polemical debates over the causes of land-use/cover change. (IGBP, 1994, p. 83)

Scale

There is widespread interest among scientists to better understand the links between and among micro-,

---

[2]An association between population growth and indices of environmental deterioration is often found in global, low-resolution data. Many have adopted a position that this lends support to Malthusian perspectives, that population growth is a driving force of environmental deterioration. Empirical studies at a finer resolution have challenged this, postulating a far more complicated relationship. There are increasing numbers of examples where population growth is associated with amelioration of environmental conditions. There is a growing literature on increased forest cover in agricultural areas where growth in population is associated with tree planting as rural populations seek to produce much of their own wood for fuel, building, fencing, and the like (Dewees, 1993; Tiffen et al., 1994; Castro, 1995; Olson et al., 1995).

meso-, and macroscale processes. The issue has been of interest to ecologists, who are concerned with the transfer of information and analyses across scales (e.g., see Meentemeyer and Box, 1987; Turner et al., 1989; Lima and Zollner, 1996), in remote sensing where the spatial resolution employed for analysis of surface conditions affects interpretation (Malingreau and Belward, 1992), and in linking information from different sources gathered at different levels of spatial resolution (Brown et al., 1996).

Ecologists recognize that patterns of environmental variability are generated by complex dynamic interactions among and between species. Analysis of environmental processes is made more complex when the anthropogenic contributions to human–environment interactions are considered. Such interactions encompass a wide range of socioeconomic and biophysical variables, each with its own spatial and temporal characteristics. These have to be assessed in an integrated analytical structure (Izac and Swift, 1994). Creativity is needed to comprehend the processes acting both within and between levels, because the processes and states of a particular level are the outcome of what is happening at that level plus the impact of influences emanating from other levels. At the microlevel, conditions are influenced by what is happening at the microlevel, together with impulses from meso- and macrolevels. Interactions between scales, from the village to the nation and the world, must therefore be recognized as determining patterns of resource use. In this hierarchical structure, each level is linked, and each level is independent to some extent (Figure 28.2).

Furthermore, a change in the scale of analysis may yield new insight as to causal processes: "Although moving to a coarser scale may result in a loss of detail, it may also result in the appearance of emergent properties due to synergisms at a higher level of system integration. These interactions were not seen at finer scales and were perhaps not acting if finer scale meant a lower level of interaction" (Meentemeyer and Box, 1987, pp. 20–21). Interpretation of such synergisms and interactions requires care to distinguish between causal interactions and mere statistical association or coincidence. Again, reference to a conceptual model is critical to valid interpretation of phenomena.

This discussion also illustrates the fact that there is no single appropriate scale at which to discuss human–environmental systems (Levin, 1992). Ultimately, a reductionist approach would yield discussion of microsettings where individual people, soil particles, and plants may be the focus. In the context of human interactions with ecosystems, this level of detail is superfluous and unmanageable. An appropriate scale requires generalizations based on these microlevel patterns and processes, as well as understanding of how they respond to and affect broader-scale processes.

The issue remains, however, that the management of natural resources implies interaction between societal and biophysical states and processes. This interaction takes place between and among different scales. In both the ecological and the human sciences, local to global interactions are recognized as contributing to patterns of resource use: "How do we know which scale and spatial pattern we should use? And how do we know when the results are meaningful and valid given the scale and resolution of the data?" (Lam and Quattrochi, 1994, pp. 89–90).

The issue has been clearly stated in the context of ecology by Levin (1992, p. 1943):

> the problem of pattern and scale is the central problem in ecology. Applied challenges, such as the prediction of the ecological causes and consequences of global climate change, require the interfacing of phenomena that occur on very different scales of space, time, and ecological organization. Furthermore, there is no single natural scale at which ecological phenomena should be studied; systems generally show characteristic variability on a range of spatial, temporal and organizational scales. . . . The key to prediction and understanding lies in the elucidation of mechanisms underlying observed patterns.

An essential component of analysis is the explication of a theoretical–conceptual framework that facilitates clarity of thought regarding the issue under consideration. Such a framework is needed to ensure an appropriate definition of the analytical model to be used. This model in turn will guide the selection of variables, of the spatial scale and unit of data collection. The conceptual framework will also inform approaches to aggregation, analysis, and interpretation and the generalization of results.

There are a number of approaches to this question. They range from case studies and strategies for linking large-scale (small-area) studies to small-scale (large-area)[3] information (McGwire et al.,

[3]There is some confusion in the published literature over the terminology to describe different scales. A number of authors have sought to clarify the issue (Meentemeyer and Box, 1987; Lam and Quattrochi, 1994). In a cartographic sense, small scale represents large areas; for example, a map of the world may be at a scale of 1 : 1,000,000, and large scale refers to small areas, for example, a map of a city may be at the scale of 1 : 5000. In this chapter we will follow this convention. Large-scale spatial units might be used to examine local case studies of village communities, field-level soil conditions, subcatchments of river basins, and, in vegetation terms, the patch. Comparable small-scale units might include national social characteristics, generalized soil maps, river basins, and continental vegetation zones.

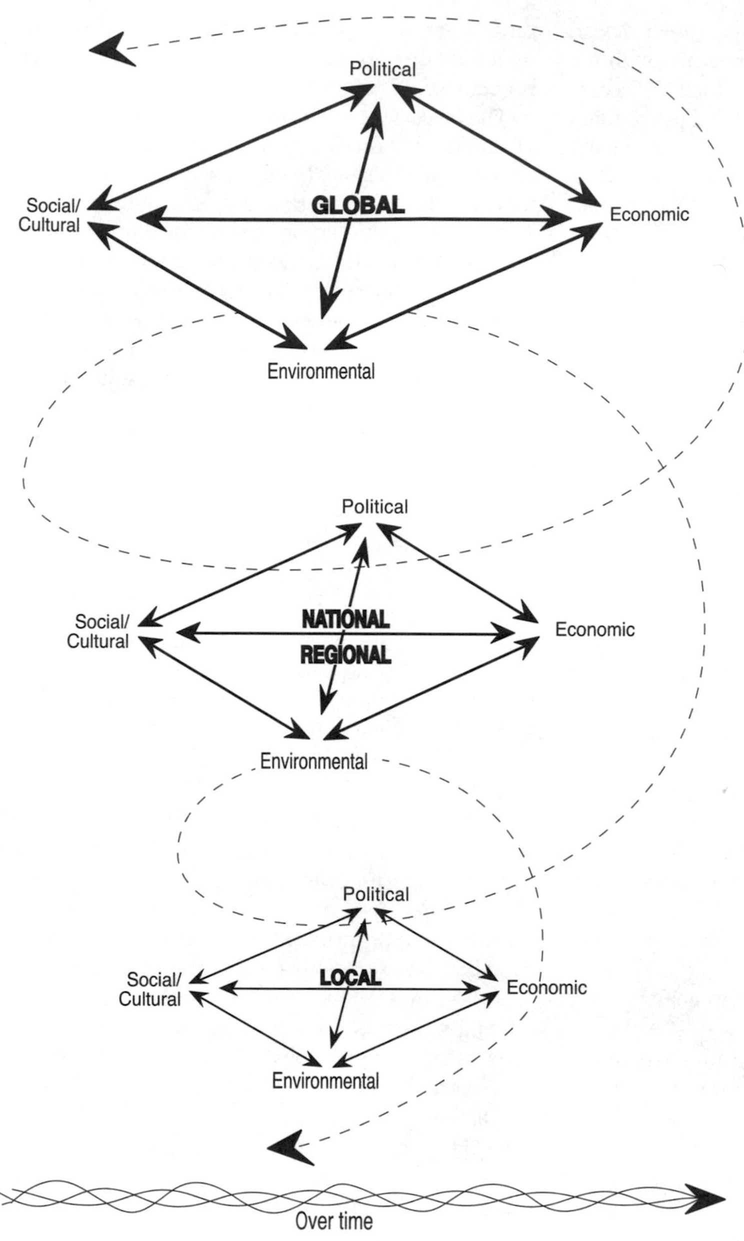

FIGURE 28.2. The Kite framework.

1993), to statistical modeling (Rastetter et al., 1992; De Cola, 1994), GIS approaches (Wong, 1996), fractal analysis (Lam and Quattrochi, 1994), and the search for a representative scale at which the multiplicity of factors and their spatial and temporal dimensions can be modeled at an average representation of these dynamics (Wood et al., 1988).

At the local level, case studies are required to understand the dynamic interactions between society and environment that result in significant changes in land use and land cover. Detailed case studies allow the dynamics of change in human and biophysical systems to be described and the causal processes of change to be assessed. A combination of methods can be used that permits examination of the critical interface between human and biophysical systems—where the hoe breaks the soil or the effluent enters the river or the air.

The most effective case studies are those for which the available time series data on both societal and environmental processes permit analysis of the patterns of change both within human and environmental systems and the interaction between them. Such information can include historical documentary material, detailed data on socioeconomic change over a 10- to 20-year period, and comparable time series information on environmental conditions using soil surveys, climatic records, and remotely sensed data and air photographs.

Aggregation

Aggregations are made to reduce complexity. This may be necessary for a variety of reasons, including parsimony in model building, cost constraints, and the practicality of working with large numbers of variables or observations.

Aggregation is a relatively uncomplicated process in which components have linear properties and it is appropriate to add together information from different units. Some additivities are unambiguous (e.g., number of people and livestock, area under different land uses, and ecological zones). When we have ratio variables or estimates from different sized populations, then weighting or some other manipulation of the information may be required to maintain representativeness.

The concept of similarity is fundamental to the process of aggregation. Similarity can be variously defined by variables, indicators, proximity, processes, and history. When the processes at a finer level are understood, then aggregation can be done on a theoretically sound basis. There remains a problem, however, in the application of models developed at the finer scale to interactions among variables that represent aggregations of finer-scale conditions. For example, "the leaves of a canopy might be lumped and treated as a single "big leaf" . . . . The fine-scale equations are often applied directly, or with minor changes, to describe the properties of this coarse-scale aggregate. The problem with this approach is that the aggregate does not generally behave the same way as the fine-scale components from which it is constituted" (Rastetter et al., 1992, p. 56). Criteria for aggregation may include optimizing objective functions (minimizing within-group variance, maximizing a regression coefficient), contiguity, and administrative convenience.

The issue of aggregation requires attention to the assumptions that have to be met to permit valid generalizations. Aggregation changes the scale of analysis. In the social and ecological sciences, it may involve sampling a number of individual cases, each with its own idiosyncrasies, to create a sample size large enough to allow generalizations to be made that are not unduly influenced by the particular behavior of individuals.

Each discipline has its own literature on the issue. Parametric and nonparametric procedures and spatial statistics all deal with the issue of aggregation and extrapolation from individual cases to populations. In single-discipline analyses, methods exist to explore interscalar issues, although they remain problematic. When dealing with environmental issues that cut across and integrate many disciplines, these methods are partially appropriate. The conceptual frameworks and methodology for such integration are emerging (see Chapters 10 through 21).

Scale and Time

The stimuli that create interaction from one scale to another are not instantaneous. Time lags occur that will vary in length depending on the characteristics of the processes involved. For example, diffusion studies have examined the dynamics of the spread of a phenomenon over space or within populations. The rate of spread may accelerate or decelerate at different times in the process and there may be factors that facilitate or inhibit diffusion (Morrill et al., 1988).

The impact of the impulses between levels is subject to time lags of different length depending on the impulse. Thus an understanding of the interactions across scales requires appreciation of temporal dynamics in the model:

> Patterns of interaction not only exist over space but through time. The outcomes of induced change in social, political or economic conditions are recognized to involve both an interaction between them, a synergism, and a recognition of time-lags such that the full impact of interactions may only become evident over a longer period of time. Child health programs, for example, may have rapid success in reducing child mortality rates but their impact upon other aspects of the production system may be long-delayed. Increased demand for land arising from subsequent population growth may only appear fifteen to twenty years later when the beneficiaries reach adulthood. . . .
>
> The question of time becomes more complicated once biophysical components are added to the developmental framework. Climatic rhythms, processes of soil formation and erosion, and vegetation growth and decay have

rates of activity that may be accelerated or delayed by human intervention. . . .

The concept of time is therefore one that changes from issue to issue and from one set of interactions to another. Political decisions and economic action may have instantaneous effects but social change may take a generation and biophysical process may take hundreds of years. . . . Time can be conceived of as made of different strands, different time scales, intertwined as a braid rather than as linear sequence of events. (Campbell and Olson, 1991, p. 21)

A second component of the temporal dimension that requires attention in studies of environmental issues over time is that variables used in these analyses may change over time in their character and in their implications for resource use. Does a variable have the same existential meaning when compared from one area to another or from one time period to another? This is particularly true of categorical variables that define the analytical structure of research. Two examples illustrate this issue.

In most society–environment analyses, the concept of economic class would be an important category. Its definition might well have different implications for modeling from time 1 to time *n*. For example, an American defined as average middle class in 1930 would have a very different material quality of life from an average middle-class American in 1996. Their expenditures on goods and services would be significantly different in terms of demands on environmental goods. In comparing across time, therefore, the definition has to be adjusted for the characteristics under investigation.

Soil type is a second example of an important environmental variable that influences human interaction with the environment. Soil maps are common and widely used in planning. It would be tempting to interpret a particular soil as having similar implications for land use over space and through time. In terms of this interaction, soil capability is a function of soil characteristics as well as the technology applied to the soil. Both may change over time and be different from one area to another. Different farmers with different economic and technological resources will view the capability of the soil differently—this is *the existential soil.* Therefore, a soil map may have a different meaning depending on the context.

What may be generalizable under such circumstances may not be the variable itself, but more the conditions of causal linkage that affect the variable, the processes that underlie the changes in the existential quality of the variable. Indicators that portray their presence and intensity may identify such causal linkages. Successful identification of such indicators is predicated on a plausible theoretical–conceptual framework.

Many contemporary projects are actively exploring frameworks that effectively link societal and biophysical processes and that are sensitive to methodological issues, such as time, scale, and uncertainty (Houghton et al., 1990; Lambin, 1992; Zimmerer, 1994; Turner et al., 1995). The similarities and differences found in the approaches of political ecology and landscape ecology provide a fertile basis for exploring such a framework.

28.7 Conclusion

The impact of land cover transformations on ecosystems is of major concern for those addressing environmental issues that relate to the long-term viability of Earth's ecosystems. These include climate change, environmental sustainability, biodiversity conservation, land degradation and desertification, air quality, and the conservation of marine ecosystems.

We must find ways to quantify patterns of variability in space and time, to understand how patterns change with scale and to understand the causes and consequences of pattern. This is a daunting task that must involve remote sensing, spatial statistics, and other methods to quantify patterns at broad scales; theoretical work to suggest mechanisms and explore relationships; and experimental work, carried out at both fine scales and through whole-system manipulations, to test hypotheses. Together, these can provide insights as to how information is transferred across scales, and hence how to simplify and aggregate models. (Levin, 1992, p. 1961)

Within such an approach, selected sets of relationships can be analyzed using numerical models that serve to predict land transformations. Others are more difficult to quantify and are most effectively analyzed using approaches such as narratives, scenario analysis, and role-playing simulations. These different approaches are not inimical, and outputs from one can serve as inputs to another.

The material presented in this chapter illustrates themes of importance from the perspective of the social sciences. The social, political, and economic dimensions of the interaction between environment and society are complex and interrelated. Issues such as the exercise of power by institutions and individuals and differentiation by socioeconomic class, age, and gender are seen as critical to an effective scientific analysis of environmental issues and their interpretation. Researchers explicitly address these in their conceptual frameworks, research design, analytical categories and methods, and interpretation of results. The complexity of conceptual frameworks and the plurality of research methods applied by the social sciences reflect these issues.

While the questions addressed, data collected, and analytical methods and models are different from those of the biophysical sciences, there are obvious common concepts that both groups contend with in their scientific endeavors. These include the following:

- *Driving forces*: the objective of identifying and assessing the relative importance of fundamental driving forces of environmental change
- *Scale*: individual versus population as the unit of analysis, the ecological fallacy, and the drawing of inferences from one scale to another
- *Time*: the notion that different, yet interacting processes operate at different time scales, nonlinearity, and uncertainty
- *Space*: the concept of spatial autocorrelation, the issue of the representativeness of case studies
- *Complexity*: the complexities implied by issues of scale, time, and space and by the interaction between driving forces
- *Conceptual frameworks*: the search for conceptual frameworks that effectively incorporate the complexity of socioenvironmental issues
- *Research methods*: methods that enable the complexity to be addressed; this includes the design of appropriate sampling procedures and defining relevant temporal and spatial contexts for the analysis

Such common concepts provide a basis for dialogue to devise scientific approaches that can integrate the social and biophysical sciences in the search for more effective conceptual frameworks and analytical approaches. Both social and biophysical scientists are concerned to understand, model, and attempt to predict patterns of interaction between society and environment. Each group of scientists has developed its own suite of conceptual and analytical approaches, and each group increasingly recognizes the importance of the insights of the other. Landscape ecology and political ecology are two contemporary frameworks that illustrate the mutual interest in the societal and the biophysical.

The dialogue between social and biophysical scientists has in the past been constrained by a variety of concerns about the scientific validity of their respective approaches, including epistemology, positivism versus more pluralistic approaches, and methodology (quantitative versus qualitative analysis). Increasingly, such divisive concerns are seen by both groups of scientists as inhibiting rather than promoting understanding of issues that cut across the social and biophysical. There is widespread recognition that collaborative research can be developed around themes and concepts of common interest and that perspectives specific to each can bring additional scientific insights. Fundamental to the success of such collaboration is that we engage in a respectful dialogue that appreciates both what are of common interest and of more specific importance to particular scientific domains.

28.8 References

Blaikie, P. M. 1985. *The political economy of soil erosion in developing countries*. London and New York: Longman.

Blaikie, P. M. 1994. *Political ecology in the 1990s: an evolving view of nature and society*. East Lansing, MI: Center for Advanced Study of International Development, Michigan State University.

Blaikie, P. M.; Brookfield, H. 1987. *Land degradation and society*. London: Methuen.

Blalock, H. M. 1982. *Conceptualization and measurement in the social sciences*. Beverly Hills, CA: Sage Publications.

Boserup, E. 1965. *The conditions of agricultural growth; the economics of agrarian change under population pressure*. London: Allen & Unwin.

Boserup, E. 1981. *Population and technological change: a study of long-term trends*. Chicago: University of Chicago Press.

Brink, A. P. 1982. *A chain of voices*. New York: Morrow.

Bromley, D. W. 1992. *Making the commons work: theory, practice and policy*. San Francisco: International Center for Self-Governance.

Brown, D. G.; Olson, J. M; Campbell, D. J.; Berry, L. 1996. A hierarchical approach to the integration of social and physical data sets: the case of the Rwanda Society–Environment project. *Human Dimensions Quart*. 1(4):14–17.

Bruce, J. W.; Migot-Adholla, S. E., editors. 1994. *Searching for land tenure security in Africa*. Dubuque, IA: Kendall/Hunt.

Caldwell, J. C., editor. 1975. *Population growth and socio-economic change in West Africa*. New York: Columbia University Press.

Campbell, D. J. 1996. *Implications of scale for the design and application of environmental information systems*. Consultant Final Report, August 17, 1996.

Nairobi: United Nations Environment Programme—Division of Environmental Information and Assessment.

Campbell, D. J. 1998. Towards an analytical framework for land use change. In: Bergstrom, L.; Kirschmann, H., eds. *Carbon and nutrient dynamics in tropical agro-ecosystems*. Wallingford, UK: CAB International: 281–301.

Campbell, D. J.; Olson, J. M. 1991. *Framework for environment and development: the Kite*. CASID Occasional Paper No. 10. East Lansing, MI: Center for Advanced Study of International Development, Michigan State University.

Castro, A. P. 1995. *Facing Kirinyaga: a social history of forest commons in southern Mount Kenya*. London: Intermediate Technology Publications.

CIESIN (Consortium for International Earth Science Information Network). 1992. *Pathways of understanding. The interactions of humanity and global environmental change*. University Center, MI: Consortium for International Earth Science Information Network.

Colby, M. E. 1990. *Environmental management in development: the evolution of paradigms*. World Bank Discussion Paper 80. Washington, DC: The World Bank.

De Cola, L. 1994. Simulating and mapping spatial complexity using multi-scale techniques. *Int. J. Geogr. Inf. Syst.* 8(5):411–427.

Dewees, P. A. 1993. *Social and economic incentives for small-holder tree growing: a case study from Murang'a District, Kenya*. Community Forestry Case Study Series 5. Rome: Food and Agriculture Organization.

Ecological Economics. 1995. 15:2.

Ehrlich, A. H.; Ehrlich, P. R. 1987. *Earth*. New York: Franklin Watts.

Ehrlich, P. R.; Ehrlich, A. H. 1990. The population explosion: why isn't everyone as scared as we are? *Amicus J.* 12(1):22–29.

Fratkin, E. 1997. Pastoralism: governance and development issues. *Ann. Rev. Anthropol.* 26:235–261.

Giddens, A. 1982. *Classes, power, and conflict: classical and contemporary debates*. Berkeley. CA: University of California Press.

Gore, C. G. 1984. *Regions in question: space, development theory, and regional policy*. London and New York: Methuen.

Haggett, P. 1983. *Geography: a modern synthesis*. New York: Harper & Row.

Hall, F. G.; Strabel, D. E.; Sellers, P. J. 1988. Linking knowledge among spatial and temporal scales: vegetation, atmosphere, climate and remote sensing. *Landscape Ecol.* 2(1):3–22.

Hardin, G. J. 1968. The tragedy of the commons. *Science* 162:1243–1248.

Hardin, G. J. 1974. Living on a lifeboat. *BioScience* 24(10):561–568.

Hardin, G. J. 1992. Cultural carrying capacity: a biological approach to human problems. *Focus* 2(3):16–24.

Hardin, G. J. 1993. *Living within limits: ecology, economics, and population taboos*. New York and Oxford: Oxford University Press.

Hardin, G. J.; Baden, J., editors. 1977. *Managing the commons*. San Francisco: W. H. Freeman.

Harvey, D. W. 1968. Processes, patterns and scale problems in geographical research. *Trans. Inst. Brit. Geogra.* 45:71–78.

Harvey, D. W. 1973. *Social justice and the city*. Baltimore, MD: Johns Hopkins University Press.

Harvey, D. W. 1974. Population, resources and the ideology of science. *Econ. Geogr.* 50:256–277.

Harvey, D. W. 1996. *Justice, nature and the geography of difference*. Cambridge, MA: Blackwell Publishers.

Houghton, J. T.; Jenkins, G. J.; Ephraums, J. J., editors. 1990. *Climate change: the IPCC scientific assessment*. Cambridge for the Intergovernmental Panel on Climate Change. Cambridge, MA: Cambridge University Press.

IGBP (International Geosphere–Biosphere Programme). 1994. *IGBP in action: work plan 1994–1998*. IGBP Global Change Report 28. Stockholm: International Geosphere–Biosphere Programme.

Izac, A. M. N.; Swift, M. J. 1994. On agricultural sustainability and its measurement in small-scale farming in sub-Saharan Africa. *Ecol. Econ.* 11:105–125.

Klijn, F.; de Haes, H. U. 1994. A hierarchical approach to ecosystems and its implications for ecological land classification. *Landscape Ecol.* 9(2):89–104.

Lam, N. S. N.; Quattrochi, D. A. 1994. On the issue of scale, resolution, and fractal analysis in the mapping sciences. *Prof. Geogr.* 44(1):88–98.

Lambin, E. 1992. *Spatial scales, desertification, and environmental perception in the Bourgouriba Region (Burkina Faso)*. Working Paper 167. Boston: African Studies Center, Boston University.

Levin, S. A. 1992. The problem of pattern and scale in ecology. *Ecology* 73(6):1943–1967.

Lima, S. L.; Zollner, P. A. 1996. Towards a behavioral ecology of ecological landscapes. *Tree* 11(3):131–135.

Lovelock, J. E. 1979. *Gaia, a new look at life on earth*. Oxford and New York: Oxford University Press.

Lovelock, J. E. 1991. *Healing Gaia: practical medicine for the planet*. New York: Harmony Books.

Malingreau, J. P.; Belward, A. S. 1992. Scale considerations in vegetation monitoring using AVHRR data. *Int. J. Remote Sens.* 13(12):2289–2307.

Malthus, T. R. 1973. *An essay on the principle of population*. London: J. M. Dent.

Manson, G. A.; Groop, R. E. 1996. Ebbs and flows in recent U.S. interstate migration. *Prof. Geogr.* 48(2): 156.

Massey, D. B. 1984. *Spatial divisions of labor: social structures and the geography of production*. New York: Methuen.

McGwire, K.; Friedl, M.; Estes, J. E. 1993. Spatial structure, sampling design and scale in remotely-sensed imagery of a California savanna woodland. *Int. J. Remote Sens.* 14(11):2137–2164.

Meentemeyer, V.; Box, E. O. 1987. Scale effects in landscape studies. In: Turner, M. G., ed. *Landscape het-*

erogeneity and disturbance. New York: Springer-Verlag: 15–34.

Morrill, R.; Gaile, G.; Thrall, G. I. 1988. *Spatial diffusion*. Newbury Park, CA: Sage Publications.

NASA (National Aeronautics and Space Administration). 1988. *Earth system science: a closer view*. Washington, DC: Author.

Norgaard, R. B. 1989. The case for methodological pluralism. *Ecol. Econ.* 1:37–57.

Nyamweru, C. 1998. *Sacred groves and environmental conservation*. Frank P. Piskor Lecture. Canton, NY: St. Lawrence University.

Olson, J. M. 1994. *Farmer responses to land degradation in Gikongoro, Rwanda*. Ph.D. dissertation. East Lansing, MI: Department of Geography, Michigan State University.

Olson, J. M. 1998. *Intensification processes and political bounds: a framework of land use change in the East African Highlands*. Paper presented at Earth's Changing Land: Joint Global Change and Terrestrial Ecosystems and Land Use and Land Cover Change Open Science Conference on Global Change; March 1998, Barcelona, Spain.

Olson, J. M., Manyara, G.; Campbell; D. J.; Lusch, D.; Hu, J. 1995. *Exploring methods for integrating data on socio-economic and environmental processes that influence land use change: a pilot project*. Report to the Biodiversity Support Program. Washington, DC: World Wildlife Fund.

Peet, R. 1985. The social origins of environmental determinism. *Ann. Assoc. Amer. Geogra.* 75(3):309–333.

Peet, R.; Thrift, N. J., editors. 1989. *New models in geography: the political-economy perspective*. London and Winchester, MA: Unwin Hyman.

Peet, R.; Watts, M., editors. 1996. *Liberation ecologies: environment, development, social movements*. London and New York: Routledge.

Rastetter, E. B.; King, A. W.; Cosby, B. J.; Hornberger, G. M.; O'Neill, R. V.; Hobbie, J. E. 1992. Aggregating fine-scale ecological knowledge to model coarser-scale attributes of ecosystems. *Ecol. Appl.* 2(1):55–70.

Rees, P. H. 1997. The second demographic transition: what does it mean for the future of Europe's population? *Environ. Plan.* A 29:381–390.

Robinson, W. S. 1950. Ecological correlations and the behavior of individuals. *Amer. Sociol. Rev.* 15:351–357.

Rocheleau, D.; Steinberg, P. E.; Benjamin, P. A. 1994. *A hundred years of crisis? Environment and development narratives in Ukambani, Kenya*. Boston: African Studies Center, Boston University.

Rocheleau, D.; Thomas-Slayter, B.; Wangari, E., editors. 1996. *Feminist political ecology: global issues and local experiences*. New York: Routledge.

Sack, R. D. 1980. *Conceptions of space in social thought: a geographic perspective*. Minneapolis: University of Minnesota Press.

Seeland, K., editor. 1997. *Nature is culture: indigenous knowledge and socio-cultural aspects of trees and forests in non-European cultures*. London: Intermediate Technology Publications.

Simon, J. L. 1986. *Theory of population and economic growth*. Oxford, UK: Basil Blackwell.

Simon, J. L. 1992. *Population and development in poor countries: selected essays*. Princeton, NJ: Princeton University Press.

Stutz, F. P.; de Souza, A. R. 1998. *The world economy: resources, location, trade, and development*. Upper Saddle River, NJ: Prentice Hall.

Thomas, W. L. 1956. *Man's role in changing the face of the earth*. Published for the Wenner–Gren Foundation for Anthropological Research and the National Science Foundation. Chicago: University of Chicago Press.

Tiffen, M.; Mortimore, M.; Gichuki F. 1994. *More people less erosion: environmental recovery in Kenya*. Nairobi: ACTS Press in association with the Overseas Development Institute, London.

Todaro, M. P. 1969. A model of labor migration and urban unemployment in less developed countries. *Amer. Econ. Rev.* 59:138–148.

Turner, M. G. 1990. Spatial and temporal analysis of landscape patterns. *Landscape Ecol.* 4(1):21–30.

Turner, M. G.; Dale, V. H.; Gardner, R. H. 1989. Predicting across scales: theory development and testing. *Landscape Ecol.* 3(3/4):245–252.

Turner, B. L., II; Clark, W. C.; Kates, R. W.; Richards, J. F.; Mathews, J. T.; Meyer, W. B., editors. 1993a. *The earth as transformed by human action: global and regional changes in the biosphere over the past 300 years*. New York: Cambridge University Press with Clark University.

Turner, B. L., II; Hyden, G.; Kates, W. R. 1993b. *Population growth and agricultural change in Africa*. Gainesville: University Press of Florida.

Turner, B. L., II; Skole, D.; Sanderson, S.; Fischer, G.; Fresco, L.; Leemans, R. 1995. *Land-use and land-cover change (LUCC): science/research plan*. IGBP Report, No. 35. HDP Report No. 7. Stockholm: International Geosphere–Biosphere Programme.

Vandermeer, J. 1994. The naiveté of the natural: a natural scientist looking at social scientists looking at natural scientists. In: Guizlo, M., ed. *Political ecology workshop: conference proceedings*; April 14–17, 1994. East Lansing, MI: Michigan State University.

Warren, D. M. 1991. *Using indigenous knowledge in agricultural development*. Washington, DC: The World Bank.

Wong, D. 1996. Aggregation effects in geo-referenced data. In: Griffith, D. A.; Arlinghaus, S., eds. *Practical handbook on spatial statistics*. Boca Raton, FL: CRC Press.

Wood, E. F., Sivapalan, M.; Beven, K.; Band, L. 1988. Effects of spatial variability and scale with implications to hydrologic modeling. *J. Hydrol.* 102:29–47.

Zimmerer, K. 1994. Human geography and the "new ecology": the prospect and promise of integration. *Ann. Assoc. Amer. Geogr.* 84(1):108–125.

29
Mapping Patterns of Human Use and Potential Resource Conflicts on Public Lands

James V. Schumacher, Roland L. Redmond, Melissa M. Hart, and Mark E. Jensen

29.1 Introduction

Large areas in western North America are publicly owned and managed by governmental agencies for a variety of uses. As the human population continues to grow, competing interests will place mounting pressures on how resources from these lands should be managed and used by people. To make sound decisions about the allocation of these resources, decision makers must consider all aspects of the ecosystems in which they are found. The human role in ecosystem function is one such topic deserving of more attention (Sheifer, 1996). Much as humans need to be included in ecological studies, current maps of human settlement and related patterns are not accurate enough for many assessments. One way to create more accurate maps is to integrate data from the U.S. Census Bureau with other information sources, often remotely sensed imagery (Lo and Faber, 1997; Yuan et al., 1997; Mesev, 1998; Ryavec and Veregin, 1998).

In this chapter (which was published previously by Schumacher et al., 2000), we present a mapping approach incorporating census data, land cover as mapped from Landsat Thematic Mapper (TM) imagery, land ownership, and a transportation network. In essence, we map areas that are inhabited or used by people. These maps can be used to point us toward areas where there are likely to be risks of conflict—whether risks for humans, resources, or both. To illustrate the approach, we present an example for the Lolo National Forest, focusing on methods rather than actual results.

29.2 Study Area

The administrative boundaries of Lolo National Forest cover 11,000 km^2 in western Montana, U.S.A. (Figure 29.1). To avoid edge effects along these boundaries, our analyses included a 10-km buffer around the entire forest, which increased the total study area to 22,000 km^2. This area is reasonably large and complex, and nearly all requisite input data were readily available, making the area a convenient choice for developing and illustrating our GIS mapping protocol.

29.3 Methods

29.3.1 Data Inputs

Human Population

The U.S. Constitution specifies that the nation's population be counted every 10 years. Since the first census in 1790, the uses of this information, as well as the methods by which it is collected, have changed dramatically. Since 1970, every household has been sampled by a mail-in questionnaire and tallied within a nested set of units ranging from the block (containing approximately 100 people) to the block group, tract, county, and state. For this study, we obtained block-level population counts and block boundaries from the 1990 Census (http://www.esri.com/data/online/tiger/).

Census data are commonly displayed as either dot density or choropleth maps (e.g., Figure 29.2a).

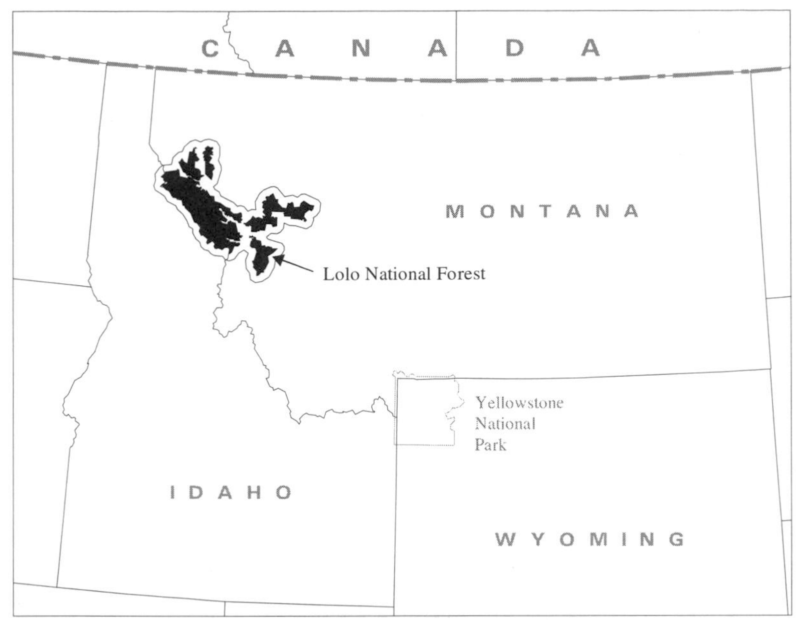

FIGURE 29.1. Administrative boundaries of the Lolo National Forest in western Montana and the surrounding 10-km buffer.

In both cases the map and sample units are the same. Although these maps may be adequate for displaying general population trends over large areas, they lack detailed information and are not suitable as input for spatial analyses using GIS. Choropleth maps do an especially poor job of representing population density across large areas of public or industrial private lands, where actual density can be zero (contrast Figure 29.2a with Figure 29.2b).

To circumvent these limitations, we created a dasymetric map of population density for the study area (Figure 29.2c) in which the numbers of people counted in each block were reassigned to smaller map units in a sequence of filtering steps (Holloway et al., 1999). First, areas of public land and industrial private land were eliminated from the blocks. Actual population counts for each block were then redistributed according to the underlying land cover, land use, and topography as follows. Using land cover data developed from Landsat TM imagery (Redmond et al., 1998), we selected four general land use/land cover classes: urban, agriculture, wooded, and open lands. All urban and agricultural lands were assumed to be populated; wooded or open cover types were assumed to be populated if their slope was less than or equal to 15%. This 15% limit was chosen based on countywide building restrictions. Although it may not be valid in all areas, it serves as an appropriate filter for steep lands in our study area. The actual population counts were allocated to spatial units (regions) on a per unit area basis: assuming all four cover types were present in a census block, 80% of the people were allocated to urban regions, 10% to open regions, and 5% each to wooded and agricultural regions. Where one or more cover types were absent, counts were assigned maintaining these relative proportions. Common sense tells us that if an area's current land use is classified as urban it will contain a higher proportion of the residents in a census unit than open or wooded areas. Although this redistribution ratio was somewhat arbitrary, it seemed to be a reasonable proportioning. If this method were applied elsewhere, or if empirical data became available for this area, the ratio could readily be modified or a different approach taken (e.g., Yuan et al., 1997).

Finally, we examined a frequency histogram of all the density values generated for the study area and, from that, defined density classes used for map display (e.g., Figure 29.2c). For the study area, we defined ten classes: nine classes with approximately equal area and an additional class representing uninhabited lands.

Land Ownership, Roads, and Trails

Land ownership data were crucial to the dasymetric mapping technique. These were obtained as digital coverages from the U.S. Bureau of Land Management, the U.S. Forest Service, and the Plum Creek Timber Company, all at the nominal scale of 1 : 100,000. Slope was calculated for each 30-m^2 grid cell from U.S. Geological Survey (USGS)

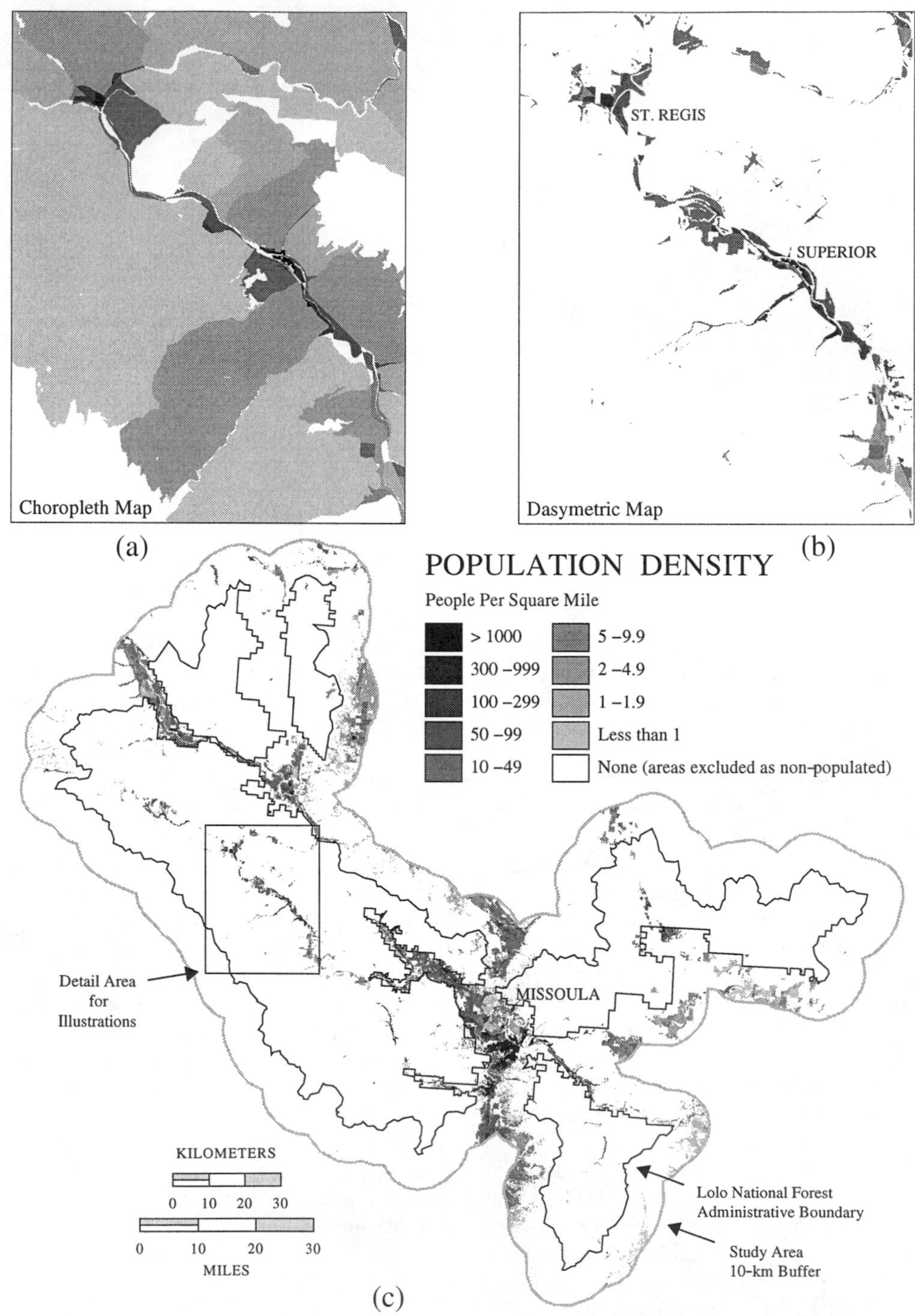

FIGURE 29.2. Human population density represented by (a) a traditional choropleth map for a small part of the study area, (b) a dasymetric map for the same area, and (c) a dasymetric map for the entire study area (note location of inset box used for parts a, b, and Figures 29.3 and 29.4).

7.5-minute digital elevation models (DEMs) that had been smoothed to minimize inherent flaws such as striping and seams, which would appear as breaks in slope. Data on roads and trails were provided in digital form and at 1 : 24,000 scale from the U.S. Forest Service, Lolo and Idaho Panhandle National Forests. All GIS data inputs were prepared at the finest resolution available and joined or appended together into seamless grids or coverages for the entire study area.

29.3.2 Mapping Intensity of Human Use

Human Population Density

To create a model of human population density, the number of people living in each 30-m^2 grid cell was calculated by dividing the number of people assigned to each of the cover-type regions in each census block by the number of 30-m^2 cells in each region. In other words, each grid cell in a region was assigned an average population density for the entire region. To estimate the broader impact of human residences on surrounding lands, we used the FOCALSUM function in ARC/INFO (ESRI, 1998) to calculate the number of people living within a 1-km radius of each 30-m^2 cell. The 1-km search radius estimates broader spatial effects; it was selected arbitrarily to describe an inverse relationship between human use and distance from homes, roads, or trails. A different search radius could be used if warranted either by empirical data or different management questions.

The resulting grid was then resampled to 90-m^2 resolution to match the terrestrial vertebrate data with which it was to be compared (described later). This output provides a relative measure of human occupancy and use. *Use* is meant in a general sense; the output model indicates areas likely to be used by people, but does not point to different types of use.

Road and Trail Density

A weighted density of roads and trails was mapped within a 1-km radius around each 30-m^2 cell using the LINEDENSITY function in ARC/INFO. Weights were assigned according to feature type (3 for improved roads, 2 for unimproved roads, and 1 for trails) to account for differential ease of access. Rather than count all roads and trails equally, when we know that some receive far heavier use, the 3–2–1 weighting was chosen as the simplest, albeit arbitrary, measure of impact. Density was calculated in units of length per unit area. This output grid was similarly resampled to 90-m^2 resolution.

Human Use Intensity

A model of intensity of human use was then created that combined the resampled population and road–trail density grids. In this case, they were weighted equally by simply adding the two together, recognizing that roads, trails, and local residences all have impacts, but lacking more detailed evidence on how to rate these highly variable impacts. Because the units differed between these two grids (numeric count versus length of lines), each of the input density grids had to be scaled to the same range of units, 0 to 100, before they could be combined (output intensity model = population grid + road and trail grid). The output grid (Figure 29.3a) shows both local intensity from human settlement and the much broader spatial patterns from roads and trails; it also clearly identifies lands that are >1 km from either.

29.3.3 Mapping Distributions of Terrestrial and Aquatic Vertebrates

Species of Special Concern

We use the term *species of special concern* to mean taxa that are rare, endemic, disjunct, threatened or endangered, either throughout their range or just in Montana, following the definition used by the Montana Natural Heritage Program. These species typically are accorded special management attention, so they offer a reasonable starting point for assessing human impacts on wildlife. Predicted habitat was modeled at a 90-m^2 cell size for each species as part of the Montana Gap Analysis Project (Redmond et al., 1998); then the outputs were overlaid to create a richness map for the Montana portion of the study area (Figure 29.3b). This was aggregated to four richness classes (0 plus three area-based classes) and then intersected with a human use intensity grid, also broken down to the same four classes (Figure 29.3c). Both species richness and human use intensity were aggregated to four classes so that individual classes could be distinguished on this two-variable map.

Bull Trout

To create a map of bull trout (*Salvelinus confluentus*) distribution for the study area, we used three inputs: (1) streams and other hydrographic features, (2) bull trout distribution data, and (3) watershed boundaries (6th-code hydrologic units, or HUCs; Seaber et al., 1987). Hydrography coverages (1 : 100,000 scale, vector format) were acquired from the USGS, National Hydrography Dataset, Pacific Northwest StreamNet River Reach Files (http://www.streamnet.org/pnwrhome.html) and clipped to the study area boundary. Stream orders (Strahler, 1957) were then assigned within ARC/INFO, and a new coverage was created that contained only higher-order streams (codes ≥ 2). A database of fish status, some known, some predicted,

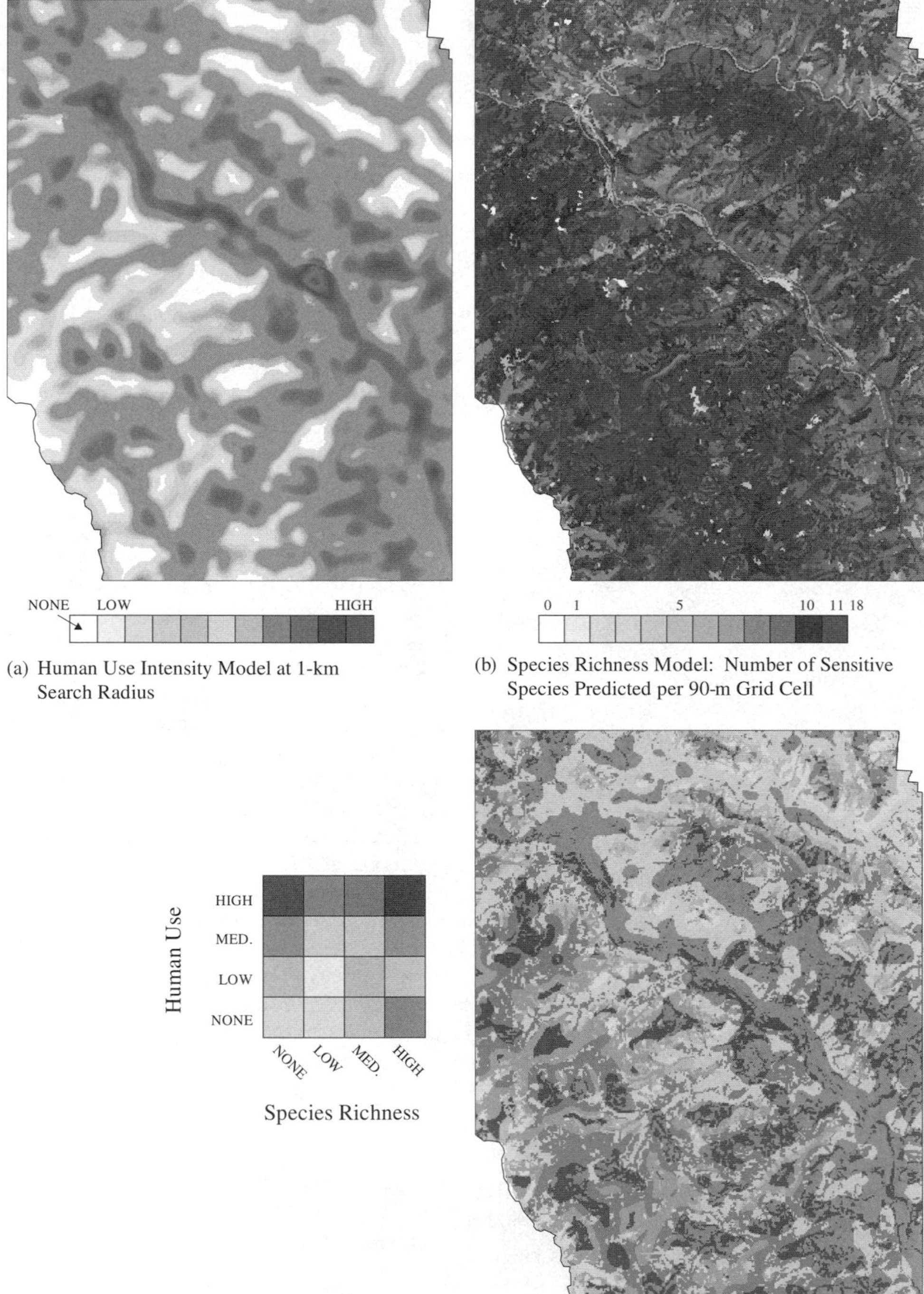

(a) Human Use Intensity Model at 1-km Search Radius

(b) Species Richness Model: Number of Sensitive Species Predicted per 90-m Grid Cell

(c) Intersection of a and b: Human Use Intensity and Predicted Species Richness

FIGURE 29.3. For those portions of inset box (Figure 29.2) located within National Forest boundaries: (a) predicted intensity of human use; (b) richness map of terrestrial vertebrate species of special concern; and (c) two-variable map showing intersection of parts a and b. (See color print of this figure in the insert.)

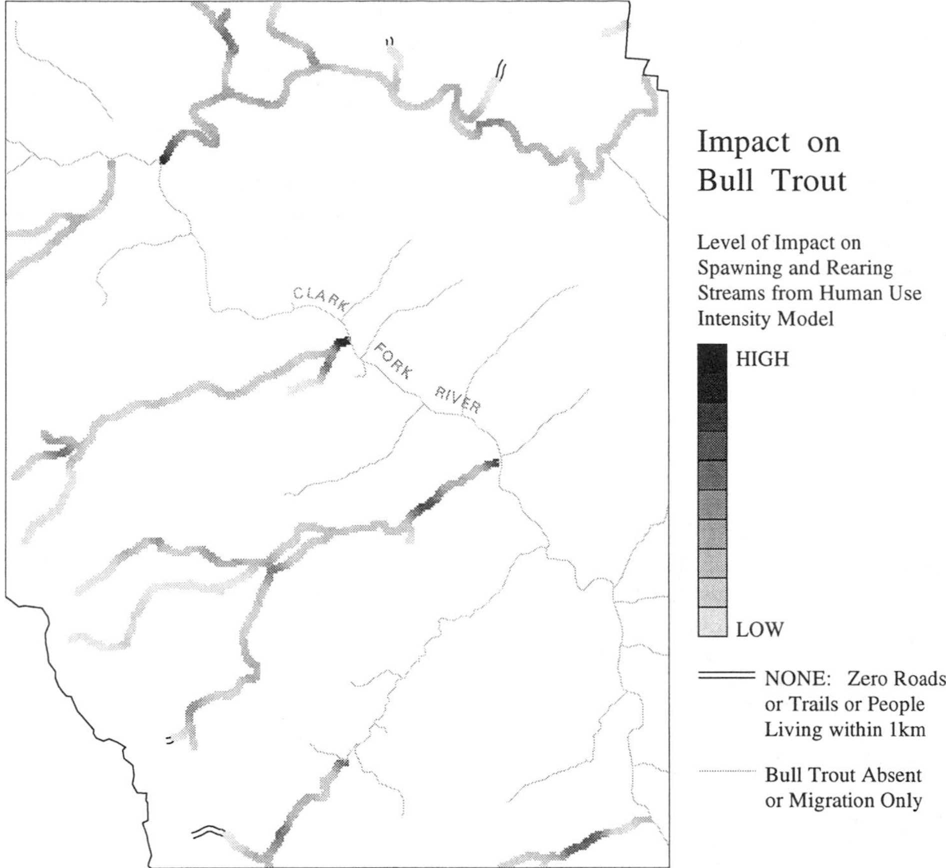

FIGURE 29.4. Intersection of predicted human use (Figure 29.3a) with distribution of bull trout spawning and brood-rearing habitat within Lolo National Forest administrative boundaries.

by 6th-code HUC had been previously compiled for the Interior Columbia Basin Ecosystem Management Project (http://www.icbemp.gov) in dBASE format. Attributes from this file were attached to a GIS coverage representing 6th-code HUCs. Fish status codes for the bull trout were combined to create a simplified coding scheme for each stream segment: present for spawning and brood-rearing, present for migration only, and absent. The higher-order stream layer was then intersected with the 6th-code HUCs to assign bull trout status codes to the streams themselves, rather than to entire watersheds. The higher-order stream coverage was then converted to grid (90 m^2) format so that it could be intersected with the human use intensity grid. Finally, a new grid was created; for each grid cell where bull trout were known or predicted to be present for spawning and brood-rearing, the corresponding human use intensity value was assigned (Figure 29.4).

29.4 Results and Discussion

29.4.1 Intensity of Human Use

The predicted intensity of human use, based only on access and where people live, for all lands administered by the Lolo National Forest shows relatively dissected, local effects for the 1-km search radius (Figure 29.3a). If this radius were increased, more distant inputs would be taken into account, and very few places would remain where low levels of human use are predicted. Conversely, a shorter search radius would have predicted more areas with low levels of human impact.

29.4.2 Impacts to Species of Special Concern

Of the 70 terrestrial vertebrates on the Montana list, 41 are predicted to occur in or around the Lolo National Forest, including 4 amphibians, 11 mammals, and 26 birds. Examples include the Coeur d'Alene salamander (*Plethodon idahoensis*), Townsend's big-eared bat (*Corynorhinus townsendii*), gray wolf (*Canis lupus*), lynx (*Lynx canadensis*), common loon (*Gavia immer*), northern goshawk (*Accipiter gentilis*), and flammulated owl (*Otus flammeolus*). When predicted habitat is overlaid for all 41 species, richness values range from 0 to 18 species per 90-m^2 cell (Figure 29.3b). Although here we focus on overall richness, each species merits individual consideration as well. In this landscape, species of special concern tend to be associated with forested areas, whereas human populations are concentrated in lower, open areas. High species richness counts and concentrated human use do, however, coincide in places (shown in dark blue in Figure 29.3c), especially where density of roads and trails is high. And even in areas where human use is highest, there is almost always at least one species of special concern to be considered. By intersecting layers of predicted species richness and human use, we are better able to identify areas where conflicts are more likely to occur.

29.4.3 Impacts to Bull Trout

This type of analysis allows public land managers to identify specific areas (stream reaches) where continued or increased levels of human use could adversely affect bull trout habitat and then to factor this information into management plans. In this example, the most important areas to examine are those with high levels of predicted human use (shown in dark gray in Figure 29.4). If predicted use is high at key points in the stream network, this could affect the accessibility of upstream habitat and thereby the future security of bull trout throughout larger watersheds. Results also show the very limited proportion of spawning and rearing streams (4.9% in the entire National Forest boundary) that fall outside a 1-km zone of predicted human impact and that these relatively undisturbed streams tend to occur in the uppermost portions of watersheds (shown as double-line segments in Figure 29.4).

29.4.4 Limitations

We are not attempting with this chapter to present a method complete with ideal parameters or specific results for a given area. Rather, our goal is to illustrate a flexible way to incorporate census data with other GIS layers, creating a far more accurate picture of human presence in ecosystems. We present the use of modifiable weighting ratios as an aid in this process. For example, the 80–10–5–5 ratio used to allocate population counts to regions is arbitrary and could be improved if revised based on actual field study or other empirical evidence. One option might be to examine correlations between census data and land cover types through multivariate regression, as Yuan et al. (1997) have done for four counties in Arkansas. The largest differences, however, between the dasymetric maps shown here and traditional choropleth maps arise from the exclusion of uninhabited lands, not the 80–10–5–5 weighting. Again, the method we present here offers a general approach; many of its steps require parameters to be defined, yet insufficient information is available to support these definitions.

Another limitation of the method is its reliance on fairly coarse scale data (e.g., census data) to predict fine-scale effects. To some extent, the use of a search radius (in this case, 1 km) in generating model output helps to compensate for this limitation by projecting results across broader areas. Nonetheless, as is always the case, users should bear in mind the nature of the data inputs when interpreting outputs.

29.5 Conclusions

In any ecosystem, humans are an integral part of the equation, but traditionally they have been neglected from an analytical standpoint. The approach described herein offers a management tool for predicting, evaluating, and visualizing human impacts in a spatial sense. But analyses need not be limited to where people live and where they might travel. Human use could be evaluated in other ways as well. For example, point data on campgrounds, mines, summer cabins, and other anthropogenic features could be incorporated or distance to urban areas measured and weighted. Although this approach is well suited to assessment of human impacts on other terrestrial and aquatic species, it could be applied to problems in all disciplines, particularly risk assessments. One potential application is mapping fire risk; both the likelihood of human-caused ignitions and potential damage to property could be assessed. Mapping risk of exposure to contaminants is another possibility; for example, human use could be analyzed with respect to pollution levels around Superfund sites. In fact, a simple protocol like this could assist land managers and planners with future decisions in any area where suitable input data are available.

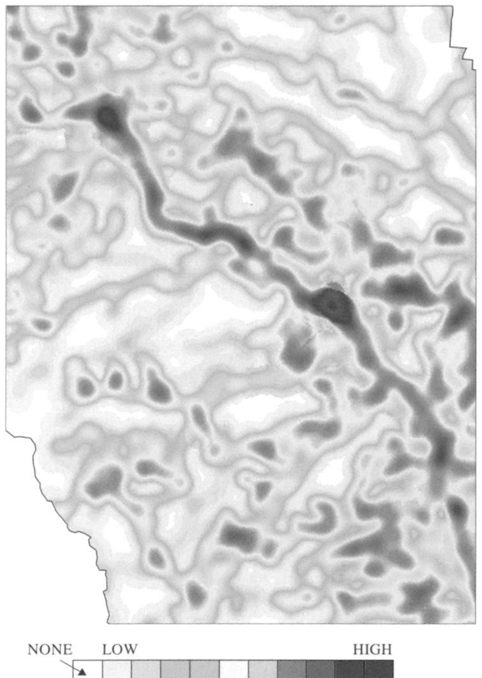

(a) Human Use Intensity Model at 1-km Search Radius

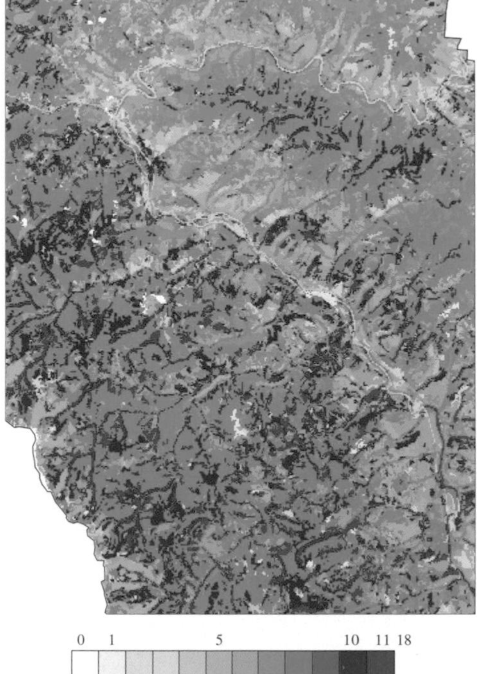

(b) Species Richness Model: Number of Sensitive Species Predicted per 90-m Grid Cell

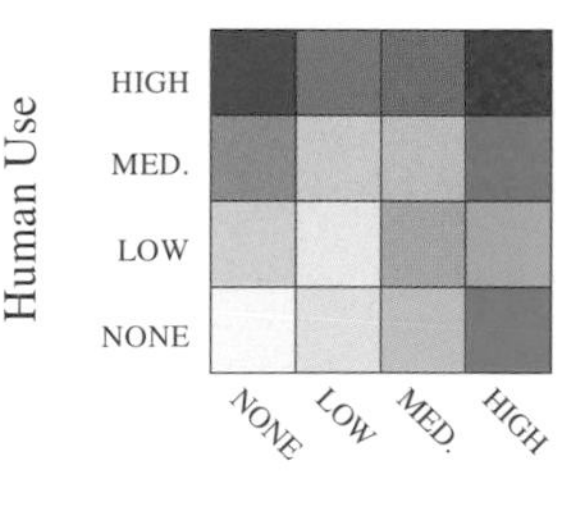

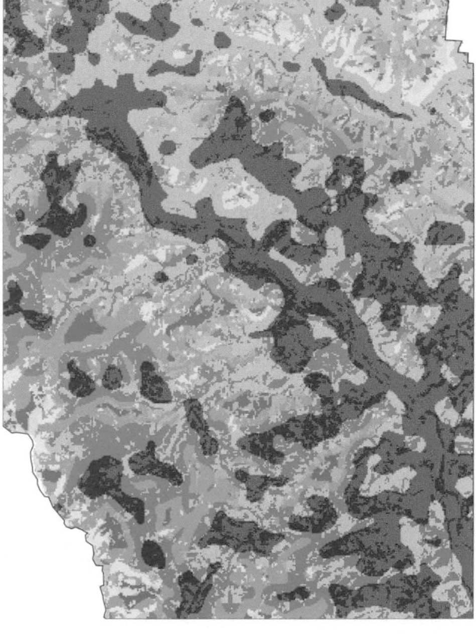

(c) Intersection of a and b: Human Use Intensity and Predicted Species Richness

Figure 29.3 For those portions of inset box (Figure 29.2) located within National Forest boundaries: (a) predicted intensity of human use; (b) richness map of terrestrial vertebrate species of special concern; and (c) two-variable map showing intersection of parts a and b.

29.6 References

ESRI. 1998. *ARC/INFO version 7.1*. Environmental Systems Research Institute, Inc., Redlands, CA.

Holloway, S. R.; Schumacher, J. V.; Redmond, R. L. 1999. People and place: dasymetric mapping using ARC/INFO. In: Morain, S., ed. *GIS solutions in natural resource management: balancing the technical–political equation*. Santa Fe, NM : OnWord Press: 283–291.

Lo, C. P.; Faber, B. J. 1997. Integration of Landsat thematic mapper and census data for quality of life assessment. *Remote Sens. Environ.* 62:143–157.

Mesev, V. 1998. The use of census data in urban image classification. *Photogramm. Eng. Remote Sensing* 64: 431–438.

Redmond, R .L.; Hart, M. M.; Winne, J. C.; Williams, W. A.; Thornton, P. C.; Ma, Z.; Tobalske, C. M.; Thornton, M. M.; McLaughlin, K. P.; Tady, T. P.; Fisher, F. B.; Running, S. W. 1998. *The Montana gap analysis project: final report*. Unpublished report. Missoula, MT: Montana Cooperative Wildlife Research Unit, University of Montana: xiii + 136p.

Ryavec, K. E.; Veregin, H. 1998. Population and rangelands in central Tibet: a GIS-based approach. *GeoJournal* 44:61–72.

Schumacher, J. V.; Hart, M. M.; Redmond, R. L.; Jensen, M. E. 2000. GIS modeling of human use of public lands in the western United States. In: *The EMAP symposium on western ecological systems: status, issues and new approaches*. April 6–8, 1999. San Francisco: J. Environ. Monit. Assess.

Seaber, P. R.; Kapinos, F. P; Knapp, G. L. 1987. *Hydrologic unit maps*. Water-Supply Pap. 2294. Corvallis, OR: U.S. Dept. of Interior, Geological Survey: 62p.

Sheifer, I. C. 1996. Integrating the human dimension in ecoregion/ecosystem studies—a view from the ecosystem management national assessments effort. *Bulletin Ecol. Soc. Amer.* 77:177–180.

Strahler, A. N. 1957. Quantitative analysis of watershed geomorphology. *Trans. Amer. Geophys. Union* 33: 913–920.

Yuan, Y.; Smith, R. M.; and Limp, W. F. 1997. Remodeling census population with spatial information from Landsat TM imagery. *Comput. Environ. Urban Syst.* 21:245–258.

Part 7
Case Studies

30
The Great Lakes Ecological Assessment

David T. Cleland, Larry A. Leefers, J. Michael Vasievich, Thomas R. Crow, and Eunice A. Padley

30.1 Introduction

French explorer Jean Nicolet led the first European expedition to the northern Great Lakes region in 1634 (Stearns, 1997). This began a period of exploration and fur trading that lasted into the early 19th century. Many small settlements were established as trading centers, army posts, and missions, which subsequently grew into towns and cities as the midwestern U.S. population grew, with an attending growth in the demand for lumber.

Although there were impacts from Native Americans dating back nearly 10,000 years (Denevan, 1992), it was the European advance that greatly altered the landscapes of the Great Lakes region (Andersen et al., 1996). Following the exploitation and collapse of furbearer populations, settlers turned their attention to the forests (Catton, 1984). White pine logging began about 1836 and, depending on location, reached a peak between 1890 and 1910, by which time virtually all merchantable pine had been cut or destroyed by fire. During the white pine era, hemlock was cut heavily as a source of tannin for processing cowhides into leather, resulting in the absence of this species in much of today's forests. Shortly before the end of the white pine era, harvesting of hardwoods began, continuing into the 1930s. The impact of near-total deforestation was amplified by frequent and often catastrophic wildfires burning through slash, as well as smaller fires that were deliberately set to clear land or that started from railroad locomotives.

While supporting the explosive growth of the Midwest, the turn of the century logging era wastefully exploited the forest resources of the three Lake States: Michigan, Minnesota, and Wisconsin. Today's forests, although generally younger and less ecologically diverse than the original forests, have exhibited remarkable resilience to early logging disturbance. As stated by Stearns (1997, pp. 8–29).

> In the immediate future, prospects for the Lake States forests appear good. Growth exceeds wood removal, and most forest managers are giving attention to sustained management. New silvicultural approaches are being investigated, and the public is becoming better informed. Climatic change is probably the major threat; although the simplification of the many forest systems may produce unforeseen declines as may infestations of exotic insects and disease. Increasing human use, especially for housing sites, also poses short- and long-term dangers. Our forests are recovering from the destruction visited on them by fire and human need and greed. To continue to do so, they must be managed wisely and conservatively.

The northern Great Lakes region is of national significance in many respects. Bordering our Canadian neighbors to the north, the region is characterized by extensive forestlands; the freshwater seas of Lakes Superior, Michigan, and Huron; tens of

We wish to acknowledge the many people and organizations who have contributed to the Great Lakes Ecological Assessment, especially Nancy Potter (NPR project champion) at the General Services Administration and Chris Risbrudt, Ecosystem Management, USDA Forest Service, Washington, DC; Jim Jordan, John Wright, Sharon Hobrla, Don Meyer, Kathleen McAllister, Jack Troyer, and Phyllis Green, USDA Forest Service; Steve Ventura, University of Wisconsin, Madison; George Host, Natural Resources Research Institute, University of Minnesota, Duluth; Daniel Fitzpatrick, U.S. Geological Service, Biological Resource Division; and many people who have contributed to the GLA at the USDA Forest Service, at the Land Information Computer Graphics Facility, University of Wisconsin at Madison, at Michigan State University, and at the Departments of Natural Resources in Michigan, Minnesota, and Wisconsin.

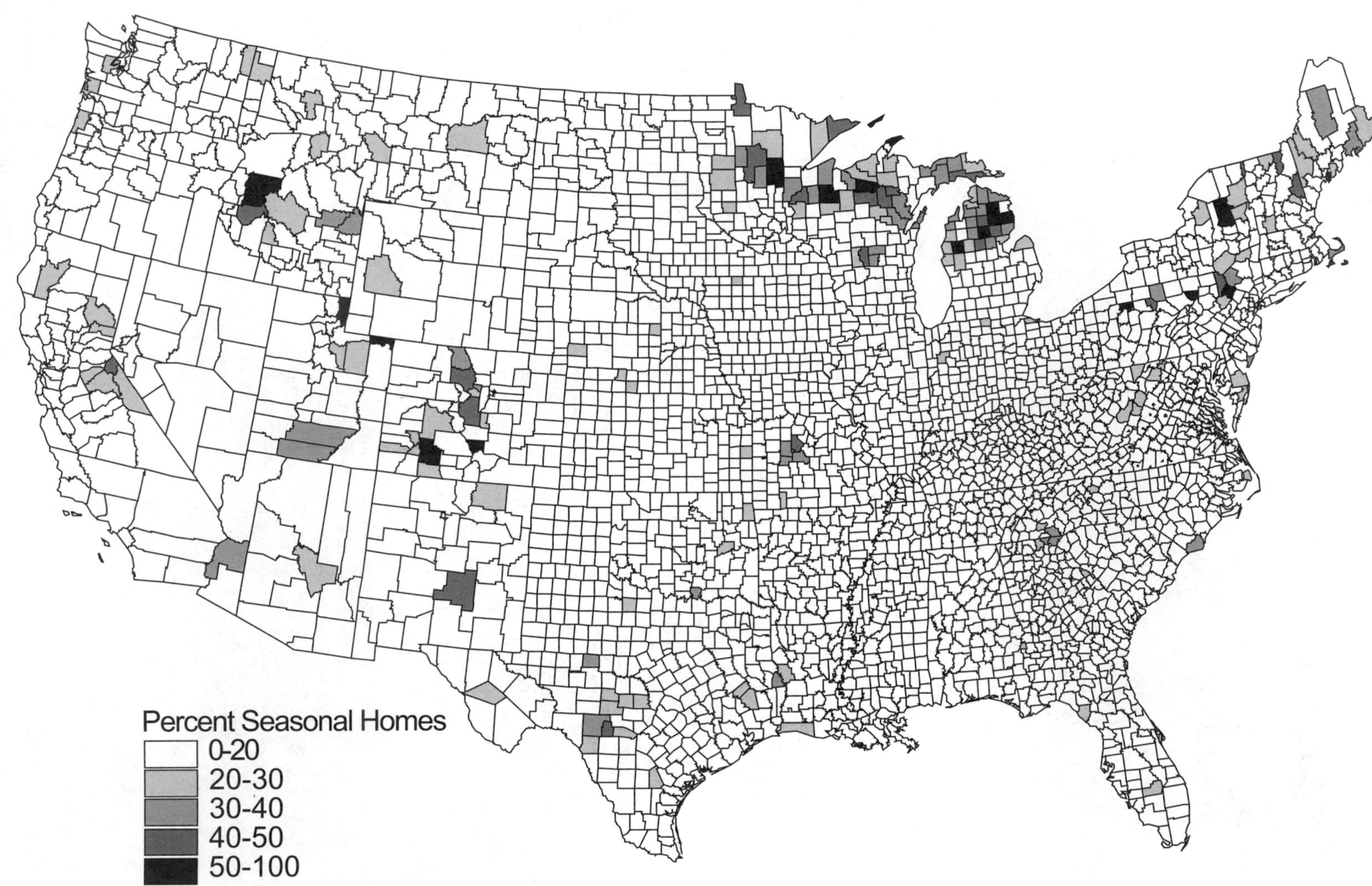

FIGURE 30.1. Seasonal homes as a percent of total housing stock, by county in the conterminous United States, 1989.

thousands of inland lakes; and vast networks of rivers and streams. Demographically, the region is within a day's drive of approximately 50 million people, leading to an enormous weekend and seasonal immigration of tourists, hunters, anglers, snowmobilers, and other outdoor enthusiasts. Furthermore, this region has one of the nation's most geographically expansive and densely concentrated areas of recreational or second-home development (Figure 30.1). Many of these homes are owned by people from nearby cities, such as Chicago, Detroit, Milwaukee, and Minneapolis, resulting in a mix of neighbors that often have differing perceptions of natural resources and related management practices. The patchwork ownership pattern of private and public lands, the high concentration of primary and seasonal homes and other developments interspersed within our forestlands, and the weekend and holiday immigration of hundreds of thousands of people create an urban–wildland interface unlike that of any other large region of the country. This widely dispersed pattern of human settlement in forested areas poses an extraordinary challenge to natural resource planners and managers.

The northern Great Lakes region is one of the most densely forested regions of the nation, with 41% of the total area, or 50.5 million acres in forested lands. About 46 million acres are considered commercial forest, and 52% of this commercial forestland is owned by the nonindustrial private sector. Interspersed with these private lands are approximately 7.3 million acres of state forests, 6.5 million acres of national forest, and 4.9 million acres of county and municipal forests (Vasievich et al., 1997). The second-growth forest is reaching commercial maturity, and substantial volume gains have been made in recent decades. Over 438,000 people are employed in the wood products and related sectors, which produce more than $39 billion in sales annually (Pedersen and Chappelle, 1997). Nationally, timber production has declined in recent years, but Lake States production has increased (Vasievich et al., 1997).

Expectations for the region's role in timber production are high, given increasing national demand, decreasing supply from the west, and the already high production from the neighboring southern and southeastern regions of the United States. Notwithstanding, surveys indicate that the majority of urban Americans perceive a need for decreased emphasis on forest commodity production and increased emphasis on noncommodity values (e.g., recreation and esthetics), yet rural users depend on forests for employment and community development. And while both standing volume and demand for forest products continue to increase in the Lake States, lands available for timber production continue to decrease due to urban and industrial expansion, development of second homes, and the resulting emergence of land-use conflicts (Resh, 1994). The regionally unique confluence of rural and urban Americans and the local and regional economic dependencies on both recreational and commodity-based production leads to competing and frequently conflicting land-use emphases. Land-use policy and practices have far-reaching economic and ecological implications, and information availability is essential for objective and publicly acceptable decision making.

30.2 Impetus and Goals

The Lake States Forestry Alliance (now the Great Lakes Forestry Alliance) held a conference in 1995 to discuss forest resource conditions and economic development opportunities in the northern Great Lakes region (Vasievich and Webster, 1997). This effort, entitled the Lake States Forest Resources Assessment, created a set of technical reports on trends and opportunities in forest resource development, with special emphasis on timber and recreational supply and demand. The Great Lakes Ecological Assessment (GLA), a separate but related effort, began during the same period. Originally, the GLA was exclusively a USDA Forest Service initiative. As it evolved, however, the GLA effort drew together many agencies to consolidate, analyze, and distribute existing environmental, biological, social, and economic information in the northern Great Lakes region. The GLA builds on the earlier Lake States Forestry Alliance effort by expanded use of spatial information and broader considerations of ecological conditions and processes.

The GLA began in earnest in June 1996 following an award from the Interagency Management Council of the National Partnership for Reinventing Government (NPR) for the project entitled Coordinated Management and Utilization of Natural Resource Information for Forest Lands in the Great Lakes Region. The request for funding to the NPR noted several conditions warranting the project. First, federal, state, and private land managers need various types of spatial information to adeptly manage and conserve our nation's natural resources. And while much information was either currently available or being collected, this information often resided at different institutions, was archived at different scales and in different formats, and was therefore typically inaccessible to government agencies or publics except on a case by case basis.

This situation led to duplicative efforts by employees in various federal and state agencies to develop information to meet their particular needs, resulting in inefficiencies and undue expenses. Furthermore, gaps in data and application methodology made data sharing among multiple agencies difficult and, in some cases, scientifically indefensible. Although many interrelated conditions and trends, such as demand for and supply of commodities and amenities, cross administrative and state boundaries, regionwide databases were not widely available. Finally, recent technological advances, including telecommunications, geographic information systems, and microcomputers, presented opportunities for assembling and distributing information that were not formerly possible. These conditions provided the impetus for the GLA.

The overall goals of the GLA are to cost effectively develop a comprehensive knowledge base for integrated resource planning and management and to provide a scientific basis for inventory, planning, management, and monitoring activities. An emphasis must be added that the GLA is not a federal decision-making process under the National Environmental Policy Act (NEPA), but rather organizes existing information and provides new information that may be considered in the NEPA process.

The GLA is an ongoing assessment and is notably different from its predecessors (see other case studies in this volume). The focus for the GLA is providing information that may be used by federal agencies in revising individual National Forest plans and environmental impact statements, state and local governments for planning purposes, and other groups or individuals needing information about the region's forests.

30.3 Scope and Location

A crucial decision for any assessment is delineation of the region, or the assessment and analysis area. In the case of the GLA, the regions of analysis were delineated within the context of the Forest Service's National Hierarchical Framework of Ecological Units (Cleland et al., 1997), a system adopted by federal and state agencies across the Lake States. The GLA encompasses 98,000 square miles within Province 212, the Laurentian mixed forest, in northern Minnesota, Wisconsin, and Michigan (Figure 30.2).

FIGURE 30.2. Province 212 with subsection boundaries and early settlement forest composition.

Most issues addressed by the GLA, such as timber supply, rural development, and sustainable forest management, can be assessed only by examining conditions and trends of land use, resource consumption, and conservation at several scales of analysis. As stated in Christensen et al. (1996, pp. 665–691).

> Boundaries defined for the study or management of one issue, process or element are often inappropriate for the study of others. For example, watersheds represent a useful unit for the study of water and nutrient fluxes driven by hydrology, but may not be ideal for studies of trophic dynamics in areas where animals move over large distances.
>
> Challenges, such as the management of wildlife populations or development of a recovery plan for an endangered species, may present themselves at one scale of organization, but a complete understanding or resolution of issues usually requires integration across several scales and levels of organization.

The GLA starts at the province level and narrows to landscape level landtype associations, thus providing an ecological basis for assessing resource conditions at multiple scales. For example, conditions such as lake densities or threatened and endangered species' distributions and processes such as past and current fire regimes have been assessed based on landtype associations. Conditions changing at broader scales, such as climatic gradients and forest biome distributions, have been assessed at section and subsection scales. One test of the National Hierarchy's validity is to compare ecological unit boundaries with information on other natural phenomena, such as surficial geology, hydrography, or forest conditions before the wholesale harvesting of the 1800s. Figure 30.2 illustrates how historical vegetative information relates to independently derived subsections: the subsections match well with the less disturbed forests encountered and recorded by the General Land Office surveyors between 1816 and 1907.

Province 212 is the ecological region of interest for the GLA, but the social and economic region is much larger, especially in the context of recreation and forest products. The region produces abundant outdoor recreation opportunities that serve people from a larger region (Stynes, 1997). All three Lake States have a similar tourist travel pattern. Most tourists travel from southern population centers (e.g., Detroit, Milwaukee, and Minneapolis) and Ohio, Indiana, Illinois, and Iowa to the northern forested counties. Although the region is a net importer of wood products, it also exports wood products to other regions (Hagenstein, 1997). Therefore, social and economic information from a three-state area, and in some cases from a seven-state area, is relevant from a planning and management perspective.

30.4 Examples of Issues

In recent years, public land managers have undertaken a number of efforts to better understand the issues of interest to various publics, including resource professionals. Extensive public involvement led to development of national forest plans in the GLA area during the mid-1980s, which listed many issues of concern to regional and local constituents. State agencies and others, through a variety of forums, identified additional issues. In 1995, a Lake States Issue Assessment Team (Seesholtz et al., 1995) identified 15 broad-scale issues for Province 212 (Table 30.1). Many of these issues are interrelated, particularly those in the biological environment. For example, old growth (issue 4) is related to biological diversity, habitat fragmentation, age-class distribution, and so on. It is also related to the social environment (e.g., to timber supply). As a result of these interrelationships, data collected and analyses performed for any single issue often address others. National Forests, one client for GLA data and information, are currently developing their own local issues; many of these correspond to the broad-scale issues identified by Seesholtz et al.

30.5 Process and Project Organization

To undertake the GLA, a small group of agency personnel and researchers was formed. This interagency effort included the U.S. Department of Agriculture's Forest Service and Natural Resource Conservation Service, the U.S. Department of Interior's Biological Resource Division of the U.S. Geological Survey, the U.S. Environmental Protection Agency, the University of Wisconsin at Madison, the University of Minnesota at Duluth, and Michigan State University. After several months of operation and meetings with the departments of natural resources in Michigan, Minnesota, and Wisconsin, employees within state agencies also contributed to and received interim products from the GLA.

The project emphasized acquiring and processing existing information from various organizations, rather than investing scarce resources in data development. Some priority data were developed, however, and thus far include regional coverages for modern wildfires, historic susceptibility of landscapes to fire and wind disturbance, and the geo-

TABLE 30.1. Regional issues identified by national forests and their interrelationships.

Biological environment	*Issue number* 1	2	3	4	5	6	7	8	9	10	11	12	13	14	15	Number of related issues
1. Biological diversity	*	×	×	×	×	×	×	×	×	–	×	×	×	×	–	12
2. Habitat fragmentation		*	×	×	×	×	×	–	–	–	–	×	×	×	×	10
3. Age-class distribution			*	×	×	×	×	–	–	–	–	×	–	×	–	8
4. Old growth				*	×	×	×	–	–	–	×	×	×	–	×	10
5. Landscape patterns					*	×	×	–	–	–	–	–	×	–	×	8
6. Rare species and communities						*	×	×	×	×	×	×	×	–	–	12
7. Forest health							*	×	×	–	×	×	–	×	–	11
Physical environment																
8. Climatic change								*	×	×	–	–	–	–	–	5
9. Atmospheric deposition									*	×	–	–	–	–	–	5
10. Great Lakes shoreline changes										*	–	–	×	–	×	5
Social environment																
11. Traditional human use (Native American)											*	–	–	–	×	5
12. Timber supply												*	–	×	–	7
13. Riparian development													*	×	×	8
14. Rural community sustainability														*	×	7
15. Recreation opportunities															*	7

Start with a column, go down to the asterisk, and then to the right to see the interrelated issues.

referencing of Forest Inventory and Analysis plots to allow evaluations and linkage with other high-resolution data (e.g., LANDSAT Thematic Mapper imagery, hydrography).

Project staffing was minimal, with only three full-time and twenty part-time staff working at various times during the first two years of the project. These employees represented five government and university organizations at nine locations. Electronic sharing of information was essential for project coordination, communications, data processing, and reporting.

The partnering arrangements were made through informal discussions, building on existing written agreements for data and information sharing. In the case of government–university arrangements, cooperative agreements were signed stipulating annual work items and attending budgets. Buy-in incentives were for the most part the recognition of mutual benefits among involved agencies and terms of cooperative agreements where universities were funded for specific work items.

Accountability was established through monthly progress reviews. Performance was measured in several ways. Semiannual reports noting accomplishments and expenditures were submitted to the Interagency Management Council of the National Partnership for Reinventing Government. Informal interagency meetings were held and reports were intermittently submitted to apprise senior managers of project status and direction. Power was exercised through negotiations. Although the project has a manager, the virtual network of universities and government agencies was not structured in a traditional supervisor and subordinate relationship. The people working on this project reported to their respective supervisors within their own agencies. A great deal of goodwill and trust was therefore required in accomplishing project objectives.

30.6 Data Compilation and Analysis

The project consists of two phases: (1) compilation of existing data and information and (2) analysis of this information. The first phase is assembling (1) environmental information such as climatic gradients; surficial and bedrock geology; ecological units at four spatial scales; soils; hydrographic data on lakes, rivers, streams, and wetlands; and contaminants; (2) biological information on current and past forest conditions; the biogeography of game, nongame, and threatened and endangered species; and outbreaks of insects and diseases; (3) ecological process information on natural and anthropogenic disturbance associated with fire and wind; and (4) social and economic information on land use and ownership, demographics, economic activity, resource supplies and demands, and infrastructure. More than 150 primary and derived data layers have been assembled thus far. Spatial information is being processed according to specified data standards in a common geographic information system data format, and metadata and accompanying documentation are being provided. All data, except for threatened and endangered species locations, are being distributed at appropriate resolution on the Internet by the Environmental Management Technical Center, USGS Biological Resource Division.

At the onset of this assessment, there was a paucity of environmental, biological, and ecological data that were consistently available across the entire northern Great Lakes region. Therefore, assessment team members set priorities on the types of data that would be most useful in addressing the issues. These data often had been constructed on a county by county or state by state basis and thus had to be further developed to provide a region-wide context for several issues. For example, maps of early European-settlement-era vegetation from General Land Office (GLO) survey records, depicting the extent and composition of forests in the mid-1800s (Figure 30.2), had been compiled independently by different scientists in each state (Marschner, 1974; Finley, 1976; Comer et al., 1995). The GLA map of historical vegetation grouped forest types into generalized classes that are commensurate with the Advanced Very High Resolution Radiometer (AVHRR) and Forest Inventory and Analysis (FIA) data to allow meaningful comparisons with current forest conditions. The metadata supporting this map are also significant in that they provide interested parties with information on the origin of the map and its strengths and weaknesses.

The General Land Office surveyors also noted natural disturbances of fire, wind, and flooding. Maps reconstructed from these notes (Comer et al., 1995) provide a historical record of the location and extent of different types of natural disturbance. This information, in conjunction with information on vegetation and the physical environment, is useful in understanding landscape patterns, forest composition, and age-class distribution existing before European settlement. The GLA is using General Land Office notes, derived maps on vegetation and disturbance, and information on surficial geology, topography, hydrography, soils, and ecological units (landtype associations and landtypes) to map the northern Great Lakes region into areas of varying susceptibility to fire and windthrow. A comprehensive review of the literature is underway and an annotated bibliography has been produced.

The following metadata, compliant with Federal Geographic Data Committee (FGDC) Standards, are being supplied for each map: (1) identification information, (2) data quality information, (3) spatial data organization information, (4) spatial reference information, (5) entity and attribute information, (6) distribution information, and (7) metadata reference information. For Michigan's early settlement vegetation (a portion of Figure 30.2), the metadata are ten single-spaced pages. A brief excerpt, the abstract, exemplifies the role of metadata:

> *1.2.1 Abstract:* Vegetative cover maps derived from General Land Office Survey notes (1816–1856), polygon interpretation included use of supplemental mapping sources such as 1 : 24k topographic quads, soil surveys, wetlands, etc. Boundary lines originally were drawn on 1 : 24k quad sheets, and digitized (by county) in Intergraph format. The GLA converted the county coverages to Arc/Info format, edge-matched the county maps, converted the individual county coverages from three state plane coordinate systems to a single Albers projection and combined them into one coverage. This was then subdivided into three smaller coverages (Upper, Northern Lower and Southern Lower Peninsula's).
>
> (GLA created metadata for historic vegetation of Michigan)

Clearly, this type of detailed record will help users better understand data and properly interpret maps. The FGDC-compliant metadata were thought to be too detailed for some users, and thus a more user-friendly summary was linked to images of early-settlement vegetation maps on the GLA Web site.

Social and economic data are plentiful in the United States. State and federal agencies compile these data at various spatial scales (e.g., the Bureau of the Census, the Bureau of Economic Analysis, and the Bureau of Labor Statistics). Our approach to defining relevant data involved surveying 150+ managers, planners, and social scientists in the northern Great Lakes region and across the United States. They identified what types and scales of data were needed to provide a social, cultural, and economic context for planning. More than 300 suggestions were received; responses reflected data and information needed to address questions on the relationships between people and natural resources. Most data identified were secondary data that can be compiled from one or more existing information sources. Although this sounds rather formidable, it is just a small subset of potential data that might be gathered.

The aforementioned issues did not constrain the types of data identified. For example, one data request was for seasonal homes (Figure 30.1). This figure shows the high concentration of seasonal

TABLE 30.2. Examples of assessments and analyses completed by issue area.

Issue	Assessment/analysis
Biological environment	
1. Biological diversity	Map biogeography of flora and fauna; map disturbance patterns.
2. Habitat fragmentation	Analyze forest patch size and distribution; analyze landscape structure.
3. Age-class distribution	Analyze forest inventory for age-class distributions.
4. Old growth	Map and attribute disturbance type and frequency.
5. Landscape patterns	Analyze forest patch structure and composition; compare with historical forest conditions and across ownerships.
6. Rare species and communities	Map species and community abundance by LTA and map presence or absence of species by county.
7. Forest health	Model spruce budworm potential in MN; fire distributions and ozone impact areas.
Physical environment	
8. Climatic change[a]	
9. Atmospheric deposition[a]	Map areas of ground-level ozone coincident with susceptible forest species.
10. Great Lakes shoreline changes[a]	
Social environment	
11. Traditional human use (Native American)[a]	
12. Timber supply	Model county-level timber harvesting and forest inventory; link FIA plot data with remotely sensed vegetation maps; evaluate growth, harvest, and mortality by ownership, state, and ecological unit.
13. Riparian development	Map hydrography and analyze lake and stream density.
14. Rural community sustainability	Map and attribute disturbance type and frequency, especially fire.
15. Recreation opportunities	Map campground location and attributes.

[a]Seesholtz et al. (1995) concluded that means for addressing several issues were beyond the scope of their report and that forest management per se would not greatly affect these issues.

homes in the northern Great Lakes region by county (finer-scale data are available, too). For managers and year-round residents in the region, seasonal homeowners and the potential expansion of seasonal homes may have important ramifications related to identified issues. For example, timber supply may be reduced and habitat fragmentation may be increased. Diverse human values and expectations associated with seasonal–urban versus permanent–rural neighbors may also need to be accommodated in resource planning and management. These types of more in-depth analysis are part of the project's second phase to examine the many dynamic interactions between people and forest resources in the region.

In the GLA, a distinction is made between the assessment and analysis phases, in which assessment is the process of gathering information and analysis is the quantitative and qualitative evaluation of the information. Assessment and analysis are independent from, but related to, subsequent planning and management activities. The second phase of the project consists of documentation and analysis of assembled information. Patterns in key conditions such as forest communities; road, river, lake, and wildfire densities; and social and economic factors are being evaluated and documented. Rather than presenting data tables or static map information, the second phase combines spatial databases with computer simulation models and data visualization tools to allow users to interact with natural resource information. A decision support system for natural resource planners and managers is also being tested. Examples of ongoing and completed analyses related to selected issues are presented in Table 30.2.

Careful documentation and objective evaluation of data are integral to the goals of the GLA. The comparison of information on current forest composition is an example of this data documentation (Figure 30.3). Vegetation classifications based on Advanced Very High Resolution Radiometer (AVHRR) and Landsat Thematic Mapper–Multispectral Scanner (TM–MSS) imagery over three regions in Wisconsin and Minnesota were compared with FIA data to determine differences due to spatial resolution of the imagery (Wolter et al., 1995; Host and Polzer, 1997). AVHRR data provide national coverage at an approximate spatial resolution of 1 km^2 (Powell et al., 1993). The classification used composite images of normalized difference vegetation index (NDVI) over an 18-

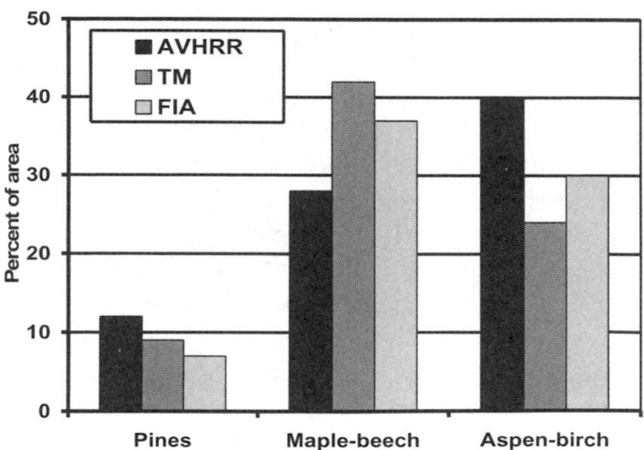

FIGURE 30.3. Comparison of AVHRR, TM, and FIA classifications of forest types in Wisconsin.

week period during the 1990 growing season. In contrast, Landsat TM–MSS imagery has a much finer spatial resolution (approximately 30 m^2), but each scene covers a much smaller geographic region. FIA data are based on a random grid of measured and modeled ground plots.

The AVHRR was found to exaggerate the amounts of certain cover types such as aspen–birch. For example, in the Wisconsin scene, 40% of the forest area was classified as aspen–birch by AVHRR, compared with 24% in the TM–MSS classification and 30% for FIA plots. Conversely, 26% of the landscape was classified as nonforest in the TM–MSS scene, compared with only 10% in the AVHRR classification. Similar patterns were observed in central Minnesota, but the TM–MSS and AVHRR classifications were quite similar for a northeastern Minnesota scene.

Numerous reasons account for these discrepancies in information on current forest conditions. In the classification of remotely sensed vegetation, forest cover types differ in their reflectance, different methods are used for interpretation, and both AVHRR and TM–MSS classification methodologies contain inherent errors. FIA data represent plot measurements of known statistical reliability, but such data are not a source of continuous spatial information. It is clear that the high level of generalization in AVHRR data, in terms of both classification accuracy and spatial resolution, restricts its use to interpretations at fairly coarse spatial scales. For example, AVHRR data would be useful for regional-scale strategic planning, but inappropriate for tactical planning at a National Forest scale. Matching the appropriate data sources to the scale of interest is critical for ecological assessments and in developing management plans.

30.7 Major Results

The Great Lakes Ecological Assessment has assembled compatible regional datasets, developed maps and analyses of resource patterns and processes, and advanced methods for displaying spatial information useful to resource planners. These products increase access to information and reduce costs for many people interested in natural resources. They also provide a new way to look at conditions across a large landscape and offer insights on interactions between natural and human processes that were not available before.

The most important result of the GLA has been to increase awareness by resource managers of ecosystem processes and patterns at broader spatial scales and the human factors that influence them. Information products are requested frequently from the GLA to help resource managers to depict areas or conditions of interest. For example, National Forest and state planners use GIS overlays produced by the Great Lakes Ecological Assessment team during public meetings. Previously, National Forests were limited to displays of conditions within National Forest boundaries or to the state boundary. The newly compiled regional maps show patterns across ownership and political boundaries, convey a broader context, and help to frame resource allocation options within a larger framework. Regional maps that show distributions across a large landscape at first surprise many viewers. Maps comparing historical and current vegetation patterns, maps of the distribution of seasonal home development in the Lake States, and combined maps showing human and natural features together were not widely available before GLA products were released. Furthermore, the GLA is providing the spatial data used to produce these maps in a form useful for subsequent analysis with common GIS software.

A major accomplishment has been the compilation of numerous spatial databases into joined, compatible regional coverages. We now have many regional maps and databases that represent patterns of the physical, biological, and socioeconomic environment important to resource managers and the publics that they serve. For example, a regional fire risk assessment framework was created with a mapped database of more than 14,000 wildfires from the past decade, a spatial factor analysis relating fire size and location to other factors, and an annotated fire bibliography (Cardille, 1998). Other useful data layers are also being compiled, such as a digital elevation model from the U.S. Geological Survey (USGS) and routed hydrography data from the USGS and Environmental Protection Agency. We are continuing to stitch other GIS databases into regional coverages, as they become available.

Analyses are being conducted to document conditions and trends of various natural and socioeconomic conditions. Basic summary analyses have been used to attribute ecological unit maps according to the density of important features that occur within them. For example, some landtype associations contain a large proportion of lakes, while most have only a few. Lakes are desirable features for recreation and seasonal home development; their presence is also correlated with occurrence of rare species. The decision-making process benefits due to increased awareness of areas where these conditions are locally abundant and regionally scarce, thus avoiding a tendency of local managers to overlook the regional significance of features that they see every day.

Other analyses used by resource managers have included those related to landscape patterns of forest patch composition, size, and arrangement. We have used vegetation maps derived from remotely sensed imagery to derive simple descriptive measures, such as patch size, perimeter–area ratios, and neighbor likelihood. The arrangement of patches in the landscape was quantified, and metrics were compared among ecological units and ownerships. Managers expect to use such measures in addressing issues of habitat fragmentation.

An important focus of the GLA has been to devise new presentations of spatial information for the region. We have evaluated data visualization techniques that allow three-dimensional landscape representations based on digital elevation models coupled with other spatial information. These methods provide a different and potentially very useful way to examine the spatial distribution of important variables. Outputs and video examples are being developed for on-line access from the GLA Web site.

Figures used in this chapter are examples of the maps and graphics that can be downloaded by the Internet. The development of a Web site has been an effective communication tool for distributing information to users. Graphics, datasets, and metadata can also be downloaded for review, additional analyses, and use in publications, plans, displays, and educational materials. Maps and data are presented and distributed at a project Web site (http://www.ncfes.umn.edu/gla/index.html) and at a linked USGS Biological Resource Division Web site.

30.8 Successes and Challenges

Many intergovernmental successes were encountered in this project. In particular, the diverse technical skills of various employees from different institutions and the increased capabilities of pooled equipment and facilities enabled successful assemblage of considerable amounts of spatial information across approximately 60 million acres for relatively low costs. Access to data was facilitated through an intergovernmental approach.

The broader goals of the GLA were accomplished by identifying mutual benefits and overlapping objectives among involved organizations. Specifically, the Forest Service's North Central Research Station and Michigan State University were heavily involved with the Lake States Forestry Alliance resource assessment and provided appreciable expertise and socioeconomic data, while receiving funds for continued data development. The National Forest System was preparing to revise forest plans by examining broad-scale issues and conditions germane to the land management planning process, and staffs were assigned to the project. The USGS Biological Resource Division has a mission of assembling and serving spatial databases to public and private agencies and assisted the project as the repository and server of completed data and metadata. The Land Information and Computer Graphics Facility (LICGF) at the University of Wisconsin housed a number of agencies, with the goal of assembling regional coverages. The LICGF provided substantial expertise, equipment, and office facilities for minimal costs and received funding for additional data assemblage. The Natural Resource Research Institute, University of Minnesota at Duluth, had been actively involved in the classification of remotely sensed vegetation and in the development of data visualization and decision support system technology, and it received funding for continued work.

Project success can also be measured as benefits to clients. This project has benefited government agencies by reducing costs of duplicate efforts in acquiring information and by developing methods for analysis, display, and application in planning and management activities. Government programs supported by the GLA include national and state forest planning and monitoring programs and, potentially, landowner-oriented programs operated through the Forest Service's State and Private Forestry organization and university extension programs. The Wisconsin and Minnesota DNR's, The Nature Conservancy, and the national forests in

Minnesota and Wisconsin have used GLA data and supporting analyses in resource assessments and in conservation and forest planning.

We believe the GLA will lead to better decisions as planners access information that provides a regional context for local issues. Developing consistent information themes on past and current conditions will also allow agencies to monitor changes in conditions more effectively. For example, as information base layers are developed, it will become easier to compare remotely sensed imagery from different dates to assess trends over time and to link other regional coverages, such as Forest Inventory and Analysis plot data, to various spatial coverages; this work is already ongoing. These developments will lead to better tracking and reporting, thereby contributing to better resource management decisions and reduced monitoring costs.

This project eventually will benefit various publics, including industrial and nonindustrial landowners, by providing ready access to information that was previously unavailable or difficult to access, by providing guidance on appropriate uses of information, and by making analytical tools available for applying the information in planning and management. Publics will benefit also by receiving better and more timely information regarding proposed actions on public lands, thereby providing the opportunity for more informed choices and comments during public participation.

30.9 Lessons Learned

As with any project involving many individuals and activities over several years, there are many lessons upon which to reflect. For brevity, we will focus on lessons that parallel material presented earlier in this paper.

Impetus and goals: To be most effective, this interagency assessment needed consensus on what was to be done and why. Through many discussions, presentations, and memoranda, collaborators communicated about many tasks. However, no matter how many times the goals were stated, perception and reality did not always match. Although it takes time, project staff must state and restate what the assessment's goals are and why they are being pursued.

Scope and location: Arguments are made for several different approaches to defining assessment areas. In dealing with ecological and socioeconomic data and information, the ecological region (e.g., Province 212 or a watershed) must first be defined; then the associated human-related region can be defined. Except for the spaceship Earth, these regions are not the same. The social and economic region is likely to be larger than the ecological region due to export of products and import of tourists and others.

Issues: Many issues are interrelated, so a variety of types of data and information is needed for a comprehensive assessment. For example, to address the timber supply issue, information on attitudes about timber harvesting may be needed, along with information on the ecological impacts of harvesting within the region. Issues may come from a variety of sources (e.g., assessment-based public involvement, ongoing forest planning efforts, and issue-oriented reports).

Process and project organization: It is possible with a modest investment, in terms of personnel and funds, to gather and disseminate information for a large geographic area. In a small group such as the GLA that has few managerial controls, it is difficult to set deadlines for collaborators, because most have many other responsibilities. Thus assessment staffing must recognize how many or how few full-time staff members are needed to achieve an assessment's goals. Because many agencies are involved, successful collaboration requires "champions" at various levels; simply providing data and analysis results is not sufficient. In addition, a mechanism to ensure task completion by collaborators is needed—for universities, task-related funding is used; a similar mechanism is needed across federal agency units to ensure accountability. Written commitment from senior agency officials is needed in cases where competitive awards are sought. The GLA is funded mostly on a year by year basis; funding uncertainty reduces the likelihood of long-term project success.

Data compilation and analysis: Formal linkages between the GLA and National Forest planning do not exist. For the GLA to more effectively enhance forest planning efforts, a more formal linkage is needed. In a broader sense, data collection is just the starting point for the GLA, and now a process is needed to prioritize future analyses. Related to this, many analyses are linked to the broad-scale issues. However, with limited resources, care must be taken to have researchers focusing on the most pressing subjects, rather than on their personal priorities. Moreover, formal written agreements on data ownership should be used to ensure data accessibility to all interested parties. Now that a fair amount of data has been compiled, comparative studies (e.g., AVHRR versus TM–MSS imagery) can be undertaken to assess data quality and applicability. Most data collected thus far have been secondary data. Future analyses may require additional data collection (and funding), both primary and secondary. Some GLA data will become obsolete or dated. A process is needed to ensure long-term data maintenance, updating, and user support. The lessons learned to date will help us to more effectively address future opportunities.

30.10 Future Work

This project is funded annually based on customer support and requested services. The NPR award was completed in 1999, and the initial goals of data assemblage and delivery have been met. The project is currently funded by the Joint Fire Science Program (USDI and USDA), and the Washington Office of the Forest Service. The Joint Fire Science project is acquiring, developing, and analyzing data on disturbance factors influencing the forests of the Lake States. The Washington Office project is quantifying select indicators of forest sustainability. Proposals to various funding agencies have been submitted, and we hope to continue to build partnerships in this new millennium.

From the onset, the Great Lakes Ecological Assessment was envisioned as one part of an overall program of integrated research and management. In this overall program, involving far more people and organizations than currently involved with the GLA, we hope to contribute to a long-term, iterative process that includes adaptive planning, management, monitoring, and research. The first step in this process identified issues at multiple scales. Next, we have been acquiring information that is already in the public domain and identifying information gaps to be filled based on need and cost. This approach is being repeated in the analysis phase, in which existing models were identified and utilized, with follow-up development of new models based on needs and priorities. Given time and resources, the GLA will augment numerical and simulation models with expert opinions or working hypotheses that serve as conceptual models for use in planning and management. Subsequent monitoring and research on subsets of these working hypotheses could either validate or refute these hypotheses. Ideally, planning, management, and research will be adjusted as appropriate and the entire process repeated as necessary to meet long-term planning and management needs.

We believe the integration of issues, plans, management, and monitoring activities becomes more feasible, given multiscaled assessments and analyses, particularly while meeting goals of sustainable forest management. For example, spatially explicit estimates of forest change could be produced by integrating remotely sensed vegetation classifications, plot data from national programs, such as the Forest Service's Forest Inventory and Analysis, and other information on forest composition, growth, and mortality with maps of constraining environmental conditions (such as the USDA Natural Resource Conservation Service's soil surveys).

A particularly useful analysis would be development of spatial estimates of timber supply linked with market demands, changing management patterns, and government planning alternatives to evaluate and monitor sustainability and multiownership trade-offs resulting from centralized decisions by public agencies. Current and potential timber supply could be analyzed to facilitate effective monitoring of the effects of management at the forest plan and project levels. Projections of potential timber supply necessary to judge long-term sustainability could be based on biological potential and social and economic factors, including growth rates, management trends, and availability of land for harvest or withdrawal from the timber base.

The Great Lakes Ecological Assessment has been unlike other assessment efforts in several ways. The project was not motivated or constrained by one or a few highly visible major issues that threaten forest sustainability. Rather, it has been a steady effort to address a set of broad issues and advance understanding of the dynamics of an entire regional forest landscape and the many natural and human processes that affect forest resources. The process has been customer driven and responsive to ongoing needs for information to improve comprehensive forest planning across a landscape composed of many owners. It has focused on gathering and analyzing information for policy makers, rather than proposing specific strategies or policies. The ultimate goal is to establish a framework of objective information and analytical methods needed to systematically identify the many forces shaping the forest resources of the region and to resolve issues of sustainability before they become crises.

30.11 References

Andersen, O.; Crow, T. R.; Lietz, S. M.; Stearns, F. W. 1996. Transformation of a landscape in the upper Midwest, USA: the history of the lower St. Croix River valley, 1830 to present. *Landscape Urban Plan*. 35: 247–267.

Cardille, J. A. 1998. *Wildfires in the northern Great Lakes region: assessment at multiple scales of factors influencing wildfires in Minnesota, Wisconsin, and Michigan.* Master's thesis. Madison, WI: Environmental Monitoring, University of Wisconsin–Madison.

Catton, B. 1984. *Michigan: a history.* New York: W.W. Norton.

Christensen, N. L.; Bartuska, A. M.; Brown, J. H.; Carpenter, S.; D'Antonio, C.; Francis, R.; Franklin, J. F.; MacMahon, J. A.; Noss, R. F.; Parsons, D. J.; Peterson, C. H.; Turner, M. G.; Woodmansee, R. G. 1996. The report of the Ecological Society of America committee on the scientific basis for ecosystem management. *Ecol. Appl.* 6(3):665–691.

Cleland, D. T.; Avers, P. E.; McNab, W. H.; Jensen, M. E.; Bailey, R. G.; King, T.; Russell, W. E. 1997. National hierarchical framework of ecological units. In: Boyce, M. S.; Haney, A., eds. *Ecosystem management: applications for sustainable forest and wildlife resources.* New Haven, CT, & London: Yale University Press: 181–200.

Comer, P. J.; Albert, D. A.; Wells, H. A.; Hart, B. L.; Raab, J. B.; Price D. L.; Kashian, D. M; Corner, R. A.; Schuen, D. W. 1995. *Michigan's native landscape, as interpreted from the General Land Office Surveys 1816–1856.* Lansing, MI: Michigan Natural Features Inventory.

Denevan, W. M. 1992. The pristine myth: the landscape of the Americas in 1492. *Ann. Assoc. Amer. Geogr.* 82(3):369–385.

Finley, R. W. 1976. *Map. Original vegetation cover of Wisconsin, compiled from U.S. General Land Office notes.* Madison, WI: University of Wisconsin Extension map (1 : 500,000).

Hagenstein, P. R. 1997. Interregional competition and the Lake States forest resources. In: Vasievich, J. M.; Webster, H. H., eds. *Lake States regional forest resources assessment, technical papers.* NC-GTR-189. St. Paul, MN: U.S. Dept. Agric., For. Serv., North Central For. Exp. Sta.: 270–289.

Host, G. E.; Polzer, P. L. 1997. *A comparison of forest cover classifications in the Great Lakes region based on AVHRR and Landsat Thematic Mapper Imagery.* Tech. Rept. NRRI/TR-97/06. Duluth, MN: Natural Resources Research Institute, University of Minnesota–Duluth.

Marschner, F. J. 1974. *The original vegetation of Minnesota, a map compiled in 1930 by F. J. Marschner under the direction of M. L. Heinselman of the U.S. Forest Service.* St Paul, MN: Cartography Laboratory of the Department of Geography, University of Minnesota–St. Paul. Map (1 : 500,000).

Pedersen, L. D.; Chappelle, D. E. 1997. Updated estimates of jobs and payrolls in tourism and forest products industries in the Lake States. In: Vasievich, J. M.; Webster, H. H., eds. *Lake States regional forest resources assessment, technical papers.* NC-GTR-189. St. Paul, MN: U.S. Dept. Agric., For. Serv., North Central For. Exp. Sta.: 179–181.

Powell, D. S.; Faulkner, J. L.; Darr, D. R.; Zhu, Z.; MacCleery, D. W. 1993. *Forest resources of the United States, 1992.* RM-GTR-234. Fort Collins, CO: U.S. Dept. Agric., For. Serv., Rocky Mountain For. Range Exp. Sta. (revised June 1994).

Resh, S. 1994. *Assessing the availability of timberland for harvest in the Lake States.* Master's thesis. East Lansing, MI: Michigan State University.

Seesholtz, D.; Freeman, P.; Sorenson, D.; Padley, E.; Jordan, J. 1995. *North woods broad-scale issue identification project.* Unnumbered report. Milwaukee, WI: U.S. Dept. Agric., For. Serv., Eastern Region.

Stearns, F. W. 1997. History of the Lake States: natural and human impacts. In: Vasievich, J. M.; Webster, H. H., eds. *Lake States regional forest resources assessment, technical papers.* NC-GTR-189. St. Paul, MN: U.S. Dept. Agric., For. Serv., North Central For. Exp. Sta.: 8–29.

Stynes, D. J. 1997. Recreation activity and tourism spending in the Lake States. In: Vasievich, J. M.; Webster, H. H., eds. *Lake States regional forest resources assessment, technical papers.* NC-GTR-189. St. Paul, MN: U.S. Dept. Agric., For. Serv., North Central For. Exp. Sta.: 139–164.

Vasievich, J. M.; Webster, H. H., editors. 1997. *Lake States regional forest resources assessment, technical papers.* NC-GTR-189. St. Paul, MN: U.S. Dept. Agric., For. Serv., North Central For. Exp. Sta.

Vasievich, J. M.; Potter-Witter, K.; Leefers, L. A. 1997. Timber supply and market trends in the Lake States. In: Vasievich, J. M.; Webster, H. H., eds. *Lake States regional forest resources assessment, technical papers.* NC-GTR-189. St. Paul, MN: U.S. Dept. Agric., For. Serv., North Central For. Exp. Sta.: 290–314.

Wolter, P. T.; Mladenoff, D.; Host, G.; Crow, T. R. 1995. Improved forest classification in the northern Lake States using multi-temporal Landsat imagery. *Photogramm. Eng. Remote Sensing* 61(9):1129–1143.

31
Upper Mississippi River Adaptive Environmental Assessment

S. S. Light

31.1 Introduction

Adaptive Environmental Assessment (AEA) employs computer simulation as a tool for understanding complex natural systems and exploring alternative management scenarios for those systems. The technique has been used in such areas as the Everglades, Columbia River, and Canadian forests, which, like the Upper Mississippi River, have competing scientific explanations for how they function and controversies regarding how they should be managed.

In 1995, an AEA process was initiated on the Upper Mississippi River (UMR). This report describes the accomplishments of the first phase of that process and summarizes the remaining work to be accomplished. (See Appendix 31.1 for a list of other reports related to phase I of the UMR AEA.)

31.2 Background

31.2.1 A Time for Decisions

The Upper Mississippi River is a national resource facing a growing number of demands as it moves into the 21st century. Shipping interests want to see the river improved as a transportation corridor. Boaters, anglers, and hunters flock to the Mighty Miss for a variety of recreational pursuits. Biologists and conservationists hope to preserve the river as a healthy and complex ecosystem. Out of these multiple demands arise differing opinions on how the river should be managed.

In the next few years a number of key decisions on river management will be made with lasting consequences for the economic and ecological future of the Upper Mississippi River (see Figure 31.1). As decision time approaches and the differences that separate people become more prominent, scientists and other experts will be called on to support people's positions and to defend the merits of various assertions. People will expect science to provide clear answers based on irrefutable evidence.

In a system as complex as the Mississippi, though, there are no easy answers, and clarity is often lost in the details, buried in volumes of analytical data. If science is to productively advise the decision-making process, it must be used in an alternative fashion that prepares us to make choices by helping us develop a more thorough understanding. It must embrace an approach that avoids the temptation of "the answer," focusing instead on the big picture and multiple views of the future.

31.2.2 Piecing the River Puzzle Together

There are two types of science: the science of parts and the science of the integration of parts. The science of parts (i.e., deduction) helps us to test an idea and explain what happens. The science of the integration of parts helps us to understand the range of ideas that might be useful in solving a problem. The two modes of science complement each other. The integrative mode attempts to make sure that we are solving the right problem, while the deductive mode tells us whether we have the right solution. Both modes are useful for different reasons.

The editor of this case study gratefully acknowledges the contributions of Mike Davis, John Duyvejonck, Bob Gaugush, Lance Gunderson, Barry Johnson, Steve Johnson, Josh Korman, Dan McGuiness, Terry Moe, Barbara Naramore, Garry Peterson, Richard Sparks, Holly Stoerker, Stan Trimble, and Carl Walters.

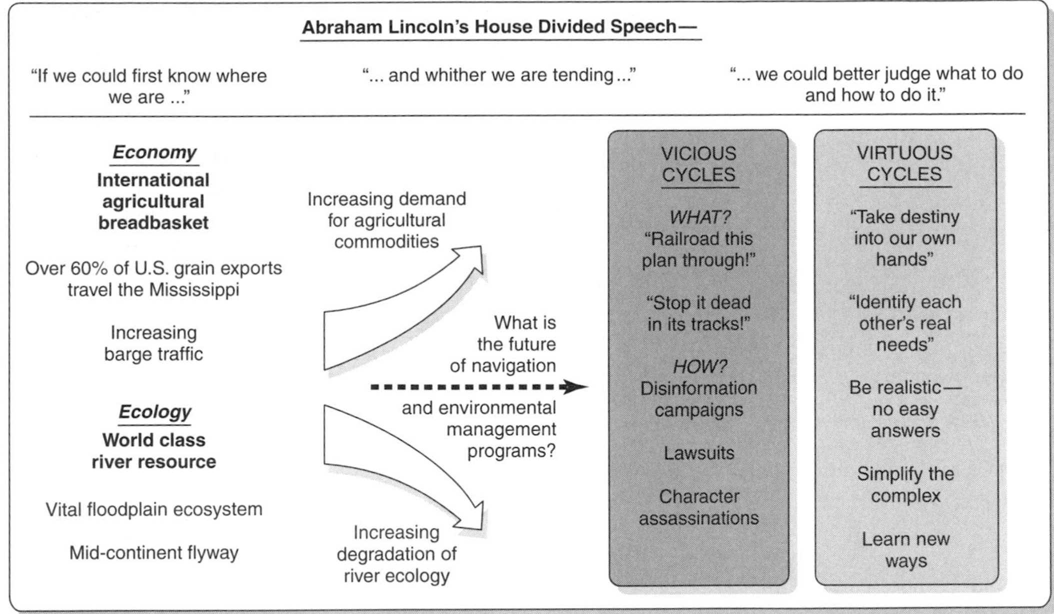

FIGURE 31.1. The path less traveled.

But we tend to rely on only the deductive mode and rarely use the integrative one.

The science of parts fails us when we need to comprehend big picture questions. Experts, for example, can tell us how many tons of commodities go up and down the river, how many ducks migrate through each year, or the number of days people spend boating or fishing on the river. But how will all these demands work together? How does the river accomplish the numerous things we expect of it and rejuvenate itself as well? And how can we be certain that future generations will continue to benefit from a vibrant river?

Even with all the information at our command, no computer can predict the river's future, because both the river and the demands we place on it are constantly interacting and changing in the process. The very act of management transforms ecosystems into new entities in order to create economic and social opportunity. To further our understanding of such a complex and changing system, we need to apply the power of integrative science.

Humans often solve problems by creating mental models of how they think the world works and then testing them. Earlier ideas that Earth was flat or that the sun revolved around it, for instance, fell apart when these views failed to account for observations. In many ways, the variety of scientific and policy perspectives on the Upper Mississippi River system constitutes a giant puzzle consisting of a collection of alternative perspectives. When these views of the river and their underlying assumptions are examined and compared, common elements emerge that can be used to build a composite framework. In this way, science can be understood and used by citizens, policy makers, and scientists alike. It can help us to identify and understand the differences that separate us and discover new ways of working together.

Even though we can never know all the puzzle pieces, by working with a combined understanding of science-based perspectives and human needs for the river, a great deal can be learned. Surprising new insights and opportunities emerge that have not been considered before. This is what AEA attempts to bring about.

31.2.3 Role of AEA

With all the current studies and reviews of river management, we might well ask why we need yet another. The shelves are full of people's perceptions and agencies' plans for the river. What could possibly be gained through another assessment? An AEA, however, is different from these other approaches (Figure 31.2). It is not a vision-led or a planning-led approach. An AEA is a cost-effective, learning-led approach to making optimum use of existing data for assessing the uncertainties, problems, and opportunities on the river. The AEA pro-

Traditional Approaches	*Adaptive Environmental* Assessment
Deductive	Inductive
Narrow focus	Broad focus
Snapshot of problem	Dynamic evolutionary perspective
Analysis of parts	Integrates understanding
Eliminate uncertainty	Highlights uncertainty
Results are conservative and unambiguous; approach perpetuates fragmented management strategies	Results are composite solutions at appropriate scales based on how resource problems define themselves

FIGURE 31.2. Traditional adaptive approaches to problem solving.

ject for the Upper Mississippi River has two objectives:

1. To develop an integrated, science-based understanding of the river as a natural system.
2. To explore alternative ways of reconciling competing demands for the river resource.

The AEA is a process that brings scientists, policy makers, river managers, stakeholder groups, and citizens together in a series of workshops to foster science-based dialogues. Within the workshops, a group of scientists with special training takes people's ideas and perceptions, together with existing data, and develops them into a computerized simulation model that attempts to portray the collective wisdom and ignorance of the group. The model becomes a representation of how people collectively think that the river works (i.e., it uses the science of the integration of parts).

An AEA does not try to ignore or avoid what we do not know about the river; in fact, people's doubt and skepticism are invited. The modeling of people's notions of how the river works makes ideas and assumptions explicit. Participants are encouraged to admit what they don't know. This is no easy task, but it leads to marvelous things happening. Instead of driving a wedge between people, acknowledgment of doubt and wonder about the river actually fosters honesty and integrity, the forerunners to building trust among people.

The AEA is not a conflict-resolution method and does not aspire to be one. Rather, it is a learning tool. The approach develops a living description of how the river works. The model produces graphs and charts that illustrate river responses to the demands and uses that we ask of it. Computer simulations become a way of establishing and building a science-based dialogue among participants. Instead of science being used as proof that one line of argument is more right than another, science helps people to structure their knowledge in ways that facilitate understanding. Participants are encouraged to challenge how the model is being assembled and what questions it will answer. The model becomes intelligible, not a black box that just spits out answers.

An AEA is a disciplined search for creative synthesis, relying on people's ingenuity. It is not a research project. The AEA relies on existing information in an attempt to learn how all the pieces of knowledge that we have fit together. The process, which takes place over a period of months or years, sharpens distinctions, builds clarity, and then begins to probe for flexibility and irreversibilities. It seeks to identify policy and management options that achieve both economic and ecological goals. Alternative ways of solving the river puzzle are proposed and examined. In the end, no single solution emerges. Instead, several composite solutions drawn from a variety of ideas are recognized, but the real product consists of shared insights and understanding.

Recognizing that the real world is very complicated, AEA attempts to simplify the complex in ways that are not simplistic. It applies science in a way that invites the complicated worlds of policy and public values to the table with scientists. An AEA helps to define problems from a management perspective, in ways such that informed citizens can comprehend the problems being addressed and participate fully in the dialogue.

31.2.4 Origins and Progress of the Upper Mississippi AEA Project

The AEA process grew out of the findings of the Upper Mississippi River Workshop and the International Large Floodplain Rivers Conference, which drew researchers from all over the globe to LaCrosse, Wisconsin, in 1994. Conference participants concluded that the system of river control structures on the Upper Mississippi might have created an initial increase in habitat diversity that was essentially unsustainable. An annual floodpulse, channel-forming floods, and infrequent droughts are major factors sustaining floodplain river ecosystems. The lock and dam system and related channel maintenance activities mute the impacts of these forces, leading many conference participants to conclude that the navigation system is an important factor in the slow but progressive degradation of the river's ecology. At the same time, it was recognized that the navigation system on the Mississippi supports considerable economic activity that is significant at regional, national, and international scales. To explore what flexibilities might exist in water control regimens and in natural systems, conferees recommended that an ecosystem assessment be completed for the Upper Mississippi River System to determine the potential for restoring the natural river hydrograph, floodplain connectivity, and energy dynamics.

Over the next several months, the Minnesota Department of Natural Resources' Mississippi River Team met with river scientists from the five-state region and assembled an ad hoc steering committee to pursue an AEA. Since then, the Upper Mississippi River AEA Steering Committee has planned and implemented phase I of a multipart effort. This first phase has focused on identifying key factors in the way the river works, describing these factors' interrelationships, and developing a computer simulation model that attempts to capture these basic interactions. The Legislative Commission on Minnesota Resources, McKnight Foundation, and National Biological Service provided funding for phase I of the UMR AEA.

The Upper Mississippi River Basin Association has administered the phase I funding. The balance of this report describes the progress that has been made through two phase I workshops and related model development efforts. The report is intended to bring closure to phase I of the UMR AEA and set the stage for phase II, which will focus on using a refined simulation model to explore various management scenarios, with an emphasis on identifying promising ways to reconcile competing demands on the river.

31.3 Determining the Scope of the Assessment: Workshop 1

An AEA usually begins with a scoping workshop in which key resource issues are defined, the types of policy actions or interventions are identified, and the critical processes and indicators of ecosystem response are developed. The scoping process leads to the next stage of assessment, when the pieces identified in the scoping session are put into a computer model. On December 5–7, 1995, a scoping workshop was held at the Alverna Center in Winona, Minnesota, to begin a series of structured conversations to assess environmental issues on the Upper Mississippi River. The workshop brought together approximately 45 people with a wide range of experiences, disciplinary backgrounds, and understanding about the river. The scoping workshop for the UMR AEA was organized around a series of plenary and small working group sessions aimed at developing a framework and ingredients for a computer model. Participants discussed resource issues, potential management actions, and the spatial and temporal scale of the analysis. A team of modeling consultants with considerable AEA experience provided technical guidance.

31.3.1 Resource Issues

Attendees at the December 1995 workshop concluded that resource issues on the Upper Mississippi could be grouped into physical, biological, and human components. The physical issues include changes to the natural hydrograph of the river and geomorphological modifications of the river channel and floodplain, including dams, wing dikes, levees, and the effects of impoundment. These changes affect the relationship between stage and discharge and the movement of water and sediments between the main channel and the floodplain and within the floodplain. The primary biological issues stem from the effects of physical modifications, including a decrease in the abundance and diversity of habitat types within the river and the riparian zone, loss of biotic diversity and of individual species, and changes in the movement patterns and abundance of fish. Human issues include the economic effects of commercial navigation, public access for recreational use, and future trends in population growth and floodplain development.

31.3.2 Management Actions to Be Simulated

Workshop participants identified a variety of potential management actions to be addressed by modeling, including those that could be applied within the river and others relating to the upland drainage area. Key physical variables that they wanted to manipulate with the model included the stage of the river, distribution of flow within the main channel and backwaters, and changes in land contours in backwater or impounded areas (e.g., building or removing levees, dredging deep holes). Participants also identified key management variables related to navigation, including changes in the channel depth and dredging policy, changes to the carrying capacity of barges, and alternative forms of transportation.

31.3.3 Spatial and Temporal Scale for Modeling

Workshop participants agreed to take a system approach, with the system defined as the Upper Mississippi River from Minneapolis to its confluence with the Ohio River. With this approach, modeling would be confined to processes occurring within the corridor of the floodplain, and processes occurring over the larger watershed or beyond would not be directly modeled. However, factors not specifically modeled can still be addressed by projecting how changes in these factors would affect the model's driving variables (e.g., a change in land use might increase demand for shipping and reduce sediment input from tributaries). Participants also determined that the minimum spatial scale for modeling within the river corridor must be small enough to simulate changes in vegetation type and to represent major structures on the floodplain (e.g., levees, dams). The time frame for modeling must allow enough time for policies to produce effects (perhaps 50 to 100 years), yet be short enough to capture seasonal effects in components such as vegetation growth, ice-up, winter habitat for fish, and shipping. It was agreed that the model would not consider short-term effects such as barge passage or wake effects from recreational craft. Even within these temporal and spatial restrictions, it was recognized that the proposed model would be too large. Participants concurred with the modeling consultants' recommendation to develop two models: a river system model that could span the entire UMR and a pool-scale model to capture finer details.

31.4 Developing the Model

A unique characteristic of the AEA process is the use of computer models to help integrate and test ideas and assumptions among a diverse set of actors with different backgrounds (Figure 31.3). As such, these models are vehicles designed primarily for facilitating communication and testing the collective grasp of a belief or idea. Only after withstanding repeated challenges and rigorous testing can the credibility of the model be sufficient to address key policy and resource issues. The modeling process is adaptive and organic; initial constructs are likely to change based on workshop participants' feedback to the modeling team (Figure 31.4).

As a starting point, the AEA modeling team constructed two models or views of the river that were linked. Both models operate by balancing inputs, outputs, and storage of sediment and water over time. One model, called the river system model, covers the Upper Mississippi River corridor and deals with processes pertinent to that scale: hydrology, sediment dynamics, land use, soil erosion, and key features of the economic system. The second model, called the pool model, focuses on ecological dynamics of the area between two successive dams (i.e., a pool). This model captures vegetation community succession in response to

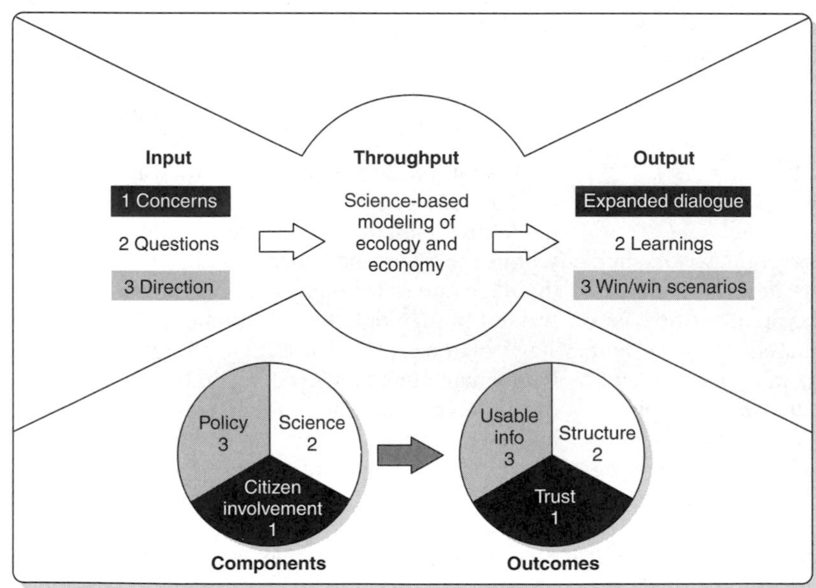

FIGURE 31.3. Upper Mississippi River AEA process.

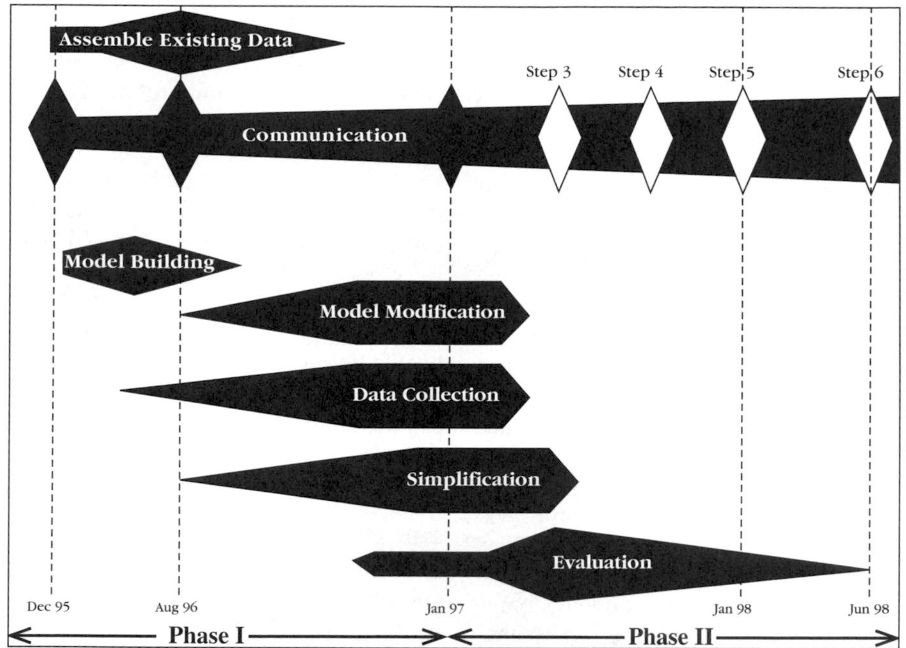

FIGURE 31.4. Stages of model development in AEA.

key physical dynamics, but maintains links to hydrologic and sediment features of the system model (Figure 31.5).

Model developers have attempted to balance mathematical computational time with the need to explore a variety of management options in a timely manner and at an appropriate level of detail. Efforts have been made to assure that the models are calibrated to real-life observations and measurements. The models are written in Visual Basic, op-

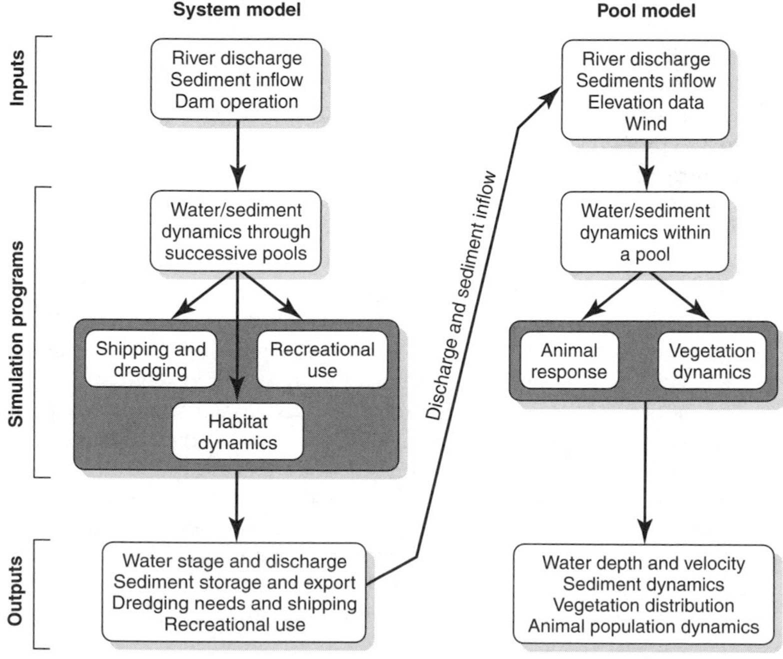

FIGURE 31.5. Components of the AEA simulation model.

erate with a user-friendly graphical interface, and require a Pentium computer using a Windows 95 or NT operating system.

31.4.1 River System Model

The river system model simulates the movement of water and sediment between navigation pools, incorporating the operating constraints of dams, natural seasonal and interannual variability in discharge, and sediment loads delivered from major tributaries. The model was designed to operate over pools 2 through 26.

However, due to constraints of time, data, and budget during development, the initial model operates only on pools 2 through 10, the pools within the Corps of Engineers' St. Paul District. Key elements of this model evaluate the movement of water, including stage and discharge, and calculate sediment dynamics. Current and planned model outputs include changes in stage, sediment storage and outflow, area of different habitat types, tonnage of barge traffic, an indicator of recreational uses, dredging requirements, and an indicator of riparian zone area flooded. Issues raised by workshop participants that are currently outside the scope of this model include exotic species, water quality, and contaminants. The model does not attempt to quantify in monetary terms the public costs and benefits provided by the environment.

The model operates on a daily time step and is driven by historical data on discharge and sediment input from 1959 to 1995. Calculation of sediment input from tributaries is based on tributary discharge and pool water levels. The model allows the user to set a stage height policy on a monthly basis over the year for one or more pools. For each day in each pool, the model calculates the daily inflow, adds that value to the current water volume of the pool, and then calculates a new water elevation for the pool. The water elevation is compared to a user-defined target elevation and water is released as needed, within the operational constraints of the dam, to reach the target level. The amount of sediment exported from a pool is based primarily on water velocity in the main channel, which is a function of discharge and water depth. The model assumes that the channel depth will be maintained at 12 feet; thus all new sediment stored in the channel is assumed to be dredged annually. A navigation–economic component predicts shipping potential in each pool on a monthly time step based on the average channel depth over the month and operating constraints of the lock and dam system.

31.4.2 Pool Model

The pool model covers the area between two successive dams, commonly referred to as a pool, with a spatial grid using 100-meter cells. The model evaluates changes in timing and duration of water levels, flows, and sediment deposition and resuspension among cells, which are then linked to distribution of floodplain topography and vegetation communities. Currently, the model simulates pool 8, for which the 100-meter grid results in about 30,000 cells. A vegetation module models changes in vegetation types, ranging from open water to upland forest. Plant growth and mortality are linked to water depth, turbidity, and duration of flooding. Regeneration and establishment of plants are related to available seed sources, soil saturation, and number of years of dry or wet stress. A stand-alone submodel was also developed that models relative abundance of various fish species based on different combinations of water depth and current velocity as determined by the pool model.

The user can modify information on land and water elevations within a pool (e.g., to incorporate levees or deep holes) by changing values on maps representing initial conditions. Water flows from upstream and tributaries can be modeled as historical flows (from 1959 to 1995), as a specific scenario defined by the user, or as natural system flows that operate as if no dams were present. The user can also modify any parameters for flow and sediment dynamics, as well as parameters for vegetation and fish response.

31.5 Assessing the Model: Workshop 2

The first-cut simulation model was developed based on the December 1995 workshop, with data provided by a variety of government agencies and guidance from the UMR AEA steering committee. The steering committee held a small technical review session in August 1996, at which an initial version of the model was assessed and the modeling team was asked to make some relatively modest modifications. A second workshop was then held on January 15–17, 1997. Approximately 45 participants, representing a wide range of organizations, disciplines, and experiences, were asked to review the first-cut simulation model, recommend refinements, and consider how they would like to use the model to explore various management scenarios. A user's guide describing the first-cut simulation model and electronic access to the model were provided to participants in the second workshop.

31.5.1 Views of the River

The January 1997 workshop served in part to clarify the very basic ways in which people perceive the Upper Mississippi River. The essence of these perspectives can be captured as caricatures, which are admittedly oversimplified, but also informative representations of how people view the current status of the river. Each caricature typifies or exaggerates a different ecological process or structural component, with economic and social implications as well. Although not explicitly stated as such, at least four different caricatures emerged during the two phase 1 UMR AEA workshops.

The Tamed River

This perspective indicates that the hydrologic character of the river has been constrained and controlled by humans. The rhythms or cycles of water flow and water depths have been dampened or tamed. Spatial and temporal variation in hydrology and other processes has decreased due to the regulated management associated with the lock and dam system. Water levels and flows are controlled, so the distribution of areas wetted over time has been changed, with some areas staying wet longer and others staying dry longer.

The Flattened River

This caricature refers to the loss of topographic diversity within the river corridor. The loss of topographic complexity is associated with changes in the hydrologic patterns and sediment movement. Sediment is accumulating in tributary deltas, many off-channel areas, and portions of the main channel. Wind and wave action further reduce topographic diversity by flattening open water areas of the river. This caricature also includes the results of manual manipulation of sediment, as cases where dredge and spoil placements from channel maintenance have led to a decrease in topographic diversity. This topographic homogenization has led to a change in vegetation patterns and animal habitats.

The Beaded River

In this view, the braided, meandering river of the past has been replaced by one with two distinct characteristics, riverine and pooled areas. The area downstream of a dam retains much of its riverine character until it meets the impounded water upstream of the next dam, which creates lakelike habitat. No longer truly riverine, the current river is like a beaded necklace, a set of pools connected by remnants of the preimpoundment river.

The Dirty River

Although not widely discussed at the two workshops, this metaphor focuses on human-induced water quality changes associated with development and watershed modifications. A recent U.S. Geological Survey report analyzes where and how the river has become more eutrophic, either from nutrient-laden runoff from land-use activities within the basin or from sewage plants along the river. Increases in a suite of other contaminants also contribute to this caricature.

Each of these caricatures suggests aspects of the complexity of the river and can be used in part to judge whether the model captures what is important about the way the river works. They also point out how the river has changed and perhaps what might be done to address these changes. As such, they can also be useful in helping to identify policy-relevant resource issues on the Upper Mississippi River.

31.5.2 Management Issues

The key issue for the ongoing AEA of the Upper Mississippi is the search for flexibility in two sectors of the system, the ecological and the economic. This assessment is searching for the flexibility both between and within these two subsystems. That is, the assessment seeks to identify policy and management options for reconciling economic and ecological goals. Nested within the range of scales of geography over time are incredible opportunities to address overriding economic and ecological concerns (Figure 31.6). The project entails a search for win–win management opportunities that resolve the issues related to river changes as described in the previous metaphors.

The management-relevant ecological issues for the Upper Mississippi appear to fall into two general categories: sediment distributions and habitat restoration. Sediment distribution has been altered in the tributaries and the mainstream river. Changes in distribution are associated with changes in sediment input and changes in water flow. In the tributaries, upland practices have increased sediment

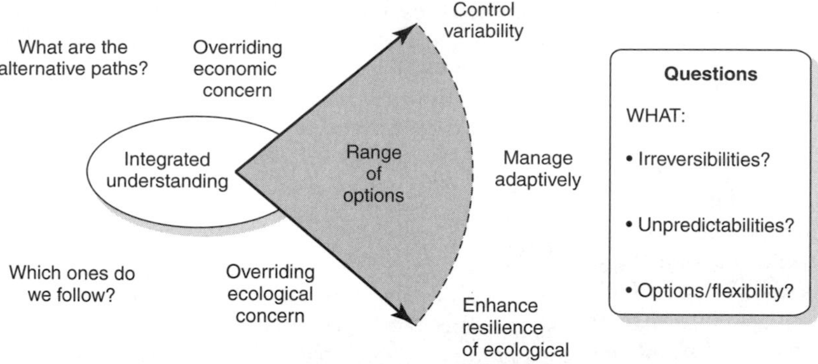

FIGURE 31.6. Exploring sustainable futures realistically.

loading. Sediments tend to fill in the main channel, creating constant dredging requirements. In the shallow areas, wind and waves resuspend sediments. Habitat changes have been observed for a suite of organisms, including many threatened and endangered species. Changes in vegetation patterns have ranged from shifts in the types of species that are dominant to wholesale loss of vegetative community types. The reversal of these unwanted changes is at the focus of ecological restoration in the Upper Mississippi Basin.

The economic issues all deal with direct use and modification of the river to meet human needs. A key issue is the maintenance of a main channel for commercial navigation. This involves channel depth and how the depth changes over time due to water control, hydrologic variation, and sedimentation. The management of sediments from dredging has large ecological and economic dimensions. Another major issue that needs to be addressed is the recreational use of the river and how management and other uses may constrain or provide opportunities for this purpose. Other significant human uses of the river, including public and industrial water supplies and wastewater assimilation, also depend on the maintenance of adequate water depth and quality.

31.5.3 Highlighting Uncertainties

Much of the January 1997 workshop focused on defining and highlighting uncertainties surrounding these management issues. The computer models provided a focus for discussions by small groups, each of which was assigned a set of topics. The groups discussed possible management scenarios, indicators and evaluators of scenarios, and gaps or weaknesses in the models. Several broad areas of uncertainty emerged from these discussions.

Hydrologic Modification

A large uncertainty has to do with how much of the historic hydrologic variability can be restored and over what time periods proposed modifications should be developed and evaluated. Alternative shapes of hydrographs were also discussed, and questions were raised regarding the time of year when drawdowns or free-flow might be attempted (winter or summer) and how these annual objectives would mesh with longer-term natural variations that occur over several years. Participants also discussed uncertainty regarding the types of structural modifications that would allow for flexibility of river uses and management options. Some of these alternatives for structural modification are identified later in the report, under the scenario development discussion.

Sediments

Uncertainties related to sediments centered on problems of modeling transport dynamics. For the river system model, a question was raised regarding potential errors in calculating tributary discharges based on reported discharges at the locks and dams and how these errors may propagate across river segments or pools. Another problem was how to distinguish among types of sediments carried in wash and bedloads and how to account for different proportions of these materials in different parts of the river. The group also identified the need to augment the model with an indicator of sediment storage in each pool.

Vegetation

The vegetation group discussed both the testing and evaluation of the existing model of vegetation dynamics and the need for alternative models. One uncertainty was how well predictions of the current model agreed with historic or observed changes in vegetation patterns. That is, the group was unable to test the current set of rules, although the model can qualitatively match current patterns. Another gap was how to model longer-term vegetation dynamics, especially how communities respond to chronic disturbances such as flooding stress and reestablishment following different levels of flooding. Another item discussed was the need for a nutrient submodel that would link hydrology, sediments, and vegetation dynamics.

Habitats

The habitat group's discussions focused on weaknesses in indicators. Questions arose as to whether it would be better to model responses of key species or to develop habitat indicators. Related to the development of habitat indicators, participants were uncertain whether enough empirical information exists to model habitat changes based only on depth and duration of inundation or if other factors need to be considered. As with the vegetation subgroup, model credibility was a key source of uncertainty, and participants cited the need to test the submodel against historic data.

Human Uses: Navigation and Recreation

Uncertainties related to human activities on the river included gaps in information and the need to determine appropriate indicators. For example, are data available to determine how different management changes would affect key commercial and recreational uses of the river? Other identified gaps include the need to model negative interactions between recreational and commercial traffic and a need to accurately identify changes in dredging requirements associated with different management schemes.

31.5.4 Confronting Uncertainties

To address the information gaps and uncertainties described, workshop participants suggested a variety of model refinement, model assessment, and scenario development activities.

Model Refinement

A number of suggestions were made regarding potential model refinements. These involved modifications to existing submodels, creation of new submodels or routines, and additions to the user interface.

One debate at the workshop was whether to continue development of two models (i.e., at the pool and system scales) or to concentrate work on a system-scale model with aggregated or broad-scale indicators for sediment, vegetation, and habitats. This was compounded by the difficulties of simulating the hydrology and sediment dynamics at a 100-meter resolution in the pool. At this spatial resolution, the model is extremely slow for somewhat reliable results (on the order of 10 minutes per simulation year). One option suggested was to use higher-powered hydraulic models, the results of which could then be plugged into the pool-scale model. The debate remains largely open, with no selection of preference made at the workshop.

A number of modifications were suggested for each of the submodels, including the following:

1. Compiling data and building better representations of sediment transport in the tributaries
2. Developing a model of sediment–landform building
3. Economic assessments for potential structural modifications (e.g., construction and operating costs of changes to dikes, levees, and channels)
4. Developing a water quality model
5. Developing stage-area relationships for habitat and vegetation responses
6. Compiling data on recreational and navigational demands
7. Developing composite dredging cost curves
8. Connecting pool models to evaluate cumulative effects
9. Adding remaining pools to the river system model (i.e., pools 11 to 26)
10. Refining vegetation models with other factors, such as flooding and water stresses, nutrients, and temperature

Workshop participants suggested modifying the user interface to permit comparisons among various output maps. One suggestion was to have the option for simultaneous display of three maps: two of the maps would represent results under different scenarios and the third would highlight differences and similarities between the other two maps. Other suggestions were to have an automatic pause at the end of a year of simulation and to have supplemental files that could be accessed to explain aspects of model code, functional relationships, or parameters. Another suggestion was to have the ability to export graphics or data files for use in other applications.

Model Assessment

A recurrent theme in the January 1997 workshop was the need to critically evaluate or assess the submodels. One example was the stated need for sensitivity analysis of the sediment components of the pool and system models. Participants also highlighted the need for the vegetation submodel to be evaluated by people who are knowledgeable about long-term dynamics and to be tested by compiling and comparing model output with historical time series data on vegetative cover. Similar statements were made about the habitat and navigation submodels. Several workshop participants said that they planned to work individually and with other colleagues to assess various aspects of the model. The U.S. Geological Survey's Environmental Management Technical Center has established a group mailing list to facilitate communication among model users: umrs-aea@emtc.nbs.gov

Scenario Development

Workshop participants identified a range of scenarios that they would like to explore using the UMR AEA simulation model. They emphasized the importance of exploring a full range of alternatives, noting that the computer model permits low-risk, low-cost experimentation because it does not require commitments to any alterations in the physical world. The scenarios described next represent general categories of management alternatives identified by participants. In all these scenarios, a suite of indicators would be examined. These indicators might include changes in pool volume, time required to reach sediment equilibrium, vegetation changes, habitat suitability plots, economic flood damage reductions, dredging costs, recreational boating, and fishing. Using the model to explore any specific scenarios would involve manipulating a few key inputs, while holding a large number of other variables constant.

1. Basin-scale modifications of land use affecting input of nutrients and sediments: an attempt to determine how surrounding land-use practices would affect aspects of water quality and sediment patterns in the river.
2. Pool drawdowns: water level manipulation to restore more natural variation in water levels through seasonal changes in stage height. Potential manipulations include a summer drawdown to dry out sediments and promote plant growth and a fall increase in water levels to flood low-lying areas so that fish and waterfowl would have access to new plant growth.
3. Reduction of flood impacts: an attempt to use a variety of floodplain management regulations and physical structures to moderate flood impacts to developed areas.
4. Extremes: remove all river regulatory structures to examine restoration options or significantly increase river regulation to support increased channel depth.
5. Physical modifications: variations include increasing spatial diversity by island construction; adding, removing, or notching training structures and levees; dredging channels deeper; partitioning pools into more management units; and creating nutrient and sediment trapping structures.
6. Improved shipping efficiency: modifications to tows and barges, alternative lock schedules, and larger locks.

31.6 Future Work and Next Steps: Phase II

As described earlier, the AEA project for the Upper Mississippi River has two objectives. Considerable progress has been made on the first objective, developing an integrated science-based understanding of the river as a natural system. Computer simulation models at both a pool and river system scale have been developed and reviewed by participants representing a wide range of perspectives and expertise. Participants in the January 1997 workshop recommended that the UMR AEA go forward to the scenario exploration phase after some additional modifications to the model are completed. Phase I, the scoping of the problem and model development, is now complete.

The next steps in the AEA process will explore alternative ways of reconciling the competing demands made of the Upper Mississippi (Figure 31.7). Development of restoration options must be articulated and explored. As testing of the river system and pool models proceeds, understanding will grow. The process must balance precision with relevance; that is, the river system is far too complex to capture entirely on any computer, so we must confine ourselves to attempting to model the most important factors in the key river processes and uses. The goal of scenario building is sustainability. Sustainability is multifaceted, and each individual has unique weighting that he or she assigns to various ecological, economic, and social issues. The AEA attempts to provide truly open access to information and devices to use information. All vested interests are asked to contribute.

In phase II, the AEA process focuses on learning that sharpens distinctions and builds clarity as it begins to probe for flexibility and irreversibili-

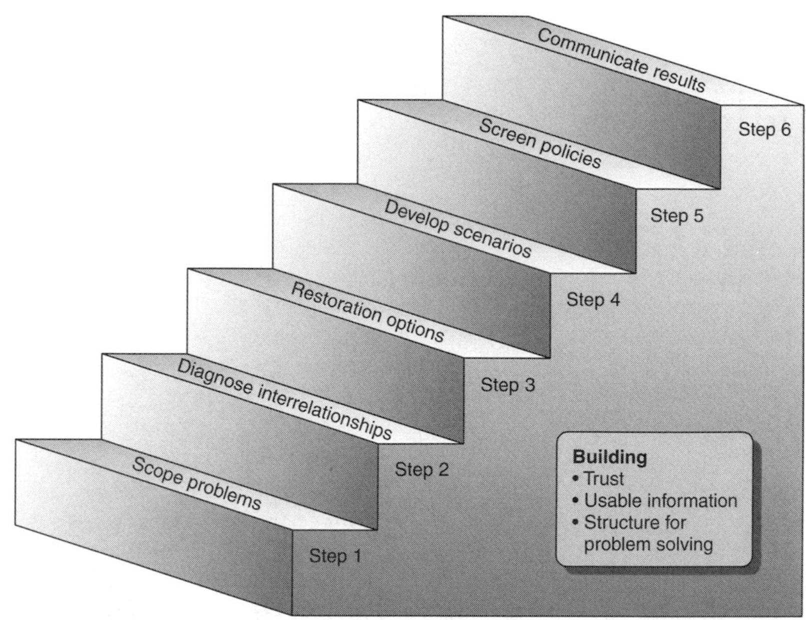

FIGURE 31.7. Steps in the process: adaptive environmental assessment.

ties. Alternative ways of solving the river puzzle are examined. Participants will identify policy and management options that achieve social, economic, and ecological goals. In the end, no single solution will emerge. Instead, several composite solutions drawn from a variety of ideas are recognized, but the real product consists of shared insights and understanding.

The phase II workshops will help build relationships and dialogue that span methods, disciplines, and institutions. Scenario development is a tool for helping the river community to take the long view in a world of considerable uncertainty, building deep and realistic confidence based on insight into possible outcomes of our choices. Key steps in phase II will include the following:

1. Identification of central economic, ecological, and social issues or decisions we face in the foreseeable future on the UMR
2. Identification of key regional factors influencing the success or failure of these decisions or issues
3. Identification of national and international driving trends that could influence key regional factors
4. Ranking key factors and driving trends on degree of importance and uncertainty
5. Selecting sound reasons for how key economic, ecological, and social variables and interrelationships will change in the future
6. Elaborating on scenarios
7. Exploring implications of alternative scenarios, including vulnerabilities and robustness of scenarios

Appendix 31.1: Additional Reports Related to UMR AEA Phase I

Adaptive Environmental Assessment, December 5–7, 1995, Scoping Workshop Evaluation Summary

Upper Mississippi River Basin Adaptive Environmental Assessment and Management, Workshop 1 Report, February 1996

User's Guide to the Adaptive Environmental Assessment Models Developed for the Upper Mississippi River, January 1997

Adaptive Environmental Assessment, January 15–17, 1997, Workshop Evaluation Summary

32 Southern Appalachian Case Study

Charles Van Sickle

32.1 Introduction

The Southern Appalachian study covers a region of 37.4 million acres. Its mountains, foothills, and valleys stretch from northern Virginia and northern West Virginia to northern Georgia and Alabama. When Native Americans came to the region, forests dominated the landscape and they still do, covering 70% of the land (Figure 32.1).

Terrain characteristics are significant in shaping the ecology of the Southern Appalachians. The region's location and its mountains produce a wide range of climatic conditions that are largely responsible for the great diversity of plants and animals found there. These ecological characteristics also influenced the cultural history and economic development of the region.

One of the most important events affecting the region was the logging that peaked around the turn of the century. Following this era, change came slowly to the Southern Appalachian Mountains. The region was considered a backwater of cultural and economic development. But, today, change is advancing at a rapid rate, especially change that affects the relationship of people to the land. The efforts of the Appalachian Regional Commission, creation of the national forests and parks, rural electrification through the establishment of the Tennessee Valley Authority, and the construction of the interstate highway system are some of the factors that have influenced development. The most important effect has been a change in the attitudes and culture of the people that live in and visit the Southern Appalachian region. Meanwhile, time has also brought change to the land and forest resources of the region.

Land managers and planners today are confronted with a baffling array of demands on the region's natural resources from a public whose diverse values and interests make it difficult to set priorities. Population growth and greater demand for recreational activities elsewhere in the Eastern United States has increased the value of the wildland recreation resources of the Appalachians. In 1988, a group of federal resource management agencies was established to meet the challenge of coordinated resource management. The Southern Appalachian Man and the Biosphere Cooperative (SAMAB) gradually expanded to include 11 federal and 3 state agencies. By 1993, SAMAB had coordinated a number of joint projects; some involved a few of the participating agencies; others, such as conferences and workshops, involved all the participants. Through these efforts, SAMAB was becoming a recognized voice for regional resource management.

Also by 1993, the Southern Region of the U.S. Department of Agriculture, Forest Service, was addressing the need for revising most of the forest plans within the Southern Appalachians. With eight national forests, some of which adjoined, there was an obvious need for close coordination and consistency among the forest plans. It was logical, therefore, to turn to a regional assessment as a framework for the planning process. It was also logical to look to SAMAB as a means of involving other federal and state agencies and to coordinate work on areas of mutual interest. The concept for a regional assessment was not well developed, but the perception was that a broad-scale description of the condition of the land and trends in the natural resources would help to identify important resource concerns and would give a better understanding of how to focus planning efforts.

In early 1994, a meeting of the Appalachian national forests' planning staffs, together with the Re-

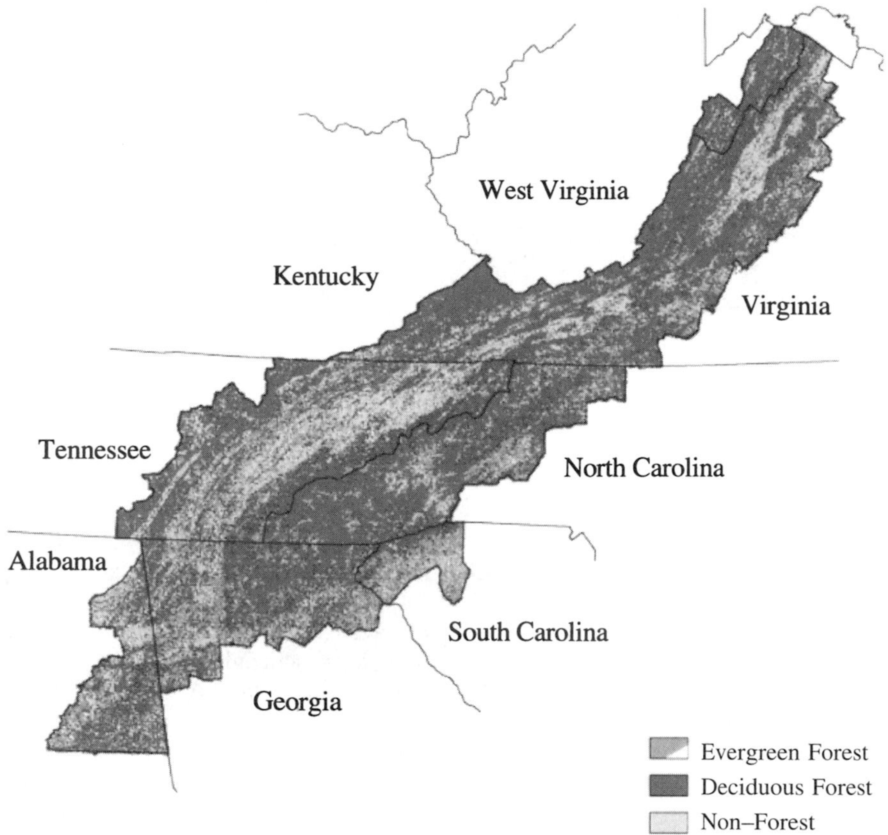

FIGURE 32.1. Forests cover 70% of the land in the Southern Appalachians. (Courtesy of the U.S. Department of Agriculture, Forest Service)

gional Office and Forest Service Research, made a commitment to conduct a regional assessment. At the same time, SAMAB agencies had decided that a regional assessment was needed to focus and coordinate interagency planning. The decision to consolidate these efforts through SAMAB was a natural one. Thus the stage was set to conduct an interagency Southern Appalachian Assessment (SAA).

32.2 Project Organization

32.2.1 Organizing the Leadership

Because the assessment was an interagency effort, an interagency leadership group was necessary to make policy, facilitate support, and provide liaison between the agencies and the assessment. The analyses would be conducted by four technical teams, and technical support would be provided by three facilitating teams: a database management and geographic information system team, a technical writing and editorial team, and a public involvement and media relations team (Figure 32.2). As planning progressed, two of the technical teams further divided their work into subgroups.

Selection of team leaders proved to be a critical step in the process. Team leaders were selected by a joint effort of the SAMAB Policy Group and the members of each team. Team leaders were expected to have some experience working with diverse groups, good communication skills, a vision of the final product of the assessment, and immense patience. Technical expertise in the assessment topic was also helpful, not only to gain the confidence of the team members, but also because the team leader was usually the spokesperson with the public.

Partitioning the work into the four technical analysis areas was a logical approach and had several advantages, but it also had some disadvantages. One clear advantage is that technical specialists communicate best with each other when they share

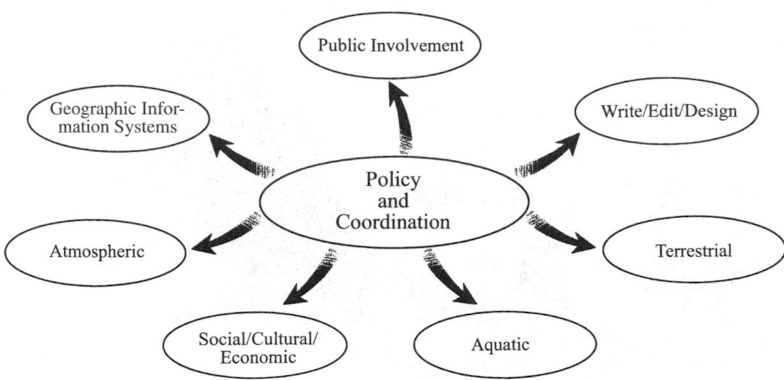

FIGURE 32.2. Interagency assessment teams organizational chart. (Courtesy of the U.S. Department of Agriculture, Forest Service)

a common or related discipline. Another obvious advantage is that the teams needed large data sets that could be shared within the teams for various analyses. A third important advantage was that each team had its own constituency among the public. The questions that arose from public discussion were easily organized around the major themes. Public involvement will be discussed in more detail in Section 32.2.3, but each team had a small and intensely interested group of public commentators who related closely to the issues being addressed.

The most significant disadvantage of the technical organization used was that the teams did not communicate well with each other. Consequently, some opportunities for coordinated or integrated analyses were not adequately exploited.

32.2.2 Communication Strategy

Eventually, more than 150 people were involved in working on the assessment, representing 11 federal agencies and three states. Cooperators from several universities worked on parts of the assessment. A large and diverse public constituency also became involved, including the U.S. Congress. Communication, both internally and externally, became extremely important in the assessment process.

The communication process was challenging because none of the assessment workers were relocated for the project. The Forest Service had effective intraagency communication through the use of the Data General network, but communication outside the agency was much more difficult. Moreover, many agencies and the public do not maintain email addresses. Communication by FAX, telephone, express mail, and other means became an essential part of getting the assessment accomplished. The real challenge was not so much how to communicate as much as what, who with, and when to communicate.

Besides those actually working on the assessment, there was a need to keep many others informed of the decisions being made, rate of progress, and plans for publication. Regular briefings were scheduled for southeastern natural resource leaders, the SAMAB Executive Committee, National Forest supervisors and planners, Washington Office staffs, and public interest groups.

In addition to briefings, a monthly newsletter was distributed to a mailing list that grew to more than 3000 names before the assessment was completed. The newsletter contained information about progress, announcements of forthcoming working-group sessions, invitations to review the assessment products, and plans for release of the final documents. The newsletter was an effective way of communicating with a limited segment of the public. When it was time to present the findings of the assessment, a much more elaborate strategy was clearly needed.

The annual SAMAB fall conference was an ideal forum for announcing the preliminary findings of the assessment. In November 1995, months before the final documents were ready for publication, a special session of the fall conference was devoted to the findings of the assessment. This gave SAMAB an opportunity to publicize the assessment, provided the public with information in advance of the published reports, and afforded the team leaders and scientists some recognition for their work.

Another part of the release strategy was to use current computer technology to make assessment products available on the Internet and on compact disks (CD-ROMs). These products offered many advantages in making data accessible to libraries, classrooms, and individuals.

The final part of the strategy was channeling information to the news media, Washington Office staffs, and members of Congress. The assessment Public Involvement Team worked with the SAMAB Public Affairs Committee to develop a detailed media release strategy. Washington briefings were made by SAA co-leaders and members of the SAMAB Executive Committee.

32.2.3 Involving the Public

One message that came through clearly from town hall meetings, which were designed to encourage public participation in the assessment process, was that people wanted to be included. This did not necessarily mean that everyone wanted to attend every meeting, but rather that they wanted to know what was going on and to decide for themselves whether to attend working sessions. One key decision was to keep the project as open as possible. Regular working meetings were scheduled for each technical team at locations throughout the assessment area. Scheduling meetings with sufficient lead time to get the announcement in the newsletter was often difficult, but a sincere effort was made.

Initially, most teams were very uneasy about the idea of having the public sit in on their working sessions. They anticipated that the public would be a disrupting influence and would make it difficult to accomplish their work. With the short time frames and overwhelming demands being placed on the teams, their concerns were certainly understandable. Public participation did not prove to be the threat that was anticipated. To the contrary, the public was often very helpful and provided the team with insight on controversial and sensitive subjects. Many issues that could have become problems later in the process were identified early and dealt with directly.

An important part of the process for working with the public was establishing the public involvement assessment team. A public involvement team member served on each of the technical teams to provide advice and consultation and to serve as liaison to the public involvement team. This interdisciplinary approach was successful at keeping communications open.

32.2.4 Document Preparation

Packaging a document the size of a regional assessment with scores of writers involved, hundreds of charts and figures, literature citations, tables, photos, and authors to acknowledge is a monumental effort. Someone was needed to worry about the details. Each technical team had the assistance of one or more technical writers plus the guidance of the writing team. Here, too, the skills necessary to present all the materials in an attractive and consistent way did not exist among the scientists and specialists doing the analyses. Getting the editors involved early saved much wasted effort once the work got underway. The Technical Writing Team helped define the overall report format, style requirements, table layouts, and the like.

Manuscript review was not adequately planned for in the initial assessment scheduling. It soon became obvious, however, that this review would become an essential part of the process. Each team leader was asked to arrange appropriate review of the assessment products. Members of the participating public argued strenuously for public comment as part of the review process. In response, it was agreed that public interest groups could nominate qualified reviewers, and names were solicited through the newsletter. The traditional peer review process familiar to most researchers was modified to include both formal peer review and comment from recognized technical experts. This review was called a technical review, rather than a peer review, to better reflect the diversity of the reviewers.

Phase 2 was a policy review of the assessment documents. Each participating agency was sent copies of the draft manuscripts and asked to review and approve the documents for policy. Copies were also sent to selected Washington Office staffs. These reviews proved to be extremely careful and thorough. In fact, they were often more technically thorough than the technical reviews. All reviews were helpful and were appropriately managed and appreciated.

32.3 Identification of Issues

An integrated interagency assessment brings many different viewpoints and priorities. For the assessment planners, issue identification was a critical starting point for organizing the project and setting priorities. Initially, however, there was no clear consensus among participating agencies regarding what an assessment was, what it should accomplish, what it should cost, how it should be funded, how long it should take, or who should lead the effort. One thing seemed clear: developing a process and a strategy for accomplishing the assessment was necessary before determining how long it would take and what resources would be needed. This process, although somewhat crude and poorly defined at first, became more refined as we proceeded and directed the outcome of the final product.

Between the spring and fall of 1994, a number of meetings were held among the assessment organizers. For the Forest Service, the assessment began with a meeting organized by the Regional Office and involved forest planners, research, and Regional Office staffs. This meeting introduced the concept of a regional assessment for the area and identified the broad objectives and expectations of the project. At that time, little or no coordination existed between the various forest planning efforts. A dominant consideration was to assure consistent development of basic data.

In April 1994, the annual SAMAB planning meeting was devoted largely to discussion of the regional assessment, issues to address, and commitments needed. Although various proposals for regional assessments had been discussed before, this meeting officially kicked off the interagency assessment. By the summer of 1994, planning was well underway. As a means of gaining public comment and support, a series of town hall meetings was conducted. The meetings were located in Roanoke, Virginia, Asheville, North Carolina, and Gainesville, Georgia, during July and August. As a result of these discussions, issues and priorities were modified and the assessment area was expanded. The annual fall conference for SAMAB draws a technically oriented audience. Technical papers are presented and scientific issues are discussed. The 1994 fall conference had a special session that explained plans for the assessment, discussed scientific issues, and honed and polished assessment issues.

Nationwide, a number of regional assessments were already underway. They differed widely in approach, scale, subject, agency involvement, and location. In an effort to learn from what had already been accomplished and bring some order to the various data collection and analysis projects, the Environmental Protection Agency held a meeting in Atlanta to review ongoing assessments. This meeting provided an exchange of ideas and approaches.

32.3.1 Assessment Questions

As a result of all these meetings, a broad framework of issues was developed, initial guidelines were formulated for public involvement, and communication links were established with the key representatives of the various agencies. The issues were expressed in the form of questions, and these questions were organized into a charter. The charter was originally intended for Forest Service use, but it was also helpful in explaining and gaining acceptance of the other agency participants.

In framing the questions, a logical approach to the assessment emerged. The assessment consisted of four major topics: aquatic; atmospheric; terrestrial; and social, cultural, and economic. These questions were used to guide the content and direction of the analyses. The key topics that guided the assessment project were the following:

Atmospheric analysis: The major atmospheric issues addressed air pollutants and exposure, concentrations of particulate matter, visibility and how it is affected by air pollution, effects on aquatic resources by acid deposition, and the impact of ground-level ozone on forests.

Terrestrial analysis: The terrestrial analysis included plant and animal resources and forest health for plants and animals. The assessment evaluated what plant and animal resources occur within the region and their habitat associations; the status, trends, and distribution of species and habitats in the region; and the habitat conditions important for maintaining the desired population levels and those needed to recover populations at risk.

Forest health area: The forest health assessment addressed occurring changes in forest vegetation or soil productivity, the potential effects of the presence or absence of fire on forest health, how forest health is affected by native or exotic pests, and how current management practices are affecting the health and integrity of forest vegetation.

Aquatic analysis: The aquatic analysis focused on the status and trends in water quality, aquatic habitat, and aquatic species; the management factors important in maintaining aquatic habitat and water quality; the extent and composition of riparian areas; the laws, policies, and programs for protection of aquatic resources and their effectiveness; how people affect aquatic resources; and the current supplies and trends in water use.

Social, cultural, and economic resources analysis: This analysis addressed several components:

1. The human dimension questions related to how the social pattern of the region had changed over the last two decades and how this change was reflected in natural resources management; how management decisions affected the economic condition of local communities and people outside the region; people's attitudes and values regarding natural resources; the important economic trends with respect to tourism and extractive and other resource-dependent industries; and why and how private and nonindustrial landowners manage land.

2. The timber economy analysis examined the supply and demand for wood products, the dependency on National Forest timber, timber production's effect on employment and income, and the National Forest land tentatively suitable for timber production.
3. The recreation resources and use component evaluated supplies and demands for recreation settings, the unique or unsatisfied recreation opportunities on public land, changes in public land use in the last 10 years, and forecasted trends; how the changing social patterns affect public land use; and how recreation opportunities modify the life-style and local culture of the area.
4. Roadless and designated wilderness areas analysis focused on identifying the location of roadless and primitive areas, the effects of Forest Service management on the integrity of roadless areas or the enhancement of natural processes in wilderness areas, the effect of the proximity to population centers on wilderness use, and the relation of wilderness and roadless areas to other assessment resources.

32.4 Scale, Extent, and Resolution of the Assessment

No discussion of scale and resolution can begin without describing the purpose, audience, and constraints imposed on the assessment. The limitations of time and budgets directly affect the level of detail and amounts of data that can be assimilated.

Two factors influenced the extent of the assessment: no special funding was allocated and the forest plan revisions imposed special requirements. Five of the seven National Forests in the region were beginning plan revision. The Regional Forester, although fully committed and supportive of the assessment, had promised that the assessment would not significantly delay plan revision and requested that the assessment be completed by January 1996. Technical teams had slightly more than 1 year from initial planning to the completion of the draft analyses. The team leaders and a handful of others worked on the assessment full time. The remainder of the 150 people on the project had competing demands on their time. Funding for the assessment, both within the Forest Service and in other agencies, came largely from the administrative units that furnished the people—the forests, research units, and agency staffs.

Because of time and budget limitations, it was decided that the assessment would be accomplished with the use of existing maps and data. Working with existing data meant that it was sometimes necessary to use whatever map scales were available, rather than designing map products to satisfy analysis applications. This did not prove to be a serious handicap and, overall, the constraints imposed on the assessment were useful in determining priorities and allocating effort.

Initially, it was assumed that the assessment would be conducted within the boundaries of the original SAMAB region, a 100-county area covering parts of Virginia, Tennessee, North Carolina, South Carolina, and Georgia. As a result of the town hall meetings, additional counties in Virginia and West Virginia were added so that the George Washington National Forest would be included. In Alabama, one county was added to include the Little River Gorge Preserve. The final 135-county study area encompassed 37.4 million acres and parts of seven states.

Counties do not always follow ecological boundaries, but county boundaries became the primary geographic reporting unit for several reasons. The demographic data were only available by county or census tract. Also, people living in the Southern Appalachian region have a very strong sense of place associated with county of residence. County and community planners were identified as potential users for the assessment data and would be interested in county data.

A number of additional boundary overlays were needed within the assessment area. Watershed and hydrological unit boundaries were needed for the aquatic assessment. Federal and state ownership boundaries were needed for the terrestrial assessment, and highway and road overlays were needed for several applications. Figure 32.3 illustrates some of the many boundary overlays that were included.

It was generally agreed that maps and locational data would be an important ingredient in the assessment. Current GIS technology, which offered new frontiers for analyses, and the maps developed for the assessment would have wide future applications for managers and planners. The first step of the GIS team was to work with the technical teams to complete an information needs assessment. Development of GIS data to support analysis was a major undertaking. Frequently, the analysts asking for the data had little understanding of the cost and effort required to supply the data, and deciding data management priorities was difficult.

A GIS team member was also assigned to each technical team. Their advice and assistance were essential to provide the maps, data, and analyses

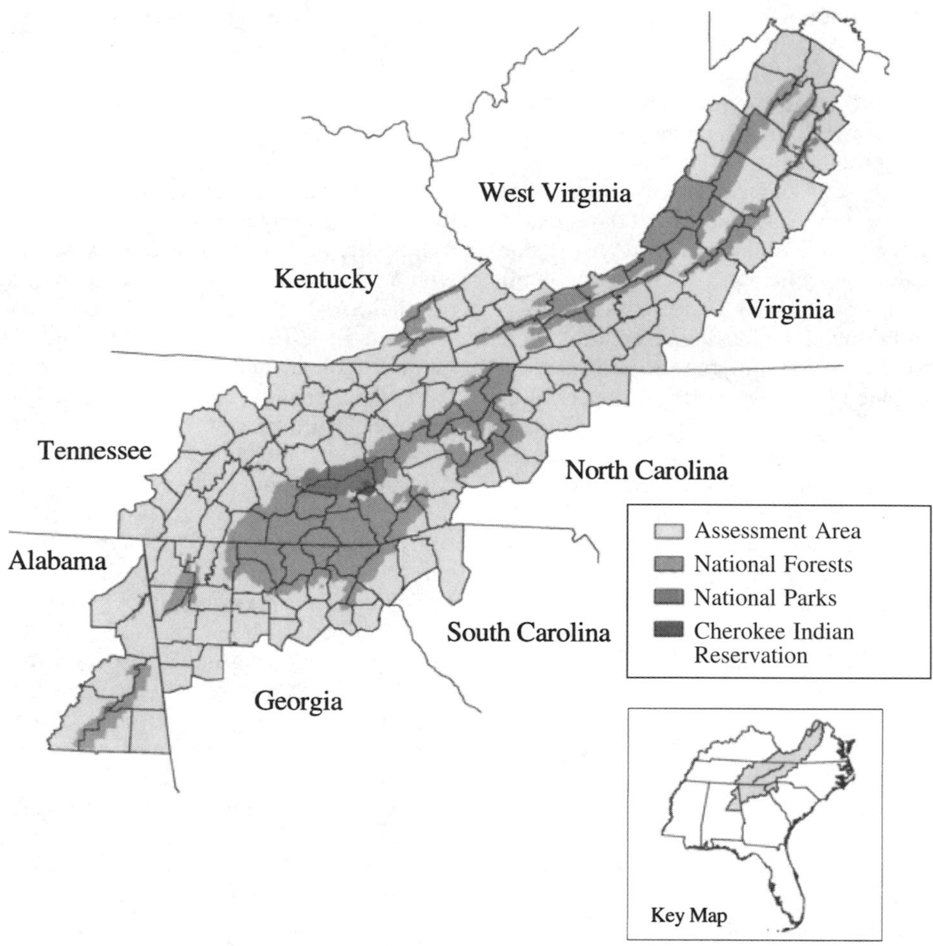

FIGURE 32.3. Boundary information provides a sense of place. (Courtesy of the U.S. Department of Agriculture, Forest Service)

that the technical teams needed. The GIS specialists were aware of techniques and cartographic protocols that added greatly to the refinement in the final reports.

It is difficult to generalize from the separate information needs assessments because they identified widely differing analytical requirements. There were, however, some things common to most of the technical applications. The maps and data assembled were at different scales depending on the intended purpose. For regional analysis, a scale of 1 : 100,000 or 1 : 250,000 was usually considered the most useful. For publication purposes, a scale of 1 : 2,000,000 was a good starting point. Certain kinds of information were needed for planning on the National Forests and for this purpose a scale of 1 : 24,000 was necessary. This was especially true for applications that required public involvement, such as the roadless areas and old-growth inventories.

Several of the teams asked for detailed ecological data. Bailey's ecoregions map was a good foundation, although the resolution at the section level was very crude. For habitat classification and other applications, a much finer level of resolution showing current conditions was needed. Consequently, a major financial commitment was made to obtain a satellite image (Landsat) of the entire area. The classified image was obtained under contract from Pacific Meridian Resources. The final product was delivered late in the process and, although it was helpful, its limited availability meant that it was not used to its full potential. Thus the factors determining scale and resolution were application, data availability, kind of analysis, and anticipated future applications.

32.5 Major Types of Data and Information Used in the Assessment

The completed assessment consists of approximately 900 pages of text, maps, tabulations, and data sets: five compact disks (CD-ROMs); and two separate home pages on the Internet. The maps and data were needed to support the analysis, the analysis supported the text, and the text was used to communicate with a variety of audiences. Thus, in describing the types of information, it is useful to make three distinctions: foundation data, analysis data, and observational data (maps are included in the category of data).

32.5.1 Foundation Data

The assessment depended almost entirely on existing data from other sources, although much of the information resided in separate and often obscure agency data files. What made the assessment unique was that the information was brought together and organized for a specific geographic region. It was a major achievement to identify, locate, and assemble this large and diverse data resource. For this reason, considerable attention was given to obtaining data that would be used by forest planners, educators, and community planners who want to conduct their own analyses. All the maps and data sets were recorded on the CD-ROMs.

There are two advantages of the CD data set (and its counterpart on the Internet). One is that it contains a huge amount of information not available anywhere else in the assessment process. The second is that it can be updated or expanded on a continuous basis. Although plans have not been completed for regularly updating these data, it is widely recognized that this could be a very important SAMAB function.

32.5.2 Analysis Data

Analysis data are the maps and models generated as part of the technical analysis phase of the assessment. These data include materials such as the habitat matrix for terrestrial wildlife, the trout range map, questionnaire survey data, and the LANDSAT land-use and land-cover map. Some of the analysis data are contained on the CD-ROM, but other data sets were not easily adapted to this form of display.

A common form of analysis data is the maps generated by combining (intersection or union) two or more GIS layers. This has been one of the most common forms of integrated analysis. An example was combining geology data with atmospheric data to show the location of streams at high risk of acidification. In some cases, maps with high resolution or precision may be combined with data sets that are low in precision. The result may be very useful, but the product is often difficult to describe in terms of reliability. To the extent possible, maps such as the black bear suitability map or the recreation settings maps need to be identified as analysis products with the criteria and assumptions clearly specified.

32.5.3 Observational Data

A third category of data covers a broad array of information in the form of economic data, demographic data, questionnaire surveys, water and air quality monitoring data, species occurrence data, and many other sources stored as numerical data sets. Observational data are usually linked to maps by geographic coordinates or place identifiers such as county boundaries. In many cases, no further locational information is possible or appropriate.

The LANDSAT Land Cover image is worthy of special mention mainly because of its large size. The ground truth component of this product was not completed at the time that the assessment was released, and users should be aware that many of the detailed classifications still require verification. Limitations of the data set are explained in the accompanying metadata. It is hoped that refining and updating this data set will be a high-priority task for those who maintain the assessment data set.

32.6 Major Types of Analysis

Each assessment team's approach to their subject was unique, but several aspects were common to each analysis area. All teams started with the set of questions from the issue identification process and were given the opportunity to revise or reorder the questions. Once agreed on, they were asked to ensure that the questions were adequately answered. The teams were also asked to focus on the status and trends affecting natural resources, but to avoid drawing conclusions and making policy or management recommendations. Each team was asked to identify missing information and to construct a list of research and monitoring needs. Finally, each team was asked to sufficiently describe their meth-

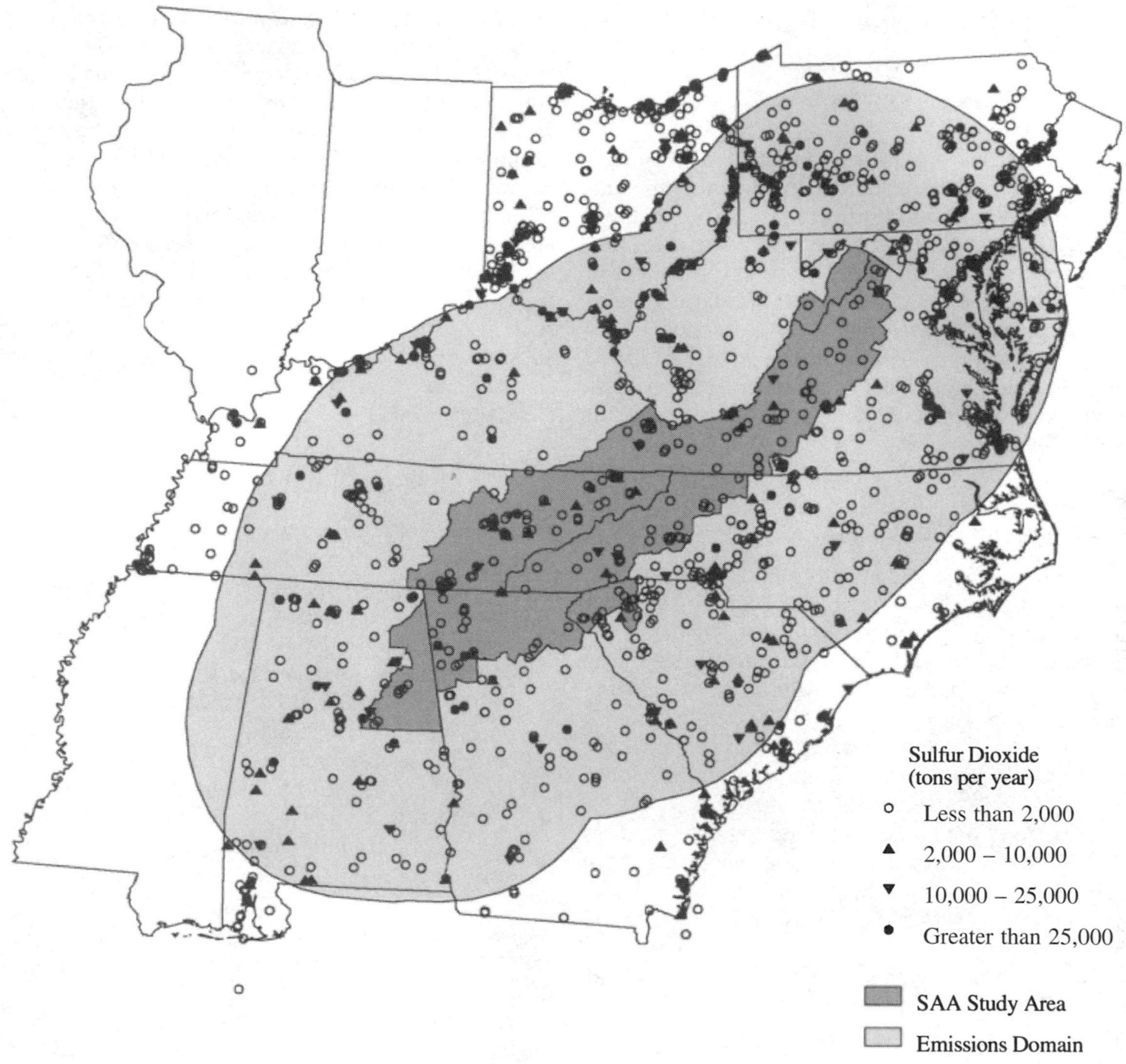

FIGURE 32.4. Location of point sources of sulfur dioxide, 1995. Nationally, point sources release the majority of sulfur dioxide. (Courtesy of the U.S. Department of Agriculture, Forest Service)

ods and procedures so that the technical reader could evaluate the appropriateness of the analysis. Where possible, the aquatic resource assessment findings were integrated with findings from the atmospheric, terrestrial, and social–cultural–economic assessments. Several additional findings emerged from these joint analyses.

32.6.1 Atmospheric Analysis

The major air pollution emissions assessed in the report were particulate matter, nitrogen oxides, volatile organic compounds, and sulfur dioxide. These pollutants are important because their secondary pollutants are suspected of causing visibility reductions, ozone impacts to vegetation, and acid deposition impacts to terrestrial and aquatic environments. They are also directly addressed by the Clean Air Act (CAA) legislation. The analysis described the location of emissions, concentrations where emissions are greatest, and likely future trends. Figure 32.4 illustrates one of several point source maps prepared as part of the analysis.

Although no violations of the national particulate matter standard (NAAQS) are presently occurring within the Southern Appalachians, regional planners need to know existing particulate concentrations to monitor for unacceptable emission levels. Common sources of particulate matter are stationary sources, such as power plants or unpaved roads, but there is considerable interest in the potential effects that prescribed fire might have on local conditions.

The Southeast has the poorest visibility in the Eastern United States, and it is usually worst in the summer months when the greatest numbers of people are viewing scenery in the mountains. Haziness has intensified since the 1940s, and sulphur dioxide emissions are believed to be primarily responsible.

In addition to being an important component of the recreation experience, visibility is specifically addressed in the CAA. The CAA established a national goal of "prevention of any future, and the remedying of any existing, impairment of visibility in mandatory Class 1 federal areas where impairment results from man-made pollution." The majority of the visibility data used in the assessment was obtained in the seven Class 1 areas within the assessment area. A description of current conditions was based on data collected through the Interagency Monitoring of Protected Visual Environments (IMPROVE) network. Historical visibility data were based on information collected at airports since the 1950s.

Acid deposition has been the cause for great concern in the assessment area since the early 1980s when some scientists began to attribute the death of high-elevation spruce and fir forests to the effects of air pollution. Moreover, media reports repeatedly proclaimed that acid rain was killing lakes and streams in the area. During the 1980s, the National Acid Precipitation Assessment Program (NAPAP) carried out the National Stream Survey to estimate the extent of stream resources affected. The assessment team used these databases and model results, together with site-specific investigations, to describe the acid-deposition threats to aquatic resources. The assessment also reviewed the EPA's findings to estimate stream reaches sensitive to acid deposition.

Ozone is potentially the most significant pollutant affecting forests in North America. Numerous surveys within the Southern Appalachians have reported ozone symptoms on sensitive plants. Whether growth loss actually occurs depends on the sensitivity of the plant species and environmental conditions such as soil moisture that affect the plants' vulnerability to ozone.

32.6.2 Terrestrial Analysis

The Southern Appalachians are a diverse and complex region. The terrestrial resources are affected by topography, geology, climate, soils, and the history of human use. The questions arising from public discussion and planning needs fell mainly into two groups: What are the status and condition of plant and animal resources, and what is the general health of the forests?

Plant and Animal Resources

To address questions that dealt with species occurrence and frequency, it was first necessary to narrow the species of concern down to a manageable number. More than 25,000 species are known to inhabit the assessment area, but 472 were selected for study based on a number of selection criteria (Figure 32.5). Slightly more than half were animals. Most of these species could be separated into 19 groups based on similar habitat requirements (associations). Species associations helped define the habitat characteristics that would be of interest in the descriptive inventory.

Many characteristics of the landscape are useful for describing habitat condition and suitability. The terrestrial team gathered data and described habitat features as part of their analysis and in response to the questions. In addition to the usual features that include land use and land cover, forest type, ownership, and ecological units, special emphasis was given to the successional stages of vegetation because of its significance for animal habitat.

About 84% of the terrestrial threatened and endangered species and 74% of the viability concern species are associated with rare communities and streamside habitats. These rare communities represent less than 3% of the assessment area, and 1% is high-quality habitat. Thus the future management for species diversity will depend on managing these habitats. Almost two-thirds is in private ownership. The remaining small area is managed under federal or state direction.

Forest Health

Forest health can be defined in many ways, but the purpose of the assessment was to give an overview of the most important factors that are affecting forest condition. These include natural and human-induced disturbance, native and exotic pests, air pollution, and changes in land use.

Historically, the most significant event affecting the Southern Appalachians was the initial logging. By 1901, the condition of the region's forests and streams had reached such a degraded state that the secretary of agriculture was asked to make an assessment of their condition. This assessment ultimately led to passing the Weeks Act, which authorized establishing the National Forests in the

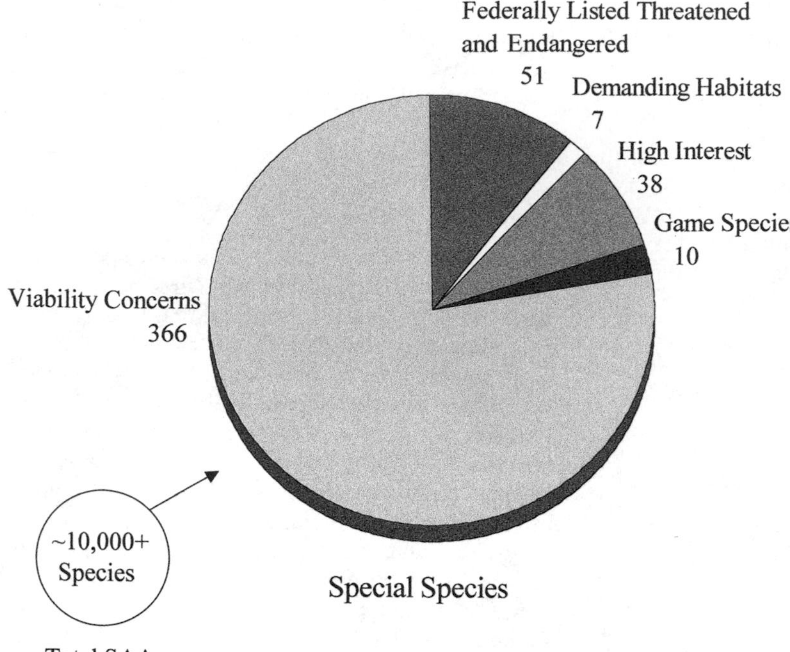

FIGURE 32.5. Selection criteria for terrestrial plant and animal species list for the Southern Appalachian Assessment. (Courtesy of the U.S. Department of Agriculture, Forest Service)

Eastern United States. Today, the region has gone through a long period of recovery and is in relatively good condition. Nevertheless, there are a number of threats to the health of the system.

Although early logging was one of the most significant disturbances to affect the region, the exclusion of fire may also be having subtle but far-reaching effects. Several vegetation communities such as Virginia pine and table mountain pine are fire dependent. Recent studies report that, prior to establishment of the national forests, fire was fairly frequent. The assessment reported on several kinds of natural disturbance, including ice and snow, wind storms, frost damage, and other weather events. Perhaps more important, however, are the insect and disease threats, most are of exotic origin.

Oak decline is a disease complex involving environmental stress (often drought), root disease, and opportunistic insect pests. Although this may be a chronic condition in the oak forests of the Southern Appalachians, it has recently become more common and severe, probably due to widespread drought in the 1950s and again in the 1980s and the advancing age of many forest stands. Recent surveys indicate a widely distributed decline that is changing forest composition and structure.

Gypsy moth is another serious threat to the occurrence of oak in Appalachian forests. The insect was introduced from Europe in the 1860s, but the rate of expansion increased alarmingly in the 1980s. During the last 10 years, the moth has defoliated more than 4 million acres in Virginia and about 1 million acres in West Virginia. Despite existing management strategies, losses are expected to continue as the moth migrates down the Appalachians. Species vary in their ability to recover from defoliation, but most will succumb after a few years of repeated attack.

Southern pine types make up only about 10% of the assessment area, but southern pine beetle is a serious threat in places where the pines occur. Other diseases that are causing concern are dogwood anthracnose, beech bark disease, butternut canker, Dutch elm disease, and hemlock wooly adelgid.

32.6.3 Aquatic Resources

The headwaters of nine major rivers lie within the boundaries of the Southern Appalachians and are the source of drinking water for much of the Southeast. Diversity of aquatic resources is high, with a rich fauna of fish, mollusks, crayfish, and aquatic insects. While human activities that impair aquatic resources have decreased, population growth and concomitant land development are increasing pressure on water resources.

The condition of water bodies is influenced by land uses within the watershed, geology, soil erosion, vegetation, and soil nutrients. Evaluation of

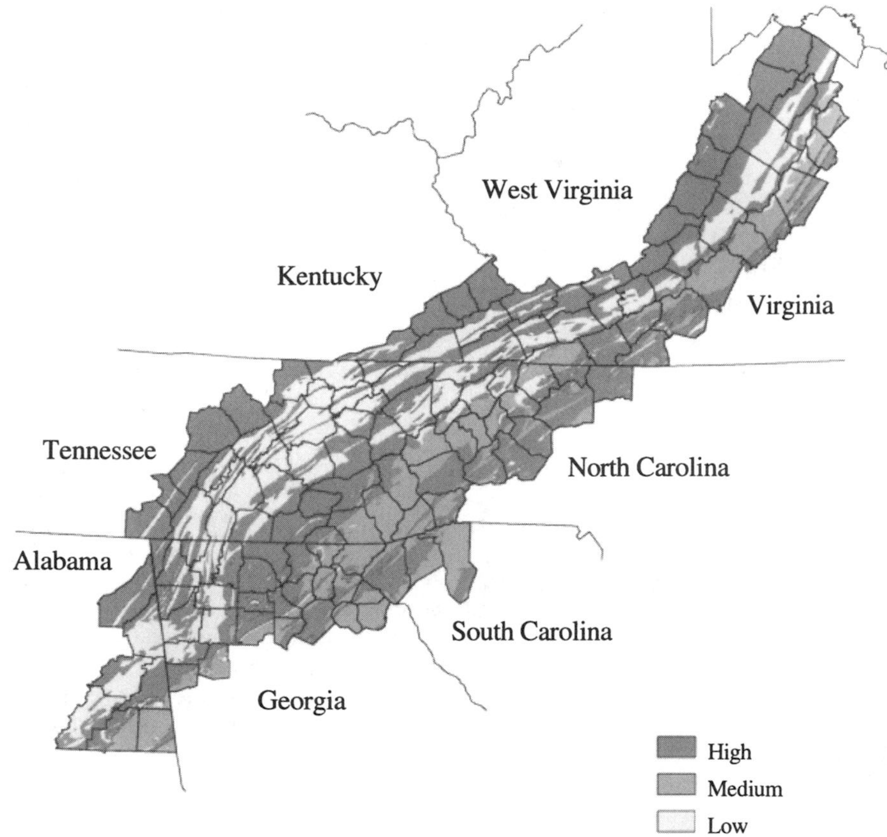

FIGURE 32.6. Soils in the SAA region that have high, medium, and low sensitivity to acid deposition. (Courtesy of the U.S. Department of Agriculture, Forest Service)

aquatic resources condition was based on how well lakes and streams support their designation uses: fishing, swimming, aquatic life, and drinking water. Trophic state of lakes was another measure of condition. Extensive data from a number of federal and state monitoring activities allowed the study team to report the condition of water bodies by county and watershed and for the major stream reaches.

For the analysis of threatened and endangered species, the team obtained the Element Occurrence Records (EOR) from seven state heritage programs. Mollusks and fish had the largest number of Fish and Wildlife Service (FWS)-listed species, reflecting greater interest and higher numbers of species at risk. In the assessment area, there are about 350 fish species, 64 of which are imperiled.

Many people fish for native or naturalized trout. Others find the native brook trout to be a beautiful fish or see trout as an indicator of high water quality. Maps were constructed that show the streams with wild trout potential and the location of stocked trout. From these maps, it was possible to predict the effects that various factors influencing water quality might have on trout habitat. For example, the maps were used to show potential impacts from acidification (Figure 32.6) and gypsy moth and hemlock wooly adelgid.

The conditions of stream habitats are highly dependent on the characteristics of natural bank and riparian zone vegetation. To assess the current condition of riparian zones over the entire region, the team combined the GIS data from the stream reach file with the LANDSAT satellite image. This provided a broad classification of riparian zone conditions and generated a GIS layer that can be used for subsequent analysis.

The last decade has been a turning point in water resource legislation and pollution control. Nevertheless, many sources of pollution still exist. Two-thirds of the water quality impacts are due to nonpoint sources such as agriculture runoff, storm water discharges, and landfill and mining leachate. Many of these sources are identified in the Natural

Resources Inventory (NRI), which is conducted periodically by the Natural Resources Conservation Service. This inventory provided an indication of soil erosion potential from agricultural sources. The universal soil loss estimates, together with the NRI data, were used to calculate an average erosion rate for each county.

32.6.4 Social, Cultural, and Economic Resources

Natural resources have value because of their utility and esthetic or intrinsic qualities that come from the human culture. Thus the importance of natural resources is directly related to people. Early in this century, the Southern Appalachian people suffered from widespread poverty and an undereducated, undernourished, and underemployed population. Today, although pockets of poverty still exist, the mountains are considered an excellent place to visit, settle and raise a family, and retire.

The federal lands that once were purchased so cheaply have now become an enormous asset to visitors and residents. The increased scarcity of these lands, with respect to the demands placed on them, is now the source of frequent controversy. The primary purposes of the assessment were to determine the condition of the natural ecosystems and to provide data that would help determine how the public land would be used, protected, and managed.

To accomplish the social, cultural, and economic analysis, the team established four subgroups. One would address the human dimensions component. A second would analyze the timber economy. The third would address recreation supply and demand. And the fourth would describe roadless areas and designated wilderness. These four groups evolved from the issues raised at the public meetings and also from the common interests among the study team.

Human Dimensions

Most of the social issues dealt with the pattern of population change and the way this is affecting people's use of natural resources. During the last 20 years, economic development was more rapid in the Southern Appalachian Mountains than in the immediately surrounding areas. As the economy grew, it became more diverse and less dependent on manufacturing. Meanwhile, the proportion of families living below the poverty level decreased from 20% to about 10%. The analysis focused heavily on the changing patterns of economic and population growth. It also examined the values and attitudes of people in the area.

In-migration is a significant component of the region's population growth, and newcomers to the region are dramatically changing the social climate for management of public land. Attitudes about the use of public lands are now influenced by retirees, resort owners, and employees in service industries who are more interested in scenery and recreation than in resource extraction. Today, the feelings of residents within the study area toward natural resources are not much different from those across the nation. Most agree that the Endangered Species Act, the Clean Water Act, and the Clean Air Act are needed to protect the region's ecosystems, but should not excessively restrict people's rights.

Increasing population in all counties, especially low-density residential development associated with retirement and second-home development, is removing forests and pastures as habitat for many species of wildlife and fish. The resulting land-use fragmentation is believed to be adversely affecting animal habitat, especially for some neotropical migratory birds. Landownership fragmentation is adversely affecting the availability of timber from private lands and the opportunities for effective forest management.

About two-thirds of the land in the assessment area is forested and, of this, almost three-fourths is privately owned. To learn more about how and why private landowners manage their land, the assessment team analyzed data from two national questionnaire surveys (the National Private Landownership Study and the National Resources Inventory). From these, they learned that most rural landowners place higher priority on the natural condition of their land than on making money from it. About a third of rural landowners derive income from their land, mainly from agriculture and timber. Recreation is the dominant noncommodity use.

Timber Economy

Logging, sawmilling, and other forms of wood processing have long been an important part of many Southern Appalachian communities. The region's timber economy has been generally stable during the last two decades, although sawlog production (mainly hardwoods) has declined somewhat and pulpwood production has increased. More than 17 species of hardwoods comprise the hardwood sawlog production, but the most important are the oaks and yellow poplar. Real prices for high-quality hardwood sawlogs have risen over the last 20 years, with northern red oak leading the group.

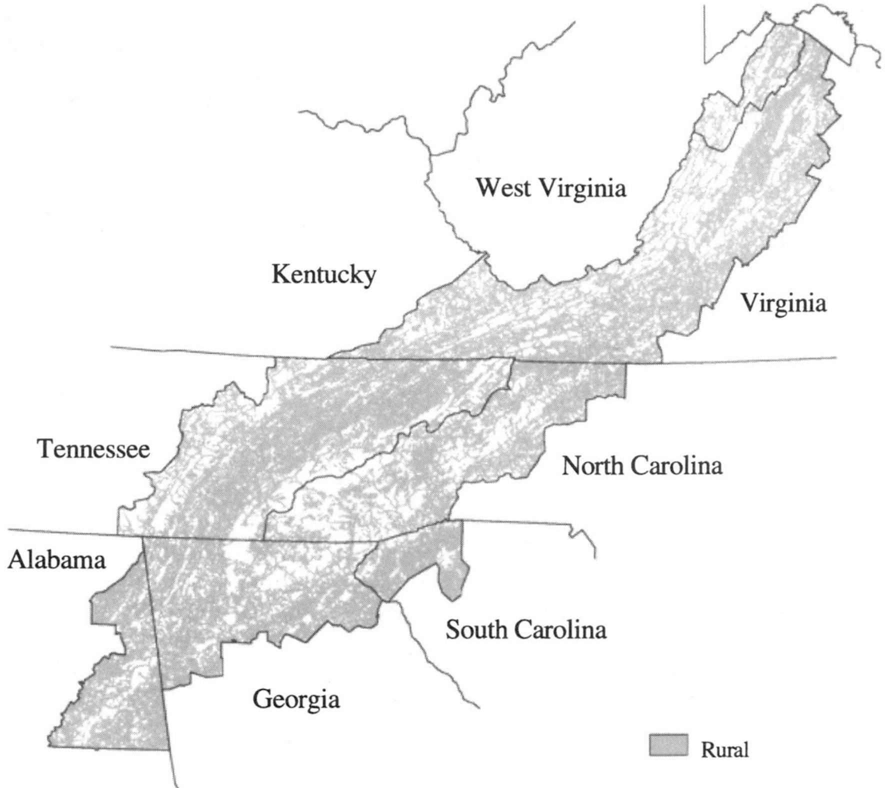

FIGURE 32.7. The recreation analysis focused mainly on rural settings. (Courtesy of the U.S. Department of Agriculture, Forest Service)

Price patterns indicate that high-quality logs are becoming more scarce, although lower quality are abundant.

Economic data for the timber industry have always been fragmentary and sometimes infrequent. The analysis team worked with a number of sources for timber prices, production, and location information. For timber supply, the survey conducted by the Forest Inventory and Analysis Unit of the Forest Service was the principal source.

The assessment analysis was conducted for the region as a whole and for four subregions. The subregions were based mainly on the economic areas defined by the Department of Commerce, Bureau of Economic Analysis. Results for the subregions contrasted important differences within the assessment area. Where possible, county or ranger district data were reported, but economic data at this level were often lacking.

Outdoor Recreation Demand and Supply

The recreation team concentrated mainly on activities in natural settings. Recreation supply was defined as the opportunity to participate in a desired recreation activity in a preferred setting. Although this would not be an economic definition, it was a realistic concept in dealing with public lands where price for participation is not obtainable. Thus the three components of supply were the settings, the activities, and the facilities available to provide support.

Settings are contained within landscapes that provide the physical and social environment that create the recreational experience. By combining the Recreation Opportunity Spectrum (which describes settings by integrating physical, social, and managerial characteristics) with the Scenery Management System (which evaluates natural and cultural influences on the landscape), landscape descriptors were developed with recreation scenery components. The descriptors allowed the team to work with the GIS data base and construct a map of the various settings. Figure 32.7 illustrates one component of the analysis.

Outdoor recreation is increasing in the assessment area partly because of increasing population and partly because rates of participation in most categories have also increased. The most avid participants are white males under 60 years of age, but in recent years participation by females has risen significantly. Several outdoor recreation surveys provided data useful for the analysis, but the data were not always available in sufficient geographic detail to be able to isolate the assessment area. Nevertheless, the trends are sufficiently clear to verify the increased demand already being felt by most land managers in the region.

Approximately 75 visitor surveys were reviewed to determine kinds of experiences that people were seeking when visiting the region's parks and forests. It was no surprise that most wanted to get away from the urban environment and enjoy the tranquility and beauty of a natural setting.

Roadless and Designated Wilderness Areas

The interest in roadless areas and wilderness focused mainly on national forests, because this topic is important during the revision of the forest plans. In previous planning efforts, the Forest Service was criticized because the criteria for roadless designation were not consistent from forest to forest and there appeared to be little coordination among forests. The majority of roadless areas are on National Forests. However, one area is in a national park and four areas are in state parks. The Great Smoky Mountains National Park is the largest single area, making up more than one-third of the total acreage.

There are 39 units of the National Wilderness Preservation System in the assessment region. The Shenandoah National Park is the largest (80,555 acres) and also the only National Park Service unit in the area. Others range in size from less than 100 acres to more than 35,000 acres on Forest Service land.

32.7 Results of the Analysis

The results of the assessment are published in four technical reports and a summary report (see Section 32.9). Each team produced a detailed report that was organized in the same general way. The questions are stated, key findings are highlighted, data methods and analyses are described, and important data gaps and research or monitoring needs are identified. The summary report attempts to capture the essential elements of the technical reports and presents them for a more general readership.

The conclusions of the assessment had both immediate and long-term implications. The key findings have been published in the *Federal Register* as part of the Notice of Intent to revise the forest plans. This is the first time that the plans of five National Forests were coordinated around a set of common issues.

32.8 Innovations Resulting from the Assessment

Analysts working on the assessment started with a broad and somewhat undefined assignment. The teams were made up of specialists representing several agencies and disciplines confronting a relatively large geographic area and a diverse body of public interests. They seemed eager for the opportunity to tackle difficult problems that sometimes challenged their discipline, but were often outside their usual sphere of responsibility.

Nationwide, several regional assessments were underway simultaneously, and we make no claim that the innovations we developed are unique. Other assessment teams probably arrived at similar solutions to problems that they faced, but there was little communication between the regional assessment efforts. The only effective means of sharing information between these regional efforts was through the Ecosystem Management Staff of the Forest Service in the Washington Office.

The widely varied initial conditions that exist for each assessment make each project fundamentally different. Consequently, every assessment must take an approach that is tailored to these conditions. In the Southern Appalachian region, the existence of the SAMAB cooperative was one of these initial conditions. SAMAB offered the Forest Service an opportunity to enlist the participation and support of an established network of cooperating federal and state agencies.

Rallying the cooperation of numerous agency partners requires compromise. Objectives must be negotiated with all the participants. The scope and time schedules need to be compatible with the human and financial resources that are being committed. During the course of the assessment, numerous opportunities emerged to expand our efforts and take advantage of data sources or partnerships that we simply did not know existed when we outlined the initial objectives. Each of these new possibilities had to be evaluated to see how it might add to the assessment and whether it would be consistent with our time and content constraints.

Projects such as the assessment involve learning as you go. The majority of the participants had no prior experience with regional assessments. Those that did, helped and encouraged the rest. Flexibility must be maintained throughout the process, because many of the problems cannot be foreseen. For example, data sources may not materialize, and analytical models sometimes do not work satisfactorily. Usually, these problems can be solved or bypassed as alternatives are considered.

The assessment teams consisted of interagency and, to a large extent, interdisciplinary specialists. The effectiveness of these groups depended on a team dynamic that encouraged participation and consensus. The Policy Team allowed the technical teams broad latitude in the hope of encouraging innovation and fostering better teamwork. The work was accomplished in a very dispersed organizational environment. That is, everyone who worked on the assessment remained at his or her established work location except for periodic working sessions. This arrangement required considerable emphasis on communication, but it also allowed people to maintain job and family obligations.

Regular briefings for agency leaders and the public were also a key part of the process. The briefings reassured line officers that the work was moving ahead as planned and gave everyone an opportunity to offer advice and suggestions.

Public involvement is more than just public relations. We were committed to include the public as much as possible in all phases of the assessment effort. A good-faith response to public involvement and customer comment meant that the project's objectives must be sufficiently flexible to allow refocus or redirection to accommodate public input. The guiding questions for the assessment were modified and embellished as a result of public involvement, agency comment, and input from the technical teams. The only factor limiting public participation was restrictions imposed by the Federal Advisory Committee Act. Even these limitations on the conduct of public meetings were largely overcome, and there was considerable benefit from the fact that any perception of restricting public access to the process was avoided. The result was an environment that facilitated open dialogue and participation between the public and the assessment teams.

The form of public participation ranged widely. On some occasions, people were only interested in satisfying their curiosity or seeing that their personal interests were not being threatened. But, more often, people were interested in the larger issues involving the welfare of the region and its resources. Meetings were held at locations throughout the region so that people could attend without undue hardship. Public participation ultimately provided many hours of very productive work in support of the assessment.

The benefits of public involvement are hard to evaluate, but were substantial. On two occasions, the assessment leaders were asked to testify before congressional subcommittee hearings regarding regional assessments. These committees were generally critical of regional assessments and the implications they might have for private landowners and public land management. At both hearings, members of the public testified in favor of the Southern Appalachian Assessment and the benefits it would provide in support of better resource management. This testimony did much to defuse hostility toward the assessment.

Data resources can be both an asset and a burden. The huge amount of data available for and requested by the assessment analytical team was an organizational challenge for the Data Management–GIS Team. A detailed listing of all maps and most data sets is available on the Internet through the SAMAB on the Forest Service home page.

The technical analysis produced many innovations. Most involved imaginative uses of GIS technology, such as the bear habitat suitability map or the map of recreation settings. Others involved constructing models that allowed extrapolation of the often limited data sources. The Atmospheric Team, for example, modeled the distribution of mean, wet sulfate loadings. In this analysis, weighted least-squares regression techniques were used to take into account the influence of regional topography on deposition.

The assessment contains numerous examples of integrated analysis, that is, joint analyses between resource elements. The analysis of atmospheric deposition effects on aquatic resources conducted in response to a specific question is an illustration. Because of the way that the teams were organized (separate teams for the major resources), other opportunities were sometimes overlooked. In an effort to facilitate more integrated analysis, the Policy Team organized a special workshop to explore the possibilities. Conducting integrated analyses involving the separate teams proved to be more difficult than anticipated. By the time this difficulty was recognized, the teams were well advanced in their separate efforts. This limited the time available to pursue integrated analysis opportunities, but it also meant that each team had well-developed databases and analyses that they could share with the other teams. The workshop facilitated several discussions of interacting resource influences that are included in the reports. In all, we believe that the integration that was accomplished was a good first step. It certainly highlighted opportunities for future work and exposed some of the limitations of the data.

32.9 Conclusion

The assessment was not the first for the region and we hope it will not be the last. As in 1901, when the first Southern Appalachian Assessment was completed, people had expressed a concern about the condition of the region's natural resources. They wondered about the health of the forest, the quality of the water, and the effects of population growth and agricultural practices. Our assessment attempted to answer these and many other questions and to show where improvements are needed. Unlike the conditions at the turn of the century, conditions today are much improved. Now the Southern Appalachians are regarded as a valuable asset that can supply the people of the United States with places to live, play, and produce essential commodities.

The Southern Appalachian Assessment began in the fall of 1994 and was completed in the spring of 1996. Although it was broad and comprehensive, it was aimed at specific questions that arose during a public outreach process. The schedule for completion was tightly maintained so that the results would be timely and relevant. This sense of urgency probably helped to sustain enthusiasm and maintain involvement. Although this approach might have limited the quality of data or the depth of analysis, we tried to make allowances for these weaknesses by highlighting opportunities for additional research or data collection.

We believe that regional assessments are not an end in themselves, but a continuing process that fosters interagency cooperation and professional networking. Maintaining a quality living environment will require hard work on the part of both public and private landowners. It will require the dedicated assistance of scientists and educators. We hope that the assessment will facilitate planning and good management and serve as the basis for continuing study.

The Southern Appalachian Assessment is presented in five separate reports as follows:

Hermann, K. A., editor. 1996. *The Southern Appalachian Assessment GIS Data Base CD ROM Set.* Southern Appalachian Man and the Biosphere Program. Norris, TN.

1. Southern Appalachian Man and the Biosphere (SAMAB). *The Southern Appalachian Assessment Summary Report. Report 1 of 5.* Atlanta, GA: U.S. Department of Agriculture, Forest Service, Southern Region.
2. Southern Appalachian Man and the Biosphere (SAMAB). *The Southern Appalachian Assessment Aquatics Technical Report. Report 2 of 5.* Atlanta, GA: U.S. Department of Agriculture, Forest Service, Southern Region.
3. Southern Appalachian Man and the Biosphere (SAMAB). *The Southern Appalachian Assessment Atmospheric Technical Report. Report 3 of 5.* Atlanta, GA: U.S. Department of Agriculture, Forest Service, Southern Region.
4. Southern Appalachian Man and the Biosphere (SAMAB). *The Southern Appalachian Assessment Social/Cultural/Economic Technical Report. Report 4 of 5.* Atlanta, GA: U.S. Department of Agriculture, Forest Service, Southern Region.
5. Southern Appalachian Man and the Biosphere (SAMAB). *The Southern Appalachian Assessment Terrestrial Technical Report. Report 5 of 5.* Atlanta, GA: U.S. Department of Agriculture, Forest Service, Southern Region.

33 Northern Forest Lands Case Study

Charles A. Levesque

33.1 Background

In 1988, a portion of the northeastern U.S. forest lands of the former Diamond International Corporation, a timberland and forest products manufacturing company, were put up for sale. Of the 1 million acres owned by Diamond in Maine, New Hampshire, Vermont, and New York, just over 200,000 acres were marketed for development purposes, while the rest went to other large timberland owners.

This act caused much consternation among those in the forestry and environmental communities in the Northeast, because the many public values associated with these private lands were perceived to be at great risk. The threat to these values caused Congress and the states to embark on a six-year, \$4.5 million effort to develop public policy recommendations to address the potential loss of forest values in the Northeast. The area of focus was the northern regions of the four states, a 26 million-acre forested region that came to be known as the Northern Forest.

The final products of this project were two reports that resulted from the two major phases of the effort: the Northern Forest Lands Study and the Northern Forest Lands Council. The first report, the *Northern Forest Lands Study* (1990), was published in 1990; the second, a report entitled *Finding Common Ground: Conserving the Northern Forest—the recommendations of the Northern Forest Lands Council* (1994), was published in late 1994. A third less distinct phase is being carried out in the four states as implementation of the recommendations from the council moves forward. To develop the public policy recommendations from these efforts, assessments were designed and completed in both phases of the effort. This case study focuses on these assessments.

33.1.1 Northern Forest Lands Study

The Study was created through congressional authorization and \$250,000 of funding provided to the USDA Forest Service in 1988 through Public Law 100-446. A second year of funding followed. Along with the study process, directed by a team of Forest Service employees, the governors of the states of Maine, New Hampshire, Vermont, and New York each appointed 3 members to a 12-member Task Force that worked in concert, as a volunteer advisory committee, with the Study effort.

The direction given to the Northern Forest Lands Study can be found in language from the Joint Explanatory Statement of the [Congressional] Committee of Conference for the law, which stated, in part:

> Changes in forest land ownership in northern New England and New York are leading to increased subdivisions, development of large tracts of lands, and a loss of traditional economic and recreational uses of these lands . . . unless prompt action is taken, there may be an irreparable loss of these forest resources. . . . The Study shall include identification and assessment of (1) forest resources including, but not limited to, timber and other forest products, fish and wildlife, lakes and rivers, and recreation; (2) historical land ownership patterns and projected future land ownership, management, and use; (3) the likely impacts of the changes in land and resource ownership, management, and use of traditional land use patterns including economic stability and employment, public use of private lands, natural integrity, and local culture and quality of life; and (4) alternative strategies to protect the long-term integrity and traditional uses of such lands. . . .

During its nearly two-year time frame, the Northern Forest Lands Study and Governors' Task Force on Northern Forest Lands conducted a natural–socioeconomic resource assessment through direct research undertaken by staff, hired contractors, and expert working groups. Resulting from this effort was a vision for the future of the Northern Forest, as well as an assessment addressing patterns of landownership; changes occurring in the ecological, economic, and social constructs of the Northern Forest; forest and related land-based natural resources; and a socioeconomic profile of the people of the Northern Forest. Also contained in the Study's final report was an identification process for land with important resources and a series of potential strategies to address the changing nature of the land ownership and uses occurring in the Northern Forest.

33.1.2 Northern Forest Lands Council

Along with the *Northern Forest Lands Study* report published in 1990, the 12-member Governors' Task Force on the Northern Forest Lands published a complementary report (Northern Forest Lands Council, 1990) that recommended a continuation of the process begun by the Study, that is, to dig much deeper into the issues uncovered during the Study process with a much more intensive assessment. The recommended four-year effort was authorized in the 1990 Congressional Farm Bill under the Forestry Title and was funded through the USDA Forest Service in federal fiscal years 1992 through 1994 for a total of just under $4.5 million. Over $3.2 million of these appropriations went directly to the four states for Geographic Information System (GIS) resource analysis work and planner staffing.

The Northern Forest Lands Council included four individuals appointed by the governors from each of the four states. These appointees represented environmental, forest industry, state government, and local community interests. A seventeenth member, from the USDA Forest Service, was appointed by the chief of the Forest Service. A staff of three managed the Council affairs in concert with a state-based forest planner in each of the four states.

During its nearly near four-year time frame, the Northern Forest Lands Council conducted its resource assessment through direct research undertaken by staff and hired contractors, through expert working groups, through citizens advisory committees, and through informal means. A number of distinct products resulted from this effort.

1. The report *Finding Common Ground: Conserving the Northern Forest*
2. Technical appendix of research and studies
3. Compilation of GIS-based natural resource and economic information by the states
4. Summary of comments on the above from thousands of people
5. New approach to consensus building among disparate views

33.2 Issues Addressed by the Assessments

33.2.1 Northern Forest Lands Study

The primary reason the Study was created by Congress and the states was due to the concern that the change of ownership and use of areas of forest land in the 26 million-acre region would adversely affect the many public values associated with these lands: recreational access, wildlife habitat, timber, clean water, and others. The assessment developed by the Study team, then, focused on three specific areas related to this original concern: (1) reasons for significant change in land ownership and use, (2) critical natural resources at risk, and (3) land conservation tools and strategies.

Reasons for Significant Change in Land Ownership and Use

Several studies were undertaken to address the perceived issue of changing land ownership and use in the Northern Forest area. A broad overview of land use from pre-European settlement to present was written by in-house staff. A series of case studies looking at land-use changes in one county in each of the four Northern Forest states was contracted to a private research firm. Finally, a series of 25 interviews was conducted by Study staff with representatives of major forest holding companies and individuals in the Northern Forest areas of the four states.

Critical Natural Resources at Risk

To determine which natural resources were at risk due to the changing ownership and use patterns occurring in the Northern Forest, a set of natural resource criteria was developed by the Governors' Task Force and Northern Forest Lands Study staff with input from interested members of the public. Using the GRANIT GIS system in New Hampshire, these criteria were translated into a graphic interpretation under a pilot project for Coos County in New Hampshire. The major resource criteria categories developed included wildlife habitat, scenic areas, river corridors, recreation opportunities, productive forest land, natural areas, lakeshore, and large contiguous blocks of forested land.

Land Conservation Tools and Strategies

After intensive review from expert work groups and the public engaged in the Study process, the Northern Forest Lands Study staff and Governors' Task Force on Northern Forest Lands identified a series of potential strategies that could protect the resources identified in the previous section. In six categories, the 28 strategies and tools covered the full range of those in use in the western world. The major subjects included the following:

1. Land-use controls and planning for conservation
2. Easements and land purchase for conservation
3. Maintaining large tracts of private forest land through incentives
4. Combining community improvement with land conservation
5. Keeping private land open to the public
6. Combining strategies in a coordinated program

33.2.2 Northern Forest Lands Council

Once the Northern Forest Lands Study completed its work, the companion Governors' Task Force on Northern Forest Lands recommended the continuation of the process through the four-year Northern Forest Lands Council process. The charge to the group was essentially the same given to the Study by Congress and the governors (outlined in a paragraph in the 1990 Farm Bill), but also to delve much deeper into the issues than the Study had the time or opportunity to do.

Second, the Council's charge ultimately was to result in a set of policy recommendations to Congress and the governors on strategies that should be employed to address the challenges that the region was facing. This was very different from the Northern Forest Lands Study process, because the Study was explicitly prohibited from recommending strategies to Congress and the states; its charge was to simply list out all possible strategies. The Council process was designed to determine which of the strategies were most important to use (and screen for ones missed in the Study process), based on the knowledge that a second-level assessment would bring to the process.

The Northern Forest Lands Council assessment focused on two levels, regional and state. At the regional level, the Council directed its assessment to seven issue areas: (1) land conversion, (2) biological resources, (3) conservation strategies, (4) local forest-based economy, (5) property taxes, (6) recreation and tourism, and (7) state and federal taxes.

At the state level, efforts were focused on a GIS data layer building process under a consistent set of standards developed jointly by the states with the assistance of a contracted GIS contractor. Called the Northern Forest Resource Inventory, these efforts were funded by Congress directly to the states, not the regional Council body. The Council served a coordinating function for the inventory. Data layers built into the state GIS systems through the process included the following:

- Land–habitat cover
- Large blocks of forest land
- Political subdivisions
- State-regulated wetlands
- Publicly owned fee or interest in lands
- Privately owned fee or interest in lands by private conservation organizations
- Transportation system: roads
- Transportation system: railroads
- Transportation system: airports
- Hydrography
- Shoreline development
- Recreational opportunities
- Historic sites
- Population
- Utility corridors and nodes
- Elevations
- Forest products manufacturing sites
- Northern Forest Resource Inventory Study
- Energy production facilities
- Boundary

33.3 Scale, Extent, and Resolution of the Assessment

Necessarily, the assessments completed for the Northern Forest Lands Study and Northern Forest Lands Council occurred at several scales and resolutions.

33.3.1 Scale

The Northern Forest Lands Study scale was a four-state 26 million-acre forested area known as the Northern Forest and more particularly described by the fact that the area's land ownership patterns were dominated by large contiguous ownerships of land, mostly privately held. The Northern Forest Lands Council efforts also focused on this multistate scale but, under the Northern Forest Resource Inventory, the assessment occurred at the substate level, concentrating on the area of the state within the Northern Forest region.

33.3.2 Extent

Northern Forest Lands Study

Many issues were studied and each was reviewed to the extent possible, given the resources available and the time frame allowed for the project. Several examples illustrate the issue of extent. The review of the patterns of land ownership, utilizing data primarily from the USDA Forest Service's Forest Inventory and Analysis and privately collected data, the assessment virtually covered every acre of the 26 million-acre region, yielding sufficient delineation to show how many acres were owned by industry and how many by nonindustrial owners and how much was in the public sector.

On another issue, that of the extent to which forest acreage in the Northern Forest was changing in use by development and other scenarios, the Study was only able to undertake a minimal case study of several counties and towns to attempt to demonstrate the changes taking place. The results here were speculative at best, but provided an insight into issues not previously analyzed due to the large multistate area involved.

Northern Forest Lands Council

The Council's charge was to build on and further the work developed through the Northern Forest Lands Study. For the base data provided by the Study, such as the patterns of land ownership, the extent of data collected was adequate. On other areas, most particularly the changing land uses study, the Council's work broadened the extent of analysis considerably. A separate town by town analysis of changing land uses for the years 1980 to 1991 of all parcels over 500 acres in size was initiated, and a less vigorous review of existing data on changing uses on smaller parcels was also included.

Additionally, the Northern Forest Lands Council furthered work begun in the Study phase once it was identified as important to the results of the effort. The extent of analysis of the following areas was broadened through the Council assessment: federal taxes, property taxes, biodiversity, recreation and tourism, and the forest-based economy of the Northern Forest.

Private Lands Issue

The Northern Forest Lands Study and Council processes' assessments looked primarily at private forest land issues. As a result, many data sources were inadequate. Good data on the hundreds of thousands of private forest owners and their lands in the Northern Forest is simply not available. In some cases, such as the work accomplished to analyze the changing land uses in the Northern Forest, primary data were ultimately collected because the issue was so fundamental to the effort. On others, such as a review of the forest harvesting practices taking place relative to water resources in the Northern Forest, a recommendation for further study was all that resulted. Private forest land information and data availability ultimately limited the extent of the assessments in both phases.

33.4 Major Types of Data and Information Used in the Assessment

Both public and private data sources were tapped for the multitude of research studies and case studies completed for the Northern Forest Lands Study and Council efforts. As can be expected, existing data and information were always the first source. For a number of issues, a simple literature search of existing data or information was as far as the process went. This was done for research under the conservation strategies issue area of the Northern Forest Lands Council, as well as others. In the other extreme, raw data were collected by a research contractor under the Northern Forest Lands Council's land conversion issue area to determine how much actual land conversion had occurred in the Northern Forest. Primary data were also collected through a written landowner survey in the current use assessment study completed by a contractor for the Northern Forest Lands Council Property Taxes issue area subcommittee.

33.4.1 Published Data Sources Utilized

Many public data sources were tapped for the Study and Council process. The most well known sources were the USDA Forest Service's Forest Inventory and Analysis (FIA); this included the multitude of related reports on landowners, the forest products industry, and other topics; state-based Natural Heritage Inventory data, threatened and endangered species information along with natural communities data; and state-based records on property taxes and current use assessment statistics.

Less known sources, such as state-based Outdoor Recreation Plans, were also tapped. Many sources were not public sources in the strictest government sense, but ranged from little-known unpublished masters degree theses to well-known studies developed by nongovernmental organizations. It is impossible here to list even a small portion of the data and information sources used, given the size of the list. The Northern Forest Lands Study's selected sources listing stretches for five pages, while the Northern Forest Lands Council Technical Appendix was a hardbound document 3 inches thick. It contained bibliographies covering well over 100 pages.

33.4.2 Case Studies Developed Specifically for the Study and Council

Given the enormity of studying a 26 million-acre, four-state area of land and public policy issues that were national in scope, it was necessary to develop case studies as an alternative means for understanding issues and developing findings relevant to the effort. The largest case study completed for the Northern Forest Lands Study was that of changing land use, for which a county and, more specifically, a town within the county were studied for each of the Northern Forest states.

Under the Council effort, a report entitled *New Directions in Conservation Strategies: A Reconnaissance of Recent Experimentation and Experience* (1993) reviewed cases of successful land conservation strategies employed throughout the United States in areas with similar forested regimes.

33.4.3 New Data Collected by National Forest Lands Council

The Northern Forest Lands Study relied primarily on existing research and commissioned a few reports for its uses. The Northern Forest Lands Council relied on very few existing sources (except the work of the Study itself) and commissioned no fewer than 29 original reports and studies for its assessment. In addition, the four state GIS offices compiled the data to create 20 data layers for the Northern Forest Resource Inventory project.

The largest new data source created by these efforts was the collection of primary data from towns and counties on the changing land use, parcel by parcel, which was accomplished through the Land Conversion Study commissioned by the Council. A private firm compiled a large database of information from this effort.

33.5 Major Types of Analyses Performed in the Assessment

33.5.1 Northern Forest Lands Study

Analyses accomplished through the Study focused on three areas: (1) reasons for significant change in land ownership and use, (2) critical natural resources at risk, and (3) land conservation tools and strategies. In all these analyses, two methods were used in a reasonably consistent fashion. Step 1 was to have research compiled or created either inhouse with staff or through outside contracted sources. In all cases, the staff of the Study, with assistance from the Governors' Task Force on Northern Forest Lands, chose the research subjects based on the charge given the project by Congress. Following completion of draft reports and compilations, expert volunteer working groups were assembled to critique the draft reports and recommend revisions to the authors. Conclusions from these analyses followed and were drawn primarily by staff of the Study.

33.5.2 Northern Forest Lands Council

Research and analyses in the Council process centered on seven major issue areas previously cited. An eighth area, the Northern Forest Resource Inventory, functioned somewhat differently from the seven major issue area efforts. The former operated with only a core group of GIS specialists as the designers of the data compilation effort.

For the work of the seven major issue areas, the analysis process used included research, expert working groups, and public dialogue. More specifically, the analysis followed the following steps.

1. Identification of study topic by issue area work group (a combination of the Council members, experts and advocates, all volunteers).
2. Staff development of request for proposals on subject or outline of topic needs, which was reviewed and approved by the work group.
3. RFPs were issued or the topic outlined was used in-house by staff (more commonly the former).
4. Contractors were selected and hired for tasks (or staff did work).

5. Draft reports on research topics were developed and reviewed and critiqued by work group and the state by state citizen advisory committees that had been formed.
6. Final reports were written on the topic.
7. With minor assistance from work group members, Northern Forest Lands Council subcommittee members developed findings from the work accomplished under the research. The findings represented the most relevant results drawn from the research.
8. All the findings were published along with a series of public policy options that were developed directly from these findings by the Council members with assistance from staff. These options represented every possible public policy change that might be made at the state and federal level to address the changing land-use trends in the Northern Forest. The resulting document was called the *Findings and Options*.
9. The public at large was invited to critique the *Findings and Options* book.
10. The findings were finalized based on public comment.
11. The draft report of the Council's recommendations was published for final public scrutiny. In addition to written comments sought on the report, a series of 26 public listening sessions (hearings) and open houses were held throughout the Northeast to gather further input on the report.

33.6 Results of Analyses

33.6.1 Northern Forest Lands Study

After nearly two years of study, the Northern Forest Lands Study's final product centered on a series of 28 potential strategies to conserve the Northern Forest. The background reports and research analyses were also included in the final report of the effort and proved as useful as the strategies themselves. The strategies resulting from the analyses of research were not recommendations. By design from the congressional charge to the Study, the 28 conservation strategies were intended to represent the full range of possible public policy changes that policy makers might look to in order to develop policies at the state and federal level conducive to the long-term conservation of the Northern Forest area.

The Governors' Task Force on Northern Forest Lands took an additional step beyond that of the Northern Forest Lands Study. Not constrained by the congressional limitations placed on the Study process, the Task Force recommended establishment of Northern Forest Lands Council with a four-year life-span to further the work begun by the Study.

33.6.2 Northern Forest Lands Council

The result of the Council analysis was a series of recommendations for action: specific public policy recommendations, some federal (8), but most at state level (29).

33.7 Innovations Resulting from the Analyses

33.7.1 New Models of Active Public Participation

Both the Northern Forest Lands Study and Council processes broke new ground in approaching natural resource assessments by developing new active venues for public participation at many levels in the processes.

1. The primary innovation in the Study process was in the invitation for active participation of landowners in defining the problem. A series of 25 in-depth interviews formed the initial contact to engage those owners whose lands were being studied.
2. The Council process for engaging the public is still being cited as *the* standard by which all other public participation processes are being judged in natural resource efforts around the Northeast. Active public participation was sought in developing research topics, reviewing draft results, and formulating recommendations.

To accomplish the enormous tasks set out in the Northern Forest Lands Study and Council charges, new approaches were needed to leverage the resources adequate to the task. The assessments of these processes necessarily used partnerships at all levels, drawing on volunteer arrangements from scientists and nonscientists alike, to develop products in an efficient and self-critiqued fashion.

33.7.2 Continuous Assessment

The immediate need for public policy answers in the Northern Forest situation resulted in assessment processes being developed that recognize that policy needs were driving the process and that timeliness was essential to realize an outcome that could be utilized in the real-time, ever-changing political world. As a result, the efforts used the following:

1. Assessment issues, scale, and extent were determined by the immediate situation.
2. Timely information provided to decision makers for the problems at hand was a paramount outcome.
3. Improved information on complex issues was developed as the needs emerged. That is, products could not be static in their design, and quickly shifting gears during the process to address immediate situations prevailed.

33.8 Conclusions

Assessments driven by immediate public policy concerns need to be responsive to both the political and scientific needs of the situation. Involving the public and outside experts in assessment processes from the early stages will likely result in greater buy-in on the results derived from these assessments.

At least one major flaw existed in each of the assessment processes developed for the Northern Forest Lands Study and Northern Forest Lands Council. In the Study, although interested publics were clamoring to become engaged in the efforts from the start, only a few formal opportunities existed whereby those interested could become involved in a way that provided ownership of the assessment and the results drawn from that assessment. More active opportunities for involvement of interested public were needed from the start.

In the Northern Forest Lands Council process, the GIS-based Northern Forest Resource Inventory process was nearly completely disjointed from the process that it was intended to serve. Several simple flaws developed early on that made it practically and politically impossible to remedy:

1. The design of the GIS process was derived from an anticipated Council process and direction that never materialized (i.e., the process went in a direction that was very different from that envisioned by those who designed the original Inventory template).
2. Congress funded the effort one year too early so that the Northern Forest Lands Council never had the opportunity to design the GIS tool functions for the effort in ways that would make it relevant to the Council effort. The states' interest in having federal dollars available for GIS data layer construction overwhelmed the specific information and data needs of the Council process. The lesson is simple: Be careful in the use of GIS for natural resource assessment purposes. Careful design based on specifically identified data and information needs must preclude any data layer development. Do not put the cart before the horse.

How do we measure the success of the output from natural resource assessments? Surely, one criterion should be whether the assessment results in continued use of the project output over time. In the case of the Northern Forest Lands Study and Council's six-year effort, the results of the efforts have stood the short test of time to date, and efforts continue at the state and federal levels to implement the recommendations for public policy changes that resulted from the effort.

33.9 References

Northern Forest Lands Council. 1990. *Report of the Governors' task force on the Northern Forest lands*. Waterbury, VT: Northern Forest Lands Council.

Northern Forest Lands Council. 1993. *New directions in conservation strategies: a reconnaissance of recent experimentation and experience*. Waterbury, VT: Northern Forest Lands Council.

Northern Forest Lands Study. 1990. *Northern forest lands study*. Waterbury, VT: Northern Forest Lands Study.

Northern Forest Lands Study. 1994. *Finding common ground: conserving the Northern Forest—the recommendations of the Northern Forest Lands Council*. Rutland, VT: U.S. Dept. Agric., For. Serv., Northeastern Region.

34 Interior Columbia Basin Ecosystem Management Project

Russell T. Graham

34.1 Introduction

The economic and cultural development of the United States is closely associated with the use and consumption of natural resources. Water, wood, and forage were the primary resources facilitating the western expansion of the United States (Malone et al., 1991; Schwantes, 1991; Robbins and Wolf, 1994). Wood was used in large quantities for shipbuilding early in the 19th century and later used to fuel and build railroads (Hutchison and Winters, 1942; Schwantes, 1991). As the West developed, shrub and grasslands supported domestic livestock grazing. Also, some lands were recognized for their crop potential and irrigation dams were constructed early in the 20th century (Wilkinson, 1992; Dietrich, 1995). In all cases, resource use required inventories, descriptions, or some other assessment that described the amount, location, or value of the resource and often how it could be utilized (Leiberg, 1897). Assessments in one form or another have always been part of natural resource use.

During the western expansion era, the intensity and size of natural resource assessments depended on the questions asked and the resource evaluated. Assessments ranged in scope from deciding where to graze a flock of sheep or harvest timber to locating proposed state boundaries (Schwantes, 1991). Similarly, today assessments are being conducted to help make decisions about natural resource use and management. Often, these assessments describe site-specific activities such as proposed timber harvests, grazing allotments, or road locations (Quigley et al., 1996; Hann et al., 1997). But frequently the effects of resource decisions transcend jurisdictional, ecosystem, or assessment boundaries. Issues and concerns about resource use are not only important locally, but often they are regionally and nationally significant (Haynes et al., 1996; Quigley et al., 1996). In addition, a wide range of laws and policies, from local to national, affect resource management (NEPA, 1970; FLPMA, 1976; NFMA, 1976). Therefore, to make more informed decisions on the use and management of natural resources, assessments need to address multiple issues across various spatial and temporal scales.

Natural resource use is often closely associated with life-styles and community stability (McCool et al., 1997). Logging, ranching, mining, and tourism are only a few of the industries that depend heavily on natural resources. Also, the resources of the western United States are not only a concern to the local people, but are also important nationally and even internationally (Noss and Cooperrider, 1994; McCool et al., 1997). At the local level, most National Forest Plans issued in 1989 and 1990 were scheduled for revision in the late 1990s and early 2000s. Nationally, the harvesting of old-growth timber was an important issue, as was the use of water. In response to concerns on the use of grass- and shrublands for grazing by domestic livestock throughout the western United States, the Bureau of Land Management started an initiative on rangeland reform. A strategy to conserve Pacific salmon throughout its range in Oregon, Washington, Idaho, and California was developed. This anadromous fish habitat and watershed conservation strategy (PACFISH, 1994) was to last 18 months. During this same era, the conservation strategy for the northern spotted owl was released and the Blue Mountains Forest health report was delivered. The President's Forest Plan was presented, as was the Forest Ecosystem Management Team (FEMAT, 1993) report. In addition, it became apparent in these efforts that cumulative effects and ecosystems transcend administrative, state, and other political boundaries.

In addition to all the above concerns, jobs, commodities, and industries were rapidly changing.

These changes were brought on by immigration of industry and people into the Northwest and the decrease in resource flows, primarily timber and forage from public lands (Haynes and Horne, 1997; McCool et al., 1997). Through interpretations of the Endangered Species Act and Forest Planning regulations, litigation and administrative bottlenecks brought gridlock to the management and use of natural resources in the Pacific and inland Northwest (Lee et al., 1997; McCool et al., 1997). Law suits challenging natural resources management were abundant; communities and industries dependent on natural resources saw their way of life threatened (Haynes and Horne, 1997; McCool et al., 1997). In addition to these political and administrative issues affecting natural resource management, the eastside forests were experiencing accelerated disease and insect epidemics, along with frequent, large stand-replacing wildfires that destroyed homes and lives and threatened communities. Also, it was apparent that many grass and shrublands on the eastside were changing because of human use and the invasion of exotic species (Hann et al., 1997).

To address these local to national issues on natural resource management, President Clinton convened a forest conference in Portland, Oregon, in the spring of 1993 to discuss the state of the forests, economy, and people of the Northwest. During the conference the president issued a statement: "Management of eastside forests will need to focus on restoring the health of forest ecosystems impacted by poor management practices of the past. . . ." The president directed the U.S. Department of Agriculture, Forest Service (FS) to develop a scientifically sound and ecosystem-based strategy for management of eastside forests (Quigley et al., 1996). The president went on to state "that the strategy should be based on the forest health study recently completed by agency scientists as well as other studies." This direction provided the impetus for the Interior Columbia Basin Ecosystem Management Project (ICBEMP), and, starting in the fall of 1993, the beginnings of the Columbia River Basin Assessment were born.

34.2 Interior Columbia Basin Ecosystem Management Project

In January 1994, following the president's direction, the chief of the Forest Service (FS) and director of the Bureau of Land Management (BLM) dictated that an ecosystem management framework and assessment be developed for lands administered by the FS and BLM on those lands east of the Cascade crest in Washington and Oregon within the Interior Columbia River basin. The effort would produce a science-based framework and assessment and a scientific evaluation of alternative management strategies. These products would be completed in the next 9 to 12 months. In addition, the charter for the ICBEMP outlined a process for using these products to develop an environmental impact statement (EIS) for the management of BLM- and FS-administered lands in eastern Oregon and Washington to be completed in spring or summer of 1994 (Haynes et al., 1996).

34.2.1 Project Organization

The executive steering committee for the ICBEMP included the regional forester from the Pacific Northwest Region, director of the Pacific Northwest Research Station, and the state BLM directors for Washington and Oregon. This group provided overall direction, as well as funding and personnel priorities. A management team composed of FS and BLM personnel was assembled that included an overall manager and assistant, science integration team (SIT) team leader and deputy, and an EIS team leader and deputy. The team was located in Walla Walla, Washington, near the border of Washington and Oregon in the heart of the eastside.

The SIT was charged to examine the historical and current ecological, economic, and social systems and discuss probable outcomes if current management practices and trends were to continue. This task would be accomplished by completing an interior Columbia River basin assessment encompassing lands in western Montana, Idaho, eastern Washington, and eastern Oregon (Figure 34.1). All lands of this broad area were to be assessed, which would provide context for a more in-depth assessment or mid-scale assessment of the lands in eastern Oregon and Washington. The basin-wide assessment would include the following:

1. Landscape, economic, cultural, and social characterization
2. Identify the probability that change may occur in the components of diversity (landscape, ecosystem processes and functions, species)
3. Identify social, cultural, and economic systems
4. Identify emerging issues that relate to ecosystem management within the basin
5. Identify the social and cultural values of natural resources
6. Identify technological gaps, research needs, and opportunities to advance the state of knowledge

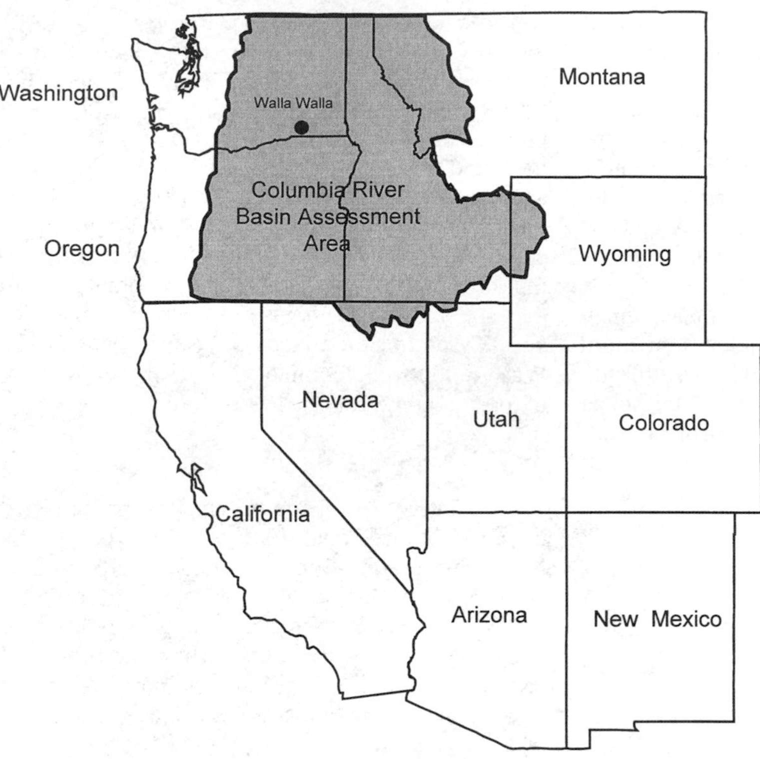

FIGURE 34.1. Interior Columbia Basin Ecosystem Management Project assessment area.

The SIT consisted of landscape ecology, terrestrial resources, aquatic resources, economics, and social sciences teams. Support for the teams included a geographic information system (GIS) team and administrative and public information staffs. An important component of the assessment was a liaison for Native American tribes.

Because the assessment would provide information for federal agency decision making, all project members were required to be federal employees according to the Federal Advisory Committee Act (all committees advising government agencies must be officially sanctioned or be federal employees). Therefore, federal employees were assembled from the USDA Forest Service, USDI Bureau of Land Management, US Bureau of Mines, USDI National Biological Service, and the US Geological Survey. In addition to the personnel in Walla Walla, Washington, detached units were located in Missoula, Montana, Boise, Idaho, Wenatchee, Washington, Seattle, Washington, and Portland, Oregon to name a few. In addition to the teams themselves, the project relied heavily on contractors to supply information about many ecosystem components.

Based on the issues facing land managers throughout the basin, a variety of policy questions were formulated to guide the individual staff areas when completing their individual assessment components. Examples of these policy questions include the following:

1. What are the effects of current and potential Forest Service and Bureau of Land Management land allocations on ecological, economic, and social systems in the basin?
2. What is required to maintain long-term productivity (in terms of various systems)?
3. What is required to maintain sustainable or harvestable or minimum viable population levels?
4. What is required to maintain and restore biological diversity (biodiversity)?
5. What management actions will restore and maintain ecosystem health?

Using these questions as a guide, the SIT was charged to conduct an integrated assessment of the basin using broad-scale information and an assessment of the lands in eastern Oregon and eastern Washington using mid-scale information. Each staff area prepared a plan to complete the assessment, framework, and evaluation of alternatives in a 12-month period.

34.2.2 Assessment Challenges

No sooner had the charter been developed, assessment plans completed, and time tables established, than additional forces started to influence the assessment. The central portions of Idaho contained tributaries of the Columbia River that were included in PACFISH. These guidelines were expiring 18 months from when issued, requiring land management agencies to adopt permanent direction before they expired. To develop these guidelines, the FS needed a mid-scale assessment of central Idaho and asked the SIT to provide this information. Just as the assessment commenced, the scope of the project was enlarged and assessment plans were modified and time lines adjusted. Even with these changes, optimism in the SIT prevailed that the products would be completed on time and within budget.

Once plans were revised to supply mid-scale information for PACFISH in central Idaho, the project changed again when Forest Plan revisions on National Forests in Idaho and western Montana became an issue. For these revisions, the forests could use the broad-scale assessment provided by the SIT, but would likely need mid-scale information similar to the data being collected for eastern Oregon and Washington. The chief of the Forest Service instructed the National Forests in western Montana and Idaho in the Columbia Basin to use the same approach as the forests of eastern Oregon and Washington to revise their Forest Plans. Also, the BLM director included all BLM-administered lands in western Montana and Idaho in the project. Once again the charter was amended and assessment plans and time lines were adjusted to include all the interior Columbia Basin in both the mid-scale and broad-scale assessments. With these modifications, over 58.3 million hectares were included in the assessment area (Figure 34.1).

34.2.3 Environmental Impact Statements

A planned component of the ICBEMP was the quick assemblage of information for the decision-making processes of the BLM and FS. In the original charter, an EIS team was assembled in Walla Walla, Washington, for completing a decision document for managing BLM- and FS-administered lands in eastern Oregon and Washington (Haynes et al., 1996). This team was separate from the SIT, but worked closely with the SIT to ensure that the information could be rapidly used in an EIS. Very quickly, members of the SIT started playing dual roles, in both the EIS and the assessment. In addition, one SIT product was a scientific evaluation of the EIS alternatives. This product could not be started by the SIT without the EIS team providing alternatives, and the EIS team could not provide the alternatives until the SIT provided the assessment. This was the first of numerous conflicts between the EIS and SIT teams trying to complete their products simultaneously.

To complete the EIS for the upper Columbia Basin (Idaho and western Montana), another team was formed and located in Boise, Idaho. This required the SIT to coordinate their efforts with both the EIS team in Walla Walla, Washington, and the EIS team in Boise, Idaho. As the project increased in scope, the size of the SIT increased, as did the executive board. It now included the FS regional foresters from the Northern and Intermountain Regions, the director of the Rocky Mountain Research Station, and the BLM state directors from Idaho and Montana. The ICBEMP, one of the largest assessment and decision-making processes ever attempted by the BLM or FS, had the potential to affect the management of large amounts of public lands in the West. The people living and working in the basin became concerned about the size of the project and its potential impacts.

34.2.4 Open Process

The ICBEMP was committed to an open process, with all interested parties having access to the information and procedures of the SIT. In concept, this was an excellent idea, but working with observers presented difficulty for many SIT members. In addition, since people throughout the basin did not have equal access to the work being conducted in Walla Walla, Washington, the project held frequent public information meetings throughout the basin. Another challenge was helping stakeholders to distinguish between the EIS (decision document) in which public involvement is required and the assessment (not a decision document).

The ICBEMP provided information and progress reports to other agencies and local, state, and national government officials. The time and effort required to complete the assessment and participate in informational meetings, open houses, and briefings were substantial. A high degree of coordination with the public information staff and the tribal liaison helped to alleviate the demands.

In addition to public involvement procedures, ICBEMP teams also needed to comply with the Federal Advisory Committee Act (FACA). This required continual consultation with federal attorneys. These attorneys advised ICBEMP teams on what processes should be used to ensure that FACA was not violated and that policies included in the National Forest Management Act and the National Environmental Policy Act were followed. Therefore, ICBEMP members carefully documented all meetings. The recorded minutes and all project documents became the official record of the project.

34.2.5 The Assessment

The changing scope, the open process, and the EISs all influenced the assessment. As the assessment increased in size, the optimism of completing the project on time and within budget continued, but time lines and budget projections changed. Large numbers of people throughout the basin were recruited to supply information and data for the assessment. Panels, committees, and work groups worked thousands of person-days preparing both a mid-scale and broad-scale assessment of the entire basin. The public, local governments, other agencies, and the EIS teams continually asked additional questions that the SIT tried to answer, while simultaneously addressing the major policy questions that it was chartered to answer.

One goal was to conduct an integrated assessment among the different disciplines. Past assessment efforts, such as the Eastside Health report (Everett et al., 1994) and FEMAT (1993), were presented as individual reports by discipline. This assessment would be integrated among disciplines throughout the process. However, this was more difficult than expected and added further to the complexity and progress of the assessment. The different staff areas were accustomed to their own data collection, analysis, terminology, and presentation of results. This was more noticeable between the social–economic sciences and the biological sciences. These differences made integration difficult and often made communication among staffs awkward. Also, completing an integrated assessment proved arduous when spatial scales used by the social and economic teams did not conform to spatial scales used by the terrestrial and landscape teams. Even within the biological sciences, inventories and procedures used to describe the animal resources did not always conform to techniques and procedures needed to describe plant communities. Not only did differing spatial scales confound the coordination among the different disciplines, but different temporal scales made integration difficult. For example, the social and economic staffs were accustomed to projecting into the future for short time periods (20 years or less), while the landscape staff used multiple decades. These differences within the SIT also led to many changes in approaches and analysis techniques.

To overcome some of these problems, the SIT divided the basin into areas that had similar biophysical characteristics (called ecological reporting units). This integration technique provided a way to rate basin condition as to ecological integrity and social resiliency. Likewise, integrated scenarios of different management strategies were developed and used to determine potential consequences and outcomes of applying these scenarios on federally administered lands in the basin. The scenarios were designed to address or bound the many possible management directions that could be used on the basin's natural resources.

34.2.6 Information Management

The GIS staff assembled, analyzed, and displayed large quantities of geographical information. The ICBEMP primarily used available information, but new data generated by all staffs required input, checking, and validation. Often these tasks were completed by teams throughout the basin and coordinated through the GIS staff in Walla Walla. The changing scope of the project and continual request for different data and analysis needs caused several false starts in data collection and analysis. This group did an excellent job of assembling the tremendous amount of data, but it was an expensive and time-consuming process.

The ICBEMP started in January 1994 with the expectation that the science products would be completed within a year. The changes in scope and analysis delayed the first product, "The Framework" (Haynes et al., 1996) until June 1996, two and one-half years after the project began. The "Integrated Scientific Assessment" was available in September 1996, and "An Assessment of Ecosystem Components" was finished in June 1997 (Quigley et al., 1996; Quigley and Arbelbide, 1997). The assessment, even with all the challenges, provided excellent products that contained substantial information and insight on the overall condition of the basin. The following are some of the highlights from the assessment.

34.3 Assessment Results

The basin covers about the 8% of the United States, an area nearly the size of Texas. Of the 58 million hectares assessed, 31 million ha are administered by 35 National Forests and 17 BLM Districts (Quigley et al., 1996) (Figure 34.1). The remain-

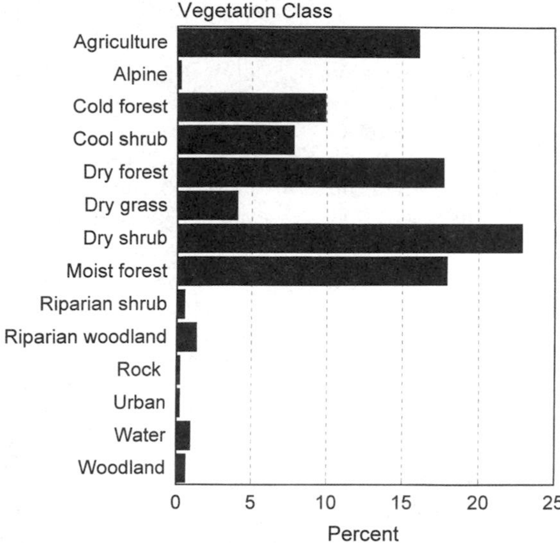

FIGURE 34.2. Proportion of the basin currently occupied by different vegetation classes.

ing area is divided among other federal, state, county, and tribal governments and private landowners. The nearly 5 million hectares of American Indian reservations represent over 20% of the reservations in the United States. Moreover, the 37 million hectares of federally designated wilderness lands that occur in the basin represent 29% of the wilderness lands within the contiguous United States. The largest proportion of the basin is dry shrub vegetation (23%), while 16% is currently used for agriculture (Figure 34.2) (Quigley et al., 1996).

The 2.9 million people in the basin, representing 1.2% of the United States population, primarily live in rural areas. Only 31% of the population lives in urban areas, and only 6 of the 100 counties in the basin could be considered metropolitan (Quigley et al., 1996). Sixty counties within the basin are in the most rural category defined by the Census Bureau. Population density on nonpublic lands in the basin averages eight people per square kilometer (km^2), and when federal lands are included, the density decreases to four people per km^2. Within these averages there is large variation, with only 0.26 people per km^2 living in Clark County, Idaho, while 132 people per km^2 live in Ada County, Idaho (Quigley et al., 1996).

Twenty-nine percent of the people living in the basin are under the age of 18 years, with 36% between the ages of 25 to 49, the prime wage-earning years. In general, the people living in the Basin are well educated, with 48% of the population having some level of higher education. The racial and ethnic composition of the basin is different than the rest of the country (Quigley et al., 1996; Haynes and Horne, 1997). Caucasians represent 92% of the people in the basin compared to 80% overall in the United States. In addition, the basin has a higher proportion of Native Americans than the national average (2.4% versus 0.8%) and smaller proportions of African Americans (0.6% versus 12.1%), Hispanic Americans (6.7% versus 9.0%), and Asian Americans (1.1% versus 2.9%) (Quigley et al., 1996). On average, each resident of the basin visits federally administered lands 32 times per year compared to a national average of 22 visits per person per year. This use totals to over 200 million recreation activity days per year made on federal lands. In addition, recreation use is increasing in the basin at a rate of 2.3% per year and will double every 31 years (Quigley et al., 1996; Haynes and Horne, 1997).

The forest area of the basin has remained relatively constant during the last two centuries, as has the volume of timber. The basin currently supplies about 10% of the annual United States timber harvest, but this proportion has been declining since the early 1960s (Figure 34.3) (Quigley et al., 1996; Haynes and Horne, 1997). Compared to the rest of the United States and the world, the basin has an abundance of forest resources for its population to use (Figure 34.4).

Cattle grazing has been an important part of the basin's economy since the late 1800s. Four percent

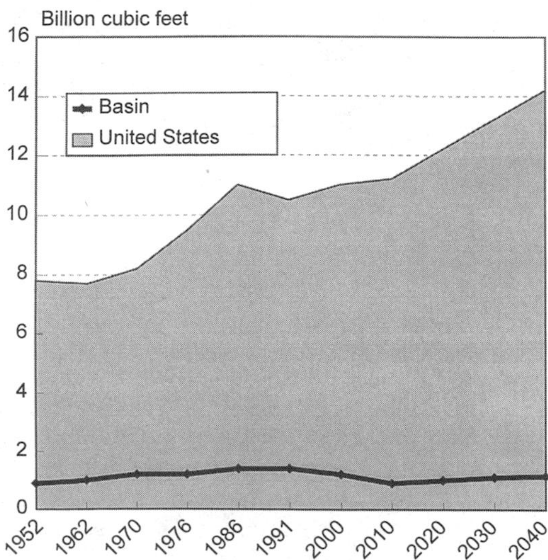

FIGURE 34.3. Historical and projected softwood timber harvest in the United States and in the basin, 1952 to 2040.

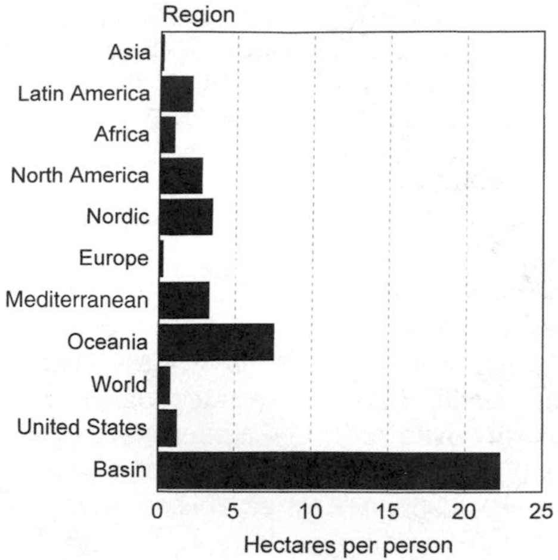

FIGURE 34.4. Amount of forest land available in different regions of the world related to the population.

of the U.S. cattle inventory occurs in the basin, with 60% of this inventory concentrated in the southern part. Depending on the location, rangeland productivity can vary from 280 kilograms (kg) per hectare of forage to over 5600 kg per hectare. Misuse and abuse degraded rangelands at the turn of the century, but they have improved since that time. Seven percent of the forage consumed by livestock (2.9 million animal unit months) in the basin is on federal rangelands (Quigley et al., 1996; Haynes and Horne, 1997).

There are 29 threatened or endangered species living in the basin. Eleven out of the 100 fishes living in the basin are listed as threatened or endangered under the Endangered Species Act (ESA, 1973) by the federal government (Lee et al., 1997). In addition to fish habitat, the waters of the basin are used heavily for irrigation and power production. The Columbia and Snake rivers represent 40% of the nation's total hydroelectric production (Quigley et al., 1996).

34.3.1 Forests and Rangelands

Native herblands, shrublands, and old single-layered forests have declined significantly since the arrival of Europeans in the basin. In addition, exotic species have increased, as has soil disturbance and roads (Figure 34.5) (Quigley et al., 1996; Hann et al., 1997). The connectivity and occurrence of early seral forests have declined, especially in areas where historical fires created conditions favorable for such species. Because of fire exclusion, selective harvesting, and grazing, forests expanded in areas that were historically woodlands and shrublands. These same factors allowed forests to become multilayered and more complex. Moreover, the forests became more dense and dominated by mid- and late-seral species. These conditions make the forests susceptible to severe fire, insect, and disease disturbances (Quigley et al., 1996; Hann et al., 1997).

Effective fire suppression decreased both very frequent and frequent fire intervals throughout the basin. Livestock grazing, timber harvesting, and the introduction of exotic plants contributed to changes in fire frequencies. Currently, fewer wildfires burn in the basin, but the proportion of high-intensity fires has doubled between the periods of 1910 to 1970 and 1970 to 1995 (Quigley et al., 1996; Hann et al., 1997) (Figure 34.6). These changes in burn frequencies and intensities also altered many of the insect and disease relations. Effective fire suppression can be partially attributed to the roads that occur in the basin. Road densities range from none to very high, 6.7 km per km^2 (Quigley et al., 1996; Hann et al., 1997).

34.3.2 Aquatic and Riparian Resources

One of the most significant issues addressed in the assessment was the status of the aquatic and riparian resources of the basin. Seven fish species, the bull trout, westslope cutthroat trout, Yellowstone cutthroat trout, redband trout, steel head, ocean-type chinook salmon, and stream-type chinook

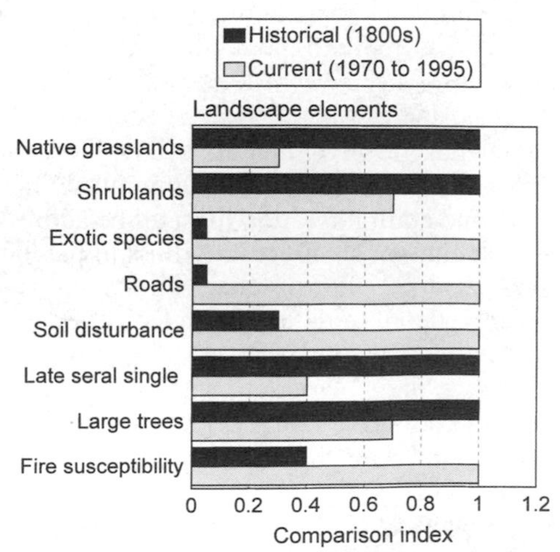

FIGURE 34.5. Comparison of historical and current selected landscape elements.

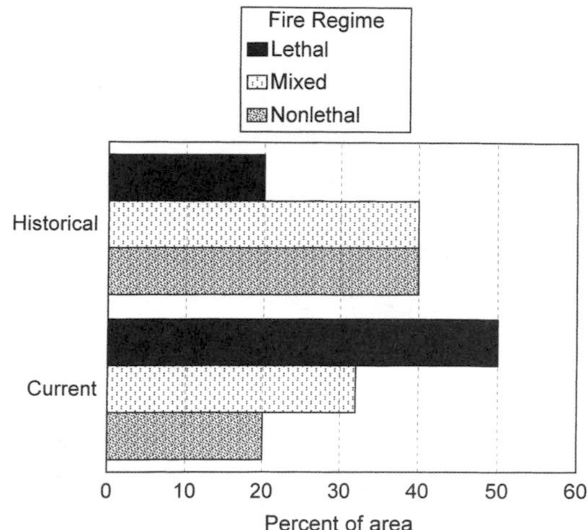

FIGURE 34.6. Changes in severity of forest fires occurring from historical (1800s) to current (1970 to 1995) time.

salmon, were selected for detailed analysis. Historically, much of the undeveloped area in central Idaho was occupied by three or more of these key species. Currently, only a fraction of this area is occupied by these species, and only a small part of this area is populated by strong populations. The strong populations are associated with higher-elevation forested landscapes and low road densities. In general, a higher proportion of the watersheds supporting strong populations of key salmonids are administered by the Forest Service compared to other lands (Quigley et al., 1996; Lee et al., 1997).

Rehabilitation of depressed populations of anadromous salmonids cannot rely on habitat improvement alone, but requires a concerted effort to address causes of mortality in all life stages (Quigley et al., 1996; Lee et al., 1997). Freshwater spawning and rearing, juvenile migration, ocean survival, and adult migration must all be addressed to ensure salmonids' survival. Salmonid habitat in the basin has changed over the last 40 to 60 years with major decreases in deep pools and pool frequency in many streams and rivers. Road construction, farming, livestock grazing, and timber harvest have all contributed to these losses. Moreover, the system of dams in the basin changed water temperatures and timing and level of peak flows, barred fish migration, and eliminated riparian vegetation. The aquatic systems of the basin no longer support the amount of native flora and fauna that they did historically. In addition to habitat changes, the introduction of nonnative Kokanee salmon, chinook salmon, lake trout, coho salmon, brown trout, Atlantic salmon, black bass, crappie, bass, and rainbow trout affected native species (Quigley et al., 1996; Lee et al., 1997).

34.3.3 Economic and Social Systems

Currently, the economies of the basin are doing quite well. The economies of the states within the basin (Oregon, Washington, Idaho, Montana) comprise 3.6 % of the U.S. economy; the economy of Washington is larger than the sum of the other three states. Per capita income is growing faster in the basin, and Washington has a higher per capita income than the U.S. average (Quigley et al., 1996; Haynes and Horne, 1997). The metropolitan centers defined by Spokane and Richland–Kenewick–Pasco, Washington, and Boise, Idaho, are the centers of economic growth in the basin. Total employment, personal income, and nonfarm labor income are growing rapidly, allowing these centers to weather national recessions better than other areas within the basin. The nonmetropolitan counties in the basin also have strong economies, and per capita income is higher than the national average in all four states. Earnings per job in nonmetropolitan areas in the basin are higher than the national average with the exception of Montana (Haynes and Horne, 1997).

The economic strengths of the basin include agriculture and related services (Quigley et al., 1996). Timber and wood products account for 2.5% of the jobs, cattle grazing accounts for 0.5%, and mining accounts for 0.5%. In contrast, it is estimated that over 14.6% of the jobs in the basin are related to recreation activities (Haynes and Horne, 1997). During the last two decades the population has grown rapidly, and the economy is evolving from one based on resources into a diverse economy oriented toward technology, service, and transportation (Haynes and Horne, 1997). Changes in FS and BLM activities will have little impact on the overall economy of the basin, but will likely affect individual communities. For example, the economies of 29 communities are sensitive to timber harvest and could be affected by changes in land management activities. Changes made by agencies in timber supply will likely affect their economies. Similarly, mining is an important economic influence in a few communities, but represents a small portion of the basin's economy (Haynes and Horne, 1997).

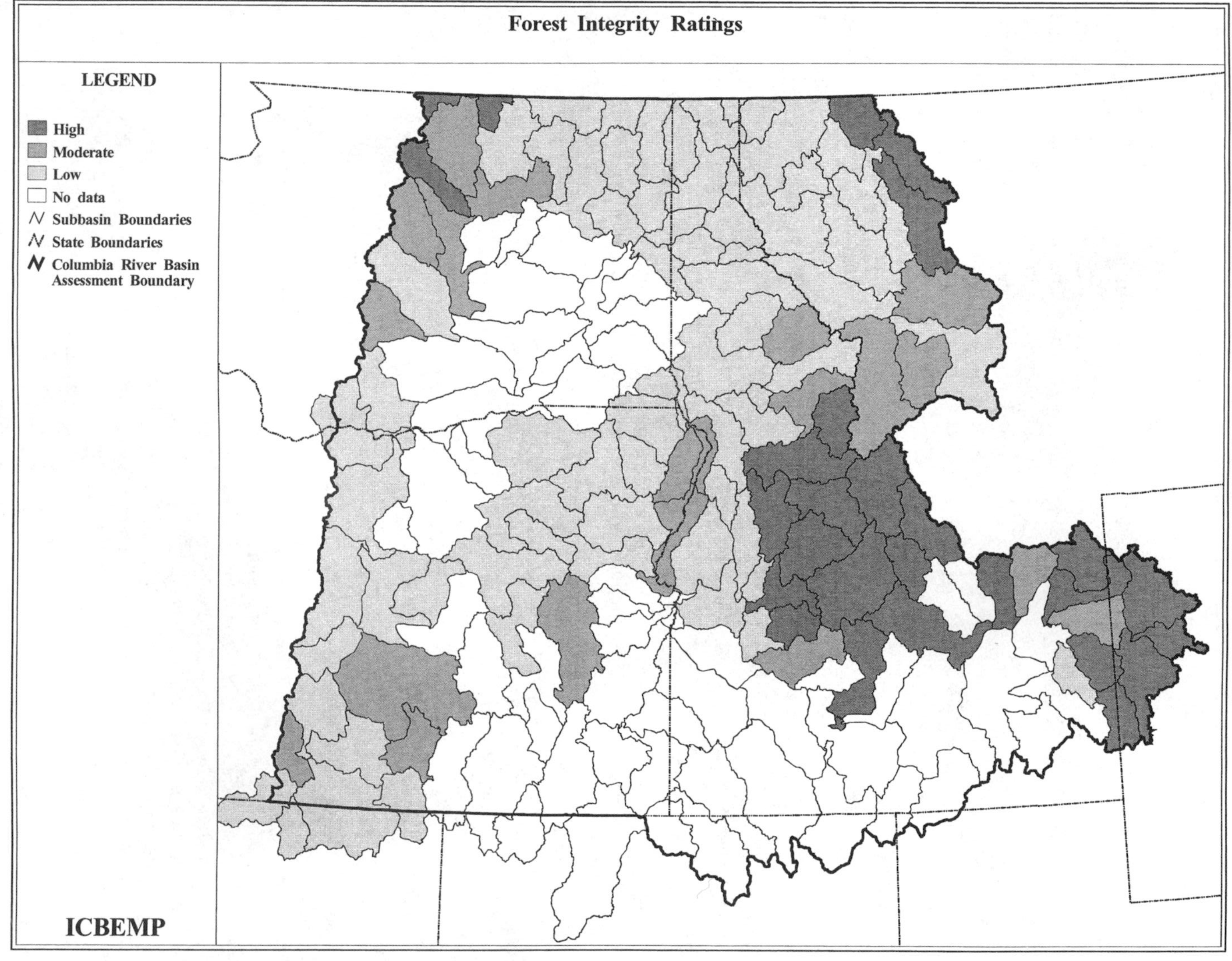

FIGURE 34.7. Forest integrity ratings for subbasins.

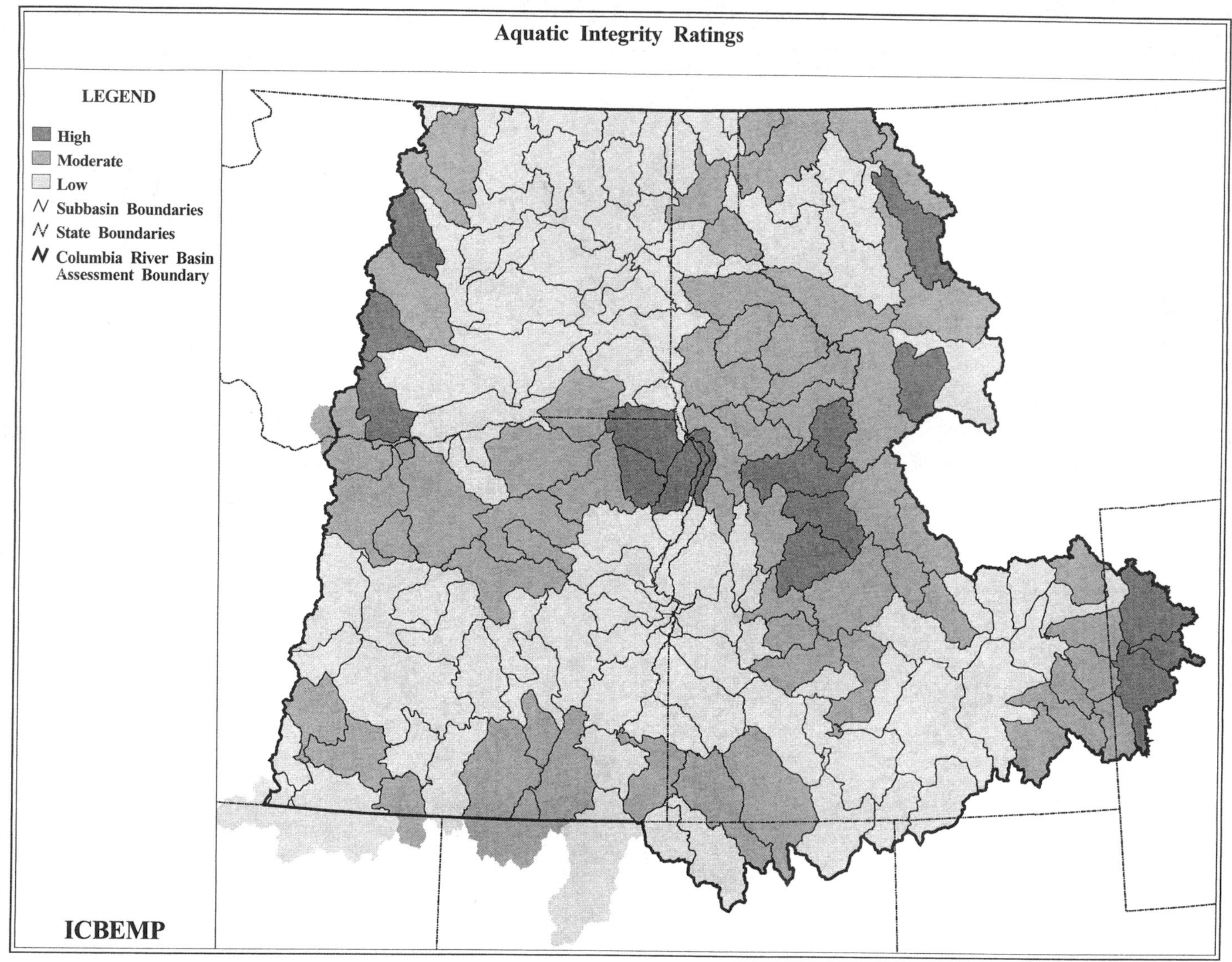

FIGURE 34.8. Aquatic integrity ratings of subbasins.

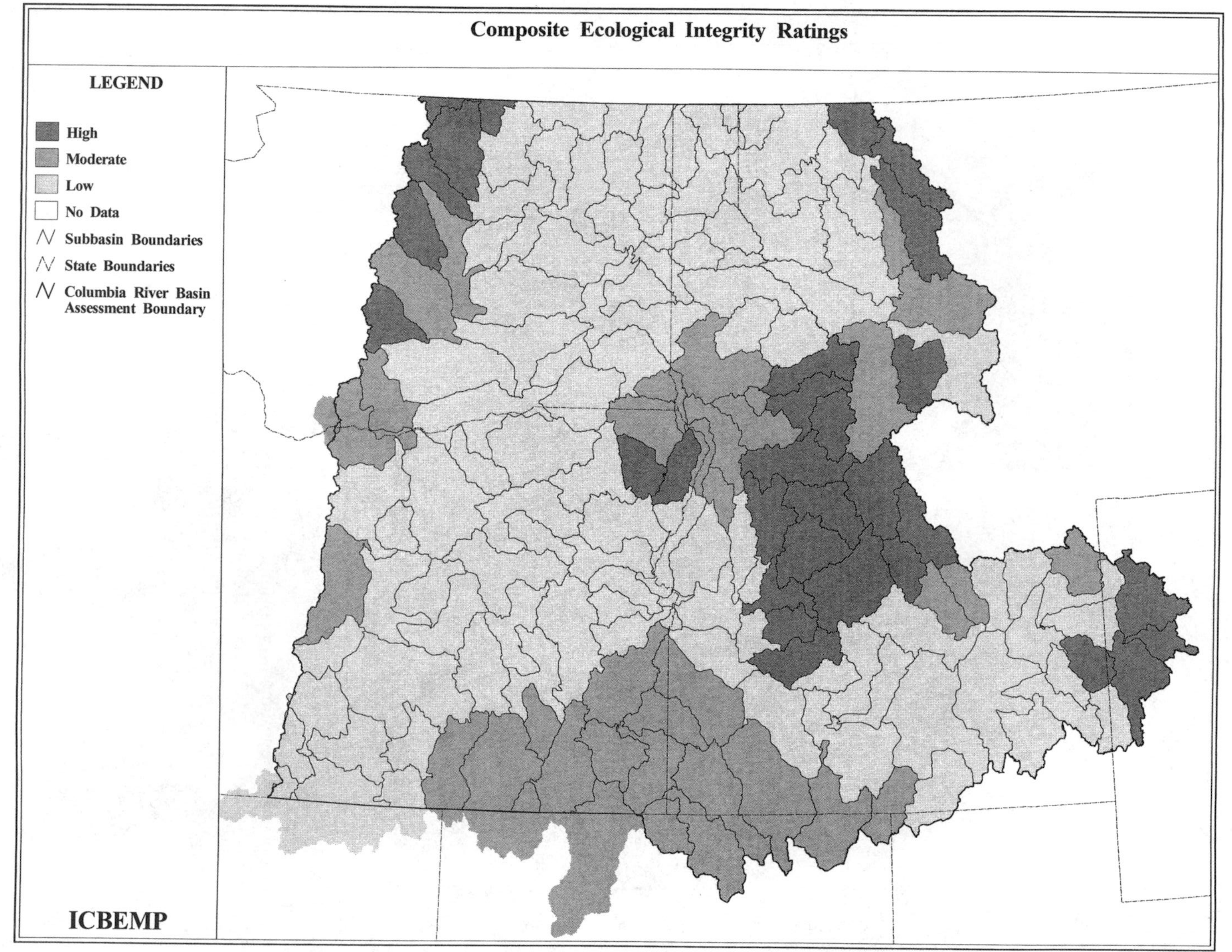

FIGURE 34.9. Composite ecological integrity ratings of subbasins synthesized from the forest and rangeland, hydrologic, and aquatic component integrity ratings.

34.3.4 Ecological Integrity and Social and Economic Resiliency

The components of the assessment produced large amounts of information describing different attributes of the basin. These separate analyses were combined to provide integrity and resiliency estimates for 164 river subbasins, ranging from 325,000 to 400,000 ha in size. The ratings were based on data and analysis, empirical models, trend analysis, and expert judgement. Forests and rangelands exhibiting high integrity were defined as having a mosaic of plant and animal communities consisting of well-connected, high-quality habitats that support a diverse assemblage of native and desired nonnative species, the full expression of potential life histories and taxonomic lineages, and the taxonomic and genetic diversity necessary for long-term persistence and adaptation in a variable environment (Quigley et al., 1996). Areas exhibiting the most elements of a system with high integrity were rated as high, and those with the fewest elements were rated low; the medium rating fell in between (Figure 34.7). Similar approaches were used to develop integrity ratings for the hydrologic and aquatic components of the basin (Figure 34.8). Combining all these ratings, a composite ecological rating was estimated for each of the subbasins (Figure 34.9). In general, 60% of the basin was rated as having low ecological integrity, with the majority of the lands having high integrity administered by the Forest Service and Bureau of Land Management (Figure 34.10).

Both social and economic resiliency measure the adaptability of human systems. Such elements as civic infrastructure, social and cultural diversity, economic diversity, and amenities were used to develop social resiliency ratings. Diversity of employment sectors was used to rate economic resiliency. Regions with high economic resiliency were described as areas in which people have ready access to a range of employment opportunities if specific business sectors experience downturns (Quigley et al., 1996). Combining both the social and economic ratings, a composite socioeconomic resiliency rating was prepared (Figure 34.11). The majority of the basin had low socioeconomic resiliency, but the majority of the population of the basin lives in areas with high resiliency (Figure 34.10).

The ICBEMP assessment described a large portion of the United States that has undergone considerable change, and the communities and economies of the basin are still rapidly changing. Old forest structures have declined, predominantly in forest types used commercially. In addition, the introduction of exotic species in both the aquatic and rangeland systems has aided in biodiversity declines in these systems. Fire regime changes have made severe fires more common (Quigley et al., 1996). The findings, data, and information assembled for this assessment should lead to more informed decisions for managing natural resources in the region and throughout the United States and North America.

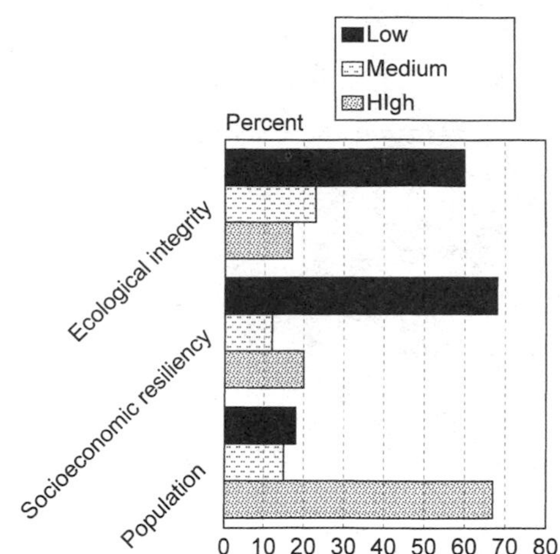

FIGURE 34.10. Proportion of the basin by ecological integrity and socioeconomic resiliency ratings and the proportion of the population with differing amounts of socioeconomic resiliency.

34.4 Conclusion

Over 150 GIS layers are available from the ICBEMP for agencies and the public to use for land management planning. No other assessment has provided this level of detailed information about ecosystems and their properties over such a wide area. Another benefit of ICBEMP was the tremendous learning accomplished by the public, agencies, and staff involved with the project. The most important lesson learned was to develop and rigorously follow an assessment process dictated by the issues. This was difficult to accomplish in the ICBEMP because of the continual changing scope and the frequent changing information needs.

The ICBEMP included the preparation of two EISs to address federal land management throughout the basin. These two decision documents were often confused with the assessment in the eyes of the public, and their constant need of information

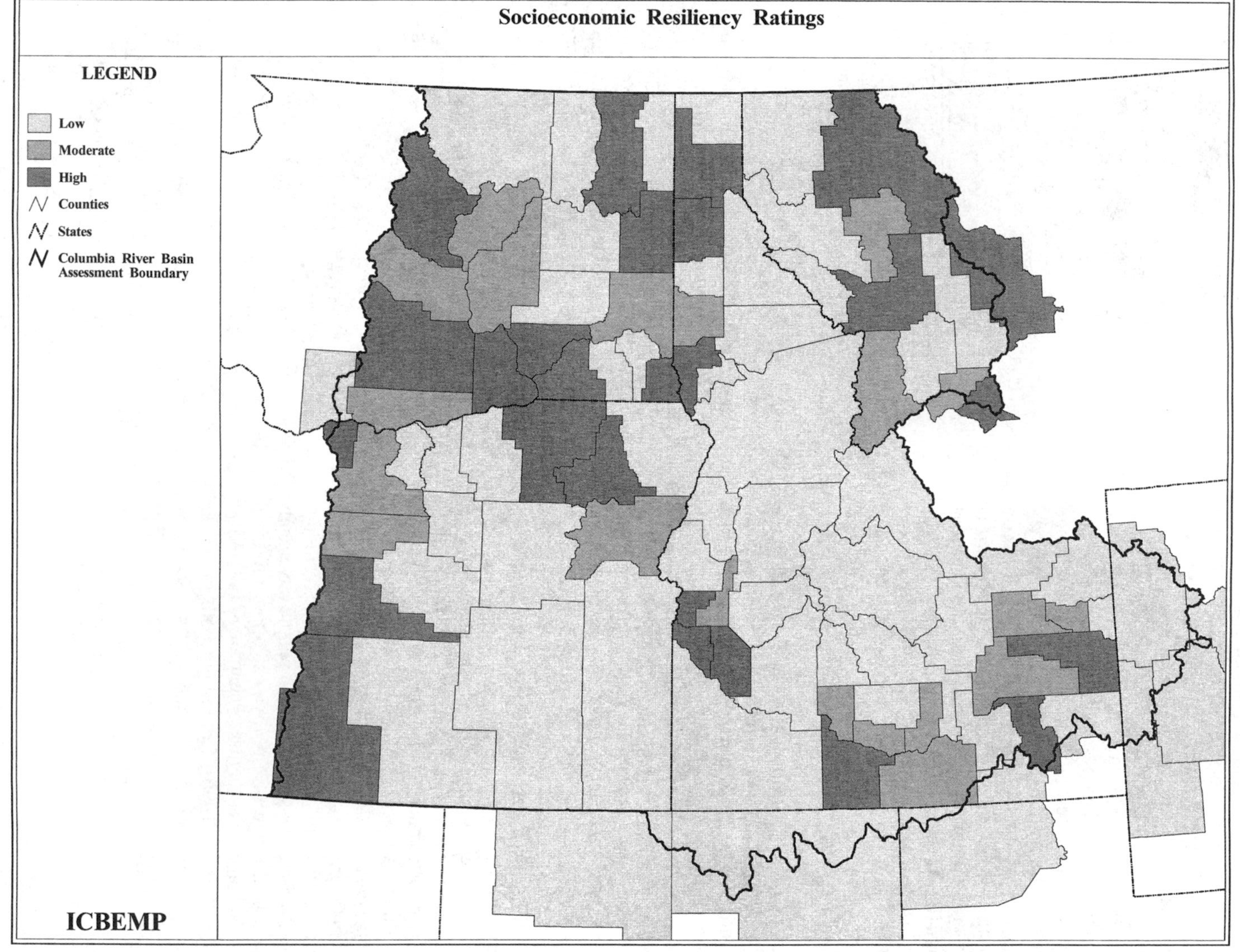

FIGURE 34.11. Socioeconomic resiliency ratings for subbasins based on an integration of economic resiliency, population density, and life-style diversity data.

and changes made the management of the assessment difficult. The success of both products may have been made easier by keeping the processes separate. An assessment needs to stand on its own, arriving at conclusions and making inferences about the data collected using appropriate analyses.

It is very easy to plan and define different scales in an assessment. But actually completing an integrated assessment using different spatial and temporal scales is difficult. The entire concept of scale is difficult to understand and as equally difficult to use, analyze, and display. The ICBEMP used different spatial and temporal scales in the assessment, but a consensus on what scales to use and what data to apply at different scales was difficult to achieve. Also, data resolution appropriate for a given scale was difficult to accomplish. In general, the comfort level for data collection and analysis is at the smallest spatial scale using the finest-resolution data. To complete an assessment on time and within budget requires good management to prevent the assessment from straying and external forces from altering the assessment plans.

Logistics for completing an assessment for an area the size of the Columbia River Basin should include a good support staff and physical plant that are located in a city with an easily accessible airport and preferably a large research library. These large-scale assessments consume large amounts of money and time. A good work environment becomes critical as long days and long work weeks are commonplace. The people working for federal agencies and in particular the Forest Service and Bureau of Land Management have excellent work ethics, which lead to can-do attitudes.

34.5 References

Dietrich, W. 1995. *Northwest passage*. New York: Simon & Schuster.

ESA. U.S. Laws, Statutes, etc., Public Law 93-205. 1973. Endangered Species Act of Dec. 28, 1973. 16 U.S.C. 1531–1536, 1538–1540.

Everett, R.; Hessburg, P.; Jensen, M.; Bormann, B. 1994. *Volume I: Executive summary*. PNW-GTR-317. Portland, OR: U.S. Dept. Agric., For. Serv., Pacific Northw. Res. Sta.

FEMAT (Forest Ecosystem Management Assessment Team). 1993. *Forest ecosystem management: an ecological, economic, and social assessment*. Portland, OR: U.S. Dept. Int., U.S. Dept. Agric., [and others].

FLPMA. U.S. Laws, Statutes, etc., and Public Law 94-579. 1976. Federal Land Policy and Management Act of 1976. Act of Oct. 21, 1976. 43 U.S.C. 1600.

Hann, W. J.; Jones, J. L.; Karl, M. G.; Hessburg, P. F.; Keane, R. E.; Long, D. G.; Menakis, J., P.; McNicoll, C. H.; Leonard, S. G.; Gravenmier, R. A.; Smith, B. G. 1997. Landscape dynamics of the Basin. In: Quigley, T. M.; Arbelbide, S. J., tech. eds. *An assessment of ecosystem components in the interior Columbia Basin and portions of the Klamath and Great Basins*. PNW-GTR-405, Vol II. Portland, OR: U.S. Dept. Agric., For. Serv., Pacific Northw. Res. Sta.: 338–1055, Chapter 3.

Haynes, R. W.; Horne, A. L. 1997. Economic assessment of the Basin. In: Quigley, T. M.; Arbelbide, S. L., tech. eds. *An assessment of ecosystem components in the Interior Columbia Basin and Portions of the Klamath and Great Basins*. PNW-GTR-405, Vol IV. Portland, OR: U.S. Dept. Agric., For. Serv., Pacific Northw. Res. Sta.: 1715–1869, Chapter 6.

Haynes, R. W.; Graham, R. T.; Quigley, T. M., technical editors. 1996. *A framework for ecosystem management in the Interior Columbia Basin*. PNW-GTR-374. Portland, OR: U.S. Dept. Agric., For. Serv., Pacific Northw. Res. Sta.

Hutchison, S. B.; Winters, R. K. 1942. *Northern Idaho forest resources and industries*. Washington, DC: U.S. Dept. Agric.

Lee, D. C.; Sedell, J. R.; Rieman, B. E.; Thurow, R. F.; Williams, J. E. 1997. Broad-scale assessment of aquatic species and habitats. In: Quigley, T. M., and Arbelbide, S. J., tech. eds. *An assessment of ecosystem components in the interior Columbia Basin and portions of the Klamath and Great Basins*. PNW-GTR-405, Vol III. Portland, OR: U.S. Dept. Agric., For. Serv., Pacific Northw. Res. Sta.: 1057–1496, Chapter 4.

Leiberg, J. B. 1897. *General report on a botanical survey of the Coeur d'Alene mountains in Idaho during the summer of 1895*. Contributions from the U.S. National herbarium Vol 5. Washington, DC: U.S. Dept. Agric., Division of Botany.

Malone, M. P.; Roeder, R. B.; Lang, W. L. 1991. *Montana: a history of two centuries*. Seattle, WA: University of Washington Press.

McCool, S. F.; Burchfield, J. A.; Allen, S. D. 1997. Social assessment. In: Quigley, T. M.; Arbelbide, S. J., tech. eds. *An assessment of ecosystem components in the interior Columbia Basin and portions of the Klamath and Great Basins*. PNW-GTR-405, Vol IV. Portland, OR: U.S. Dept. Agric., For. Serv., Pacific Northw. Res. Sta.: 1873–2012, Chapter 7.

NEPA. U.S. Laws, Statutes, etc., and Public Law 91-190 [S. 1075]. 1970. National Environmental Policy Act of 1969. [An act to establish a national policy for the environment, to provide for the establishment of a Council of Environmental Quality, and for other purposes.] In this: United States statutes at large, 1969. 42 U.S.C. sec. 4231, et seq. (1970). Washington, DC: Govt. Printing Office.

NFMA. U.S. Laws, Statutes, etc., and Public Law 94-588. 1976. National Forest Management Act of 1976. Act of Oct. 22, 1976. 16 U.S.C. 1600.

Noss, R. F.; Cooperrider, A. Y. 1994. *Saving nature's legacy: protecting and restoring biodiversity*. Washington, DC: Island Press.

PACFISH. U.S. Department of Agriculture, Forest Service and U.S. Department of Interior, Bureau of Land Management. 1994. *Draft environmental assessment. Interim strategies for managing anadromous fish-producing watersheds on federal lands in eastern Oregon and Washington, Idaho, and portions of California*. Washington, DC: U.S. Dept. Agric., For. Serv. and U.S. Dept. Int., Bureau Land Manage.

Quigley, T. M.; Arbelbide, S. J., technical editors. 1997. *An assessment of ecosystem components in the interior Columbia Basin and portions of the Klamath and Great Basins*. PNW-GTR-405, 4 Vols. Portland, OR: U.S. Dept. Agric., For. Serv., Pacific Northw. Res. Sta.

Quigley, T.; Haynes, R. W.; Graham, R. T. 1996. *Integrated scientific assessment for ecosystem management in the interior Columbia Basin*. PNW-GTR-382. Portland, OR: U.S. Dept. Agric., For. Serv., Pacific Northw. Res. Sta.

Robbins, W. D.; Wolf, D. W. 1994. *Landscape and the intermountain Northwest: an environmental history*. PNW-GTR-319. Portland, OR: U.S. Dept. Agric., For. Serv., Pacific Northw. Res. Sta.

Schwantes, C. A. 1991. *In mountain shadows: a history of Idaho*. Lincoln, NE: University of Nebraska Press.

Wilkinson, C. F. 1992. *Crossing the next meridian*. Washington, DC: Island Press.

Part 8
Implementation Strategies

35
Ecological Assessments and Implementing Ecosystem Management: Challenges, Opportunities, and the Road Ahead

Richard A. Haeuber

35.1 Introduction

Ecosystem management is a recent policy alternative proposed to address a new generation of environmental issues characterized by greater sociocultural, political, economic, and biophysical complexity. For most ecosystem management efforts, scientific uncertainty and lack of data commonly are listed among the most significant stumbling blocks to success (Yaffee et al., 1996). As we move beyond the boundaries of the safe and known into the uncharted territory of ecosystem management, at both grand and small scales, the compass that science can provide to assist decision making is frequently the missing element (Lee, 1993). Ecological assessments are fundamental to ecosystem management precisely because they provide a tool pointing the way and giving direction to difficult decisions made in the face of great uncertainty (Lackey, 1998).

At the most basic level, the goal of ecological assessments is to provide functional data to guide land-use planning and management decisions. Success in fulfilling this goal demands that ecological assessments collect accurate and useful data and information, be relevant to critical land-use and management decisions, and directly support planning and management processes and institutions. Yet, if ecological assessments are to fulfill these requirements, providing the knowledge base necessary to guide goal setting and decision making in the face of uncertainty, substantial challenges must be faced and resolved. In this chapter, we explore examples of challenges facing ecological assessment efforts in three broad categories: knowledge based, sociopolitical, and institutional. Certainly, these challenges are not insurmountable, and we briefly examine general and specific developments currently underway in environmental policy and law that may provide opportunities for the further evolution of both ecosystem management and ecological assessment tools. Inevitably, however, for ecosystem management to be a widely accepted land management approach—and, more specifically, for ecological assessment to become a useful planning and management tool—the research community must be willing to assume an active role. As the Ecological Society of America observes in its white paper on ecosystem management, " 'Science as a model for ecosystem management' is unlikely to become reality unless scientists are involved with managers and the public in adaptive management processes" (Christensen et al., 1996). Yet the question remains: Are the natural, physical, and social sciences communities ready and able to take part in this endeavor? We weigh this prospect in considering the road ahead for ecosystem management.

35.2 Implementing Ecosystem Management: Challenges for Ecological Assessment

35.2.1 Knowledge-based Challenges to Ecological Assessment

The basic function of ecological assessments is to characterize the affected environment, including species and community patterns, habitat factors, biotic–abiotic relations, and ecosystem properties (Bourgeron et al., 1994; also see Chapter 1). However, significant natural science challenges make ecological assessments more than an exercise in data gathering and reporting. For example, the difficulties involved in characterizing ecological systems to inform management typify a more general challenge confronting ecological science in the face of movement to a nonequilibrium or dynamic equilibrium conception of ecological systems (Pickett et al., 1992; Fiedler et al., 1997). It is not enough simply to describe biophysical characteristics and patterns. Because ecological systems are inherently dynamic, characterization must be dynamic as well, including an understanding of the processes that create patterns on the land and the spatial and temporal scales at which such processes operate (see Chapter 1).

If systems are constantly in motion through the influence of processes such as fire and flooding, how can an ecological assessment create a portrait that is helpful both now and in the future? Holling and Meffe (1996) propose as a golden rule that "natural resource management should strive to retain critical types and ranges of natural variation in ecosystems." Historical range of variability is certainly a useful concept for guiding ecological assessments and answering ecosystem management questions (see Chapter 19). Yet questions remain as to what a critical type or range of variation is for a specific system (Holling and Meffe, 1996) and what factors are most usefully measured to determine historic range of variability in a management context (Botkin et al., 1997). Furthermore, applying the concept in some instances is hampered by lack of data and difficulties in interpreting the historical record (Morgan et al., 1994).

Because ecological systems are dynamic in both space and time, issues of scale pose a significant challenge to ecological assessment. Patterns and processes are manifest at many spatial scales, from a stand of trees to a watershed to a drainage basin, and temporal scales may range from a single event to centuries or more. In the past, many ecologists were content either to ignore spatial and temporal heterogeneity altogether or at least to make their work easier by pretending it did not exist (Wiens, 1997). However, it is increasingly apparent that understanding and interpreting ecosystem structure and process depend on the scale at which measurements are made. There is no single "correct" scale at which to describe ecological patterns and processes. In fact, the issue of scale in understanding ecological patterns and processes has been referred to as the "fundamental conceptual problem in ecology" (Levin, 1992; Christensen et al., 1996). As far as relevance to ecosystem management is concerned, the appropriate scale depends largely on the management or policy issue in question. Scaling from the leaf to the ecosystem to the landscape to the region, as well as understanding the transfer of information, matter, and energy between these scales, is not only a central question for ecology, but poses a fundamental issue for ecological assessments.

As difficult as the task may be, simply describing, understanding, and evaluating biophysical patterns and processes are insufficient. Ecosystem management involves people, their economic and other activities, and the shared and individual beliefs and behavioral patterns that define their relationship with the land (Slocombe, 1993b). Thus ecological assessments must create a dynamic portrait that integrates human systems and activities (social, political, economic, and cultural) with biological and physical systems, replete with patterns, processes, relationships, and trajectories of change between and among the elements. The picture that emerges from an ecological assessment must portray an "ecosocial complex" (Haila, 1998) or "sociobiophysical system" (Grzybowski and Slocombe, 1988).

The integration task faced by ecological assessments is one of the most difficult challenges confronting both the natural and social sciences and is at the forefront of the scientific endeavor (Clark et al., 1999). The challenge, of course, is not new. Over the last few decades, some have attempted to bridge the gap between C.P. Snow's "two cultures" (Snow, 1993). Such efforts have met with varying levels of success, although examples of truly integrated interdisciplinary research are rare (Miller, 1994). In the last decade, however, integration efforts have become more intense as both the natural and social sciences communities realize the necessity of building bridges if they are to answer challenging intellectual questions and address pressing societal problems (Lubchenco et al., 1991; Stern et al., 1992). Much of the impetus for integration issued from the global change research community, particularly due to greater recognition that factors such as land use and land cover change are relevant to all components of global change, from biodiversity loss to altered biogeochemical cycles (CIESIN, 1992; Riebsame et al., 1994; Vitousek et al., 1997). However, the integration challenge presents itself wherever researchers try to understand the interaction between ecological and human dimensions of systems at all scales, from local-level urban systems (Pickett et al., 1997) to subcontinental regional systems (Slocombe, 1993a; Blood, 1994).

Efforts to integrate across natural and physical sciences in developing a comprehensive understanding of how systems function confront difficulties that are epistemological, technical, substantive, and process-oriented in nature. Perhaps the most basic obstacle is that individuals from different disciplines often possess very different cognitive maps or schema. As one scholar of cognitive theory observes, "Every individual acquires during the course of his development a set of beliefs and personal constructs about the physical and social environment. Beliefs provide a relatively coherent way of organizing and making sense of what would otherwise be a confusing and overwhelming array of signals and cues picked up from the environment from his senses" (George, 1980). These maps provide mental constructs that assist individuals in absorbing, processing, and interpreting new information and making sense of the world around them (Steinbruner, 1974; Lau and Sears, 1986; Fiske and Taylor, 1991; Rosati, 2001). To the extent that disciplinary training constitutes an important influence on intellectual development, it provides a paradigmatic and perceptual superstructure through which researchers view, understand, and interpret the world. Disciplinary superstructures consist of basic concepts, modes of inquiry, observational categories, representation techniques, standards of proof, and types of explanation (Vedeld, 1994; Petrie, 1976). As a result, individual members of different disciplines often observe, understand, and communicate about phenomena in ways that are alien to one another, creating distinct and difficult barriers to interaction.

Given the different ways in which members of different disciplines ask questions about phenomena, and even the questions asked, they collect and measure very different types of data and undertake such collection and measurement activities in different ways. Consequently, technical difficulties are posed by issues of comparability, exchange, and integration of data that may be presented in very different formats and secured from very different sources (see Chapter 9). In addition, members of different disciplines often question the validity and reliability of specific data, as well as collection and measurement methods (Chen, 1981).

Such epistemological and technical issues are exacerbated by substantive gaps in our knowledge within fields and discrepancies among fields. For example, the study of scaling issues illustrates that this is an area of evolving knowledge in understanding biological, physical, and social systems. In recent years, the ecological science community has concentrated a great deal of attention on understanding ecological patterns and processes at different scales and the relationship between pattern and process at different scales. In many ways, though, the social sciences are much less advanced in their understanding of relationships between types of human activities at different spatial, temporal, or institutional scales (Jacobson and Price, 1990). For example, social scientists still puzzle over the often paradoxical relationship between individual preferences and community, state, or national-level decisions (Gibson et al., 1998). Similarly, the ways in which higher-level institutions in a nested hierarchy of governance constrain decisions made at lower levels is an active area of research (Gibson et al., 1998). Overshadowing the struggle of discrete disciplines to understand cross-scalar phenomena and relationships is the fact that we simply do not understand very well the connections, at any scale, between the biological, physical, and human components of the systems studied in ecological assessments (Blood, 1994).

Personnel and process obstacles are perhaps the most practical challenges to achieving truly integrated ecological assessments. The litany is familiar. Researchers qualified and experienced in interdisciplinary research are in short supply because universities typically do not produce such people, and interdisciplinary programs often sacrifice disciplinary depth and rigor in the search for substantive breadth. Many institutions, particularly universities, do not reward interdisciplinary undertakings and may actually penalize scholars for publishing in interdisciplinary journals. A productive, high-quality integrative project demands a process to clearly identify goals and explicitly frame research questions. Pitfalls exist on the road to laying this foundation, such as failure to invest enough time in setting goals and conceptualizing research issues, or one disciplinary perspective may monopolize the act of framing research questions (Miller, 1994).

In the end, though, the most concrete obstacles are money, time, and expectations. As with all research endeavors, money is a critical limiting factor, with funding for integrative, interdisciplinary efforts often in shorter supply than funding for more traditional projects. Constrained funding means that time is in short supply as well, even though time is possibly the most critical element in interdisciplinary efforts, because it takes time simply to understand differing cognitive maps, much less develop good working relationships. And, finally, expectations of both researchers and funding institutions often exceed results. In the short to medium term, the most important result may entail forging a strong interdisciplinary team that can clearly define goals, develop well-framed research questions based on these goals, and create a research strategy with reasonable expectations of fulfilling the goals. In the face of tight budgets, however, such projects are often doomed to failure.

35.2.2 Sociopolitical Challenges to Ecological Assessments

One of the most significant barriers for assessment and implementation stems from the fact that, while scientific information is essential, land-use and management decisions are firmly embedded in a sociopolitical context. As Cortner and Shannon (1993) observe, "planning is an inherently political exercise." By extension, ecological assessments take place in such a context as well. The sociopolitical context for decision making is characterized by individual preferences and values and the institutional arrangements and organizational structures that shape individual choice. In regard to individual preferences, one key task for ecological assessments entails the delineation of ideas, perceptions, and values as they pertain to ecological systems. In this area, the most important sociopolitical challenge for ecological assessment involves the creation of mechanisms to facilitate broad public participation in defining human ecological systems, elaborating desired future conditions, and establishing broad goals and specific objectives for ecosystem management (Slocombe, 1998).

Debate over public participation in environmental decision making extends back over the last two to three decades (Reich, 1985), with many observers critical of traditional public participation processes (Shannon, 1992). Common criticisms include the failure to provide an adequate forum for representing public interests, exclusion of the general public in favor of activists, and faulty mechanisms to enable information exchange between the public and professional staff or decision makers (Moote et al., 1997). The consequences frequently include conflict, litigation, policy and management gridlock, and failure to address serious environmental and natural resource management issues. For example, an analysis of public involvement in comprehensive planning processes for 13 national forests undertaken in the mid-1980s found that the process failed to develop a partnership between forests and the public, instead creating latent adversarial relationships (Blahna and Yonts-Shepard, 1989). The lesson from years of conflict has been that decisions lacking legitimacy—meaning public acceptability—are decisions that will not stand.

Deficiencies in traditional mechanisms used to solicit public input into land-use and management decisions prompted calls for new approaches to include the public in decision making, referred to by labels such as "stakeholder involvement" (Peelle, 1995), "consensus building" (Innes, 1996), and "participatory democracy" (Moote et al., 1997). An essential task of such processes involves differentiating between categories of motivation driving human actions: needs, or motivations that are universal in all humans; values, consisting of ideas, habits, customs, and beliefs characteristic of particular social groups; and interests, or the occupational, social, political, and economic aspirations of individual members of the public (Burton, 1990). These categories of human motivation differ in their malleability, with needs being least changeable and interests the most malleable, and play a central role in defining public perceptions of desired future conditions for any ecological system. Significant hallmarks of consensus building or participatory democracy processes include the following (RTEEC, 1993; Innes, 1996; Moote et al., 1997):

1. Inclusive, wide representation of diverse stakeholders and viewpoints.
2. Public interest is jointly discovered and willed.
3. Groups are self-organizing.
4. Participants move toward strategy for action in qualitative, discursive manner.
5. Group deliberation and choice determine appropriate ends and actions.
6. Process relies on experiential knowledge of stakeholders.
7. Coordination is horizontal and self-managing, rather than a top-down exercise of power.
8. Decision-making authority is shared among all participants.
9. Groups create social, intellectual, and political capital.

Of course, the creation and successful use of such processes to elicit broad stakeholder participation is fraught with difficulties. At the most basic level, questions arise as to representation. Who represents the public, and how can we be certain that participation processes actually tap the broadest current of public perceptions and values? On one level, the challenge involves defining the set of interested publics or stakeholders that may be affected by a policy or management decision. Novel ways have been proposed to deal with this question; for example, the Malpai Borderlands project used the historic fire regime to define geographically the set of affected stakeholders. The issue of representation has more subtle dimensions, however. For instance, participation processes are often self-selecting in the sense that the loudest, most active interests are included in the process. Strong evidence exists that limited segments of the public participate in organized interests, a fact that may well introduce social, class, ethnic, and other types of bias into participatory processes (Verba and Nie, 1972; Kingdon, 1984). The frequent result is that the most widely held values or perspectives are not articulated, because the majority of unorganized, inactive segments of the public fails to engage in the process. Including only the squeakiest wheels can create polarized, adversarial processes. More importantly, perhaps, failure to identify and include the broadest cross section of the public produces decisions lacking legitimacy and increases the likelihood of conflict during implementation. Thus ecological assessments face the challenge of identifying broad segments of the interested and affected public, deciding who realistically represents these interests, and then actively facilitating their participation over the duration of the assessment process.

Another difficulty involves the fact that membership of interest groups or segments of the public is not exclusive. Whether we recognize it or not, we all belong to many groups, defined according to various interests, values, and characteristics, whose goals often are not directly compatible. For example, the person who enjoys the outdoors may simultaneously be a hunter, hiker, and snowmobile enthusiast, and the organized groups representing each of these interests may well have very different agendas. In effect, we all belong to fuzzy sets of interests with overlapping memberships and potentially contradictory value sets. Under such circumstances, the elaboration of values and desired future conditions can be a very messy process. Recent advances in social assessment techniques provide useful approaches for confronting this challenge, such as identifying "functional communities" in which people share a common perception of their relationship to the natural resources and ecological systems in a particular geographical area (Jakes et al., 1998). Still another facet of the issue of elaborating values involves the difficulty of aggregating values from one scale to another and the potential contradiction between values articulated at different scales of human organization. One of the most frequent manifestations of the disjunction between values at different scales involves the NIMBY ("not in my backyard") phenomenon in which decisions embraced by the general public, such as use of landfills or incinerators, are opposed by individuals, groups, or communities in the face of site-specific implementation.

Assuming that an ecological assessment is able to identify and secure broad, representative, and equitable participation from all affected interests, other difficulties remain. The struggle over ideas is the essence of policy and management decision making in a political environment (Stone, 1988). With regard to ecosystem management, for example, the very definition of a landscape is fundamentally a function of human perceptions and values (Rapport et al., 1998). In this regard, the challenge involves communicating complex ecological information and overcoming closely held, but often incorrect, conceptual models. As discussed earlier, important changes are occurring in the way that ecological science understands the structure and functioning of natural systems. One significant trend involves the shift from community succession toward climax as a theoretical centerpiece of ecological science to a view in which ecological systems are dynamic, exhibiting equilibrial and nonequilibrial phenomena depending on the scale of observation (Fiedler et al., 1997). Such a shift necessitates discarding the classic balance of nature approach to conservation that has been dominant for decades, supplying the mental model for resource managers, environmental advocates, conservationists, policy makers, and the public.

Perhaps more important than conceptual models, however, are the conservation statutes, regulations, policies, and organizational structures crafted over decades. The existing policy, institutional, and management framework reflects the belief that, if properly delineated and successfully protected, ecological systems can be cordoned off from the surrounding landscape and maintained in a virtually perpetual steady state. This belief in well-defined and static endpoints for protected systems is inconsistent with current understanding, which sees ecological systems as dynamic, hierarchical, nonlinear, and nonequilibrium and embedded in a nested hierarchy of systems functioning at various spatial scales (see Chapter 2).

While discovering and understanding individual preferences and values is important, therefore, it is equally essential to understand the influence of institutional arrangements and structures that shape the choices and behaviors of individuals, communities, organizations, and agencies. In this sense, ecological assessments must focus not only on the individual, but also on institutional arrangements (e.g., land ownership patterns, jurisdictional arrangements, legal frameworks) and social organization (e.g., demographic patterns and trends, community capacity). Focusing attention on this level of analysis is essential, because many opportunities for change may be found in the structure of incentives and rules governing institutional arrangements and organizations. Changing the understanding of groups and individuals, as well as the legal, policy, and administrative framework underpinning conservation in the United States, entails a sociopolitical struggle over ideas and perceptions. Because the ultimate goal of all ecological assessments is the elaboration of technically, ecologically, and socially viable options, as well as the costs and trade-offs involved in selecting one option over another, assessments are a central tool in the struggle over ideas and perceptions.

35.2.3 Institutional Challenges to Ecological Assessments

Broadly defined, institutions are "constellations of rights, rules, and relationships that define social practices and guide interactions among those who participate in them" (Young and Underdal, 1997). Institutions can be formal prescriptions or constraints, such as constitutions, statutes, and regulations, or they may affect behavior informally through norms, customs, and traditions. The mechanism for imposing behavioral constraints may be an individual organization, a network of organizations, or a group of individuals defining themselves as a community through a set of shared values or interests. Institutions are the cement that binds societies, defining how individuals act toward one another and the groups of which they are a part. Institutions also determine how individuals and groups define their relations with their physical and biological surroundings, linking society and nature. Implementing ecosystem management and successfully employing ecological assessments as a tool in this endeavor necessitates far-reaching changes in the institutional framework within which conservation is undertaken in this country. The institutional challenges include addressing the mismatch between existing institutions and conservation needs, promoting evolution of current institutions to better implement ecosystem management, and creating new institutions that are flexible, adaptive, and inclusive.

The discussion of scales earlier in this chapter highlights a set of general mismatches between ecosystems and institutions that commonly creates difficulties for conservation and resource management. Fundamental differences exist between the scale of resource issues and the institutions designed to manage them (Meidinger, 1997; Folke et al., 1998). Spatial scale issues are among the most frequently cited problems. For example, regional-scale ecological systems cross established administrative and jurisdictional boundaries, but must be managed in the context of a governance system that seldom empowers regional-scale institutions (Wuichet, 1995). Regarding temporal scales, mismatches between resources and institutions are frequently described as a function of the short-term time horizons of planners and politicians. However, a more basic temporal mismatch involves the underlying foundation of traditional resource management philosophies, which aim to reduce variability in natural systems over short time frames. Recent experiences with both fire and flooding illustrate that reduction of variability through suppression of endogenous disturbances leads to intense long-term disruptions of natural and social systems (Holling and Meffe, 1996; Haeuber and Michener, 1998). The spatial and temporal incongruities between organizations, institutions, and ecological systems are as problematic for assessment efforts as they are for management initiatives.

Consideration of specific institutions for understanding and managing ecological systems usually focuses on governmental organizations, particularly federal natural resource or environmental protection agencies. For various reasons, government agencies at all levels appear ill-prepared to design and implement successful ecosystem management strategies. For example, the fundamental incentives driving organizations within bureaucratic systems are budgets (money) and turf (power), making it difficult to create functional inter- and intraagency ecosystem management institutional relationships. Such organizational imperatives are diametrically opposed to the qualities required for successful ecosystem management, such as compromise, cooperation, and sharing of financial and human resources.

An exemplary consequence of this mismatch involves the human resources dedicated to ecosystem management efforts. The staff detailed to ecosystem management initiatives frequently is of uneven quality; most often, they are assigned simply to represent their home agency in a situation that is expected to be either very contentious or entirely meaningless. Agency representatives that successfully integrate within an ecosystem management "team" may find that the transition severs connections to their home agency support base, either due to changes within themselves or political fallout from decisions by the ecosystem management project team. Another difficult situation arises in cases where agency representatives find that positions at their home agencies are no longer available when their detail with an ecosystem management project is complete. From the level of broad organizational incentives to the individual personnel decisions necessary for designing and implementing successful ecological assessments, these types of issues require changes in agency and institutional subcultures that are both evolutionary and revolutionary (Kennedy and Quigley, 1994; Shannon, 1994).

Another basic requirement to implement ecosystem management is flexible institutions. Functionally, the requirement for flexibility translates into institutions that can adopt an adaptive approach to management. According to Walters (1997), adaptive management refers to "learning by doing" and entails more than simply undertaking better ecological monitoring and responding to unexpected management impacts. Two hallmarks of adaptive management are (1) integration of interdisciplinary experience and information into models that make predictions about the impact of alternative policies and (2) the design and implementation of management actions that are, in fact, experiments themselves. Walters (1997) identifies four common institutional obstacles that account for low success rates in implementing adaptive management: (1) modeling for adaptive management planning is frequently supplanted by ongoing modeling exercises; (2) adaptive management experiments are frequently seen as excessively expensive or ecologically risky; (3) vested interests within management agencies often oppose experimental policies; and (4) conflict over trade-offs between ecological values (e.g., restoration of a periodic flooding regime at the expense of a nonnative rare species). More fundamentally, existing institutions are rarely designed and funded to undertake the long-term monitoring and evaluation necessary to implement adaptive management (Ringold et al., 1996). Barriers such as these must be addressed as institutions strive to design ecological assessments and implement ecosystem management.

Perhaps the central institutional difficulty, however, involves envisioning and creating institutions that tackle aspects of each of the foregoing challenges (Meidinger, 1997). From the discussion of spatial scales, we can see that institutions must fit the scale of issues to be addressed, with some concerns requiring centralized management, while others are uniquely local in scope. Thus a strong argument can be made that no single agency or organization will be sufficient for undertaking ecological assessments or implementing ecosystem management. Similarly, the issues of temporal scale presented by ecosystem management point to institutions that must be capable of confronting equally short-term decisions of great immediacy and long-term issues that may unfold over decades. Such temporal adaptability must be combined with the ability to make human resource and other decisions that are flexible as well. With respect to agencies, for example, this requires transparency and openness to the broad world, whether it be another agency, a local government, or a loosely knit community coalition. Overcoming such basic obstacles as staffing decisions must be coupled with the ability to undertake management actions that can be changed with the input of new information and experience. The ability to collect, process, and act on values and interests must be added to this mix as well.

Grumbine (1997) observes that a groundswell of grassroots support for ecosystem management has commenced among federal agencies, and at least part of this change involves increased willingness to engage in collaborative, multistakeholder approaches. However, even assuming drastic changes in the aspects of institutional subcultures currently impeding implementation of ecosystem management, mere evolution of existing organizations and institutions is insufficient. The changes must range far beyond the ways in which government agencies conduct business. A fundamental element of institutions successfully implementing ecosystem management will be partnerships and networks, including agencies and decision-making bodies at all levels, nongovernmental organizations of all types, loosely affiliated groups of issue-oriented citizens, and interested individuals. Such partnerships are the building blocks for a new type of institution; they must be flexible and, in a sense, ephemeral, remaining dormant when not necessary and recreating themselves as the need arises.

Some observers maintain that we currently are witnessing such changes. They point to the emergence of an entirely new mode of conservation and environmental protection, characterized by a reinvigoration and reinvolvement of civil society in conservation, resource management, and environmental policy (John, 1994). Frequently referred to as "community-based environmental protection" (EPA, 1997; NAPA, 1997), such partnerships are widely hailed as the next generation of conservation and environmental protection (Mann and Plummer, 1998; Sawhill, 1998). Literally hundreds of instances exist in which diverse groups of people have taken matters into their own hands

(Yaffee et al., 1996; EPA, 1997). From the Yampa River Basin in Colorado to the Catskills–Delaware watershed in upstate New York to the Tensas River basin in Louisiana, people have joined together in exploring new ways to resolve thorny sociopolitical and biophysical problems. Some long-standing examples of ecosystem management do exist, such as 20-year-old efforts in the New Jersey Pinelands (Collins and Russell, 1988). However, most ecosystem management projects are much more recent (Yaffee, 1996). Given that creating, institutionalizing, and sustaining such partnerships over the long term is a difficult task, success in this endeavor is still far from secure.

35.3 Opportunities

Thus far we have focused on the challenges faced in designing and carrying out successful ecological assessments and using the results of assessments to implement ecosystem management. We should not conclude from this discussion that the future is bleak for ecological assessment as a tool or ecosystem management as a conservation approach. In fact, current and emerging opportunities exist for undertaking, refining, and employing ecological assessment as a tool in support of broad land-use, resource management, and conservation decisions. Examples of such opportunities include the gradually unfolding application of ecosystem management to conservation issues and specific developments in the use of environmental assessment tools and approaches in fulfilling NEPA requirements.

35.3.1 Evolution of Ecosystem Management

Over the past decade, ecosystem management has been the subject of hundreds of books, journal articles, seminars, conferences, workshops, and even congressional hearings. Ecosystem management has not developed without controversy. Some critics believe that the goals of ecosystem management disregard socially approved multiple-use resource management objectives, as codified in existing statutes, by dictating that ecosystems be returned to some vaguely specified condition existing at a time predating intense human impact (Sedjo, 1995). For others, the problems involve undue restrictions on economic use of public lands, increased government control over private lands, and the general subordination of private property rights and human welfare to environmental protection (Fitzsimmons, 1994, 1998). Still others question the efficacy of ecosystem management from a process perspective, expressing doubt that a few neighbors sitting down together can adequately resolve complex cultural, social, economic, political, and biophysical questions (Coggins, 1998). From a very different perspective, the prevalent version of ecosystem management is faulted for being inherently anthropocentric, resting on assumptions that are presumptuous and false, and doomed to failure because it ignores the root causes of environmental problems (Stanley, 1995). And, from the scientific community, critics raise concerns about the ability, willingness, and preparation of science and scientists to contribute meaningfully to developing and implementing ecosystem management (Hilborn and Ludwig, 1993; Cooperrider, 1996).

After so much scrutiny, where do we stand in developing and implementing ecological approaches to land and resource management? High-level federal environmental and natural resource management policy reviews have endorsed ecosystem approaches to managing the nation's land and natural resources (IEMTF, 1995; PCSD, 1996). In addition, each of the major regulatory, land management, and natural resources agencies has drafted policy guidance regarding ecosystem management (Haeuber, 1996). At the state level, ecosystem management is being adopted as an approach for addressing natural resource management and environmental protection issues (Brown and Marshall, 1996). Various sectors of business and industry, such as forestry and utility companies, are also exploring ecosystem approaches for managing land and resources that they own or control (Heissenbuttel, 1996; Mattice et al., 1996). And, as discussed above, the many local- to regional-scale ecosystem management efforts underway around the country bear witness to the outpouring of creative energy unleashed in recent years (see Chapters 1 through 4).

Perhaps most importantly, however, it appears that the legal and policy framework is changing in ways that support greater use of ecological approaches to land management in the United States. With regard to public lands, for example, laws such as the National Forest Management Act, Federal Land Policy Management Act, Endangered Species Act, National Environmental Policy Act, and the National Park Service Organic Act collectively provide a strong base from which to make a legal case supporting ecosystem management (Keiter, 1994). True, these statutes are not new, harking back to the 1960s, 1970s, and even the early years of the twentieth century in the case of the National Park

Service Organic Act. However, these laws are now employed by federal agencies to build support for ecosystem management, a distinct departure from past practice (Keiter, 1998). Another important development involves the fact that courts utilize relevant clauses of such statutes not only to support ecosystem management, but also to encourage agency use of comprehensive ecological assessments in natural resources management and planning (Bada, 1995).

Although such developments are relatively recent, the next real hurdle is for the body of law governing private lands to absorb the lessons of ecology. Even this development seems possible, with private land increasingly regarded as a vital component of an interconnected ecological system that confers important benefits for the public good (Keiter, 1998). The November 1998 elections provide significant evidence of advances in this direction, as voters in state and local jurisdictions around the nation approved over 200 initiatives involving ecologically based planning laws and open-space preservation efforts. As these changes unfold, ecological assessment will gain in importance as a tool in support of practical conservation decisions at multiple spatial scales.

35.3.2 NEPA Evolution

Enacted in 1970, the National Environmental Policy Act of 1969 is our "basic national charter for protection of the environment" (42 U.S.C.). At the heart of the statute is the requirement that federal agencies prepare an environmental assessment (EA) or detailed statement of the environmental effects of any proposed action that may significantly affect the quality of the human environment. If the agency finds no potential significant impacts, no further action is required; if significant impacts are anticipated as a consequence of the action, a full environmental impact statement is required. Although federal agencies are required to take a thorough look at environmental consequences of a proposed action, they are not required to choose the least environmentally harmful alternative based on this information (Keiter and Alder, 1997). Throughout its first 20 years of existence, this absence of substantive provisions prompted many to criticize NEPA as a primarily procedural law with little real power to remedy environmental ills (Keiter, 1990; Fogleman, 1993). Over the last several years, however, public understanding of NEPA has evolved. Application of the policy has entered a new stage, leading many to perceive it as an important tool for implementing ecosystem management.

One element of the case for NEPA's relevance to implementing ecosystem management rests on the language and procedural provisions contained in the original statute and its implementing regulations (Keiter, 1990). For example, the purpose of the original act (as amended in 1975) is to "prevent or eliminate damage to the environment and biosphere" and "enrich the understanding of the ecological systems and natural resources important to the Nation," language that may be interpreted as being consistent with ecological approaches to biodiversity protection. Similarly, NEPA regulations mandate consultation among federal agencies; consideration of possible conflicts between proposed actions and the objectives of federal, state, regional, and local land-use plans and policies; and recognition of the requirements of other laws governing public lands and resources. Another supportive aspect entails the specific requirement that environmental assessments and impact statements undertaken in the context of NEPA "utilize a systemic, interdisciplinary approach which will insure the integrated use of the natural and social sciences and the environmental design arts in planning and in decisionmaking which may have an impact on man's environment" [NEPA, section 102(2)(A)]. Finally, one hallmark of NEPA implementation has been its public participation requirement and its use as an opening for inserting public input into dialogue over environmental policy implementation. In some measure, these procedural requirements may facilitate the interorganizational cooperation, integrated interdisciplinary analysis, and public participation necessary for undertaking ecological assessments and implementing ecosystem management.

For many observers, the real breakthrough in NEPA implementation is just emerging and involves advances in the types of situations to which NEPA and its EA requirements can be applied. Some areas, such as cumulative impacts assessment, are specifically identified in the original statutory language. Cumulative impacts generally refer to effects arising from multiple activities in a given region that overlap spatially or multiple perturbations from a single, repeated activity over time (Lane, 1998; Beanlands, 1995). In the NEPA context, cumulative impacts refer to "impact on the environment which results from the incremental impact of the action when added to other past, present, and reasonably foreseeable future actions" (40 CFR sec. 1508.7). Whether it be through atmospheric nutrient deposition from multiple sources, "leapfrog development" patterns in a forested landscape, or repetitive cutting of a forest plantation over time, cumulative impacts are one of the most pervasive avenues of human-induced stress on ecological systems. Whenever many small decisions affecting a particular area may be made independently over long time periods, the incremental impacts are rarely recognized, seldom addressed, and difficult to assess. As a consequence, cumulative impacts may be identified as an area of concern in NEPA, but they have not figured prominently in most EAs undertaken to fulfill NEPA requirements (Clark, 1994; McCold and Holman, 1995). For ecological assessments, the challenge lies in providing a tool to understand the complexity of interconnected causal pathways, the feedbacks among and between human activities and environmental responses, and the additive effects of human actions over large spatial scales and long temporal scales.

In contrast to cumulative impacts assessment, some point to entirely new classes of actions that are open to scrutiny and evaluation for environmental impacts. From this perspective, areas such as strategic environmental assessment are at the cutting edge of environmental assessment practice. Strategic EA is defined as the "formalized, systematic and comprehensive process of evaluating the environmental impacts of a policy, plan, or program and its alternatives" (Therivel et al., 1992). Traditional EA practice basically has been reactive, focused primarily on evaluating potential impacts at the individual project level or, to a much lesser extent, over larger (e.g., regional) spatial scales. In contrast, strategic EAs involve evaluating potential environmental consequences of actions above the project level and integrating those into decision-making processes. For example, a strategic EA might concentrate on a suggested policy and associated legislative proposals that, if adopted, might establish a development trajectory responsible for multiple individual projects with numerous environmental impacts (Buckley, 1998). Similarly, sectoral EA involves evaluating the potential environmental impacts of proposed actions for an entire sector (e.g., agriculture), with implications for specific projects within that sector (e.g., large-scale animal feedlot operations). Policy and sectoral EAs enable influence over project selection, a level of input almost certainly precluded by project-level EAs. Other EAs in the same vein include issue-oriented analyses and evaluations at regional or broader geographical scales (Buckley, 1998). Even more ambitious, some propose undertaking EAs to examine the implications of international treaties (e.g., North American Free Trade Agreement, General Agreement on Tariffs and Trade) or national budgets (e.g., subsidies for grazing, funding for environmental research and development) (Goodland, 1998).

35.4 The Road Ahead

It is clear that widespread adoption and implementation of ecosystem management is far from assured. Similarly, many challenges confront those who envision the common use of ecological assessments as a tool in the service of ecosystem management. Yet these challenges are far from insurmountable. We can readily observe changes afoot that provide opportunities for adopting ecological approaches to land and resource management decisions and refinement of ecological assessment as a tool in support of such decisions. What is necessary for this evolution to progress?

For Keiter (1995), the enduring "problem" of ecosystem management is that "continuing confusion and uncertainty persist over how ecosystem management should be defined and what it means in practice." Brunner and Clark (1997) posit that the best way to resolve this dilemma is through a practice-based approach to ecosystem management that relies on the development and diffusion of "prototypes" of successful ecosystem management experience. They suggest a national process designed to accelerate improvements in ecosystem management principles for practical purposes. In their model, innovative ecosystem management approaches are appraised, the information gained is communicated to the broader community, and then the innovations are adapted by groups around the country to fit their own unique circumstances.

It is unclear whether a national prototyping process will, or even should, be created. However, we can argue that a de facto prototyping process is well underway. A tremendous amount of creative energy has been unleashed around the nation as individuals, groups, and organizations experiment with novel ways to resolve complex environmental and natural resource issues. Yaffee et al. (1996) describe over 100 such experiments around the country, involving diverse organizational partners, institutional frameworks, and spatial scales. These examples constitute only a handful of the many experiments underway around the nation. As these experiences are described, shared, and evaluated in the context of formal and informal gatherings and publications, this diffusion process becomes the evolution of ecosystem management and ecological assessment in action. In essence, the millions of participants in this drama represent the "invisible hands" propelling the transformation of environmental policy in this country (Howes et al., 1998).

Scientists and disciplinary communities are important players in this evolutionary process. This chapter describes the challenges facing ecological assessments and the implementation of ecosystem management. These challenges confront scientists, researchers, and the professional communities to which they belong. The willingness and ability of ecological scientists to involve themselves in ecosystem management issues, and sustainability in general, have been a matter of debate in recent years (e.g., see *Ecological Applications*, November 1993; Cooperrider, 1996). As Holling (1993) describes the issue, it is one of choosing between "traditional" science, which is disciplinary, reductionist, and detached from policy, and science that is integrative and engaged in addressing the critical, though often messy, issues confronting society. Funtowicz and Ravetz (1993, 1994) characterize this "post-normal" science as being issue-driven, comfortable with ethical and epistemological uncertainty, useful in circumstances of conflict among stakeholders, and supportive of participatory democracy by making the scientific enterprise accountable to an "extended peer community." For Lubchenco (1998), the choice to actively facilitate and support the pursuit of such inquiry, in concert with traditional basic science, involves nothing less than formulating a new "social contract" between science and society.

Policy making involves the struggle over ideas and perceptions. In this struggle, science can provide new ideas, arguments, and metaphors for understanding issues and relationships among seemingly unconnected problems. However, scientific information is only one among many inputs into the decision-making process. It is essential to understand that decisions about the environment and natural resources finally are choices among competing values and the degree to which society desires these values. In other words, these decisions are political. Scientists alone cannot and should not decide which policy and management decisions are appropriate.

Instead, scientists can define, clearly communicate, and interpret the implications of different policy options and management scenarios. Their participation should include elaborating the trade-offs entailed in policy and management choices, the degree of uncertainty involved in these trade-offs, and an assessment of postimplementation performance. In providing these services, researchers of all disciplines not only will contribute in resolving difficult conservation issues, but also assist in defining decision support tools, disseminating the prototypes of such tools, and furthering the evolution of ecosystem management. Fulfilling these needs is both the role and the challenge of ecological assessments. Ultimately, the success of ecological assessments will be measured in terms of their ability to collect and provide useful information, develop a slate of viable policy and management options based on this information, and communicate these options to decision makers in a clear and meaningful fashion. If the scientific community is able to accept and address this challenge, we all benefit; if not, decisions will still be made, business will go on as usual, and the cry for more and better science to aid decision making will remain unanswered.

35.5 References

Bada, C. 1995. Federal agency management plans are "ongoing" actions under Endangered Species Act's section 7: Pacific Rivers Council vs. Thomas and Northwest Forest Resources Council. *Natur. Resour. J.* 35:981–996.

Beanlands, G. 1995. Cumulative effects and sustainable development. In: Munasinghe, M.; Shearer, W., eds. *Defining and measuring sustainability: the biogeophysical foundations*. Washington, DC: The World Bank: 77–87.

Blahna, D. J.; Yonts-Shepard, S. 1989. Public involvement in resource planning: toward bridging the gap between policy and implementation. *Soc. Natur. Resour.* 2:209–227.

Blood, E. 1994. Prospects for the development of integrated regional models. In: Groffman, P. M.; Likens, G. E., eds. *Integrated regional models: interactions between humans and their environment*. New York: Chapman and Hall.

Botkin, D. B.; Megonigal, P.; Sampson, R. N. 1997. *Considerations of the state of ecosystem science and the art of ecosystem management*. Santa Barbara, CA: Center for the Study of the Environment.

Bourgeron, P. S.; Humphries, H. C.; DeVelice, R. L.; Jensen, M. E. 1994. Ecological theory in relation to landscape evaluation and ecosystem characterization. In: Jensen, M. E.; Bourgeron, P. S., tech. eds. *Volume II: ecosystem management: principles and applications*. PNW-GTR-318. Portland, OR: U.S. Dept. Agric., For. Serv., Pacific Northw. Res. Sta.: 58–72.

Brown, R. S.; Marshall, K. 1996. Ecosystem management in state governments. *Ecol. Appl.* 6(3):721–723.

Brunner, R. D.; Clark, T. W. 1997. A practice-based approach to ecosystem management. *Conserv. Biol.* 11(1):48–58.

Buckley, R. 1998. Strategic environmental assessment. In: Porter, A. L.; Fittipaldi, J. J., eds. *Environmental methods review: retooling impact assessment for the new century*. Fargo, ND: The Press Club: 77–86.

Burton, J. W. 1990. *Conflict: resolution and prevention*. New York: St. Martin's Press.

Chen, R. S. 1981. Interdisciplinary research and integration: the case of CO_2 and climate. *Climatic Change* 3:429–447.

Christensen, N. L.; Bartuska, A. M.; Brown, J. H.; Carpenter, S.; D'Antonio, C.; Francis, R.; Franklin, J. F.; MacMahon, J. A.; Noss, R. F.; Parsons, D. J.; Peterson, C. H.; Turner, M. G.; Woodmansee, R. G. 1996. The report of the Ecological Society of America committee on the scientific basis of ecosystem management. *Ecol. Appl.* 6(3):665–691.

CIESIN (Center for International Earth Science Information Network). 1992. *Pathways of understanding: the interactions of humanity and global environmental change*. University Center, MI: CIESIN.

Clark, R. 1994. Cumulative effects assessment: a tool for sustainable development. *Impact Assess.* 12(3):319–331.

Clark, R. N.; Stankey, G. H.; Brown, P. J.; Burchfield, J. A.; Haynes, R. W.; McCool, S. F. 1999. Toward an ecological approach: integrating social, economic, cultural, biological, and physical considerations. In: Sexton, W. T.; Szaro, R.; Johnson, N. C.; Malk, A. J.; ed. *Ecological stewardship: a common reference for ecosystem management*. Oxford, UK: Elsevier Science: 297–318.

Coggins, G. C. 1998. Of Californicators, quislings and crazies. *Chronicle Community* 2(2):27–33.

Collins, B. R.; Russell, E. W. B., editors. 1988. *Protecting the New Jersey Pinelands*. New Brunswick, NJ: Rutgers University Press.

Cooperrider, A. Y. 1996. Science as a model for ecosystem management—panacea or problem? *Ecol. Appl.* 6(3):736–737.

Cortner, H. J.; Shannon, M. A. 1993. Embedding public participation. *J. For.* 91:14–16.

EPA (Environmental Protection Agency). 1997. *Community-based environmental protection: a resource book for protecting ecosystems and communities*. Washington, DC: U.S. Environ. Protection Agency.

Fiedler, P. L.; White, P. S.; Leidy, R. A. 1997. The paradigm shift in ecology and its implications for conservation. In: Pickett, S. T. A.; Ostfeld, R. S.; Shachak, M.; Likens, G. E., eds. *The ecological basis of conservation*. New York: Chapman and Hall: 83–92.

Fiske, S. T.; Taylor, S. E. 1991. *Social cognition*. New York: McGraw-Hill.

Fitzsimmons, A. K. 1994. Federal ecosystem management: a train wreck in the making. *Cato Inst. Policy Analysis* 217:1–23.

Fitzsimmons, A. K. 1998. Why a policy of federal management and protection of ecosystems is a bad idea. *Landscape Urban Plan.* 40:195–202.

Fogleman, V. M. 1993. Toward a stronger national policy on environment. *Forum Appl. Res. Public Policy* 8(2):79–84.

Folke, C.; Pritchard, L., Jr.; Berkes, F.; Calding, J.; Svedin, U. 1998. *The problem of fit between ecosystems and institutions*. IHDP Working Paper No. 2. Bonn, Germany: International Human Dimensions Programme.

Funtowicz, S. O.; Ravetz, J. R. 1993. Science for the post-normal age. *Futures* 25(7):739–755.

Funtowicz, S. O.; Ravetz, J. R. 1994. The worth of a songbird: ecological economics as a post-normal science. *Ecol. Econ.* 10:197–207.

George, A. L. 1980. *Presidential decisionmaking in foreign policy: the effective use of information and advice*. Boulder, CO: Westview Press.

Gibson, C.; Ostrom, E.; Ahn, T.-K. 1998. *Scaling issues in the social sciences*. IHDP Working Paper No. 1. Bonn, Germany: International Human Dimensions Programme.

Goodland, R. 1998. Strategic environmental assessment. In: Porter, A. L.; Fittipaldi, J. J., eds. *Environmental methods review: retooling impact assessment for the new century*. Fargo, ND: The Press Club: 87–94.

Grumbine, R. E. 1997. Reflections on "What is ecosystem management?" *Conserv. Biol.* 11(1):41–47.

Grzybowski, A. G. S.; Slocombe, D. S. 1988. Self-organization theories and environmental management: the case of South Moresby. *Environ. Manage.* 12(4):463–78.

Haeuber, R. A.; Michener, W. K. 1998. Policy implications of recent natural and managed floods. *Bioscience* 48(9):765–772.

Haeuber, R. A. 1996. Setting the environmental policy agenda: the case of ecosystem management. *Natur. Resour. J.* 36:1–28.

Haila, Y. 1998. Assessing ecosystem health across spatial scales. In: Rapport, D. J.; Constanza, R.; Epstein, P. R.; Gaudet, C.; Levins, R., eds. *Ecosystem health*. Oxford, UK: Blackwell Science: 81–102.

Heissenbuttel, A. E. 1996. Ecosystem management—principles for practical application. *Ecol. Appl.* 6(3): 730–732.

Hilborn, R.; Ludwig, D. 1993. The limits of applied ecological research. *Ecol. Appl.* 3(4):550–552.

Holling, C. S. 1993. Investing in research for sustainability. *Ecol. Appl.* 3(4):552–555.

Holling, C. S.; Meffe, G. K. 1996. Command and control and the pathology of natural resource management. *Conserv. Biol.* 10(2):328–337.

Howes, J.; DeWitt, J.; Minard, R. A., Jr. 1998. Resolving the paradox of environmental protection. *Issues Sci. Technol.* 14(4):57–64.

IEMTF (Interagency Ecosystem Management Task Force). 1995. *The ecosystem approach: healthy ecosystems and sustainable communities. Volume I. Overview*. Washington, DC: IEMTF.

Innes, J. E. 1996. Planning through consensus building. *J. Amer. Plan. Assoc.* 62(4):460–472.

Jacobson, H. K.; Price, M. F. 1990. *A framework for research on the human dimensions of global environ-*

mental change. ISSC/UNESCO Series: 3. Paris: International Social Science Council.

Jakes, P.; Fish, T.; Carr, D.; Blahna, D. 1998. Functional communities: a tool for national forest planning. *J. For.* 96(3):33–36.

John, D. 1994. *Civic environmentalism*. Washington, DC: Congressional Quarterly Press.

Keiter, R. B. 1990. NEPA and the emerging concept of ecosystem management on the public lands. *Land Water Law Rev.* 25(1):43–60.

Keiter, R. B. 1994. Beyond the boundary line: constructing a law of ecosystem management. *Univ. Colorado Law Rev.* 65(2):293–333.

Keiter, R. B. 1995. Greater Yellowstone: managing a charismatic ecosystem. *Nat. Resour. Environ. Issues* 3:75–85.

Keiter, R. B. 1998. Ecosystems and the law: toward an integrated approach. *Ecol. Appl.* 8(2):332–341.

Keiter, R. B.; Adler, R. 1997. NEPA and ecological management: an analysis with reference to military base lands. In: Fittipaldi, J. J.; Wuichet, J. W., eds. *Army ecosystem management policy study*. Atlanta, GA: Army Environ. Policy Inst.: 6–1 to 6–23.

Kennedy, J. J.; Quigley, T. M. 1994. Evolution of Forest Service organizational culture and adaptation issues in embracing ecosystem management. In: Jensen, M. E.; Bourgeron, P. S., tech. eds. *Volume II: ecosystem management: principles and applications*. PNW-GTR-318. Portland, OR: U.S. Dept. Agric., For. Serv., Pacific Northw. Res. Sta.: 16–26.

Kingdon, J. W. 1984. *Agendas, alternatives, and public policies*. New York: Harper Collins.

Lackey, R. T. 1998. Seven pillars of ecosystem management. *Landscape Urban Plan.* 40(1/3):21–30.

Lane, P. 1998. Assessing cumulative health effects in ecosystems. In: Rapport, D. J.; Constanza, R.; Epstein, P. R.; Gaudet, C.; Levins, R., eds. *Ecosystem health*. Oxford, UK: Blackwell Science: 129–153.

Lau, R. R.; Sears, D. O., editors. 1986. *Political cognition*. Hillsdale, NJ: Lawrence Erlbaum.

Lee, K. N. 1993. *Compass and gyroscope*. Washington, DC: Island Press.

Levin, S. A. 1992. The problem of pattern and scale in ecology. *Ecology* 73(6):1943–1967.

Lubchenco, J. 1998. Entering the century of the environment: a new contract for science. *Science* 279:491–497.

Lubchenco, J.; Olson, A. M.; Brubaker, L. B.; Carpenter, S. R.; Holland, M. M.; Hubbell, S. P.; Levin, S. A.; MacMahon, J. A.; Matson, P. A.; Melillo, J. M.; Mooney, H. A.; Peterson, C. H.; Pulliam, H. R.; Real, L. A.; Regal, P. J.; Risser, P. G. 1991. The sustainable biosphere initiative: an ecological research agenda. *Ecology* 72(2):371–412.

Mann, C. C.; Plummer, M. L. 1998. Grass-roots seeds of compromise. *Washington Post*, October 11, C3.

Mattice, J.; Fraser, M.; Ragone, S.; Daugherty, D.; Wisniewski, J. 1996. Managing for biodiversity: emerging ideas for the electric utility industry—summary statement. *Environ. Manage.* 20(6):781–788.

McCold, L.; Holman, J. 1995. Cumulative impact assessments: how well are they considered? *Environ. Professional* 17(1):2–8.

Meidinger, E. E. 1997. Organizational and legal challenges for ecosystem management. In: Kohm, K. A.; Franklin, J. F., eds. *Creating a forestry for the 21st Century: the science of ecosystem management*. Washington, DC: Island Press.

Miller, R. B. 1994. Interactions and collaboration in global change across the social and natural sciences. *Ambio* 23(1):19–42.

Moote, M. A.; McClaran, M. P.; Chickering, D. K. 1997. Theory in practice: applying participatory democracy theory to public land planning. *Environ. Manage.* 21(6):877–889.

Morgan, P.; Aplet, G. H.; Haufler, J. B.; Humphries, H. C.; Moore, M. M.; Wilson, W. D. 1994. Historical range of variability: a useful tool for evaluating ecosystem change. *J. Sustain. For.* 2:87–111.

NAPA (National Academy of Public Administration). 1997. *Principles for federal managers of community-based programs*. Washington, DC: NAPA.

Peelle, E. 1995. From public participation to stakeholder involvement: the rocky road to more inclusiveness. In: *Proceedings, National Association of Environmental Professionals Annual Meeting*, June 1995.

Petrie, H. G. 1976. Do you see what I see? The epistemology of interdisciplinary inquiry. *J. Aesthet. Educ.* 10(1):29–43.

PCSD (President's Council on Sustainable Development). 1996. *Sustainable America: a new consensus*. Washington, DC: PCSD.

Pickett, S. T. A.; Parker, V. T.; Fiedler. P. L. 1992. The new paradigm in ecology: implications for conservation biology above the species level. In: Fiedler, P. L.; Jain, S. K., eds. *Conservation biology: the theory and practice of nature conservation preservation and management*. New York: Chapman and Hall: 65–88.

Pickett, S. T. A.; Burch, W. R., Jr.; Dalton, S. E.; Foresman, T. W.; Grove, J. M.; Rowntree, R. 1997. A conceptual framework for the study of human ecosystems in urban areas. *Urban Ecol.* 1(4):185–199.

Rapport, D. J.; Gaudet, C.; Karr, J. R.; Baron, J. S.; Bohlen, C.; Jackson, W.; Jones, B.; Naiman, R. J.; Norton, B.; Pollock, M. M. 1998. Evaluating landscape health: integrating societal goals and biophysical processes. *J. Environ. Manage.* 53:1–15.

Reich, R. 1985. Public administration and public deliberation: an interpretive essay. *Yale Law J.* 94: 1617–1641.

Riebsame, W. E.; Parton, W. J.; Galvin, K. A.; Burke, I. C.; Bohren, L.; Young, R.; Knop, E. 1994. Integrated modeling of land use and cover change. *Bioscience* 44(5):350–356.

Ringold, P. L.; Alegria, J.; Czaplewski, R.; Mulder, B. S.; Tolle, T.; Burnett, K. 1996. Adaptive monitoring design for ecosystem management. *Ecol. Appl.* 6(3):745–747.

Rosati, J. 2001. The power of human images and cognition in foreign policy. *Mershon Intern. Studies Rev.*

RTEEC. 1993. *Building consensus for a sustainable future: guiding principles.* Ottawa, Ontario: National Roundtable on the Environment and the Economy.

Sawhill, J. C. 1998. *Your tree or mine? A blueprint of choices. Washington Post,* October 11, C3.

Sedjo, R. A. 1995. Ecosystem management: an uncharted path for public forests. *Resources* 121:10, 18–20.

Shannon, M. A. 1992. Building public decisions: learning through planning. In: *U.S. Congress, Office of Technology Assessment, Forest Service planning: accommodating uses, producing outputs, and sustaining ecosystems. Volume II, Part A: Contractor's documents.* Washington, DC: U.S. Government Printing Office: 227–338.

Shannon, M. A. 1994. Coordination among federal agencies: cultures, budgets, and policies. In: *Committee on environment and public works. U.S. Senate. Ecosystem management: status and potential.* S. Prt. 103–98. Washington, DC: U.S. Government Printing Office.

Slocombe, D. S. 1993a. Environmental planning, ecosystem science, and ecosystem approaches for integrating environment and development. *Environ. Manage.* 17(3):289–303.

Slocombe, D. S. 1993b. Implementing ecosystem-based management. *Bioscience* 43(9):612–622.

Slocombe, D. S. 1998. Defining goals and criteria for ecosystem-based management. *Environ. Manage.* 22(4):483–493.

Snow, C. P. 1993. *The two cultures.* Canto edition. Cambridge, MA: Cambridge University Press.

Stanley, T. R., Jr. 1995. Ecosystem management and the arrogance of humanism. *Conserv. Biol.* 9(2):255–262.

Steinbruner, J. D. 1974. *The cybernetic theory of decision.* Princeton, NJ: Princeton University Press.

Stern, P. C.; Young, O. R.; Druckman, D. 1992. Global environmental change: understanding the human dimensions. Washington, DC: National Academy Press.

Stone, D. A. 1988. *Policy paradox and political reason.* Boston: Scott, Foresman.

Therivel, R.; Thompson, S.; Wilson, E.; Heaney, D.; Pritchard, D. 1992. *Strategic environmental assessment.* London: Earthscan.

Vedeld, P. O. 1994. The environment and interdisciplinarity: ecological and neoclassical economical approaches to the use of natural resources. *Ecol. Econ.* 10:1–13.

Verba, S.; Nie, N. 1972. *Participation in America: political democracy and social equality.* New York: Harper & Row.

Vitousek, P. M.; Mooney, H. A.; Lubchenco, J.; Melillo, J. M. 1997. Human domination of Earth's ecosystems. *Science* 277:494–499.

Walters, C. 1997. Challenges in adaptive management of riparian and coastal ecosystems. *Conserv. Ecol.* 2(1): 1–19.

Wiens, J. A. 1997. The emerging role of patchiness in conservation biology. In: Pickett, S. T. A.; Ostfeld, R. S.; Shachak, M.; Likens, G. E., eds. *The ecological basis of conservation.* New York: Chapman and Hall: 93–107.

Wuichet, J. W. 1995. Toward an ecosystem management policy grounded in hierarchy theory. *Ecosystem Health* 1(3):161–169.

Yaffee, S. L. 1996. Ecosystem management in practice: the importance of institutions. *Ecol. Appl.* 6(3):724–726.

Yaffee, S. L.; Philips, A. F.; Frentz, I. C.; Hardy, P. W.; Maleki, S. M.; Thorpe, B. E. 1996. *Ecosystem management in the United States: an assessment of current experience.* Washington, DC: Island Press.

Young, O. R.; Underdal, A. 1997. *Institutional dimensions of global change.* IHDP Scoping Report. Bonn, Germany: International Human Dimensions Programme.

Index